Springer Proceedings in Energy

The series Springer Proceedings in Energy covers a broad range of multidisciplinary subjects in those research fields closely related to present and future forms of energy as a resource for human societies. Typically based on material presented at conferences, workshops and similar scientific meetings, volumes published in this series will constitute comprehensive state-of-the-art references on energy-related science and technology studies. The subjects of these conferences will fall typically within these broad categories:

- Energy Efficiency
- Fossil Fuels
- Nuclear Energy
- Policy, Economics, Management & Transport
- Renewable and Green Energy
- Systems, Storage and Harvesting
- Materials for Energy

eBook Volumes in the Springer Proceedings in Energy will be available online in the world's most extensive eBook collection, as part of the Springer Energy eBook Collection. Please send your proposals/inquiry to Dr. Loyola DSilva, Senior Publishing Editor, Springer (loyola.dsilva@springer.com)

More information about this series at http://www.springer.com/series/13370

Ahmed Belasri · Sid Ahmed Beldjilali
Editors

ICREEC 2019

Proceedings of the 1st International Conference on Renewable Energy and Energy Conversion

Springer

Editors
Ahmed Belasri
Laboratory of Plasma Physics, Conductive Materials and their Applications
Department of Energy Physics
Faculty of Physics, University of Sciences and Technology of Oran Mohamed Boudiaf USTO-MB
Oran, Algeria

Sid Ahmed Beldjilali
Laboratory of Plasma Physics, Conductive Materials and their Applications
Department of Energy Physics
Faculty of Physics, University of Sciences and Technology of Oran Mohamed Boudiaf USTO-MB
Oran, Algeria

ISSN 2352-2534 ISSN 2352-2542 (electronic)
Springer Proceedings in Energy
ISBN 978-981-15-5443-8 ISBN 978-981-15-5444-5 (eBook)
https://doi.org/10.1007/978-981-15-5444-5

This Springer imprint is published by the registered company Springer Nature Singapore Pte Ltd.
The registered company address is: 152 Beach Road, #21-01/04 Gateway East, Singapore 189721, Singapore

Preface

Renewable energies are types of green energies collected from renewable resources. They include mainly solar power and wind power. Geothermal power, hydropower, biomass, and tidal power are also additional forms of renewable energy that produce electric power. In contrast to the fossil energy, the renewable energy releases advantageously fewer potentially harmful emission into the atmosphere. However, the renewable energy can inconveniently be costly. Although wind power and solar power have become cost-competitive with coal-fired and nuclear powers, some forms are not cost-competitive globally yet.

The First International Conference on Renewable Energy and Energy Conversion (ICREEC'2019) was held at the University of Sciences and Technology of Oran USTO-MB (Oran, Algeria) during 11–13 November 2019, to cover the topics related to these new renewable sustainable energies. This conference has attracted a large number of researchers, research scholars, industrials, delegates, and students to interact and share their experience and knowledge in technology application. ICREEC'2019 has provided an excellent international forum for sharing knowledge and results in renewable energies and energy conversion. The aim of the conference was to provide a platform to the researchers and practitioners from both academia as well as industry to meet the share cutting-edge development in the field.

During this scientific event, contributions of researchers of different experimental, theoretical, and numerical investigations were focused in different topics, in particular photovoltaic energy, bioenergy, and biomass, new materials for photovoltaic cells, energy conversion techniques, plasma processes, and impact of energy on environment and climate change. Aiming to promote an international exchange of scientific knowledge and experience in the different fields, the scientific program of the conference included plenary/keynote lectures and session talks as well as poster presentations.

These proceedings contain full research papers submitted by participants (keynote speakers, research scholar, and PhD students). All of these submissions went through a rigorous peer-reviewed process. In overall, 200 research papers were submitted by participants. Each paper was reviewed by two referees, members of the Scientific Committee of the conference. An ensemble of 85 papers were accepted, yielding an acceptance rate of 42 %. The papers presented herein represent a set of high-quality contributions to the technological literature on renewable energy research and experience addressing a wide range of topics. Materials from all of the sessions are available on the conference website at https://icreec2019.wixsite.com/icreec2019.

We would like, therefore, to thank everyone who contributed to this effort including paper authors, session presenters, program committee members, partners, volunteers, and sponsors. Without their support, the event would not have been as successful.

Oran, Algeria
December 2019

Prof. Ahmed Belasri
Editor & Conference Chair

Dr. Sid Ahmed Beldjilali
Editor & Conference Co-Chair

Organization

Organizers

University of Sciences and Technology of Oran Mohamed Boudiaf, USTO-MB,
Faculty of Physics - Department of Energy Physics,
Laboratory of Plasma Physics, Conductive Materials and their Applications,
El M'naouar, BP 1505, 31000, Oran, Algeria.
www.univ-usto.dz

Université des Sciences et de la Technologie d'Oran Mohamed Boudiaf USTO-MB,
Faculté de Physique, Département de Physique Energétique,
Laboratoire de Physique des Plasmas, Matériaux Conducteurs et leurs Applications,
El M'naouar, BP 1505, 31000, Oran, Algeria.
www.univ-usto.dz

Scientific Committee

Hafid Aourag, University of Tlemcen, Algeria
Ahmed Belasri, USTO-MB, Algeria
Nouredinne Zekri, USTO-MB, Algeria
El-hachemi Amara, CDTA, Algeria
Jean Pierre Bœuf, LAPLACE-CNRS, Toulouse, France
Mostefa Kameche, USTO-MB, Algeria
Nassira Benharrats, USTO-MB, Algeria
Jörg Hermann, LP3-CNRS, Aix-Marseille Université, France
Francisco Pontiga, University of Sevilla, Spain
Mokhtar Kamel Inal, University of Tlemcen, Algeria
Christophe Innocent, IEM-Montpellier, France
Abdelghani Tebboune, USTO-MB, Algeria
Fatima Hamdache, USTO-MB, Algeria
Sid Ahmed Beldjilali, USTO-MB, Algeria

Saad Hamzaoui, USTO-MB, Algeria
Ahmed Hafid Belbachir, USTO-MB, Algeria
Zahir Harrache, USTO-MB, Algeria
Ali Hennad, USTO-MB, Algeria
Mokhtar El Cheikh, USTO-MB, Algeria
Mohammed Noureddine Belkaid, USTO-MB, Algeria
Said Haidsi, USTO-MB, Algeria
Mokhtar Zerdali, USTO-MB, Algeria
Samia Bahlouli, USTO-MB, Algeria
Ahmed Guessab, ENPO, Algeria
Yacine Laghaouti, USTO-MB, Algeria
Djaaffar Rached, USTO-MB, Algeria
Dris Aid Amir, USTO-MB, Algeria
Khadija Khodja, URAER, Ghardaïa, Algeria
Sid Ahmed Amine Sabeur, USTO-MB, Algeria
Anis Samy Amine Dib, USTO-MB, Algeria
Abbas Belfar, USTO-MB, Algeria
Samira Sali, CRTSE, Algeria
Sabrina Messaoud Aberkane, CDTA, Algeria
Emanuel Axente, INFLPR–CETAL, Romania
Abed Bouadi, C. U. Relizane, Algeria
Bachir Liani, University of Tlemcen, Algeria
Itaka Kenji, Hirosaki University, Japan
Leanne Pitchford, LAPLACE-CNRS, Toulouse, France
Zeyneb Bedrane, University of Tlemcen, Algeria
Tobias Schrag, InES, Ingolstadt, Germany
Ali Cheknane, University of Laghouat, Algeria
Philippe Belenguer, LAPLACE-CNRS, Toulouse, France
Andrei Popescu, INFLPR–CETAL, Romania
Djoudi Bouhafs, CRTSE, Algeria
Youcef Bouhadda, URAER, Ghardaïa, Algeria
Bruno Caillier, INU Champollion, France
Djelloul Djafer, URAER, Ghardaia, Algeria
Noureddine Yassaa, CDER, Algiers, Algeria
Hocine Aît Kaci, USTO-MB, Algeria
Abdelatif Hassini, University Oran 2, Algeria
Tewfik Baba-Hamed, USTO-MB, Algeria
Bouziane Amine Hammou, USTO-MB, Algeria
Amine Boudghene Stambouli, USTO-MB, Algeria
Ahcène Bouabdallah, USTHB, Algeria
Mostefa Benhaliliba, USTO-MB, Algeria
Mohamed Abdelouahab, USTO-MB, Algeria
Soumia Bendella, USTO-MB, Algeria
Mohammed Sadok, URER/MS-ADRAR, Algeria
Mohammed Adnane, USTO-MB, Algeria

Nourdinne Benabadji, USTO-MB, Algeria
Aissa Keffous, CRTSE, Algeria
Lahouari Ghalouci, USTO-MB, Algeria
Khelifa Yanallah, University of Tiaret, Algeria
Lahouari Temimi, USTO-MB, Algeria
Abdelkader Aris, ENPO, Algeria
Khaled Touafek, URAER, Ghardaia, Algeria

Organizing Committee

Ahmed Belasri
Sid Ahmed Beldjilali
Fatiha Ghaleb
Abdelghani Tebboune
Bouziane Amine Hammou
Fatima Hamdache
Nouredinne Zekri
Lahouari Ghalouci
Mohammed Noureddine Belkaid
Aid Driss Amir
Asma Saim
Naima Bessai
Khadija Khodja
Mostefa Kameche
Hocine Ait Kaci
Soumia Bendella
Zoheir Harrache
Mohammed Adnane
Samia Bahlouli
Mokhtar Zerdali
Salim Baadj
Nadjet Larbi Dahou
Djaaffar Rached
Larbi Saadia
Mokhtar El Cheikh
Wafa Benstaali
Saadia Benhalouche
Tayeb Belaroussi
Tewfik Baba Hamed
Halima Loukil
Tewfik Sahraoui
Abdelwehed Hamed
Amar Benmoussa
Barkahoum Larouci
Farouk Boutaiba
Zineb Cherrak
Mohamed Amine Benelmouaz
Abbas Belfar
Sid Ahmed Amine Sabeur

Keynotes Speakers

Full Name	Title of Communication
Pr. Tobias Schrag, Full Professor, Institute of new Energy Systems (InES), Technische Hochschule Ingolstadt/Germany	Solar Energy Use in District Heating Networks
Pr. Kenji Itaka, Full Professor, North Japan Research Institute for Sustainable Energy, University of Hirosaki/Japan	Flux-mediated Carbothermic Reduction process from high purity silica to solar-grade silicon
Pr. Jean-Pierre Bœuf, Research Director, Laplace-CNRS, Université de Toulouse/France	Negative Ion Source For The Neutral Beam Injection System Of ITER
Pr. Jörg Hermann, Research Director, LP3-CNRS, AixMarseille Université/France	Ideal radiation source generated by laser ablation for a revolution in analytics
Pr. Christophe Innocent, Researcher -Chargé de Recherche, IEM-Montpellier –CNRS, Université de Montpellier/France	Bioelectrochemical devices for energy production
Pr. Francisco PONTIGA, Full Professor, University of Sevilla/Spain	Carbon dioxide capture in fluidized beds of nanosilica/$Ca(OH)_2$
Mr. Mohamed Abbés BOURASSI, Director ENIE Solar, ENIE/Algeria	Le Développement des Applications Solaires: Expérience de l'Eclairage Public en Algérie
Mr. Yvon Pellegrin, CSO Serytec, Montpellier/France	Adaptation of the PV Industry to Not Being a Pain for Investors and EPCs
Pr. Amine Boudghene STAMBOULI, Full Professor, USTO-MB, Oran/Algeria	SSB and Beyond - Options of a Global Energy Transition
Pr. Thomas Heiser, Full Professor, ICube-MaCEPV-CNRS, Université de Strasbourg/France	Organic solar cells: State-of-the-art and future perspectives
Dr. Christian Carme, CEO, C3 SAS company/France	ActiveWaveTM, the new wave energy recovery system
Pr. Ali Cheknane, Full Professor, Université de Laghouat/Algeria	Novel Structure of Organic Sensitizer for Dye-Sensitized Solar Cell (DSSC)
Pr. Andrei C. Popescu, Scientific Researcher II, INFLPR– CETAL/Romania	Additive technologies for production and processing of metallic implants
Pr. Leanne Pitchford, Research Director, Laplace-CNRS, Université de Toulouse/France	Modélisation des plasmas froids pour synthese de fertiliser utilisant les energies renouvelables
Dr. Frederic Pelascini, Scientific Leader, Cetim Grand Est, Strasbourg/France	LIBS imaging: a breakthrough in material and biomedical sciences?
Pr. El-hachemi Amara, Research Director, CDTA, Algiers/Algeria	Laser Additive Manufacturing
Mr. Youcef Khanfar, Director, Direction Resources Nouvelles SONATRACH, Algiers/Algeria	La transition énergétique de SONATRACH: Enjeux et perspectives
Pr. Saad Hamzaoui, Full Professor, Université USTO-MB, Oran/Algeria	Jalons vers un Slicium Algérien

(continued)

(continued)

Full Name	Title of Communication
Pr. Emanuel Axente, Scientific Researcher I, INFLPR– CETAL/Romania	High speed picosecond laser 3D processing of photosensitive glasses: A new approach for the fabrication of complex large area microfluidics
Dr. Rabie Benioub, Scientific Researcher, SEAVAC, Inc/Japan	Zero Carbon-Emission Process Concept for Solar Cells Manufacturing
Pr. Bruno Caillier, Full Professor, DPHE, INU Champollion, Albi/France	From power supply to light emission: a dielectric barrier discharge insight
Pr. Djoudi Bouhafs, Research Director, CRTSE, Algiers/Algeria	Boron-Oxygen metastable defects degradation of carrier lifetime by illumination in Cz and multicrystalline p-type silicon wafers
Dr. Cristina Muja, Research Engineer, DPHE, INU Champollion, Albi/France	Seeds treatment by low pressure microwave plasma source
Pr. Ahcène Bouabdallah, Full Professor, Université USTHB, Algiers/Algeria	On the Thermodynamic Approach of the Crystal Growth in the Czochralski Technique

Editing Committee Editorial

Sid Ahmed Beldjilali
Ahmed Belasri
Mostefa Kameche
Fatiha Ghaleb
Saadia Benhalouche
Mohamed Amine Benelmouaz

Partners

- Ministry of Higher Education and Scientific Research MESRS
- Directorate General for Scientific Research and Technological Development DGRSDT
- Thematic Agency of Research in Sciences and Technology ATRST

Sponsors

- SONATRACH
- SONELGAZ
- AIR ALGÉRIE
- ENIE & ENIE Solar

Contents

About the Editors

Ahmed Belasri is a full Professor at the University of Sciences and Technology of Oran USTO-MB and was the head of the Plasma Physics Laboratory LPPMCA during 17 years. He was also the president of the Scientific Council of the physics department. Since 2000, Pr. Ahmed Belasri is the Head of the Research group "Plasma and energy conversion", and responsible of the research team "Electric discharge lamps" at the Plasma Physics Laboratory LPPMCA. His research interests are on treatment of gaseous discharges, pollution and climate change, plasma physics for energy and photovoltaic system.

Sid Ahmed Beldjilali is an Associate Professor at the University of Sciences and Technology of Oran USTO-MB. He is also a Member of the Scientific Council of Energy Physics Department. Dr. Beldjilali is currently a member of the Research Group "Plasma and energy conversion", and responsible of the research team "Laser-induced breakdown spectroscopy LIBS" at the Plasma Physics Laboratory LPPMCA. His research interests are on plasma and laser processes, calibration-free LIBS, quality control of materials for energy.

Renewable and Sustainable Energy

Chapter 1
Solar Energy Use in District Heating Networks

T. Schrag, M. Ehrenwirth, T. Ramm, A. Vannahme, and C. Trinkl

Abstract District Heating Networks are an economically efficient way to decarbonise our heat supply by integrating a higher share of renewables into these networks. This paper provides an overview of utilizing solar energy in district heating networks. In particular solar thermal as well as photovoltaics and central options as well as decentral ones are discussed. Although the current use of photovoltaics in district heating networks is limited, the decrease of module prizes and the increasing need for coupling the electricity with the heat markets make it an actual field of research and application. A showcase of an existing district heating network combining all four options for integrating of solar energy is presented.

Keywords District heating systems · Solar energy · Heat pumps

1.1 Introduction

District heating networks (DHNs) have a long history in many countries facing northern and central European climates. Since the installation of the first DHNs tremendous progress has been made concerning the efficiency of DHNs. Nowadays the development is driven by improving the integration of renewable energy sources into DHNs as well as the integration of DHNs into the overall energy system and cost reduction. DHNs were identified as an integral part of the transition towards a climate-friendly overall energy system. To promote this transition, the European Union formulated their respective targets [12]. In order to achieve the German governments aims of covering 60% of gross final energy consumption by renewable energy sources in 2050 [5], it is necessary to expand the overall number of DHNs, increase the economic efficiency of existing DHNs and to integrate a higher share of renewables into these networks. Especially in rural areas, where biomass is a widely available option for renewable heat generation, the economical

---

T. Schrag (✉) · M. Ehrenwirth · T. Ramm · A. Vannahme · C. Trinkl
Institute of New Energy Systems, Technische Hochschule Ingolstadt, Ingolstadt, Germany
e-mail: tobias.schrag@thi.de

A. Belasri and S. A. Beldjilali (eds.), *ICREEC 2019*, Springer Proceedings in Energy,
https://doi.org/10.1007/978-981-15-5444-5_1

operation of DHNs can be challenging due to the low specific heat demand density. Therefore, an efficient integration of solar energy of all forms into DHNs is an urgent task inside the German energy transition (i.e. decarbonisation) program.

1.2 Possibilities of Solar Integration into District Heating

Solar assisted DHNs are an established technology since the first plants in the late 1980s in Sweden [6]. These systems are characterized by large central solar thermal (ST) collector arrays in combination with a central heat storage, aiming for shifting the solar gains from the summer into the heating period. Compared to domestic solar thermal systems, these large-scale systems both benefit from the economy of scale and the better surface-to-volume-ratio of large heat storages. On the other hand, large-scale solar assisted DHNs suffer from large investment costs of the combined solar system and network as well as from the heat losses of both the storage and the network. Current system designs vary from low cost designs without storage and reduced solar shares to systems with high solar shares such as Drake Landing in Canada with a 98% solar fraction [11]. If space is not an issue, the ground-based installation of a collector array is most economical. This kind of installations is widespread in Denmark and other Scandinavian countries, yielding low heat generation costs (c.f. Fig. 1.1, lower left corner).

However, in case of newly build DHNs, available roof areas are often used for the installation of solar thermal collectors. In this context, a variety of solutions regarding the integration of the solar thermal both into the buildings as well as into the central storage system are available. Solar thermal heat can either directly be transferred into the central storage or to bidirectional substations, which can both function as conventional substations and decentralized producers. The major advantage of such a setup is the direct and decentralized heat consumption of the single buildings within the DHN, allowing for a heat loss reduction of the network. Distributed solar thermal collector arrays and heat storages are also an interesting

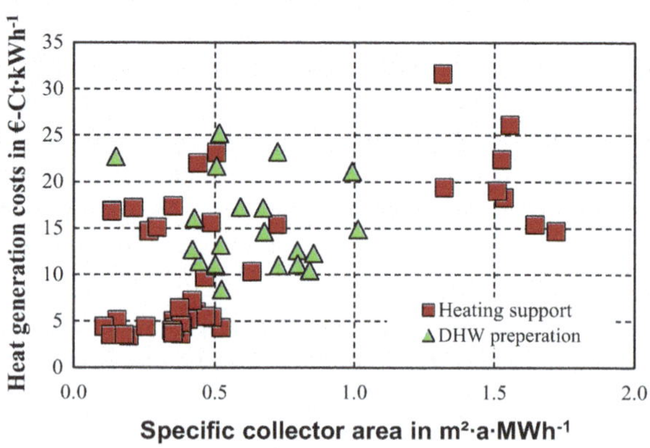

Fig. 1.1 Correlation between specific collector area and heat generation costs according to [1]. The lower left corner (i.e. specific collector area ranging between 0–0.6 m^2 a MWh^{-1}, heat generation costs ranging between 0–10 €-Ct kWh^{-1}) represent typical Danish systems

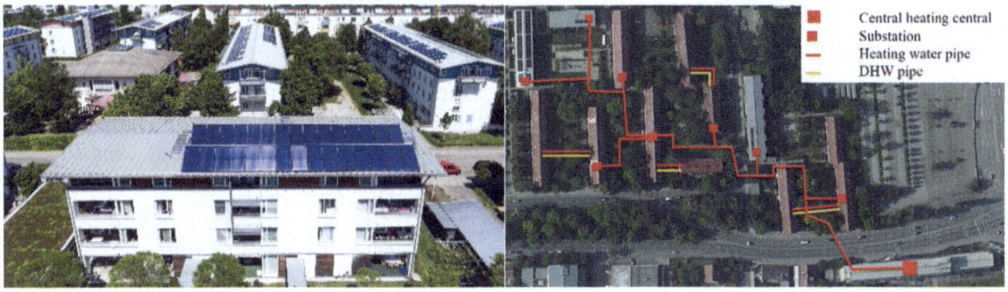

Fig. 1.2 Solar retrofitting of an existing DHN. Left: Typical multi-family building within the DHN. Right: Layout plan of the investigated DHN [2]

option for integrating renewable energy into densely built urban areas (both in case of newly build DHNs or retrofitting purposes). [2] investigated different hydraulic circuits within a bi-directional DHN. By either directly contributing to the domestic hot water (DHW) production of an individual building, feeding solar heat into the DHN or combining the aforementioned concepts (i.e. direct consumption of decentralized generated solar heat and feeding the excess heat into the DHN), the authors could successfully demonstrate a 'solar retrofitting' of an existing DHN (c.f. Fig. 1.2).

Another widespread approach is the coexistence of a DHN and domestic solar thermal systems. As this coexistence was not intended originally, the DHN and the decentralized solar thermal systems do rarely harmonize.

One aim of ongoing projects is to optimize the coordination and interaction of these two heat sources and to improve the overall efficiency of the system.

In addition to the solar thermal systems, also PV systems can be integrated into a DHN, either connected with a direct electrical heater or via a heat pump (HP). The integration of a PV array into DHN systems allows for omitting cost-intensive electrical battery or seasonal heat storages. The integration of PV into DHN can either be realized by integrating a central PV array or decentralized installations on the available roof areas. Up to now, PV is rarely integrated into DHN in Germany due to the feed-in tariffs, which make feeding into the electricity grid the only economic reasonable solution. But as PV systems have to reduce their production in times of grid overload heat production is more economic than reducing the production. Some decentral PV systems are already connected to heating rods and cause similar problems for DHNs as decentral solar thermal systems. As the feed in tariffs will be stopped after 20 years from 2020 onwards, large PV systems have to sell their energy on the market, enabling an economic operation t of central power to heat (P2H) installation.

In Austria, where the economic boundary conditions deviate from the German ones, the first central PV systems are already contribution to the heat generation of DHN.

Table 1.1 Possibilities of integration of solar energy into DHNs

	Solar thermal (ST)	PV
Central	Dominant existing form of the combination of DHN and solar energy	Hardly existing yet, but with higher shares of renewable electricity production, central P2H becomes a very economic sector coupling unit
Decentral	Existing, but often no harmonization with DHN	Similar like decentral ST, likely to increase with renewable shares in future

The possibilities of solar energy integrated into DHNs are summarized in Table 1.1.

The DHN in Dollnstein (Germany) was, as far as known by the authors, the first DHN in the world that can be operated both as warm and cold DHN denoted as DHN with variable temperatures. This systems allows for an efficient integration of ST heat; further information regarding the system was presented in [10].

The hydraulic scheme of the system is illustrated in Fig. 1.3, showing the full system with additional opportunities of solar integration which are not implemented yet. The central heating plant consists of different heat producers (combined heat and power (CHP) plant, ground water heat pump, peak boiler and ST system) as well as two thermal energy storages (TESs) with different temperature levels to decouple production and consumption. The heating plant feeds the heating network from the stratified storage on different temperature levels (30–75 °C) to deliver the heat to the consumers substations. To use the heat within this wide temperature range, the substations include heat exchangers, HPs as well as TESs. While the substations are standardized, the consumers strongly vary regarding their heat demand as well as the installed (heating) systems. Some of the consumers additionally have solar thermal systems and/or PV on their homes. The decentralized ST systems supply renewable heat to the single homes and reduce the demand from the heating network but do not actively interact with the DHN. During winter, the DHN usually operates on high temperatures up to 75 °C to deliver higher amounts of heat, up to 8 MWh per day. During summer, the losses would exceed the demand, which is not viable, since the heat production of the most heat generators within the system is related to fuel costs.

The central ST system on the roof (south-facing, 22° tilt) of the central heating system consists of 40 flat-plate collectors with a gross collector area of 100.4 m^2 and is described subsequently. Depending on the operation status of the entire system (e.g. the temperature levels of the TESs), the system feeds into the low-temperature storage or the stratified storage tank. During winter, the ST supplies the low temperature storage until the TES achieved a certain temperature level. During summer, the stratified storage has a lower temperature level and the ST system mostly charges this storage tank.

The analysis of the measurement data (year 2018) confirms the efficient integration of the central ST system. The solar thermal coverage of the central heat production fed to the stratified storage reached 73% in July and 78% in August. Operating the ST system on low average temperatures enables high solar yields (796 kWh $m^{-2}a^{-1}$). Due to the feeding of the low temperature storage, also in winter low average temperatures and high efficiencies within the ST system could be observed. In addition, higher source temperatures improve the efficiency of the central HP. The comparatively high specific solar heat yield shows that the ST system is very efficiently integrated into the system and that positive flow conditions for the ST system were established with the flow temperature reduction in the summer and the different storage tank temperature levels.

The electric yields from the central PV system were planned to be utilised within the central and decentralized HPs, especially during summer period, when high solar electric yields can be achieved. Due to legislation and the PV supporting scheme, the central PV field is until now directly feeding into the public electricity grid. Central as well as decentralized storages could help to utilise the electricity from PV by decoupling heat production and consumption. In the same way, the decentralized PV systems could be utilised.

As stated earlier, the ST systems which are installed on individual buildings are currently working independent from the DHN. Although they are lowering the heat demand of each single family home (SFH) and therefore positively contribute to a CO_2 reduction of the overall system, the DHN is currently not responding to these decentralized heat producers in an interactive manner. The electric yields from both the central and decentralized distributed PV systems are currently fed into the public electricity grid due to legal issues. However, self-consuming this electricity

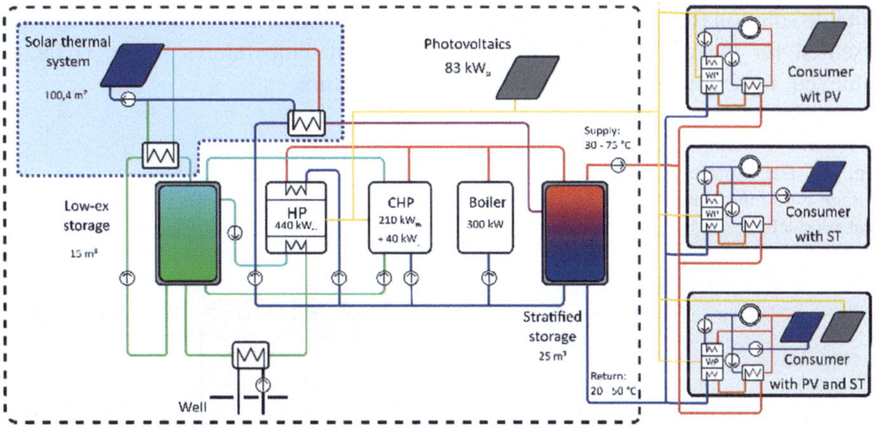

Fig. 1.3 Solar Integration into District heating networks has four entry points: Solar thermal or PV, both central and decentral. Central heating plant ––; solar thermal system … (*Source* extended from [9])

for both the central and decentralized HPs—especially during summer period—would be a meaningful measure both economically as well as ecologically. It is therefore intended to adopt this approach in case of a revision of the German Renewable Energies Act (EEG).

1.3 Integration of Decentral Solar Systems in DHN

To better integrate potential heat producers such as ST systems, PV modules or tiled stoves, current research aims for improving the standard control algorithms of DHNs. As described in the previous section, the central heat producers of a DHN are typically unable to adopt to such decentralized heat producers and are therefore demanding additional information (such as weather forecast, behavioral patterns etc.).

ST and PV systems are very popular amongst homeowners in rural areas in Germany. In combination with heat supply through a DHN, the optimal integration of this decentral sources is important for the DHN's efficiency. A large number of smaller DHNs operated in rural areas are supplied by biomass or -gas heating plants. With the expiry of the EEG government [4] subsidy, the economic efficiency of biogas operated DHN becomes more urgent. Thus, for increasing the energetic and economic efficiency of the overall system, a design and control-optimized integration of decentralized heat sources is of particular importance, because then the return temperatures and consequently the heat losses are lower. SFH, which account for a comparatively high share in rural areas, are connected with the DHN via heat exchanger or a storage tank for water heating with a typical capacity of approximately 200 l [3]. In case of a second heat source, a district heating substation (DHS), which is a storage tank itself, with a capacity typically ranging between 500 l and 1000 l is a suitable solution (c.f. Fig. 1.4). A good stratification within the storage tank, which is essential for the energy efficiency and low return temperatures in the DHN, is achieved, for example, by stratifying the heating return.

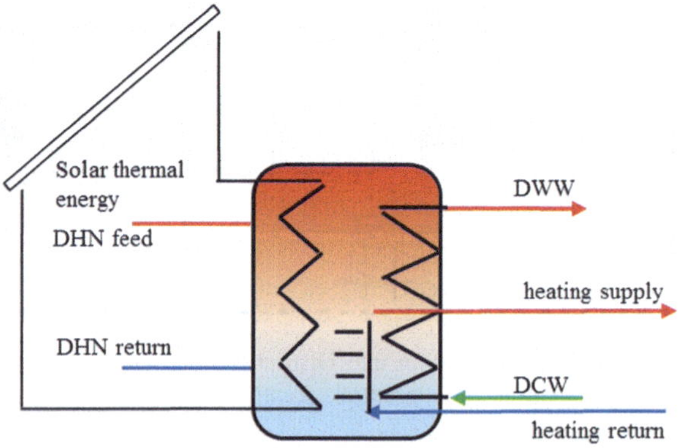

Fig. 1.4 DHS stratified storage with integrated ST heat production

Due to the low heat demand during summer, this season represents a particular challenge in terms of relative DHN losses. Therefore, a seasonally adjustable control strategy is required, preferably also accounting for different branches of the DHN individually. In case of distributed ST collector arrays on individual SFHs, the corresponding branch may be decoupled from the remaining DHN as it can produce and distribute the necessary heat independently. Furthermore, a central monitoring of each DHS enables an adjustment of the de- and central control strategy.

Previous research primarily deals with the integration of decentralized sources in the context of low-temperature district heating. [13, 14] investigated possibilities of water heating at heat supply feed temperatures of around 40 °C. According to the authors, a combination of PV system and a heating rod is very cost-effective [13, 14]. They furthermore observed that it is particularly energy-efficient if the heating element is installed directly in the feed pipe of the instantaneous DHW heater. [8] carried out studies on low-temperature DHNs for passive and low-energy houses, investigating e.g. the possibility of integrating waste heat and ST as well as the development of a DHS with booster heat pump. As a result, the authors argued for the further development of the low-temperature DHS for bi-directional heat utilization. In [7], this mode of operation for a 50 m^2 ST system on a SFH in a DHN with few consumers was investigated. The bi-directional use of energy from the ST reduces heat generation costs by 60% compared to the stand-alone solution. A test rig for space-heating systems in residential buildings is currently under development to carry out the optimal integration of heating rod connect with a photovoltaic array and solar thermal collectors in a TES and DHS, respectively.

1.4 Outlook

As the possibilities of integration of solar energy have been shown, it has to be mentioned that an even better integration of decentral solar application is possible in a new kind of DHN, so called "cold heating network" or DHN of the 5th generation, that supply only 10–25 °C. These temperatures are neither sufficient for heating purposes nor for water heating and therefore always require the installation of decentralized HPs which use the network as low temperature source. Decentralized PV is often a prerequisite to make these heating systems economically competitive and decentralized ST does not increase the return temperatures but can be integrated to increase the supply temperature. In addition, these networks can serve as a supply for cooling energy, which becomes increasingly important in central European countries, too. In hotter climates such as in Algeria, the heating demand is lower compared to European countries, but also the solar irradiation is higher. Therefore, in urban areas with a heat demand district heating systems based on waste heat and central ST plants seem to be a very economic solutions to decarbonise the energy system. Furthermore district cooling networks can have similar advantages as DHN and can be operated with less primary energy if temperature differences to the environment are kept as small as possible.

References

1. D. Beckenbauer, S. Brandmayr, W. Zörner, Erfassung und Vergleich von Kennzahlen großer europäischer Solaranlagen für Wohngebäude – Ge-schosswohnungsbau, in *Gleisdorf Solar 2014. 11. Internationale Konferenz für Solares Heizen und Kühlen* (AEE INTEC, Gleisdorf, 2014)
2. D. Beckenbauer, M. Klärner, W. Zörner, V. Cheng, Solar retrofitting in urban district heating networks - influence of the collector type on the solar yield, in *3rd Solar District Heating Conference* (2015)
3. Buderus Heiztechnik GmbH, *Planungsunterlage Größenbestimmung und Auswahl von Speicher-Wassererwärmern*, https://www.buderus.dk/. Accessed 24 Oct 2019
4. Bundesministerium für Justiz, *Gesetz für den Ausbau erneuerbarer Energien (Erneuerbare-Energien-Gesetz - EEG 2014)/Act on the Development of Renewable Energy Sources (Renewable Energy Sources Act - RES Act 2014)* (2014)
5. Bundesministerium für Umwelt, Naturschutz, Bau und Reaktorsicherheit, *Klimaschutz in Zahlen: Sektorenziele 2030*
6. J.O. Dalebaeck, *Large-Scale Swedish Solar-Heating Technology System Design and Rating* (Chalmers University of Technology, Goeteborg, 1988)
7. B. Di Pietra, F. Zanghirella, G. Puglisi, An evaluation of distributed solar thermal "net metering" in small-scale district heating systems. Energy Procedia **78**, 1859–1864 (2015)
8. M. Köfinger, R.-R. Schmidt, D. Basciotti et al., *NextGenerationHeat. Niedertemperaturfern wärme am Beispiel unterschiedlicher Regionen Österreichs mit niedriger Wärmebedarfsdichte. NEUE ENERGIEN 2020*, Endbericht (2015), http://www.get.ac.at/Downloads.html?file=tl_files/Download/Endberichte/nextgenerationheat_publizierbarer_endbericht_final.pdf. Accessed 16 Mar 2018
9. T. Ramm, M. Ehrenwirth, T. Schrag, Modelling of the central heating station within a district heating system with variable temperatures, in *Proceedings of the 13th International Modelica Conference, Linköping Electronic Conference Proceedings*, Regensburg, Germany, 4–6 March 2019 (Linköping University Electronic Press, 2019), pp. 567–576. https://doi.org/10.3384/ecp19157567
10. T. Ramm, C. Hammel, M. Klärner, A. Kruck, T. Schrag, Energy storage and integrated energy approach for district heating systems. Energy Procedia **135**, 391–397 (2017)
11. B. Sibbitt, D. McClenahan, R. Djebbar, J. Thornton, B. Wong, J. Carriere, J. Kokko, The performance of a high solar fraction seasonal storage district heating system – five years of operation. Energy Procedia **30**, 856–865 (2012)
12. The European parliament and the Council of the European Union. Regulation (EU) 2018/842 of the European Parliament and of the Council of 30 May 2018. On binding annual greenhouse gas emission reductions by Member States from 2021 to 2030 contributing to climate action to meet commitments under the Paris Agreement and amending Regulation (EU) No 525/2013. Official Journal of the European Union
13. X. Yang, H. Li, S. Svendsen, Decentralized substations for low-temperature district heating with no Legionella risk, and low return temperatures. Energy **110**, 65–74 (2016)
14. X. Yang, H. Li, S. Svendsen, Evaluations of different domestic hot water preparing methods with ultra-low-temperature district heating. Energy **109**, 248–259 (2016)

Chapter 2
Influence of Heat Absorber $HDPE_{BIO}$ on the Performance of Solar Collector

Nassim Baba Ahmed and Khaled Aliane

Abstract Energy saving and environment protection constitute fundamentals of the modern energy efficiency concept. Following the Algeria program on renewable energy and energy efficiency dictated by its energy transition policy, it is expected that 40% of electricity production will be from renewable energy source by 2030. This paper accompanies this policy with respect the energy efficiency concept. This is why this experimental work deals with design, build and study of a friendly environmental solar collector based on the high-density polyethylene ($HDPE_{BIO}$) from vegetal origin as heat absorber. This material is commonly used in a variety of industrial applications due to its low cost, ease of use, water barrier properties and chemical resistance. However, this material is sensitive to oxidative aging and investigations must be achieved to predict its durability. In this context, the $HDPE_{BIO}$ used in the experiment is characterized to confirm its role as a heat absorber. The objective of this work is to contribute to a better understanding of the mechanisms involved in the influence of $PEHD_{BIO}$ that leads to its use as heat absorber in the solar collector. Comments on this innovative design and performances of the studied solar collector are presented and the obtained results are encouraging: Use of biodegradable materials in the solar collector is an original idea, a solar collector is performed and improved by 50% using chicanes.

Keywords $HDPE_{BIO}$ · Solar collector · Heat absorber · Chicanes

2.1 Introduction

The development of solar collectors requires the reduction of thermal losses of the absorber by a rigorous design of its collectors with respect to the energy efficiency protocol and by a relevant choice of the components feeding its solar collectors.

N. Baba Ahmed (✉) · K. Aliane
Laboratory of Computational Mechanics, Abou Bekr Belkaïd University of Tlemcen, Tlemcen, Algeria
e-mail: nassimbaba@yahoo.fr

A. Belasri and S. A. Beldjilali (eds.), *ICREEC 2019*, Springer Proceedings in Energy,
https://doi.org/10.1007/978-981-15-5444-5_2

In order to be able to hope to make consequent improvements on the solar collector while respecting sustainable development, we propose a synthesis of some works already carried out in the field of solar collectors. Duffie et al. [1] have realized a planar solar collector by measuring the temperatures of the absorber, the back plate and the glazing in a stationary one-dimensional regime. The influence of environmental conditions on the solar collector (wind speed and direction, ageing of the collector surface, convective heat loss, thermal inertia and angle of incidence) was numerically processed and experimentally validated by Rodriguez et al. [2, 3]. The study of the thermal behaviour of solar collectors was carried out by Molero et al. [4] by developing an unsteady three-dimensional mathematical model for design testing. Minn et al. [5] were interested in the phenomenon of interaction of the absorber temperature with the heat transfer fluid. Letz et al. [6] have made a comprehensive analysis of the behavior of an air collector under the effect of natural and artificial sunlight using the nodal method by solving thermal equilibrium equations.

The orientation of the solar collector has a significant impact on its overall efficiency. Thus, Dang et al. [7] took an interest in this problem by developing an experimental study. The effect of the inclination of the whole solar collector and thus the Rayleigh number on the flow regime is also important. This allowed Hollands et al. [8] to determine the critical Rayleigh number for horizontal, vertical and tilted positions. In 2012, D'Antoni and colleagues [9] analysed the performance of solar thermal collectors in order to put forward the different possible variants already proposed to help researchers better express their future developments. In 2017, Suresh and colleagues [10] point out that the efficiency of solar conversion is much higher for industrial heating than for electricity production, and that industrial heating applications account for a significant share of industrial energy consumption. In this sense, in Algeria, studies have been carried out by Ihaddadene et al. [11], Kabeel et al. [12] by varying the flow rate of the working fluid and automatically control the position of the solar collector in relation to the sun, but remains the problem of heat transfer and the shape of the solar collector to be solved as well as refrigeration.

This work deals with the experimental study of an innovative material as absorber for the solar collector design. This material is the HDPE (High density polyethylene) friendly environmental, because synthetized from vegetal origin. The aging study of this material and the global solar collector performances have been investigated. Obtained results are presented and some comments have been done.

2.2 Characterization Study

The built solar collector has served as a test bench at the University of Tlemcen, Algeria. The Fig. 2.1 showed a view of the designed solar collector. The experimental tests were carried out during the summer period from 20th to 30th June 2016 under a clear sky and a low or moderate wind speed. The solar collector was

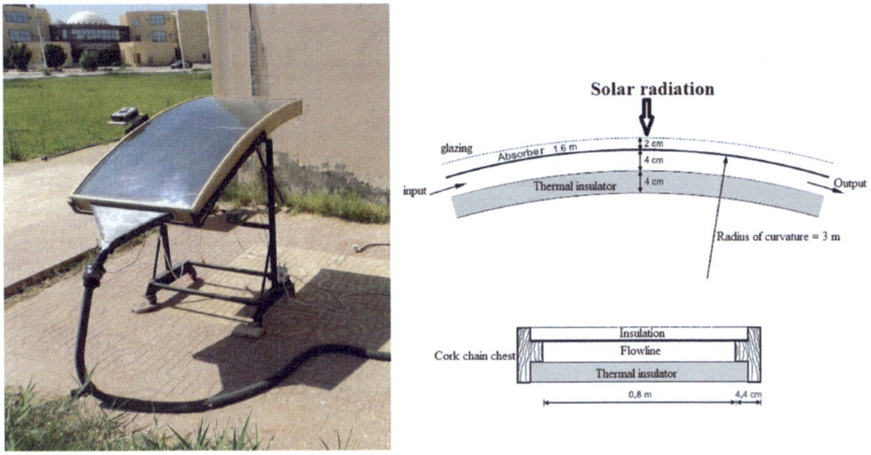

Fig. 2.1 Curved solar collector

oriented towards the south and inclined by 15° relating to the horizontal plane, which is considered optimal position for Tlemcen city during the summer season. The airflow through the solar collector was assisted by a regulated exhaust fan system. There is a row of 20 holes with 1 cm diameter used for ambient air circulation into the inner space of the solar collector.

The heated air is subsequently evacuated through a converging section at the outlet of the flow manifold. The solar collector structure is built with galvanized sheet (thickness 1.5 mm, length 1.6 m, width 0.8 m). The thickness of the air free space is 2 cm located between a transparent cover with 4 mm thickness flexible glazing at the upper side and the absorber in HDPEBIO material with dimensions: thickness 3 mm, length 1.6 m, width 0.8 m at the bottom side. The working fluid circulates through a 4 cm channel thickness located between the absorber and the bottom insulating wall.

Two superposed layers of glass wool and cork-oak with respectively 1.6 and 2.4 cm cover the insulating wall. Different experimental techniques for the thermomechanical characterization as well as the aging investigation of the selected PEHDBIO samples have been achieved in order to analyze the structural and textural properties of our material. The Table 2.1 shows the measured properties as results of HDPEBIO characterization.

Table 2.1 Measured properties of $HDPE_{BIO}$

Properties	Unity	Value
Volume density	g cm^{-3}	0.974 ± 0.01
Thermal capacity	J/Kg K	1900 ± 0.012
Thermal conductivity	W/m K	0.43 ± 0.01
Total emissivity		0.958 ± 0.005
Thermal diffusivity	m^2/s	(0.314 ± 0.02) 10^{-6}
Density	g/cm	0.945 ± 0.001
Melting temperature	°C	202.34 ± 0.02
Crystallinity	%	55 ± 2
Weight-average molar mass	g/mol	498.54 ± 0.44
Glass transition temperature	°C	−127.22 ± 0.02
Coefficient of expansion	C^{-1}	19 10^{-5}

2.3 Solar Collector Energy Study

2.3.1 Efficiency Result

This study compares the performances of two solar collectors. The reference solar collector (R) with air free space thickness of 2 cm and the solar collector (R-3) with 1.7 cm thickness. The working fluid is 1.8 l/min flowrate water. From these measurements, the efficiency of the solar collector [13] is determined using Eq. (2.1):

$$\eta = \frac{P_u}{P_{abs}} = \frac{P_u}{A_c \cdot G} = \frac{\dot{m} \cdot C_p \cdot \Delta T}{A_c \cdot G} \tag{2.1}$$

Where P_u is the useful power recovered by the working fluid, P_{abs} is the solar power arriving on the solar collector, A_c is a solar collector surface, G is a global incident solar flux, $\dot{m}$ is the mass flow rate of working fluid and C_p is the specific heat at constant pressure.

2.3.2 Solar Collector Improvement

The optimization of the air space thickness is unfortunately not possible due to the solar collector structure which is not variable geometry. To overcome this problem, we added chicanes like shown on Fig. 2.2. These chicanes are transparent to solar radiation and have a thickness of 0.1 mm and a height of 18 mm. They are placed with regular interval and respecting a gap of 1 mm relating to sides of the absorber and the glazing. The effect of chicanes addition on the global efficiency of the solar collector is evaluated. We still have two solar collectors, the first one is without chicanes and serves as a reference case. The second one is equipped with five crossed horizontal chicanes and four longitudinal ones in Fig. 2.3. Respectively show the evolution of the global solar radiation (RSGI) and the temperatures T_{inlet}, T_{outlet} (respectively temperatures of inlet working fluid and outlet working fluid) of working fluid and $T_{absorber}$ (temperature of absorber) for reference solar collector corresponding to date 21st June 2019 and for solar collector with chicanes corresponding to date 25th June 2019.

Fig. 2.2 Solar collector with chicanes

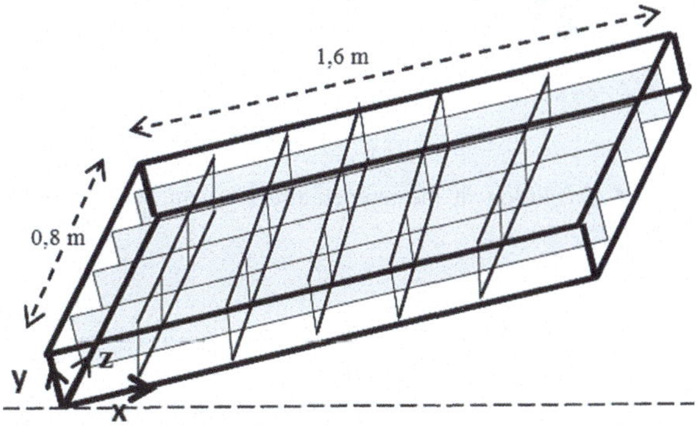

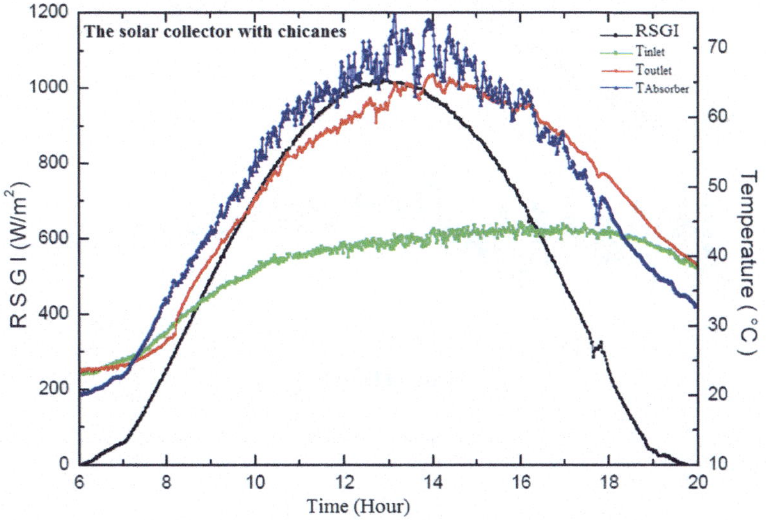

Fig. 2.3 Evolution of the global solar radiation (RSGI) and the temperatures T_{inlet}, T_{outlet} of working fluid and $T_{absorber}$ during one day

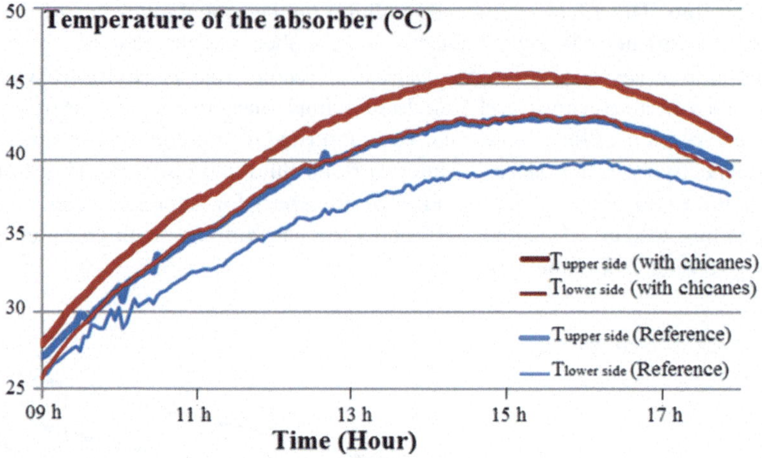

Fig. 2.4 Evolution of the temperature gradient between the absorber faces for the two studied solar collectors

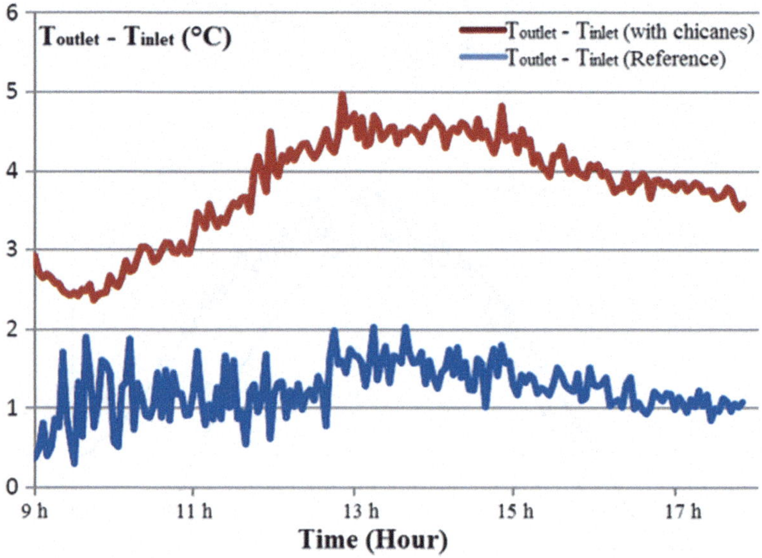

Fig. 2.5 Evolution of the temperature difference between the input and output of the two studied solar collectors

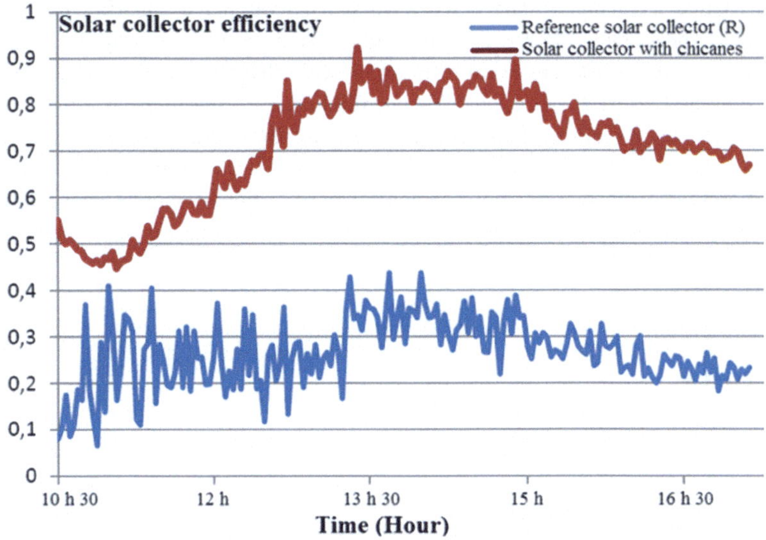

Fig. 2.6 Comparison of the efficiency of the two solar collectors

For these operating conditions, the performances of the solar collector are showed on Figs. 2.4, 2.5 and 2.6. It appears from the figures that the modified solar collector with chicanes presents a better global efficiency due to the higher gradient temperature level evolution as explained by Eq. (2.1). Hence, chicanes have greatly improved the efficiency of the solar collector because they limit the convection effect by various confinements. The efficiency increased with 50%, in Fig. 2.6. Consequently, it seems than the increase in the number of obstacles would intensify the solar collector improvement. An optimization work will be numerically performed and then experimentally validated to determine the best chicane configuration to add in the air free space of the solar collector. This work is currently under elaboration.

2.4 Conclusion

The conception of the solar collector and then the thermomechanical characterization of the heat absorber used, the $HDPE_{BIO}$, allowed us to undertake two studies, the first to treat the energy aspect of the solar collector and then the second considers the aging of our material Heat absorber. The results obtained from the energy study on the solar collector show thickness effect of the air space on the flow regime; the latter may be stationary or unsteady. Thus, the increase in the thickness favors the intensification of the convection and therefore the thermal losses by the front face. The introduction of the partitions in the cavity allowed the reduction of the convective exchanges. The tests were carried out by evaluating the thermal behavior of the solar collector as a function of the external conditions (radiance, temperature, …) and of course, the flow rate of the heat transfer fluid (water). The results obtained show the influence of the thickness on the thermal efficiency and the need to find an optimum thickness for a better yield. Indeed, the more the thickness increases, the more intense the convection becomes, which leads to forward losses of the sensor and a reduction in the efficiency. The introduction of obstacles contributes to the reduction of convection in the air space and allows improving the efficiency of the sensor.

References

1. J.A. Duffie, W.A. Beckman, *Solar Engineering of Thermal Processes*, 3rd edn. (Wiley, Hoboken, 2006)
2. M.C. Rodriguez, Flat plate thermal solar collector efficiency: transient behavior under working conditions part II: model application and design contributions. Appl. Therm. Eng. **31**, 2385–2393 (2011)
3. M.C. Rodríguez-Hidalgo, P.A. Rodríguez-Aumente, A. Lecuona, G.L. Gutiérrez-Urueta, R. Ventas, Flat plate thermal solar collector efficiency: transient behavior under working conditions. Part I: model description and experimental validation. Appl. Therm. Eng. **31**, 2394–2404 (2011)
4. N. Molero Villar, J.M. Cejudo Lopez, F.D. Dominguez Munoz, A. Rodriguez Garcia, A. Carrillo Andres, Numerical 3-D heat flux simulations on flat plate solar collectors. Sol. Energy **83**, 1086–1092 (2009)
5. M.A. Minn, K.C. Ng, W.H. Khong, T. Melvin, A distributed model for a tedlar-foil flat plate solar collector. Renewable Energy **27**, 507–523 (2002)
6. T. Letz, M. Lallemand, Etude théorique et expérimentale d'un capteur solaire plan à air en régime dynamique. Revue de Physique Appliquée **21**, 727–734 (1986)
7. A. Dang, K.J. Shama, Performance of flat plate solar collectors in off-south orientation in India. Energy Convers. Manag. **23**, 125–130 (1983)
8. K.G.T. Hollands, L. Konicek, Experimental study of the stability of differentially heated inclined air layers. Int. J. Heat Mass Transfer. **16**, 1467–1476 (1973)
9. M. D'Antoni, O. Saro, Massive solar-thermal collectors: a critical literature review. Renew. Sustain. Energy Rev. **16**(6), 3666–3679 (2012)
10. N.S. Suresh, B.S. Rao, Solar energy for process heating: a case study of select Indian industries. J. Clean. Prod. **151**, 439–451 (2017)
11. R. Ihaddadene, N. Ihaddadene, N. Madani, F.Z. Hamdibacha, The effects of volumetric flow rate and inclination angle on the performance of a solar thermalcollector. Energy Convers. Manag. **78**, 931–937 (2014)
12. A.E. Kabeel, A. Khalil, S.S. Elsayed, A.M. Alatyar, Modified mathematical model for evaluating the performance of water-in-glass evacuated tube solar collector considering tube shading effect. Energy **89**, 24–34 (2015)
13. A. Guedira, A. Benallou, Procédure de qualification des capteurs plans sous ensoleillement naturel au Maroc. Revue des Energies Renouvelables : Journées de Thermique JITH, 79–84 (2001)

Chapter 3
Experimental Contribution to the Design of a Microprocessor Cooling System by Thermoelectric Module

Abdelillah Abed Belarbi and M'hamed Beriache

Abstract In this study, experiments are performed to investigate the coupling of thermoelectric module and rectangular fin heat sink subjected to an impacting air jet for cooling desktop microprocessors. A controlled thermoelectric test system was conceived and performed for this purpose. The control of the thermoelectric forced air combined cooling system was designed on the basis of electronic Arduino card. The proposed thermoelectric forced convection cooling combined system was compared with the conventional forced air cooling technique. Three electrical powers for the heat source (CPU) were adopted and compared in this experimental study: 60, 87 and 95 W. Performance of thermoelectric cooling module with three preset temperature was experimentally investigated below diverse working conditions. Effects of thermoelectric input current and air jet velocity on the case temperature (T_{case}) were analysed. The thermoelectric cooler had a considerable effect on the cooling of the CPU. However, the consumption of the energy was also augmented. Experimental results indicated that the cooling effect improved with increasing of thermoelectric operating current. However, Temperature of the heat source increased with high power input of the CPU. For a power input of 95 W of the CPU, the T_{case} was maintained under 50 °C with thermoelectric input power of 45 and 5.8 W for the fan, giving improvement around 15% comparing to conventional forced air cooling.

Keywords Thermoelectric · Heat sink · Impacting air jet · Electronics cooling

A. A. Belarbi (✉) · M. Beriache
Laboratory of Rheology and Mechanics, Department of Mechanical Engineering,
Faculty of Technology, Hassiba Benbouali University of Chlef, B.P 78C,
02180 Ouled Fares Chlef, Algeria
e-mail: a.belarbi@univ-chlef.dz

A. Belasri and S. A. Beldjilali (eds.), *ICREEC 2019*, Springer Proceedings in Energy,
https://doi.org/10.1007/978-981-15-5444-5_3

3.1 Introduction

Modern progresses in the electronic manufacturing industry have resulted in a brusque augment in power dissipation which led to severe working conditions, especially for microprocessor Desktop. To function properly, the microprocessor must have an operating case temperature below than a working boundary limited by the manufacturer. The increase in the operating frequency and the explosion of miniaturization projects has caused several problems due to the considerable increase in the heat flux generated by the electronic components. If the heat is not dissipated in a timely manner, the case temperature will be higher than the maximum operating temperature mentioned by the manufacturer, so the thermal control becomes more and more complex. investigation of the thermal performance of CPUs is now a key part of the design of electronic functions subjected to a severe environment. Overheating of the CPU decrease its performance and its lifespan also can provoke failure. Necessity to enhance the cooling systems of CPU has expanded research on heat transfer afterwards. Cooling the microprocessor by Thermoelectric module have benefit over standard cooling systems, counting lightest, small dimension, great dependability, no mechanical process, powered by direct current [1]. Heat sink coupled with thermoelectric cooler could be one appropriate technique to improve the cooling on CPU [2]. Researches on enhance performance of thermoelectric cooling system of CPUs Were reported in several publications. Chein and Huang [3] studied thermoelectric cooler used in the electronic cooling, results showed that the use of a heat sink with air and water as a cooler enhance cooling performances of thermoelectric cooler. Naphon et al. [4] carry out a study on cooling with liquid and rectangular fin heat sink with and without thermoelectric module for cooling CPU. The parameters taken into consideration during this study are: manufacturing materials of heat sink, width of the heat sink channel, flow rate of the liquid coolant, run condition of PC. The results obtained reveal that the cooling with thermoelectric module has an important result on cooling CPU of PC. However, the consumption of energy increases. Chang et al. [5] and Huang et al. [6] designed two thermoelectric cooling systems with air and with water respectively for cooling electronic components. The efficient working parameters determined were the supply current and cooling capacity. Gould et al. [7] designed and constructed a thermoelectric test system for cooling in a standard personal computer for different computer executing applications. Results revealed that effect of the type of heat sink greatly affects the overall cooling performances of the thermoelectric system.

This study presents a comprehensive experimental investigation into the benefits of cooling system with thermoelectric module cooler and heat sink designed for cooling desktop central processing unit (CPU). In a first time, effects of thermoelectric input current and air jet velocity on the case temperature are studied. In second time, a thermoelectric cooling control system is designed. The Arduino card is used to control the input current of the thermoelectric module cooler and the jet velocity in the aim to enhance the cooling performance of the cooling system and optimize electrical energy consumption.

3.2 Experimental Equipment

Figure 3.1 illustrate the experimental test bench designed and constructed for the purpose to study under steady state the aerodynamic and the thermal and performances of the heat sink submitted to an impinging air jet intended for cooling of the desktop CPUs. The bench is appropriate for studying CPU cooling performance with and without thermoelectric cooling module (TEC). The experimental test rig consists of a heat sink and experimental equipments intended to measure the temperature and the air jet velocity. The impacting air jet (1) contains an axial fan 70 mm × 70 mm × 15 mm, type 12VDC 0.7 A, delivering 42 m^3/h of air, and air jet diameter D = 66 mm, powered by a DC voltage source (2). A Plexiglas duct (3) is used to route the airflow from the fan to the heat sink. The heat sink (4) with plate fin and mini channel used in this study is made of pure extruded aluminum (Fig. 3.2), density of the heat sink $\rho = 2702$ kg/m^3 and its thermal conductivity around 237 W/m K. The dimensions of the heat sink are illustrating in Table 3.1 and Fig. 3.2a below.

The dimensions of the heat source used in the present study (5) are 35 mm × 35 mm × 15 mm, emitting a three different power of 60, 87 and 95 W simulating the central processing unit (CPU), attached on the heat sink base by its top side and bounded by insulation on the other sides, to minimize thermal losses. A thermal paste is used to ensure good contact (Fig. 3.2). The air jet is focused over the heat sink.

A thermoelectric cooling module type TEC-12706, the dimension of which is 40 mm × 40 mm × 3.8 mm, mounted in-between the heat source and heat sink in the aim to assess the efficiency of thermoelectric cooling solution. The thermoelectric module is powered from 0 to 4 A with a step of 1 A. The cold surface of the thermoelectric module is that fixed to the heat source, and the hot one is that attached to the heat sink (Fig. 3.2). The heat source temperature is considered the same temperature of the cold surface of the thermoelectric module. It is referred to as the upper surface attached to the cold surface of the thermoelectric cooler and

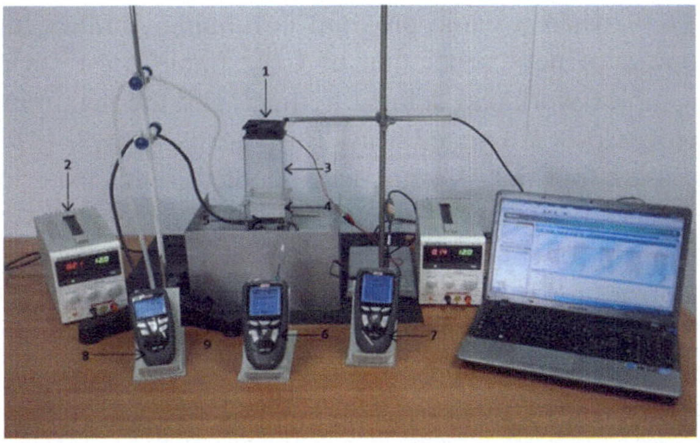

Fig. 3.1 Experimental test bench

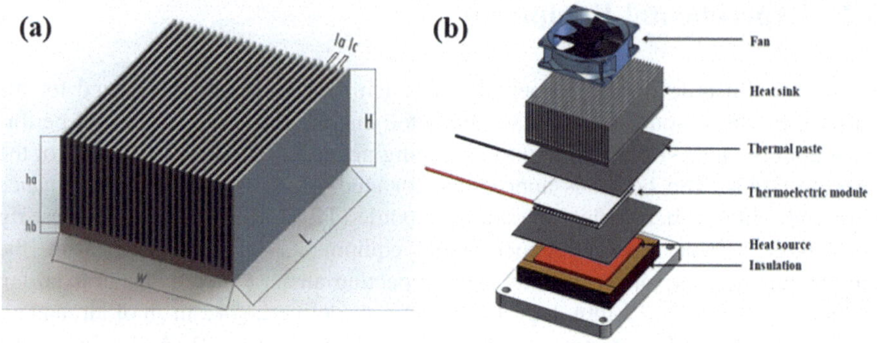

Fig. 3.2 **a** Heat sink. **b** Assemblage of heat sink, fan with the thermoelectric cooling module on the heat source (CPU)

Table 3.1 Heat sink dimensions

N	W (mm)	ha (mm)	L (mm)	h_b (mm)	l_c (mm)	l_a (mm)	H (mm)
27	66	32	82	4	1.5	1	36

designed in order to allows the fixation of the thermocouple, and then measure the temperature (T_{case}).

3.3 Experimental Methodology

Using heat sink with airflow impingement coupled with thermoelectric module (TEC) has been considered in the purpose to improve cooling performances of desktop computer CPU. Referring to [7], electrical power used by the computer under steady state conditions is approximately 60 W, this electrical power rise to 87 W when a game program is running. Further, the maximum heat dissipation power of new generation of CPU Intel-Core-i7 is around 95 W under a highly complex workload defined by Intel, and the maximum temperature allowed on the processor's integrated heat distributor is 72.2 °C [8]. On this basis, three electrical powers for the heat source are adopted in this experimental study, to know: 60, 87 and 95 W. In this study, several mean velocities are indeed adopted, under a laminar flow in the heat sink channels, ranging from 1 to 5 m/s. The air jet velocity is measured a hot wire anemometer probe moving perpendicularly (7) displaying ±0.3% of accuracy. The cold surface of the thermoelectric module was attached to the heat source, and the hot surface attached to the base of the heat sink. Input current of the thermoelectric module was varied from 0.0 A to 4 A by 1 A steps. The temperatures are measured using K-type thermocouples Ø 0.2 mm, connected to a thermometer Digital converter (6) with an accuracy of ±0.4%. The heat source is designed in order to penetrate the thermocouple to the centre of the surface of the heat source in order to measure the case temperature, which define the temperature of the chassis designated by the manufacturer as the maximum permitted temperature on processor. The hot surface of the heat source is taken to be the identical temperature as the cold surface of thermoelectric module. The tests lead on the coupled cooling system have been conducted using the Arduino electronic card, in order to realize a temperature control system for maintaining the temperature of the CPU at a preset level of 30 °C, 40 °C and 50 °C, according to the power consumption of the CPU. A temperature sensor LM35 placed on the heat source is connected to the Arduino card is used to measure the temperature.

3.4 Results and Discussion

The T_{case} is measured for three different heat sources 60, 87 and 95 W, in a first stage the test bench is implemented without a thermoelectric module, the T_{case} obtained is that of the direct contact between the heat sink base and the heat source. In a second time, the thermoelectric module (TEC) is mounted between the base of the heat sink and the heat source by varying the velocity of air jet then thermoelectric input current. Figure 3.3 highlights the effect of the input current of thermoelectric module and the impinging air jet on the CPU temperature (T_{case}). Results are plotted with respect to a reference critical temperature of 72.2 °C. Thermoelectric input current is varied from 1 A to 4 A with a step of 1 A. Figure 3.3a illustrates the CPU temperature (T_{case}) variations for a 60 W heat source power simulating steady state of a CPU. The (T_{case}) temperature decreases when the air jet velocity increases, further, the increase of thermoelectric input current until 3 A implying decrease in the T_{case} around 15.7 °C margining capability comparing to cooling without TEC. Typically the T_{case} decreases below ambient temperature for a jet velocity superior than 4 m/s, the enhancement is around 38%. In Fig. 3.3b, for a heat source of 87 W simulating power consumption of a CPU when game programme is running, the increase of thermoelectric input current up to 4A decreases the T_{case} giving a 16% improvement. For Fig. 3.3c, the heat source dissipates a power of 95 W simulating the maximum consumption of an Intel-Core-i7. Cooling by TEC provides 6 °C wider margining capability.

The results for the thermoelectric cooling control system are shown in Fig. 3.4. For heat source of 60 W, the case temperature can be maintained at approximately 30 °C, with a power consumption of thermoelectric module around 22 and 2.8 W for the fan. For a set-point of 40 °C, the input power of the thermoelectric module is typically 14 and 0.68 W for the fan. For 50 °C, the input power is around 4 and 0.68 W for the fan. The improvement is around 32% comparing to cooling without TEC.

For heat source of 87 W, and set-point of 30 °C, the case temperature (Tc_{ase}) can be maintained at 36.2 °C with thermoelectric input power of 42 and 5.8 W for the

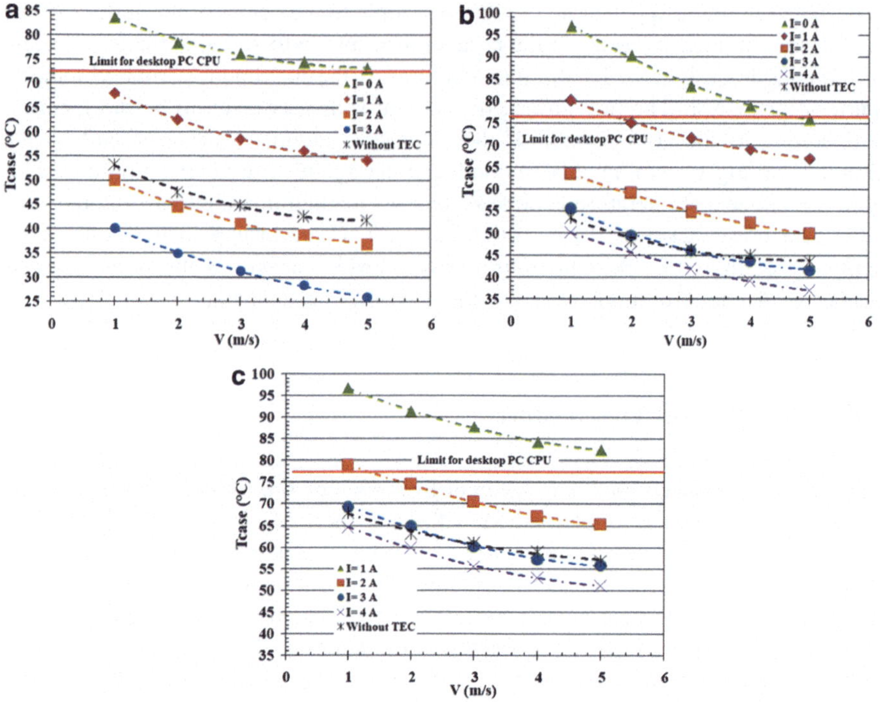

Fig. 3.3 Effect of thermoelectric input current on heat source temperature **a** Q_{elec} = 60 W, **b** Q_{elec} = 87 W, **c** Q_{elec} = 95 W

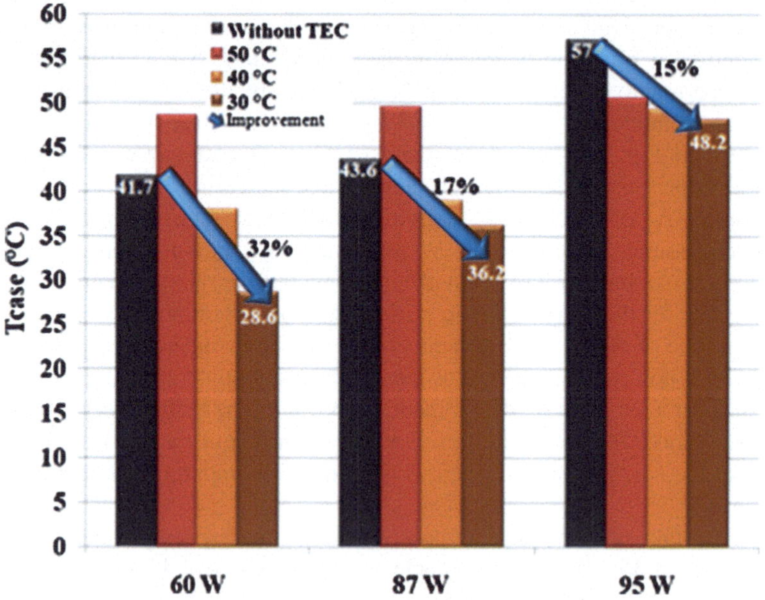

Fig. 3.4 Case temperature based thermoelectric cooling control system

fan, the case temperature can be maintained for a set-point of 40 °C for a thermoelectric input power of 26.1 and 2.3 W for the fan, also for a set point of 50 °C thermoelectric input power of 10.2 and 5.8 W for the fan. The improvement is around 17% comparing to without TEC. For heat source of 95 W, the case temperature can be maintained only for a set point of 50 °C with thermoelectric input power of 45 and 5.8 W for the fan, the case temperature can be decrease at 48 °C. The improvement is around 15% comparing to that without TEC.

3.5 Conclusion

In this article, experimental investigation on coupling of thermoelectric module and a rectangular plate fins heat sink with mini channel subjected to an impingement air jet for desktop CPU have been conducted by the design of a thermoelectric cooling control system by mean of electronic Arduino card. The results obtained with and without thermoelectric cooling are compared. Experimental results have indicated that both CPU temperature and heat sink thermal resistance are significantly improved according to 60, 87 and 95 W heat source power respectively by at least 31%, 17% and 15% when thermoelectric cooling system is installed, compared to that of the air impingement jet. This increase is more significant when the TEC's input courant is higher and less significant for low input courant.

References

1. D.L. Zhao, G. Tan, A review of thermoelectric cooling: materials, modelling and applications. Appl. Therm. Eng. **66**(1–2), 15–24 (2014)
2. S. Twaha, J. Zhu, Y. Yan, B. Li, A comprehensive review of thermoelectric technology: materials, applications, modelling and performance improvement. Renew. Sustain. Energy Rev. **65**, 698–726 (2016)
3. R. Chein, G. Huang, Thermoelectric cooler application in electronic cooling. Appl. Therm. Eng. **24**, 2207–2217 (2004)
4. P. Naphon, S. Wiriyasart, Liquid cooling in the mini-rectangular fin heat sink with and without thermoelectric for CPU. Int. Commun. Heat Mass Transfer **36**, 166–171 (2009)
5. Y.-W. Chang, C.-C. Chang, M.-T. Ke, S.-L. Chen, Thermoelectric air-cooling module for electronic devices. Appl. Therm. Eng. **29**, 2731–2737 (2009)
6. H.-S. Huang, Y.-C. Weng, Y.-W. Chang, S.-L. Chen, M.-T. Ke, Thermoelectric water-cooling device applied to electronic equipment. Int. Commun. Heat Mass Transfer **37**, 140–146 (2010)
7. C.A. Gould, N.Y.A. Shammas, S. Grainger, I. Taylor, Thermoelectric cooling of microelectronic circuits and waste heat electrical power generation in a desktop personal computer. Mater. Sci. Eng. B **176**, 316–325 (2011)
8. https://ark.intel.com/fr/products/41315/Intel-Core-i7-870-Processor-8M-Cache-2_93-GHz

Chapter 4
Experimental Study of a Thermoelectric Refrigerator

M. M. Hadjiat, K. Salhi, S. Ouali, A. Ait-Ouali, M. Hazmoune, and K. Imessad

Abstract In this work, a thermoelectric refrigerator is experimented at economic mode working on reduced voltage. The aim of this study is show the performance of a refrigerator alimented with tension lower than 12 V. In this study, we realized a thermoelectric refrigerator in the center of renewable energy CDER at Algiers. Multiple test was done regarding the optimization of energy consumption and achievement of lower temperature possible. The thermoelectric module TEC1-12706 is used for cooling. This TE module can achieve the maximum of efficiency when it function with a tension of 15 V. In our experimental study, we aliment the TE module with a tension of 9 V, the result show that the temperature of 0.5 L in the refrigerator was reduced from 26 to 18 in two hours and the coefficient of performance COP is equal to 0.45. In this work, a thermoelectric refrigerator was manufactured and tested in the laboratory. The experimental study shows that its efficiency is relatively acceptable and that the prototype can operate even at low voltage. At a voltage of 9 V, the COP remains above 0.4 and the temperature in the cooler box is slightly below 14 °C.

Keywords Thermoelectric · COP · Refrigeration · Efficiency

4.1 Introduction

Nowadays, refreshments are no longer a luxury but a necessity. In fact, greenhouse gas emissions increase from one year to the next. As a result, the natural climatic equilibrium is modified and the climate is readjusted by a warming of the surface of the earth. So unusual heat waves have appeared all over the world, killing dozens of people and achieving record temperatures. In medical field, it is urgent to find a new refrigeration system, which facilitates the transport of medicines, especially in desert areas and allows precise control of the interior temperature. Currently, most

M. M. Hadjiat (✉) · K. Salhi · S. Ouali · A. Ait-Ouali · M. Hazmoune · K. Imessad
Centre de Développement des Energies Renouvelables, CDER, 16340 Algiers, Algeria
e-mail: mhadjiat@gmail.com

A. Belasri and S. A. Beldjilali (eds.), *ICREEC 2019*, Springer Proceedings in Energy,
https://doi.org/10.1007/978-981-15-5444-5_4

refrigerators use refrigerant compression systems, e.g. domestic refrigerators. Undoubtedly, the compression system has a relatively high COP (coefficient of performance), however, the system is heavy and bulky, it requires a lot of space, and it consumes a lot of energy and does not support displacement. Therefore, the compression system is not suitable for medical needs. Currently cooling by thermoelectricity represents a promising alternative to the compression system and it is perhaps the best possible choice. Indeed, the researchers have made enormous progress in the field of cooling by thermoelectricity. The thermoelectric cooler is a recent refrigeration technology based on the Peltier effect. Recently, low cost thermoelectric cells have been rapidly developed by the industry and are produced in several sizes. Although the efficiency of the thermoelectric module is relatively low compared to traditional refrigeration systems, thermoelectric cooling remains irreplaceable in certain applications. The cooling system by thermoelectricity has several advantages; it is a silent device, does not consume a lot of energy therefore environmentally friendly, it allows controlling the temperature with better precision and offers the possibility of product transport medical or food in difficult conditions. Therefore, it is necessary to use this device for cooling and in the medical field. Abdul-Wahab et al. [1] designed and studied a thermoelectric solar refrigerator for desert areas where there is no electricity. They showed that the interior temperature of the cooler dropped from 27 to 5 °C in 44 min, but the COP was approximately 0.16. Vián and Astrain [2] have improved a heat exchanger for the cold side of Peltier thermoelectric modules based on the principle of a phase change thermosiphon with capillary action. The experimental study has shown that the thermal resistance of the device has been improved by 37% and that the COP has increased by 32% (from 0.297 to 0.393). Astrain et al. [3] presents the results of an experimental study of certain heat exchangers for the hot side of the thermoelectric module: a water-air system, a finned heat sink with fan; a pump and fan coil and a heat pipe with fan. This experience has shown that high COP values can be reached for thermoelectric refrigerators if the heat exchangers are optimized correctly, thus exceeding 0.7 for COP. Mirmanto et al. [4] study the cooler box by varying the positions of the thermoelectric module as follows: above, below and on the wall of the cooler. The results showed that the maximum COP is obtained when the position of the thermoelectric module is on the wall. In this paper, the study of the performance of Peltier cells at low voltage is presented. The thermoelectric module used for cooling is TEC1-12706. The maximum efficiency of this module is at 12 V, in our experimental study, we used 9 V in alimentation, the aim is to experiment the refrigerator at economic mode, when there is not sufficient power to aliment the refrigerator.

4.2 Experimental Setup

4.2.1 Prototype Description

The experimental prototype is shown in Fig. 4.1. It is composed of a cooler box and a heatsink-fan. The cooler box was made of Polystyrene with a thickness of 40 mm. The inner cooler box size was 20 cm × 20 cm × 25 cm. All temperatures were measured using K-type thermocouples. The voltage and current supplied to the thermoelectric module were measured with a multimeter (Professional Extech EX542 Multimeter model).

The thermoelectric module used was TEC1-12706, see Fig. 4.2, with a power of 38.08 W. The power of the fan installed on the external heat sink was 0.712 W. An ExtechSDL200 datalogger is used to record the experimental data. the duration of the data recording taken into account is 8 h, a container containing 0.5 L of water has been installed in the space of the cooler as an additional cooling charge. The temperatures measured were, ambient temperature, water temperatures, thermoelectric temperature on the hot side, thermoelectric temperature on the cold side and the space temperature of the cooler. The characteristics of the thermoelectric module used in the experiment are provided in Table 4.1.

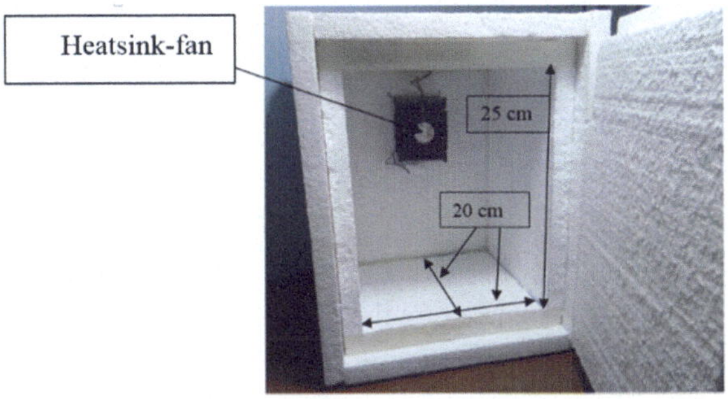

Fig. 4.1 Experimental prototype

Fig. 4.2 Peltier cell TEC1-12706

Table 4.1 Characteristics of TEC1-12706

Type	N	I_{max} (A)	V_{max} (V)	Q_{max} (W)	ΔT (K)
TEC1-12706	127	6	15.2	56.5	68

4.2.2 *Fundamental Equations of TE Module*

The behavior of a thermoelectric couple is determined by three essential parameters: K the thermal resistance, R the electrical resistance and α Seebeck coefficient. Knowing that in a module there is N pair and assuming that all couples are the same, the parameters of a Peltier module are:

$$\alpha_m = \alpha N \tag{4.1}$$

$$R_m = RN \tag{4.2}$$

$$K_m = KN \tag{4.3}$$

However, these coefficients can be calculated using parameters Vmax, Imax and ΔTmax which are given by the manufacturer in Table 4.1.

$$\alpha_m = \frac{V_{max}}{T_h} \tag{4.4}$$

$$R_m = \frac{V_{max}(T_h - \Delta T_{max})}{I_{max} T_h} \tag{4.5}$$

$$\frac{1}{K_m} = \frac{\Delta T_{max}}{I_{max} V_{max}} \frac{2T_h}{(T_h - \Delta T_{max})} \tag{4.6}$$

The performance of a thermoelectric material is called the figure of merit (Z) and is expressed as

$$Z = \frac{\alpha_m^2}{K_m R_m} \tag{4.7}$$

TE is characterized by numerous performance expressions, including heat absorbed on the cold side, heat rejected on the hot side. Expressions of heat flow through the cold and hot junctions for N semiconductor thermocouples can therefore be expressed as follows:

$$Q_c = \alpha_m I T_c + \frac{1}{2} R_m I^2 + K_m (T_h - T_c) \tag{4.8}$$

$$Q_h = \alpha_m I T_h - \frac{1}{2} R_m I^2 + K_m (T_h - T_c) \quad (4.9)$$

The experimental COP of a TE module is expressed as:

$$COP = \frac{Q_c}{P} \quad (4.10)$$

where P is, the electric power supplied to the thermoelectric (W) expressed as:

$$P = VI = \alpha_m I (T_h - T_c) + I^2 R_m \quad (4.11)$$

I is the current (A) and V is the voltage (V). Moreover, the COP of Carnot (COPc) is expressed in this study as:

$$COP_c = \frac{T_c}{T_h - T_c} \quad (4.12)$$

4.3 Result and Discussion

The results obtained from experimental study are illustrated in the figures below. Figure 4.3 show thermal behavior of cooler box room. The temperatures Tin, Tw and Tc decreases strongly and then gets stable. In (a) it show the evolution of water temperature, it decrease from 26 to 16 °C in three hours whereas the ambient temperature is 28.5 °C. In (b) it show the temperature profile of the hot side Th and the cold side Tc of the thermoelectric module, we can note that the difference ΔT = Th − Tc is equal to 30 °C.

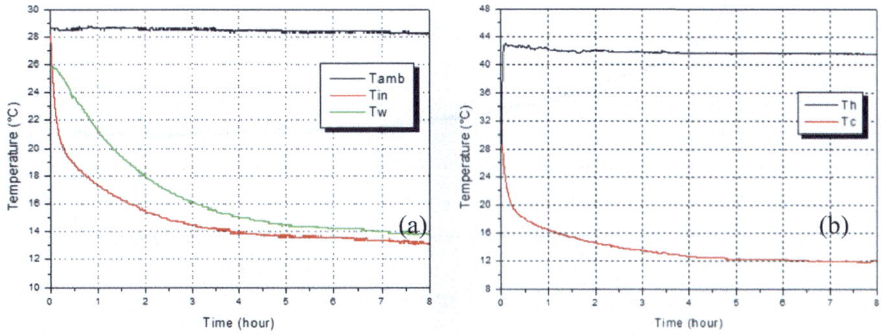

Fig. 4.3 Temperature variation: in **a** of water (Tw), interior of refrigerator (Tin) and ambient Temperature (Tamb). In **b** of hot side (Th) and cold side (Tc) of TE module

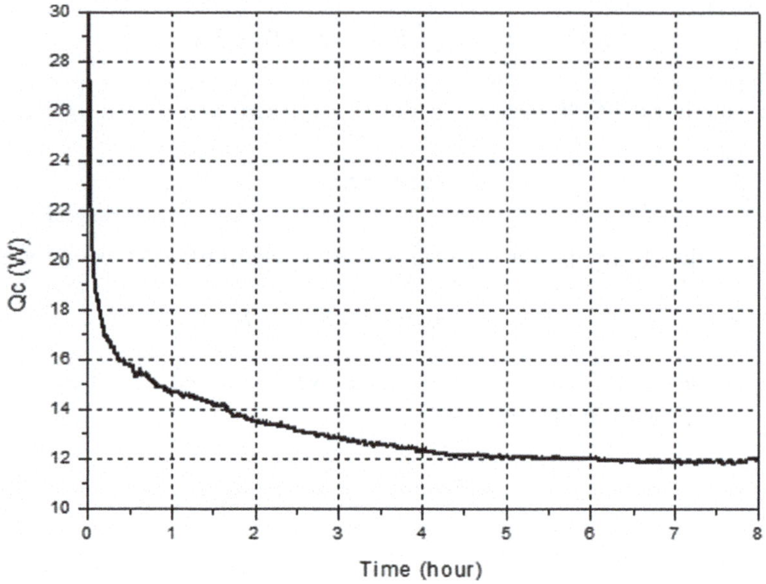

Fig. 4.4 Variation of Qc

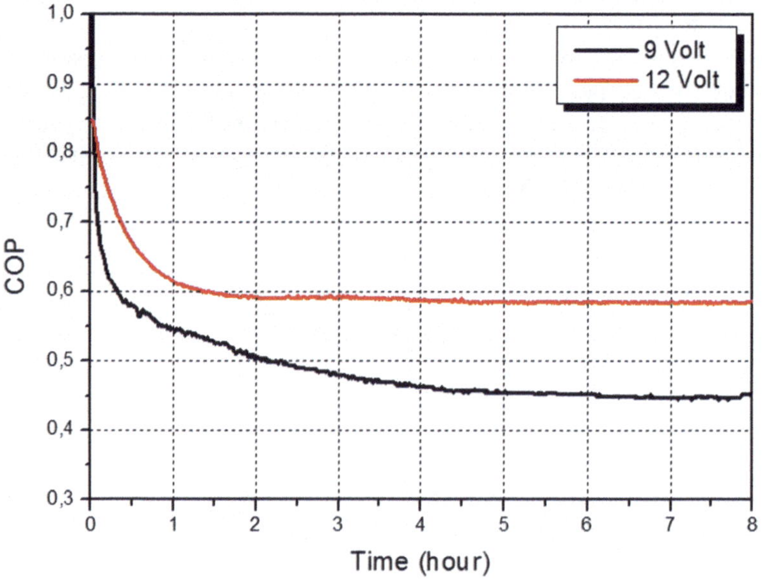

Fig. 4.5 Variation of experimental COP

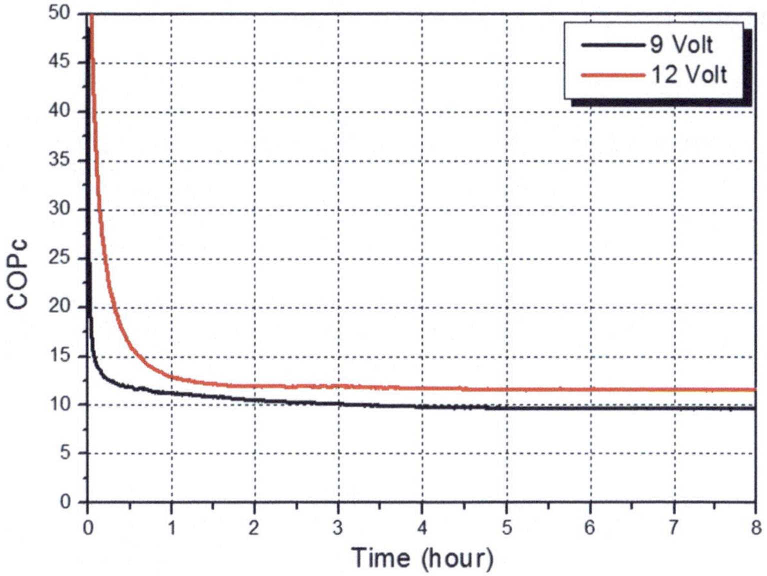

Fig. 4.6 Variation of Carnot COPc

Figure 4.4 show the cooling capacity Qc given by Eq. (4.8). Figures 4.5 and 4.6 show the experimental COP and Carnot COP respectively for the prototype alimented with 9 V and 12 V, The comparison between the two curves (9 V and 12 V) shows that the performance of the device decreases in decreasing voltage. The figure of merit is $Z = 2.2 \times 10^{-3} K^{-1}$.

4.4 Conclusion

In this work, a thermoelectric refrigerator was manufactured and tested in the laboratory. The experimental study shows that its efficiency is relatively acceptable and that the prototype can operate even at low voltage. At a voltage of 9 V, the COP remains above 0.4 and the temperature in the cooler box is slightly below 14 °C.

References

1. S.A. Abdul-Wahab, A. Elkamel, A.M. Al-Damkhi, I.A. Al-Habsi, H.S. Al-Rubai'ey, A.K. Al-Battashi, A.R. Al-Tamimi, K.H. Al-Mamari, M.U. Chutani, Design and experimental investigation of the portable solar thermoelectric refrigerator. Renew. Energy **34**, 30–34 (2009)
2. J.G. Vián, D. Astrain, Development of a thermoelectric refrigerator with twophase thermosyphons and capillary lift. Appl. Therm. Eng. **29**, 1935–1940 (2009)
3. D. Astrain, P. Aranguren, A. Martínez, A. Rodríguez, M.G. Pérez, A comparative study of different heat exchange systems in a thermoelectric refrigerator and their influence on the efficiency. Appl. Therm. Eng. **103**, 1289–1298 (2016)
4. M. Mirmanto, S. Syahrul, Y. Wirdan, Experimental performances of a thermoelectric cooler box with thermoelectric position variations. Eng. Sci. Technol. Int. J. **22**, 177–184 (2019)

Chapter 5
A Numerical Study of the Effect of Natural Ventilation Displacement on the Buoyant Thermal Plume Evolution in an Enclosure

Mokhtar Khrissat, Yassine Laghouati, Hamid Oualli, and Ahcène Bouabdallah

Abstract This numerical study deals with the evolution of a buoyant axisymmetric thermal plume issued from localized source in enclosure. The geometry configuration and the physical conditions in the confined space are the same as those chosen by Abdalla et al. (Eng. Appl. Comput. Fluid Mech. 3(4):608–630, [1]). It is aimed to analyze the influence of the natural displacement on development of thermal buoyant plume in a room connected to the exterior via high and low openings. It is sought to find the critical height at which stratification disappears when varying the airflow rate value. The buoyancy driven natural ventilation in an enclosure is modeled using the ICEM-CFD software11. The numerical results are obtained for two turbulence models based on Reynolds-Averaged Navier-Stokes (RANS): standard k-ε model and SST model. The results consistency is checked by comparisons to experimental and theoretical data. It is found that before equilibrium, ventilation velocity causes a transitional and turbulent mixing. The steady state is reached when air stratification is fully established leading to interface appearance separating the dense fresh air from the light warm buoyant air in lower and upper layers respectively. Furthermore, it is noted that the interface height is independent from the thermal source strength but depends strongly on the supplying fresh air velocity.

Keywords Heat source · Buoyant thermal plume · Natural ventilation displacement · Confined area · Turbulent RANS model

---

M. Khrissat · Y. Laghouati (✉)
Laboratory LAMOSI, University of Sciences and Technology of Oran USTO MB, PB 1505 El Mnaouar, 31000 Oran, Algeria
e-mail: laghouatiy@yahoo.fr

H. Oualli · A. Bouabdallah
Laboratory of Thermodynamics and Energetical Systems, Faculty of Physics, USTHB, PB 32 El Alia, Bab Ezzouar, 16111 Algiers, Algeria

A. Belasri and S. A. Beldjilali (eds.), *ICREEC 2019*, Springer Proceedings in Energy,
https://doi.org/10.1007/978-981-15-5444-5_5

5.1 Introduction

The buoyant plumes in both uniform and stratified media and particularly fire plumes are often encountered in a large variety of natural and built environment.

Thermal buoyant plumes have been widely studied to understand their evolution using analytical and empirical models. Among references in literature, we can cite Hunt and van Der Bremen [2], where an extensive review of theoretical and experimental works carried out on turbulent buoyant plumes is presented.

The most important research work of Morton et al. (MTT) [3] remains up to date a reference work describing the dynamics of turbulent buoyant plumes growing from the point and the line sources into homogeneous and stably stratified medium. Baines and Turner [4] confirmed experimentally MTT theory research.

In order to understand the draining flows mechanism by natural ventilation in a space containing a buoyant fluid, Linden et al. [5], elaborated experiences on emptying-filling-box concept based upon the buoyancy-driven displacement ventilation model produced by the internal and external fluids (air). They noticed that the interface height between the two hot and fresh fluid layers depends on the source geometry, the openings size and the confined space height. It is found also that the source buoyancy flux has no effect on the interface position while the stratification stability and the plume velocities amplitudes significantly increase if the convective power source is important.

In the light of all these theoretical and experimental works, the main objective is to better understand the behavior of plumes under specific conditions and provide solutions to ensure people safety and health. Particularly, this numerical work is devoted to make assessment of the influence of natural displacement ventilation on the buoyant thermal plume behavior in an enclosure. The working conditions retained are similar to those adopted by Abdalla et al. [1]. In addition, we test the sensitivity limit of the stratification for different velocities of inlet fresh air.

5.2 Theoretical Plume Model and Displacement Ventilation

5.2.1 Plume Model

We consider a thermal Boussinesq axisymmetric plume moving upwardly in a homogeneous confined medium initially at room temperature. This plume is produced by heat source of convective power PC localized at ground level. The plume buoyancy flux B is given by:

$$B = \frac{g\beta P_c}{\rho c_p} \tag{5.1}$$

ρ, β, cp and g designed the plume density, isobaric thermal expansion, specific heat and the gravity.

According to Morton et al. [3], the buoyancy flux, volume flux and reduced gravity are expressed as:

$$B = Q_p.G' \tag{5.2}$$

$$Q_p(y, B) = C(B.y^5)^{1/3} \tag{5.3}$$

$$G'(y, B) = (B^2.y^{-5})^{1/3}/C \tag{5.4}$$

where y denotes the vertical distance of the plume from the source.

The constant $C = \left(\frac{6\alpha}{5}\right)\left[\frac{9\alpha}{10}\right]^{1/3}\pi^{2/3}$ is depending on the entrainment coefficient α.

Linden et al. [5] proposed the experimental value of the parameter α = 0.1 for the model «Top-hat» profiles. The relationships between the «Top-hat» and «Gaussian» quantities (plume radius and vertical velocity) are: $b_T = \sqrt{2}.b_G$, and $V_T = V_G/2$.

5.2.2 *Natural Displacement Ventilation*

The theory of buoyancy driven displacement ventilation has been largely studied by Linden et al. [5], Nielsen [6], Cook and Hunter [7]. The confined space is subject to a natural circulation of ambient air entering through the lower openings at wind velocity UN. The wind pressure drop exerted at the fresh air inlet opening is:

$$\Delta P_{\mathrm{Vent.}} = -(1/2)f.\rho_\infty U_N^2 \tag{5.5}$$

where, f and UN denote respectively the pressure loss coefficient and ventilation velocity normal to the front surface of openings. The ventilation flow rate Q_{vent} can also be calculated from the pressure difference [8]:

$$Q_{\mathrm{Vent.}} = C_D.\frac{C_D a_t a_b}{\left(\frac{1}{2}\left((C_D^2/C_e)a_t^2 + a_b^2\right)\right)^{1/2}}\sqrt{\frac{2\Delta P}{\rho_\infty}} \tag{5.6}$$

where ab and at are respectively the bottom and top openings areas. Ce and C_D are respectively the expansion and discharge coefficients.

5.2.3 *Working Conditions*

Geometrical Configuration. As reported in Fig. 5.1, the model is a rectangular box with L = 1 m, l = 0.5 m and H = 1 m are length, width and height respectively.

The velocity and temperature fields are evaluated at two monitor points P1 and P2 and two vertical stations within the confined space near the walls. The coordinates of the monitor points are represented in Table 5.1.

Conditions of Thermal Source. The problem is approached considering the plume as an incompressible ideal gas. The power convective heat produced from the source is PC = 2.104 W/m^2. The source temperature is fixed at T0 = 400 °C for the all simulation tests. The heat transfer by radiation is neglected. The walls and ceiling are assumed to be adiabatic. The temperature of the ambient air is 18 °C. The simulations are performed for three different ventilation velocities: $U_N = 0.35$ m/s, $U_N = 0.5$ m/s and $U_N = 0.70$ m/s.

Main Initial and Boundary Conditions. The Main Conditions Are:

- Initial conditions: For t = 0, $T = T_0$, $B = B_0$, and $M_0 = 0$.
- The boundary conditions: For y = 0 and y = H, $p = p_0$ (pressure acting on the upper and lower openings).

M_0, B_0 are respectively the momentum and initial buoyancy fluxes.

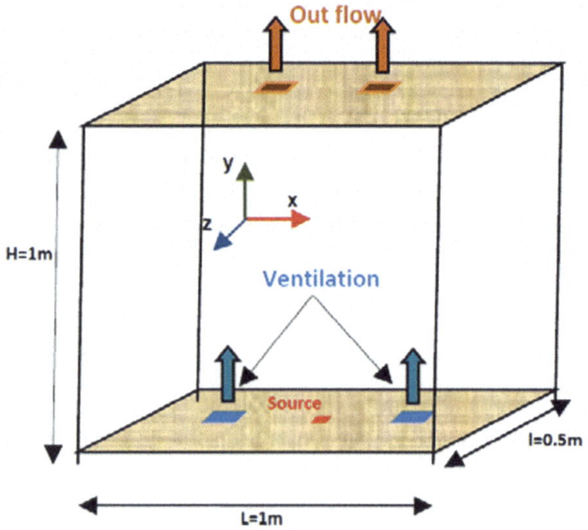

Fig. 5.1 Illustration of the naturally ventilated enclosure

Table 5.1 Monitor points and vertical stations coordinate

Monitor point	X (m)	Y (m)	Z (m)
P1	0.5	0.5	0.25
P2	0.5	0.95	0.25
Vertical station 1	0.975 0.975	0 1	0.245 0.245
Vertical station 2	0.5 0.5	0 1	0.245 0.245

5.3 Numerical Procedure

The software package ICEM-CFD11 is used for the geometrical configuration and grid generation. Spatial discretization of the computational domain into a set of 3D unstructured mesh of 874,408 cells is generated. Simulation is realized with the CFX 11 software [9]. The adopted convergence criterion is about 0.0001%. The time step is fixed to $\Delta t = 0.1$ s, corresponding to optimal parameter CFL values between 0.35 and 0.7.

We consider an Initial turbulent kinetic energy $k = 0.01\ m^2s^{-2}$, turbulent energy dissipation $\varepsilon = 0.01\ m^2s^{-3}$, Turbulent length scale $\ell = 0.03$ m and turbulent intensities at the inlet ventilation and the source are about 10%.

Two turbulence models are tested, namely standard k-ε and SST models. One advantage of this study is relevant to the SST model known to encompass the k-epsilon model away from the walls and the k-omega model near the walls. Predictions of the flow presented in this paper are compared with Abdalla et al. [1] numerical data.

5.4 Results and Discussions

5.4.1 Compared Values for k-ε and SST Models

The convective instability is controlled by the dimensionless Rayleigh number Ra expressed by:

$$Ra = \frac{g\beta H^3 \Delta T}{\kappa \nu} \tag{5.7}$$

where g, β and H denote respectively gravity, isobaric thermal expansion coefficient and the box height respectively. ΔT, κ and ν are temperature difference between hot and fresh ambient layers, thermal diffusivity and the kinematic viscosity of the air.

The calculated values of the Rayleigh number characterizing laminar-turbulent regime transition are between 7.39 × 108 and 1.39 × 109 as shown in Table 5.2.

Table 5.2 Compared values for k-ε and SST models

Velocity Un (m s^{-1})	0.35	0.5	0.7
(SST)k-omega	Ra = 1.39×10^9 h = 0.42 m ΔT = 13.15 °C	Ra = 9.12×10^8 h = 0.59 m ΔT = 9.4 °C	Ra = 7.39×10^8 h = 0.71 m ΔT = 7 °C
Standard k-epsilon	Ra = 1.31×10^9 h = 0.46 m ΔT = 14 °C	Ra = 9.99×10^8 h = 0.63 m ΔT = 10.4 °C	Ra = 8.75×10^8 h = 0.75 m ΔT = 9 °C

5.4.2 Spatiotemporal Evolution of the Velocity and Temperature of the Plume in the Confined Space

As illustrated in Fig. 5.2, a qualitative view of the development of hot plume in the chamber is shown.

Initially the plume has an elongated shape with head and then evolves to a mantle just before becoming a cone.

During the transitional period not exceeding 3 s, the plume front moves upwards to impact the ceiling. After a period, two layers' stratification and un intermediate thick interface are observed.

5.4.3 Distribution of Velocity and Temperature Mean Fields

Vertical velocity and temperature profiles for different dimensionless plume height positions (y/d) are reported in Fig. 5.3.

The rather sharp Gaussian curves of velocity and temperature become progressively flattened with the increasing ratio (y/d). It seems that when moving away from the source, the plume loses its ascent velocity and expands radially. The results show reasonable self-similarity of profiles far from the source as described by MTT [3].

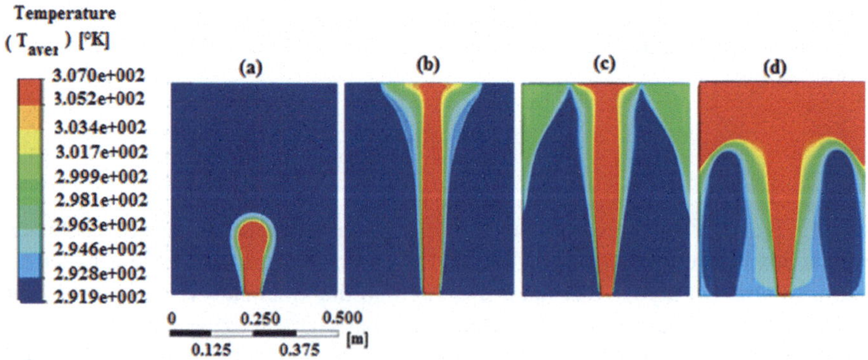

Fig. 5.2 Time sequence of the developed turbulent buoyant plume for U_N = 0.35 m/s (SST model) **a** t = 1 s, **b** t = 3 s, **c** t = 5 s and **d** t = 55 s

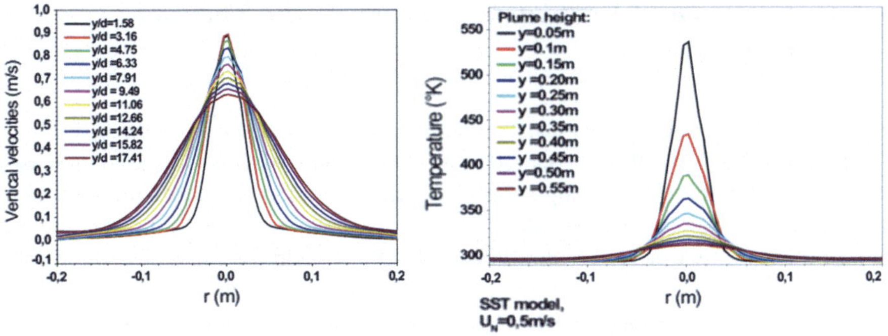

Fig. 5.3 Vertical velocity and temperature profiles for different height positions

Figure 5.4 represents the results of the evolution of the volume flux in the plume versus the ascending plume heights for the ventilation velocity UN = 0.35 m/s (turbulence model SST). Comparing our values with those of Abdalla et al. [1], for the volume flux Q less than 0.002 (m^3 s^{-1}), the precision of calculus is about 4%.

Compared to other experimental and numerical works, our results are in line with those of Linden et al. [5]. The growth of the volume flux of the plume depends on the entrainment parameter α. The assumed value of the entrainment coefficient α is 0.1.

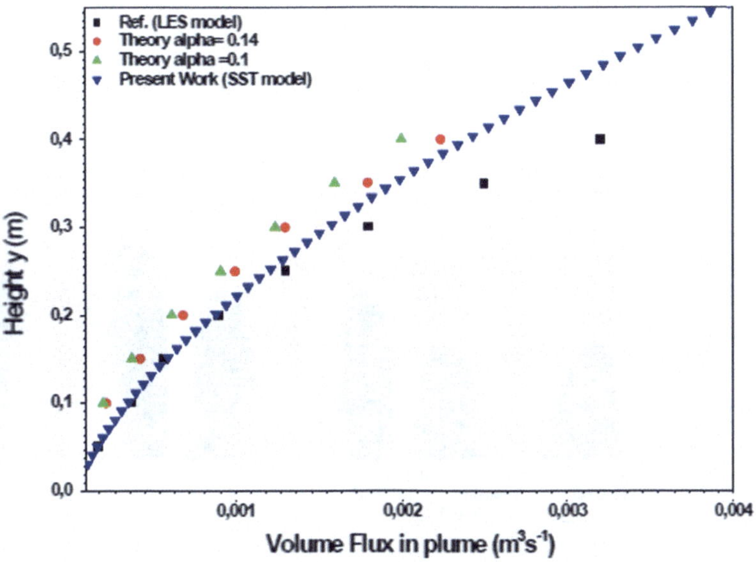

Fig. 5.4 Variation of the volume flux in plume versus the height

5.4.4 *Influence of Ventilation Velocity on the Thermal Interface*

Figure 5.5 shows the contours of temperature in the enclosure for three different ventilation velocities. It is observed that the amount of fresh airflow increases stratification height. For a strong supply air jet, the interface becomes more instable and reaches the ceiling. This result is confirmed by Webster et al. [10], the amount of air brought down increases stratification height. In addition, at higher air supply volumes, the stratification is minimum and the interface reaches the ceiling.

In our case, the interface attains the ceiling for ventilation velocities exceeding the critical velocity $U_N = 0.7$ m/s. The hot layer volume is reduced and the volume of fresh air in the enclosure becomes important and stratification eventually disappears.

Figure 5.6 shows evolution of the interface height for SST and standard k-epsilon models. For the different values of natural ventilation velocity at vertical station 1 near the wall, it shows that the temperature differences between the two layers decrease and consequently the interface height increases if the fresh air intake flow increases.

In the case of SST model and for natural ventilation velocity $U_N = 0.35$ m/s, the affected temperature difference between the two distinct layers and the reached interface height are respectively $\Delta T = 13.9$ °C and h = 0.42 m. However, for the critical velocity $U_N = 0.7$ m/s, the temperature difference is evaluated as $\Delta T = 7$ °C and the interface height becomes h = 0.71 m. Unlike the obtained values with the SST, those of the standard k-epsilon model are overestimated as represented in Fig. 5.6 in the cases of low ventilation rate.

Recently Gilani et al. [11] tested RANS turbulence models for the buoyancy driven displacement ventilation. They confirm that the SST model can accurately predict the temperature distribution and stratification.

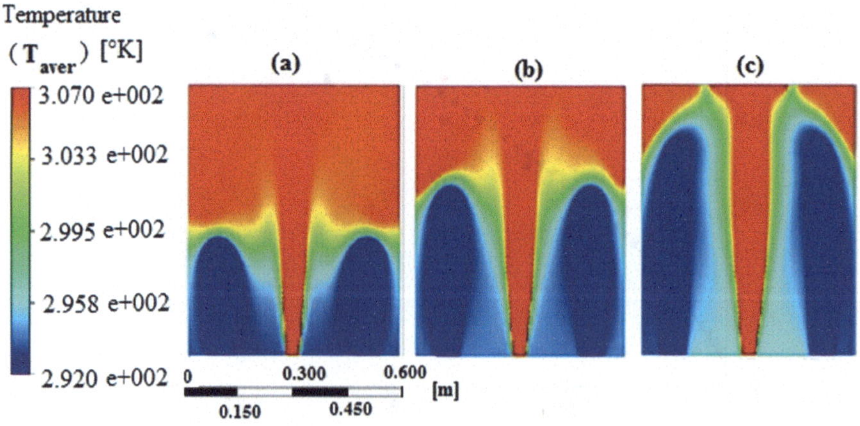

Fig. 5.5 Influence of ventilation velocity on the thermal interface (SST model): **a** $U_N = 0.35$ m s^{-1}, **b** $U_N = 0.5$ m s^{-1} and **c** $U_N = 0.7$ m s^{-1}

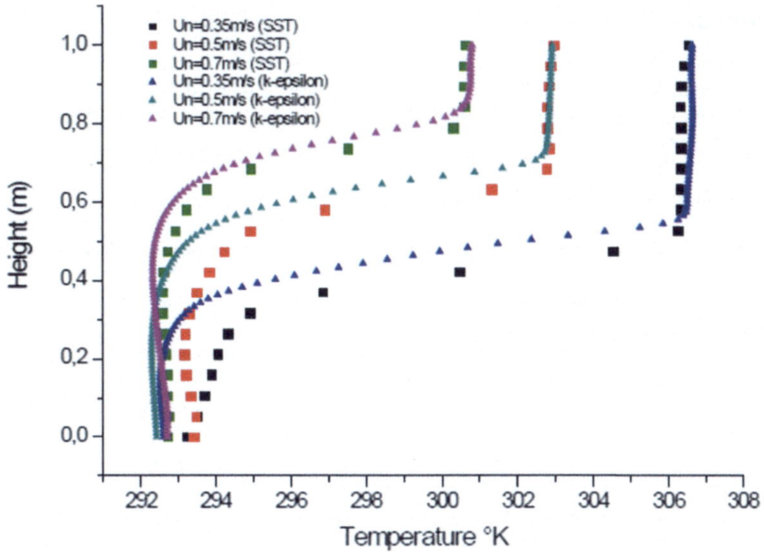

Fig. 5.6 Evolution of the interface height for different ventilation velocities near de wall (SST and k-epsilon model)

The present numerical results agreed with those of the experimental works of Hunt and Linden [8] showing that the mean temperatures of the plume are affected by the volume flux of entrained fresh air and consequently the temperature and the interface heights of the warm upper layer.

5.5 Conclusion

The influence of the natural displacement ventilation on the development of thermal buoyant plume in an enclosure is investigated using two RANS turbulence models: the standard k-epsilon and SST. The results are consistent and in good agreement with those of Linden et al. [5] and Abdalla et al. [1].

The SST turbulence model seems to be better suited to plume flow problems. It is noted that the variation of the supplying fresh air velocity between the cool and the warm layers changes the position of the interface. For the critical ventilation velocity $U_N = 0.7$ m/s, the difference of temperature of the two layers is $\Delta T = 7$ °C, and the recorded interface height is h = 0.71 m. If the critical velocity value is exceeded, the interface becomes more instable and the stratification even disappears.

References

1. I.E. Abdalla, M.J. Cook, G.R. Hunt, Numerical study of thermal plume characteristics and entrainment in an enclosure with a point heat source. Eng. Appl. Comput. Fluid Mech. **3**(4), 608–630 (2009)
2. G.R. Hunt, T.S. Van Den Bremer, Classical plume theory: 1937–2010 and beyond. IMA J. Appl. Math. **76**(3), 424–448 (2011)
3. B.R. Morton, G.I. Taylor, J.S. Turner, Buoyant gravitational convection from instantaneous and maintained point sources instantaneous and maintained point sources. Proc. R. Soc. Lond. A **234**, 1–23 (1956)
4. W.D. Baines, J.S. Turner, Turbulent buoyant convection from a source in a confined region. J. Fluid Mech. **37**, 51–80 (1969)
5. P.F. Linden, G.F. Lane-Serff, D.A. Smeed, Emptying filling boxes: the fluid mechanics of natural ventilation. J. Fluid Mech. **212**, 309–335 (1990)
6. P.V. Nielsen, Displacement ventilation: theory and design. Indoor Envirn. Eng. **R0038**(18), 1–42 (1993)
7. M.J. Cook, Y. Ji, G.R. Hunt, CFD modelling of natural ventilation: combined wind and buoyancy forces. Int. J. Vent. **1**(3), 169–179 (2003)
8. G.R. Hunt, P.F. Linden, Steady-state flows in an enclosure ventilated by buoyancy forces assisted by wind. J. Fluid Mech. **426**, 355–386 (2001)
9. Ansys CFX version 11, www.ansys.com/cfx
10. T. Webster, F. Bauman, T. Reese, M. Shi, Thermal stratification permanence of underfloor air distribution systems. Proc. Indoor Air CA USA **4**, 260–265 (2002)
11. S. Gilani, H.B. Montazeri, J. Blocken, CFD simulation of stratified indoor environment in displacement ventilation: validation and sensitivity analysis. Build. Environ. **95**, 299–313 (2016)

Chapter 6
Empirical Mathematical Model to Evaluate the Effective Diffusivity of Pumpkin Slice During Forced Air Convection Dryer

A. Benseddik, A. Azzi, D. Lalmi, H. Bensaha, M. N. Zidoune, and K. Allaf

Abstract In this work, through central composite design, the Response Surface Methodology (RSM) has been adopted to develop an empirical model to describe the effective moisture diffusivity versus the drying conditions (air temperature, velocity) and slices thickness. The instantaneous effective moisture diffusivity estimated by the inverse method based on experimental drying kinetics data for thin-layer of pumpkin slice during hot air forced convection. A range of air temperature (40–80 °C), air velocity (2–15 m/s) and slices thickness (0.4–1.4 cm) were tested. The relationships between drying conditions, slices thickness and effective moisture diffusivity were determined. The results obtained following of the efficient moisture diffusivity tests show good agreement with the predicted values. The mentioned thin-layer drying empirical model comprise the drying conditions and slices thickness can be applied in drying process simulation of pumpkin slices in a forced air convection dryer.

Using diffusion model included the drying conditions and slices thickness for modeling moisture diffusivity of pumpkin slices during convective airflow drying is very significant for a better understanding development of the drying process

A. Benseddik (✉) · D. Lalmi · H. Bensaha
Unité de Recherche Appliquée en Energies Renouvelables, URAER,
Centre de Développement des Energies Renouvelables, CDER, 47133 Ghardaïa, Algeria
e-mail: a_benseddik2008@yahoo.fr

A. Benseddik · K. Allaf
Laboratoire des Sciences de l'Ingénieur pour l'Environnement : LaSIE FRE 3474
CNRS - Université de La Rochelle, La Rochelle, France

A. Benseddik · A. Azzi
Unité de Recherche Matériaux et Energies Renouvelables (URMER), Faculté des Sciences,
Université de Tlemcen, BP 119, 13000 Tlemcen, Algeria

M. N. Zidoune
Laboratoire de nutrition et technologies alimentaires (L.N.T.A), équipe Transformation et Elaboration des Produits Agro-alimentaires (T.E.P.A.), INATAA,
Université Constantine 1, Constantine, Algeria

A. Belasri and S. A. Beldjilali (eds.), *ICREEC 2019*, Springer Proceedings in Energy,
https://doi.org/10.1007/978-981-15-5444-5_6

through the obtained results on 3D, 4D surfaces and contours. An experimental design according to the central composite design (CCD) has been constructed for investigating the air temperature, velocity and thickness influences on effective moisture diffusivity. Results illustrated by the plots show that the second-order polynomial model provide an adequate description of the responses of effective moisture diffusivity in hot airflow drying processes of diffusivity of pumpkin slices in the experimental ranges.

Keywords Hot airflow drying · Multiple regression analysis · Response surface methodology

6.1 Introduction

The drying process has specific particularities in the food industry; essentially, in removing water from a material. Drying helps in extending the shelf-life of food and agricultural products and makes the product smaller and lighter. Furthermore, it does not require any cooling, that makes storage and transportation processes easier while the related costs are dramatically decreased [1]. On the other hand, utilization of high amount of energy in drying industry, makes drying among the most energy-intensive operations with major industrial prominence [2–6]. The simulation and theoretical data of each product's behavior during moisture removal process are very considerable for drying systems study, equipment development, optimization, as well as, commercial viability determination [7].

Comprehension of effective diffusivity in relationship with the drying conditions, slices thickness is essential for estimating the drying time, required for reducing the moisture content at the desired level. It consists the main issue for scientists and engineers who are focused on the design study of the dryers of pumpkin slice processing. Therefore, many researchers attempted in developing new model by considering the drying conditions of some fruits and vegetables. Moreover, these models provided good results predicting in food process engineering application. However, they require additional effort in the aim to find a new relationship between the effective moisture diffusivity, the drying conditions and slices thickness. The effective moisture diffusivity previously obtained is not defined versus the drying conditions. Therefore it is important to find the relationship between effective moisture diffusivity and operating variables.

The objective of this work is to determine the effective diffusivity of pumpkin slice during forced air convection dryer, as well as its relationship with the drying conditions and slices thickness. The inverse problem of the effective moisture diffusivity has been employed based on drying kinetic data through the use of an experimental pilot training airflow dryer [8]. We conclude that the described work offers a more complete approach by introducing the intrinsic variables (material).

In the present study, experimental data for the drying pumpkin slices are used in order to (i) determine the instantaneous effective moisture diffusivity of pumpkin slices (ii) investigate the effects of drying conditions and slices thickness on the effective moisture diffusivity.

6.2 Materials and Methods

6.2.1 Biological Materials

The pumpkin, used in the test was brought from a local market of La Rochelle, France. Pumpkins are kept in a refrigerator of the laboratory at 4 °C, before the experiments start. Previous any experiment, the pieces of pumpkins were cut into cubes according to the central composite design using a culinary aid device. The initial moisture content (wet basis) of the products is on average (87%). The measurements were carried out by an adequate drying oven at 105 °C for 24 h (AOAC 1990) [9].

6.2.2 Determination of Moisture Ratio

The moisture content of the pumpkin slice during drying process is calculated according to the following formula:

$$M_t = \frac{w_t - w_d}{w_d} \tag{6.1}$$

The moisture loss during drying process was calculated through form [10]:

$$MR = \frac{M_t - M_e}{M_0 - M_e} \tag{6.2}$$

where M_t = moisture (dry basis) at time t, M_e = equilibrium moisture content (dry basis), M_0 = initial moisture content (dry basis), w_t is the dry matter at time t and w_d is the dry matter.

6.2.3 Instantaneous Effective Moisture Diffusivity

The mechanisms of moisture transport during drying process can be modeled mathematically from the Fick's second law diffusion (Eq. (6.3)):

$$\frac{dMR}{dt} = D_{eff}\nabla^2 M \tag{6.3}$$

Supposing that the transfer of moisture during drying carried out by diffusion, the temperature and diffusion coefficient have constant values and the shrinkage is neglected, the analytical solution of Eq. (6.3) for an infinite plate is as Follows [11]:

$$\frac{dMR}{dt} = D_{eff}\nabla^2 M \tag{6.4}$$

$$MR = \frac{8}{\pi^2}\sum\nolimits_{n=0}^{\infty}\frac{1}{(2n+1)}exp\left[-(2n+1)^2\frac{\pi^2 D_{eff}t}{4L^2}\right] \tag{6.5}$$

The second and all higher terms in Eq. (6.5) are negligible compared to the first at a long process. Thus, Eq. (6.6) can be written as follows:

$$MR = \frac{8}{\pi^2}exp\left[-\frac{\pi^2 D_{eff}t}{4L^2}\right] \tag{6.6}$$

where MR is the moisture ratio [-], t is the drying time in [s], D_{eff} is the moisture diffusivity in [m^2/s] and L is the half thickness of samples in [m].

The instantaneous moisture diffusivity is estimated by the inverse method, from experimental data obtained during drying process.

6.2.4 Experimental Design

Response surface methodology was applied for evaluating the relationship between the air velocity (m/s), temperature (°C), and slices thickness (cm) using Centurion Version XVI statistical software. The independent variables, used in the RSM design are listed in Table 6.1.

Table 6.1 Central composite design for 5 levels and 3 variables

Variables	Label	Coded levels				
		−1.682	−1	0	1	1.682
x_1	Temperature [°C]	40	48	60	72	80
x_2	Velocity [m/s]	2	4.5	8.5	12.5	15
x_3	Thickness [cm]	0.2	0.4	0.8	1.2	1.4

The tests were designed following the central composite design (CCD), as shown in Table 6.2. The effective moisture diffusivity of pumpkin slice was taken as the response during the designed experiments. Thus, the data recorded following the central composite design model experiments can be described as follows:

$$y = a_0 + \sum_{i=1}^{3} a_i x_i + \sum_{i=1}^{3} a_i x_i^2 + \sum_{i=1}^{3} \sum_{j=1}^{3} a_{ij} x_i x_j \qquad (6.7)$$

where y is the predicted responses used as dependent variables; $x_i (i = 1, 2 \text{ and } 3)$ are the independent variables; and a_0, $x_i (i = 1, 2 \text{ and } 3)$ and $a_{ij} (i = 1, 2, 3;\ j = i, \ldots, 3)$ were the model coefficient parameters.

Table 6.2 Design and responses of the central composition design

Run number	Tree factors						Response value
	Coded values			Uncoded values			Effective moisture diffusivity
	x_1	x_2	x_3	T [°C]	V [ms^{-1}]	L [cm]	[%]
1	0	−1.682	0	60	2	0.8	4.0356e−08
2	0	0	0	60	9	0.8	9.1425e−08
3	0	0	0	60	9	0.8	9.3807e−08
4	−1	−1	−1	48	5	0.4	1.4692e−07
5	−1	−1	+1	48	5	1.2	4.2740e−08
6	−1	+1	+1	48	12	1.2	7.5711e−08
7	+1	−1	+1	72	5	1.2	6.4505e−08
8	+1	+1	+1	72	12	1.2	9.5404e−08
9	+1	+1	−1	72	12	0.4	2.9728e−07
10	+1	−1	−1	72	5	0.4	2.4963e−07
11	0	0	−1.682	60	9	0.2	——
12	−1	+1	−1	48	12	0.4	1.6609e−07
13	−1.682	0	0	40	9	0.8	6.4274e−08
14	+1.682	0	0	80	9	0.8	1.4574e−07
15	0	+1.682	0	60	15	0.8	9.2417e−08
16	0	0	0	60	9	0.8	1.0517e−07
17	0	0	+1.682	60	9	1.4	4.3542e−08
18	0	0	0	60	9	0.8	4.3542e−08

6.3 Results and Discussion

The effective moisture diffusivity response obtained following the experimentation is listed in Table 6.2. The experimental data obtained were used to estimate the quadratic equation coefficients. It has been assumed that a good-fitting model should have $R^2 > 80\%$. When $R^2 = 1$, the empirical model is appropriate for fitting the real data. A lower R^2 at 80% or less indicates that the model is inapt to interpret the relationship between independent variables [12]. The response surface analysis of the data shows that the relationship between dependent (Y) and independent variables (X1, X2, X3) is quadratic, with a good-fitting have R2 = 0.946. Equation (6.8) gives the relationship between dependent (effective moisture diffusivity **D_eff**) and independent variables (air temperature **T**, velocity **V** and slices thickness **E**) and the independent variables () at time t = 90 min.

$$\mathbf{D\_eff} = \big(8.247 + 3.019 * \boldsymbol{T} + 1.598 * \boldsymbol{V} - 7.767 * \boldsymbol{E} + 1.215 * \boldsymbol{T}^2 + 33.01 * \boldsymbol{T} * \boldsymbol{V} - 2.405 * \boldsymbol{T} * \boldsymbol{E} - 15.015 * \boldsymbol{V}^2 - 368.75 * \boldsymbol{V} * \boldsymbol{E} + 4.097 * \boldsymbol{E}^2\big) \times 10^{-8} \tag{6.8}$$

Figures 6.1 and 6.2 presented the plots of the response surface and isopleths explaining the effects of air velocity, temperature and slices thickness, on the effective moisture diffusivity. We keep one independent variable constant X3 = 0 at its central level and varying the other independent variables X1 and X2. It seems that the D_eff is increased versus the air velocity increase. Which revealed that higher effective moisture diffusivity is favorable for getting high effective moisture diffusivity at a high air temperature.

As a final point, the effective moisture diffusivity modeled by (Eq. (6.8)) is presented under different values of air temperature (X1), velocity (X2) and thickness (X3) for a given time in Fig. 6.3.

MATLAB® was used for 3D response surface curve and 2D contour plot of the model described by (Eq. (6.8)).

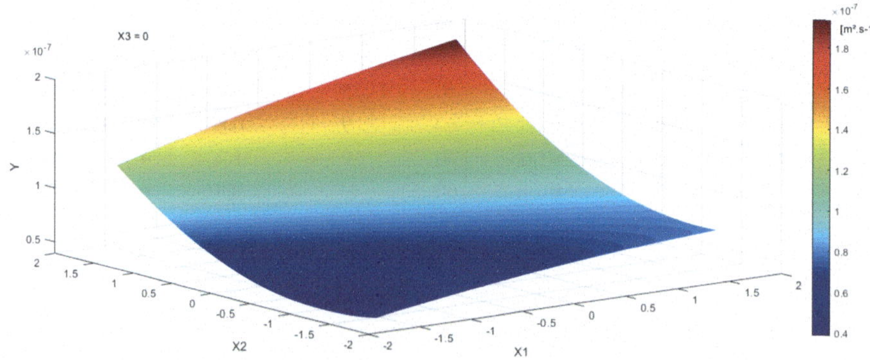

Fig. 6.1 Variations in moisture diffusivity with air temperature and velocity at 90 min

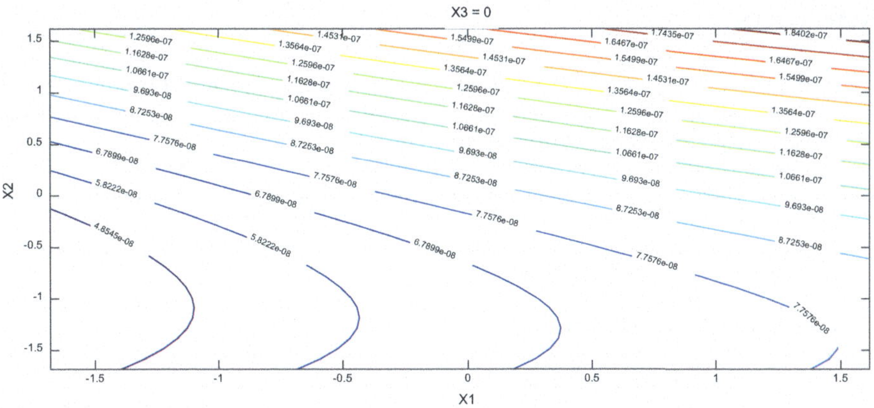

Fig. 6.2 Isopleths of effective moisture diffusivity with air temperature and velocity at 90 min

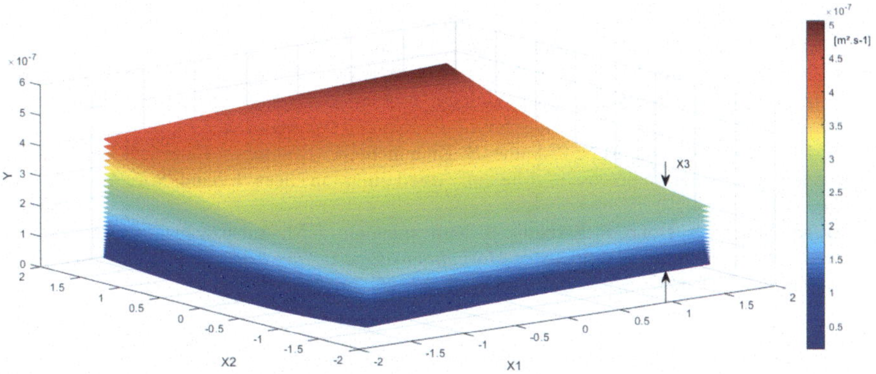

Fig. 6.3 Plot of 3D effective moisture diffusivity evaluated by Eq. (6.8) at 90 min

6.4 Conclusion

Using diffusion model included the drying conditions and slices thickness for modelling the effective moisture diffusivity of pumpkin slices. This obtained model is very important for developing a better drying process through the results obtained on 3D, 4D surfaces and contours. An experimental design has been adopted, according to the central composite design (CCD) for studying the influence of air temperature, air velocity and slices thickness on effective moisture diffusivity. Results averred that the second-order polynomial model provide an adequate description of the responses of effective moisture diffusivity in hot airflow drying processes of diffusivity of pumpkin slices in the experimental ranges.

References

1. H. Kocabiyik, Combined infrared radiation and hot air drying, in *Infrared Heating for Food and Agricultural Processing*, Chap. 06, ed. by Z. Pan, G.G. Atungulu (CRC Press Taylor & Francis Group, Boca Raton, 2010), pp. 101–116
2. M. Carsky, Design of a dryer for citrus peels. J. Food Eng. **87**, 40–44 (2008)
3. I. Dincer, Thermodynamics, exergy and environmental impact. Energy Sources **22**(8), 723–732 (2000)
4. I. Dincer, Y.A. Cengel, Energy, entropy and exergy concepts and their roles in thermal engineering. Entropy Int. J. **3**(3), 116–149 (2001)
5. I. Dincer, A.Z. Sahin, A new model for thermodynamic analysis of a drying process. Int. J. Heat Mass Transf. **47**, 645–652 (2004)
6. Q. Shi, C. Xue, Y. Zhao, Z. Li, X. Wang, Drying characteristics of horse mackerel (*Trachurus japonicus*) dried in a heat pump dehumidifier. J. Food Eng. **84**, 12–20 (2008)
7. P.C. Corrêa, O. Resende, A.P. Martinazzo, A.L.D.G. Goneli, F.M. Botelho, Modelagem matemática para a descrição do processo de secagem do feijão (*Phaseolus vulgaris* L.) em camadas delgadas. Engenharia Agrícola **27**(2), 501–510 (2007)
8. A. Benseddik, A. Azzi, M.N. Zidoune, K. Allaf, Mathematical empirical models of thin-layer airflow drying kinetics of pumpkin slice. Eng. Agric. Environ. Food **11**(4), 220–231 (2018)
9. A.O.A.C. Official methods of analysis. Edition: Association of Official Analytical Chemist (AOAC), Washington, D.C., 16ème edition (1990)
10. H. Niu, Y. Li, Y. Lei, L. Zhang, J. Peng, S. Guo, Microwave drying of anthracite: a parameter optimized by response surface methodology. Journal Arabe pour la Science & Engineering **37**, 65–73 (2012)
11. A. Benseddik, A. Azzi, M.N. Zidoune, R. Khanniche, C. Besombes, Empirical and diffusion models of rehydration process of differently dried pumpkin slices. J. Saudi Soc. Agric. Sci. **18** (4), 401–410 (2019)
12. D. Krishnaiah, A. Bono, R. Sarbatly, R. Nithyanandam, S.M. Anisuzzaman, Optimisation of spray drying operating conditions of *Morinda citrifolia* L. fruit extract using response surface methodology. J. King Saud Univ. Eng. Sci. **27**, 26–36 (2015)

Chapter 7
Experimental Study of Thermal and Hygrometric Behavior of Earth to Air Heat Exchanger with Two Working Regimes in Arid Region

Nasreddine Sakhri, Belkacem Draoui, Younes Menni, Ebrahim Elkhali Lairedj, Soufiane Merabti, and Noureddine Kaid

Abstract Natural ventilation is essential for buildings, offices and dwellings occupants' thermal comfort and good living conditions. Also, for good production in greenhouses. Earth to air heat exchanger is a renewable source based on geothermal energy of the earth. Many working parameters affect earth to air heat exchanger thermal performance were studied: pipe burial depths, length, diameter, inlet air velocity. In the present paper, an experimental investigation of EAHE hygrometric-thermal performance with two working regimes: first work naturally with zero energy consumption and the second regime with fan (33 W). EAHE with zero energy consumption was able to create 2 working modes: heating by rising the outlet air temperature by 8 °C (between from 17 to 7 h), reduce 12 °C in cooling regime (From 8 to 16 h), and rises relative humidity by 45%. EAHE with fan provide only a rise in outlet air temperature by 12 °C. In addition, relative humidity rises by 20%. The Obtained results confirm the capacities of EAHE to contribute in thermal comfort assessment not only for cold season but also across the year with less or no energy consumption.

Keywords Renewable energy · Natural ventilation · Earth to air heat exchanger

N. Sakhri (✉) · B. Draoui · S. Merabti
Laboratory of Energy in Arid Areas (ENERGARID), University of Bechar, Bechar, Algeria
e-mail: sakhrinasreddine@gmail.com

Y. Menni
Unit of Research on Materials and Renewable Energies, Tlemcen, Algeria

E. E. Lairedj
Department of Electronics, University of Sidi-Bel-Abbès, Sidi Bel Abbès, Algeria

N. Kaid
Department of Technology, University Center Salhi Ahmed of Naâma, Naâma, Algeria

A. Belasri and S. A. Beldjilali (eds.), *ICREEC 2019*, Springer Proceedings in Energy,
https://doi.org/10.1007/978-981-15-5444-5_7

7.1 Introduction

Earth to air heat exchangers technique has aroused the intention of researchers who have tried to model it numerically and study it experimentally. Benhamou et al. [1] studied the cooling capacity of underground pipes. Misra et al. [2] have studied the parameters that affect the performance of the underground heat exchanger technique for passive cooling. Several works [3] discuss numerical, parametric, experimental and economic aspects of this technique. It is also a natural ventilation technique, used for centuries (3000 B.C) until today. Many researchers around the world have conducted numerical and experimental research to mount the great potentials that they represent this technique. All begin with the geothermal source of the location. The good knowledge of the vertical profile of the temperature is first step to know at which depth the tube of the exchanger will be buried (Fig. 7.1).

Fig. 7.1 Experimental setup of air heat exchanger

7.2 Methodology

The experience was carried out between December 2018 and February 2019. Winter season in the Southern Algeria is characterized by severe conditions where the temperature reaches negative values.

An open loop horizontal earth to air heat exchanger was made of 60 m PVC pipes (thicknesses: 0.002 m, diameter: 0.11 m). Thermal conductivity coefficient of PVC is 0.2 W/m K. The pipe was buried in homogeneous sandy-loam soil at 1.5 m underground where annual unchangeable temperature of soil was around 28 °C [4]. The inlet and the outlet were made also of 3 m PVC pipe for each side with 1.5 m underground and 1.5 m on the ground equipped with thermal insulation for the inlet and outlet side (exposed to outside weather conditions). The air inlet and the outlet were made of PVC elbows with solar protection on the top. ETAHE inlet was directed in the direction of prevailing winds (North) in the first case and equipped with an exterior fan of 33 W for the second case with 3 m/s airflow speed. The EAHE outlet was directed to the South. Figure 7.2 shows experimental set-up of studied case.

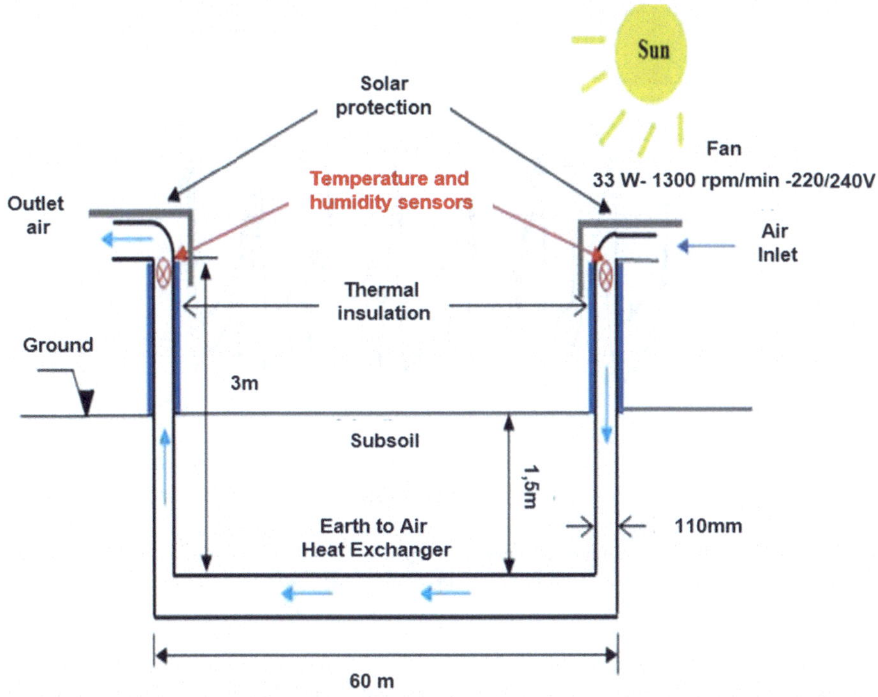

Fig. 7.2 Stand-alone EAHE with exterior fan

7.3 Results and Discussion

7.3.1 EAHE with Solar Protection

Between December 2018 and February 2019, measured inlet and outlet air temperature are presented in Fig. 7.2. For the inlet, temperature was situated between 38–39 °C for maximum and 6 °C for the minimum. The MAX-MIN temperature difference was 33 °C. Outlet air temperature maximum was 27.3 °C at 12 o'clock and 35 °C at 17:00 due to sun path effect. The minimum temperature was 11 °C. MAX-MIN outlet temperature difference was 26 °C. ETAHE reduce this MAX-MIN between the inlet and outlet by 6 °C. Even with solar protection, temperature reaches 38 °C because of the reflected radiance and indirected sun radiance. Also, $T_{min\text{-}outlet}$ was higher than $T_{min\text{-}intlet}$ which indicates that air submits a preheating during its passage through the device.

In comparison between inlet and outlet relative humidity, we observe a differences in maximum RH values. For inlet maximum RH, values were in the range of 30 to 80%. Relative humidity for ETAHE outlet exceeds 90% for higher values between 19^h and 09^h. The hygrometric regime of the device outlet was more stable in comparison with the inlet where EAHE plays a humidification role due to

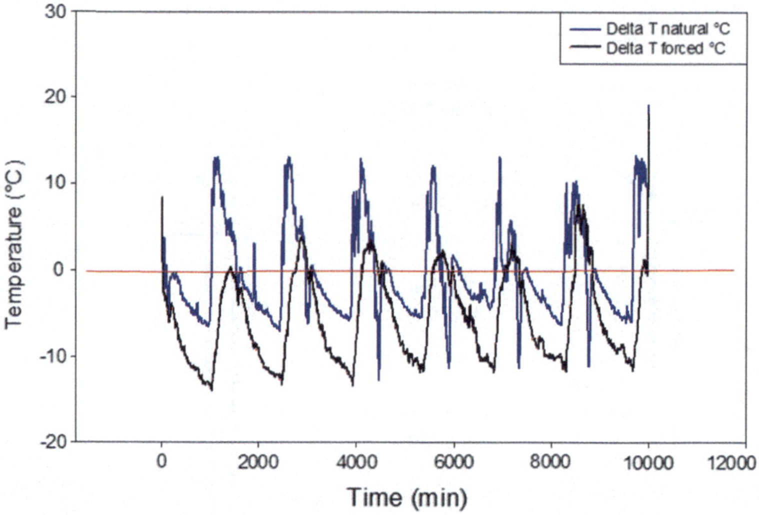

Fig. 7.3 ETAHE thermal regime expressed in [Delta T = $T_{inlet} - T_{outlet}$] in blue: EAHE stand-alone with zero energy consumption and in black: EAHE with electric fan

condensation phenomena. Inside EAHE pipe, the water vapour contained in air circulating inside ETAHE condensate to form water and this accumulation of water rise the outlet relative humidity to reach and exceeds 90% (and sometimes get 98%).

ETAHE with solar protection inlet and outlet hourly mean temperature difference was calculated $\Delta T = T_{inlet} - T_{outlet}$ and presented in Fig. 7.3. The obtained results show the presence of two major regimes:

- $\Delta T = T_{inlet} - T_{outlet} > 0 \rightarrow (T_{inlet} > T_{outlet})$: In this case T_{outlet} was lower than T_{inlet} so the ETAHE with solar protection work as cooling device.
- $\Delta T = T_{inlet} - T_{outlet} < 0 \rightarrow (T_{inlet} < T_{outlet})$: A heating regime took place and the outlet air temperature was higher than the inlet which is perfect for preheating.

From 17^h to 7^h: in this part of the day $T_{outlet} > T_{intlet}$, outside air temperature was lower and the ETAHE rise the outlet air temperature by 6 °C. The device was able to create a heating regime.

From 8^h to 16^h: the sun heats up the outside air and even with the solar protection air inlet reach 38 °C. ETAHE T_{outlet} < ETAHE T_{intlet} where underground pipe work as a heat sink by reducing this temperature from 38 to 27 °C. A cooling regime took place and the difference reaches its higher value at 11^h (between 11 to 12 °C).

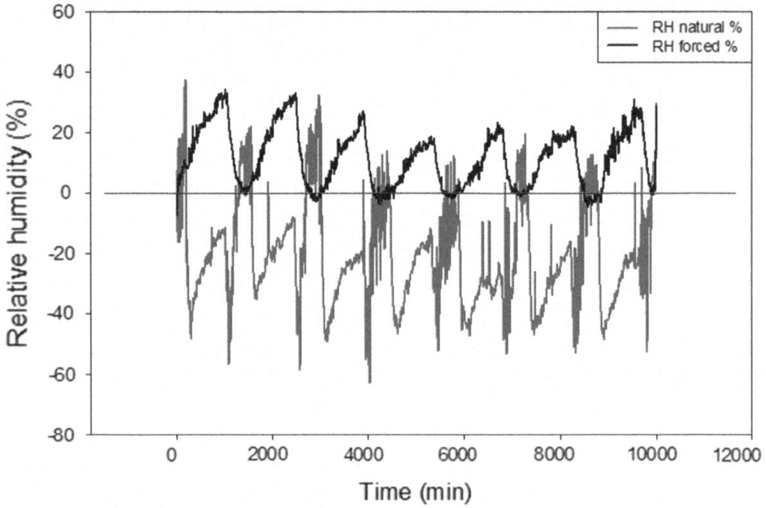

Fig. 7.4 ETAHE hygrometry regime expressed in [Delta T = Tinlet − Toutlet] in green: EAHE stand-alone with zero energy consumption and black: EAHE with electric fan

For relative humidity, the ΔRH ($RH_{inlet} - RH_{outlet}$) shows also that the big part of this difference is situated in negative zone where $RH_{outlet} > RH_{inlet}$. ETAHE with solar protection of upper parts rise the relative humidity for most of the day by 10 to 46%. We remark also that relative humidity decrease between 13:00 and 16:00 by 1 to 15%. The solar protection contributes in rising relative humidity of air passing through the pipe. For 24 h, we can divide daily hygrometric regime of the EAHE as follow: 17% (from 13^h to 16^h) of a dehumidification regime, and 83% (17^h to 12^h) of a humidification regime. ETAHE with solar protection contribute in solving a big thermal comfort problem in arid regions which is the low values of relative humidity inside buildings. The ETAHE outlet will be connected directly to the inside dwellings and plays a humidification role but air relative humidity and quality must be controlled.

Hygro-thermal analysis of the obtained results shows that the thermal regime of the ETAHE is influenced directly by outside climatic conditions which explain the presence of two thermal regimes: heating in time of outside air low temperature values (present 70% of daily regime) and cooling in time of high outside air temperature (which present 30% of daily regime).

Hygrometric regime shows less dependence to weather conditions where ETAHE outlet relative humidity was higher most of the time with the presence of water surface and green cover in the site of the study.

7.3.2 EAHE Equipped with Fan and Solar Protection

$T_{outside}$ was between −1 and 27 °C. In the same time, 12.2 °C was lowest outlet air temperature at EAHE outlet in comparison with the highest 22.5 °C by more of 10 °C difference. EAHE was able to rise temperature from (−1) to 12.2 °C with more than 13 °C difference. At least, half of heating demand can be reduced by natural resources for winter season. The device plays the role of thermal regulation to reduce difference between highest and lowest air temperature for EAHE inlet and outlet. These results ensure steadiness even with irregular weather conditions which is not the case with the first system without any exterior fan. The first system depend greatly to outside climate conditions. RH was between 13 and 59% for EAHE inlet and 15 to 31% for the outlet. Two thermal regimes are present even with the use of the fan:

From 17^h to 11^h: Heating (ΔT_{max} = 13 °C). This thermal regime induces a Dehumidification regime ($HR_{inlet} > HR_{outlet}$) with 30%.

From 12^h to 16^h: Cooling regime where it gets maximum at 14^h to 15^h with 6.8 °C associated to a humidification regime where $HR_{inlet} < HR_{outlet}$ but the rise doesn't exceed 3%.

Earth to air heat exchanger efficiency or thermal efficiency is calculated as follow:

$$\varepsilon = T_{inlet} - T_{outlet}/(T_{inlet} - T_{soil}) \tag{7.1}$$

T_{soil}: Soil temperature equal to 28 °C in our case.

Figure 7.4 present EAHE efficiency. For stand-alone EAHE with solar protection, Thermal efficiency was situated between 0.2 and 6 with a heterogeneous distribution and sometimes become negative. The second case (EAHE + exterior electric fan), values were positive and negative. Positives (with 0.4) between midnight to 10 h and 17 h to 23 h. Negatives values (−0.2 to −0.4) between 4 and 5 pm. Positive EAHE thermal efficiency was related to the heating regime and the negative values were related to the cooling regime which also confirms the great dependence of this type of earth to air heat exchangers with exterior parts exposed to local climatic conditions despite these effects are reduced through thermal insulation.

7.4 Conclusion

Earth to air heat exchanger EAHE based on geothermal source was used for many centuries by arid regions population for natural ventilation.

In the present study, a comparison between two working regimes of an EAHE. The first work naturally and the second used an exterior fan. Obtained results for the stand-alone EAHE system shows that thermal and hygrometric analysis of the outlet air leaving the system gives the following conclusions:

- Heating regime (T soil > T outside air): almost by 62.5% where ETAHE with solar protection serve as a heat source and was able to heat up air passing through the buried pipe by 7 to 8 °C.
- Cooling regime (T soil < T outside air): almost by 37.5% where the system works as a heat sink to reduce by 11 to 12 °C the outlet air.
- ETAHE with solar protection raises the relative humidity by 45% and serve as a condensation device.

For earth to air heat exchanger with electric fan, heating regime was the dominant with more than 85% where air temperature leaving the tube rise by 13 °C. Hygrometric regime was dehumidification mode with 30% of relative humidity reduction. Dehumidification can solve the problem of micro-organisms and lead to good internal air quality IAQ inside living space connected to EAHE outlet. For cold season, hating needs are the most needed which improve indoor thermal comfort in arid regions like the South of Algeria. Great potentials for EAHE contribute in the improvement of thermal comfort in arid regions naturally by stand-alone system without any electrical or energy consumptions or associated to exterior fan to give more control to the user.

References

1. M. Benhamou, B. Draoui, M. Hamouda, Improvement of the summer cooling induced by an earth-to-air heat exchanger integrated in a residential building under hot and arid climate. Appl. Energy **208**, 428–445 (2017)
2. R. Misra, S. Jakhar, K.K. Agrawal, S. Sharma, D.K. Jamuwa, M.S. Soni, G. Das Agrawal, Field investigations to determine the thermal performance of earth air tunnel heat exchanger with dry and wet soil: energy and exergetic analysis. Energy Build. **171**, 107–115 (2018)
3. N. Sakhri, Y. Menni, A. Chamkha, E. Lorenzini, N. Kaid, H. Ameur, M. Bensafi, Study of heat and mass transfer through an earth to air heat exchanger equipped with fan in South West of Algeria. Int. J. Heat Technol. **37**(3), 689–695 (2019)
4. N. Sakhri, B. Draoui, Y. Menni, Experimental study of earth to air heat exchanger performance in arid regions. First step: in-situ measurement of ground vertical temperature profile for different depths. J. Adv. Res. Fluid Mech. Therm. Sci. **56**(2), 183–194 (2019). Vol. 37, No. 3, 689–695, September (2019)

Chapter 8
Water-Cooled Photovoltaic Panel Efficiency

Sonia Ait Saada, Idir Kecili, and Rezki Nebbali

Abstract The purpose of this work is to improve the efficiency of a photovoltaic solar panel with water cooling system circulating along the back side of a PV panel. The numerical simulation was done on CFD code, the effect of water flow rate and the ambient air temperature on the conversion efficiency of the cooled PV panel were discussed. The results showed that at a flow rate of 100 g/s or more, the average temperature of the PV panel stabilizes, the distribution of the temperature field on the cooled solar panel with a water flow rate of 100 g/s is almost homogeneous over the entire solar panel, with the exception of the fixing zone of the electrical box which prevents the evacuation of the heat absorbed by the silicon, which raises the panel temperature locally. The results show that is enough to use 100 g/s of water flow rate to ensure a sufficient cooling. For air temperatures of 25, 35 and 45 °C, this cooling technique has improved the efficiency of the solar panel by 24.25, 28.92 and 33.92%. Concerning the junction box, which is often neglected, it has been shown that it affects the distribution of the temperature field of the PV panel and constitutes a localized area that can significantly alter the performance of the solar panel.

Keywords Junction box · Photovoltaic · Cooling

8.1 Introduction

In order to minimize the negative effect of the PV panel's temperature increase on its efficiency, we propose to cool it by water. Several cooling techniques were proposed, [1–3] have reviewed the different existing techniques. A distinction was made between active water cooling. [4] showed the influence of the heterogeneity of the temperature field distribution on a PV panel cooled by the circulation of

S. Ait Saada (✉) · I. Kecili · R. Nebbali
Laboratory of Energy, Mechanics and Materials (LEMM), Mouloud Mammeri University, Tizi-Ouzou, Algeria
e-mail: atsonia@outlook.com

A. Belasri and S. A. Beldjilali (eds.), *ICREEC 2019*, Springer Proceedings in Energy,
https://doi.org/10.1007/978-981-15-5444-5_8

water through pipes mounted on the back side of the PV panel. [5] proposed to cool a PV panel by water spray on its front side to reduce reflectivity and ensure the cleaning of the glass surface. This process improved the efficiency of the PV panel by 11.7% against 9% for the uncooled one. In the same way, [6] further improves this efficiency to 14% by simultaneously spraying water on both sides of a PV panel. [7] studied the effect of a water jet on a set of solar cells. They show that the PV panel cooled from 69.7 to 36.6 °C and 47.6 to 31.1 °C, which correspond to efficiency improvement of 17.9% and 15.5%, respectively, in June and December.

In this work we studied a PV panel (1580 × 808 × 45 mm) cooled by water which flows underside of the PV panel through a cavity of about 4 cm thickness. The influence of air temperature and water flow on the electrical efficiency of this PV panel is then studied.

8.2 Uncooled PV Panel

The temperature T_P of the uncooled PV panel is determined by solving the single equation of heat balance, expressed by:

$$\alpha R_G S - \Phi_c - \Phi_r = 0 \quad (8.1)$$

$\Phi_c = h\, S\, (T_P - T_{air})$ Convective heat flux.
$\Phi r = \sigma\, S\, (T_P^4 - T_V^4)$ Radiative flux exchanged between the panel and the sky.
$T_v = 0.0552\ Tair^{1.5}$ Temperature of the sky [8].
In situation of no wind, the convective exchange coefficient between the PV panel surface and the ambient air is evaluated by correlations [8]:

$$N_U = 0.54 \times (Ra)^{0.25} \text{ for } 10^4 < Ra < 10^6 \quad (8.2)$$

$$N_U = 0.15 \times (Ra)^{0.33} \text{ for } 10^6 < Ra < 10^{11} \quad (8.3)$$

8.3 Water-Cooled PV Panel

Temperature field of the glass and silicon media were determined by the thermal balance expressed by:

$$\Delta T + Q/\lambda = 0 \tag{8.4}$$

where the internal heat sources Q were deduced by the radiative heat balance between the PV panel and the sky, expressed by:

$$Q = \left(\varepsilon R_G S - \sigma \varepsilon \left(T_p^4 - T_v^4\right)\right)/e\left(W.m^{-3}\right) \tag{8.5}$$

Moreover, velocity and temperature field distribution of the water flowing underside the PV panel were determined by solving the coupled equations of continuity, momentum and energy. To do this, CFD-Fluent code was used.

8.4 PV Panel Efficiency

The electrical efficiency of the PV panel is calculated by the following relationship [9]:

$$\eta = \eta_{ref}\left[1 - \beta_{ref}\left(T_p - T_{ref}\right)\right] \tag{8.6}$$

β_{ref} = 0.37%/°C/Power temperature coefficient.
η_{ref} = 14.9% Electrical efficiency of the PV panel under reference conditions T_{PV} = 25 °C and R_G = 1000 W/m^2.
The comparison between the cooled PV panel and the uncooled one is given by the relative efficiency (Table 8.1):

$$\eta_r = 100 \times |\eta - \eta_0|/\eta_0 \tag{8.7}$$

where η_0 denotes the efficiency of the uncooled PV panel (Table 8.2).

Table 8.1 Properties of the PV panel [10]

Layers	e(mm)	λ (W/mK)	ρ (kg/m^2)	C_P(J/kg K)	ε
PV cell	0.3	148	2330	677	0.7
Glass	3.2	1.8	3000	500	

Table 8.2 Temperatures and efficiencies of the uncooled PV panel at R_G = 1000 W/m^2

T_{air} (°C)	45 °C	35 °C	25 °C
T_{pv} (°C)	101.84	93.76	85.65
η_0 (%)	10.66	11.1	11.55

8.5 Results and Discussion

Figure 8.1 shows the evolution of the average panel temperature for different air temperatures and water flow rates. It's shown that from a flow rate of 100 g/s the temperature of the PV panel is stabilized and reaches values of 34.80, 35.45 and 36.21 °C for air temperatures, respectively 25, 35 and 45 °C. Thus, the efficiencies of the PV panel increase (Fig. 8.2) from 11.6 to 14.35%, from 11.10 to 14.31% and from 10.70 to 14.27%, which corresponds, respectively, to the efficiency improvement of 24.25, 28.92 and 33.92% (Fig. 8.3).

Figure 8.4 illustrates the evolution of the average temperature of the cooled PV panel with 100 g/s of water flow rate under various radiations and air temperatures. We observe that the temperature of the PV panel increases with radiation. Indeed, at air temperature of 45 °C, the PV panel temperature reaches 31.80 and 36.21 °C, respectively, at radiations of 200 and 1000 W/m^2. However, the PV panel efficiency were not affected significantly. It vary from 14.52% to 14.27% Figs. 8.4 and 8.5.

Figures 8.6 and 8.7 show the distribution of the temperature field on the cooled PV panel with 100 g/s of water flow rate and air temperatures of 25, 35 and 45 °C. There is an almost homogeneous temperature distribution over the whole PV panel. Temperature field vary overall between 33 and 45 °C for T_{air} = 25 °C, T_{air} = 35 °C and T_{air} = 45 °C. However, it can be seen that the highest value of the temperature were observed at the region of the box electrical wires (Fig. 8.7).

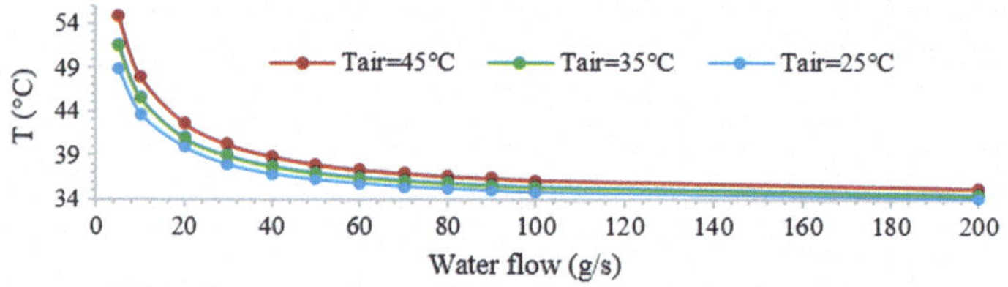

Fig. 8.1 Evolution of the average panel temperature for different air temperatures as a function of water flow

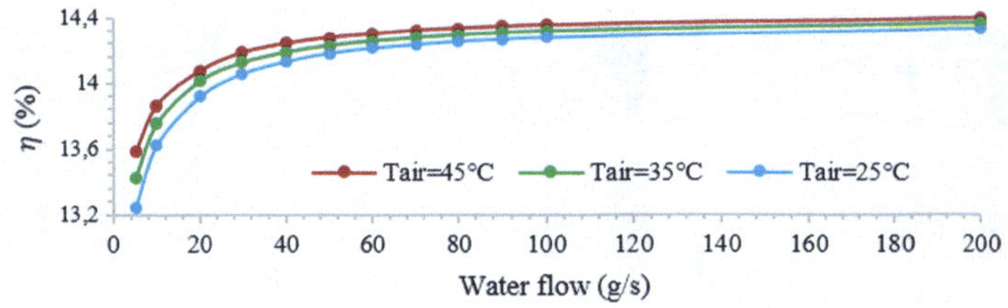

Fig. 8.2 Evolution of the efficiency of the solar panel with the water flow, for different air temperatures

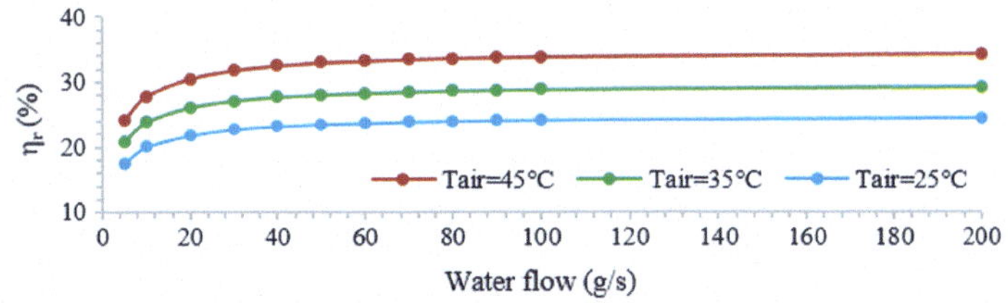

Fig. 8.3 Evolution of the relative difference between the yields of the cooled and uncooled solar panel with the water flow, for different air temperatures

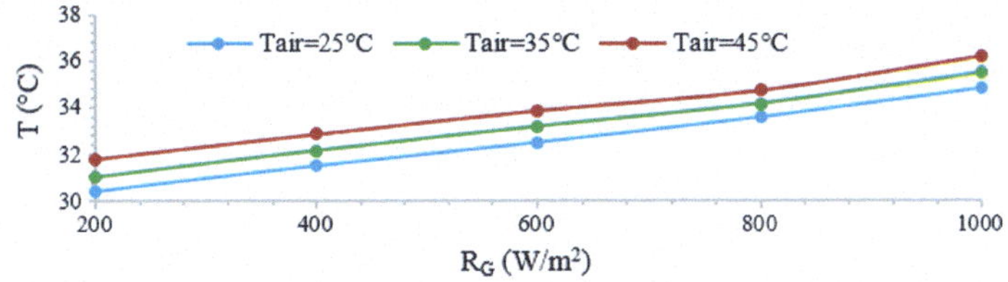

Fig. 8.4 Evolution of the average temperature of the cooled PV panel with a flow rate of 100 g/s for different air temperatures according to solar radiation

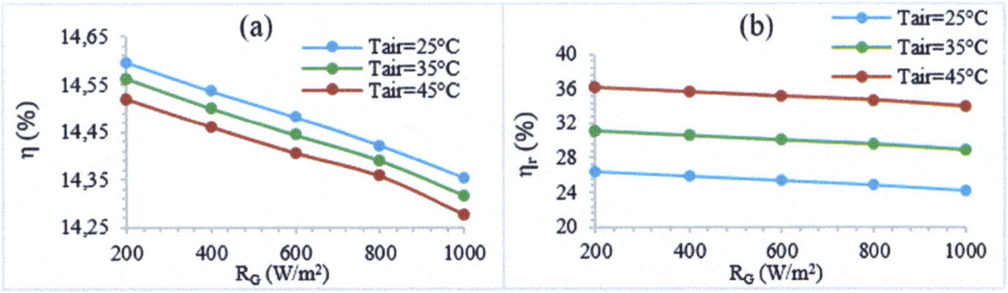

Fig. 8.5 Evolution of the efficiency (a) and relative deviation (b) of the cooled PV panel with a flow rate of 100 g/s for different air temperatures as a function of solar radiation

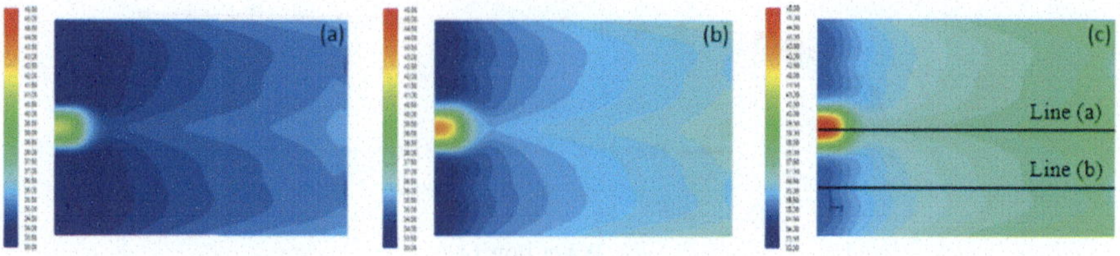

Fig. 8.6 Temperature field distribution on the cooled solar panel for a water flow rate of 100 g/s and an air temperature of 25 °C, 35 °C and 45 °C with RG = 1000 W/m^2

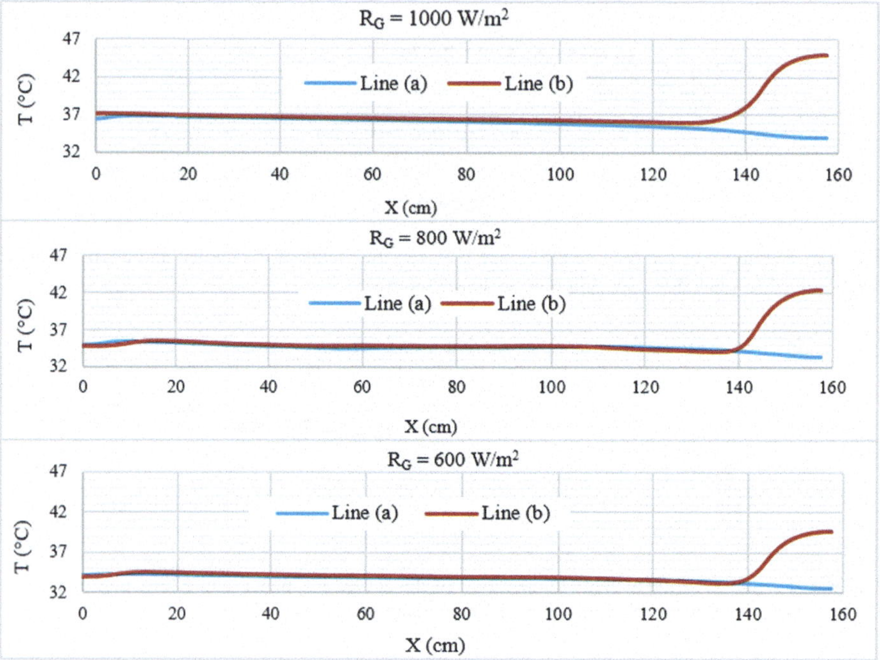

Fig. 8.7 Temperature of the cooled PV panel with a water flow rate of 100 g/s along lines (a) and (b) (Fig. 8.6c) under radiation of 1000, 800, 600 W/m^2 and an air temperature of 45 °C

8.6 Conclusion

The purpose of this work was to improve the electrical conversion performance of a PV panel cooled by water. Moreover, we highlight the influence of the box electrical wires on temperature homogenization.

The results show that is enough to use 100 g/s of water flow rate to ensure a sufficient cooling. For air temperatures of 25, 35 and 45 °C, this cooling technique has improved the efficiency of the solar panel by 24.25, 28.92 and 33.92%. Concerning the junction box, which is often neglected, it has been shown that it affects the distribution of the temperature field of the PV panel and constitutes a localized area that can significantly alter the performance of the solar panel.

References

1. A. Shukla, K. Kanta, A. Sharma, P.H. Biwole, Cooling methodologies of photovoltaic module for enhancing electrical efficiency: a review. Sol. Energy Mater. Sol. Cells **160**, 275–286 (2017)
2. S. Sargunanathan, A. Elango, S. Tharves Mohideen, Performance enhancement of solar photovoltaic cells using effective cooling methods: a review. Renew. Sustain. Energy Rev. **64**, 382–393 (2016)
3. M. Hasanuzzaman, A.B.M.A. Malek, M.M. Islam, A.K. Pandey, N.A. Rahim, Global advancement of cooling technologies for PV systems: a review. Solar Energy **137**, 25–45 (2016)
4. A. Al Tarabsheh, S. Voutetakis, A.I. Papadopoulos, P. Seferlis, I. Etier, O. Saraereh, Investigation of temperature effects in efficiency. Improvement of non-uniformly cooled photovoltaic cells (2013)
5. A. Elnozahy, A.K. Abdel Rahman, A.H.H. Alib, M. Abdel-Salam, S. Ookawara, Performance of a PV module integrated with standalone building in hot arid areas as enhanced by surface cooling and cleaning. Energy Build. **88**, 100–109 (2015)
6. S. Nizetic, D. Coko, A. Yadav, F. Grubišić-Čaboa, Water spray cooling technique applied on a photovoltaic panel: the performance response. Energy Convers. Manag. **108**, 287–296 (2015)
7. H. Bahaidarah, A. Subhan, P. Gandhidasan, S. Rehman, Experimental performance evaluation and modeling of jet impingement cooling for thermal management of photovoltaics. Solar Energy **135**, 605–617 (2016)
8. J.P. Holman, *Heat Transfer*, 8th edn. (McGraw-Hill, New York, 1997)
9. E. Skoplaki, J.A. Palyvos, On the temperature dependence of photovoltaic module electrical performance: a review of efficiency/power correlations. Solar Energy **83**, 614–624 (2009)
10. S. Armstrong, W.G. Hurley, A thermal model for photovoltaic panels under varying atmos-pheric conditions. Appl. Therm. Eng. **30**, 1488–1495 (2010)

Chapter 9
A Fast Low-Cost Automatization Process for Solar Cell Spectral Response Measurement System

M. A. Zafrane, L. Filali, and A. Boudghene Stanbouli

Abstract Response spectral analysis is obviously the most considerable experience in solar cells characterizations. However, it's proportional to the photonics luminosity function, also known as «Quantum efficiency». In process characterization system, the monochromatic light source is considered the crucial element of the spectral response measurement system. In this work, an adapted and improved automatic system, to measure the external quantum efficiency (EQE) of solar cells based on an ATMEGA328, has been designed. For this purpose and in order to perform different electro-optical measurements to assess very low current signals, a monochromator has been automated and synchronized. Moreover, a Graphical User Interface has also been developed in order to make all the procedure easy to operate. Spectral response of different samples of silicon has therefore been characterized and assessed.

Keywords Monochromator · ATMEGA328 · Solar cell · Sensors

9.1 Introduction

Photovoltaic (PV) energy is a new energy source, have became extremely indispensable part of electrical production in many more countries all around the globe [1–6].

Moreover, photovoltaic solar energy is a form of renewable energy comes from the conversion of sunlight irradiation into electricity within semiconductor materials (photovoltaic solar modules, comprising several interconnected photovoltaic cells)

M. A. Zafrane (✉) · A. Boudghene Stanbouli
Département d'électronique, Université des Sciences et de la Technologies d'Oran Mohamed Boudiaf, USTO-MB, BP 1505, El M'naouer, 31000 Oran, Algeria
e-mail: mohammedamine.zafrane@univ-usto.dz

L. Filali
Laboratoire de Physique des Couches Minces et Matériaux pour l'Electronique, Université d'Oran 1, Ahmed Ben Bella, BP 1524, El M'naouar, 31100 Oran, Algeria

A. Belasri and S. A. Beldjilali (eds.), *ICREEC 2019*, Springer Proceedings in Energy,
https://doi.org/10.1007/978-981-15-5444-5_9

such as silicon or coated with a thin metallic layer [7]. Characterization has been essential to the advancement and realization of photovoltaic. Thus, is characterized by more parameters that are determined by the manufacturers under the Standard Test Conditions (STC) [8]. However, the spectral response, also known as External Quantum Efficiency is one of more important characterization, this characterization provides a complete overview of the current of the solar cell relative to the number of photons incident on the panel. Moreover, the spectral response of solar cell is defined by the ratio of the current generated to the power absorbed.

Furthermore, in the characterization of photovoltaic materials, the excellent instruments give more ability to analyze information (optical properties) and experiences in an objective manner. Furthermore, most and large characterization of solar cells using monochromator system. Checking the motor for rotation can be accomplished through the rotation of the grating. However, an only with one wavelength is selected.

Thus the, majority of monochromator have already been used a commutable gratings, in orders to efficiently cover, an immense range of wavelengths. In regards to the solar cells capability of respond [9].

This paper highlights the development and implementation of electronics card, this latter is used for the restarting and eventually the control of the monochromator system, monochromator has been in case of failure at "Laboratory of Thin Film Physics and Materials for Electronics", University of Oran, Ahmed Ben Bella. However, a thorough study of the structure and operation modes of the monochromator system of all those that exist in the country has been accomplished.

The development involves the design and manufacturing of a low cost electronic control boards, based on microcontrollers has been adapted accordingly on monochromator system. Therefore, an effective implementation of resolution ATMEGA328 is also part of the critical work being done. In addition, an I2C [10] current intensity and voltage sensor has been implemented, present a height precision and performance to perform a height reality to characterization process, these new solutions, when we added to the characterization line, will help you to increase productivity through quality control processes that do not require human intervention. Finally, a Graphical User Interface has been designed to clearly instruct you in the use and of the application.

9.2 Methodology

9.2.1 Microcontroller Selection

In this project, an ATMEGA328 has been selected [11]. In particular, it integrates all the electronic functionalities which make it possible to carry out programming works without difficulty; also we use an integrated development card. For this, it is just necessary to connect the card to a PC using a USB cable.

Furthermore, is a reliable element and capable of running a complete program. It represents the solution adapted to control the monochromator and data acquisition. Also it is possible to solder it on electronic boards.

9.2.2 *Sensors Selection*

The sensors used in this development are based on MEMS technology in Fig. 9.1, present height precision and performance in data measurement.

The current sensor used in this project and in order to optimize the acquisition system, is the INA219 model. it is a digital sensor with an I2C interface. It provides digital values of current, voltage and power necessary for precise decision-making in precisely controlled systems [12].

We added in this study a TSL 2561 luminosity sensor, This light sensor uses advanced technology, known as "MEMS", in order to respond to measurement requirements; its measurement range is fairly professional to detect light ranges from up to 0.1–40,000+ Lux. this sensor has advantages, such as precision, low cost and easy configuration in the measurement chain [13].

9.3 Results and Validation

However, once the design, and the suitable complements for control has been selected. Before making this architecture in PCB board, it is advisable to develop a simulator tool, in order to ensure that our motor (monochromator) operate correctly

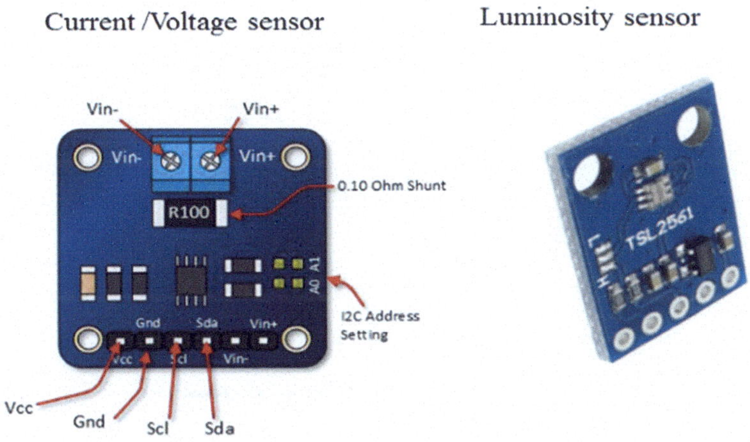

Fig. 9.1 Current/voltage/light sensors

and safely. Moreover, we recommend the use of software's simulation tools. In we present the corresponded schematic for monochromator control (Fig. 9.2).

The following figure shows the test results of the motor control. Therefore, the code program developed is running correctly (Fig. 9.3).

The following figure shows the simulation results of the sensors. Furthermore, in the simulation environment, it was noted that the sensors works nicely. Furthermore, we present in the Fig. 9.4, the implementation of the system for data acquisition on bread board. Firstly, we present the result of light luminosity sensors.

Thereafter the light luminosity sensor test, samples measures was transmitted to the computer. Moreover, a high efficiency has been noted. This experience will also be useful in the second sensor; it is a Current/Voltage sensor, INA 219. In Fig. 9.5, we illustrate the measuring bench.

A series of successful measures have been introduced, in order to ensure an effective functioning of the integrated system, as shown in previous figures.

In addition to these, an electronic card for control and acquiring electrical measurements from the system has been designed with a focus on the motor step control. The electronic board realized.

Finally, an integration and operational test has been carried at laboratory. An encouraging result can be seen in this regard in a number of characterizations. However, the results provide a highly flexible and adaptable facility. Figure 9.6, above presents a photograph of the experimental set up for characterization of solar cells using an electronic development in this study.

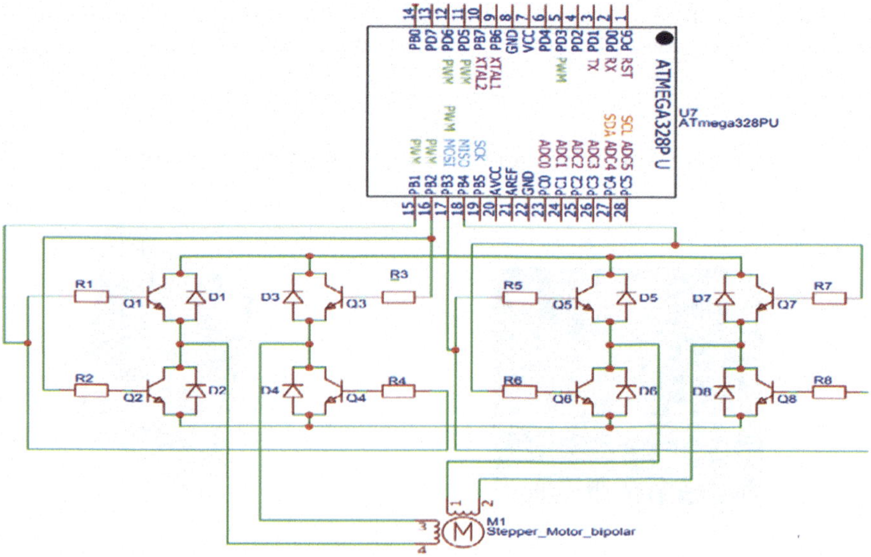

Fig. 9.2 Control card diagram

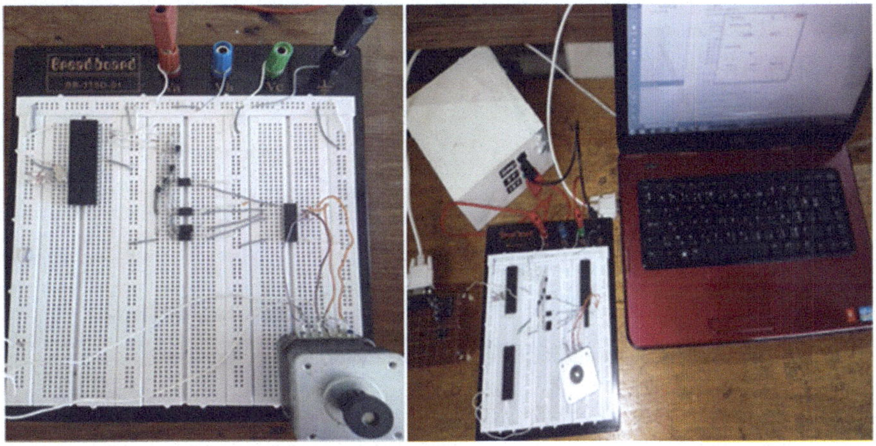

Fig. 9.3 Test results with graphical user interface

Fig. 9.4 Light luminosity sensors test

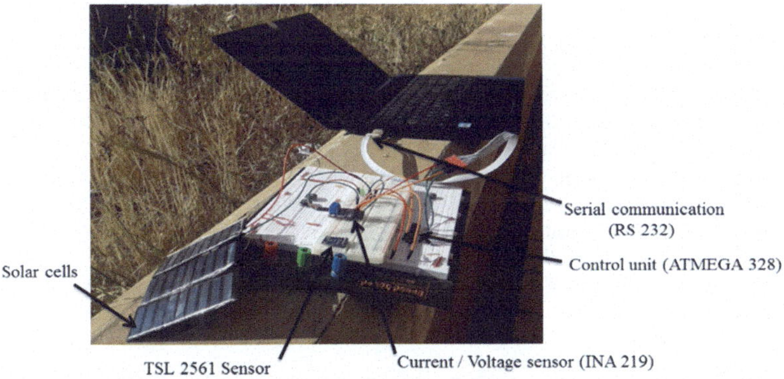

Fig. 9.5 Current/Voltage sensor test

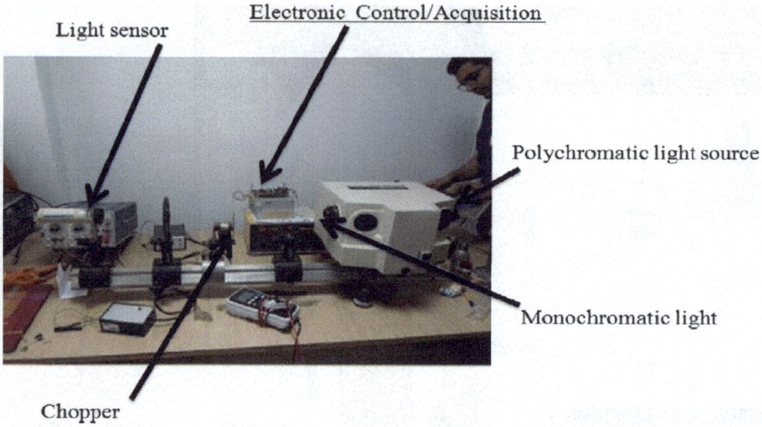

Fig. 9.6 Photograph of the experimental set up for characterization of solar cells

9.4 Conclusion

This paper is mainly focused on development of a novel approach (measurement and control) to efficiently monitor the spectral response of solar cells; enable easy to perform the standard current voltage characterization. The overall cost of making this instrument is favorable, much reduced to that of standard instruments used for the characterization of solar cells. Moreover, the instrument produced during this study provides a complete assessment of the area of solar cells and modules with high reliability and which does not require any additional hardware.

References

1. J. Peng, L. Lu, H. Yang, Review on life cycle assessment of energy payback and greenhouse gas emission of solar photovoltaic systems. Renew. Sustain. Energy Rev. **19**, 255–274 (2013)
2. T. Bhattacharya, A.K. Chakraborty, K. Pal, Effects of ambient temperature and wind speed on performance of monocrystalline solar photovoltaic module in Tripura, India. J. Sol. Energy **2014**, 1–5 (2014)
3. V.V. Tyagi, N.A.A. Rahim, N.A. Rahim, J.A.L. Selvaraj, Progress in solar PV technology: research and achievement. Renew. Sustain. Energy Rev. **20**, 443–461 (2013)
4. G.V.S. Priscila, O.A.G. Mario, Renew. Sustain. Energy Rev. **74**, 590–601 (2017)
5. A. Reinders, P. Verlinden, W. van Sark, A. Freundlich (eds.), *Photovoltaic Solar Energy: From Fundamentals to Applications*, 1st edn. (Wiley, Hoboken, 2017). Part Seven Characterization and Measurements Methods
6. L. Phillips, in *Managing Global Warming*. Solar energy (2019)
7. B. Estibals, Contribution à l'amélioration des chaines de conversion photovoltaïques par l'introduction d'architectures, https://tel.archives-ouvertes.fr/tel-00546390

8. L.L. Kazmerski, Photovoltaics characterization: a survey of diagnostic measurements. Center for Measurements and Characterization, National Renewable Energy Laboratory, Golden, Colorado 80401
9. H. Mäckel, A. Cuevas, Capturing the spectral response of solar cells with a quasi-steady-state, large-signal technique. Prog. Photovoltaics **14**(3), 203–212 (2006)
10. R. Wei, K. Ouyang, X. Bao, X. Gao, C. Chen, High precision smart calibration system for temperature sensors. Sens. Actuators A Phys. **297**, 111561 (2019)
11. https://www.microchip.com/wwwproducts/en/ATmega328
12. http://www.ti.com/lit/ds/symlink/ina219.pdf
13. https://www.mouser.com/ds/2/737/tsl2561-932888.pdf

Chapter 10
Optimal Intelligent Energy Management to Integrate a Photovoltaic Park into Electricity Grid Using a Real-Time Objective Function—Application to the Naâma Park

A. A. Tadjeddine, A. Chaker, F. Zekkour, A. Belaghit, and H. Hamiani

Abstract In order to improve the performance and solve the problem of integration of renewable electrical powers from a photovoltaic park at the peak time, optimal production management as well as synchronization with the increase in normal consumption, not only solve the droop voltage but also minimizes the cost of producing electricity from natural gas. Using software designed and developed, the load curve for a duration of 16 h gives the objective function the set point to find the improved voltage solution and the optimal location for reactive compensation in a real electrical network. This planned solution supports the sustainable development on the radial network 225 kV of Saida-Bechar region, and increases the stability margin for the five substations positions to study in N and N-1 situations. Analysis techniques, modeling and control system are studied and confirmed by exploiting the advanced numerical simulation. In this contribution, we studied the voltage drop in remote stations due to the high electrical load, we have solved the problem of voltage drop by local reactive compensation and the use of active power from photovoltaic farms in order to relieve and reduce the production costs of the gas turbine in bus 15. Finally, a reduction of 19.23% in the total active losses of the electrical system and an optimal location of the bank of capacitors from 5 Mvar to bus 12.

Keywords SMART grid · Power integration · PV · Dispatching · Decentralized production · Real-time

---

A. A. Tadjeddine (✉) · A. Chaker · F. Zekkour
SCAMRE Laboratory, ENPOran - MA, Oran, Algeria
e-mail: a.taj.ali31@gmail.com

A. Belaghit
LABAB Laboratory, ENPOran - MA, Oran, Algeria

A. A. Tadjeddine · A. Chaker · F. Zekkour · A. Belaghit · H. Hamiani
LAAS Laboratory, ENPOran - MA, Oran, Algeria

A. Belasri and S. A. Beldjilali (eds.), *ICREEC 2019*, Springer Proceedings in Energy,
https://doi.org/10.1007/978-981-15-5444-5_10

10.1 Introduction

The Algerian progress society in different sectors; industrial, agriculture, domestic, health, need more energy consumption and an optimal management to avoid the electricity power losses. However, the industrial and demographic development requires the grid manager to investigate, monitoring and planning optimal structuring production service, transmission, repartition and distribution of electricity, constantly respecting the economic and ecological aspect.

The energy transition towards renewable energy is the perfect tool for integrating the power produced by the photovoltaic system, which may have several advantages, but there are still technical problems with the stability of the electricity grid [1].

Allowed these disadvantages, there may be mentioned the low installed power to be injected, especially acute change in the electrical charge that has fluctuations in the sinusoidal shape factor and the difficulty in forecasting production of PV. As a result, several problems will be induced, for example phase imbalance and voltage; the change of power flows; uncontrollable overshoot of network capabilities; the dysfunction of the protections … etc. [2].

The optimal management of integration of renewable energy resources (photovoltaic and wind) and reactive power control devices in electrical grids call for real-time status information to synchronize between all energy sources in the grid, to improve energy effectiveness and profiles voltage substations in different operating conditions [3]. The good quality of the power and the continuity of service play the essential part in the optimal management of the electrical network, then, the compensation of the reactive power transited is the means that ensures the stability of the voltage and increasing the active power flow. The reactive power compensation, capacitive TSC (Thyristor switched capacitor) or inductive TCR (Thyristor-Controlled Reactor) requirement installed for transport grids on an electrical interconnection or compensation substation, according to the desired objectives. The capacitive compensation for the distribution grid frequently installed upstream of consumers to optimize the power factor [3, 4].

The transit of reactive power in enormous capacities through transmission and repartition lines affects the disturbance of the electricity grid, in particular the overall stability of the electricity system. So this difficulty increases the total active and reactive losses, and consequently the voltage drop in the substations which obliges the electricity distribution companies to increase the regulation range of the autotransformers and use a many primary sources to produce electricity, that is to say an increase in the costs of managing electrical energy. The voltage drop of a transmission line AC is given by the Eq. (10.1) [5]:

$$U_{3ph} = \sqrt{3} \cdot I \cdot (R \cdot \cos \phi 1 + \omega L \cdot \sin \phi 1) \tag{10.1}$$

The maximum voltage drop in an AC system conventional is given by the Eq. (10.2) [5]:

$$u = \frac{100\% \cdot P \cdot l}{\kappa \cdot A \cdot U2} = \frac{100\% \cdot \sqrt{3} \cdot I \cdot \cos \phi \cdot l}{U \cdot A \cdot \kappa} \quad (10.2)$$

Subsequently, the maximum transferable power will be (10.3) [5]:

$$P_{max} = \frac{u \cdot U2 \cdot A \cdot \kappa}{100\% \cdot l} \quad (10.3)$$

So, the relative loss of transmissible power Δp is (10.4) [5]:

$$\Delta p = \frac{u}{(cos\,\phi 1)^2} \quad (10.4)$$

where, P: active power, U: complex voltage, I: line current, u: relative voltage (%), cos(ϕ1): power factor, A: line section, l: line length, κ: conductivity.

The objective of this work is to study the impact of the integration of photovoltaic parks (PV) and reactive compensation on the power generated by the gas turbine and the correction of the voltage in remote bus in electrical grid of Saida-Bechar 225/63 kV.

10.2 Methodologies and Devices

For this study, we opted for the choice of a strategic site, collection of electrical data, and finally, implementation and realization of advanced calculations by the computer.

10.2.1 System Description

The studied system is composed as indicated in Table 10.1 below:

Figure 10.1 shows the model in the simulator of study.

Table 10.1 Electrical system data

Generator	Central TG/CC	Renewable energy sources
		Parcs photovoltaic
	02	03
Buses	Load buses	Generator buses
	13	07
SVC	02	

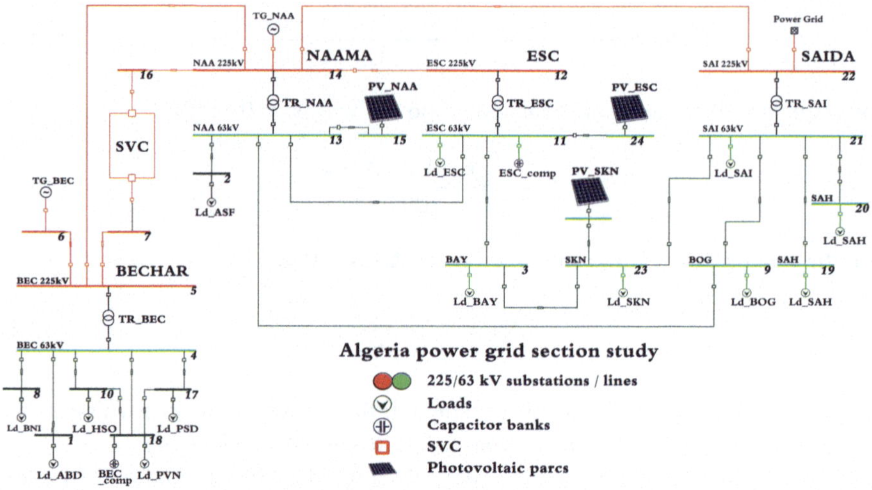

Fig. 10.1 Study model with 225/63 kV autotransformers

10.3 Results and Discussion

The power distribution study for the permanent case was performed by the Newton Raphson Adapted method and for the transient stability analysis; we used the Gauss Seidel method with the acceleration factor of 1.48 and a precision from 10^{-5}.

10.3.1 Static Regime

In Fig. 10.2 shows the nodal voltage violations in the initial case without using the three photovoltaic parks and without compensating reactive energy in the two nodes of 225 kV (bus 6, 14).

In Fig. 10.3 shows the calculated and optimal corrections for nodes that have voltage violations.

The results obtained for the distribution of loads without PV and OEC (Optimal Location Compensator) show the different critical nodes. Note that 09 nodes have critical voltage violations, two of which have violations above the upper limit (bus 13, 20).

After the calculation and the correction of the voltage by the reactive compensation and the integration of the active power by the three photovoltaic parks one notices well that all the tensions go within the admissible limits.

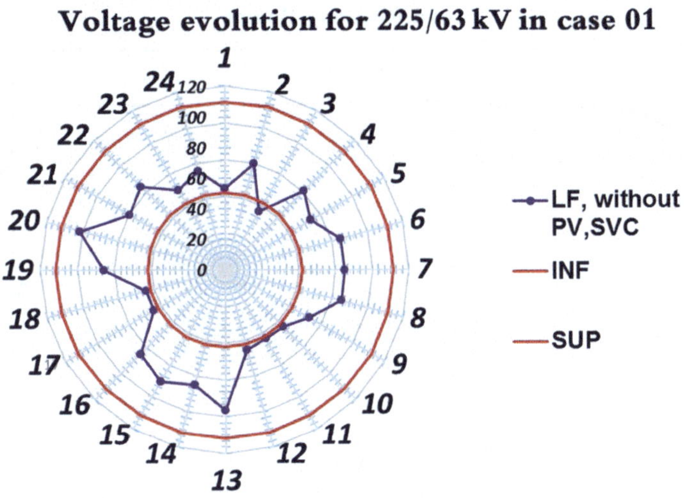

Fig. 10.2 Node voltages without PV and without compensation

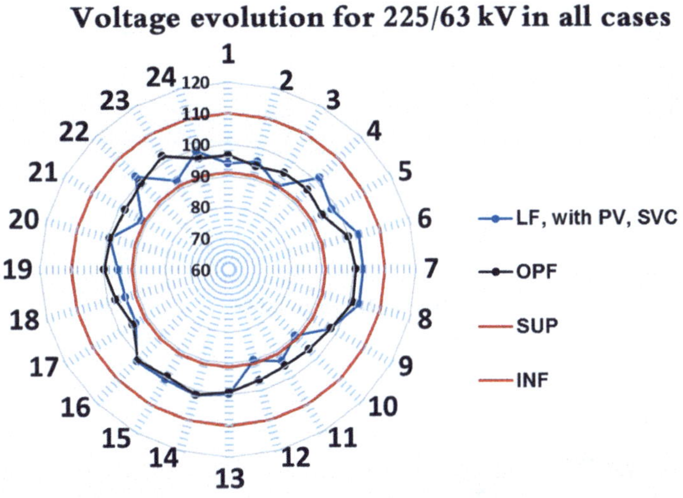

Fig. 10.3 Tension results with integration of the 03 PVs and OEC

10.3.2 Dynamic Regime

Figures 10.4, 10.5 and 10.6 show the reactive power curve, the voltage variation and the voltage evolution by frequency in the 05 nodes of the single 63 kV substation.

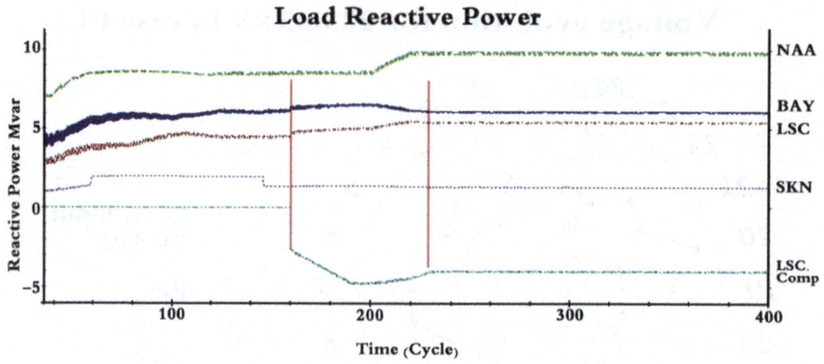

Fig. 10.4 Reactive charge in the nodes

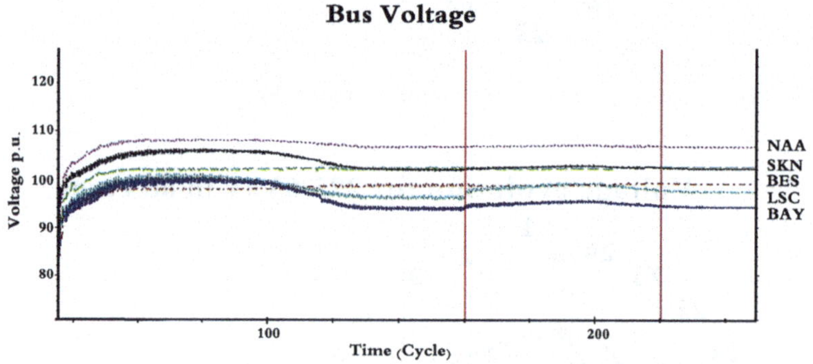

Fig. 10.5 Nodal voltages evolution

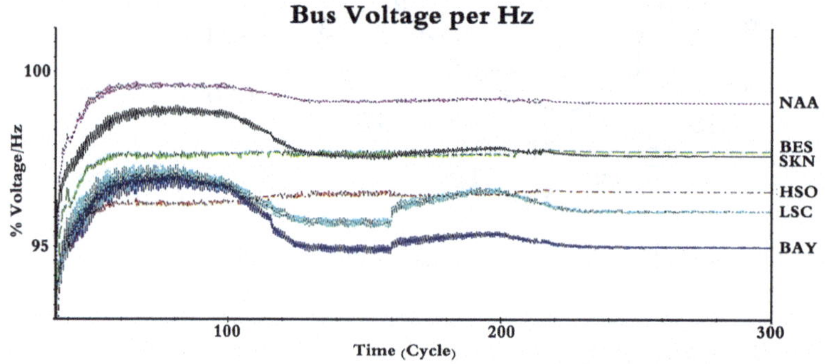

Fig. 10.6 Nodal voltage per Hz

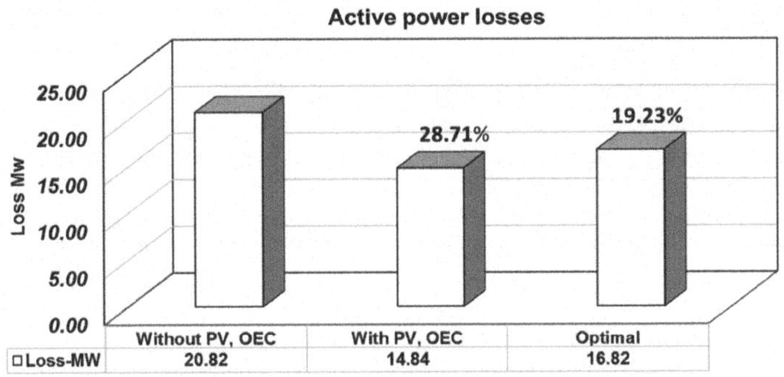

Fig. 10.7 Evolution of the total active losses

In order to solve the problem of the voltage collapse in the target nodes, the algorithm of optimal location of the capacitor banks gives a capacity of 5 Mvar to the node 12 with the use of the SVCs of the nodes 8 and 16.

In Fig. 10.4, the interval between vertical red lines presents the integration of the reactive power of 40% of the maximum capacity, for this reason, it is noted that the voltage is corrected in Fig. 10.6, however there are alternations in frequency. The algorithm of our objective function follows these changes so that the frequency and voltage remain within the permissible limits.

The active losses of the system are reduced thanks to the reactive compensation and the integration of photovoltaics on the network (see Fig. 10.7).

The loss reduction rate is estimated at 28.71% compared to the initial case for the second case. We distinguish that there is a small increase in the optimal case with a reduction of 19.23%. This result confirms that the objective function to correct the voltages taking into account the other factors that react on the network.

10.4 Conclusion

In order to follow the load curve change guideline, the optimal management by an objective algorithm seeks to minimize the active, reactive losses and to correct the voltage collapse in remote stations of the 225 kV Saida-Bechar radial network.

In this contribution, we studied the voltage drop in remote stations due to the high electrical load, we have solved the problem of voltage drop by local reactive compensation and the use of active power from photovoltaic farms in order to relieve and reduce the production costs of the gas turbine in bus 15.

Finally, a reduction of 19.23% in the total active losses of the electrical system and an optimal location of the bank of capacitors from 5 Mvar to bus 12.

References

1. L.L. Grigsby, *Electric Power Generation, Transmission, and Distribution*, 3rd edn. (CRC Press edn., Electric Power Generation, 2012), pp. 2.1−2.9
2. VECHIU, Modeling, control and integration of decentralized production in networks, Dissertation, Grenoble INP, chap. 01,04, 2013
3. W. Zhang, W. Liu, X. Wang, L. Liu, F. Ferrese, Distributed multiple agent system based on optimal reactive power control for smart grids. IEEE Trans. Smart Grid **5**(5), 2421–2431 (2014)
4. R.R. Bayliss, B.J. Hardy, *Transmission and Distribution of Electrical Engineering*, 4th edn. (Newnes edn., 2012), pp. 1124–1131. Chap. 28 Fundamentals
5. H. Wolfgang, J. Jurgen, J. Wolfgang, *Reactive Power Compensation - A Practical Guide* (Wiley edn., 2012), pp. 22–27. Chapter 3 Effect of Reactive Power

Chapter 11
Direct Power Control Improvement Using Fuzzy Switching Table of Grid Connected PV System

Sofia Lalouni Belaid, Zoubir Boudries, and Amel Aoumer

Abstract Photovoltaic (PV) energy system has considerable focus during the last decade. PV is a very attractive source of renewable energy since it's an inexhaustible source, free and clear. A lot of paper has been focused to the optimum exploitation of the PV energy in order to maximize its efficiency. MPPT based on Fuzzy logic controller is applied in this paper in order to accelerate the reaching of maximum PV power. The principal purpose of the paper is the improvement of conventional direct power control (CDPC) performance applied to grid connected photovoltaic system. In this technique, the selection of the switching state vectors of the inverter is based on the errors between the instantaneous reactive and active powers and the respective reference quantities applied to hysteresis comparators. The output of these controllers in combination with the located sector, constitute the three inputs of CDPC table. In this paper CDPC is improved using fuzzy switching table for the selection of the optimal voltage vector position namely fuzzy direct power control (FDPC). The objectives are the reduction of harmonic of grid current, the reactive power compensation and the injection of maximum PV power into the grid. In order to evaluate the fuzzy DPC performances, the global studied system is modeled and designed in MATLAB/Simulink; the obtained results show that the active power is perfectly tracked using MPPT fuzzy controller and the reactive power tends toward zero with power factor close to unity. The grid phase current waveform is sinusoidal with THD of 0.25%.

Keywords PV system · Direct Power Control (DPC) · Fuzzy DPC

S. L. Belaid (✉) · Z. Boudries · A. Aoumer
Laboratoire de Technologie Industrielle et de l'Information (LTII), Faculté de Technologie, Université de Bejaia, Bejaia, Algeria
e-mail: lalouni_sofia@yahoo.fr

A. Belasri and S. A. Beldjilali (eds.), *ICREEC 2019*, Springer Proceedings in Energy,
https://doi.org/10.1007/978-981-15-5444-5_11

11.1 Introduction

Solar energy is the kind of renewable power generation technologies. It is being widely used, since it is free, abundant, clean, and its technology is now well developed [1]. Photovoltaic installations are classified according to their operating requirements named standalone and grid connected PV systems. In standalone application, isolate area is supplied by DC or AC power with converters and energy storage equipment [2]. In grid connected application, the generated power supply to the utility services without any energy storage devices. Industrial and domestic equipment use more and more electronic circuits with a nonlinear behavior responsible for grid pollution, they cause power quality deterioration and generate harmonics [3]. These last decades, PWM converters have become the most popular electronic devices used as an interface between the electricity production system and utility grid due to their inherent merits: the power flows bi-directionally, the line current presents low THD, the size of the filter capacitor is reduced and the DC side voltage is well stabilized. Many researchers have been done on control strategies of such system such as Voltage-oriented control (V_{OC}) [4] and p–q theory [5]. Recently, direct power control (DPC), proposed by Akagi et al. in 1983 is widely used [6, 7]. The DPC has become popular as an alternative to the conventional V_{OC} strategy for grid connected PWM converters [8]. DPC method has many advantages over conventional control techniques: easy to implement, quick response and robustness [7–9]. The major inconvenience of DPC is located in its variable switching frequency which mainly depends on the hysteresis comparator bands and the DPC table structure; that introduces a wide band of harmonics into the grid current. This problem can be minimized by averted hysteresis comparators [7, 10]. This paper is focused on the studied of a grid connected photovoltaic system with a new approach namely Fuzzy DPC (FDPC), based upon a replacement of both hysteresis comparators and switching table by fuzzy logic controller, for the purpose of the grid injection of maximum produced PV power, the control of reactive power to provide unit power factor and minimize harmonics.

11.2 Control System Description

The circuit topology of the studied grid PV system includes PV generator connected through controlled DC/DC converter and inverter to an AC grid as shown in Fig. 11.1. The DC/DC converter is a boost one, it achieves the MPPT function using fuzzy logic controller, for the maximization of power coming from PV generator. Whereas, the inverter is used to transfer the PV power to the grid, it's controlled using fuzzy direct power controller with the view to improve the performances of the injected powers.

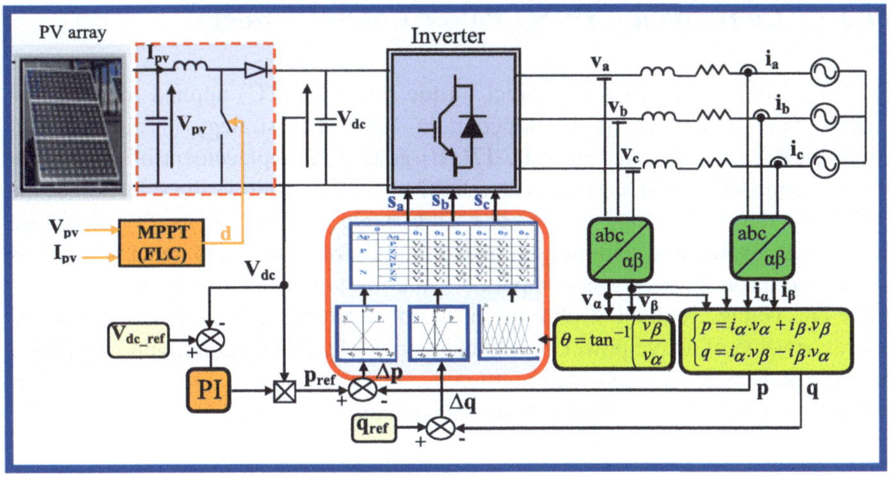

Fig. 11.1 FDPC strategy of grid connected PV system

11.2.1 Modeling of PV Array and MPPT Control

Different models of photovoltaic array are presented in the literature [1–4, 9]. The model chosen in this paper is a five parameters model (I_{sc}, V_{oc}, I_{mpp}, V_{mpp} and R_s) of PV array whose electrical parameters are given in [2]. The photovoltaic current is calculated using the following equation:

$$I_{pv} = I_{sc}.\{1 - K_1[\exp(K_2 V_{pv}^{K_3}) - 1]\} \tag{1}$$

Where, I_{sc} the short circuit current, $K_1 = 0.01175$ is determined experimentally and the coefficients K_2 and K_3 are given in [2]. The PV array $P_{pv}(V_{pv})$ and $I_{pv}(V_{pv})$ characteristics have non-linear nature, then, an MPPT control must be introduced for maximum operating of PV generator. Fuzzy logic control (FLC) is used in this paper [1, 2, 4].

11.2.2 Fuzzy Logic MPPT Controller

The fuzzy MPPT controller (FLC) has better advantages over conventional methods; it is robust, adaptive and doesn't need exact mathematical model of the system but experience of designer is required to choose fuzzy parameters [4]. The FLC uses as measurement the PV voltage and current then calculates PV power, it disturbs the PV voltage with either positive or negative increment with small or large value in the direction that can enhance the PV power. The MPPT FLC blocs and rules are indicated in [2, 4].

11.2.3 Conventional Direct Power Control Strategy

In 1991, Ohnishi proposes the direct torque control (DTC) applied to induction motors, after that Noguchi developed it in 1998. The strategy of direct power control (DPC) was inspired by DTC [7]. Its goal is the replacement of the current control loop and the PWM block with a DPC table which allows the selection of the optimal inverter switching states. The DPC table uses as inputs the difference between the reference and the estimated values of active and reactive powers digitalized through hysteresis comparators and the voltage source vectors position. After DC voltage regulation, through a proportional-integrator (PI) controller, the reference value of active power p_{ref} is achieved while the operating with unit power factor is performed by imposing the reference reactive power q_{ref} at zero [7, 9]. The estimation of powers is obtained from this equation:

$$\begin{cases} p = i_\alpha.v_\alpha + i_\beta.v_\beta \\ q = i_\alpha.v_\beta - i_\beta.v_\alpha \end{cases} \quad (2)$$

The comparison results between both reference value and estimate one of either reactive or active powers represent the inputs of the two levels hysteresis comparators. The voltage source vectors position (θ) is determined as follow:

$$\theta = \arctan\frac{v_\beta}{v_\alpha} \quad (3)$$

The conventional DPC table is presented in Table 11.1, it has three inputs: the two digitized variable (*d*p, *d*q) and the line voltage position θ. The output corresponds to the switching vector V_i (S_a, S_b, S_c) that can be generated by the voltage source inverter.

11.3 DPC Using Fuzzy Logic Controller (FDPC)

The major inconvenience of DPC is located in its variable switching frequency which mainly depends on the hysteresis bands and the structure of switching table, it introduces a wide band of harmonics into the grid current. This problem can be

Table 11.1 Switching table of classical DPC

θ		θ_1	θ_2	θ_3	θ_4	θ_5	θ_6
d_p	d_q						
1	1	V_3	V_4	V_5	V_6	V_1	V_2
	0	V_5	V_6	V_1	V_2	V_3	V_4
0	1	V_2	V_3	V_4	V_5	V_6	V_1
	0	V_1	V_2	V_3	V_4	V_5	V_6

Table 11.2 Switching table of FDPC [10]

θ		θ_1	θ_2	θ_3	θ_4	θ_5	θ_6
Δp	Δq						
P	P	V_3	V_4	V_5	V_6	V_1	V_2
	Z	V_7	V_0	V_7	V_0	V_7	V_0
	N	V_5	V_6	V_1	V_2	V_3	V_4
N	P	V_1	V_2	V_3	V_4	V_5	V_6
	Z	V_0	V_7	V_0	V_7	V_0	V_7
	N	V_6	V_1	V_2	V_3	V_4	V_5

minimized by replacing hysteresis comparators and classical DPC table by intelligent controller. By exploiting Fuzzy DPC controller (FDPC) we can divide the active and reactive power errors into more subsections instead of two states (0, 1) [10, 11]. Then, the generated switching vector will be based on the different levels of power errors. The fuzzy inference system has three inputs: the two power (active and reactive) errors and the voltage source vectors position the voltage space vector represent the controlled output. The membership function of active power error is characterized by (02) linguistic terms: Positive (P) and Negative (N) while the membership function of reactive power error is depicted by (03) linguistic terms: Positive (P), Zero (Z) and Negative (N). The voltage vectors position (θ_n) is divided into six 06 sectors and represented by six fuzzy sets ($\theta_1-\theta_6$) distributed between 0° and 360°. The grid PV system controlled by FDPC is illustrated in Fig. 11.1. The FDPC strategy rule table is given in Table 11.2 [10]. The inference method of Mamdani is used and the defuzzification is made using the criterion of maximum. The outputs can't be directly applied to switches S_a, S_b and S_c.

11.4 Simulation Results and Analysis

The grid connected PV system consists of photovoltaic array of 3.3 kW (03 strings with 10 PV panels connected in series). The MPPT is controlled by fuzzy logic controller. The proposed fuzzy direct power controller FDPC is used for the control of the inverter. The simulation of the controlled system of Fig. 11.1 is done using MATLAB/Simulink. To show the performances of the FDPC strategy and to test its robustness, different level of irradiance are applied (Fig. 11.2). The DC voltage waveform plotted in Fig. 11.3 is well controlled to its reference value of 550 V and reached its steady state in short time with negligible oscillation. The grid phase voltage and current are synchronized with unit power factor as presented in Fig. 11.4. Figure 11.5 illustrations the waveform of grid powers (active and reactive). The powers are decoupled and tracked their reference values, where the reactive power tends toward zero. As we can see in Figs. 11.6 and 11.7, which present a zoom of active and reactive power, the conventional DPC presents significant ripples and it's evident that the FDPC offers an improved waveform with

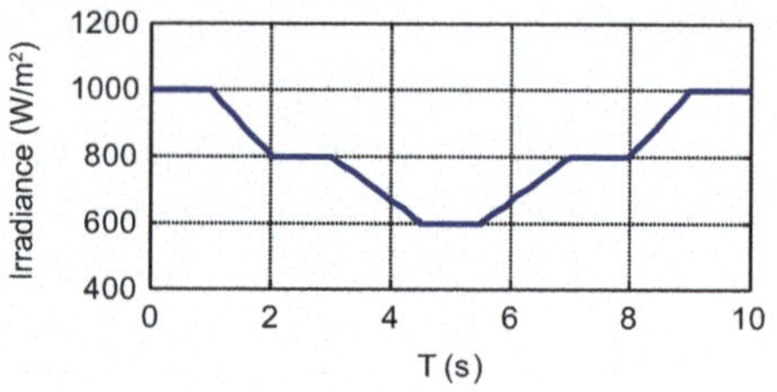

Fig. 11.2 Solar irradiance

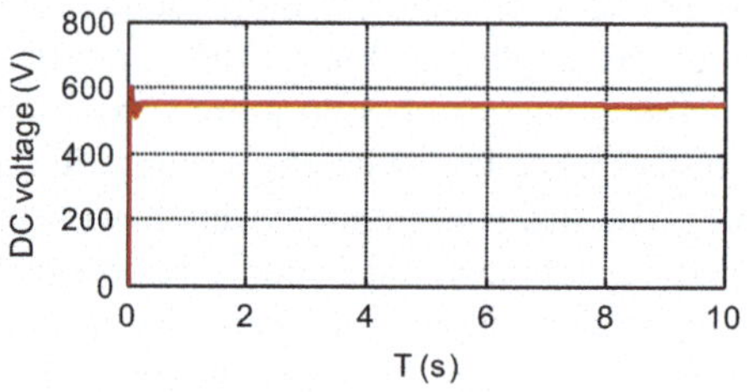

Fig. 11.3 DC bus voltage

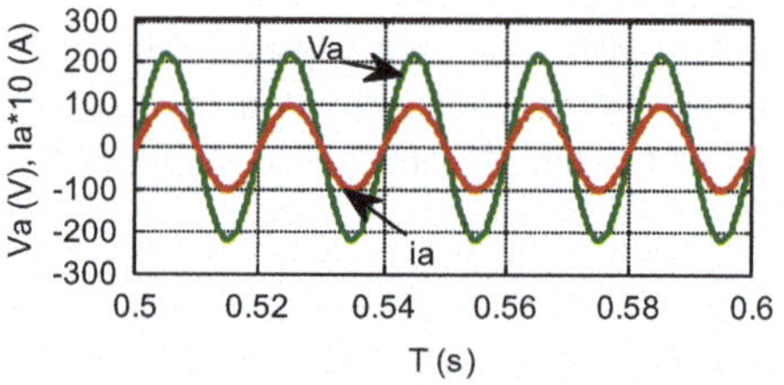

Fig. 11.4 Generator PV voltage

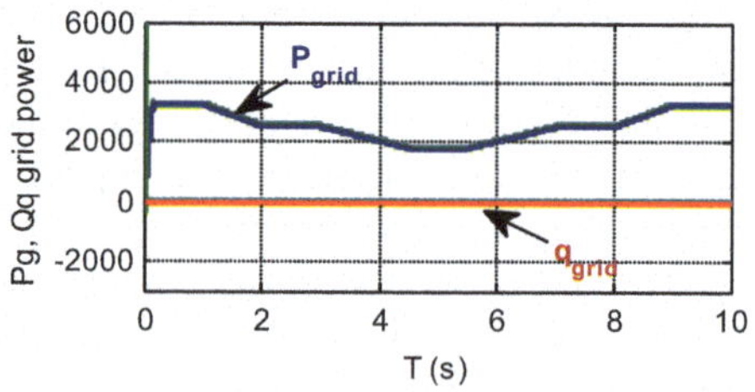

Fig. 11.5 Grid active and reactive powers

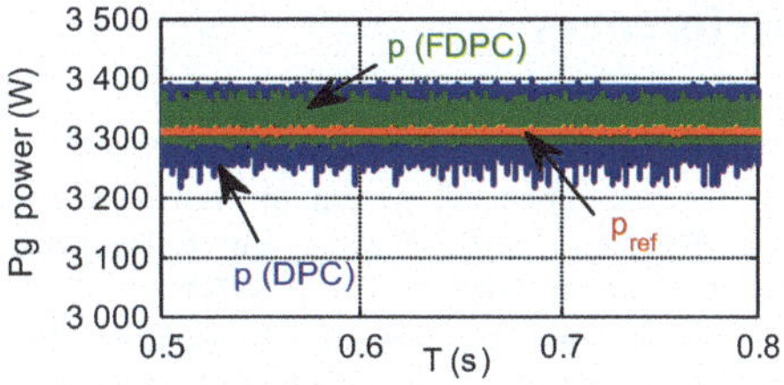

Fig. 11.6 Zoom of grid active power

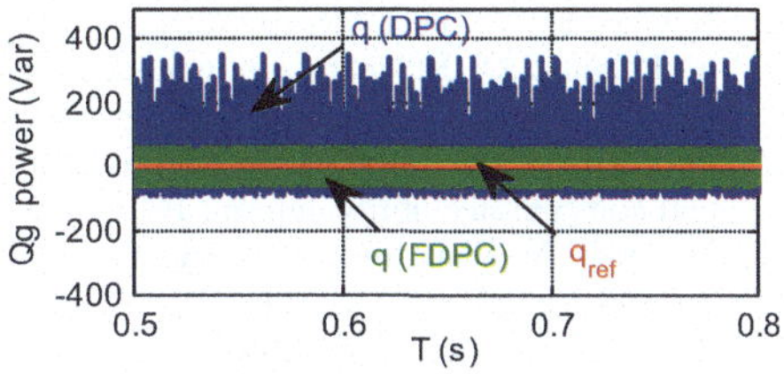

Fig. 11.7 Zoom of grid reactive power

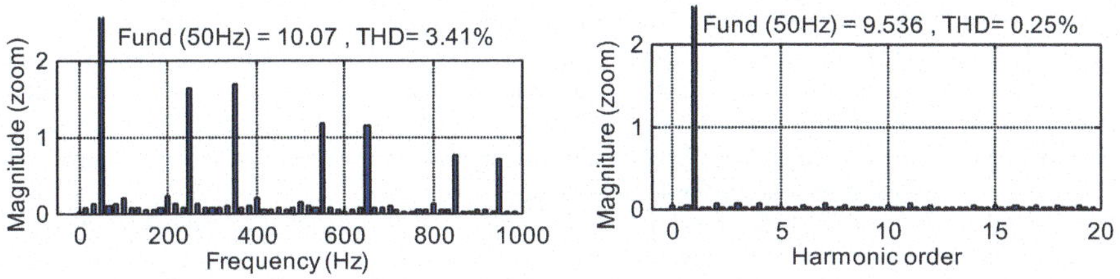

Fig. 11.8 Grid current harmonic spectrum with (DPC) (in the left side) and (FDPC) (in the right side)

fewer ripples. The grid current waveforms for both control techniques DPC and FDPC is sinusoidal with a THD respectively of 3.41% and 0.25% as shown in Fig. 11.8.

11.5 Conclusion

The FDPC strategy is proposed to counter the drawbacks of the conventional DPC related to high ripples that introduces a wide band of harmonics into the grid current. With the FDPC strategy, hysteresis comparators and DPC table are replaced by an intelligent method. The simulation has been done with MATLAB's software for comparing the fuzzy control strategy with the conventional one. From the obtained results, it can be concluded that the principal goals have been attained: the injection of the PV power tracked using FLC into the grid, the reactive power compensation, the unit power factor and low THD (0.25%) of grid current.

References

1. S. Lalouni Belaid, N. Benyahia, Power management of a stand-alone hybrid (photovoltaic/hydrogen/battery) system, in *Fourth International Conference on Energy, Materials, Applied Energetics and Pollution*, vol. 1 (Publisher: Applied Energetics Laboratory, Constantine 1 University, 2018), pp. 1–6. ISBN 978-9931-9229-4-0
2. S. Lalouni, D. Rekoua, T. Rekoua, E. Matagne, Fuzzy logic control of stand-alone photovoltaic system with battery storage. J. Power Sources **193**, 899–907 (2009)
3. M.R. Bengourina, M. Rahli, L. Hassaine, Direct power control of a grid connected photovoltaic system, associated with an active power filter. Revue des Energies Renouvelables **20**(1), 99–109 (2017)
4. S. Lalouni, D. Rekioua, Optimal control of a grid connected photovoltaic system with constant switching frequency. Energy Procedia **36**, 189–199 (2013)
5. N.D. Tuyen, G. Fujita, PV-active power filter combination supplies power to nonlinear load and compensates utility current. IEEE Power Energy Technol. Syst. J. **2**, 1–11 (2015)
6. H. Akagi, Y. Kanazawa, A. Nabae, Generalized theory of the instantaneous reactive power in three-phase circuits, in *IPEC 1983-International Power Electronics Conference, Tokyo, Japan* (1983), pp. 1375−1386
7. A. Chaoui, J.-P. Gaubert, A. Bouafia, Direct power control switching table concept and analysis for three-phase shunt active power filter. J. Electr. Syst. **9**(1), 52–65 (2013)
8. T.A. Trivedi, R. Jadeja, P. Bhatt, A review on direct power control for applications to grid connected PWM converters. Eng. Technol. Appl. Sci. Res. **5**(4), 841–849 (2015)
9. A.M. Nader, D. Abderrahmane, Direct power control for a photovoltaic conversion chain connected to a grid. Rev. Roum. Sci. Techn. Électrotechn. et Énerg. **61**(4), 378–382 (2016)
10. H. Hamdaoui, A. Semmah, A. Masssoum, P. Wira, Aa. Ayad, A. Meroufel, Elaboration d'une table de commutation floue Pour la commande d'un filtre actif de puissance. Rev. Roum. Sci. Techn. Électrotechn. et Énerg. **58**(4), 405–414 (2013)
11. H. Sudheer, S.F. Kodad, B. Sarvesh, Direct torque and flux control of induction machine using fuzzy logic controller, in *International Conference on Advances in Electrical, Electronics, Information, Communication and Bio-Informatics (AEEICB 2016)* (2016), pp. 1–5

Chapter 12
Numerical and Experimental Study of Parabolic Trough Solar Collector

Belkacem Agagna and Arezki Smaili

Abstract The Solar power plants with parabolic trough collectors (PTCs) represent more than 92% of the solar thermal power plants currently in operation. It is also the most suitable technology for most plants under construction and planned [1]. The present study evaluates the thermal performance of LS-2 parabolic trough collector of Solar Electric Generating System (SEGS VI) plant. A thermal model is developed for the solar receiver based on comprehensive energy analysis. The model has been validated with the experimental data from Sandia National Laboratory. Besides, the predictions from the model were compared with the published modeling studies (both on sun and off sun tests have been considered). In all cases, the results of the developed model show a good agreement with the experimental tests and the prediction results of the other models. Particular interest has been given for the impact of the glass cover diameter on the efficiency of the solar receiver.

Keywords Parabolic trough solar collector · Solar receiver · Heat transfer analysis · Modeling

12.1 Introduction

The solar receiver is the main component in the PTC, it absorbs the incident solar radiation and converts it into useful heat. The improvement of the performance of this component requires an accurate analysis of the thermal losses. This can be done through a comprehensive heat transfer analysis. In this work, we are interested in the solar collector of the power plant (SEGS VI) [2, 3]. The collector used in this plant (LS-2) has been tested in SNL laboratories (Sandia National Laboratories), Albuquerque (New Mexico) site. The results of the experiments were published by

B. Agagna (✉) · A. Smaili
Laboratoire de Génie mécanique et Développement, Ecole Nationale Polytechnique, B.P. 182, El-Harrach, Algeria
e-mail: belkacem.agagna@g.enp.edu.dz

A. Belasri and S. A. Beldjilali (eds.), *ICREEC 2019*, Springer Proceedings in Energy,
https://doi.org/10.1007/978-981-15-5444-5_12

Dudley [2]. Forristall [3] developed a 1D model implemented in EES software (Engineering Equation Solver) and validated it through a detailed comparison with SNL experimental data. García-Valladares et al. [4] have developed a numerical model for a single-pass receiver. They are also proposed a double pass type absorber. Cheng et al. [5] combined the optical analysis with CFD to simulate the complex heat transfer problem of a PTC. Kalogirou [6] has established the energy balance of the receiver and developed a model written in SEA language. The model has been validated on the basis of Dudley's experiments [2].

The present paper presents comprehensive modeling of the solar receiver. It is organized in two parts. The first part present the model and its validation by comparison with experimental data from Dudley [2]. In the second part, a comparative study with previous work is presented, with a focus on the impact of variation in the diameter of the glass cover on system performance.

12.2 Mathematical Modeling

The absorber tube is often coated with a selective layer and surrounded by a transparent glass envelope. It is placed on the focal line of the PTC, as shown in Fig. 12.1a. The incident solar energy absorbed, is not entirely transmitted to the heat transfer fluid because a part is dissipated in the form of heat losses. Figure 12.1b illustrates the heat transfer in LS2 solar receiver. To analysis the heat transfer, we have developed a model, which allows energy analysis of the parabolic trough concentrator and the prediction of the thermal performance.

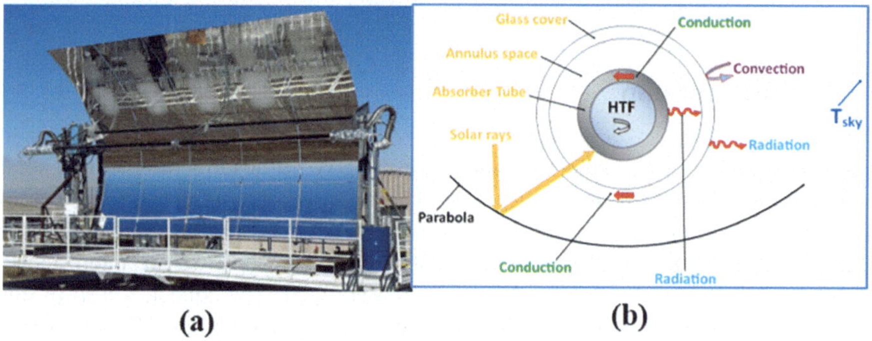

Fig. 12.1 LS2 PTC tested at SNL (left). Heat transfer model of the PTC (right)

12.2.1 Hypothesis

When calculating the energy balance for a turbulent regime, the following assumptions have been taken into account: (i) The regime is permanent; (ii) The thicknesses of the glass and the absorber are negligible [2, 4, 6]; (iii) The solar flux at the absorber is uniform; (iv) For a vacuum annulus, convective exchanges are negligible.

12.2.2 Thermal Model

Convection Heat Transfer Between the HTF and the Absorber Tube. The convection heat transfer from the receiver tube to the HTF can be given by:

$$q_{abs} = h_i.\pi D_{r,i}\left(T_r - T_f\right).L \tag{12.1}$$

With:

$$h_i = \frac{Nu.k_f}{D_{r,i}} \tag{12.2}$$

where h_i is the HTF convection heat transfer and is evaluated as a function of the Nusselt number N_u, and k_f represents the thermal conductance of the HTF [6].

Thermal Radiation Heat Transfer in the Annulus Space. The radiation heat transfer between the receiver tube and glass envelope is estimated:

$$q_{loss\, r,an} = h_{r,an}.A_{r,o}.\left(T_r - T_g\right) \tag{12.3}$$

With

$$h_{r,an} = \frac{\sigma\left(T_r^2 + T_g^2\right)\left(T_r + T_g\right)}{\frac{1}{\varepsilon_r} + \left(\frac{A_{r,o}}{A_{gl,i}}\right)\left(\frac{1}{\varepsilon_g} - 1\right)} \tag{12.4}$$

where, Tg is the glass envelope temperature and $h_{r,an}$ is the radiative heat transfer coefficient between the receiver tube and the glass envelope.

With: ε_r, ε_g and σ are the emissivity of the receiver, the emissivity of the glass envelope and the Stéfan-Boltzmann constant, respectively. $A_{gl,i}$, $A_{r,o}$ are the inner surface of the glass cover (m^2), the outer surface of the absorber tube (m^2), respectively.

Convection Heat Transfer from the Glass-Cover to the Ambient. The convection heat transfer from the glass envelope to the ambient is the largest source of heat loss, especially if there is wind. From Newton's law of cooling [6]:

$$q_{loss\,conv,ext} = h_w.A_{gl,o}.\left(T_g - T_a\right) \tag{12.5}$$

With

$$h_w = (Nu.k)/D_{gl,o} \tag{12.6}$$

Nusselt number can be estimated as function of Reynolds number [3, 6].

Radiation Heat Transfer from Glass Cover to the Sky. The net radiation transfer between the glass and the sky can be given by [3, 6]

$$q_{loss\,r,ext} = h_{rca}.A_{gl,o}.\left(T_g - T_a\right) \tag{12.7}$$

With

$$h_{rca} = \sigma\varepsilon_g\left(T_g + T_a\right)\left(T_g^2 + T_a^2\right) \tag{12.8}$$

12.3 Results and Discussion

12.3.1 Estimation of the Thermal Efficiency

Table 12.1 illustrates the operating conditions [2] and the results obtained.

Table 12.1 Operating conditions and obtained results

Case	Gb (W/m^2)	Wind (m/s)	Flow (l/mn)	T_{amb} (°C)	T_{in} (°C) HTF	Delta Air (°C)	η_{Dudley} (%) Exp [2]
1	933.7	2.6	47.7	21.2	102.2	91.9	72.51
2	968.2	3.7	47.8	22.4	151.0	139.8	70.90
3	982.3	2.5	49.1	24.3	197.5	184.3	70.17
4	909.5	3.3	54.7	26.2	250.7	233.9	70.25
5	937.9	1.0	55.5	28.8	297.8	278.6	67.98
6	880.6	2.9	55.6	27.5	299.0	280.7	68.92
7	903.2	4.2	56.3	31.1	355.9	334.1	63.82
8	920.9	2.6	56.8	29.5	379.5	359.4	62.34

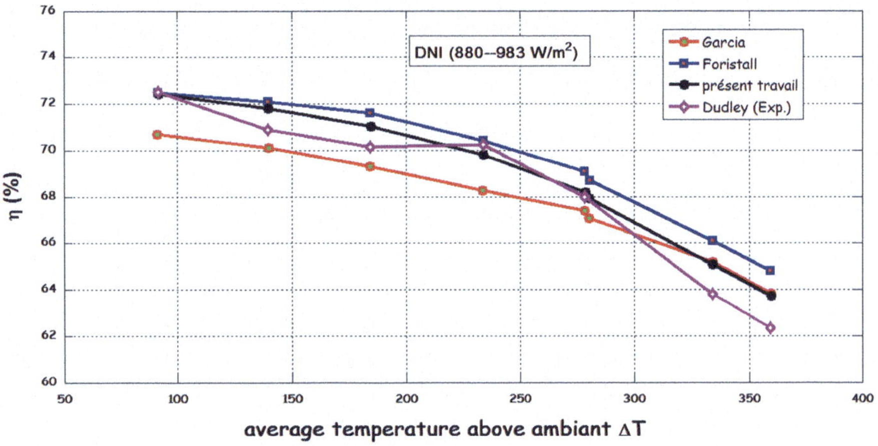

Fig. 12.2 Thermal efficiency predictions of the PTC vs. ΔT

From Fig. 12.2, it is noted that the efficiency of the system is maximum where the temperature difference ΔT is low. Besides, the efficiency of the solar concentrator decreases remarkably with the increase of ΔT. This is because heat losses increase with the increase in the operating temperature.

12.3.2 Estimation of Heat Losses

Table 12.2 shows the operating conditions [2] with solar irradiation of zero W/m^2. The results show an increase in heat losses with the rise of the temperature difference ΔT. High operating temperature increases exponentially the radiation heat losses to the ambient, which explains clearly the decrease of the yield in the preceding cases.

Table 12.2 Operating conditions and obtained results

Case	Wind (m/s)	Flow (l/mn)	T_{amb} (°C)	T_{in} (°C)	T_{out} (°C) Exp	q''_{Dudley} (W/m^2) Exp [2]	$q''_{present\ model}$ (W/m^2)	q''_{Garcia} (W/m^2) [4]
1	3.2	27.4	26.3	99.55	99.54	0.3	3.75	3.81
2	2.9	27.4	25.4	100	99.97	0.85	3.82	3.91
3	1.1	53.6	19.9	153.4	153.3	5.3	8.55	8.68
4	0.1	54.7	22.5	199.4	199.0	14.04	13.71	13.57
5	1.5	55.6	24.2	253.8	229.2	23.4	22.91	23.04
6	2.0	56.0	26.7	299.0	297.9	36.7	32.68	32.6
7	0.6	56.8	27.6	348.3	319.9	55.8	45.57	45.65

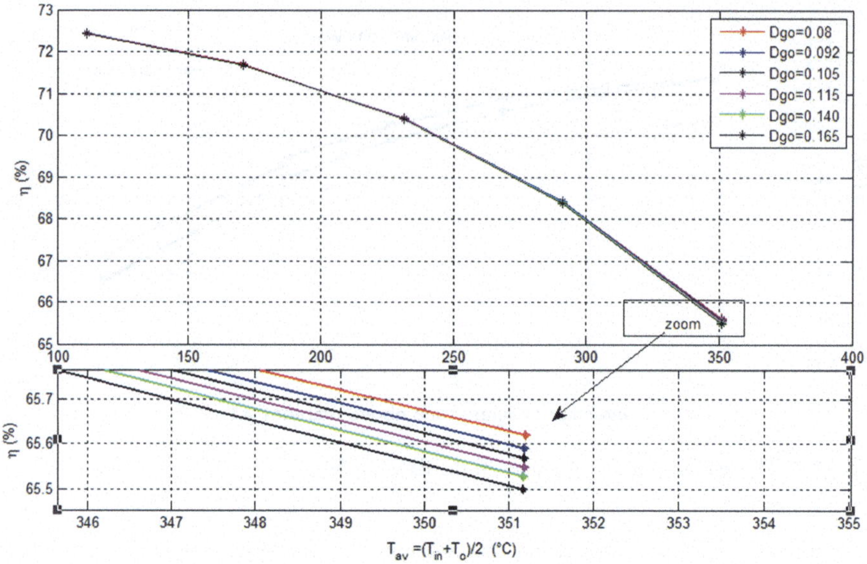

Fig. 12.3 Variation of the efficiency as a function of ΔT for different cover glass diameters

12.3.3 The Sensitivity of the Performance to the Diameter of the Glass Cover

The operating conditions including constant solar irradiation of 933.7 W/m^2 [2], are summarized with the following parameters: $V_w = 2.6$ m/s, V = 47.7 l/mn and $T_{air} = 21.2$ °C. For all the considered cases, a thickness of 0.003 m is used. Figure 12.3 illustrates the variation of the efficiency as a function of ΔT. It is clear that the efficiency decreases with the increase of the temperature difference. It has been observed that the efficiency decreases slightly with increasing the outer diameter of the envelope.

Figure 12.4 shows the variation of the thermal losses as a function of ΔT for different cover glass diameters. The increase in the outer diameter induces a slight increase in thermal losses. It can be seen from Figs. 12.3 and 12.4 that the variation of the outer diameters of the glass envelope does not have much effect on the performance of the solar receiver.

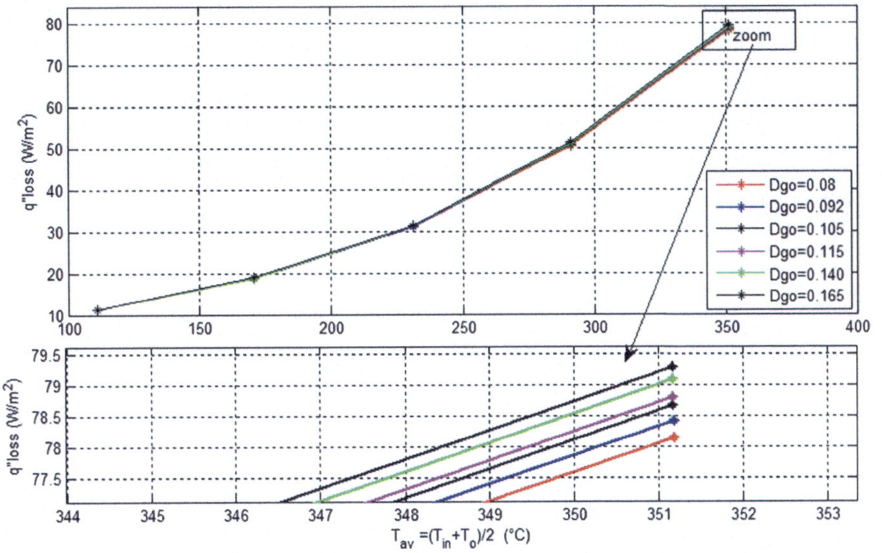

Fig. 12.4 Variation of the thermal losses as a function of ΔT for different cover glass diameters

12.4 Conclusion

The present study focused on the heat transfer analysis of the receiver of PTC. A simplified one-dimensional model has been established and used to analyze the thermal performance. The predictions of the model have been successively compared with the experimental results of Dudley et al. [2]. Besides, good agreement has been observed when comparing the predictions with the previously published studies. The sensitivity analysis indicates that the increase in the outer diameter of the glass cover does not influence the performance of the PTC.

References

1. D.A. Baharoon, H.A. Rahman, W.Z.W. Omar, S.O. Fadhl, Historical development of concentrating solar power technologies to generate clean electricity efficiently. Renew. Sustain. Energy Rev. **41**, 996–1027 (2015)
2. V.E. Dudley, G.J. Kolb, M. Sloan, D. Kearney, Test results: SEGS LS-2 solar collector. Report of Sandia National Laboratories (SAND94-1884) (1994)
3. R. Forristall, Heat transfer analysis and modeling of a parabolic trough solar receiver implemented in Engineering Solver Equation. Technical Report (NREL, Colorado, 2003)

4. O. García-Valladares, N. Velázquez, Numerical simulation of parabolic trough solar collector: improvement using counter flow concentric circular heat exchangers. Int. J. Heat Mass Trans. **52**(3–4), 597–609 (2009)
5. Z.D. Cheng, Y.L. He, J. Xiao, Y.B. Tao, R.J. Xu, Three-dimensional numerical study of heat transfer characteristics in the receiver tube of parabolic trough solar collector. Int. Commun. Heat Mass Transf. **37**, 782–787 (2010)
6. S.A. Kalogirou, A detailed thermal model of a parabolic trough collector receiver. Energy **48**, 298–306 (2012)

Chapter 13
Grid-Connected Photovoltaic System

Nasreddine Attou, Sid-Ahmed Zidi, Mohamed Khatir, and Samir Hadjeri

Abstract As energy needs increase and fossil resources decrease, the development of grid-connected photovoltaic energy is becoming an important part of the energy mix in the majority of countries. In this article, our attention has been concentrated on a strategy to control and interface photovoltaic power injection systems to the grid without batteries in order to make a significant and reliable contribution to sustainable energy supply and further improve the performance and stability of the power system. Different control mechanisms are considered in power flow management, maximum power point tracking (MPPT) for a three-phase photovoltaic inverter connected to the grid, PLL design standards (Phase Locked Loop), the current and voltage regulator VDC are also presented. The simulation results demonstrate the ability of the proposed control systems to control the energy flow, ensuring a good transfer of all maximum power to the grid.

Keywords PV system · Grid · Power injection · MPPT

13.1 Introduction

A PV system connected to the grid without batteries is the simplest and most economical solar energy installation available and since it does not require batteries, it is more cost-effective and requires less maintenance and reinvestment than stand-alone systems. It should be noted that a grid-connected solar energy system

N. Attou (✉) · S.-A. Zidi · M. Khatir · S. Hadjeri
Intelligent Control and Electrical Power System Laboratory, University Djillali Liabes, Sidi-Bel-Abbes, Algeria
e-mail: attioun12@gmail.com

A. Belasri and S. A. Beldjilali (eds.), *ICREEC 2019*, Springer Proceedings in Energy,
https://doi.org/10.1007/978-981-15-5444-5_13

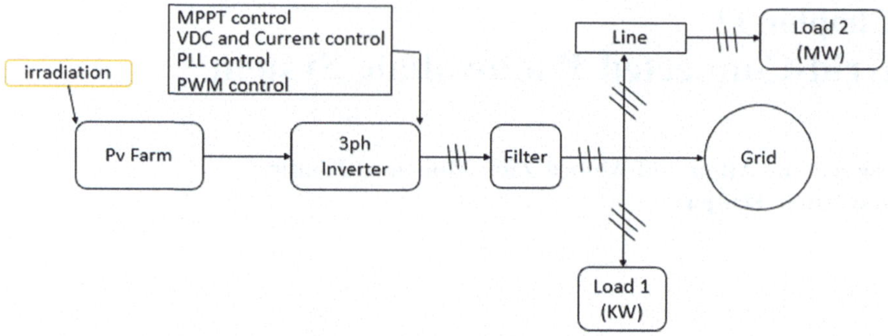

Fig. 13.1 Grid connected PV system block diagram

feeds their solar energy directly back into the grid. If, on a sunny day, the photovoltaic solar system produces more electricity, this solar energy is instantly reintroduced into the grid based on insolation conditions and actual electricity demand [1, 2].

Grid-connected photovoltaic systems are composed of photovoltaic panels connected to the grid via a DC-AC inverter with a maximum power tracker (MPPT) and a permanent controller of the power injected, a bidirectional interface between the AC output circuits of the PV system and the grid, the main electricity grid and the DC and AC loads as well as the control system necessary to ensure the safe operation of the group (see Fig. 13.1) [2].

To effectively control the power flow in the electrical system, multiple parameters and specific conditions are taken into consideration when connecting PV energy to the grid. The impact of PV modules on power grids cannot be ignored. Grid-connected PV systems can cause problems with the grid's flow control and stability.

In this article, Our work focused on the design of a photovoltaic grid-connected system using a controller for monitoring the maximum power point of the PV farm (MPPT) of type (P&O), a phase locked loop (PLL) in order to ensure synchronization with the grid and thus ensuring correct generation of the reference, two voltage and current regulation loops necessary for inverter operation using MATLAB Simulink [2, 3].

13.2 Methods

A grid-connected inverter's control system is responsible for managing a distributed generator's power injection into the grid. Most of the time, a control structure based on two loops but the most widely used strategy is the one that uses a slower external voltage regulation loop and a faster internal current regulation loop.

In such a way that the grid-connected photovoltaic system will operate flexibly, reliably and efficiently, it is necessary to comply with the following instructions, namely the order of the phases for the solar installation and the grid, frequency and voltage adaptation, and the sequence of the phases for the solar installation and the grid [3].

To overcome these constraints, a control strategy for the PV system injected into the grid is implemented based on a control interface that includes 4 types of control:

13.2.1 MPPT Control (Maximum Power Point Tracking)

This control is associated to an inverter that allows an adaptation between the PV and the load so that the power generated corresponds to its maximum value and is transferred directly to the grid. The MPPT control is established in our study using the algorithm (Perturb and Observe (P&O)). This Controller automatically varies the VDC reference signal of the inverter's DC voltage regulator to obtain a DC voltage that will obtain the maximum power from the PV generator [4].

13.2.2 Current and Voltage Control

VDC Regulator. A linear voltage regulator generates a fixed output voltage of a predetermined amplitude that remains constant whatever variations in its input voltage or load conditions (under normal conditions or rapidly changing atmospheric conditions).

From the PV operation standpoint, The VDC voltage control is used to monitor the PV panel's output voltage and to drive it to the maximum power point value based on the solar radiation and cell temperature parameters.

The control of the direct voltage VDC provides a constant real power flow to the inverter. This is usually achieved by an external loop control by comparing the reference voltage to the current state of the DC link voltage via a PI controller [4, 5].

Current Regulator. The current control loop regulates the Id current and the Iq current in the dq_0 reference by means of a PI controller. The main objective is to control the current output of the converter by generating an appropriate reference output voltage. This direct current is normally intended to extract as much PV production power as possible [5].

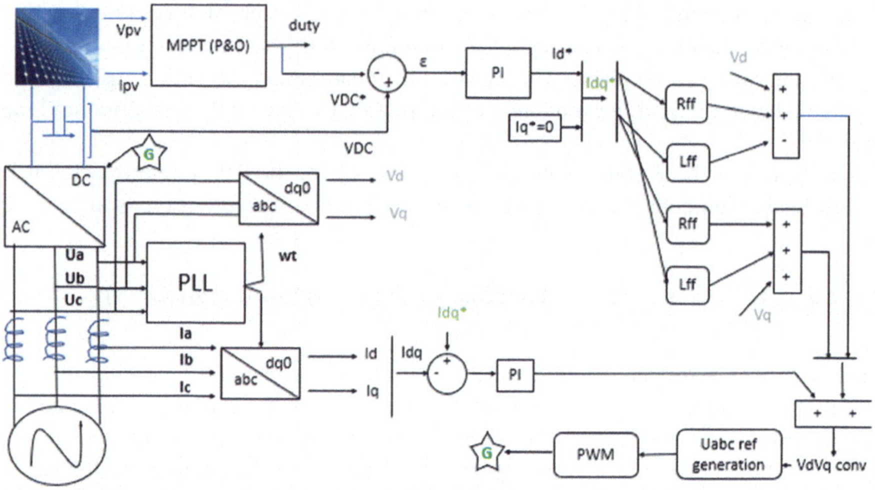

Fig. 13.2 Control structure of PV connected grid [1]

Control of the two current of the d-axis and the q-axis allows the regulation of active and reactive energy respectively.

As shown in Fig. 13.2, in the control loop, the terms and cross-coupling voltage are forwarded so that the injection of active and reactive energy into the grid can be controlled individually [6, 7].

13.2.3 Grid Synchronization (Phase Locked Loop-PLL)

The PLL performs phase and frequency tracking to provide a reference signal to synchronize it with the grid voltage thus reducing undesirable frequency variations. At each moment when this loop detects a variation in the angle between the two signals (i.e.) a frequency variation it acts quickly and precisely to resynchronize the two signals. The voltage outputs *Vd* and *Vq* of the current regulator are converted into three $Uabc - ref$ modulation signals used by the PWM generator [7].

The modelled PLL and d_q0 transformer block is shown in Fig. 13.2.

13.2.4 System Description

The three level *VSC* converter regulates the DC bus voltage at 500 V and maintains the unit power factor. The control system uses two control loops: an external control loop that regulates the voltage of the DC link at ± 250 V and an internal control loop that regulates the *Id* and *Iq* grid currents (active and reactive components). The current reference Id is the output of the external DC voltage controller. To preserve the unit power factor, the present reference *Iq* is set to zero. The voltage outputs *Vd* and *Vq* of the current regulator are converted into three modulation signals *Uref _abc* used by the three-level pulse generator with pulse width modulation (*PWM*). The block of control is shown in Fig. 13.2. The selected frequency for the *PWM* generator is 1975 Hz. A phase-locked loop is used to orient the frame *dq*. Then, the PLL calculates the angular velocity of the charge voltage vector *wl* [8].

13.3 Results and Discussion

The grid-connected photovoltaic system was simulated according to different irradiance values and tested with two three-phase loads of 5 MW and 200 KW.

The power of the PV generator, the voltages between phase and current at the output of the grid and PV systems, The VDC voltage supplied by PV systems is shown under varying solar radiation systems (Figs. 13.3 and 13.4).

The Fig. 13.3 shows a fluctuation in the current injected by the PV system during the day and this is due to changes in solar irradiation, the proportional-integral current regulator (PI) is used to maintain the current injected into the sinusoidal grid and to have high dynamic performances under rapidly changing atmospheric conditions. It is also important to keep in mind that the voltage and frequency of the signal provided by the PV installation is similar to that of the grid, This shows the role of the current and voltage control loop regulations used, PWM generator and phase lock loop in adjusting the power flow and improving the waveform fed into the grid [8].

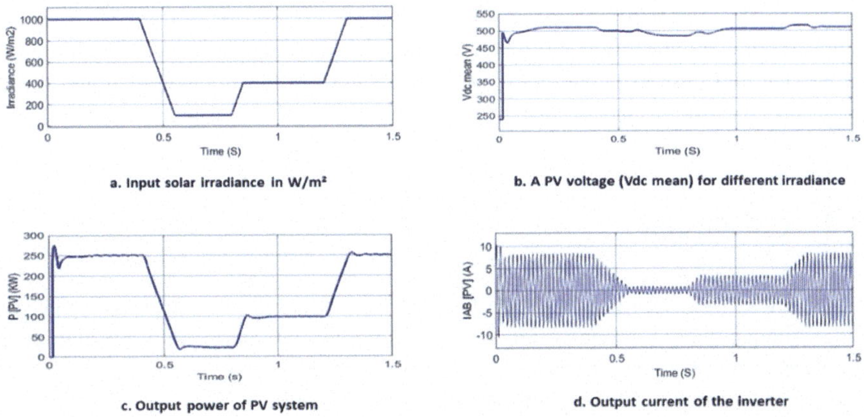

Fig. 13.3 Current and Voltage variation of the PV systems under different solar insolation

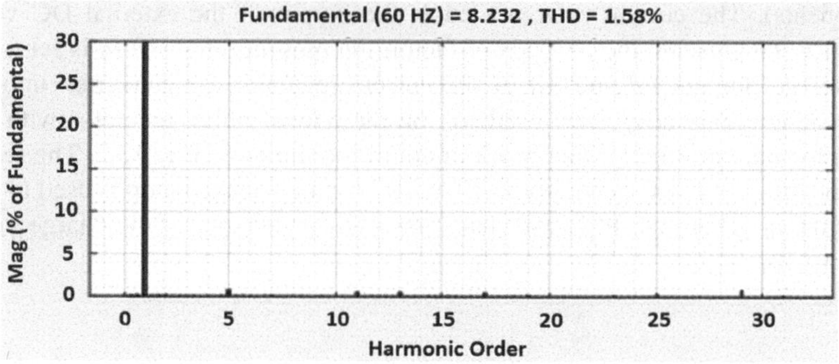

Fig. 13.4 The harmonics spectrum of output current inverter

The analysis of the insolation variation results at different times showed that:

The initial input irradiance of the PV generator model is 1000 W/m^2 and the operating temperature is 25°(°C). When the equilibrium state is reached at $t = 0.15$ s a PV voltage (Vdc_mean) of 507 V is obtained and the MPPT regulator extracts the maximum power (Pdc_mean) from the field is 255 kW.

At $t = 0.6$ s, solar irradiation is rapidly decreased to 100 W/m^2 due to the variation in sunlight during the day (passing cloudy). Due to the MPPT operation, the control system reduces the VDC reference to 480 V in order to extract the maximum power from the PV system (24 kW).

At $t = 0.95$ s, an increase in irradiation has a value of 400 W/m^2, the DC control loop regulate its value to 503 V in order to inject 100 kW of power from the PV to the grid.

Finally, a $t = 1.25$ s, the illumination reaches its maximum value again, which allows the PV to inject this maximum power.

It can be seen that each time the irradiance changes the MPPT regulator starts to follow the maximum power by regulating the PV voltage in order to extract the maximum power. The PWM control with the control blocks improves the quality of the voltage and current wave quality of the PV system fed into the grid and has a low THD (see Fig. 13.4).

13.4 Conclusion

In this article, our attention has been focused on a strategy to control and interface photovoltaic power injection systems to the grid without batteries in order to make a significant and reliable contribution to sustainable energy supply and further improve the behavior and stability of the power system.

The simulation results demonstrated the capability of the control systems implemented in the control of energy flow during power injection to the grid such as power system stability and power quality and in improving system performance, including low THD, high power factor, and fast dynamic response.

Moreover, the use of an algorithm capable of monitoring the maximum power point (MPPT) in grid-connected photovoltaic inverters and the design of the phase-locked loop (PLL) ensuring fast and accurate phase angle detection, amplitude and operating voltage frequency ensuring correct reference signal generation. The VDC current and voltage regulator is therefore also required to ensure that all maximum power is transmitted to the grid at a higher power quality.

References

1. A. Reaz Reisi, A. Alidousti, *Optimal Designing Grid-Connected PV Systems* (IntechOpen, Rijeka, 2019)
2. Y. Abdalla, I. Farog, Y. Mamoun, Grid connected photovoltaic system, in *International Conference on Communication, Control, Computing and Electronics Engineering (ICCCCEE)* (IEEE, 2017), pp. 1–5
3. R. Kadri, J. Gaubert, G. Champenois, An improved maximum power point tracking for photovoltaic grid-connected inverter based on voltage-oriented control. IEEE Trans. Industr. Electron. **58**(1), 66–75 (2010)
4. A. Deswal, V.K. Garg, Voltage source converter (VSC) control of grid connected PV system. Adv. Res. Electr. Electron. Eng. **3**(4), 257–259 (2016)
5. S. Zhou, J. Liu, L. Zhou, DQ current control of voltage source converters with a decoupling method based on preprocessed reference current feed-forward. IEEE Trans. Power Electron. **32** (11), 8904–8921 (2017)
6. B.Z. Li, S. Huang, X. Chen, Enhanced dq current control for single-phase voltage-source inverters. IET Power Electron. **11**(9), 1537–1546 (2018)
7. M. Lakshmi, S. Hemamalini, Decoupled control of grid connected photovoltaic system using fractional order controller. Ain Shams Eng. J. **9**, 2090–4479 (2016)
8. N. Hamrouni, M. Jraidi, A. Cherif, New control strategy for 2-stage grid-connected photovoltaic power system. Renew. Energy **33**(10), 2212–2221 (2008)

Chapter 14
The Confinement Effect on a Supercritical Fluid Mixture Enclosed in a Micro-cavity

H. Smahi, D. Ameur, J. Dib, and I. Raspo

Abstract In this article, we study a supercritical mixture model enclosed in a rectangular cavity. The mixture is composed of two elements: the first is supercritical CO_2 and the other is naphthalene. This article aims to study the cavity confinement effect on the mixture and the mass transfer adsorption with a heating plate and an adsorbent at the lower limit. The mathematical model used is based on the solution of the Navier-Stokes equations, associated with energy and mass diffusion equations, including the additional Peng-Robinson equation of state. These equations are solved under the assumption of a low Mach number. Under high heating of the lower plate, the results show an influence of the cavity confinement on the adsorption of the mixture and the mass transfer. Thus, an important cavity confinement for heights up to 0.25 mm leads to an increase in mass transfer. The results also showed that the convection flow is less intense and leads to an adsorption profile less homogeneous.

Keywords Adsorption · Mass transfer · Mixture fluid

14.1 Introduction

Supercritical fluid stands for a fluid whose pressure and temperature are higher than those corresponding to the critical point. Thus, the fluid physical properties (density, viscosity, and diffuseness) are considered to be intermediate between those of gases and liquids [1, 2] in this phase diagram area. Supercritical Fluid Extraction (SFE) can be associated with adsorption on activated carbon for soil remediation, since adsorption allows species selective transfer from a fluid phase to a solid

H. Smahi (✉) · D. Ameur · J. Dib
Laboratory of Theoretical Physics, Faculty of Sciences, University of Tlemcen, Tlemcen, Algeria
e-mail: housseynsmahi@gmail.com

I. Raspo
Aix-Marseille Université, CNRS, Centrale Marseille, M2P2 UMR 7340, Marseille, France

A. Belasri and S. A. Beldjilali (eds.), *ICREEC 2019*, Springer Proceedings in Energy,
https://doi.org/10.1007/978-981-15-5444-5_14

adsorbed. The SFE process has been already employed in soil remediation using propane and CO_2 as solvents, and widely applied to different chemical nature organic compounds, mostly high molecular weight hydrocarbons [3]. Moreover, various organic products extraction such as hexachlorobenzene, naphthalene, anthracene, and phenol employing supercritical carbon dioxide has been reported [4, 5]. It is to say that when supercritical solvent is recirculated, the process must be almost contaminant-free due to environmental legislation requiring low concentration of soil contaminants. When employing volatile pollutants, the solvent and solute separation by decompression is not conducive enough to attain the required low concentration in the soil, hence a second adsorption step on activated carbon is required. In this context, Madraș et al. [6] proposed a process aiming to soils remediation soils contaminated with heavy molecular weight organic compounds [7], combining both adsorption and supercritical extraction. While it has been found that in the case of organic pollutants, continuously extracted using supercritical carbon dioxide and then deposited on activated carbon, that the process would operate in the optimal mode only if the adsorption at high temperatures was carried preceded by a soil desorption at low temperature [8].

Few numerical and theoretical studies were carried out on adsorption from supercritical fluids. In [9, 10], Raspo et al. exhibited that SFC hydrodynamic behavior may be coupled with the solid solubility critical behavior in order to trigger the reaction in diluted supercritical binary mixtures, for a generic heterogeneous reaction relevant for adsorption/desorption. More precisely, by heating the mixture on one side, the Piston effect causes heterogeneous reaction liberation for the opposite plate which has been maintained at the initial temperature. Furthermore in [11], Wannassi and Raspo investigated adsorption in a binary model mixture, the naphthalene-CO_2 mixture, near the solvent critical point in a laterally heated cavity. The equilibrium constant divergent behavior and the Piston effect were taken into consideration and their influence on the adsorption process was highlighted. In order to design an adsorption plant, it is indispensable to acquire reliable mass transfer models necessary to determine optimal operating conditions and scaling. Nevertheless, few attempts have been made to model the supercritical adsorption process [8–11] and most studies have focused on the SFC extraction. In the present paper, a solute model adsorption from supercritical CO_2 is investigated in a small lateral heated cavity employing 2D numerical simulations. Naphthalene was chosen as a solute model since its phase equilibrium with CO_2 has been carefully studied in [9–12].

14.2 Physical and Mathematical Model

As shown in Fig. 14.1, the physical model consists of a dilute mixture (Naphthalene-CO_2) enclosed in a rectangular cavity with a heated and adsorbent plate on a part of the bottom boundary. The cavity aspect ratio is $L/H = 10$ and three different cavity heights H are considered: 1, 0.5 and 0.25 mm. On the

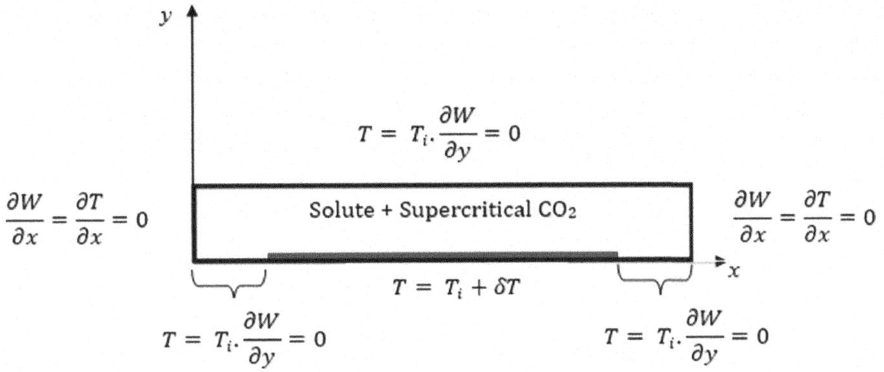

Fig. 14.1 Sketch of the physical configuration

Table 14.1 Pure component properties

	T_c (K)	ρ_c (kg m^{-3})	M (kg mol^{-1})	ω
CO_2 (1)	304.21	467.8	4.401×10^{-2}	0.225
Naphthalene (2)	748.40	314.9	1.282×10^{-1}	0.302

non-reactive vertical walls, an adiabatic boundary condition is applied. At initial time, the fluid is considered in thermodynamic equilibrium at a constant temperature T_i slightly above a particular mixture critical temperature T_{cm} = 307.65 K (T_i = (1 + ε)) and stratified with the mean density equal to the mixture critical density ρ_{cm} = 470 kg m^{-3}. Parameter $\varepsilon = (T_i - T_{cm})/T_{cm}$ stands for the dimensionless proximity to this mixture critical point named the Lower Critical End Point (LCEP) slightly above the critical point of CO_2. A weak heating is then gradually applied at the solid bottom plate defined by $H < x < L - H$. The temperature ΔT is fixed at 10 mK.

The Naphthalene-CO_2 mixture was selected as a reference system due to wide experimental data on solubility that can be found in literature, allowing an easier thermodynamic model validation [11]. Each pure compound physical properties are given in Table 14.1. We consider a dilute mixture of Naphthalene (the minority species) and supercritical CO_2. Naphthalene is the species adsorbed on the bottom plate.

The mathematical model is based on the 2D time-dependent and compressible Navier–Stokes equations, coupled with the Peng–Robinson equation of state and the energy and mass diffusion equations. These equations are solved in the framework of low Mach number approximation allowing reduced computational costs. Hence, the total pressure P is split into two parts: a non-homogeneous dynamic part $P_{dyn}(x, y, t)$, appearing in the momentum equation and which varies with time and space, and a homogeneous thermodynamic part $P_{th}(t)$, which appears in the equation of state and in the energy equation and only depends on time t. In order to account for the strong stratification in fluids near the critical point, we employ low Mach number approximation modification. T_{cm} stands for characteristic temperature, ρ_{cm} for characteristic density, H for characteristic length, the Piston effect time scale t_{PE} [9–12] for characteristic time and the corresponding velocity $V_{PE} = H/t_{PE}$ for characteristic velocity. The dimensionless governing equations read as follows:

$$\frac{\partial \rho}{\partial t} + \frac{\partial}{\partial x}(\rho V) = 0 \tag{14.1}$$

$$\rho\frac{\partial V}{\partial t} + \rho V.\nabla V = -\nabla P_{dyn} + \frac{1}{Re}\Delta V + \frac{1}{3Re}\nabla(\nabla.V) + \frac{1}{Fr}(\rho - \rho_i) \tag{14.2}$$

$$\begin{aligned}\rho\frac{\partial T}{\partial t} + \rho V.\Delta T &= -\frac{C_{V0}}{C_{Vi}}(\gamma_0 - 1)T\left(\frac{\partial P_{th}}{\partial T}\right)_{\rho,w}(\nabla.V) + \frac{\gamma}{RePr}\nabla.(\lambda^*\nabla T)\\ &+ \frac{1}{(-1)^2 Le}\nabla.\left[\left((\bar{U}_2^* - \bar{U}_1^*) + \frac{C_{V0}}{C_{Vi}}(\gamma_0 - 1)P_{th}(\bar{V}_2^* - \bar{V}_1^*)\right)\rho D_{21}^*\nabla w\right] - \frac{1}{(\gamma - 1)^2 Le}\\ &\left[(\bar{U}_2^* - \bar{U}_1^*) + \frac{C_{V0}}{C_{Vi}}(\gamma_0 - 1)\left(P_{th} - T\left(\frac{\partial P_{th}}{\partial T}\right)\right)_{\rho,\omega}(\bar{V}_2^* - \bar{V}_1^*)\right]\nabla.(\rho D_{21}^*\nabla w)\end{aligned} \tag{14.3}$$

$$\rho\frac{\partial w}{\partial t} + \rho V.\nabla w = \frac{1}{(\gamma - 1)^2 Le}\nabla.(\rho D_{21}^*\nabla w) \tag{14.4}$$

In above equations, V defines components velocity u and v in the x- and y-directions respectively. w is the mass fraction, γ is the ratio of the isobaric C_p and isochoric C_v specific heats calculated from the equation of state, γ_0 and C_{vo} correspond to the values for a perfect gas ($\gamma_0 = 1.3$, $C_{vo} = 3R/M_1$) and the subscript i refers to values at the initial condition. Dimensionless parameters are the Mach number Ma, the Reynolds number Re, the Froude number Fr, the Prandtl number Pr and the Lewis number Le and are respectively given by:

$$Ma = \frac{V_{PE}}{C_0},\quad Re = \frac{\rho_i V_{PE} H}{\mu_i},\quad Fr = \frac{V_{PE}^2}{gH},\quad Pr = \frac{\mu_i C_{Vi}}{\lambda_i},\quad Le = \frac{\lambda_i}{\rho_i C_{Vi}(D_{21})_i}$$

14.3 Results

For all performed simulations, the distance to the mixture critical point is fixed to 0.5 K for an initial temperature Ti = 308.15 K and one value of the bottom plate heating ΔT = 10 mK. The cavity height varies from 0.5 mm to 0.25 mm. The value of the initial mass fraction $w_i = 7.6751 \times 10^{-3}$ corresponds to the solubility of Naphthalene in CO_2 at 308.15 K. For this value of temperature, the adsorption constant derivatives are $(\partial K_a/\partial T) = 0.295$ and $(\partial K_a/\partial P) = -2.90 \times 10^{-6}$. Concerning the adsorption reaction, $Da = HK_{ai}/(D_{21})_i = 10^{-5}$. The binary mixture is very dilute because of low solute solubility, which may suggests that the transport coefficients of the mixture are those of the pure solvent and are: $\mu_i = 3.3402 \times 10^{-5}$ Pa s^{-1}, $\lambda_i = 9.6172 \times 10^{-2}$ W m^{-1} K^{-1}, $C_{Vi} = 1306.27$ J kg^{-1} K^{-1}, $(D_{21})_i = 2.1969 \times 10^{-8}$ m^2 s^{-1}, $\beta_i = 2.3134 \times 10^{-1}$ K^{-1}, $\gamma = 16.75$.

Figure 14.2 shows that more we descend in the cavity height *H*, more the thermal and mass diffusion boundary layers become thickness. This is explained by the fact that only mass diffusion coefficient and thermal diffusivity varie with the distance to the critical point. The comparison of the temperature fields between the different heights shows that confinement multiplies and increases the number of thermal plumes.

For all three height cases showed in Fig. 14.3, profiles reveal that the solute adsorption is strongly inhomogeneous. The relative mass fraction perturbation profiles magnitude $(w - w_i)/w_i$ shows that the maximum of adsorbed is observed for $H = 0.25$ mm. Therefore, in a smaller cavity, it is possible to increase the amount of solute adsorbed, which is very interesting in the framework of chemical engineering processes. Moreover, for all cases, the effect of the less intense convection is also visible on the temporal evolution of the mean Sherwood number (Fig. 14.4).

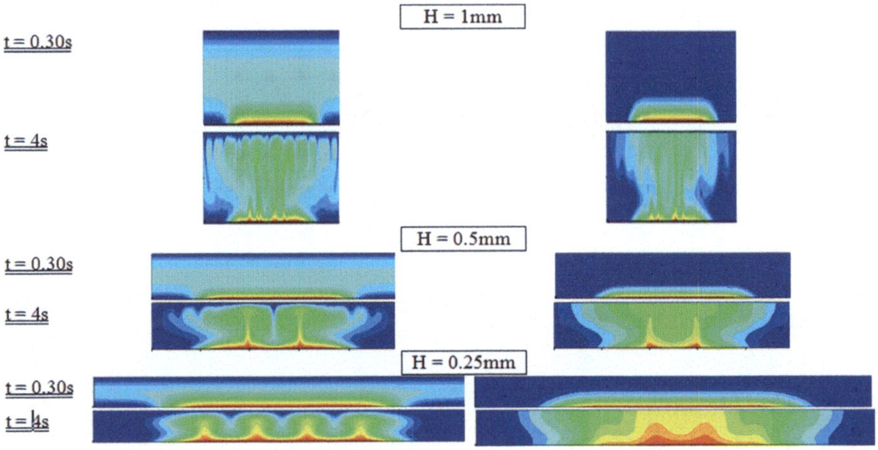

Fig. 14.2 Mass fraction perturbation $w - w_i$ (right column) and isolines of temperature (left column) for the heated 100 mK in the same $T_i = 308.15$ K and for three values of height H = 1 mm, 0.5 mm and 0.25 mm

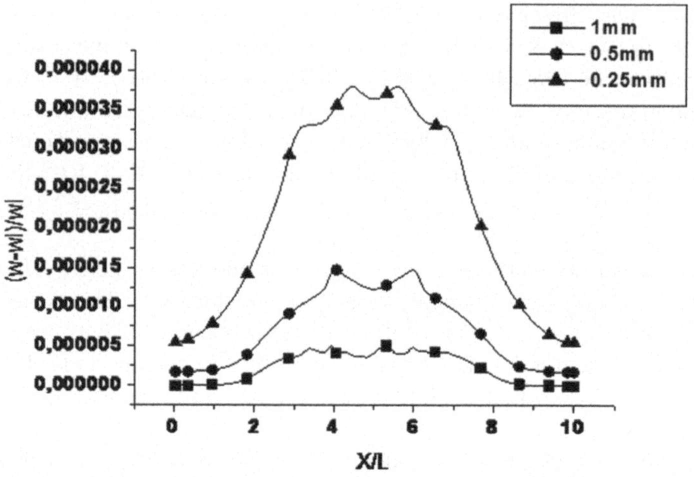

Fig. 14.3 Profiles of the relative mass fraction perturbation at t = 4 s on the bottom boundary for ΔT = 100 mK and three values of height H = 1 mm, 0.5 mm and 0.25 mm

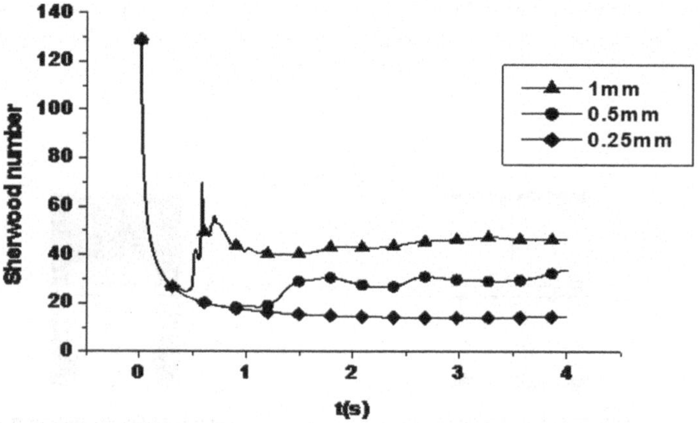

Fig. 14.4 Temporal evolution of the mean Sherwood number for ΔT = 100 mK and three values of height H = 1 mm, 0.5 mm and 0.25 mm

14.4 Conclusion

A numerical study of the heat and mass transfer in an almost critical binary mixture diluted, in a macro/micro-cavity heated from below, is proposed. The effect of confinement on the adsorption rate is also analyzed. The results showed that decreasing the height of the cavity improves the absorption process. The results also showed that confinement generates a less intense convection flow. Finally, the confinement of the cavity leads to an increase in mass transfer to the bottom, very interesting in the context of chemical engineering processes.

References

1. D.D. Do, H.D. Do, Adsorption of supercritical fluid in non-porous and porous carbons; analysis of adsorbed phase volume and density. Carbon **41**, 1777–1791 (2003)
2. D.W. Hall, J.A. Sandrin, R.E. McBride, An overview of solvent extraction treatment technologies. Environ. Prog. **9**(2), 98–105 (1990)
3. J.M. Becnel, K.M. Dooley, Supercritical fluid extraction of polycyclic aromatic hydrocarbon mixtures from contaminated soils. Ind. Eng. Chem. Res. **37**, 584–594 (1998)
4. A. Akgerman, C. Erkey, S.M. Ghoerishi, Supercritical extraction of hexachlorobenzene from soil. Ind. Eng. Chem. Res. **31**, 333–339 (1992)
5. K.M. Dooley, D. Ghonasgi, F.C. Knopf, Environ. Prog. **9**, 197 (1990)
6. G. Madras, C. Erkey, A. Akgerman, Supercritical extraction of organic contaminants from soil combined with adsorption onto activated carbon. Environ. Prog. **13**(1), 45–50 (1994)
7. E. Alonso, F.J. Cantero, J. Garcia, M.J. Cocero, Scale-up for a process of supercritical extraction with adsorption of solute onto active carbon. Application to soil remediation. J. Super. Fluids **24**, 123–135 (2002)
8. E. Reverchon, G. Lamberti, P. Subra, Modelling and simulation of the supercritical adsorption of complex terpene mixtures. Chem. Eng. Sci. **53**, 3537–3544 (1998)
9. I. Raspo, S. Meradji, B. Zappoli, Heterogeneous reaction induced by piston effect in supercritical binary mixtures. Chem. Eng. Sci. **62**, 182–4192 (2007)
10. I. Raspo, S. Meradji, B. Zappoli, Fast mass transfer at a solid-supercritical fluid interface by piston effect, in *Progress in Computational Heat and Mass Transfer (Proceedings of the 4th ICCHMT), Lavoisier*, ed. by R. Bennacer, A.A. Mohamad, M. El Ganaoui, J. Sicard (2005), pp. 132–137
11. M. Wannassi, I. Raspo, Numerical study of non-isothermal adsorption of Naphthalene in supercritical CO_2: behavior near critical point. J. Super. Fluids **117**, 203–218 (2016)
12. H. Smahi, D. Ameur, J. Dib, I. Raspo, Numerical study of adsorption for fluids mixture near critical point in a micro-cavity, in *CFM 2018* (2018)

Chapter 15
Effect of Parameters on the Stratification of a Solar Water Heater

Mohamed Lazreg, Touhami Baki, and Driss Nehari

Abstract In this paper, the effect of stratification and integration of the storage tank of an SDHW, a storage tank containing the system behavior is analyzed with the storage tank under different thermal stratifications. There are many different types of solar collectors, but all of them are constructed with the same basic premise in mind. In general. Get simulated solar collector results By simulating (collector with surface of 2 m^2 and a storage tank of 300 L), operated in Oran (Algeria), A solar collector is a device that collects and/or concentrates solar radiation from the Sun. These devices are primarily used for active solar heating and allow for the heating of water for personal use. These collectors are generally mounted on the roof and must be very sturdy as they are exposed to a variety of different weather conditions. The use of these solar collectors provides an alternative for traditional domestic water heating using a water heater, potentially reducing energy costs over time. As well as in domestic settings, a large number of these collectors can be combined in an array and used to generate electricity in solar thermal power plants. What we do in this research is to analyze the temperature inside the tank and this taking into account the changes in the temperature that come in touching the stratification of the tank. This study gives us the ideal way to obtain good results for the system by studying nodes.

Keywords Solar water heater · Balloon · Stratification · Temperature

---

M. Lazreg (✉)
Mechanical Department, ENPO Maurice Audin,
Ex ENSET, Rue Oran, Es Senia, BP 1523 Oran El M'naouer, Oran, Algeria
e-mail: lazreg20007@yahoo.fr

T. Baki
Faculty of Mechanics, Gaseous Fuels and Environment Laboratory, USTO-MB, El Mnaouer, BP 1505, Bir El Djir, 31000 Oran, Algeria

D. Nehari
Laboratory of Intelligent Structures, Belhadj Bouchaïb University Center, Sidi Bel Abbes Road, BP 284, 46000 Aïn Témouchent, Algeria

A. Belasri and S. A. Beldjilali (eds.), *ICREEC 2019*, Springer Proceedings in Energy,
https://doi.org/10.1007/978-981-15-5444-5_15

15.1 Introduction

With the growth of energy demand and in order to reduce greenhouse gas emissions including CO_2 in the atmosphere, in addition to a solar field exceeding 3,000 h of sunshine per year reference [3], Algeria has decided to develop different applications of renewable energies. Among the most promising applications are solar water heaters. Indeed, national programs, awareness campaigns, investments and several international partnership and cooperation projects have been carried out; the targeted sectors are housing, transportation, industry. A solar water heater is a device that can meet up to 80% of the needs of large consumers of hot water such as hotels, buildings, and steam rooms. There are several types of solar water heaters. Example the diagrams (Fig. 15.1).

The most used solar water heater is with heat exchanger inside. National program for the development of ESCs the will of the public authorities is translated by national programs that target the two tertiary and residential sectors. These programs include: Horizon 2011 program to power 5,500 solar hot water homes (funded by UNDP), ALSOL program: 1000 individual water heaters in the housing and 1000 in industry funded by the National Fund for the control of energy (FNME) Installation, program of 16,000 m^2 of solar water heater for isolated sites (funded under the PNME).

At this level we will look at this work to study simulation by TRNSYS.

Simple designs include a simple glass-topped insulated box with a flat solar absorber made of sheet metal, attached to copper heat exchanger pipes and dark-colored, or a set of metal tubes surrounded by an evacuated (near vacuum) glass cylinder. In industrial cases a parabolic mirror can concentrate sunlight on the tube. Heat is stored in a hot water storage tank. The volume of this tank needs to be larger with solar heating systems to compensate for bad weather. What our simulation system does is to address temperature changes within days within months of the year Using the simulation program TRNSYS, and this by analyzing the temperature within the tank and consequently the stratification (Fig. 15.2) of the tank.

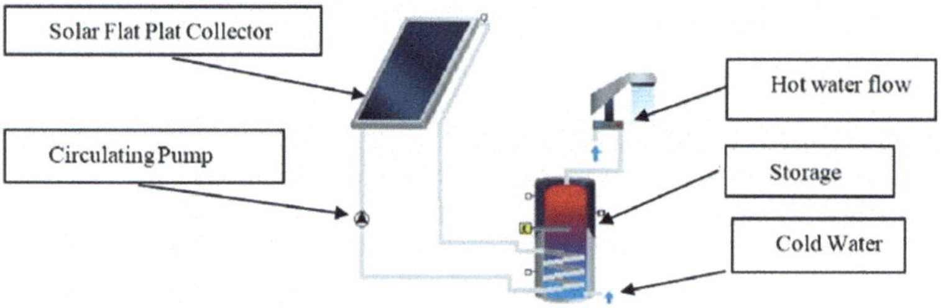

Fig. 15.1 Studied system as SDHW (TSOL show)

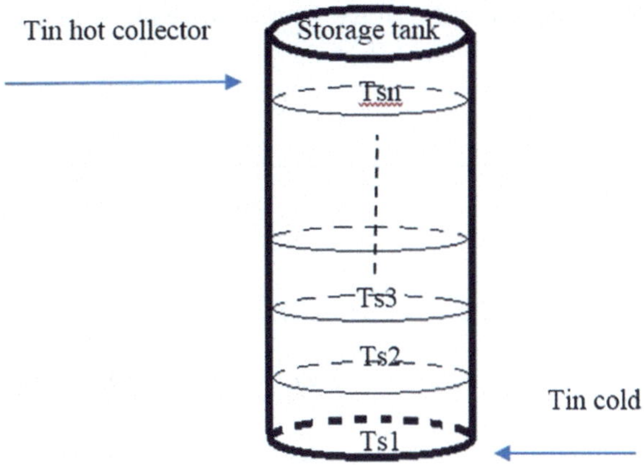

Fig. 15.2 Stratified tank schematic diagram

15.2 Methods

The heat storage and discharge system in the tank works. We can divide the stages of the process into two basic stages, representing the conversion of solar energy into thermal energy and this is what we see in solar heating systems. T1 is the first stage or what is called the storage stage and is the stage in which the heat is contained and stored until reaching T2 and here is the stage called exothermic process reference [2]. Here, the role of the auxiliary system that works to meet the difference in temperature is lost in the system. And Fig. 15.3 illustrates the two stages and they are storage and thermal expulsion.

Energy calculation and mass conservation, it is in the storage stage. This is within the following calculation hierarchy:

$$Q_u^{stor} = Q_b^{stor} + Q_s^{stor} + Q_l^{stor} \tag{15.1}$$

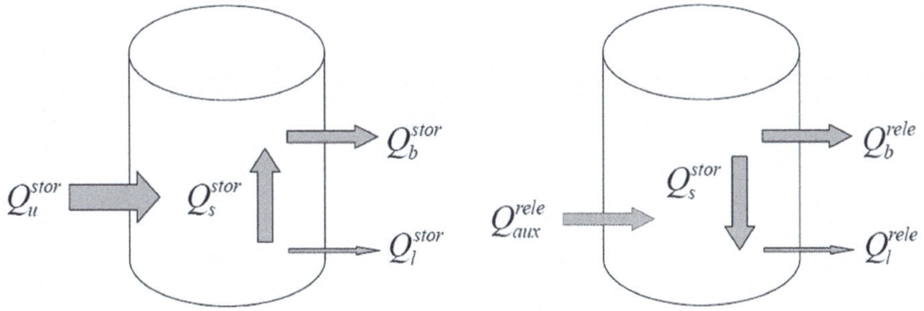

Fig. 15.3 (**a**) Heat storage phase; (**b**) thermal expulsion phase

We can express the stored heat in the tank as follows:

$$Q_u^{stor} = \int_{\tau_0}^{\tau_3} I_{\theta(\tau)}.A_c.n_{cd}d\tau \quad (15.2)$$

$$Q_s^{stor} = \int_{\tau_1}^{\tau_2} cpV\Delta Tdt \quad (15.3)$$

What we see here is thermal expulsion:

$$Q_s^{stor} + Q_{aux}^{rele} = Q_b^{rele} + Q_l^{rele} \quad (15.4)$$

In order to obtain the required volume of the tank we make some hypotheses:

1-Neglecting to change the density of water and heat to avoid changes in water temperature
2-Give the tank a medium temperature and overlook the stratification.

The tank is located in an outdoor environment, and certainly there is a heat exchange that we can express as follows:

$$Q_l^{all} = \int_{0}^{24} UA(T_s - T_a)dt \quad (15.5)$$

We can express the volume and area of the tank in a mathematical way:

$$A = 2.5 \times (4\pi)^{\frac{1}{3}} V^{\frac{2}{3}} \quad (15.6)$$

Note: In the system we are going to build, we will dispense with the auxiliary system of electric water heater. And dependence on the volume of the tank and this is shown by the third equation of clarifying the relationship between the maximum temperature and the role of the volume of the tank in it:

$$V = \frac{Q_s^{stor}}{c_p(T_2 - T_1)} \quad (15.7)$$

What worries us is the loss of temperature, and this is what it shows in:

$$Q_l^{all} = \int_0^{24} 2.5U(4\pi)^{\frac{1}{3}}V^{\frac{2}{3}}(T_s - T_a)dt \tag{15.8}$$

S is the area of the tank, m^3/m^2:

$$S = \frac{A_c}{V} \tag{15.9}$$

15.3 Simulation

What we see in Fig. 15.4 is building a simulation model TRNSYS. Installation of an individual solar water heater under the climate of Oran, a particular follow-up was carried out to show the influence of certain parameters on the stratification of the balloon; our choice fell on a panel of 2 m^2 and a balloon of 300 L, the chosen parameters are the drawing of the consumption of water, the dimensioning of the exchanger and the isolation of the balloon. The temperature curves obtained showed a variability according to the chosen parameters.

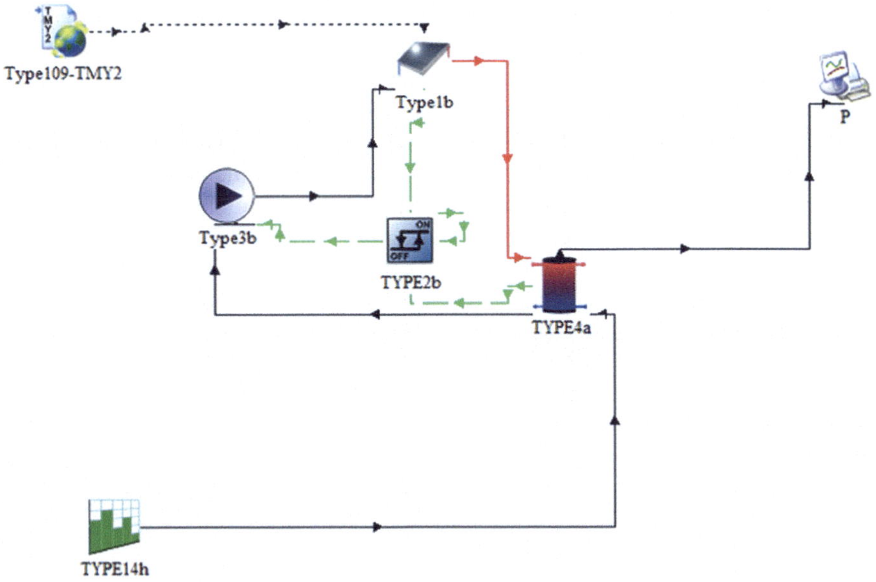

Fig. 15.4 Simulation system picture

15.4 Results and Discussion

Our approach is based on the simulation of an installation of an individual solar water heater with the code TRNSYS, under the weather conditions of May in Oran, see Fig. 15.5, showing the average daily temperature, Fig. 15.6 shows the stratification of the temperature at different levels of the balloon, the latter has been divided into four nodes of equal height, during the night there is no energy input the difference in temperature is consistent between the nodes, when the system starts the variation is a few degrees and the temperature is increasing from the bottom to the top of the balloon; from 10 to 15 h stratification goes through a transitional period before having equilibrium. What we observe is the consistency of the results obtained with those of the reference [1], This is in the norm of behavior of the change of temperature at the level of the nodes. What we observe from the simulation results for May the beginning of the first two weeks the temperature is approximately the difference between the first node and the eighth node between 25 °C to 1 °C.

What we observe in the results obtained from the program TRNSYS This is to show the temperature change with the division that touches the tank 6 nodes from January, June and September. The lowest month temperature is January and the strongest temperature known in the study in June.

Simulation results for the six nodes, we found it difficult to present the results with a single graph of 1 to 6 nodes plus the three months required January, June and September so we decided to divide the results into several graphs, several curves touch the most desired results. 6 nodes from January, June and September these months were randomly selected giving an overview of the days of the year.

The temperature in January reaches a maximum of 50 °C approx., this is the biggest temperature record for this month for nodes (1, 3 et 6) (Figs. 15.7, 15.8 and 15.9). The maximum temperature in June is about 95 °C, the maximum recorded in September 80 °C most temperature records were at peak time of the day.

After comparing the results obtained, we find temperature affinity for nodes, so that there is a slight difference for midday times. It does not exceed one or two degrees of temperature, because the tank's internal system maintains the temperature even in the late evening. This certifies the success of research in this field for the development of hot water tanks.

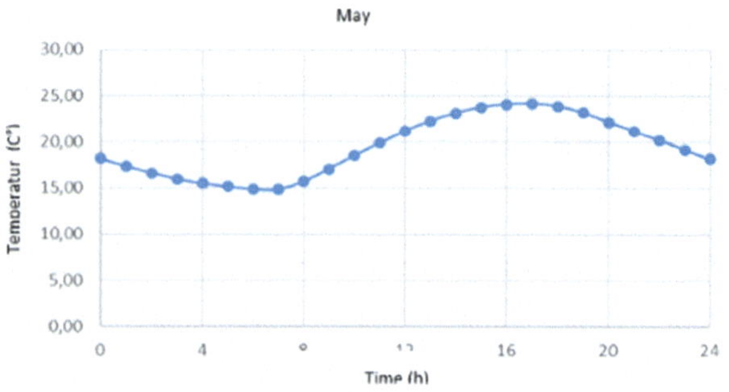

Fig. 15.5 The average daily temperature (May)

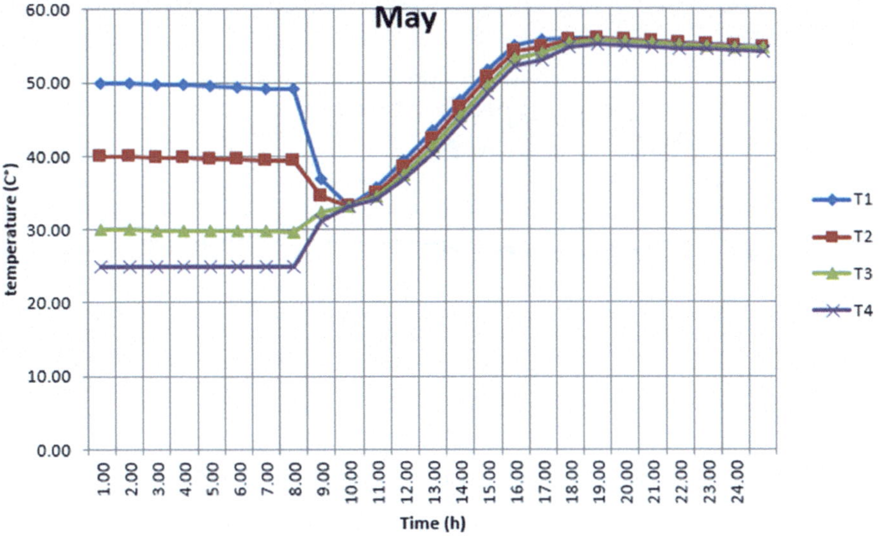

Fig. 15.6 The stratification of the different levels of the solar flask (May)

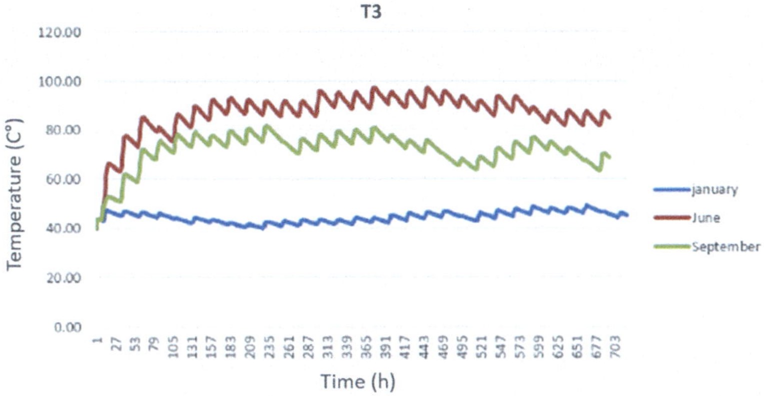

Fig. 15.7 Temperature change for months (January, Jun and September) for node T3

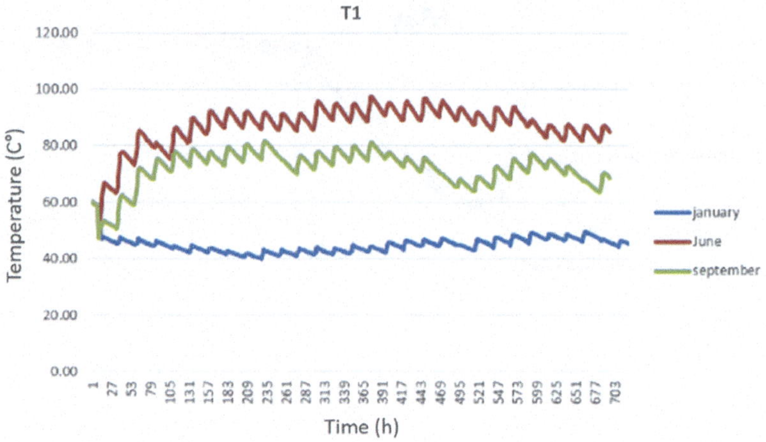

Fig. 15.8 Temperature change for months (January, Jun and September) for node T1

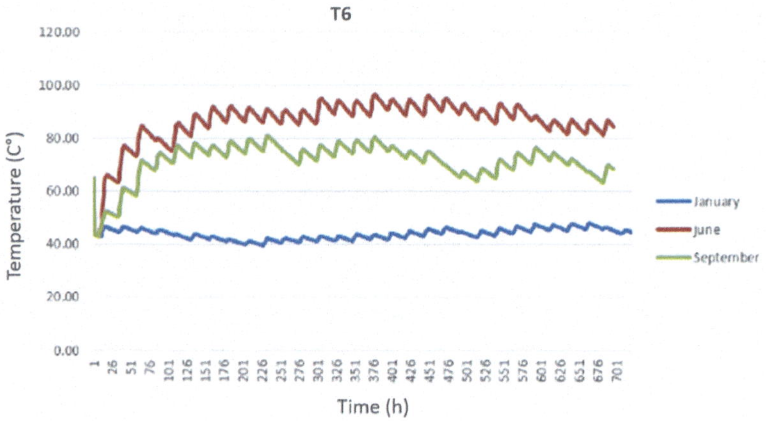

Fig. 15.9 Temperature change for months (January, Jun and September) for node T6

15.5 Conclusion

In this paper, the effect of stratification and integration of the storage tank of an SDHW. storage tank containing the system behavior is analyzed with the storage tank under different thermal stratifications.

System performance is designed within 3 months of the year, representing the average conditions experienced by the months of the year. We divide the results into two parts with four nodes and six nodes. The results indicate that nodes 4 are capable of describing tank, and this stratification give accuracy in the desired results for May for Oran, at peak time we get the maximum temperature 55 °C Approximately.

What we notice is a drop-in temperature for the morning except January, which touches a high temperature in the morning. The temperature spacing between the 6 nodes is also low in the 3 months. The results obtained are acceptable for May compared to the experimental results as well as the simulation results for the remaining months, it can be concluded that the stratification a water tank for a solar heater helps in obtaining more accurate results this is what the results say.

Our country, by launching the reforms, is aware that the development of renewable energies corresponds to the option of local development that values existing resources, promotes employment and meets a social expectation for sustainable development. Despite the constraints that are the price of gas, lack of subsidies, lack of solar equipment manufacturing plant, customs taxes.... Economic, political and cultural trends are conducive to the development and growth of the Solar Water Heater market with total respect for the environment.

An exchange of experiences between Algeria and its neighbors, as well as with European countries that have gained experience in the field is essential.

References

1. S. Li, Y. Zhang, K. Zhang, X. Li, Y. Li, X. Zhang, Study on performance of storage tanks in solar water heater system in charge and discharge progress. Energy Procedia **48**, 384–393 (2014)
2. T. Li, Y. Liu, D. Wang, K. Shang, J. Liu, Optimization analysis on storage tank volume in solar heating system. Procedia Eng. **121**, 1356–1364 (2015)
3. F. Bouhired, Développement des Chauffe Eau Solaires en Algérie. Recherche et Développement e-mail: fbouhired@cder.dz division solaire thermique et géothermique N° 17 (2010)

Chapter 16
Validation of an Analytical Calculation Computer Programming on a Flat Plate Thermal Solar Collector

A. Seddaoui, M. Z. Dar Ramdane, and R. Noureddine

Abstract This work aims to validate an analytical calculation computer programming for a flat plate thermal solar collector with experimental results. For this reason, an experiment is carried out on a flat plate thermosiphon solar water collector with closed circulation considering the weather conditions of the Renewable Energy Applied Research Unit (URAER) of Ghardaïa located at the south of Algeria, where the outdoor test is done. The analytical calculation computer programming is based on iterative loops by introducing experimental data of the global solar irradiation and the different measured temperatures of the collector as initial conditions. The calculated results show a good agreement with the measured results. The average error associated with the calculated absorber temperatures for both the clear and partly cloudy day is less than 10%. The proposed computer program may be a useful tool for the optimisation of thermal performance of a flat plate collector under several climatic situations.

Keywords Thermal solar energy · Experiment · Flat plate collector · Theoretical analysis

16.1 Introduction

The intensive use of fossil fuels (oil, gas, coal … etc.) for nearly 150 years has made the world in front of several threats such as the expected depletion of these energy sources as well as the increase of the global warming, along with the catastrophic climatic consequences [1]. Furthermore, energy needs have strongly

A. Seddaoui (✉) · M. Z. Dar Ramdane · R. Noureddine
Institute of Maintenance and Industrial Safety, University of Oran 2 Mohamed Ben Ahmed, B.P. 170 Es-Senia, 31000 Oran, Algeria
e-mail: seddaoui.ali@univ-oran2.dz

A. Belasri and S. A. Beldjilali (eds.), *ICREEC 2019*, Springer Proceedings in Energy,
https://doi.org/10.1007/978-981-15-5444-5_16

increased due to the industrial development and the high consumption populations, leading to an energy shortage [2]. Thus, the current situation is characterised by an urgent case since no alternative solutions are found for both energy consumption and generation, considering that energy is one of the most important prerequisites for continuing the industrial development [3]. For several years, all the world has been concerned with saving energy and limiting greenhouse gas emissions. Especially in Algeria, where a large part of energy production depends on fossil energy sources [4]. Although, the research is going on the way towards the development of efficient techniques using solar energy.

The growing demand of the users towards hot water makes this service requires a large part of the consumed energy. To avoid this problem, it is necessary to highlight new solutions. In this context, solar thermal systems which are under permanent development offer clean alternative energy. The flat plate solar collectors for hot water production provide an effective solution and gives the opportunity to reduce the consumption of gas and electricity. In order to investigate experimentally and theoretically the thermal performance of solar collectors, several studies have been carried out making it possible to better understand these systems [5–8].

In this work, an experiment is done on a closed thermosyphon solar water collector under different weather conditions at URAER in order to validate the proposed analytical modelling approach. This approach is based on the calculation of heat balances in the system elements using thermal resistance method. A computer programming is developed under FORTRAN environment. The analytical calculation results show a good agreement with the experimental results for a clear and partly cloudy day in terms of the absorber plate temperature.

16.2 Experimental Work

The experimental work is carried out on a flat plate thermal solar water collector with closed thermosiphon circulation. The collector is covered by a single flat plate ordinary glass and an aluminium absorber plate coated with matte black. Ten vertical pipes are made of copper. The back is insulated using a 50 mm rock wool material.

Thermocouples type K are installed to measure different temperatures such as ambient temperature, absorber temperature and inlet fluid temperature. A pyranometer type Kipp & Zonen is used to measure solar radiation. Measured temperatures and solar radiation are recorded for each half hour.

16.3 Analytical Method

Solar collectors' performance may be investigated by an experimental testing under controlled environmental conditions as well as the theoretical analysis of the heat transfer and energy balance involved in solar collector system. The energy balance allows characterizing the mathematical model describing the thermal behaviour of a collector. This model is translated into a functional algorithm (Computer Programming) of different subsets, in order to solve the equations of the system that are based on the variations of input and output parameters.

In general, the thermal balance of most flat plate thermal collectors is expressed as:

$$Qab = Qu + Qp \tag{16.1}$$

Where *Qab* is the absorbed energy rate; *Qu* is the useful energy rate and *Qp* heat loss rate.

The heat losses are analysed through each part of the collector, from the top to the back side as follows: The top heat loss from the cover to the ambient environment (Qp_{t1}) is expressed by the top heat transfer coefficient U_1, ambient temperature T_a and cover temperature T_c as followed:

$$Qp_{t1} = U_1(T_c - T_a) \tag{16.2}$$

$$U_1 = 2{,}8 + 3v_{wind} + \frac{\sigma\varepsilon_c\left(T_c^4 - T_{sky}^4\right)}{T_c - T_a} \tag{16.3}$$

The top heat loss from the absorber to the cover (Qp_{t2}) is expressed by the top heat transfer coefficient U_2, cover temperature T_c and absorber temperature T_p as followed:

$$Qp_{t2} = U_2\left(T_p - T_c\right) \tag{16.4}$$

$$U_2 = \frac{N_u K_{air}}{L_{air}} + \frac{\sigma\left(T_p^2 + T_c^2\right)\left(T_p + T_c\right)}{\frac{1}{\varepsilon_p} + \frac{1}{\varepsilon_c} - 1} \tag{16.5}$$

The back heat loss through the back insulation Qp_{b3} is expressed by the back heat transfer coefficient U_3, back insulation temperature T_b and absorber temperature T_p as followed:

$$Qp_{b3} = U_3\left(T_p - T_b\right) \tag{16.6}$$

$$U_3 = \left(\frac{e_{is}}{\lambda_{is}}\right)^{-1} \tag{16.7}$$

The back heat loss from the back insulation to the surrounding environment is generally negligible, because the back insulation temperature is almost the same as the ambient temperature.

After determining top and back heat losses, the global heat loss Qp is expressed as followed:

$$Qp = U_t(T_p - T_a) \tag{16.8}$$

$$U_t = \left(\frac{1}{U_1} + \frac{1}{U_2}\right)^{-1} + U_3 \tag{16.9}$$

Here, U_t is the overall heat transfer coefficient.

The heat removal factor F_R of collector is defined as:

$$F_R = \frac{\dot{m}_f C_f}{AU_t}\left(1 - exp\left(-\frac{nF'WL_{tb}U_t}{\dot{m}_f c_f}\right)\right) \tag{16.10}$$

The useful energy gain Qu is formulated as:

$$Qu = AF_R\left[(\tau\alpha)G^* - U_t(T_{fi} - T_a)\right] \tag{16.11}$$

And, the absorber plate temperature T_p is calculated as:

$$T_p = T_{fi} + \frac{Qu}{AF_R U_t}(1 - F_R) \tag{16.12}$$

An iterative loop is used to calculate the absorber temperature T_p. Initial values of the cover and absorbing plate temperature are given in which U_t, F_R and Qu are calculated. A new absorber temperature T_p is obtained and used to recalculate U_t, F_R and Qu. The iterative loop process is repeated until getting an insignificant difference between two consecutive iterations.

16.4 Result and Discussion

In order to study the thermal behaviour of a solar collector, a detailed theoretical analysis of each part of the collector is required and a computer program is performed. The results obtained are validated by experimental measurement in terms of the absorber plate temperature. Where, the mean error expressed in percentage and associated with measured and computed absorber temperature is calculated as follows:

$$MPE = \frac{100\%}{n} \sum_{i=1}^{n} \left| \frac{T_{cal} - T_{mes}}{T_{mes}} \right| \quad (16.13)$$

Figure 16.1 represents the evolution of the global solar irradiation and the ambient temperature as a function of the local time for a clear (a) and partly cloudy day (b). In the case of the clear day (Fig. 16.1a), a regular evolution of solar irradiation is observed over time. As this type of day is preferred by most researchers who use modelling to help them predict results with the smallest margin of error. However, during the partially cloudy day (Fig. 16.1b), perturbations were observed starting in the afternoon. These climatic perturbations have been manifested by the presence of clouds that induce large variations in solar irradiation making it difficult to simulate the results of this type of day.

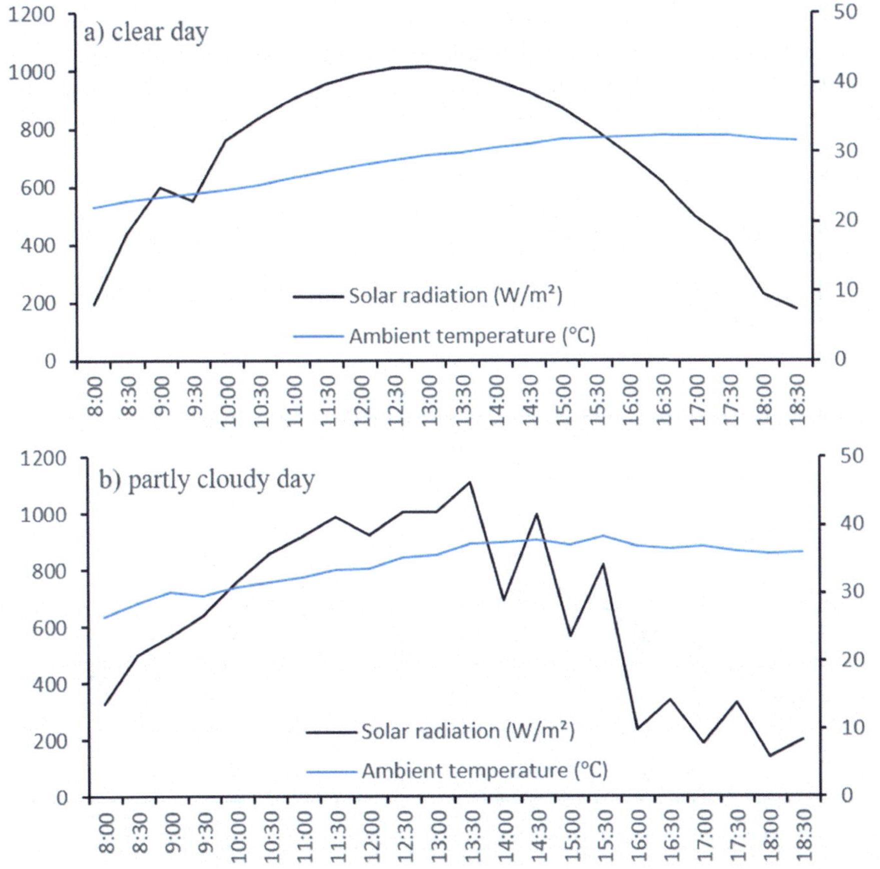

Fig. 16.1 Evolution of the global solar radiation and ambient temperature on a clear day (a) and partly cloudy day (b)

Figure 16.2 show measured and calculated temperature profiles of the absorber plate for clear (a) and partially cloudy day (b).

It is clearly shown that the measured and calculated absorber temperatures of the clear day are related to the evolution of the solar irradiation. The solar irradiation and the temperatures increase simultaneously from the morning to the noon, then the temperatures decrease according to the normal daily decrease of the solar irradiation. It is noted that the experimental results agree well with the calculated results and the mean error averaged for the whole day using Eq. 16.13 is equal to 8.4%.

For a partially cloudy day, the measured solar irradiation is considered as input data in the computer program. As mentioned above, this day is marked by the passage of a few perturbing clouds throughout the afternoon. Due to the unexpected decrease and increase in solar irradiation, the calculated absorber temperature shows significant fluctuating slopes similar to solar radiation. However, the measured absorber temperatures do not act at the same time with the solar radiation. This may occur for several reasons such as the experimental conditions, where the measured data are averaged each half hour and difficulties appear while fixing of the absorber thermocouples that may result to a damping of the absorber temperature

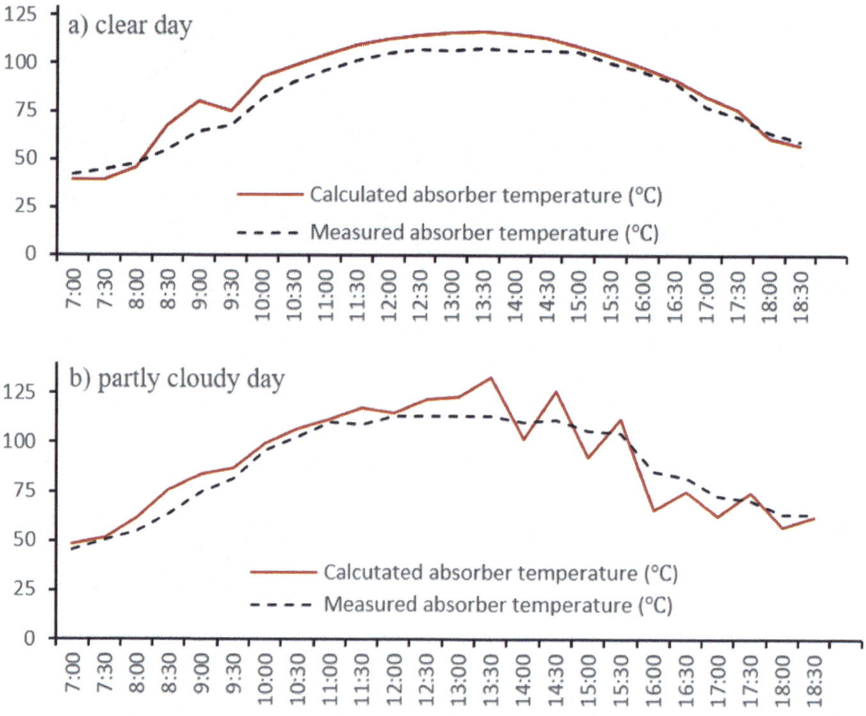

Fig. 16.2 Variation of measured and calculated absorber temperatures on a clear day (a) and partly cloudy day (b)

fluctuation. The mean error averaged for the whole day calculated with Eq. 16.13 is equal to 9.1%. It is shown that the mean error of the partly cloudy day increase compared to the clear day due to instability of the inputs in solar radiation.

16.5 Conclusion

This work is an experimental and theoretical investigation of a solar flat plate collector. A computer program that predicts the performance of a solar collector is performed. The comparison of the calculated results with the experimental data is done in terms of absorber plate temperature.

The calculated results are in good agreement with the measured data and the mean error averaged for the whole day is equal to 8.4% and 9.1% for a clear and partly cloudy day, respectively. The presented theoretical model used in the calculation of solar collector offers a promising tool to predict temperature profiles under different climatic conditions.

References

1. S. Shafiee, E. Topal, When will fossil fuel reserves be diminished? Energy Policy **37**(1), 181–189 (2009)
2. L. Pérez-Lombard, J. Ortiz, C. Pout, A review on buildings energy consumption information. Energy Build. **40**(3), 394–398 (2008)
3. N. Armaroli, V. Balzani, The future of energy supply: challenges and opportunities. Angew. Chem. Int. Ed. **46**(1–2), 52–66 (2007)
4. A.B. Stambouli, Z. Khiat, S. Flazi, Y. Kitamura, A review on the renewable energy development in Algeria: current perspective, energy scenario and sustainability issues. Renew. Sustain. Energy Rev. **16**(7), 4445–4460 (2012)
5. Y.W. Koholé, G. Tchuen, Experimental and numerical investigation of a thermosyphon solar water heater. Int. J. Ambient Energy **41**, 1–11 (2018)
6. B.O. Bolaji, I.O. Abiala, Theoretical and experimental analyses of heat transfer in a flat-plate solar collector. Walailak J. Sci. Technol. **9**, 239–248 (2012)
7. I.E. Maree, A.G. Ismaeel, Experimental and theoretical calculation of efficiency for flat plate solar collectors in Erbil City
8. Y. Menni, A. Azzi, A.J. Chamkha, A review of solar energy collectors: models and applications. J. Appl. Comput. Mech. **4**(4), 375–401 (2018)

Chapter 17
Solar Chimney Power Generation in the South of Algeria: Experimental Study

A. Azizi, T. Tahri, M. H. Sellami, L. Segni, R. Belakroum, and K. Loudiyie

Abstract Based of the difference of climatic conditions, the efficiency of solar power plant varies from one region to another; thus, a specific design is required for different locations. Solar chimney power plant (SCPP) consists of a vertical cylinder surrounded by transparent collector. The air trapped between the collector and the ground is heated by solar radiation. Hydrostatic pressure difference between the base and the top of the chimney draws air upwards with a speed that can launch a turbine driven generator located at the base of chimney. A prototype of (SCPP) in which the air temperature and velocity were recorded as a function of meteorological parameters was installed at Ouargla University. The chimney's height is 6 m with a diameter of 0.16 m; the collector of 16.54 m^2 of area is covered with plastic with four air inputs and 0.2 m high at the entrance with tilt angle of about 6° compared to the ground. The Best results are obtained on August 15th, 2016 at 13:00 h: the air velocity was 3.1 m/s; the temperature at the collector outlet was 63 °C. Where the climatic parameters were 994 W/m^2, 40.5 °C, for solar irradiance and ambient temperature respectively.

Keywords Solar chimney · Solar irradiance · Electrical power · Experimental · Air velocity · Ouargla

---

A. Azizi (✉)
Mechanical and Energetic Laboratory, Mechanical Engineering Department,
Faculty of Technology, Hassiba Benbouali University, Chlef, Algeria
e-mail: abdelghani.azizi1@gmail.com

T. Tahri
Laboratory of Electrical Engineering and Renewable Energy LGEER,
Faculty of Technology, Electrical Department, Hassiba Benbouali University, Chlef, Algeria

M. H. Sellami · L. Segni
Process Engineering Laboratory, Kasdi Merbah University, 30000 Ouargla, Algeria

R. Belakroum
Laboratory of Dynamics, Interactions and Reactivity of Systems,
Faculty of Applied Sciences, Kasdi Merbah University, 30000 Ouargla, Algeria

K. Loudiyie
Laboratory of Renewables Energies, Al Akhawayne University, Ifrane, Morocco

A. Belasri and S. A. Beldjilali (eds.), *ICREEC 2019*, Springer Proceedings in Energy,
https://doi.org/10.1007/978-981-15-5444-5_17

17.1 Introduction

In our times, become the development of environmentally friendly and sustainable energy policies worldwide important subject. In front of the increase of environmental temperature and climate change, and in aim to reduce the detrimental effects of greenhouse gases as CO_2 emission problems because of the power generation by fossil fuels, humanity must find new methods and techniques using renewable and clean energies such as solar, wind and geothermal energy to generate clean and low cost power. One of these techniques whose convert solar thermal energy to electrical power is the solar chimney power plant; this device can be used in sunny areas especially in arid and semi-arid lands such as southern Algeria where the solar radiation rate is high [1].

The town of Ouargla situated on (latitude 31.95′ N, longitude 5.40′ E and altitude 141 m), where the solar irradiance period is around 3500 h per year, delivering some 2650 kWh/m^2 y of solar irradiance on the horizontal surface [2]; thus, Ouargla city is an important place to produce electricity using the solar chimney power plant [3].

Since 1981 was built in Manzanares, Spain a solar chimney prototype of 50 kW and 200 m height. The plant functioned from 1982 until 1989 and was connected to the urban power grid between 1986 and 1989 [4]. Since, many energy specialists, researchers and investigators focused their studies and researches in this field. Some of them carried out a mathematical model by computer simulation in aim to determine firstly, thermodynamic relationships of the airflow within the solar chimney; secondly, to optimize the chimney geometric design and optimal meteorological parameters to predict the system performance.

Hamdan [5] implemented and solved static fluid and ideal gas combined with Bernoulli equations in order to predict solar chimney power plant performance; he developed a model evaluating geometric parameters' effects on chimney power plant generation. Finally, the author studied the feasibility of solar chimney especially for UAE and Gulf climatic conditions.

Dai et al. [6] carried out a theoretical model in order to install solar chimney power plant in northwest sunny region of China, which the diameter and height are respectively 10 m and 200 m; the diameter of the solar collector is 500 m. They predicted to produce monthly from 110 to 190 kW of electrical power. Their theoretical model studied and analyzed the effect of several key parameters namely: chimney geometry (height and diameter), collector diameter, turbine efficiency and meteorological parameters such as ambient temperature and solar irradiance on the performance of power generation.

Alibakhsh et al. [7] has been used a mathematical model optimizing geometric parameters of solar chimney power plant in Teheran. They based their simulating study on a fundamental flow description; the study revealed that the height and the diameter are the key parameters of the solar chimney, and the best pilot unit that will be set up has the following geometric parameters: 3 m of height, 10 cm of diameter, 6 cm of collector inlet and 4 m of collector diameter.

Solar chimney power plant efficiency varies from one location to another because of the difference of weather conditions. In this context, Fei et al. [8] carried out a simulation program based on TRNSYS, which can identify the major meteorological parameter that influences the solar chimney performance. In this study, a techno-economic analysis and configuration geometric design of commercial solar chimney are also carried out for regions with different solar irradiation abundance. Other researchers and investigators used the simulated results, which were obtained from mathematical models as a base to build small pilot plants.

In this article, we wish to present an experimental study conducted at the University of Ouargla on environmentally friendly energy (SCPP), and discuss the effects of metrological parameters such as solar radiation, ambient temperature on air velocity. Inside the SCPP.

17.2 Materials and Methods

In aim to perform an investigation into the measured temperature field and air velocity for the solar chimney power plant, we have built in Ouargla University Algeria a small scale prototype presented in Fig. 17.1 composed of a (4.60 m × 3.60 m) collector area and 6 m of chimney's tall using a (PVC) pipe with 16 cm diameter. The pipe was further isolated by glass wool. The collector is constructed using a transparent plastic and the collector height from the ground is 0.2 m near its edge and 0.4 m at the center. The main geometric parameters of the solar chimney power plant are listed in the Table 17.1.

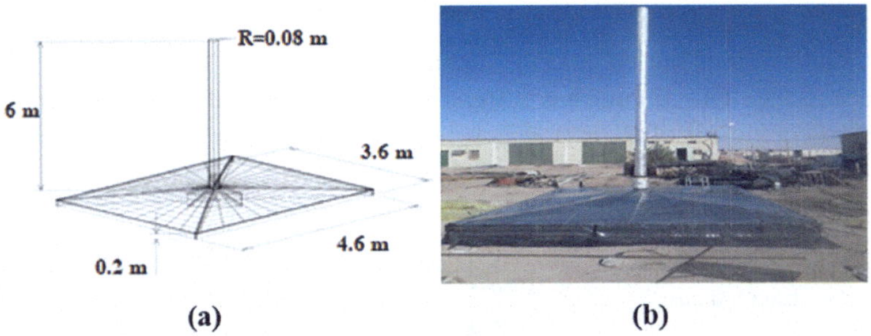

Fig. 17.1 (**a**) Schematic diagram of the solar chimney prototype. (**b**) A front view of experimental solar chimney prototype

Table 17.1 Main geometric parameters of the solar chimney prototype

Property	Value
Collector area	4.6 × 3.6 m^2
Chimney height	6 m
Height from collector inlet to the ground	0.2 m
Chimney diameter	0.16 m
Height from collector outlet to the ground	0.4 m
Collector mean tilt angle	≈6°

17.3 Results and Discussion

17.3.1 Observation and Measurement of Environmental Parameter

Figure 17.2 displays the variation of measured solar irradiance G and ambient temperature Ta versus local time for our tests location where the solar chimney was set up for the day of August 15, 2016. During experiments, ambient temperature varied between 31.4 °C at 8:00 h and 40.5 °C at 13:00 h; also the solar irradiance increased and reached the maximum value of 994 W/m^2 at 13:00 h and begin to decrease until reached the value of 944.45 W/m^2. The resultants presented show that Ouargla city has very potential of solar energy that can be used for production of electricity such as solar chimney power plant.

17.3.2 Collector Outlet Velocity

Figure 17.3 presents the variation of the air velocity inside the chimney versus the solar irradiance. This Figure has a trend curve of a third-degree polynomial with regression coefficient of R^2 = 0.982. It is clear that the air velocity increases with

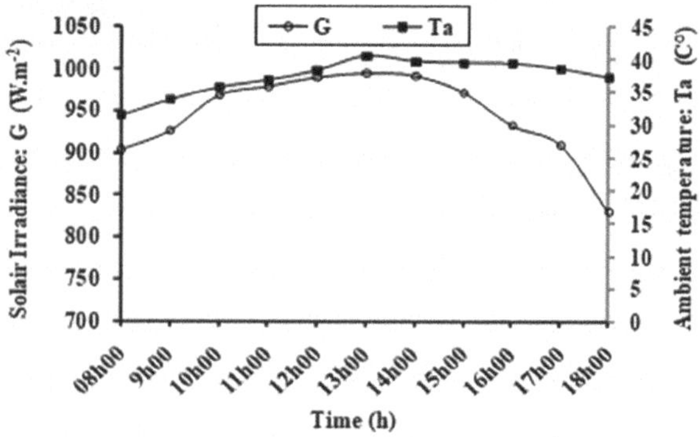

Fig. 17.2 The measured ambient temperature and solar irradiance vs. time (August 15, 2016)

Fig. 17.3 The variation of air velocity at the collector outlet vs. solar irradiance

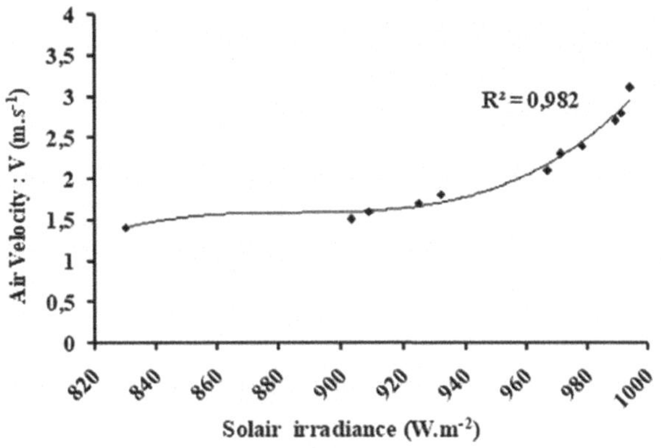

the increase of solar irradiance but according to our results, the proportionality between them is not linear.

In Fig. 17.4, we can see the variation of the air velocity at the collector versus the ambient temperature. Generally, by rising up the ambient temperature decreases slightly as a function of altitude; therefore, the air temperature near the ground is slightly higher. According to the last criterion, the air temperature before it enters the collector and the ambient temperature are the same and since the temperature of the air is in some way the motor of its upward movement, so logically, the air velocity increases with the increase of ambient temperature.

Figure 17.5 shows the relation between the air velocity and the air temperature at the collector outlet (chimney inlet). According to this Figure we can say that the air velocity is proportional non-linear with the air temperature at chimney inlet. Therefore, increasing the air temperature at chimney inlet leads directly to an increase of the air velocity and hence leads logically to an improvement of power generation efficiency.

Fig. 17.4 The variation of air velocity at the collector outlet vs. ambient temperature

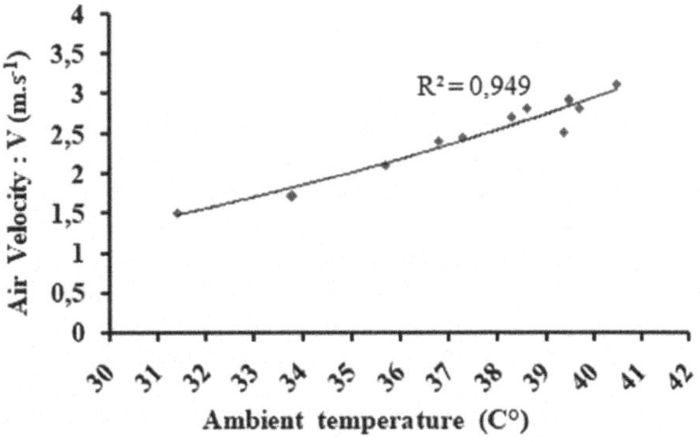

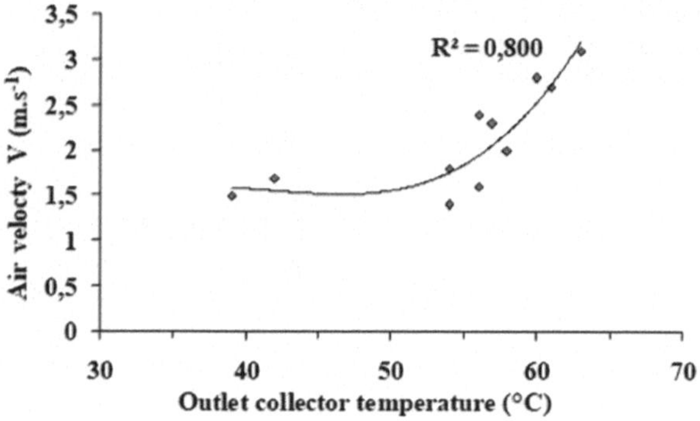

Fig. 17.5 The variation of air velocity at the chimney entrance vs. temperature at the collector outlet

17.4 Conclusion

A small prototype of SCPP was installed in Ouargla University in aim to collect results for the subsequent extraction of electrical energy with perfect design.

According to experiments, the air velocity inside the solar chimney increases proportionally by increase of solar irradiance, ambient temperature, and the best results were obtained in the chosen warm and sunny day of August 15th, 2016. The results revealed that: 3.1 m/s of air velocity was recorded; the temperature at the chimney entrance was 63 °C at 13:00 h for the ambient temperature and the solar irradiance values respectively of 40.5 °C and 994 W/m^2.

References

1. R.S. Amir, I. Hadi, Experimental and numerical investigation on an innovative solar chimney. Energy Convers. Manag. **95**, 446–452 (2015)
2. M.H. Sellami, S. Guemari, R. Touahir, K. Loudiyi, Solar distillation using a blackened mixture of Portland cement and alluvial sand as a heat storage medium. Desalination **394**, 155–161 (2016)
3. A. Azizi, T. Tahri, M.H. Sellami, L. Segni, R. Belakroum, K. Loudiyie, Experimental and CFD investigation of small-scale solar chimney for power generation. Case study: southeast of Algeria. Desalin. Water Treat. **160**, 1–8 (2019)
4. A. Koonsrisuk, S. Lorente, A. Bejan, Constructal solar chimney configuration. Int. J. Heat Mass Transf. **53**(1–3), 327–333 (2010)
5. M.O. Hamdan, Analysis of solar chimney power plant in the Arabian Gulf region. Renew. Energy **36**, 2593–2598 (2011)
6. Y.J. Dai, H.B. Huang, R.Z. Wang, Case study of solar chimney power plants in Northwestern regions of China. Renew. Energy. **28**, 1295–1304 (2003)
7. K. Alibakhsh, G. Mehran, G. Mehrdad, Simulation and optimization of geometric parameters of a solar chimney in Tehran. Energy Convers. Manag. **83**, 28–34 (2014)
8. C. Fei, L. Huashan, Z. Liang, B. Tianyang, G. Liejin, Design and simulation of the solar chimney power plants with TRNSYS. Sol. Energy **98**, 23–33 (2013)

Chapter 18
A Preliminary Study on Blade Sweep to Improve Performance of an OWC Turbine

Sidi Mohamed Karim Benslimane, Antonio Mañino Ferrando, María Clavero Gilabert, and Ali Nemdili

Abstract The energy of sea waves is one of the Renewable Energies. The Wells turbine uses the airflow produced by the pressure change inside the oscillating water column. A visible radial drift from the main flow occurs downstream of the rotor is noticed in the turbines wells with a constant chord length of the blade. Indeed, a blade chord linearly increasing with a radius is proposed in order to minimize this phenomenon and to increase the local coefficient of lift of the blade. In this context an investigated experimental and computational method are carried out to analyze the performances of two types of Wells turbines, one with a constant chord blade (rectangular-shaped blade) and the other with the blade's agreement increasing linearly (Swept blade). The results obtained show an improvement in the performance of the Wells turbine with a swept blade. The analysis of the obtained results shows clearly that the characteristics of the Wells turbine with sweep blades are highest.

Keywords Energy conversion · Wells turbine · Performance · NACA0018 · Blade with a rectangular plan form · Swept blade

---

S. M. K. Benslimane (✉)
Département de Génie Maritime, Université des Sciences et de la Technologie d'Oran Mohamed Boudiaf USTO-MB, BP 1505 El M'naouar, 31000 Oran, Algeria
e-mail: smbens@yahoo.fr

A. M. Ferrando
Department of Structural Mechanics and Hydraulic Engineering, University of Granada, Granada, Spain

M. C. Gilabert
Department of Civil Engineering, University of Granada, Granada, Spain

A. Nemdili
Département d'Hydraulique, Université des Sciences et de la Technologie d'Oran Mohamed Boudiaf USTO-MB, BP 1505 El M'naouar, 31000 Oran, Algeria

A. Belasri and S. A. Beldjilali (eds.), *ICREEC 2019*, Springer Proceedings in Energy,
https://doi.org/10.1007/978-981-15-5444-5_18

18.1 Introduction

The burning of fuels has a great influence on the environment, which obliges us to use renewable energies. One of the high potential renewable energy sources is certainly energy from sea waves and could be exploited by coastal countries [1]. Interesting devices designed for the extraction of wave energy are those based on the oscillating water column (OWC) principle.

The converter is the Wells turbine, introduced by Wells DR.AA in 1976 [2], is a self-rectifying air turbine, with a constant blade chord length of symmetrical cross-section, arranged in a 90° stagger angle cascade, and has the specificity to turn in the same direction, when flow take two opposite directions. Thus, the mechanical energy produce, is converted into electrical energy with generator at the end of the device [2, 3].

Several versions of the Wells turbine were carried out to improve the performance [4], e.g., the contra-rotating, the biplane, the variable pitch, and with a variable atmospheric conditions [5–7]. Most of these research have used a turbine with airfoil blades NACA 00XX with rectangular plan form. Experimental investigations, and CFD analysis of the flow-field through the conventional Wells turbine blades have shown a stall zone at tip blade and becomes important when the flow increases, which significantly affects the local torque and therefore performance, [8].

This study presents an experimental and numerical investigation showing the effect of sweeping the blade of a Wells turbine on the aerodynamic performance. The turbine consists of a hub of 0.192 m in diameter, designed for fixing once, eight removable blades with a constant chord (blade with a rectangular plan form), and a second time, eight blades whose chord increases linearly with the radius (swept blade) In both configurations, diameter of turbine is 0.288 mm, and profile is NACA 0018.

The power output is predicted numerically, by solving the 3-D Navier-Stokes equations with computational Fluid Dynamics (CFD) using a commercial Fluent Code.

18.2 Materials and Methods

A combined system of the experimental and numerical simulation has been developed to carry this study In order to analyze the performance of new design turbine, two types of blades based on the NACA 0018 profile changing the length of the cord was tested in the wind tunnel. While, the difficulty of finding a suitable coupler for the new design of the turbine, it was appropriate to numerically calculate the power produced, observing the experimental model, the boundary conditions and the RPMs resulting from the tests.

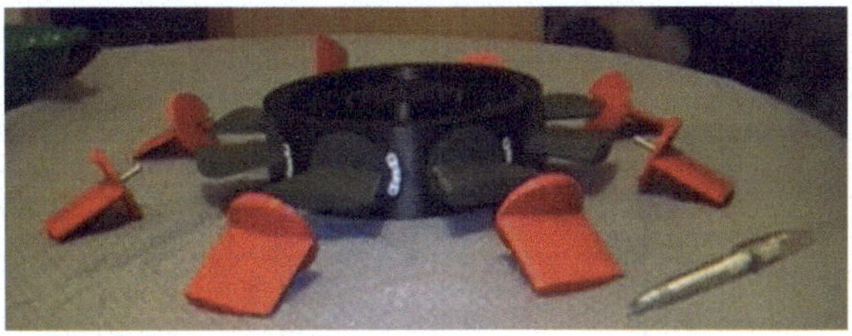

Fig. 18.1 Turbine design

Table 18.1 Geometric characteristics of the turbine

Wells turbine with rectangular plan form blades	Wells turbine with sweep blades
Hub radius Rh = 96 mm Tip radius Rt = 144 mm Hub-tip ratio Rh/Rt = 0.67 Number of blades Z = 8	
NACA0018 With constant chord length C = 52.20 mm	NACA0018 With sweep chord length C (mid) = 59.7 mm
Solidity S = 0.55	Solidity S = 0.63

18.3 Turbine Design

The turbine experimental model used for tests is presented in Fig. 18.1, mainly consists of a hub whose diameter is 0.192 m on which will set perpendicular and around its axis 08 removable blades. First blade with rectangular plan and second with a linear Swept with angle of $\lambda = 29.20°$.

The Table 18.1 exposes the geometric characteristics of both blades.

18.4 Experimental Setup

Tests are conducted with the same conditions on the flow velocities generated by the wind tunnel of the IISTA Fluid Dynamics Laboratory (University of Granada). The wind tunnel has a test section 2.15—1.8—15 m (width, height, length), the wind speed, up to 200 km/h, is controlled by frequency converter controlled by automaton, the power installed is 160 KW.

As shown in Fig. 18.2, the turbine was fitted within a uniform duct section, which had a length of 160 cm in the flow direction. However, the system was mounted using standard ball type bearings. The flow duct was tapped with ports for pressure measurements at 03 locations in the axial direction, at each of which there were up to 04 ports equally distributed around the circumference of the duct.

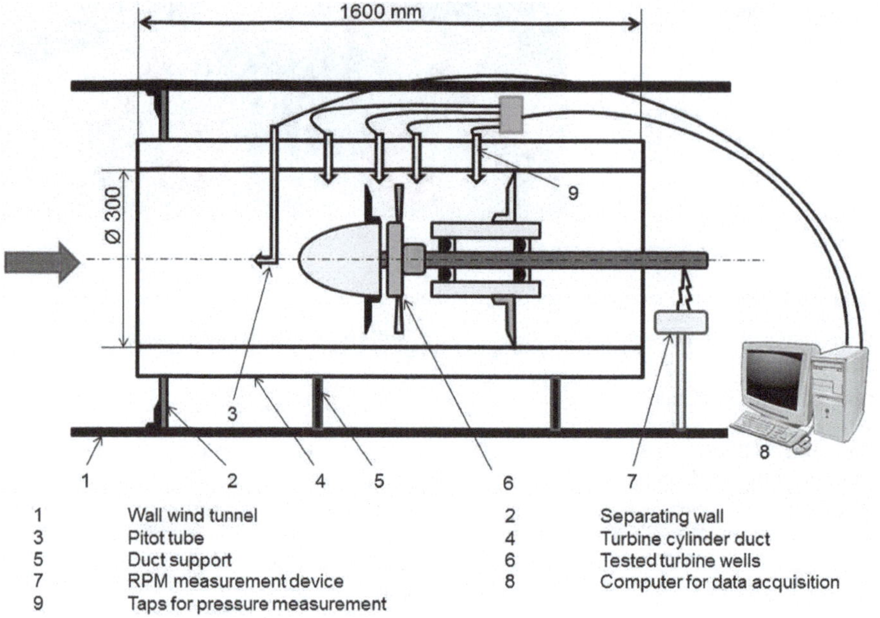

Fig. 18.2 Device in wind tunnel

Fig. 18.3 Pressure scanner connected to the taps

The circumferential ports were physically connected by air conduits, and were therefore combined into a single pressure signal, at each axial location.

Indeed, a pressure scanner (i.e. piezo resistive sensor) has been used over study area in order to characterize the pressure field (Fig. 18.3) and to transform this pressure value into an electrical signal on a computer. Moreover, a Pitot-tube velocity measurement device was positioned just inside the hub of turbine.

The tachometer is setting at end of device, to achieve the revolutions per minutes (RPM).

18.5 Numerical Method Setup

A combined Fluent-Gambit models system is considered to calculate the power output.

The Geometry of the numerical model in three dimensional, is made according to the experimental model presented in Table 18.1. Moreover, the condition of the asymmetry of the turbine is used to simplify the mechanisms and allows us to mesh only a quarter of the overall domain of the turbine. the power output of turbine are got from the numerical simulation by solving the incompressible Navier-Stokes equations in a non-inertial reference frame rotating with the turbine using FLUENT 6.3.26. The determination of the torque generated by the rotor of the turbine, is directly related to the tangential forces acting on the blades of the turbine. The turbulence model used is "k-epsillon SST".

A series of simulations were carried out, each time with an input speed change, and the angular velocity of the rotor (same conditions of the wind tunnel tests).

The geometry and grids are generated with the preprocessing Software Gambit. As shown in Fig. 18.4, a hybrid mesh is created with Map-Hex mesh near the rotor and Tri Pave mesh in the other regions.

18.6 Experimental and Numerical Analysis

The turbine performance under steady flow conditions is evaluated by turbine efficiency η, torque coefficient CT and input coefficient CA against flow coefficient Φ. The definitions of these parameters are as follows:

Pressure drop Δp^* across the rotor,

$$\Delta p^* = \frac{\Delta p_0}{\rho\,\omega^2 R_t^2} \tag{18.1}$$

where, Δp^0 is the stagnation pressure drop across the turbine, taking the pressure values at the inlet and the outlet on the turbine tests, and ρ density of air.

Fig. 18.4 Geometry and grid of the 1/8 Wells turbine model

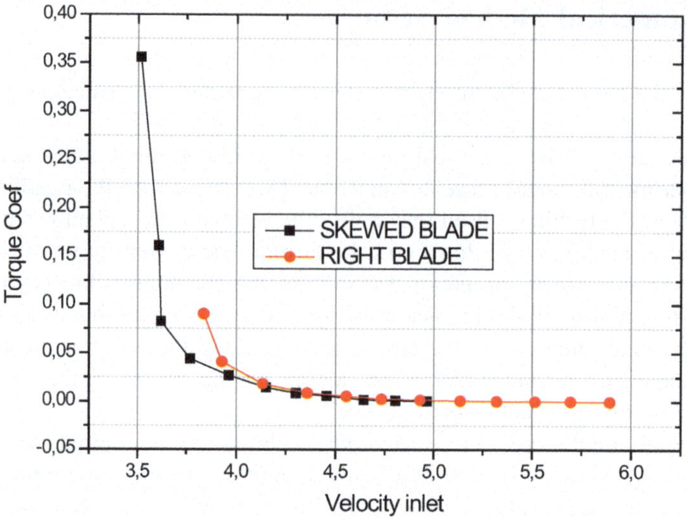

Fig. 18.5 Torque coefficient for the both blades used

Torque coefficient C_T, obtained from the numerical analysis for each inlet axial velocity V_A used in test is shown in Fig. 18.5.

$$C_T = \frac{T}{\frac{1}{2}\rho\left(V_A^2 + U_t^2\right) R_t Z b C} \tag{18.2}$$

Input coefficient CA:

$$C_T = \frac{\Delta p_0 \, \pi R_t^2}{\frac{1}{2}\rho\left(V_A^2 + U_t^2\right) R_t Z b C} \tag{18.3}$$

Flow coefficient Φ:

$$\Phi \qquad \Phi = \frac{V_A}{U_I} \tag{18.4}$$

And efficiency η:

$$\eta = \frac{T\omega}{\Delta p_0 Q} = \frac{C_T}{C_A}\left(\frac{1}{\Phi}\right) \tag{18.5}$$

where:
U_t is the peripheral velocity, R_t is the tip radius. V_A is the axial velocity normal to the plane of rotation, Z is the number of blades, b is the blade span, C is the blade chord and the medium blade chord in sweep blade, and Q is the volume flow rate.

The analysis of pressure drop across flow rate turbine function for the right blade and the sweep blade indicate that when the value of Φ (Φ < 0.1) decreases, Δp* is the same in the two turbine, and when Φ increase the Δp* of sweep is higher than the right one. The graph shape for the both turbines is parabolic. Figure 18.6b–d present the experimental results of the sweep blade effect on the turbine performance. We note from the Fig. 18.5, it is clear that the value of CT in the case of the turbine with sweep blade is higher than that in the case of the turbine with right blade in the whole range of α (α = arctg Φ). Regarding input coefficient CA characteristics in Fig. 18.6d, although the CA-α also increases with sweep blade. Furthermore, Fig. 18.7 its peak efficiency increases with sweep blade than right blade from the value of Φ = 0.05, the value in the case of the maximum efficiency of turbine sweep blade is approximately 62%. When compared to the turbine with right blade, it increases by approximately 10%. From Fig. 18.6c, it can be noted that the CT slightly improved by using the swept blade.

According to the above facts, it is clear that the increase of peak efficiency and the postponement of stall point are achieved by swept blade. Consequently, the swept blade than square one is effective in enhancing the turbine performance.

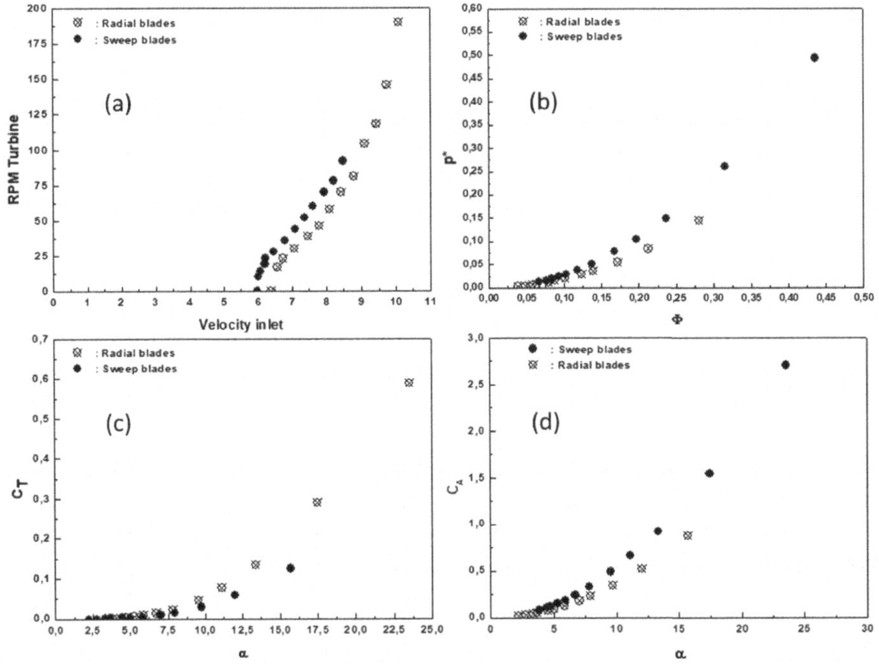

Fig. 18.6 Characteristics comparison between turbine right blades and sweep blade **a** RPM's, **b** pressure drop across turbines, **c** torque coefficient, **d** input coefficient

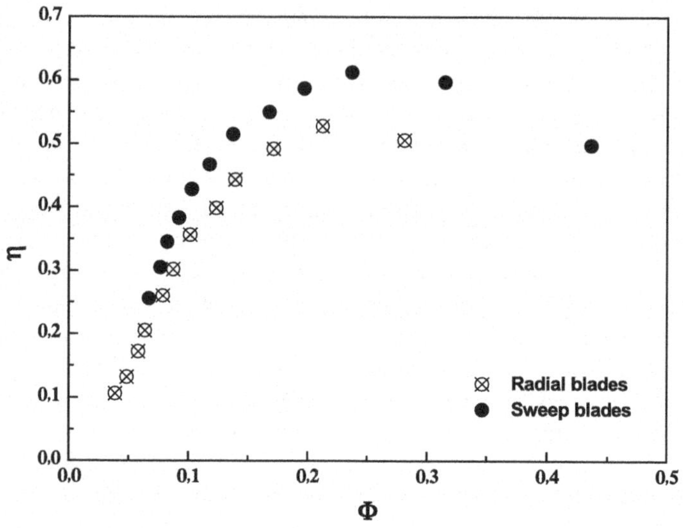

Fig. 18.7 Efficiency (η = f (Φ))

18.7 Conclusion

In this study, an experimental investigation is carried out with the aid of combined system experimental-Fluent in order to improve the performance of the Wells turbine for wave energy conversion. In this context, two form of blades turbines rectangular plan form blades and sweep blades are tested in the Wind Tunnel of the IISTA Fluid Dynamics Laboratory (University of Granada) using unidirectional airflow.

The analysis of the obtained results shows clearly that the characteristics of the Wells turbine with sweep blades are highest. The flow-field on the blades showed a great influence on the separation of the boundary layer. The distribution of the flow pressure around the blade changes significantly at tip, where the flow-field on the rectangular plan form blade have shown a stall zone at tip blade and becomes important when the flow increases, which significantly affects the local torque and therefore performance.

The sweep of blade turbine is a promising design factor for improved performance.

References

1. S. H. Salter MA, Wave energy: some questions and answers. Int. J. Ambient Energy **14**(1), 17–23 (1993). https://doi.org/10.1080/01430750.1993.9675587

2. T. Setoguchi, M. Takao, Current status of self rectifying air turbines for wave energy conversion. Energy Convers. Manag. **47**(15–16), 2382–2396 (2006)
3. T. Setoguchi, K. Kaneko, M. Inoue, Determine of optimum geometry of wells turbine rotor for wave power generator, in *Proceedings of 3rd Symposium on Ocean Wave Utilization*, Japan Agency for Marine-Earth Science and Technology (JAMSTEC), Tokyo (1991), pp. 141–149. (in Japanese)
4. S. Raghunathan, C.P. Abtan, Aerodynamic performance of a Wells air turbine. J. Energy **7**(3), 226–230 (1983)
5. K. Kaneko, T. Setoguchi, S. Raghunathan, Self-rectifying turbine for wave energy conversion, in *Proceedings of 1st Offshore and Polar Engineering Conference*, Edinburgh, UK, 11–16 August 1991, vol. 1 (1991), pp. 385–392
6. E. Medina-López, A. Moñino Ferrando, M. Clavero Gilabert, C. del Pino, M. Losada Rodríguez, Note on a real gas model for OWC performance. Renew. Energy **85**, 588–597 (2016)
7. K. Takasaki, M. Takao, T. Setoguchi, Effect of blade shape on the performance of wells turbine for wave energy conversion. Int. J. Mech. Aerosp. Ind. Mechatron. Manuf. Eng. **8**(12), 133–136 (2014)
8. A. Moñino, E. Medina-López, M. Clavero, S. Benslimane, Numerical simulation of a simple OWC problem for turbine performance. Int. J. Mar. Energy **20**, 17–32 (2017)

Chapter 19
Thermal Behaviour Study of a Bare Plate Thermal Solar Air Collector with One Pass Coupled with Compound Parabolic Concentrator (CPC)

S. Djedoui, S. Bouhassoun, B. Benameur, and R. Saim

Abstract The objective of this work is to analyze a thermal behavior of a bare plate sun-powered collector coupled with Compound Parabolic Concentrator (CPC), considering solar data of Tlemcen. To carry out this work, a program in a MATLAB environment has been developed. The nonlinear mathematical equations coupled with the boundary conditions of this problem are solved numerically. A discretization of the systems of nonlinear equations with finite difference, as well as a solution by the method of Newton-Raphson was adopted. A parametric study was performed, taking into account the effect of: CPC concentration coefficient, mass flow rate and air passage depth on collector performance coupled with a CPC either glazed or unglazed.

Keywords Flat plate solar air heaters · Unsteady · CPC · Finite difference · Newton-Raphson algorithm

19.1 Introduction

Several studies being done to improve thermal effectiveness of solar thermal collectors, mainly efficiency, the amount of heat extracted from the air from the collector's inlet to the outlet, the temperature difference from the overall radiation received…etc. as follow this conception sustains a one canal for the air flow

S. Djedoui (✉)
Smart Structures Laboratory, Belhadj Bouchaib University Center, Aïn Témouchent, Algeria
e-mail: sara.djedoui@gmail.com

S. Bouhassoun
AbouBakr Belkaid University, Tlemcen, Algeria

B. Benameur · R. Saim
Energy and Applied Thermal Laboratory, AbouBakr Belkaid University, Tlemcen, Algeria

A. Belasri and S. A. Beldjilali (eds.), *ICREEC 2019*, Springer Proceedings in Energy,
https://doi.org/10.1007/978-981-15-5444-5_19

between the absorbing plate and the insulation bottom plates. Choudhury et al. [1] investigated the thermal efficiency of this conception (Fig. 19.1).

The thermal efficiency of this conception was foreseen again by ONG [2]. The mathematic model suggested by ONG is the same to that of Choudhury et al. [1], with a limited modification. In this mathematical model, the collector was theoretically pretended to be short so the hypothesis were reasonable.

Njomo [3] and Njomo and Daguenet [4] studied the heat transferring character in the aforementioned configuration, with hypothesis in heat transfer modelization.

There are two fundamental categories of theses thermal collectors, the non-focusing and focusing collectors. The focusing ones can moreover classified built on the tracking system namely only one axis, double axis and non-tracking. Moving collectors have superior maintenance demand, Kalogirou [5].

Since the invention of parabolic concentrators (CPCs) in 1974, numerous articles have been released treating a broad series of this configuration and its analysis. Some of them are illustrated in Hsieh et al. [6].

However, after a thorough review of every work show that the largest part of them are focused on the optical, geometric and thermal character of the CPC with a tubular shaped absorber. Among these researches that has been launched in this context, in particular the work of Harmim et al. [7] have developed solar cooking in the Saharan environment. Tchinda [8] introduced a mathematical model to determine the thermal efficiency of solar air heater with a truncated concentrator with a planar unilateral absorbing plate.

Tchinda and Ngos [9] developed mathematical equations examining thermal interactions in CPC air collectors with a planar absorbing plate. Pramuang and Exell [10] mentioned the results of an experimented test whereby the Chungpaibulpatana and Exell [11] method lead to define the parameters of sun-powered collectors with a plane receptor with a CPC to heat air instead of water.

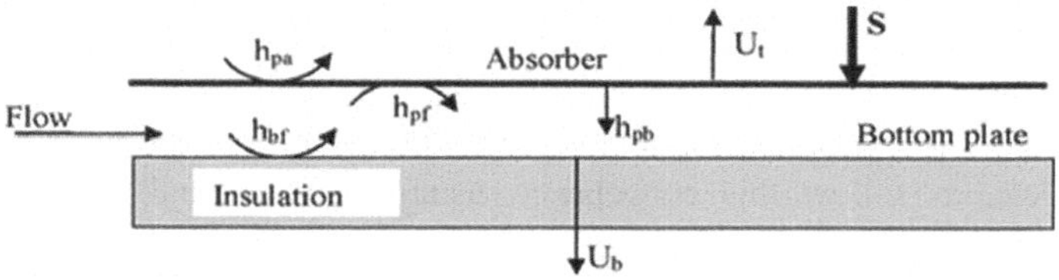

Fig. 19.1 A bare plate solar collector

19.2 Methods

19.2.1 Geometric Model and Scheme of Thermal Exchanges

See Fig. 19.2.

19.2.2 Energy Balances

The proposed model is built through the energy balance of each element identified in Fig. 19.2.

Bare Plate Air Heater Coupled with Glazed CPC
Cover:

$$S_{cpc} + h_{nc}(T_{ab} - T_c) + U_t(T_c - T_a) + h_{rcab}(T_{ab} - T_c) = \rho_c Cp_c e_c \frac{\partial T_c}{\partial t} \quad (19.1)$$

Absorber:

$$S_{1c} - h_{nc}(T_{\text{ab}} - T_c) - h_{rcab}(T_{\text{ab}} - T_c) - h_1\left(T_{\text{ab}} - T_f\right) - h_{rab-p}\left(T_{\text{ab}} - T_{\text{p}}\right) = \rho_{ab} Cp_{ab} e_{ab} \frac{\partial T_{ab}}{\partial t} \quad (19.2)$$

Fluid:

$$h_1\left(T_{\text{ab}} - T_f\right) - h_2\left(T_f - T_{\text{p}}\right) - \Gamma\left(T_f - T_{fe}\right) = \rho_f Cp_f e_f \frac{\partial T_f}{\partial t} \quad (19.3)$$

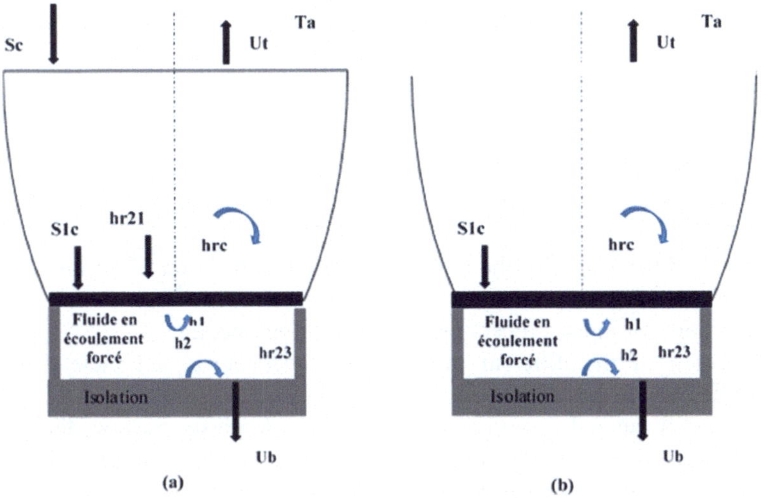

Fig. 19.2 Uncovered plate air collector coupled with: **a** Glazed CPC. **b** Unglazed CPC

Plate:

$$h_{rab\_p}(T_{ab} - T_p) + h_4(T_f - T_p) - U_b(T_p - T_a) = \rho_p C p_p e_p \frac{\partial T_p}{\partial t} \tag{19.4}$$

Bare Plate Air Heater Coupled with Unglazed CPC
Absorber:

$$-h_{r12}(T_1 - T_2) - h_1(T_1 - T_f) - U_t(T_1 - T_a) = \rho_{ab} C p_{ab} e_{ab} \frac{\partial T_{ab}}{\partial t} \tag{19.5}$$

Fluid:

$$h_1(T_1 - T_f) - h_2(T_f - T_2) - \Gamma(T_f - T_{fe}) = \rho_f C p_f e_f \frac{\partial T_f}{\partial t} \tag{19.6}$$

Plate:

$$h_{rab\_p}(T_{ab} - T_p) + h_2(T_f - T_p) - U_b(T_p - T_a) = \rho_p C p_p e_p \frac{\partial T_p}{\partial t} \tag{19.7}$$

Boundary and Initials Conditions

$$\begin{cases} T_f(0,t) = Ta \\ T_{ab}(x,0) = Ta \\ \mathrm{T_f(x,0) = Ta} \\ \mathrm{T_p(x,0) = Ta} \end{cases} \tag{19.8}$$

The different heat transfer coefficients are taken from Ong [2].

19.2.3 Algorithm

- Division of the length of the collector where the direction of air circulation in specific number of elementary sections.
- Initialization of input temperatures as ambient temperature.
- Calculation of the different heat transfer coefficients and their derivatives according to the initialized temperatures.
- Using the linearized systems with the Newton-Raphson method, the problem will be solved by returning the result of each iteration to the next phase, until reaching the Iteration condition of shutdown to obtain the converged solution.

- The temperature of the elemental section output will be placed as the input temperature for the next section.
- This calculation process is repeated until all elementary sections are finished.

19.3 Results and Interpretations

To show the effect of the CPC on the air heater, three cases of radiation received by the absorber were selected: without CPC, with unglazed CPC and with glazed CPC. Results are presented in the following figure (Fig. 19.3).

As expected, the introduction of CPC provides a significant improvement in the output temperature in the heater, because of CPC's ability to focus solar radiation using these parabolic reflectors.

19.3.1 Parametric Study

The results are obtained by considering the day of 21 June at 12: 00 pm, where they are compared each time for both cases (glazed and unglazed CPC) through curves using the same length of the sensor (L = 2 m). The air inlet temperature is considered T_{fe} = Ta = 22 °C. The mass flow is m_f = 0.01 kg/s.

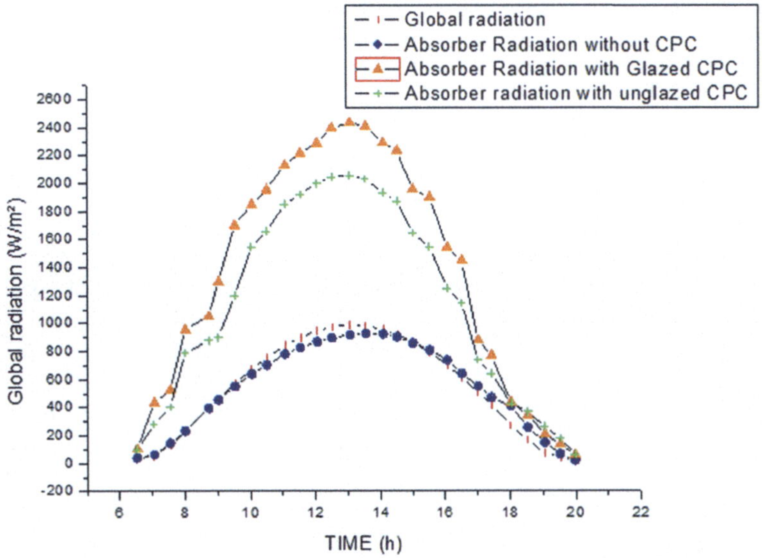

Fig. 19.3 Daily radiation behavior absorbed at the absorber level for the three cases (June 21)

Table 19.1 Influence of concentration coefficient on the heater performance in the presence and absence of CPC

T_{fe} = Ta = 22 °C L = 2 m m_f = 0.01 kg/s e = 0.0254 m	(Ca)	Uncovered plate air collector		
		Without CPC	Unglazed CPC	Glazed CPC
T_{fs} (°C)	4.5	32.867	72.454	78.944
	7	32.867	96.038	101.756
	9	32.867	112.603	117.626
Q_{perte} (watts)	4.5	352.495	1300.265	1043.547
	7	352.495	2008.660	1660.524
	9	352.495	2567.559	2150.120
η_{th}	4.5	0.2207	1.0250	1.1569
	7	0.2207	1.5041	1.6203
	9	0.2207	1.8406	1.9427
ΔP (Pas)	4.5	0.1977	0.1983	0.1988
	7	0.1977	0.1987	0.1993
	9	0.1977	0.1990	0.1996
η_{eff}	4.5	0.2207	1.0250	1.1569
	7	0.2207	1.5041	1.6203
	9	0.2207	1.8406	1.9427

Table 19.2 Influence of mass flow on the heater performance in the presence and absence of CPC

T_{fe} = Ta = 22 °C L = 2 m **Ca** = 3 $e_1 = e_2$ = 0.0254 m	Air flow (kg/s)	Uncovered plate air collector		
		Without CPC	CPC unglazed	Glazed CPC
T_{fs} (°C)	0.02	30.057	46.538	50.874
	0.07	25.849	32.970	34.003
	0.12	24.590	29.277	29.598
Q_{perte} (watts)	0.02	302.045	731.466	517.689
	0.07	196.214	457.375	267.364
	0.12	155.505	351.886	199.663
η_{th}	0.02	0.3274	0.9970	1.1732
	0.07	0.5474	1.5602	1.7071
	0.12	0.6316	1.7743	1.8524
ΔP (Pas)	0.02	0.7380	0.7386	0.7395
	0.07	8.1933	8.1953	8.1986
	0.12	23.3057	23.3090	23.3149

Influence of Concentration Coefficient. It is also noted from Table 19.1 that the heater with glazed CPC has higher performance, because of the many covers that have concentrated radiation resistance transmitted to the absorber and thus reduce the radiation absorbed. Therefore, we can see that using CPC coverage is useful for a bare air heater.

Influence of Mass Flow. We have chosen a flow interval m_f = [0.01–0.2] kg/s which corresponds to a Reynolds interval Re = [3627–72532] which starts from a transient regime and ends with a fully turbulent regime. The length and the channel depth are respectively L = 2 m and $e_1 = e_2 = 0.0254$ m (Table 19.2).

The results are obtained by considering the day of June 21 at 12 h: 00 where they have as following.

19.4 Conclusions

A MATLAB calculation code has been developed. This code allows us to simulate the thermal behavior for flat air solar collectors in the presence and absence of CPC. We have validated our calculation program for a single-pass unglazed air heater in unsteady mode. The results achieved show good accord with the reference [2].

The solar collector coupled with CPC, have been studied and compared to other configurations glazed/unglazed CPC, in unsteady state considering the solar field of Tlemcen.

This work allowed us to reach the following conclusions:

- The increase in the CPC concentration coefficient considerably improves the air temperature at the outlet, as well as the two thermal and effective efficiencies.
- The increase in mass flow of air causes a decrease in the outlet temperature and an increase in thermal efficiency, effective efficiency and pressure losses in all cases either with or without CPC.
- The increase in mass flow of air causes a decrease in the outlet temperature and an increase in thermal efficiency, effective efficiency and pressure losses in all cases either with or without CPC.
- Increasing the airflow depth (e) causes a decrease in sensor performance either without or with CPC.
- The pressure drops are better for low mass airflows and small depths of passage either in cases without or with CPC.
- The daily behavior of the sensors either without or with CPC is strongly influenced by the solar radiation and the ambient temperature where it presents higher temperatures for the days of high solar intensity.

References

1. C. Choudhury, H.P. Garg, Design analysis of corrugated and flat plate solar air heaters. Renew. Energy **1**(5–6), 595–607 (1991)
2. K.S. Ong, Thermal performance of solar air heaters: mathematical model and solution procedure. Sol. Energy **55**, 93–109 (2005)
3. D. Njomo, Unglazed selective absorber solar air collector: heat exchange analysis. Heat Mass Transf. **36**(4), 313–317 (2000)
4. D. Njomo, M. Daguenet, Sensitivity analysis of thermal performances of flat plate solar air heaters. Heat Mass Transf. **42**(12), 1065–1081 (2006)
5. S.A. Kalogirou, Solar thermal collectors and applications. Prog. Energy Combust. Sci. **30**(3), 231–295 (2004)
6. C.K. Hsieh, Thermal analysis of CPC collectors. Sol. Energy **27**(1), 19–29 (1981)
7. A. Harmim, M. Merzouk, M. Boukar, M. Amar, Design et test expérimental d'un concentrateur parabolique composé non-symétrique pour la cuisson solaire. Revue des Energies Renouvelables **14**(4), 591–600 (2011)
8. R. Tchinda, Thermal behavior of solar air heaters with compound parabolic concentrator. Energy Convers. Manag. **49**, 529–540 (2008)
9. R. Tchinda, N. Ngos, A theoretical evaluation of the thermal performance of CPC with flat one-sided absorber. Int. Commun. Heat Mass Transf. **33**, 709–718 (2006)
10. S. Pramuang, R.H.B. Exell, Transient test of a solar air heater with a compound parabolic concentrator. Renew. Energy **30**(5), 715–728 (2005)
11. S. Chungpaibulpatana, R.H.B. Exell, The effects of using a one-node heat capacitance model for determining solar collector performance parameters by transient test methods. Sol. Wind Technol. **5**(4), 411–421 (1988)

Chapter 20
Performance Optimization of Thin Film Triple-Junction for Photovoltaic Applications

Fatiha Benbekhti, Souad Tahraoui, Habiba Houari, and Hayat Khouani

Abstract Since the advent of thin layers, the efficiency of manufactured photovoltaic cells continues to increase. In addition, these new materials have many advantages over crystalline silicon, such as absorption, malleability and lightness. Our study focused on a model of photovoltaic cells with multijunction. Our goal is to model, simulate and optimize this cell, in order to derive the best characteristics for the manufacture of solar cells. We have studied and modeled the triple junction (a-Si:H/μc-SiGe:H/μc-Si:H) by the Silvaco Atlas simulator. This cell was created using the actual solar cell parameters documented in different papers. To reduce manufacturing cost and improuve conversion efficiency.

Keywords Silicon · Germanium · Thin films · Optimization · Simulation

20.1 Introduction

Improved performance photovoltaic conversation is one of the major and most important issues for solar cells. These cells used today consist essentially of inorganic semiconductors. The dominant semiconductor material on the photovoltaic market is undoubtedly silicon [1–3]. This domination of silicon in its crystalline and amorphous forms occupies 99% of the total photovoltaic market [2]. In terms of physical and optoelectronic properties, crystalline silicon is not among the most

F. Benbekhti (✉)
Technology Department, Khemis Miliana University,
Thniet El Had Street, Khemis Miliana, Ain Defla, Algeria
e-mail: Fatiha.benbekhti@univ-dbkm.dz

S. Tahraoui
Electronic Engineering Department, Chlef University, N19, Ouled Fares, Chlef, Algeria

H. Houari · H. Khouani
Electric Engineering and Electronic Department, Tlemcen University,
BP 119, Fg Pasteur, Tlemcen, Algeria

A. Belasri and S. A. Beldjilali (eds.), *ICREEC 2019*, Springer Proceedings in Energy,
https://doi.org/10.1007/978-981-15-5444-5_20

suitable materials in photovoltaic solar radiation conversion [4, 5]. However, its abundance, its low cost and the good control of the silicon technology propels it to the first place in the photovoltaic market.

Thin-film solar cells are promising candidates for future photovoltaic generations because they offer several advantages: first, the material often used for semiconductor direct band gap, including their absorption coefficient is very high [6–8]. Second, a thin absorbent layer (several micrometers) is sufficient to absorb the entire incident light, which reduces the cost of materials. Third, due to the small size of the active material of the solar cell, rare and expensive materials can be used. Fourth, although the conversion efficiency of thin film solar cells is lower than the crystalline silicon cells, but their manufacture is less costly because of the various vacuum deposition techniques used, which further reduces costs treatment [5, 8]. Finally, the deposition of thin films on a variety of substrates (flexible substrates, light substrates such as polymeric sheets, etc.), expands the range of applications of these solar cells [8–11].

20.2 Materials and Methods

A triple junction based solar cell was created and simulated with the following parameters, a thin layer of Aluminum *Al* (1 μm)/*ZnO* (100 nm), Bottom Cell composed of three microcrystalline Silicon thin films deposited as follows *n-μc-Si:H* (10 nm)/*i-μc-Si:H* (1.7 μm)/*p-μc-Si:H* (15 nm) with a high doping (10^{20} at/cm^3). Middle Cell consisting of three thin films of microcrystalline silicon-germanium deposited as follows *n-μc-SiGe:H* (15 nm)/*i-μc-SiGe:H* (500 nm)/*p-μc-SiGe:H* (20 nm) with a high doping (10^{20} at/cm^3). Top Cell consisting of three thin layers of amorphous silicon deposited as follows *n-a-Si:H* (10 nm)/*i-a-Si:H* (200 nm)/*p-a-Si:H* (15 nm) with a high doping (10^{20} at/cm^3) and a layer oxide of zinc doped with aluminum *ZnO:Al* (100 nm) (Fig. 20.1).

Numerical simulation by Silvaco is based on the numerical solution of three fundamental equations of charge transport in semiconductors, which are respectively the Poisson equation and the continuity equation for the electrons and holes and current equation.

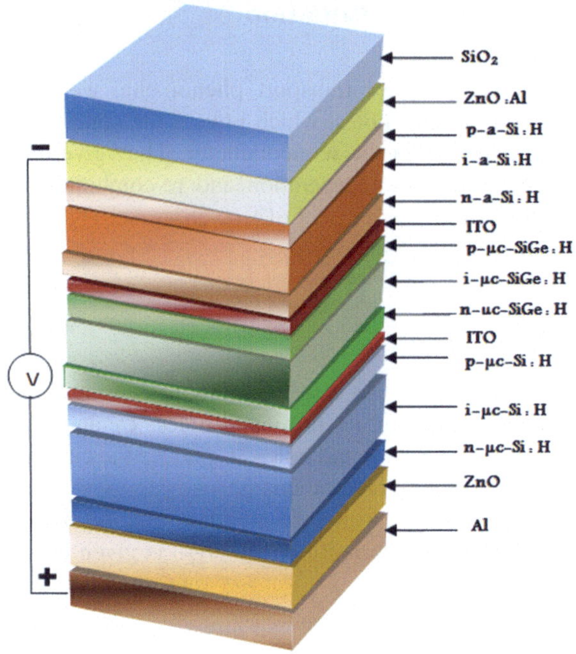

Fig. 20.1 Schematic of triple junction solar cell: *a-Si:H/μc-SiGe:H/μc-Si:H*

20.3 Poisson's Equation

$$\Delta\psi = -\frac{e}{\varepsilon}(p - n + N_d - N_a) \tag{20.1}$$

where ψ is the electrostatic potential, e: is the elementary electric charge. ε: is the dielectric permittivity of the semiconductor. N_d and N_a are ionized donor and acceptor concentrations, n and p are the electron and free hole densities.

20.4 Continuity Equation of Electrons and Holes

The currents of electrons and holes result from the sum of two terms: a concentration gradient and an electrostatic potential gradient:

$$\begin{cases} J_n = ne\mu_n E + e\,D_n \overrightarrow{gradn} \\ J_p = pe\mu_p E - e\,D_p \overrightarrow{gradp} \end{cases} \tag{20.2}$$

where n and p are the densities of free electrons and holes, e: is the elementary electric charge, μ_n and μ_p: the mobility of electrons and holes. D_n and D_p: the diffusion coefficient of electrons and holes.

20.5 Current Equation

To describe carrier transport phenomena, we use the continuity equations for electrons and for holes, which govern the condition of dynamic equilibrium of the charge carriers in the semiconductor. They give the relation between the currents, the mechanisms of generation and re-combination and the spatial and temporal distribution of the carriers of free charges:

$$\begin{cases} \frac{\delta n}{\delta t} = G_n - U_n + \frac{1}{e} div\,\vec{J}_n \\ \frac{\delta p}{\delta t} = G_p - U_p + \frac{1}{e} div\,\vec{J}_p \end{cases} \tag{20.3}$$

where n and p are the densities of free electrons and holes. G_n and G_p: represent the generation rates of electrons and holes. U_n and U_p: represent the rates of recombination of electrons and holes. J_n and J_p: the current density of electrons and holes, e: the elementary electric charge.

In photovoltaic cells, the standard relation between open-circuit voltage (V_{OC}) and short-circuit current density (J_{SC}) is given by

$$V_{OC} = \frac{kT}{q} \ln\left(\frac{J_{SC}}{J_0} + 1\right) \tag{20.4}$$

where k is the Boltzmann constant, q the unit charge, T the temperature and J_0 the dark saturation current density. The values of V_{OC} and J_{SC} can be extracted from an IV characteristic.

From the IV characteristic 4 different parameters can be extracted: J_{max}, V_{max}, V_{OC} and Jsc. The definition of the *fill factor* (FF) is given in equation:

$$FF = \frac{J_{\max} V_{\max}}{J_{SC} V_{OC}} \tag{20.5}$$

The efficiency (η) of a solar cell refers to the power conversion efficiency, defined as the ratio between the maximum power delivered by the cell (P_{max}) and the incident light power (P_{in}) [6]. It is given by Eq. (20.6), with.

$$\eta = \frac{P_{\max}}{P_{in}} = FF \frac{J_{SC} V_{OC}}{P_{in}} \tag{20.6}$$

20.6 Results and Discussions

The IV characteristics of the triple-junction solar cell previously modeled are studied separately for each sub-cell and then for the global cell, for different parameters (thickness, doping, dimensions, etc.).

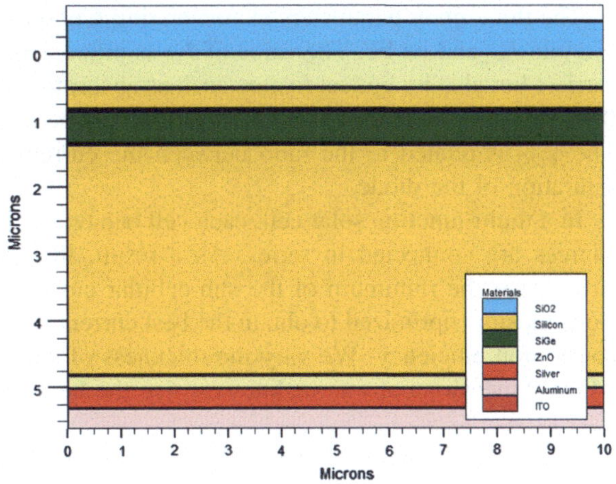

Fig. 20.2 Structure of triple Junction based solar cell simulated: *a-Si:H/μc-SiGe:H/μc-i:H*

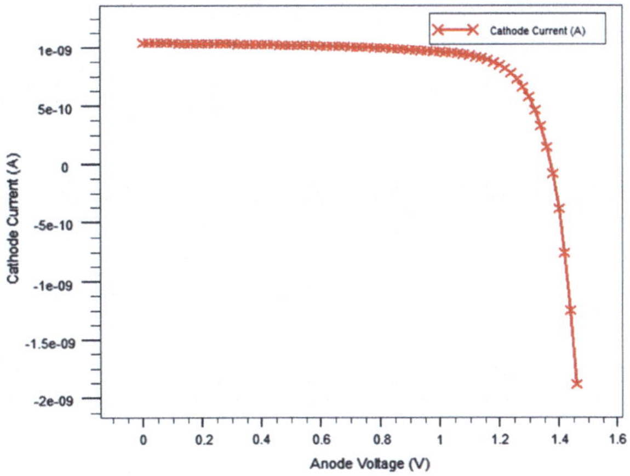

Fig. 20.3 IV characteristic obtained after simulation of triple junction *a-Si:H/μc-SiGe:H/μc-Si:H*

The dimensions are then optimized to obtain the maximum of J_{SC} possible and the composition of the global cell is modified, in order to analyze its effect on the overall performance of the cell. Finally, an optimal structure is selected with the highest efficiency and its additional layers are further optimized to achieve the highest cell efficiency (Fig. 20.2).

Figure 20.3, shows the IV characteristic obtained after simulation of the a-Si:H/μc-SiGe:H/μc-Si:H. we obtain a voltage value V_{OC} equal to 1.3966 V and a value of the short circuit current J_{SC} equal to 11.477 mA/cm^2 with a form factor equal to 77.1403 and an efficiency equal to 12.3647%.

The three main parameters characterizing a photovoltaic device are therefore its V_{OC}, its J_{SC} and its FF. The value of J_{SC} is influenced by the diffusion length of the carriers but also by optical factors such as absorption and reflection of the luminous flux. The form factor is limited in most cases by parasitic resistors and short circuits. The V_{OC} is related to the ratio between the current densities of short circuit and saturation of the diode.

In a multi-junction solar cell, each cell behaves like a current source. All current sources are connected in series. As a result, the total current generated by the structure is the minimum of the sub-cellular currents. For this reason, the dimensions must be optimized to obtain the best current possible in order to maximize the conversion efficiency. We vary the thickness of each cell individually and we see which dimensions of the thicknesses give the best results.

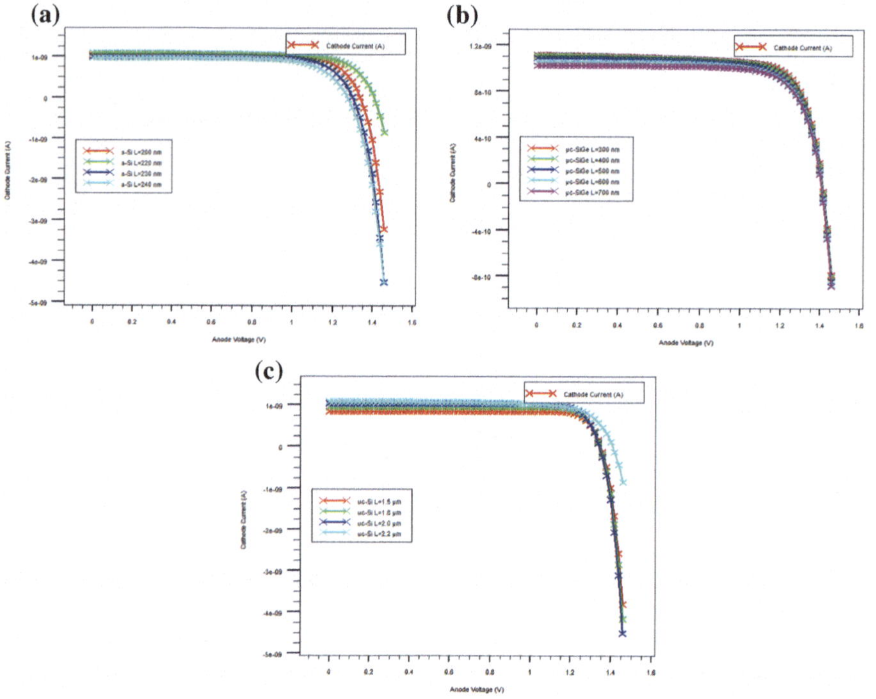

Fig. 20.4 **a** IV characteristics by varying the top cell thickness a-Si:H, **b** IV characteristics by varying the middle cell thickness, **c** IV characteristics IV by varying the bottom cell thickness

Figure 20.4a shows the variation of J_{SC} with the thickness of the top cell from 200 to 260 nm. Figure 20.4b shows the variation of J_{SC} with the thickness of the Middle Cell from 300 to 700 nm. And Fig. 20.4c shows the variation of J_{SC} with the thickness of the Bottom Cell from 1.5 to 3 μm.

From the figure Fig. 20.4a, we note that maximum J_{SC} is given for Top cell thickness of 220 nm, which corresponds to the best efficiency with a value of 12.3033%. And for Middle cell thickness of 500 nm, which corresponds to the best return with a value of 12.1952%. And for Bottom cell thickness of 2.2 μm, which corresponds to the best yield with a value of 12.3703%.

From the results obtained in the three figures above (Fig. 20.4), we thus come to the conclusion that for the structure a-Si/μc-SiGe/μc-Si. The highest J_{SC} is obtained when the Bottom cell thickness is 2.2 μm, the Middle cell thickness is 500 μm and the Top cell thickness is 220 nm. The best return is 12.36%.

20.7 Conclusion

The aim of our work was to be able to make significant improvements to the performance of silicon-based solar cells by the inclusion of thin films silicon on the one hand, and on the other hand, by incorporating germanium into the structure silicon. The total thickness of the multi-junction solar cell is similar to a solar cell with conventional single junction, but each component cell is thinner and therefore less sensitive to defects induced by light. An additional advantage of multi-junction cell structure is that each component cell can be adapted to a specific part of the solar spectrum, thereby extending a usable part of the spectrum and the increase of solar cell conversion efficiency.

However, due to the reduction in the thickness of the absorber material and due to lower material quality, the efficiency of thin-film crystalline silicon solar cells is much lower than the efficiency of silicon wafer solar cells.

Renewable sources of energy are the ideal choice and solar power is by far the most prominent energy source owing to its versatility, inexhaustible and environmental friendly features.

References

1. M. Sever, J. Krč, M. Topič, Optimisation of periodic surface textures in thin-film silicon solar cells using rigorous optical modelling by considering realistic layer Growth, in *E-MRS Spring Meeting 2013 Symposium D - Advanced Inorganic Materials and Structures for Photovoltaics*, Strasbourg, France, 27–31 May 2013 (2014). Energy Procedia **44**, 138–144
2. N. Jain, Design of III–V multijunction solar cells on silicon substrate. Ph.D. thesis, Polytechnic Institute and State University (2013)

3. V. Von, W. Tress, Device physics of organic solar cells drift-diffusion simulation in comparison with experimental data of solar cells based on small molecules. Dissertation (2011)
4. L. Hong, Rusli, X. Wang, H. Wang, X. Xu, L. He, H. Zheng, Towards perfect anti-reflection and absorption for nanodome-array thin film silicon solar cell, in *PV Asia Pacific Conference 2012* (2013). Energy Procedia **33**, 150–156
5. M. Grau, Réalisation de nouvelles structures de cellules solaires photovoltaïques à partir de couches minces de silicium cristallin sur substrat de silicium préparé par frittage de poudres. Ph.D. thesis, INSA–Lyon (2012)
6. F. Abdo, Croissance de Couches Minces de Silicium par Epitaxie en Phase Liquide à basse Température pour Applications Photovoltaïques. Ph.D. thesis (2007)
7. J. Rao, S. Varlamov, Light trapping in thin film polycrystalline silicon solar cell using diffractive gratings. Energy Procedia **33**, 129–136 (2013)
8. M. Zeman, O. Isabella, K. Jäger, R. Santbergen, S. Solntsev, M. Topic, J. Krc, Advanced light management approaches for thin-film silicon solar cells, in *International Conference on Materials for Advanced Technologies 2011 Symposium O* (2012). Energy Procedia **15**, 189–199
9. F. Oehler, Croissance et Caractérisation de Nanofils de Silicium. Ph.D. thesis, Claude Bernad Lyon 1 University (2007)
10. K.v. Maydell, K. Grunewald, M. Kellermann, O. Sergeev, P. Klement, N. Reininghaus, T. Kilper, Microcrystalline SiGe absorber layers in thin-film silicon solar cells, in *E-MRS Spring Meeting 2013 Symposium D - Advanced Inorganic Materials and Structures for Photovoltaics*, Strasbourg, France, 27–31 May 2013 (2014). Energy Procedia **44**, 209–215
11. L.W. Veldhuizen, C.H.M. van derWerf, Y. Kuang, N.J. Bakker, S.J. Yun, R.E.I. Schropp, Optimization of hydrogenated amorphous silicon germanium thin films and solar cells deposited by hot wire chemical vapor deposition. Thin Solid Films **595**, 226–230 (2015)

Chapter 21
Design and Size Optimization of the PV Modules for Stand-Alone Photovoltaic Systems

A. Bensaha, F. Benkouider, and A. Cheknane

Abstract In this paper, we have presented a proposed size optimization of the PV modules by selection various types of the PV modules with their cost, finally for choosing PV array at lower cost by using the MATLAB software. Mathematical approach was presented for optimal sizing of PV system components in addition to the total capital cost of the system. As a result, the system composed of 8 polycrystalline solar modules that yields the most cost-effective system among the 9 considered systems, so the optimized PV array size is 2.24 KWc with the cost of 1984$. Our results are very encouraging as they show us the significant improvements in photovoltaic performance.

Keywords Size optimization · Standalone PV system · Cost

21.1 Introduction

The accurate sizing of a stand-alone photovoltaic system is a fundamental procedure to optimize system operation in terms of both energy consumption and costs. The sizing optimization of standalone photovoltaic system components is a real problem, which consists of obtaining an acceptable energy and an economic cost for the consumer.

A. Bensaha (✉)
Unité de Recherche Appliquée en Energies Renouvelables, URAER,
Centre de Développement des Energies Renouvelables, CDER, 47133 Ghardaïa, Algeria
e-mail: Bensaha.madjid@gmail.com

F. Benkouider
Laboratoire de Communication, signaux et systèmes, Université Amar Telidji de Laghouat,
BP G37, route de Ghardaia, 03000 Laghouat, Algeria

A. Cheknane
Laboratoire des matériauxfonctionnel, Université Amar Telidji de Laghouat,
BP G37, route de Ghardaia, 03000 Laghouat, Algeria

A. Belasri and S. A. Beldjilali (eds.), *ICREEC 2019*, Springer Proceedings in Energy,
https://doi.org/10.1007/978-981-15-5444-5_21

Many works are carried out focusing on optimization of PV system components. So that the number of PV modules, battery storage capacity, inverter capacity can be optimally selected. The size of the PV system and its performance strongly depend on metrological data variables such as solar irradiance and ambient temperature, therefore, to optimize a PV system, extensive studies related to the metrological variables have been done in [1].

Belmili et al. [2] have used the fuzzy logic control to extract the maximum power point of the photovoltaic generator (PVG) and the wind generator (WG), he designed the optimal sizing of the system for minimizing between 3 to 9% from the installing powers. Javed et al. [3] have proposed four types of load profiles for a stand-alone PV system which has been designed and simulated using the PVSyst simulation in rural areas in Indian weather conditions (Jodhpur location). The cost analysis shows that the optimized load profile type system is the most economical design for a standalone PV system. El Shenawy et al. [4] have been presenting the use of solar photovoltaic energy to supply the electrical energy of a household of about 50 m^2 in a rural area situated in Shalateen (Egypt), they followed a methodology for sizing all the necessary components that they needed for a standalone PV system. Moreover, they studied the economic analysis of the system by using the terms of life cycle cost, and they estimated the cost percentage of each component compared with the total cost, the results show that the PV array has the highest cost.

Following the previous studies, the present study have been performed to a studio located in Ghardaia (Algeria) a semi arid area located at 450 m above sea level and at a latitude 32°38′ North and a longitude of 2°32′ Est. We have presented a proposed sizing optimization for the PV system components especially for the PV array by presented a various types of PV modules with their power output and cost to choose a PV installation at lower cost.

21.2 Materials and Methods

21.2.1 Presentation of the Standalone PV System Components

A standalone PV system is an installation that produces electricity from the PV array, which consists of photovoltaic modules interconnected between them. They transform the sunlight into direct current (DC), the modules are connected to a charge controller, then to batteries that will store the electrical energy, we have to add an inverter to convert the DC current of the modules and batteries in alternating current (Fig. 21.1).

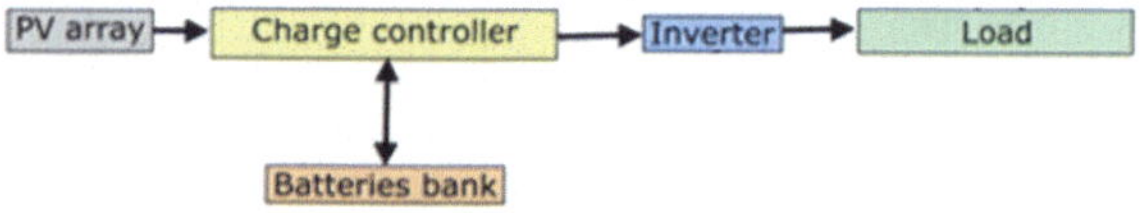

Fig. 21.1 Standalone PV system components

21.2.2 System Description

The standalone PV system consists mainly of the following systems:

- The collection system.
- The storage system.
- The system of regulation.
- The system of conversion.

The collection system is the solar panels consists of photovoltaic modules interconnected in series and/or in parallel in order to produce the required power. These modules are mounted on a metal frame that supports the solar field with a specific inclination angle of the site.

The storage system is an essential element in a standalone PV system, its role is to store the photovoltaic energy produced during the day to use it at night and days without sun.

The system of regulation is considered as the heart of a PV system; it also serves to protect the storage system. Its main role is to manage the energy from the PV array to the inverter.

The system of conversion is an equipment that is usually available between the storage system and the load, its role is to transform the direct current into alternating current generally required by the user.

21.2.3 Sizing Methodology

The sizing methodology adopted to the present study is explained systematically as follows:

Optimization of the Number Photovoltaic Modules. To calculate the optimal number of solar photovoltaic panels, we need to know the peak sun hours (PSH) of the region where the installation is located. In this study we were doing the sizing optimization about the data of Ghardaïa site so that it varies from 6.2 to 8 h in the day. The average number of sunshine hours per day on the optimum tilted surface is equals as 7.1 h/day. The PV array size and the number of solar modules can be calculated as follows [5]:

$$P_{PV-array} = \frac{E_{load}}{PSH} \tag{21.1}$$

$$N_{PV\_modules} = \frac{P_{PV-array}}{P_{PV\_module}} \tag{21.2}$$

The cost of PV array varies with the variation of the PV array capacity and it is determined by knowing the cost of each solar PV module and the number of modules required by the PV array and that can be obtained using the following equation:

$$C_{PV\_array\_cost} = \alpha_{PV\_module\_cost} \times N_{PV\_module} \quad (21.3)$$

Figure 21.2 shows the algorithm for optimizing the PV generator size.

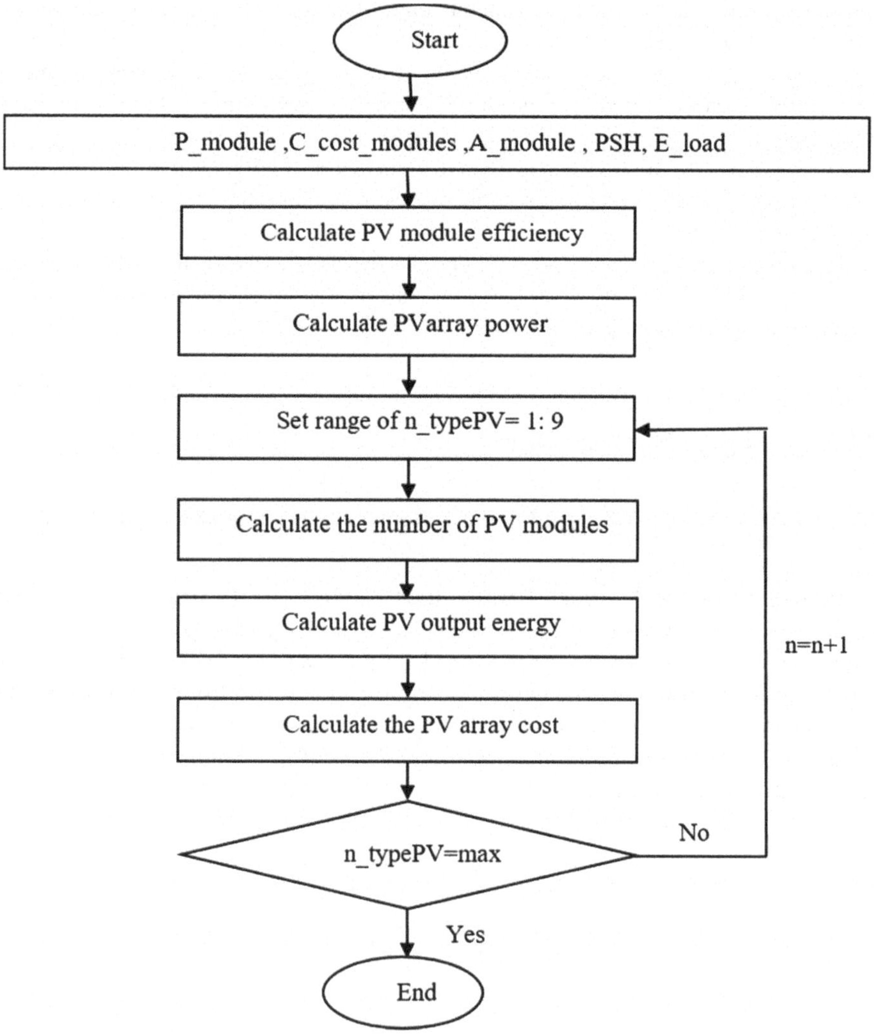

Fig. 21.2 Algorithm of the PV generator optimization

Determining the Number of Batteries. To determine the number of required batteries of the PV system, we must know the storage capacity of the batteries that could be calculated by the following equation [5]:

$$C_{storage} = \frac{E_{load} \times n_{autonomy\_days}}{P_{depth\_decharge} \times U_{pv\_system\_voltage}} \quad (21.4)$$

where E_{load} is daily energy consumption, $n_{Autonomy\ days}$ is the number of autonomy days, $P_{depth\ of\ discharge}$ is the percentage of depth of batteries discharge and $U_{PV\ system\ voltage}$ is the system voltage utilized in our sizing of the PV system components.

So, the number of batteries are determined as follows:

$$N_{batteries} = \frac{C_{storage}}{C_{capacity\_battery}} \quad (21.5)$$

Calculating the Charge Controller Capacity. The charge controller is the heart of the photovoltaic system. It allows monitoring and protecting the batteries bank. The charge controller has two main functions: protecting batteries against overloads and deep discharges and optimizing the energy transfer from PV array to load. To choose the capacity of charge controller, we must know the maximum current produced from the PV modules and it is calculated as follows:

$$I_{charge\_controller} = \frac{P_{pv\_array}}{U_{pv\_system\_voltage}} \quad (21.6)$$

Determining the Inverter Size (W). An inverter is an electronic device that converts a direct current into a sinusoidal alternating current; it receives the current produced by the PV array to convert them to an alternating current of a voltage of 220 volts and 50 Hz frequency, so the size of the inverter has been determined by this formula [5]:

$$P_{inverter} = P_{PV\_array} \times 1.1 \quad (21.7)$$

21.3 Results and Discussions

This study aims to optimize the size of PV generator, Table 21.1 outline simulation results obtained using MATLAB environment. It can be seen from Table 21.1 that the system composed of 8 polycrystalline solar modules that yields the most cost-effective system among the 9 considered systems. The maximum current entered to the charge controlled which was calculated by the Eq. (21.6) is equal to 46.66 A. So, the utilized capacity of the charge controller must be higher than determined capacity. That's why a charge controller at capacity of 55 A was chosen.

Table 21.1 Simulation results of the optimization cost of the PV array

N	Type of PV module	PV module peak power (Wp)	PV module cost ($)	Number of PV modules obtained	PV array cost obtained ($)
1	Monocrystalline	90	84	24	2016
2	Monocrystalline	100	93	24	2232
3	Monocrystalline	200	187	12	2244
4	Polycrystalline	70	69	32	2208
5	Polycrystalline	145	128	16	2048
6	Polycrystalline	235	208	10	2080
7	Polycrystalline	240	212	10	2120
8	*Polycrystalline*	*280*	*248*	*8*	*1984*
9	Polycrystalline	285	252	8	2016

The inverter size required for the present study was determined as 2464 W based on the Eq. (21.7) so we must search for an inverter at capacity superior of this value as well as we must choose an inverter at a capacity of 3 kVA.

21.4 Conclusion

In this paper a case study applied to a studio located in Ghardaia region was conducted to optimize the sizing of the PV modules for standalone PV systems. The optimization was conducted by using MATLAB environment based on various solar PV types with their output power and cost to choose PV panels at lower cost. The results showed that the type eight of the polycrystalline PV module has the best choice and we have also selected the appropriate charge controller and size of the inverter required for the PV installation.

References

1. S. KacemGairaa, Analysis of solar radiation measurements at Ghardaïa area, south Algeria. Energy Procedia **6**, 122–129 (2011)
2. H. Belmili, S. Boulouma, B. Boualem, A.M. Fayçal, Optimized control and sizing of standalone PV-wind energy conversion system. Energy Procedia **107**, 76–84 (2017)
3. K. Javed, H. Ashfaq, R. Singh, Optimized load profile & cost analysis of standalone photovoltaic system for rural power applications in Indian scenario. Smart Sci. **6**, 245–255 (2018)
4. E.T. El Shenawy, A.H. Hegazy, M. Abdellatef, Design and optimization of stand-alone PV system for Egyptian rural communities. Int. J. Appl. Eng. Res. **12**, 10433–10446 (2017)
5. A. Chel, G.N. Tiwari, A. Chandra, Simplified method of sizing and life cycle cost assessment of building integrated photovoltaic system. Energy Build. **41**, 1172–1180 (2009)

Bioenergy & Biomass

Chapter 22
A Modeling and Optimization of the Transport Phenomena of Water in a Fuel Cell H_2/O_2

Khalid Yacoubi

Abstract Electrolysis of alkaline water is an electrochemical process that produces hydrogen and oxygen by oxidation of water at anode and reduction of proton at cathode. These two gases are used as fuel in fuel cells, in particular Alkaline Fuel Cell (AFC). Potassium hydroxide solution is used as electrolyte and platinum as catalyst in anode and cathode. The chemical reaction taking place inside the cell, is the recombination of both gases to yield water, electric energy and heat energy. One of the major challenges for the proper functioning of the fuel cell is discharging water and heat that can damage the electrolytic solution. Therefore, it is essential to master the phenomena of mass transfer of water that operates within the fuel cell, and the optimum temperature of the inlet gas, knowing that fuel cells typically operate in isothermal conditions. The modeling of these phenomena and the real-time simulation of the main factors involved in the process can be investigated in order to ensure the good cell efficiency. The influences of current density, moisture and gas initial concentration of potassium, were observed the amount of water produced and evaporated within the cell.

Keywords Water transport · Alkaline fuel cell · Mathematical model

22.1 Introduction

During the 20ThC, the world has experienced a surprising proliferation of unusual weather events (heat waves, persistent droughts, floods, rising sea level, hurricanes). At that environmental concerns must be added the one that posed by the decline of world reserves of fossil resources. For these reasons, we are witnessing the emergence of cleaner energy technology, and the coming years will be those of

K. Yacoubi (✉)
Laboratoire Ingénierie des Procédées de L'Environnement (LIPE), Faculté de Chimie, Université des Sciences et de la Technologie d'Oran Mohamed Boudiaf USTO-MB, BP 1505, El M'naouer, Oran, Algeria
e-mail: yacoubi_khalid@yahoo.com

A. Belasri and S. A. Beldjilali (eds.), *ICREEC 2019*, Springer Proceedings in Energy,
https://doi.org/10.1007/978-981-15-5444-5_22

the diversification of renewable energy sources. The hydrogen carrier is a realistic solution in terms of alternative energy source, especially for embedded or mobile systems. The fuel cell uses chemical energy of hydrogen and oxygen to generate electricity without pollution. Other products are simply pure water and heat. The system is efficient and clean because it produces no emission of toxic gases [1, 2]. The AFC batteries are the oldest variants of fuel cells. They were developed in the early sixties, and they had a renewed interest with the European space program.

Also A. Khalidi et al. based their study on the implementation of an alkaline with H_2-O_2 immobilized electrolyte, which is intended in space applications. After optimizing the flow of hydrogen to the accuracy of the removal of water produced by the overall reaction of the cell, another problem has appeared to them after a long operating time of the cell, which is the accumulation of OH^- ions on the cathode side of the electrolyte [3].

Jang-Ho Jo and Sung-Chul Yi presented a mathematical model for a single cell of an alkaline fuel cell AFC. Parameters and operating conditions of the model are based on fuel cells Obiter, which are used as an energy source for space shuttles from NASA. Simulated result is obtained, which shows good agreement with some experimental data. The profiles of the variables (local density and the overvoltage current) are also obtained based on the cell voltage. A study of the influence of the electrolytes indicates that the initial concentration of the performance of the AFC is maximized at a concentration of 35% [4].

Ivan Verhaert et al. have also worked on the AFC. They give a thermodynamics model. The alkaline fuel cells are the ones of static low-temperature applications, such as cogeneration in buildings, is a promising market. To ensure a long life for the fuel cell, water and thermal management must be done carefully [5].

S. Rowshanzamir et al. also conducted a study on the performance of hydrogen-air fuel cells in the fuel cell along with the immobility of the alkaline electrolyte. They developed a one-dimensional mathematical model of partial based on the principles of transport phenomena in the gas phase, for alkaline fuel cell that checks the operating conditions resulting in water balance, and that determines a set of optimum values for the current density, the gas flow rate, the rate of humidity, temperature and pressure to ensure that the cell operate without a water deficit or excess [6].

Betty Y.S. Lin and al worked on the alkaline fuel cell (AFC), whose technologies are among the most mature. These fuel cell technologies have been in use since the mid-1960s by NASA in the Apollo program. The process of AFC has been revisited in order to define the characteristics of the current operation of the baseline for the development of future technologies in the AFC that can take advantage of advances in alkaline water electrolysers. The purpose is to advance a design that connects the actions of many devices in the AFC with AWE (Alkaline Water electrolyzer). According to the Betty et al. work a good performance can be achieved by AFC while temperatures around 70 °C. The model we developed predicts an ideal temperature of 80 °C [7].

According to previous work, the major problems in a fuel cell H_2-O_2 are the production of water and heat that undermine the proper behavior of the cell. That is why we have developed a one-dimensional transient model for this type of battery, resulting mathematical equations Nernst-Planck Fick and conservation of mass, which will be solved by numerical methods, making varying different criteria such as the current density, the weight percentage of KOH, the temperature of the hydrogen input of the fuel cell, the temperature inside the fuel cell and the specific mass flow of the steam. The simulation model will allow us to choose the optimal solution of the operation of the fuel cell every time.

22.2 Alkaline Fuel Cell AFC

Alkaline fuel cells are developed in the early 60s to power the Apollo spacecraft. They use an electrolyte liquid, usually potassium hydroxide (KOH), which has the advantage of accelerating the reduction of oxygen. However, they work properly only when the use of pure hydrogen and oxygen. Indeed, when using air or reformed hydrogen, carbon dioxide content in the latter reacts with the KOH electrolyte and form the potassium carbonate, which reduces the mobility of ions [4]. This fuel cell has the advantage of using a variety of catalysts and presents the best performance of all fuel cells. The catalysts are usually a combination of nickel and a metal such as aluminum inactive to reduce the overall manufacturing cost of this technology. For the bipolar plates, metallic magnesium or graphite compounds is used [8]. The electrolyte is an aqueous potassium hydroxide solution. The electrodes consist of macro-pores and micro-pores between the gas chambers (chamber hydrogen and oxygen chamber) of the electrolyte.

Each electrode consists of two parts: a gas filled macro-porous and micro-porous another, supposed very thin, and filled with the electrolytic solution. The electrochemical reactions take place in the micro-porous area. To represent the electrodes, the single pore model is used. This model assumes that the electrode is formed of a number of uniform and parallel cylindrical pores. The reaction is:

$$H_2 + \frac{1}{2}O_2 \rightarrow H_2O \tag{22.1}$$

22.3 Development of the Mathematical Model

22.3.1 Region of the Electrode

FICK's Law. For alkaline fuel cell one mole of water is formed at the anode for each mole of hydrogen consumed by the reaction. Steady-state mass transfer is defined as a process of equimolar diffusion against the current. For a binary gas mixture at low pressure and at constant temperature, the diffusion through the electrode is substantially independent of the composition of the mixture.

Considering the constant molar concentration of the mixture through the thickness of the macro-pore region of the anode in the steady state, the mass of water vapor is carried by the given corrected Fick law, the correction factor being the porosity ε electrode [9].

$$m_{H_2O} = \frac{\varepsilon D_{H_2O-H_2} M_2}{tg} \frac{(P_K - P_V)}{RT} \tag{22.2}$$

22.3.2 Region of Electrolyte

The model of the electrolytic solution. The cell is considered isolated at both ends in the x direction and at both ends in the y direction. The temperature inside the cell is maintained constant by a control system, so the fuel cell operates isothermally.

Nernst-Planck Equation. The flow of "i" species flow, subjected to a gradient of concentration and potential can be written [10];

$$J_i = -D_i \nabla c_i - c_i \mu_i \nabla \Phi + c_i v \tag{22.3}$$

This formula is valid for an isothermal system consisting of "n" neutral or ionized species.

From the equation above, it is easy to obtain the NERNST-PLANK formula-applying to a K^+ and OH^- ions binary system [11].

$$J^+ = -D^+ \nabla C^+ - C^+ \frac{FD^+}{RT} \nabla \Phi + VC^+ \tag{22.4}$$

$$J^- = -D^- \nabla C^- + C^- \frac{FD^-}{RT} \nabla \Phi + VC^- \tag{22.5}$$

22.3.3 Matter Outcome on Solute KOH

The equation of continuity applied to solute transient can be written as [12];

$$\frac{\partial ci}{\partial t} = (-\nabla J_i + \delta i) \tag{22.6}$$

where as in a concentration gradient along an axis perpendicular to the surface of the electrode in the y direction, and then passing to the mass concentration can be written

$$\frac{\partial \rho_1}{\partial t} = \frac{\partial^2 \rho_1}{\partial y^2} - \frac{\partial (V\rho_1)}{\partial y} \tag{22.7}$$

The passage of two moles of electrons through the cell uses two moles of OH-ion sand two moles of water to the anode against the cathode produces two moles of OH-and consumes one mole of water. If *i* amperes per unit electrode surface flow in the cell, the average flow of volume related to this unit area of anode and cathode are expressed as follows:

$$Va = \left(2r - \frac{m_{2a}}{M_2}\right) v_{2a} - 2r v_{-,a} \tag{22.8}$$

$$Vc = \left(r + \frac{m_{2c}}{M_2}\right) v_{2a} - 2r v_{-,c} \tag{22.9}$$

$r = \frac{i}{2F}$: Number of moles of water produced per unit of time and area.

A difference between Va and Vc leads to an increase in the volume of electrolyte solution equal to:

$$\rho_1 [Va - Vc] = \rho_1 \Delta v \tag{22.10}$$

This increase is according to they-axis. In introducing in Eq. (22.7), we can write:

$$\frac{\partial \rho_1}{\partial t} = D \frac{\partial^2 \rho_1}{\partial y^2} - V \frac{\partial (\rho_1)}{\partial y} \tag{22.11}$$

Equation of the Distribution of the Water Density. By solving the Eq. (22.11), we must ensure that in the calculations, the following relation is always verified. To one mole of KOH-H_2O solution:

$$\rho_1 \frac{v_1}{M_1} + \rho_2 \frac{v_2}{M_2} = 1 \tag{22.12}$$

Boundary Conditions. They are the following

At the anode:

$$D \frac{\partial \rho_1}{\partial y} = 2rM_1(1 - t) + \rho_1 V_a \tag{22.13}$$

At the cathode:

$$D\frac{\partial \rho_1}{\partial y} = 2rM_1(1-t) + \rho_1 V_c \tag{22.14}$$

To solve numerically [13, 14] the equation of diffusion-convection:

$$\frac{\partial \rho}{\partial t} = D\frac{\partial^2 \rho}{\partial y^2} - V\frac{\partial \rho}{\partial y} \tag{22.15}$$

22.4 Calculation of Parameters

22.4.1 Diffusion Coefficient

For the calculation of the diffusion coefficients in electrolytic solutions, there are formulations established in the case of diluted solutions. To extend the use of these formulas in concentrated solutions are introduced into these equations simple methods of empirical correction. We propose a simple formula of correction for the effect of temperature:

Case of Liquids. The diffusion coefficient is given by the Wilke, Chang and TynCalus formula [15]:

$$\frac{D_{AB}\mu}{T} = \frac{D^0_{AB}\mu_0}{T^0} = Cste \tag{22.16}$$

Along With:

$$D_{AB} = D^0_{AB}\frac{T}{334\,\mu} \tag{22.17}$$

Case of Gas. In the case of binary mixtures of hydrogen-steam and oxygen-steam Gilliland [9] gives the relationship as:

$$D^{H20}_{H2} = \frac{0{,}0043\,T^{1{,}5}\sqrt{\frac{1}{M_{H2}} + \frac{1}{M_{H2O}}}}{P\left[\left(\frac{1}{V_{H2}}\right)^{0{,}33} + \left(\frac{1}{V_{H20}}\right)^{0{,}33}\right]^2} = 1{,}21\,cm^2/s \tag{22.18}$$

$$D^{H20}_{O2} = \frac{0{,}0043\,T^{1{,}5}\sqrt{\frac{1}{M_{O2}} + \frac{1}{M_{H2O}}}}{P\left[\left(\frac{1}{V_{O2}}\right)^{0{,}33} + \left(\frac{1}{V_{H20}}\right)^{0{,}33}\right]^2} = 0{,}351\,cm^2/s \tag{22.19}$$

22.4.2 *Volume of the Electrolytic Solution*

The mass fraction of KOH electrolyte is expressed by taking a mass of 1000 g water.

$$V_S = \frac{1000\,f}{\rho_S(1 - f)} \tag{22.20}$$

This last equation shows that V_S depends on the mass fraction of KOH and the density of the solution that it is a function of temperature and composition (Fig. 22.1).

22.5 Results and Discussion

The resolution of the balance Eq. (22.11) allows us to draw a number of curves that we present on Figs. 22.2 and 22.3. In these figures, we will try to find the optimum conditions for operation of the cell such as the temperature of moisture from hydrogen, the initial concentration and the current density (i), the mass percentage of KOH in the potassium hydroxide solution (f%), the temperature of the hydrogen at the entrance of the cell (Tb), the temperature inside the cell (T) and the specific mass flow of water vapor through the cathode (m2c). But for now temperature of the fuel cell has set, which most authors have adopted, is that of 80 °C. Then in a second time to refine our research we draw the first curve that covers a wide field of hydrogen inlet temperature in the cell.

In Figs. 22.2a–c, it is considered that there is the water vapor which passes through the cathode. Figure 22.2a shows the evolution of the production and evaporation of the water curve (through the anode and the cathode) over time for different hydrogen temperature values at the entry of the fuel cell. Curve of production and evaporation are superimposed, hence the amount of water produced is

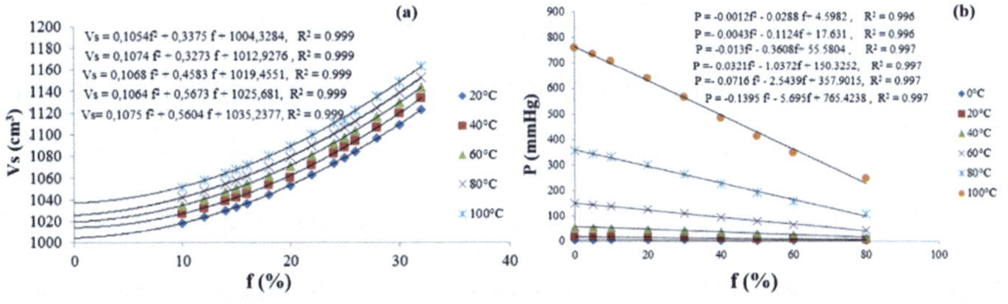

Fig. 22.1 **a** Volume of the potassium hydroxide solution with a mass fraction of KOH at different temperatures. **b** Variation of the water vapor pressure above a potassium hydroxide solution as a function of the mass fraction of KOH at different temperatures

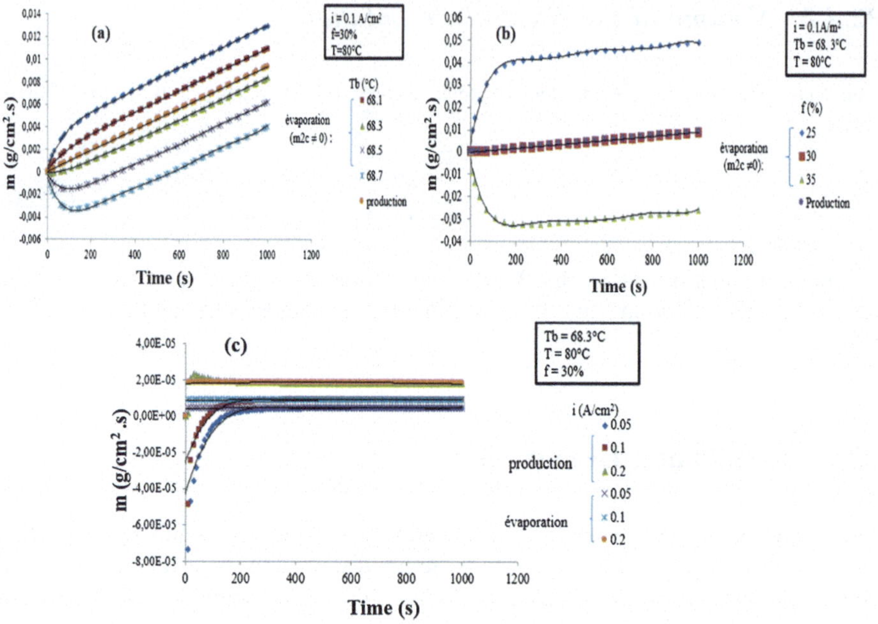

Fig. 22.2 **a** Influence of the moisture from the hydrogen on evaporation (m2c $\neq$ 0). **b** Influence of the initial concentration on the evaporation of water (m2c $\neq$ 0). **c** Effect of current density on the evaporation of water (m2c $\neq$ 0)

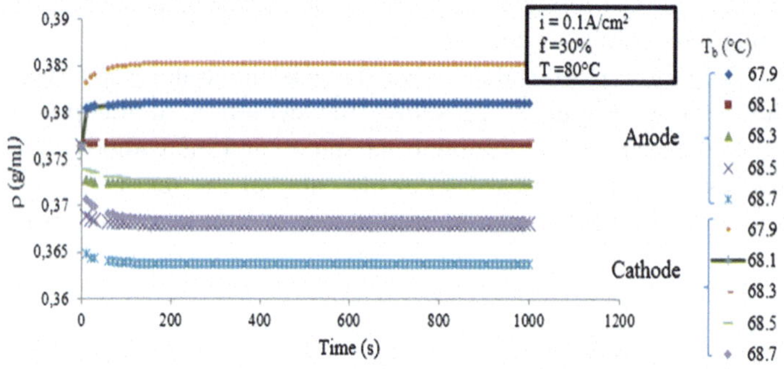

Fig. 22.3 Influence of the moisture from the hydrogen on the evolution of the density of KOH to the electrodes (m2c $\neq$ 0)

equal to the amount of water evaporated. Finally, it will be noted that the best operation is obtained by a temperature of the hydrogen gas of 68.3 °C.

Figure 22.2b shows the influence of the initial concentration of the potassium hydroxide to the quantity of water produced and evaporated. It is noted that the optimum operation is given to a KOH mass percentage of 30%. For percentages below this value was water which diffuses into hydrogen and otherwise water will accumulate in the cell. This is due to the fact that the vapor pressure of water in the potassium hydroxide solution varies inversely with the concentration of KOH.

Figure 22.2c shows the specific mass flow of the produced water and evaporated over time for different current density values.

In working conditions imposed, it is preferable to operate at 0.1 A/cm^2 because the current density value introduces a transitional regime of relatively short duration during which the flow of water evaporated does not vary significantly only during the first moments.

Figure 22.3 shows the influence of the production and evaporation of the water on the density of KOH to the electrodes as a function of time for different values of temperatures of the hydrogen at the entrance of the fuel cell, the curves of the current density of the anode and cathode superimposed. In fact, in this case, the water vapor pressure in the hydrogen and oxygen are initially equal. During operation of the fuel cell, the consumption of water at the cathode causes a diffusion of water vapor from the oxygen chamber to the interior of the cell, while at the anode the water vapor diffuses to the hydrogen chamber.

22.6 Conclusion

In the work that has been presented, we have developed a model that reflects the water transport phenomenon in a fuel cell. The proposed model combines the general laws of electrochemistry and mass transfer. For its development the following hypothesizes were issued: unidimensional problem, using the model of single pore (uniform parallel pores and cylindrical), use of an amount of hydrogen in excess of what is consumed by the electrochemical reaction rate large enough to consider gas constant moisture between the input and the output, isothermal regime, linear speed profile, constant concentration of gas mixture through the electrodes, constant volume between the electrodes. To solve the problem describing the model, we have established, with the help of correlations giving some parameters (diffusion coefficient, water vapor pressure in the potassium hydroxide solution and volume electrolyte solution) involved in the calculation. The equation of the model is a nonlinear differential equation of second order; we opted for a numerical resolution, after discretization of the problem based on the finite differences method. After resolving the equation, we have drawn a number of graphs by varying several parameters (current density, moisture and gas initial concentration of potassium) to see their influence on the amount of water produced and evaporated in the cell. This study allowed us to define the optimal values of these parameters are the cell temperature T = 80 °C, hydrogen at the entrance Tb = 68.3 °C, KOH mass fraction f = 30% and current density i = 0.1 A/cm^2.

Outlooks

In the immediate future, it would be interesting to conceive an experimental set up to test the degree of compatibility of this model with experience. The major problem mounting this device will identify the materials of each element of the fuel cell which is a simple thing as the industrial secrecy surrounding the manufacture of such technology. Then the optimal parameters that we are able to find through simulation should be checked such as hydrogen temperature at the entrance to the cell, the mass fraction of the KOH solution, the current density, temperature of the cell and the humidity of the hydrogen are actually transposable to experience. However, there is one final aspect that could not explain of an AFC and more precisely at the electrodes or there is a dry area surrounding the hydrogen entry point that diminish over time and alters the functioning of the fuel cell by letting leakage of KOH gas. We believe this is due to the assumption that the gas flow is large enough to be considered its constant moisture between the input and the output of the hydrogen. It encourages us to think complete the model of the electrolyte solution by a second bi-dimensional model of the hydrogen compartment because it is clear that there is pressure drop between the inlet and the outlet of the gas compartment and therefore the partial pressure of water vapor as well. Nevertheless, this will be mentioned in future research.

References

1. E. Giilzow, Alkaline fuel cells: a critical view. J. Power Sources **61**, 99 (1996)
2. D.M. Lu, N. Djilali, T. Berning, Three-dimensional computational analysis of transport phenomena in a PEM fuel cell. J. Power Sources **106**, 284 (2002)
3. A. Khalidi, B. Lafage, P. Taxil, G. Gave, M.J. Clifton, P. Cezac, Electrolyte and water transfer through the porous electrodes of an immobilised-alkali hydrogen-oxygen fuel cell. Int. J. Hydrogen Energy **21**(1), 25 (1996)
4. J. Jo, S. Yi, A computational simulation of an alkaline fuel cell. J. Power Sources **84**, 87 (1999)
5. I. Verhaert, M. De Paepea, G. Mulder, Thermodynamic model for an alkaline fuel cell. J. Power Sources **193**(1), 233 (2009)
6. S. Rowshanzamir, M. Kazemeini, M.K. Isfahani, Mass balance and water management for hedrogen-air fuel cells of alkaline matrix type. Int. J. Hydrogen Energy **23**(6), 499 (1998)
7. B.Y.S. Lin, D.W. Kirk, S.J. Thorpe, Performance of alkaline fuel cells: a possible future energy system. J. Power Sources **161**(1), 474 (2006)
8. E. De Geeter, M. Mangan, S. Spaepen, W. Stinissen, G. Vennekens, Alkaline fuel cells for road traction. J. Power Sources **80**, 207 (1999)
9. M. Fogiel, *The Transport Phenomena Problem Solver: Momentum, Energy, Mass* (Research and Education Association, New York, 1986)
10. H.H. Girault, *Electrochimie physique et analytique* (Presses Polytechniques et Universitaires Romandes, Lausanne, 2007)
11. M. Liang, C.S. Kao, L.H. Ou, S.W. Cho, The Nerst-Plank equation in cylindrical space and time coordinates for the chloride-ion transport phenomenon in concrete materials using theatrical study. J. Sci. Eng. **13**, 257 (2010)
12. D. Fives, A. Rojey, P.de P. Legoff, *Transfert de matière, Efficacité des Opérations de Séparation du Génie Chimique* (Editions Technip, Paris, 1976)

13. J.P. Nougier, *Méthodes de calcul numérique*, 3rd edn. (Masson, Paris, 1987)
14. S.V. Patankar, *Numerical Heat Transfer and Fluid Flow* (McGraw-Hill Book Company, New York, 1980)
15. R.C. Reid, J.M. Praunitz, T.K. Sherwood, *The Properties of Gases and Liquid*, 3rd edn. (McGraw-Hill Book Company, New York, 1977)

Chapter 23
Cellulose Degradation Methods for Biofuel Production

A. Allouache, M. A. Aziza, and T. Ahmed Zaid

Abstract Transport sector is highly dependent on oil and it is responsible for the emission of significant quantities of greenhouse gases. In order to reduce the dependency on fossil energies and reduce greenhouse gas emissions, Algeria has embarked on an ambitious program for the production of renewable energies. In this context, bioethanol represents an excellent economic and ecological solution. It is produced by alcoholic fermentation from different substrates rich in simple or complex sugar. The production from biomass rich in simple sugars is the easiest method since it does not need any pre-treatment. However, this process is excluded because all used substrates are basic food products. The second way of using non-food substrates rich in complex sugars would be the most suitable, but the production process is more complex because it requires pre-treatment and hydrolysis to release sugars. This paper describes the different methods of degradation of cellulose contained in waste paper. Office paper was collected and hydrolysed according to two methods, the first biological using a fungal enzyme, the second thermochemical using diluted sulfuric acid at high temperature, the two hydrolysed solutions were analysed by HPLC to measure released glucose, and the results showed that the thermochemical method is better by releasing twice the sugar released during biological hydrolysis.

Keywords Biofuel · Bioethanol · Cellulose · Glucose

A. Allouache (✉) · M. A. Aziza
Centre de Développement des Énergies Renouvelables (CDER),
Route de l'Observatoire, B.P. 62, 16340 Bouzareah, Alger, Algeria
e-mail: a.allouache@cder.dz

A. Allouache · T. Ahmed Zaid
Laboratoire de Valorisation des énergies Fossiles, Département Génie Chimique,
Ecole Nationale Supérieure Polytechnique, El Harrach, Algeria

A. Belasri and S. A. Beldjilali (eds.), *ICREEC 2019*, Springer Proceedings in Energy,
https://doi.org/10.1007/978-981-15-5444-5_23

23.1 Introduction

The exploitation of renewable energy sources has become a necessity to move from a fossil fuel-dependent economy to a greener economy based on renewable resources. Biomass can be used to efficiently produce renewable liquids or gaseous fuels, providing alternatives to fossil fuels. Ethanol, is already produced from sugar and starch crops, as corn or sugarcane, it is used worldwide as an alternative to gasoline. However, the controversy over the production of ethanol from food crops has led to the development of technologies using lignocellulosic biomass substances as feedstock allowing the changeover from the first to the second generation that uses lignocellulosic biomass to produce ethanol [1].

However, lignocellulosic ethanol production requires a complex process including pretreatment and enzymes and onerous costs leading to reduced competitiveness compared to sugar-based ethanol [2].

To reduce cellulosic bioethanol costs and simplify the process, using a coproduct or a waste like paper waste seems to be a good solution. The Algerian national consumption of paper and cardboard is estimated at 600 000 tons/year, which corresponds to 15 kg/person/year generating 335,000 tons of waste annually, while the recycling capacity does not exceed 10% of all the waste generated annually offering a huge quantities of pure cellulosic waste that can be used as bioethanol substrate [3]. Using waste paper to produce bioethanol is the best way to reduce production cost and help to eliminate environment problems caused by this waste. The cellulose, hemicellulose, and lignin in waste paper are less complicated, because lignin is already removed during paper production, so cellulose and hemicellulose are almost free and accessible to enzymes [2].

The composition of office waste paper was reported in several works, (Table 23.1) the majority noted high cellulose content (35.7–78.6%) and a very low hemicellulose and lignin content (4.7–13% and 0.93–5.78%, respectively) [4].

Cellulose is a plant polymer composed of chains of D-glucoses monomers linked together by bonds α (1-4). The macromolecules are linked together by hydrogen bonds, making the cellulose very compact, hydrophobic and resistant to various hydrolysis treatments. The release of the glucose units passes through the hydrolysis of the acetate functions of the cellulose. Hydrolysis can be carried out in several ways, for example with enzymes or with acid (Fig. 23.1).

Table 23.1 Literature data of composition of waste paper in sugars (%)

	Cellulose	Hemicellulose	Lignin	Reference
Office paper	78.6	4.7	1.2	[4]
	87.4	8.4	2.3	[5]
	64.7	13	0.93	[6]
	35.7	13	5.78	[7]

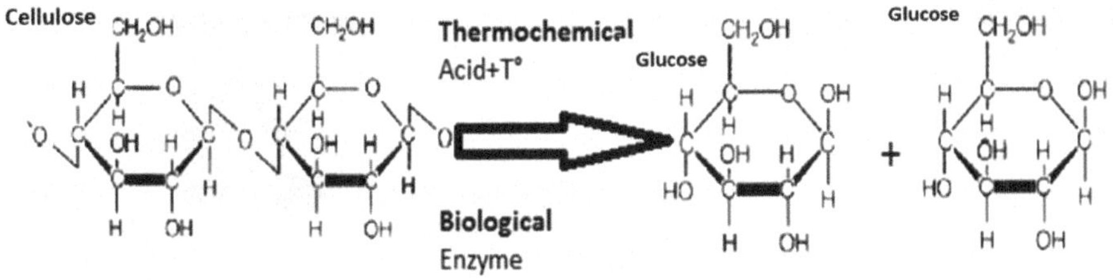

Fig. 23.1 Cellulose α (1-4) bond rupture by hydrolysis

After hydrolysis, glucose molecules are released, Bergeron [8] reported the composition of hydrolyzed waste office paper; (Table 23.2) glucose presented the most important fraction by 53% on dry basis.

Following the hydrolysis of the cellulose, the glucose solution obtained and then purified is added to a bioreactor containing yeasts that metabolize glucose and produce ethanol. Like cellulose, hemicelluloses can also be depolymerized into sugars and fermented to ethanol [9].

23.2 Material and Method

Office paper was collected and treated by enzymes and acid under the following conditions: (Fig. 23.2).

23.2.1 Biological Hydrolysis

Waste office paper are added to a citrate buffer, a commercial fungal cellulose is added. The solution is incubated during 60 min at 50 °C. This experience was carried out during 4 days; glucose was daily measured by HPLC.

Table 23.2 Compositions of hydrolyzed wastepaper

	Weight percent [wt%] on dry basis	Reference
Glucose	53.0	[8]
Xylose	8.0	
Other sugars	2.6	
Lignin	24.7	
Ash	9.6	
Total	97.9	

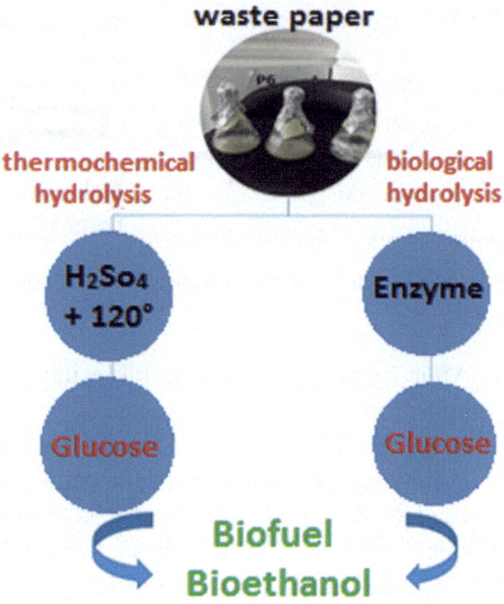

Fig. 23.2 Biological and thermochemical hydrolysis of waste paper

23.2.2 *Thermochemical Hydrolysis*

1 g of waste office paper are weighed and placed in tubes, in each tube 1, 4, 8 and 12 ml of 72% H_2SO_4 are added and mixed to impregnate paper with acid. The tubes are put in a bath at 30 °C for 1 h; water is added in each tube then autoclaved for 1 h at 120 °C.

After 1 h the tubes are cooled in an ice bath, the solution is centrifuged and filtered for analysis at HPLC.

Samples for glucose analysis were collected, before injection samples were centrifuged and filtered.

23.2.3 *Saccharification Rate*

Saccharification rate corresponds to the amount of cellulose that have been hydrolysed by enzymatic or thermochemical hydrolysis, it was evaluated using the following equation [10]:

$$\text{Saccharification rate}\,\% = \frac{\textit{the amount of glucose produced} \times 0.9}{\text{the amount of cellulose included in waste paper}} \times 100 \tag{23.1}$$

23.2.4 *Theoretical Ethanol Production*

Based on a theoretical yield of 0.51 g ethanol/g sugars, [11] the ethanol yield of biological and thermochemical from this experiment were calculated and then compared.

23.3 Results and Discussion

The results of the thermochemical and biological hydrolysis of wastepaper are presented in Tables 23.3 and Fig. 23.3.

23.3.1 *Biological Hydrolysis*

Results showed a progressive increase of glucose release Fig. 23.3, with a slow hydrolysis the first day and a maximum of 19.59 g/l recorded in the fourth day, this represent a good results even if the hydrolysis still incomplete. Guerfali [4] reported a close result 22.8 g/l done under the same conditions.

23.3.2 *Thermochemical Hydrolysis*

Results showed good yields in glucose release from cellulose hydrolysis. Acid volume added to wastepaper is a determining factor for a complete cellulose hydrolysis, indeed using a ratio of 2, 4 or even 8 allow us to have an incomplete hydrolysis, acid-waste paper ratio should be at least 12 to obtain good yield in glucose.

Compared with enzymatic hydrolysis, thermochemical hydrolysis presents a higher yield; it represents twice glucose released under enzymatic hydrolysis.

Table 23.3 Glucose release under thermochemical hydrolysis effect

H_2SO_4 volume (ml)	Glucose (g/l)
2	5.3
4	9.05
8	26.7
12	41.3

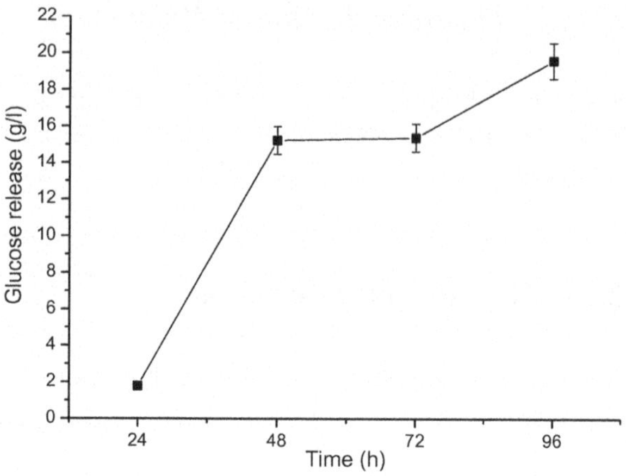

Fig. 23.3 Glucose release under enzymatic hydrolysis effect

23.3.3 Saccharification Rate

Considering a maximal cellulose content of 87.6 g/l and using Eq. (23.1), saccharification rate were calculated to be:

Table 23.4 Saccharification rate (%)

	Glucose released (g/l)	Saccharification rate (%)
Biological hydrolysis	19.59	20.12
Thermochemical hydrolysis	41.3	42.43

As expected thermochemical hydrolysis presents the best saccharification rate, which is proportional to glucose released during Cellulose hydrolysis.

23.3.4 Theoretical Ethanol Production

Considering the maximum theoretical yield of 0.51 g of ethanol produced for each gram of released glucose, ethanol conversion rate was calculated to be: (Table 23.5)

Table 23.5 Office paper and newspaper composition (%)

	Glucose released (g/l)	Theoretical Ethanol production (g/l)
Biological hydrolysis	19.59	9.99
Thermochemical hydrolysis	41.3	21.06

Results showed that theoretical ethanol production from released glucose after a thermochemical hydrolysis is higher than that produced after a biological hydrolysis, it reached 21.06 g/l.

23.4 Conclusion

Office paper represents an excellent feedstock for bioethanol production; it is rich in cellulose and available in large quantities. However, the conversion of its carbohydrates to fermentable sugars cannot be done without hydrolysis. This paper presented two methods of cellulose hydrolysis to glucose; the results showed that acid hydrolysis method had a higher yield in fermentable sugars compared to the enzymatic one. Even if this latter presented an acceptable yield and is more environmentally friendly.

However, there is still a lot of work to be done to improve the acid hydrolysis, notably reducing acid volume and its onerous cost while keeping an optimal glucose release. In addition, the reduction in environmental impact is due to acid spills on the sewage system.

References

1. B. Frankó, M. Galbe, O. Wallberg, Influence of bark on fuel ethanol production from steam-pretreated spruce. Biotechnol. Biofuels **8**, 15 (2015). https://doi.org/10.1186/s13068-015-0199-x
2. A. Al Azkawi, A. Elliston, S. Al-Bahry, S. Nallusamy, Waste paper to bioethanol: current and future prospective. Biofuels Bioprod. Bioref. (2019). https://doi.org/10.1002/bbb.1983
3. B. Djemaci, La gestion des déchets municipaux en Algérie: analyse prospective et éléments d'efficacité, Sciences de l'environnement, Université de Rouen, Français (2012). fftel-00804063f
4. M. Guerfali, H. Belghith, Enhanced enzymatic hydrolysis of waste paper for ethanol production using separate saccharification and fermentation. Appl. Biochem. Biotechnol. (2014). https://doi.org/10.1007/s12010-014-1243-1
5. M.A. Barlaz, Forest products decomposition in municipal solid waste landfills. Waste Manag. **26**(4), 321–333 (2006)
6. K.H. Chu, X. Feng, Process Saf. Environ. Prot. **91**, 123–130 (2012)
7. W.E. Eleazer, W.S. Odle, Y.S. Wang, M.A. Barlaz, Environ. Sci. Technol. **31**, 911–917 (1997)
8. P.W. Bergeron, C.J. Riley, Wastepaper as a feedstock for ethanol production, National Renewable Energy Laboratory Prepared under task no. D0041 010, November 1991
9. F. Élie, Hydrolyse du sucre, octobre 2009 ©Frédéric Élie, octobre 2004, octobre 2009, p. 7, http://fred.elie.free.fr
10. S.K. Sharma, K.L. Kalra, G.S. Kocher, Fermentation of enzymatic hydrolysate of sunflower hulls for ethanol production and its scale-up. Biomass Bioenergy **27**(4), 399–402 (2004)
11. M.S. Krishnan, N.W. Ho, G.T. Tsao, Fermentation kinetics of ethanol production from glucose and xylose by recombinant Saccharomyces 1400 (pLNH33). Appl. Biochem. Biotechnol. **77–79**, 373–388 (1999). Spring

Chapter 24
Numerical Calculation of Turbulent Reacting Flow in a Gas Turbine Combustion Chamber

Ahmed Guessab and Abdelkader Aris

Abstract In this numerical study, the effect of hydrogen on methane combustion for swirled, turbulent diffusion flames (100% CH_4, 70% CH_4 + 30% H_2, 50% CH_4 + 50% H_2, 30% CH_4 + 70% H_2 and 100% H_2) is presented. Local mean gas properties (NO and temperature) are predicted by solving the appropriate conservation equations in the finite volume form with the corresponding boundary conditions. The k-ε two equations turbulence model is employed to describe the turbulent nature of the flow. A five-step kinetic model is assumed to govern the reaction mechanism for pure methane and pure hydrogen using the eddy dissipation concept model for description of the combustion process. The principal objective of present study is to predict the chemical compositions and then examine the validity of the mathematical models for gas turbine combustor problems. The second objective is to discuss the consequences of hydrogen addition on the emission of gas turbine can-combustor chamber. Numerical investigation offers an improvement in the analysis and design of the gas turbine can-type combustors.

Keywords Gas turbine · K-ε model · NO · Finite volume

24.1 Introduction

Hydrogen and Methane are important fuels for particle application. In addition, the physical-chemical description of the oxidation is the basis for more complex fuels. In practice, hydrogen is mainly used as fuel rocket motors for space launchers with high specific impulse and a large cooling power. Hydrogen is also considered as a potential fuel projects launchers powered by aerobic hypersonic engines during flight in the atmosphere. Methane is an important practical fuel, constituting approximately 90% of the composition of natural gas. Methane is not currently used

---

A. Guessab (✉) · A. Aris
Research Laboratory in Mechanical Manufacturing Technology,
ENP d'Oran (MuriceAudin), BP 1523, El M'Naouer, Oran, Algeria
e-mail: ahmed.guessab@enp-oran.dz

A. Belasri and S. A. Beldjilali (eds.), *ICREEC 2019*, Springer Proceedings in Energy,
https://doi.org/10.1007/978-981-15-5444-5_24

for aerospace propulsion, but it is considered an alternative to hydrogen as its calorific value per unit volume is two and a half times greater in the cryogenic state. The characteristics of ignition and combustion of methane can be improved by the addition of hydrogen [1, 2]. Interest in methane-hydrogen mixture as fuel has increased significantly over the past five years. Since methane is the simplest hydrocarbon fuel available, several studies have been focused on methane-air flames. The development of the detailed chemical kinetic mechanism for the natural gas combustion began in the 70s, when several high temperature kinetic models for the hydrogen, carbon monoxide and methane oxidation were constructed under the support of large quantity of experimental data. In this a brief description of the combustor used in the present numerical study is given and the combustor CFD grid model. Then, the kinetic approach described hydrogen, methane and mixture methane-hydrogen combustion and NOx formation mechanism, respectively. Pure hydrogen combustion has been simulated using five-step reaction scheme. Furthermore, simulation results of the pure hydrogen, mixture methane-hydrogen and pure methane combustion are reported. The aim of this work was to evaluate the performances of a cylindrical combustor chamber fed by methane riche mixtures with hydrogen.

24.2 Combustor Description

Figure 24.1 shows a schematic of the combustor simulated in this study. The inlet air is guided by vanes to give the air a swirling velocity component. The swirl air passes through swirl of turn to the flow of (30°, 45° and 60°). For the results reported here, the fuels are burned at an overall equivalence ration of (0.6, 0.9 and 1.2) for the different combustor geometry studies in this study. In the air stream, the uniform inlet primary air velocity is 12 m/s, with a preheated temperature of 300 K. The fuel is injected through six fuels inlets in the swirling primary air flow. There are six small fuel inlets, each with a surface area of 0.1 cm^2. In the fuel stream, the uniform inlet gas velocity is 40 m/s, with a temperature of 300 K. The combustor grid was realized using Gambit [3, 4] software. Figure 24.2 shows the combustor grid used in CFD calculations.

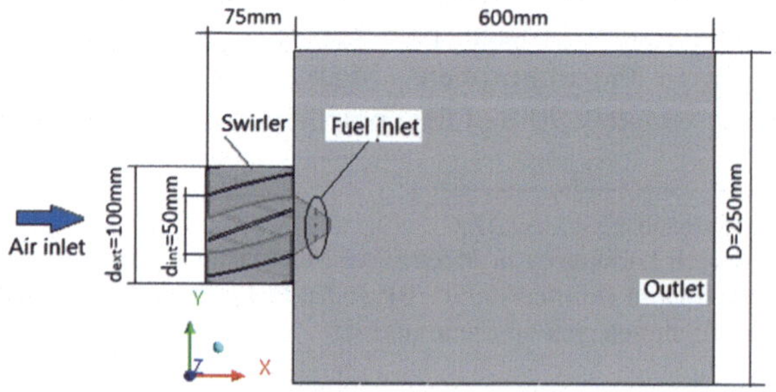

Fig. 24.1 Schematic of the combustor

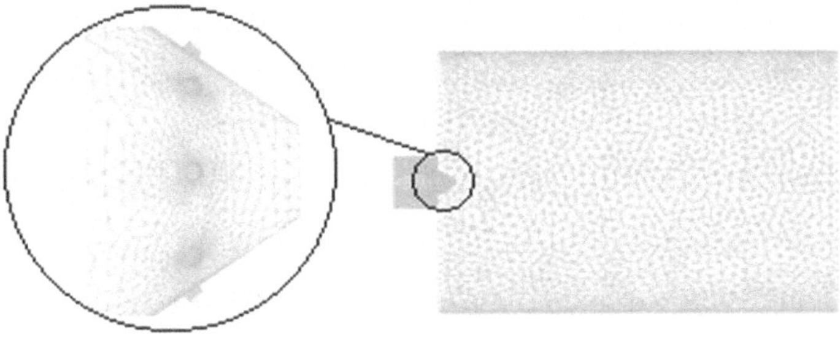

Fig. 24.2 Computational grid used in CFD simulation

24.3 Chemical Kinetics Mechanism

24.3.1 Five-Step for Hydrogen and Methane Combustion Kinetic Mechanism

In this model, reaction rates for the chemical species involved in the combustion process were calculated using a hybrid kinetic scheme based on Arrhenius model and the eddy dissipation concept. Where a, b and β are numerical coefficients. A_0 is the pre-exponential factor, T is the temperature (K), E_a is the activation energy (J/mole) and R is the universal gas constant (J/mole/K). According to Refs. [5–8], the following reduced reactions scheme and Arrhenius kinetic parameters for the above reactions are reported in Tables 24.1 and 24.2.

24.3.2 Solution Procedure

The governing equations and the associated boundary conditions are solved by code using the Finite volume method (FVM) [4] with k-ε turbulence model and standard wall functions. The pressure distribution is estimated by the Pressure-Implicit Split Operation (PISO) predictor-correction technique [9]. Calculations are performed with a no uniform grid distribution. The convergence criteria were set to 10^{-4} for

Table 24.1 Five-step hydrogen combustion and Arrhenius kinetic parameters [5, 6]

Reaction	A_k	E_k	α	β	β_k
$H_2 + O_2 \Leftrightarrow 2OH$	1.0e + 12	1.63e + 08	1	1	0
$OH + H_2 \Leftrightarrow H_2O + H$	6.3e + 14	2.5e + 07	1	1	0
$H + O_2 \Leftrightarrow OH + O$	2.2e + 14	6.9e + 07	1	1	0
$O + H_2 \Leftrightarrow OH + H$	1.1e013	3.9e + 07	1	1	0
$H + OH + M \Leftrightarrow HO_2 + M$	1.0e + 17	0	1	1	0

Table 24.2 The original Jones-Lindstedt global multi-step combustion mechanism for CH_4 with the kinetic rate data; units in [m], [s], [kmol], [J] and [K] [7, 8]

Reactions	Rate equations [Kmol/(m^3.s)]	A	b	E
$CH_4 + 0.5O_2 \Longrightarrow CO + 2H_2$	$\frac{d[CH_4]}{dt} = AT^b e^{-E/(RT)}[CH_4]^{0.5}[O_2]^{1.25}$	4.4×10^{11}	0	1.26×10^8
$CH4 + H_2O \Longrightarrow CO + 3H_2$	$\frac{d[CH_4]}{dt} = AT^b e^{-E/(RT)}[CH_4][H_2O]$	3.0×10^8	0	1.26×10^8
$H_2 + 0.5\ O_2 \Longrightarrow H_2O$	$\frac{d[H_2]}{dt} = AT^b e^{-E/(RT)}[H_2]^{0.25}[O_2]^{1.5}$	6.8×10^{15}	−1	1.67×10^8
$H_2O \Longrightarrow H_2 + 0.5O_2$	$\frac{d[H_2O]}{dt} = AT^b e^{-E/(RT)}[H_2]^{-0.75}[O_2][H_2O]$	1.25×10^{17}	−0.877	4.095×10^8
$CO + H2O \Longleftrightarrow CO_2 + H_2$	$\frac{d[CO]}{dt} = AT^b e^{-E/(RT)}[CO][H_2O](\text{forward})$	2.75×10^9	0	8.4×10^7

the mass, momentum, turbulent kinetic energy and the dissipation rate of the turbulent kinetic energy and the chemical species conservation equations. For the energy and the pollution equations, the convergence criteria were set to 10^{-6}. Geometry and grid generation is done using Gambit which is the preprocessor bundled with Fluent. The mesh is generated by automatic method. This mesh has 15,100 elements. The numerical calculations are performed on a DELL computer with CPU time of around 45 min. About 5248 iterations are required to obtain convergent solutions for the present calculation.

24.4 Results and Discussion

This part is concerned with the effect of different parameters on the flow. The parameters considered are: air swirl and fuel mixture of methane-hydrogen. These parameters are frequently employed in combustion chambers for reduce pollutant emissions. In this study, the fundamentals theory of combustion such as chemical mechanisms, velocity-composition eddy dissipation concept (EDC) model coupled with a k-ε turbulence model based mean flow computational fluid dynamics model are used here to describe the effect of such parameters on the characteristics of the combustion systems. The aim of this study is to investigate the effects of the dilution of the flame by the combustion products on NO emissions, the hydrogen content in fuel from 0%, 30%, 50% and 70%. Swirling jets are used as a mean of controlling flames in combustion chambers. Figure 24.3 shows an evolution of the laminar flame speed (LFS) depending upon % N_2 dilution in methane gas fuel. This figure confirm the capability of the Jones-Lindstedt kinetics mechanisms to capture the effect of hydrogen addition on flame speed. Moreover, the effect of nitrogen dilution is also correctly described. Figure 24.4 there is a considerable increasing of NO emission while increasing the percentage of hydrogen in the fuel mixture. This

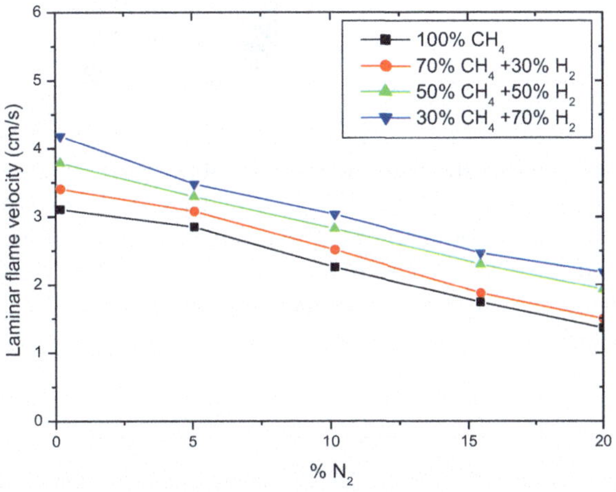

Fig. 24.3 Effect of fuel composition and nitrogen dilution on laminar flame speed

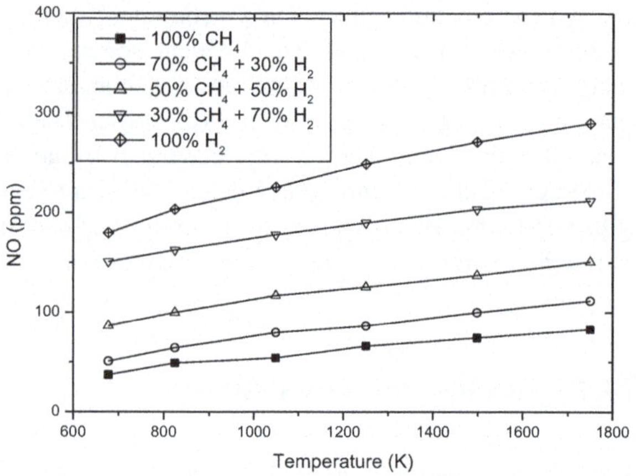

Fig. 24.4 NO_x emission at the combustor exit with different fuel mixtures

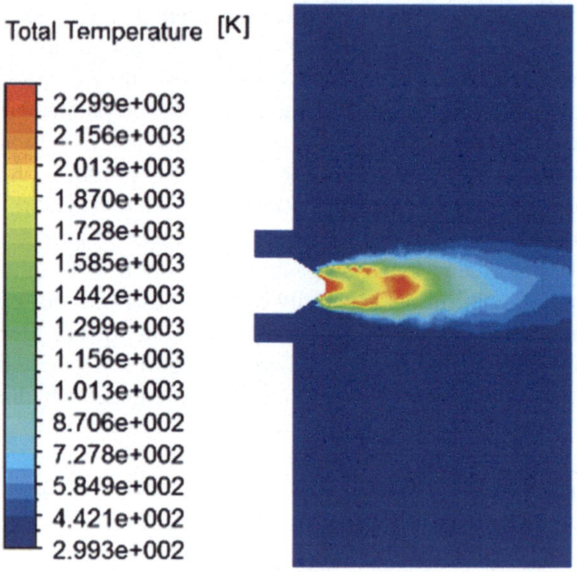

Fig. 24.5 The contours of predicted total temperature for pure hydrogen (100% H_2) combustion in can-type combustor

is caused by the increased flame temperature but also by the worse mixing quality. Figures (24.5, 24.6, 24.7, 24.8 and 24.9) shows temperature distributions in the can-combustor chamber. The calculated results of total temperature for natural gas and mixtures to pure hydrogen are shown. The values of the highest temperature while burning pure hydrogen were about 2299 K and while using natural gas about 2115 K.

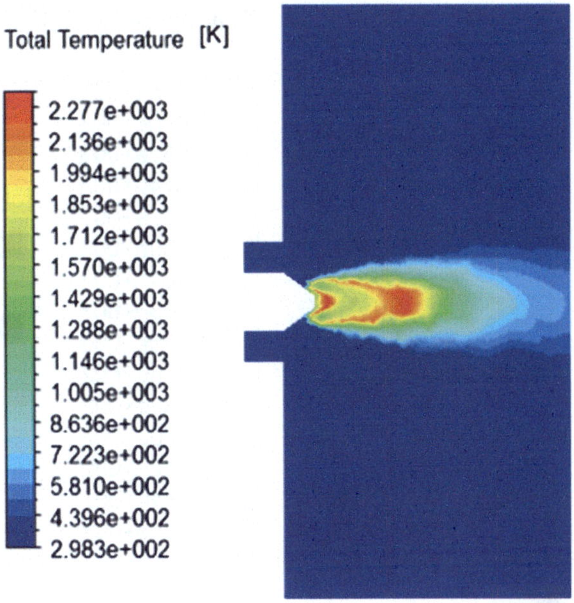

Fig. 24.6 The contours of predicted total temperature for mixture (30% CH4 + 70% H_2) combustion in can-type combustor

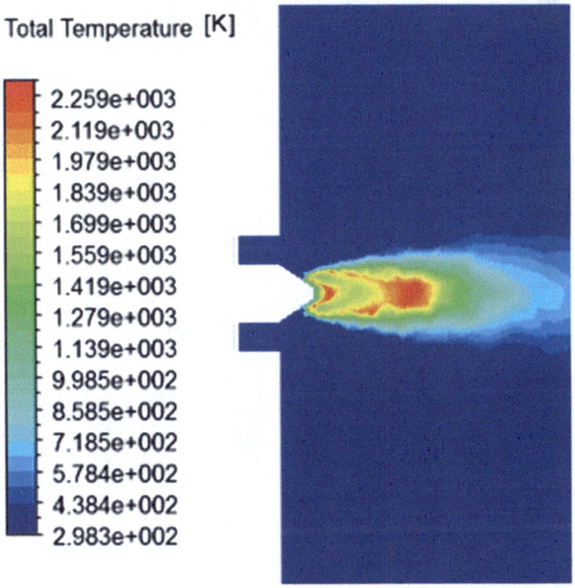

Fig. 24.7 The contours of predicted total temperature for mixture (50% CH_4 + 50%H_2) combustion in can-type combustor

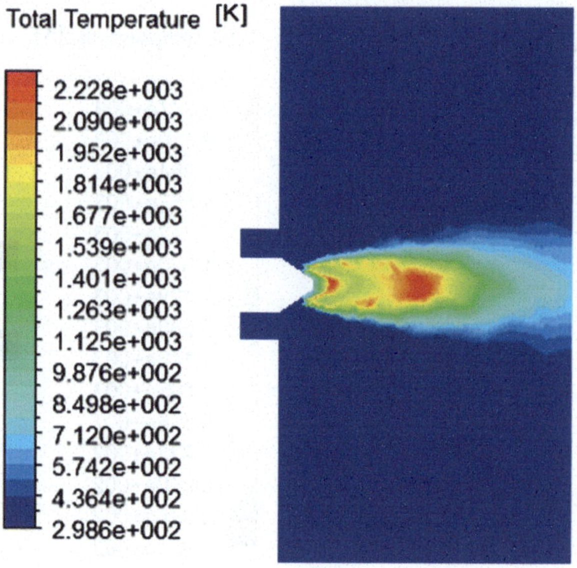

Fig. 24.8 The contours of predicted total temperature for mixture (70% CH_4 + 30% H_2) combustion in can-type combustor

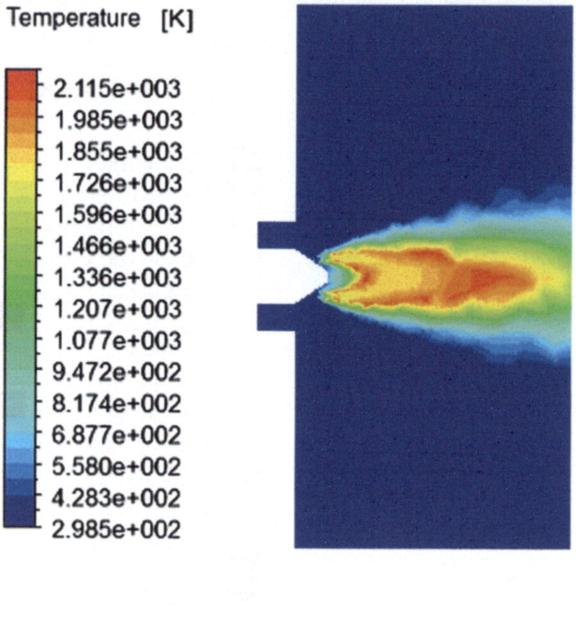

Fig. 24.9 The contours of predicted total temperature for pure methane combustion in can-type combustor

24.5 Conclusion

In the present study, a numerical calculation for turbulent swirl combusting flow in a gas-turbine combustion chamber has been investigated. The Numerical results showed that the addition of hydrogen increased the temperature and reduced the length of the flame. The results show that there is a considerable increase in NO emissions while increasing the percentage of hydrogen in the fuel. This is due to the increase in the temperature of the flame but also to the poor quality of the mixture. Finally, without any modification of a traditional axisymmetric Can-combustor geometry, hydrogen rich mixtures, until pure hydrogen have been successfully used as an alternative fuel.

References

1. K.L. Kaufman, Effect of hydrogen addition and burner diameter on the stability and structure of lean, premixed flames. Master of Science thesis, University of Iowa (2014)
2. http://ir.uiowa.edu/etd/4661
3. R.W. Schefer, Reduced turbine emissions using hydrogen-enriched fuels, in *14th World Energy Conference*, Montreal, Canada, 9–13 June (2002)
4. FLUENT Incorporated, GAMBIT. Tutorial Guide, October (1999)
5. ANSYS-Fluent, Theory Guide: Release 14.0 (2014). Last modified 23 January 2009
6. K.K. Kuo, *Principles of Combustion* (Wiley, New York, 1986)
7. J. Riccardi, P. Gheri, G. Giorgiani, M. Schiavetti, G. Gigliucci, CFD simulation for the development of gas turbine low-NO_x hydrogen combustors, in *WHEC 2016*, 13–16 June, Lyon, France (2006)
8. D.G. Nicol, A chemical and numerical study of NO_x and pollutant formation in low-emissions combustion. Ph.D. Dissertation, University of Washington (1995)
9. W.P. Jones, R.P. Lindstedt, Global reaction schemes for hydrocarbon combustion. Combust. Flame **73**(3), 233–249 (1988)

Chapter 25
Growth of Electroactive Biofilm onto Carbon Felt Bioanode in Microbial Fuel Cell: Enhancement of Bioenergy Production

Hakima Kebaili, Mostefa Kameche, Christophe Innocent, Widya Ernayati Kosimaningrum, and Tewfik Sahraoui

Abstract A microbial fuel cell (MFC) is a device that converts organic matter to electricity using microorganisms as the biocatalyst [1]. It was demonstrated that it was possible to produce electricity in a MFC from domestic wastewater, while at the same time accomplishing biological wastewater treatment (removal of chemical oxygen demand; COD) [2]. It has been recently proved by Ketep et al., that the use of secondary biofilms has slightly improved performance of microbial fuel cells and confirms the good results obtained with paper effluents, in terms of current densities and COD abatement [3]. In our experiments, we showed also the thicker the biofilm layer, the less time and inefficient the electronic transfer to the anode, hence the need to replicate the electro-active biofilm on a new electrode and in a new leachate. This has increased bioenergy significantly and without much latency. The electrochemical characterization in terms of cyclic voltammetry and impedance spectroscopy confirms this phenomenon with oxidation current peaks and charge transfer resistance.

Keywords Re-inoculation · Electro-active biofilm · Wastewater treatment

H. Kebaili (✉) · M. Kameche
Laboratory of Physico-Chemistry of Materials, Catalysis and Environment, University of Sciences and Technology of Oran Mohamed Boudiaf, M'Nouar, BP 1505, Oran, Algeria
e-mail: kebailihakima@gmail.com

C. Innocent · W. E. Kosimaningrum
European Membrane Institute, UMR 5635, University of Montpellier, CC 047, Place Eugene Battalion, 34095 Montpellier Cedex 5, France

W. E. Kosimaningrum
Analytical Chemistry Division, Faculty of Mathematics and Natural Science, Institute of Technology Bandung, Bandung, Indonesia

T. Sahraoui
Laboratory of Electronic Microscopy and Materials Sciences, University of Sciences and Technology of Oran Mohamed Boudiaf, M'Nouar, 1505, Oran, Algeria

A. Belasri and S. A. Beldjilali (eds.), *ICREEC 2019*, Springer Proceedings in Energy,
https://doi.org/10.1007/978-981-15-5444-5_25

25.1 Introduction

Many years ago, generation of electricity has been achieved from substrate wastes, by using bacteria as catalyst [4]. On this basis, a Microbial Fuel Cell (MFC) can be conceived to be a good tool for conversion of organic matter into electric energy [5]. In this biological cell, the oxidation of organic matter by bacteria at the anode, along with reduction of oxygen at the cathode generate electric power energy [6]. Specific biochemical activities of anaerobes drive the MFC chemical process. The organic substrate is oxidized into CO_2, protons and electrons [5]. The protons are transferred into the electrolyte inside the cell, whilst the electrons flow outside in the external circuit to feed an electric device [7]. Currently, efforts are being made to improve the performance of the cell by reducing its operating and construction costs. However, its application is limited because of its lower power density produced, generally few mW/m^2 [4]. At present, a number of experiments mainly focused on the electricity producing bacterium, electrode material, battery configuration and the effect of medium on the electricity generation performance of MFC. Although many major bottlenecks exist, MFCs promise sustainable energy generation in the future [8]. Instead of requiring metal catalysts, MFCs utilize a wide microbial diversity that can convert a great deal of organic matter components into sustainable and renewable energy [9]. In the present investigation, a MFC using garden soil leachate as biofilm catalyst, was conceived. Then, the bioanode (bio-film carrier) has been characterized by cyclic voltammetry to study the electro biochemical process by measuring anode oxidation peaks [10]. Moreover, the electrochemical spectroscopy has been also used to optimize the electron transfer from the biofilm to the bioanode by measuring charge resistance transfer [11].

25.2 Experimental

25.2.1 Inoculum and Microbial Fuel Cell

Effluent samples were prepared from Garden Soil (GS) mixed with Potassium Chloride solution. They were used without further filtration as inoculum source for microbial reactor. As illustrated by Fig. 25.1, the MFC consisted of two glass spherical half-cells, connected to each other by a cation exchange membrane. The bio-anode made of Carbon Felt (CF) and cathode of stainless steel, were placed inside the anolyte and the catholyte respectively.

The anolyte contained the leachate fed continuously with Sodium Acetate solution, whereas the cathodic compartment contained ferrycyanide solution. The MFC was discharged to external resistor (1000 Ω), to generate electricity. The electro-active biofilm was therefore formed on the surface of the electrode within 10 days and the evolution of the cell voltage with time, was then measured with a multi-meter in parallel.

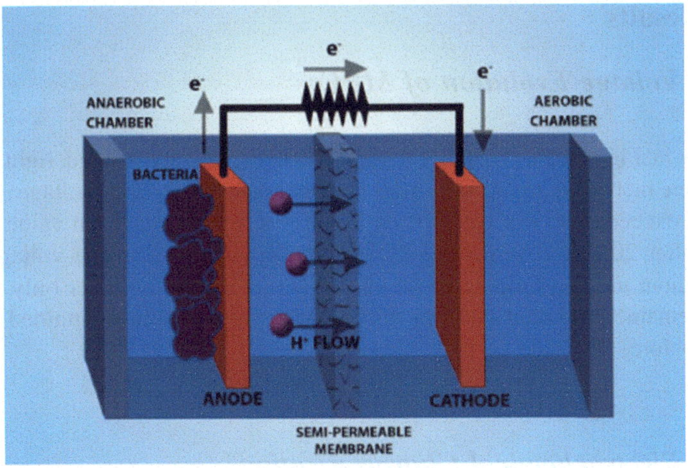

Fig. 25.1 Scheme of microbial fuel cell [12]

25.2.2 Scratching/Transplanting of Electro-Active Biofilm

In order to exploit the electroactive property of the biofilm in a new MFC, we scratch it and translate it according to the methods reported in the literature. In the present investigation, we adopt the method used by Ketep et al. [13] and Yates et al. [14]. In effect, the primary electroactive biofilm (1st G) was scrapped-off from bio-anode by sonication into a sterilized solution which was later added to sterilized leachate, in order to be re-inoculated onto a new electrode.

The biofilm formed from the following re-inoculation was then performed in identical experimental conditions (2nd G).

25.2.3 Electrochemical Characterisation

Cyclic voltammetry (CV) voltammograms and Electrochemical Impedance Spectroscopy (EIS) diagrams were obtained with the potentiostat-galvanostat (PGZ-301) using three-electrode glass cell. Working electrode was flat Carbon Felt, saturated Ag/AgCl reference electrode and Platinum wire auxiliary electrode.

25.3 Results

25.3.1 Voltage Evolution of MFCs

As shown in Fig. 25.2a, for the 1st generation, the voltage started right up from 0 mV, then increased to 30 mV after a lag time period. The voltage increased gradually, exceeds 100 mV after 10 days and reaches its maximum value less than 200 mV after 20 days. As for the 2nd generation (Fig. 25.2b), the voltage started earlier without any lag time and rose up to the same value but after only two days where it remained constant during a while. However, the voltage attained the same value as before.

25.3.2 Electrochemical Characterization

Impedance Spectroscopy. The impedance spectra of the different MFCs are represented according to the Nyquist diagram (Figs. 25.3). They are represented in the figures given below, for the different electrodes (pristine, 1st generation and 2nd generation). Obviously, the variations of the curves are different, reflecting the duration of operation of the MFC, in the presence of the bacterial biofilm. However, the MFC without biofilm has the highest resistance, the pristine work electrode gives a relatively higher resistance, and finally that of the 2nd generation has the lowest resistance due to the electro activity of the bacterial biofilm where the electronic conduction is better.

Cyclic Voltammetry (CV). The voltammograms recorded in the absence of substrate confirmed the electric current due to acetate oxidation. It is clear that the voltammogram of the 1st generation which is the more pronounced due to the presence of biofilm both onto the bioanade and in solution. Whilst, the

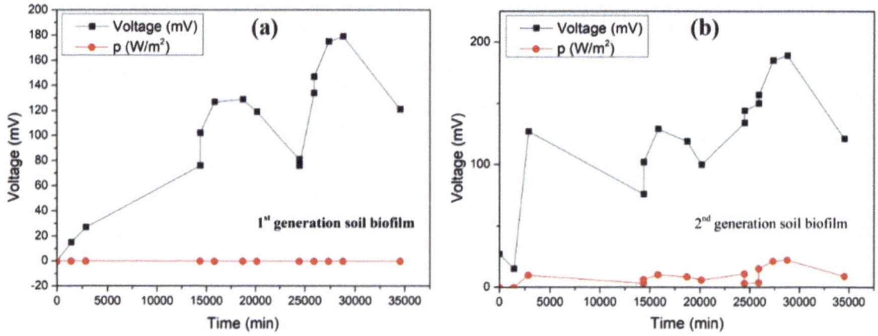

Fig. 25.2 Evolution voltage curves of 1st (Fig. 25.2a) and 2nd (Fig. 25.2b) generations of microbial fuel cells using garden soil as inoculum

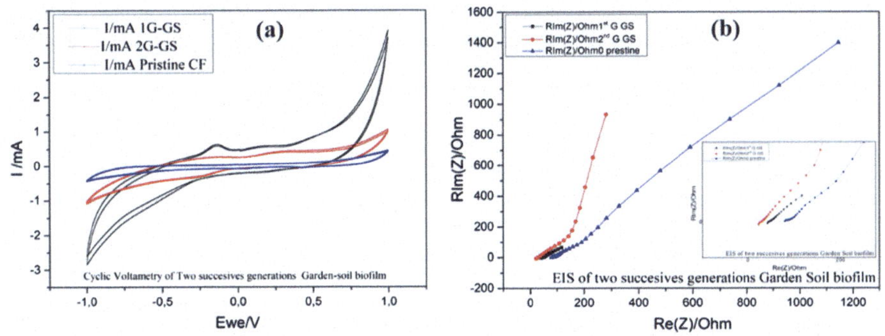

Fig. 25.3 Carbon felt electrode voltamogram (Fig. 25.3a) and Nyquist diagram (Fig. 25.3b) of pristine, 1st generation and 2nd generation garden soil biofilm

voltammograms of pristine and 2nd generations are less pronounced because on one hand the absence of elctroactive biofilm on the pristine electrode, and on another hand the solution of the 2nd generation does not contain enough bacterial electroactive cells.

25.4 Conclusion

We demonstrated the possibility to improve the energy production in a microbial fuel cell by doping the new bacterial medium with the biofilm of the 1st generation. Indeed, the MFC of the 2nd generation started functioning without any lag time. These interesting observations are explained in terms of electrochemical characterizations i.e. cyclic voltammetry and electrochemical impedance spectroscopy.

Acknowledgements The authors are grateful to the financial support from the Tassili scheme programmme14 MDU 912 in partnership between Algeria and France.

References

1. B. Min, B.E. Logan, Continuous electricity generation from domestic wastewater and organic substrates in a flat plate microbial fuel cell. Environ. Sci. Technol. **38**(21), 5809–5814 (2004)
2. H. Liu, R. Ramanathan, B.E. Logan, Production of electricity during wastewater treatment using a single chamber microbial fuel cell. Environ. Sci. Technol. **38**(7), 2281–2285 (2004)
3. F.S. Ketep, Piles à combustible microbiennes pour la production d'électricité couplée au traitement des eaux de l'industrie papetière. Thèse de doctorat, Université de Grenoble (2012)
4. Z. Du, H. Li, T. Gu, A state of the art review on microbial fuel cells: a promising technology for wastewater treatment and bioenergy. Biotechnol. Adv. **25**, 464–482 (2007). https://doi.org/10.1016/j.biotechadv.2007.05.004

5. B. Midyurova, V. Nenov, M. Ates, N. Uludag, Improvements of MFC's proton exchange membranes and cathodes. J. Mater. Environ. Sci. **7**(1), 85–95 (2016). ISSN 2028-2508 CODEN: JMESCN
6. B.E. Logan, C. Murano, K. Scott, N.D. Gray, I.M. Head, Electricity generation from cysteine in a microbial fuel cell. Water Res. **39**(5), 942–952 (2005). https://doi.org/10.1016/j.watres.2004.11.019
7. A. Capitaine, T. Chailloux, G. Pillonnet, B. Allard, Microbial fuel cells for energy harvesting: electrical characterization and modeling, in *Symposium de Genie Electrique (SGE 2016): EF-EPF-MGE 2016*, 7–9 June 2016, Grenoble, France (2016)
8. K. Rabaey, W. Verstraete, Microbial fuel cells: novel biotechnology for energy generation. Trends Biotechnol. **23**(6), 291–298 (2005)
9. N. Chabert, O.A. Ali, W. Achouak, All ecosystems potentially host electrogenic bacteria. Bioelectrochemistry **106**(Part A), 88–96 (2015). https://doi.org/10.1016/j.bibelechem.2015.07.004
10. K. Fricke, F. Harnisch, U. Schröder, On the use of cyclic voltammetry for the study of anodic electron transfer in microbial fuel cells. Energy Environ. Sci. **1**(1), 144–147 (2008). https://doi.org/10.1039/b802363h
11. N. Sekar, R.P. Ramasamy, Electrochemical impedance spectroscopy for microbial fuel cell characterization. J. Microb. Biochem. Technol. **S6** (2013). http://dx.doi.org/10.4172/1948-5948.S6-004
12. https://www.researchgate.net/profile/Alexandra_Perebikovsky/publication/263733888/figure/fig6/AS:296534060093446@1447710637729/Microbial-fuel-cell-schematic-Microbial-fuel-cells-utilize-microbes-to-oxidize-biofuels.png.12/03/2019
13. S.F. Ketep, A. Bergel, M. Bertrand, W. Achouak, E. Fourest, Lowering the applied potential during successive scratching/re-inoculation improves the performance of microbial anodes for microbial fuel cells. Bioresour. Technol. **127**, 448–455 (2013). https://doi.org/10.1016/j.biortech.2012.09.008
14. M.D. Yates, L. Ma, J. Sack, J.P. Golden, S.M. Strycharz-Glaven, S.R. Yates, L.M. Tender, Microbial electrochemical energy storage and recovery in a combined electrotrophic and electrogenic biofilm. Environ. Sci. Technol. Lett. **4**(9), 374–379 (2017). https://doi.org/10.1021/acs.estlett.7b00335

Chapter 26
Electro-Catalytic Electrodes and Ionic Exchange Membranes in Microbial Fuel Cell

Aicha Zerrouki, Mostefa Kameche, Hakima Kebaili, Ahcene Ait Amer, and Christophe Innocent

Abstract Microbial Fuel Cell (MFC), an emerging energy device that produces sustainable bio-energy and treats wastewater and industrial effluent. The MFC consists of two compartments separated by an ionic exchange membrane. In the anolyte, the electro-catalytic biofilm formed on the anode, degrades the organic matter into electrons and proton. Whilst in the catholyte, the cathode reduces dissolved oxygen. The carbon based material usually used at the anode because it is biocompatible and chemically stable in the electrolyte. In order to increase its active surface, its modification with conducting polymer can increase the current density produced from MFC. Besides, the materials based on carbon and graphite are widely used as cathodes, but they should be enriched with noble metals necessary for the catalysis of the electrochemical reaction. In MFC, the polymer ionic membrane can affect significantly its performance. In effect, various parameters such as membrane internal resistance, pH splitting, oxygen diffusion, and substrate loss and bio-fouling, have been investigated. In addition, the slow transfer of proton at neutral pH, creates pH gradient at electrodes. Thus, the accumulation of protons at the anode causes acidification slowing oxidation activity. Conversely, the weak proton availability near cathode reduces the rate of cathode reaction. The use of proton membrane can replaced by an anionic membrane. The adjustment of pH in anolyte and cathode compartments, influence the cell voltage. The highest value was obtained with anionic membrane.

Keywords Microbial Fuel Cells · Electrodes · Membranes

A. Zerrouki (✉) · A. A. Amer
Laboratory of Chemistry and Electrochemistry of Metallic Complexes, University of Sciences and Technology of Oran-Mohammed Boudiaf, M'Nouar BP 1505, Oran, Algeria
e-mail: aichazerrouki98@gmail.com

M. Kameche · H. Kebaili
Laboratory of Physico-Chemistry of Materials, Catalysis and Environment, University of Sciences and Technology of Oran-Mohammed Boudiaf, M'Nouar BP 1505, Oran, Algeria

C. Innocent
European Membrane Institute, UMR 5635, University of Montpellier, CC 047, Place Eugene Bataillon, 34095 Montpellier Cedex 5, France

A. Belasri and S. A. Beldjilali (eds.), *ICREEC 2019*, Springer Proceedings in Energy,
https://doi.org/10.1007/978-981-15-5444-5_26

26.1 Introduction

The conventional fuel cells are currently employed to substitute fossil toxic energies. This new technology converts generally energy from hydrogen or methanol into electricity by oxidizing the fuel at the anode and the reduction of the oxygen at the cathode [1]. By inspiring from this novel technology, nowadays, MFCs are becoming good alternative systems to treat wastewaters and to produce low energy for functioning small electrical devices [2]. They employ the bio-electrochemical catalysis of microbes to generate electric current from the oxidation of organic and inorganic substrates. They are thus new energy sources, renewable and environmentally friendly. As mentioned by Gude [3], the substrates such as simple chemical compounds, urban sewage, agricultural, brewery, food, and dairy farm and various industrial wastewaters can be metabolized by microorganisms can be served as potential feedstock in MFCs.

A MFC consists of two electrodes anode (carrier of electro-active biofilm) and cathode, separated by an ionic exchange membrane. As a result of the oxidation of the organic matter, the electrons are produced and flow in the external circuit to produce electric current. The protons are transferred through the membrane to compensate the electron transfer, ensuring electro-neutrality. In general, the material used at the anode must be biocompatible and chemically stable in the electrolyte [4]. Although its conductivity electron transfer is lower than that of metals, it is stable in microbial cultures, cheap, easy to use, but it has a wide active surface [5]. On the other hand, the reaction efficiency at the cathode depends on electrons acceptor, availability of produced protons from the anode and the performance of the catalyst. As the anode, the cathode uses low cost carbon or graphite. But it should be enriched with noble metals necessary for the catalysis of the electrochemical reaction.

The ion exchange membrane separates the biological anode from the cathode reactions, and facilitating the transport of ions in order to maintain electro-neutrality in the system [6]. At the cathode of the MFC, protons are consumed in equal amounts as electrons, making in evidence that only protons are ideally transported through the membrane. The electro-neutrality is consequently observed without any pH changes in cathode compartment. Nevertheless, bio-electrochemical for the treatment of wastewater requires the utilization of a Cation Exchange Membrane (CEM) which allow the transport metal cations and ammonium in the internal circuit of the MFC. This results in an increase of the pH in the cathode compartment, which unfortunately reduces the cell performance [7, 8].

The objective of this present investigation, was to study both the effects of electro-catalytic electrodes and ionic exchange membranes on the performance of the MFC. The power curves of MFCs utilizing CEM and Anion Exchange Membrane (AEM) separately, have been plotted. Besides, the effect of Raney-Nickel and material doped with platinum nanoparticles, have been highlighted as potential materials used as cathode for potential reduction.

26.2 Materials and Methods

26.2.1 MFC Configuration

The MFC consisted of two compartments: anodic and cathodic was adopted in this study. The anode consisted of thick carbon bar in contact with carbon granules, both were placed inside an anodic chamber filled with pre-treated wastewater samples taken from industrial Wastewater Treatment Plant of Elkerma (Oran, Algeria). The samples contained a substantial amount of bacterial species which have been grown under polarization to form the electroactive biofilm. A Raney-Nickel carbon cloth was used as cathode. The two electrodes anode and cathode were discharged to an external electric resistance, to allow the flow of electrons. A CEM was used as separator, to allow the proton transfer, ensuring the electro-neutrality of the electrochemical device. For the study of the effect of the type of separator, the same MFC design was used for AEM.

26.2.2 Growth of Biofilm

The growth of anodic electro-active microbial biofilms from wastewater inoculums is necessary for the MFC to produce bioenergy by oxidizing organic matter. There are direct and indirect electron transfers. Indeed, the utilization of the galvano-potentiostat permits adjustment of the electrode potential and ensures reproducible microbial culturing. The electric current density and the Coulombic efficiency obtained from this bio-electrochemical process are necessary to assess the bio-electrocatalytic activity. Cyclic voltammetry is useful to study the extracellular electron transfer of electro-active bio-film [9]. The three mechanisms can be involved in electron transfer from the biofilm to the electrodes: (A) Indirect transfer via mediators or fermentation products; (B) direct transfer via cytochrome proteins; (C) direct transfer via conductive pili [10].

26.2.3 Cathode Used in MFC

The novel material Raney-Nickel carbon cloth was tested in the MFC. It used carbon cloth as support which was doped with Raney-Nickel particles. The voltage of MFC using this electrode, started with the highest voltage of about 0.5 V, a plausible value commonly obtained for well functioning MFCs. It then decreased steeply by loosing 90% of its efficiency after 1 day of functioning. Then, it rose steadily a little bit leveled out during 3 days. But upon the addition of fuel, it rose up sharply by recovering 20% of its initial efficiency. However, the voltage ended with slight decrease, making in evidence the slow electrochemical process of the electro-active biofilm towards the end of the biological cell. Indeed, the high potential observed using this electrode may be explained by its great activity towards the reduction of both oxygen and protons. In comparing this novel material to a platinum metal, certainly the energy efficiency is really much less; however this latter is very expensive. A current study is being undertaken by doping carbon cloth by nanoparticles of platinum, making the process much inexpensive.

26.2.4 Ionic Exchange Membranes Used as Separators in MFC

In order to elucidate the effect of the ionic exchange membrane upon the functioning of the MFC, two membranes were used: CEM and AEM. The schemes of the MFC using these membranes are illustrated by Figs. 26.1, 26.2 and 26.3. In effect, in Fig. 26.1, upon the oxidation of the organic matter contained in the wastewater, both protons and metal cations can be transferred through the CEM. Whilst in Fig. 26.2, only the protons can be diffused through the AEM. Thus, as shown in Fig. 26.4, the curve of MFC using AEM harvests higher electric energy than that using CEM. Indeed, upon the addition of substrate, the difference is remarkably clear. At the cathode of the MFC with CEM, the reduction reaction consumed fewer protons than MFC with AEM, which resulted in an increase of pH. Therefore, the electro-neutrality is no longer fulfilled and as a consequence, the efficiency of the MFC became low. These finding is in good agreement, when the pHs of both compartments were deliberately changed, as it has been already noticed in our previous study [11]. Nevertheless, we think that the scheme of Fig. 26.3, with three compartments, seems to give a MFC more efficient since, on one hand the cation metals are blocked by AEM and kept in the central compartment. Therefore, the pH of the catholyte cannot change too much and the voltage drop can be avoided. On the other hand, the anolyte can be efficiently purified from metallic species which may hinder the activity of the biofilm. This experiment is planned to be undertaken in the future, to elucidate definitively this problem.

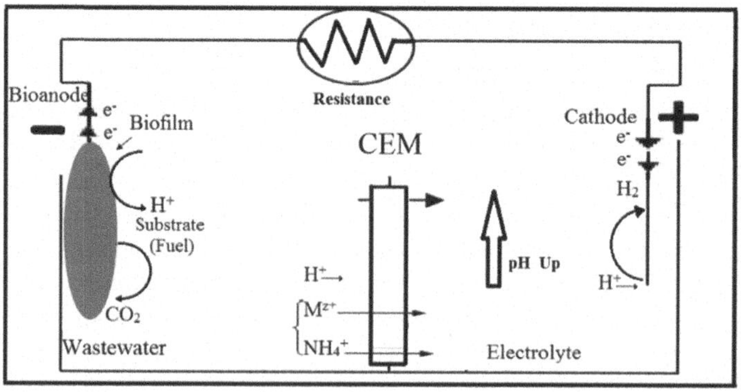

Fig. 26.1 Scheme of MFC using Cation Exchange Membrane

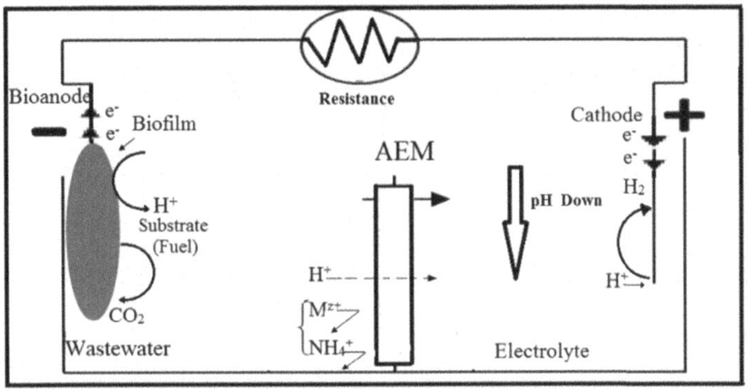

Fig. 26.2 Scheme of MFC using Anion Exchange Membrane

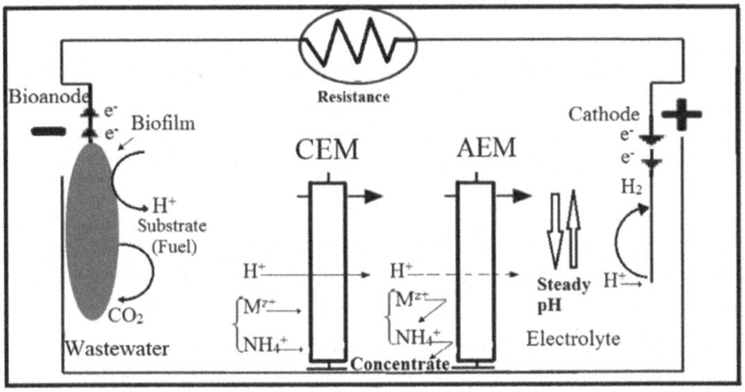

Fig. 26.3 Scheme of MFC using Cation and Anion Exchange Membranes

26.3 Effect of PH Compartment on MFC

Within 1 h of functioning, the pHs of the anode and cathode compartments were 4 and 10 respectively. This means that a great deal of metal cations contained in the wastewater diffused through the membrane, causing an increase of the pH and therefore a loss in voltage. As shown in Fig. 26.4, the power curves obtained with MFC using CEM and AEM highlight and confirm this effect. In a MFC used for the treatment of wastewater containing metal cation, the effect of the type of the membrane on the ion transport and pH, should be studied before going further in the exploitation of the MFC. Thus, a MFC allowing the transport of cationic metals, implies an increase of the pH of the catholyte and therefore a loss in the value of the voltage. So, in order to prevent the metal ions through the membrane, the modification of the membrane is necessary.

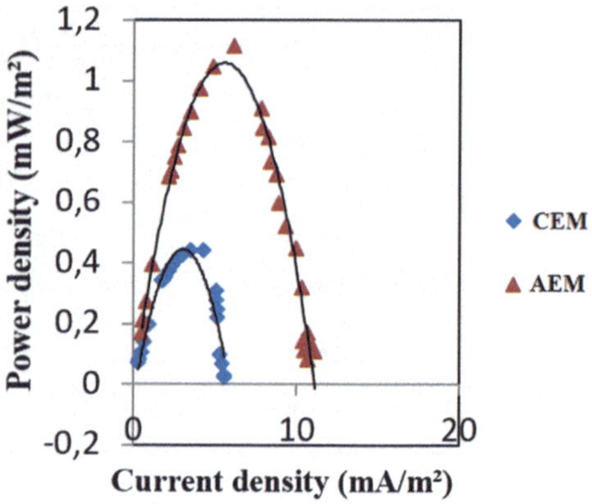

Fig. 26.4 Power curves of MFC using the Cation and Anion Exchange Membranes [11]

26.4 Conclusion

This paper showed the possibility to treat wastewater by means of Fuel Cells using bio-catalysts at electrodes. The effects of these latter and polymer ionic membranes were discussed, in terms of energy efficiency yielded by the MFC. Indeed, the use of carbon cloth doped with nickel can be a good alternative with respect to platinum extremely very expensive. Besides, it was revealed that in contrast to hydrogen fuel cell, the use of the AEM was beneficial for the treatment of wastewater containing cation metals. The easy transfer of these metals using CEM improved substantially the pH of the cathode compartment and therefore reducing the performance of the cell. Thus, the utilization of the AEM as separator in MFC for the treatment of wastewater is required. At the end of this study, we come up to design in future a suitable 3 compartment cell, to improve the efficiency of the MFC along its conception as a suitable desalination process for removing heavy metals contained in wastewater.

Acknowledgements The authors would like to acknowledge the responsible authorities of the Wastewater Treatment Plant of El kerma (Oran, Algeria) for their permission in using pre-treated wastewater samples for bio-film growth in microbial fuel cell.

References

1. S. Bent, *Hydrogen and Fuel Cells: Emerging Technologies and Applications*, 2nd edn. (Elsevier, Amsterdam, 2011)
2. S.M. Haile, Materials for fuel cells. Mater. Today **6**(3), 24–29 (2003)
3. V.G. Gude, Wastewater treatment in microbial fuel cells - an overview. J. Clean. Prod. **122**, 287–307 (2016)
4. F.S. Ketep, Piles à combustible microbiennes pour la production d'électricité couplée au traitement des eaux de l'industrie papetière, pp. 60–61. Doctoral thesis, University of Grenoble, France (2012)

5. K. Watanabe, Recent developments in microbial fuel cell technologies for sustainable bioenergy. J. Biosci. Bioeng. **106**(6), 528–536 (2008)
6. J.X. Leong et al., Ion exchange membranes as separators in microbial fuel cells for bioenergy conversion: a comprehensive review. Adv. Mater. Res.-Switz **28**, 575–587 (2013)
7. R.A. Rozandal et al., Effect of the type of ion exchange membrane on performance, ion transport, and pH in biocatalyzedelectrolysis of wastewater. Water Sci. Technol. **57**, 1757–1762 (2008)
8. F. Harnish et al., Modeling the ion transfer and polarization of ion exchange membranes in bioelectrochemical systems. Bioelectrochemistry **75**(2), 136–141 (2009)
9. C. Gimkiewicz, F. Harnisch, Waste water derived electroactive microbial biofilms: growth, maintenance, and basic characterization. J. Vis. Exp. **82** (2013). https://doi.org/10.3791/50800
10. C. Santoro, C. Arbizzani, B. Erable, I. Ieropoulos, Microbial fuel cells: fundamentals to applications. J. Power Sources **356**, 225–244 (2017)
11. A. Zerrouki, M. Kameche, H. Kebaili, I.S. Boukoussa, M.A. Flitti, H. Ilikti, C. Innocent, An investigation on polymer ion exchange membranes used as separators in low-energy microbial fuel cells. Polym. Bull. **75**(11), 4947–4965 (2018)

Chapter 27
Effect of $SrAl_2H_2$ on the Stability of MgH_2 for Hydrogen Storage Application

Khadidja Khodja, Youcef Bouhadda, and Kamel Benyelloul

Abstract High-density hydrogen storage is a significant challenge for stationary, mobile and transportation applications, so the storage in solid materials remains the safest way. By performing the first-principles calculations within the generalized gradient approximation (GGA) and using the pseudo-potentials and plane waves based on the density-functional theory (DFT), the effect of zintl phase hydride $SrAl_2H_2$ on the stability of the magnesium dihydride compound (MgH_2) is studied for an application in the field of hydrogen storage in solid materials. Their crystal structures are known and occurs in a trigonal and tetragonal respectively. Therefore; we investigated the enthalpy of formation for three reactions between MgH_2 and $SrAl_2H_2$ compounds by calculating the total energies of MgSr and $SrAl_2$ compounds. These last compounds crystalize in cubic and orthorhombic structure respectively. We also investigated the structural and the electronic properties of these compounds and discuss the chemical bond nature using total and partial density of electronic states. Our calculated results are generally in good agreement with theoretical and experimental data.

Keywords Hydrogen storage · Zintl phase hydride · Enthalpy formation · Density of state · Density functional theory

27.1 Introduction

Facing the question of the long-term energy fossil exhaustion and the global warming caused by human activities that no serious scientist could dispute, it become necessary to develop a new energy carrier, which is free of carbon. For that, two solutions must be considered. First, the energy expenditure must be mastered and the second the energy produced from renewable sources must be less expensive

K. Khodja (✉) · Y. Bouhadda · K. Benyelloul
Unité de Recherche Appliquée en Energies Renouvelables, URAER,
Centre de Développement des Energies Renouvelables, CDER, 47133 Ghardaïa, Algeria
e-mail: khodja_kh@yahoo.fr

A. Belasri and S. A. Beldjilali (eds.), *ICREEC 2019*, Springer Proceedings in Energy,
https://doi.org/10.1007/978-981-15-5444-5_27

and accessible to everyone. The hydrogen is the future energy carrier for mobile applications [1]. The United States Department of Energy (DOE) has determined that 6 wt% of the total weight of energy system storage must contain hydrogen to obtain a range comparable to that of standard gasoline vehicles. Realizing that this mass corresponds to 50 m^3 at standard conditions, one can easily imagine the technological challenge represented by the storage of hydrogen for mobile applications. Two storage methods exist, the first is based on liquefied hydrogen (at 20.3 K) and the second, on pure gas under very high pressure (200 bar to 700 bar) are the best known and the simplest to design.

Several research have been done to study the stability of the MgH_2 compound, Song et al. [2] investigated the effect of the elements Cu, Ni, Al, Nb, and Fe to Ti. It is found that the instability is due to a weakened bonding between magnesium and hydrogen atoms. Wan et al. [3] studied the alloying effect on the thermodynamic stability. The obtained results predict structural transition and properties of MgH_2 based hydrides for hydrogen storage. Gong and Shao [4] studied the stability of pristine and doped 2D MgH_2 by the first principles calculations and the results show that all the systems are dynamically stable.

In our work, we focus on the storage of hydrogen in solid materials [5–7]. This paper is devoted to the study of the effect of $SrAl_2H_2$ compound on the stability of MgH_2 using the first principle calculation.

27.2 Method and Computational Details

The electronic structure calculations are based on the density-functional theory (DFT) [8] using generalized gradient approximation (GGA) performed by ABINIT code [9] based on pseudo-potentials and plane waves [10]. The electronic wave functions were expanded in plane waves up to a kinetic energy cutoff of 50 Hartree and 80 Hartree for MgSr and MgH_2 respectively. Integrals over the Brillouin zone were approximated by sums of $8 \times 8 \times 8$ mesh of special k-point for both compound.

27.3 Results and Discussion

27.3.1 Structural Features

The structures of the compounds MgSr et MgH_2 have been determined experimentally by Nowotny [11] et Bortz et al. [12] respectively. The structures of these compounds are known and occurs in a body-centered cubic (bcc) and tetragonal for MgSr and MgH_2 respectively (see Fig. 27.1). The relaxed structural parameters and atomic coordinates compared with experimental and theoretical work are detailed in Table 27.1.

Table 27.1 Relaxed structural parameters and atomic coordinates compared with experimental and theoretical work

Structure	Our work	Experimental	Theoretical	Atomic coordinates			
					x	y	z
MgSr (221, Pm-3 m) a = b = c(Å)	4.1349	3.9 [11]	4.18 [13]	Mg Sr	0 0.5	0 0.5	0 0.5
MgH_2 (136, P4 2/m n m) a = b ≠ c(Å)	4.5726 3.0578	4.501 [12] 3.01 [12]	4.519 [14] 3.022 [14]	Mg H	0 0.304	0 0.304	0 0

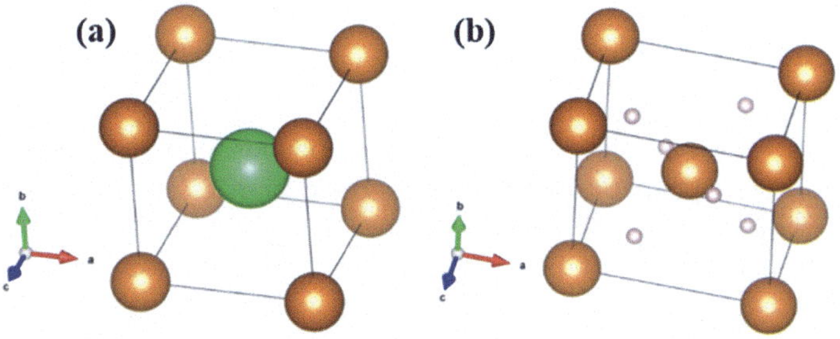

Fig. 27.1 Cristal structure of **a** MgSr and **b** MgH_2, where Sr, Mg and H are the big, medium and small sphere respectively

27.3.2 Enthalpy of Formation

For the enthalpy of formation calculation, we considered the following reactions:

$$MgH_2 + SrAl_2H_2 \rightarrow MgSr + 2Al + 2H_2 \quad (27.1)$$

$$MgH_2 + SrAl_2H_2 \rightarrow SrAl_2 + Mg + 2H_2 \quad (27.2)$$

$$MgH_2 + SrAl_2H_2 \rightarrow Mg + Sr + 2Al + 2H_2 \quad (27.3)$$

In Table 27.2, we reported the total energy and the formation enthalpy computed for the three reactions cited below with GGA approximation. The results show that the formation enthalpy for reaction (27.1) is negative which means that this reaction is exothermic (stable), but the reactions (27.2) and (27.3), whose formation enthalpy is positive, are endothermic (not stable) and need lot of energy to be realized.

Table 27.2 Calculated formation enthalpy and total energy

Elements	Total energy (eV)	Reaction	Enthalpy of formation (kJ/mol)
Mg	−79.29	(1)	−307.79
Sr	−36.50		
Al	−228.68		
H_2	−31.40	(2)	3724.3
$SrAl_2$	−101.85		
MgSr	−75.82		
MgH_2	−71.31	(3)	3670
$SrAl_2H_2$	−184.96		

27.3.3 Electronic Density of States (TDOS and PDOS)

To investigate the electronic structure of the compounds MgSr and MgH2, we used the calculated lattice parameters reported in Table 27.1.

Figure 27.2 shows the total density of states. For MgSr the DOS reaches a value of 1.56 (states/eV/cell) at the Fermi level. That is why, the electronic structure of this compound is metallic (see Fig. 27.2a). In Fig. 27.2b there is a wide band gap (about 3.83 eV) between the conduction band and valence band, revealing the MgH_2 insulating nature and the DOS reaches a maximum value of 4.33 (states/eV/cell). The energy gap obtained is in good agreement with the one reported in Reshak work [15] which is 3.735 eV.

The partial density of states is plotted in Fig. 27.3. In the lowest energy section (from −5 eV to 0), the first two peaks are produced by the Mg-s and Mg-p binding interactions, i.e. an orbital contribution is present, while the partial density of Sr has a small contribution (see Fig. 27.3a). From Fig. 27.3b, it is revealed that there is hybridization between the H and the Mg states in the valence bands, which have a dominant hydrogen character.

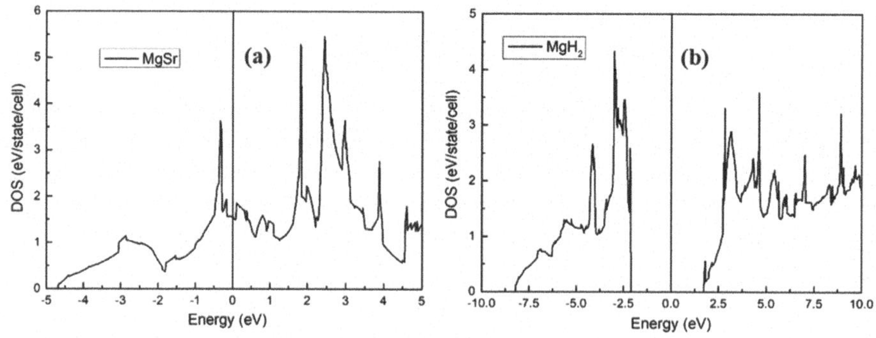

Fig. 27.2 Total density of state: **a** MgSr, **b** MgH_2

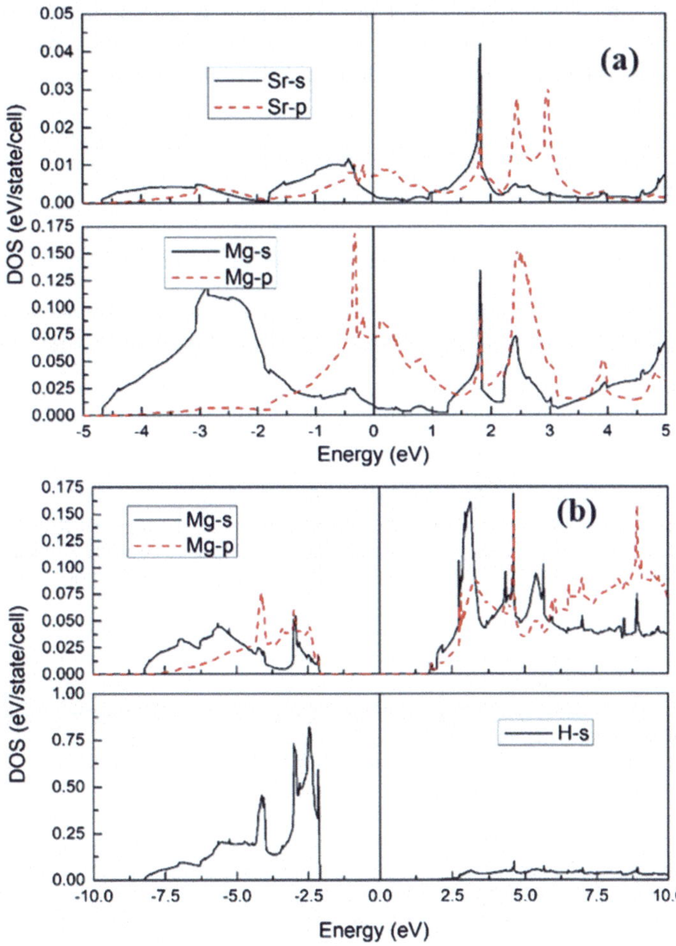

Fig. 27.3 Partial density of state: **a** MgSr, **b** MgH_2

27.4 Conclusion

In this work, we investigate the electronic structure calculations for MgSr and MgH_2 compound using the ABINIT code to investigate the effect of $SrAl_2H_2$ on the stability of MgH_2. The electronic structure is obtained from GGA aproximation. The enthalpy of formation for three reactions is investigated after the calculation of relaxed structural parameters. This allowed us to better analyze our compounds by studying the total and partial density of states. The investigation of the formation enthalpy leads us to conclude that the reactions (27.2) and (27.3) are endothermic unlike the reaction (27.1) which is exothermic. Also from the plot of the DOS, we noted that,

MgSr has no gap, which means that the compound is a metallic. MgH_2 has an isolating nature with a band gap value of 3.83 eV and this is in good agreement with a theoretical work. The hydrogen atoms dominate the valence band.

References

1. A. Züttel, A. Borgschulte, L. Schlapbach, *Hydrogen as a Future Energy Carrier* (Wiley, Hoboken, 2008)
2. Y. Song, Z.X. Guo, R. Yang, Influence of selected alloying elements on the stability of magnesium dihydride for hydrogen storage applications: a first-principles investigation. Phys. Rev. B **69**, 094205 (2004)
3. Z.-Z. Wan, Z.-M. Wang, D.-H. Wang, Y. Zhong, J.-Q. Deng, H.-Y. Zhou, C.-H. Hu, Alloying effect study on thermodynamic stability of MgH_2 by first-principles calculation. Chin. J. Chem. Phys. **29**, 545–548 (2016)
4. X. Gong, X. Shao, Electronic structure, and dehydrogenation properties of pristine and doped 2D MgH_2 by the first principles study. Metals **8**, 482 (2018)
5. K. Benyelloul, L. Seddik, Y. Bouhadda, M. Bououdina, H. Aourag, K. Khodja, Effect of pressure on structural, elastic and mechanical properties of transition metal hydrides Mg7TMH16 (TM = Sc, Ti, V, Y, Zr and Nb): first-principles investigation. J. Phys. Chem. Solids **111**, 229–237 (2017)
6. K. Khodja, Y. Bouhadda, L. Seddik, K. Benyelloul, Solid-state structures and properties of scandium hydride; hydrogen storage and switchable mirrors application. Eur. Phys. J. Appl. Phys. **74**, 24614 (2016)
7. K. Khodja, Y. Bouhadda, Electronic properties of $SrAl_2H_2$ for hydrogen storage. Models Optim. Math. Anal. J. **2**, 54–58 (2014)
8. W. Kohn, L.J. Sham, Self-consistent equations including exchange and correlation effects. Phys. Rev. **140**, A1133–A1138 (1965)
9. X. Gonze, J.-M. Beuken, R. Caracas, F. Detraux, M. Fuchs, G.M. Rignanese, L. Sindic, M. Verstraete, G. Zerah, F. Jollet, M. Torrent, A. Roy, M. Mikami, P. Ghosez, J.-Y. Raty, D.C. Allan, First-principles computation of material properties: the ABINIT software project. Comput. Mater. Sci. **25**, 478–492 (2002)
10. N. Troullier, J.L. Martins, Efficient pseudopotentials for plane-wave calculations. Phys. Rev. B **43**, 1993–2006 (1991)
11. H. Nowotny, Die Kristallstrukturen von Ni5Ce, Ni5La, Ni5Ca, Cu5La, Cu5Ca, Zn5La, Zn5Ca, Ni2Ce, MgCe, MgLa und MgSr. Z. Metallkd. **34**, 247–253 (1942)
12. M. Bortz, B. Bertheville, G. Böttger, K. Yvon, Structure of the high pressure phase γ-MgH_2 by neutron powder diffraction. J. Alloy. Compd. **287**, L4–L6 (1999)
13. D. Zhou, J. Liu, S. Xu, P. Peng, First-principles investigation of the binary intermetallics in Mg–Al–Sr alloy: Stability, elastic properties and electronic structure. Comput. Mater. Sci. **86**, 24–29 (2014)
14. Y. Bouhadda, A. Rabehi, S. Bezzari-Tahar-Chaouche, First-principle calculation of MgH_2 and LiH for hydrogen storage. Revue des Energies Renouvelables **10**, 545–550 (2007)
15. A.H. Reshak, MgH_2 and LiH metal hydrides crystals as novel hydrogen storage material: electronic structure and optical properties. Int. J. Hydrogen Energy **38**, 11946–11954 (2013)

Chapter 28
Production of Bio-Energy Using Biological Fuel Cell: Application to Electro-Dialysis for Recovery of Heavy Metal Traces from Treated Wastewater

Chahinez Yahiaoui, Mostefa Kameche, and Christophe Innocent

Abstract Biological Fuel Cell (BFC) is a device that converts chemical energy into electric energy by using a biological catalyst such as enzyme, microbe, bacteria, etc. …, and a fuel. Although enzymes and microorganisms are highly efficient biocatalysts. The microbial desalination cell (MDC) a newly developed technology, integrated the Microbial Fuel Cell (MFC) process and electro-dialysis (ED) for wastewater treatment, water desalination and production of renewable energy. The cell developed herein, was composed of two compartments: anode and cathode compartments separated by a cation exchange membrane. The biocatalyst was either on the electrode and suspended in solution. The anode was in general a carbon-based electrode, while the cathode was stainless steel plate. The connection between the two electrodes via an electrical resistance allowed the flow of electrons from the anode to the cathode through the external circuit. This flow was compensated by the flow of ions through the internal circuit. Though it was relatively weak, the produced bio-energy was always enough to feed low energy devices. We have demonstrated herein the potential energy for the treatment of a wastewater. We have proved the degradation of organic matter using the bio-film, by measuring the abatement of Chemical Oxygen Demand (COD). Besides, we have eliminated the residual heavy metal traces using by electro-dialysis. In effect, we used the bio-energy produced by the (BFC) to feed the three-compartment electro-dialysis cell (EDC). Compared to electro-dialysis, our results proved the possibility to recover metal traces at approximately the same rate of abatement, using this novel biological device.

C. Yahiaoui (✉) · M. Kameche
Laboratory of Physico-Chemistry of Materials, Catalysis and Environment, University of Sciences and Technology of Oran-Mohammed Boudiaf, M'Nouar, BP 1505, Oran, Algeria
e-mail: cyahiaoui31@gmail.com

C. Yahiaoui · M. Kameche · C. Innocent
European Membrane Institute, UMR 5635, University of Montpellier, CC 047, Place Eugene Battalion, 34095 Montpellier Cedex 5, France

A. Belasri and S. A. Beldjilali (eds.), *ICREEC 2019*, Springer Proceedings in Energy,
https://doi.org/10.1007/978-981-15-5444-5_28

Keywords Environment · Heavy metals · Bacterial desalination cell

28.1 Introduction

The growing demand of our modern societies for less polluting energy sources has led the researchers in the field to move towards the development of new means of energy production [9]. The new alternative sources are for example the conversion of energy by electrochemical techniques such as fuel cells [2]. Amongst these fuel cells; they are those called biological fuel cells because of the biological nature of their catalysts [1, 11]. Instead of using metal catalysts, we look nowadays for processes implemented by bacteria to find solutions necessary for their growth, to be used as biocatalysts towards the conversion of fuels (sugars, alcohols, hydrogen, etc.) into low electric energy. This physiological device functions identically as a fuel cell in which the catalysts would be replaced by enzymes. What advantages besides their bioavailability and biodegradability, the enzymes are more effective and specific for the transformation of their substrate than chemical catalysts like platinum. They also allow a simplification of the battery by avoiding the chemical membrane that separates the anode and cathode compartments [8].

The principle of the bio pile is directly derived from the operation of fuel cells; two electrodes ensure the operation of the generator: an anode, place of the oxidation of a fuel, and a cathode, place of reduction of the oxidant (usually oxygen). The electrodes collect the electrons between the electrodes and thus allow the generation of an electric current. These electrode reactions require catalysts. In the case of bio piles, catalysis is provided by bio-elements: redox enzymes for enzymatic bio piles, or microorganisms (bacteria, algae, yeasts ...) for microbial bio piles [3, 5].

In this study, bacterial fuel cell based on yeast as biocatalyst will be developed as energy production system (YFC). The polarization and power curves make it possible to check the performance of this fuel cell [4].

Finally, in order to eliminate the content in a dilute aqueous solution, we used bioenergy from YFC. For this reason, we have used it for the desalination of aqueous solutions containing traces of heavy metals such lead [6].

28.2 Experimental

28.2.1 Materials and Chemical Compounds

Owing to its Ox/Red properties, Methylene blue was used as mediator for electron transfer in YFC using yeast as biocatalyst for oxidation of glucose. Carbon Graphite (CG) with dimensions 5 cm × 0.8 cm and 0.5 cm and stainless-steel plate (5.0 cm × 1.0 cm × 0.2 cm) were used as anode and cathode respectively. The electrodes were initially washed with HCl and then rinsed with ultrapure water before using in experiment.

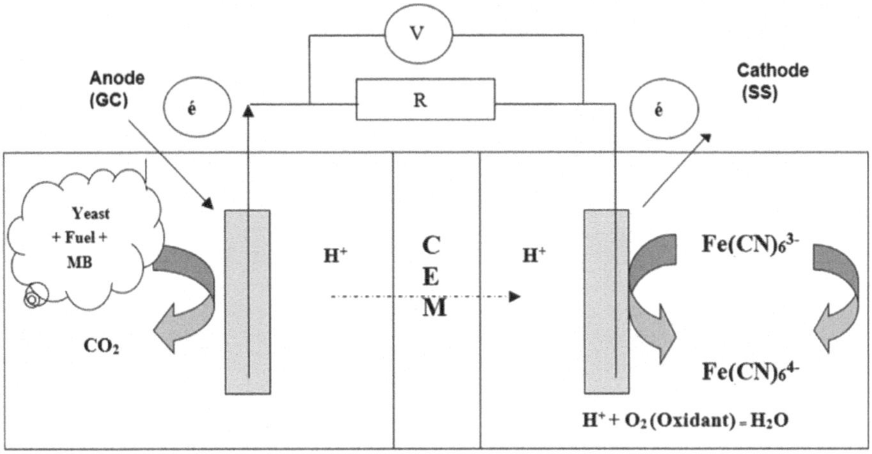

Fig. 28.1 Scheme of YFC using MB as catalyst

28.2.2 YFC Design

The YFC was built in a two chamber cell with anode and cathode compartments 70 mL each, separated by proton exchange membrane. CG and stainless-steel plate were used as anode and cathode electrodes, respectively. A scheme of the YFC is shown in Fig. 28.1. The anode compartment was filled with glucose solution 0.1 M in phosphate buffer solution 0.1 M allowing pH 7. It contained a solution of the commercial yeast used as biocatalyst. The experiments were conducted with commercial textile impregnated with Methylene Blue in the yeast solution as molecule facilitating electron transfer (i.e. mediator). The cathode compartment was filled with potassium ferricyanide solution, acting as Oxy/Red agent and allowing reduction of oxygen.

28.2.2.1 Evolution Curve

In the YFC, commercial textile impregnated with methylene blue was used as potential mediator for Ox/Red reactions at bioanode. When the biological cell became mature, it was then discharged on the electric resistance 10 kΩ, to allow the flow of current obtained from the oxidation organic substrate (glucose). The X-axis of Fig. 28.2 shows the working days of the cell, while the electric voltage in (mV) appears in the Y-axis. It may be seen clearly that the electric voltage rises steadily in the first days and reaches the peak value (140 mV) in the 28th day. After 30 days, the voltage drops off down to 80 mV and remains stable up to 36 days [7].

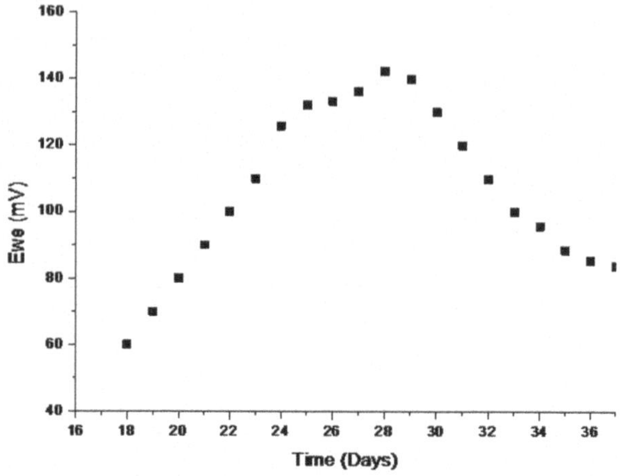

Fig. 28.2 Evolution curve of YFC during 36 days

28.2.2.2 Electrochemical Characterization

The voltammograms obtained with CG electrode with different mixtures of yeast, were carried out. During the electrochemical analysis, in the forward scan, when the oxidation potential was applied, the current augmented rapidly to attain the anodic peak and then decreased as soon as the concentration of analyte dropped off. Similarly, in the backward scan, the cathodic peak appeared due to the reduction of the oxidized electro active species, which generated in the forward scan. The dye MB plays therefore the role of mediator, since it raises up the faraday current. Thus, the values obtained make in evidence the current produced as a result of the oxidation of the glucose in the presence of the bacterial yeast.

28.2.2.3 YFC Used as Desalination Cell

The YFC was used as a desalination cell called Yeast Desalination Cell (YDC), to remove traces of lead from an aqueous solution. It consisted of anodic, central and cathodic compartments, which contained yeast solution (bio-catalyst), heavy metal solution and ferricyanide solution respectively. The anodic and central was separated by an anionic exchange membrane, while the central and cathodic compartments were separated by a cationic exchange membrane (Fig. 28.3). As soon as the YDC started working, an electric field applied and acted on the ionic species of the central compartment containing cation metals Mz+ and anions A− [10]. In practice, the lead cations moved towards the cationic exchange membrane and crossed it, whilst the nitrate anions migrated in opposite direction towards the anionic membrane crossed it as well. Obviously, the concentration of Mz+ decreased in the anodic feed compartment and in contrary it increased in the cathode receiving compartment (Fig. 28.4). The depletion of lead ions was optimum after a while, yielding an abatement rate more than 50%.

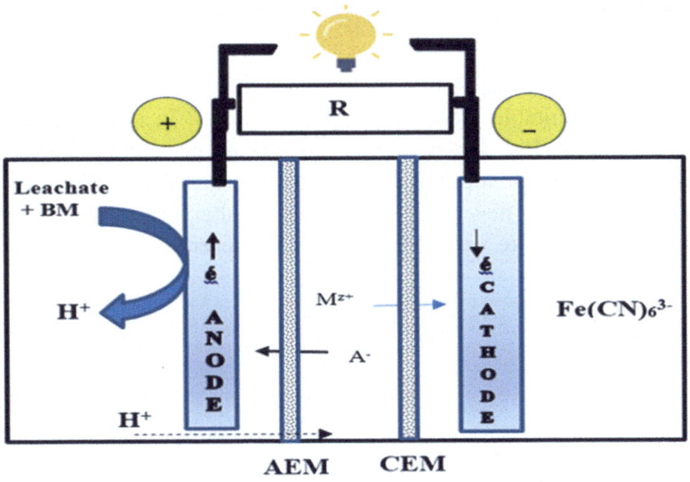

Fig. 28.3 Scheme of YDC; AEM (Anion Exchange Membrane); CEM (Cation Exchange Membrane)

Fig. 28.4 Desalination evolution curves using YFC as power source

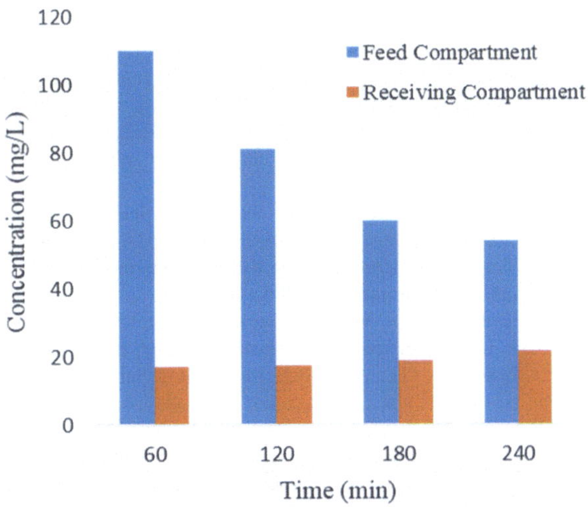

28.3 Conclusion

In the present investigation, it has been shown that the YFC produces electric energy, provided a suitable mediator has been chosen. In our case, the MB has proved this mediation. Besides, owing to low electric current generated, our YFC has been used as a biological power source, to remove traces of heavy metals from an aqueous solution. This new device could a good alternative renewable energy to fossil energy; it is therefore very beneficial to the preservation of the environment.

References

1. S. Cheng, H. Liu, B.E. Logan, Increased power generation in a continuous flow MFC with advective flow through the porous anode and reduced electrode spacing. Environ. Sci. Technol. **40**, 2426–2432 (2006)
2. A.E. Franks, K.P. Nevin, Microbial fuel cells. A current review. Energies **3**, 899–919 (2010)
3. B. Logan, S. Cheng, V. Watson, G. Estadt, Microbial fuel cell with non-Pt cathode catalysts. J. Power Sources **171**, 275–281 (2007)
4. U. Mardiana, C. Innocent, H. Jarrar, M. Cretin, Buchari, S. Gandasasmita, Electropolymerized neutral redox mediator for yeast fuel cell. Int. J. Electrochem. Sci. **10**, 8886–8898 (2015)
5. U. Mardiana, C. Innocent, H. Jarrar, M. Cretin, Buchari, S. Gandasasmita, Yeast fuel cell: application for desalination, in *IOP Conference Series: Materials Science and Engineering*, p. 107012049 (2016)
6. M. Mehanna, T. Saito, J. Yan, M. Hickner, C. Xiaoxin, X. Huang, B.E. Logan, Using microbial desalination cells to reduce water salinity prior to reverse osmosis. Energy Environ. Sci. **3**(8), 1114–1120 (2010)
7. R. Rossi, L. Setti, Effect of methylene blue on electron mediated microbial fuel cell by saccharomyces cerevisiae. Environ. Eng. Manag. J. **16**(9), 2011–2018 (2016)
8. S. Tingry, M. Cretin, C. Innocent, Enzymatic biofuel cells as electrical power source. L'Actualité Chimique No. 373 (2013)
9. K. Scott, E.H. Yu, M.M. Ghangrekar, B. Erable, N.M. Duteanu, Biological and microbial fuel cells, in *Comprehensive Renewable Energy*. Elsevier (2012)
10. N.A. Shehab, G.L. Amy, B.E. Logan, P.E. Saikaly, Enhanced water desalination efficiency in an air-cathode stacked microbial electrode ionization cell (SMEDIC). J. Membrane Sci. **469**, 364–370 (2014)
11. L. Zhao, J. Brouwer, J. Naviaux, A. Hochbaum, Modeling of polarization losses of a microbial fuel cell, in *Proceedings of the ASME 12th International Conference on Fuel Cell Science Engineering and Technology* (2014)

Chapter 29
Hydrogen Storage for Mobile Applications: First-Principles Study of $SrAlH_3$

Y. Bouhadda, K. Benyelloul, K. Khodja, N. Fenineche, and M. Bououdina

Abstract Hydrogen storage is a great challenge for material scientists to overcome for on-board applications. Current storage methods (e.g. gas/liquid form) and materials (e.g. metal or complex hydrides) are far from being practical which requires exploration of new materials. Computational methods such as density functional theory has been proven to provide extensive structural, physical and mechanical properties as well as analyzing stabilities of compounds without complexity and cost of any experiments. In this sense, this study adopts density functional theory in order to suggest and thoroughly investigate new type of perovskite materials for solid state storage of hydrogen. $SrAlH_3$ perovskite hydride is chosen and investigated using density functional theory in terms of ground state properties, electronic, chemical bonding properties for solid state storage of hydrogen. Electronic band structures and their corresponding density of states of compounds are obtained. The results indicate that the compound is mechanically stable and has semiconductor nature with gap energy equal to 0.6 eV. The bond chemical nature was analyzed using ELF combined with total and projected DOS.

Keywords Hydrogen storage · DFT · Perovskite · $SrAlH_3$

Y. Bouhadda (✉) · K. Benyelloul · K. Khodja
Unité de Recherche Appliquée en Energies Renouvelables, URAER,
Centre de Développement des énergies Renouvelables, CDER, 47133 Ghardaïa, Algeria
e-mail: bouhadda@yahoo.com

N. Fenineche
FR FCLAB, UTBM Bât. F, Rue Thierry Mieg, 90010 Belfort Cedex, France

M. Bououdina
Department of Physics, College of Science, University of Bahrain, PO Box 32038, Zallaq, Bahrain

A. Belasri and S. A. Beldjilali (eds.), *ICREEC 2019*, Springer Proceedings in Energy,
https://doi.org/10.1007/978-981-15-5444-5_29

29.1 Introduction

Nowadays, energy demand is high and mostly satisfied by fossil fuels. On the other hand, this latter are responsible for the climate imbalance. In fact, their combustions release greenhouse gases into the atmosphere, which leads to global warming. An alternative considered to meet our energy needs is hydrogen [1, 2].

Hydrogen is the most environmentally friendly and clean energy. Indeed, the hydrogen can be produced from renewable energy resources (such as photovoltaic electrolyzers), then can be stored and used in many applications such as fuel cell vehicles [3, 4]. The widely most used techniques to store hydrogen are either in gaseous or liquid form but they are expensive and unable to ensure greater safety [5, 6]. In the last few decades, hydrides have been studied for hydrogen storage.

Among them, ABH_3 family has been intensively studied for their attractive perovskite crystal structure. The perovskite hydrides are important for hydrogen storage and can have cubic, orthorhombic, hexagonal or tetragonal phases depending on pressure and temperature. The known perovskite hydrides with the ideal Pm3m structure (cubic) are $LiBeH_3$, $SrLiH_3$, $CsCaH_3$, $BaLiH_3$, $KMgH_3$, and $RbCaH_3$ [7–11].

We have previously studied the thermodynamic, electronic, dynamic, and mechanic properties of both $KMgH_3$ and $KMgH_3$ [12–15]. Indeed, we have examined the optical response and the theoretical optical spectrum [12] of $NaMgH_3$. However, the study of the thermodynamic functions of $NaMgH_3$ was done for the first time [13] and show that $NaMgH_3$ is dynamically stable.

We have also investigated the mechanical properties of $NaMgH_3$ and we have found that is mechanically stable at ambient pressure and can be classified as brittle material [14]. Also, we have found that the calculated linear bulk moduli are in good agreement with the theoretical value reported in the literature. We have determined Debye temperature (648 K°) using theoretical elastic constants. This value is higher compared to that of $KMgH_3$, which reveals that $NaMgH_3$ is harder than $KMgH_3$. Moreover, we have calculated and presented the dynamical and thermodynamic properties of $KMgH_3$, such as thermodynamic functions and formation energy that was calculated for different possible reaction pathways [15].

In this work, we investigate the structural, electronic and chemical bond properties of ideal perovskite $SrAlH_3$ material using density functional theory. In the best of our knowledge, no experimental or theoretical investigation has been reported in the literature on perovskite $SrAlH_3$ hydride.

29.2 Computational Methods

In this study, we have used for all electronic structure computations the ABINIT code [16], which based on plane-waves and pseudopotentials within density functional theory (DFT) [17]. Indeed, an efficient fast Fourier transform algorithm

[18] is employed for the conversion of wave functions between real and reciprocal space. In addition, we used an adaptation to a fixed potential of the band-by-band conjugate-gradient method [19] with a potential-based conjugate-gradient algorithm for the determination of the self-consistent potential [20]. To represent atomic cores we used the Generalized Gradient Approximation (GGA–PBE) [21] to describe exchange energy with Fritz–Haber–Institute GGA pseudopotentials [22].

The atoms are progressively relaxed to equilibrium until the Hellmann–Feynman forces on all atoms were less than 0.005 meV/Å. The electronic wave functions were extended in plane waves up to a kinetic energy of cut-off 60 hartree. Integrals over the Brillouin zone were done with similar k-grid densities in all calculations: $6 \times 6 \times 6$ mesh of special k-point. A tolerance in the total energy of 0.01 meV/atom was chosen for the self-consistency.

29.3 Results and Discussion

$SrAlH_3$ crystallizes in an ideal perovskite-type (Pm3m; see Fig. 29.1) structure (space group 221) which contains one formula with the Wyckoff positions of the atoms are: Al 1a (0, 0, 0), Sr 1b (0.5, 0.5, 0.5) and H 3c (0, 0.5, 0.5). Therefore we have relaxed only the lattice constant a. After minimization, we obtained for lattice constant a value of 4.515 Å. In addition, we have obtained the bulk modulus (B_0) and its pressure derivative (B') (Table 29.1) using the Birch-Murnhagan equation of state by fitting the total energy as a function of cell volume (Fig. 29.2).

The electronic structure is investigated for better understanding of the chemical bonding properties of hydrides and further proposes a mechanism about structural stability. The total and the atomic partial density of states (DOS and PDOS) of $SrAlH_3$ are plotted in Fig. 29.3.

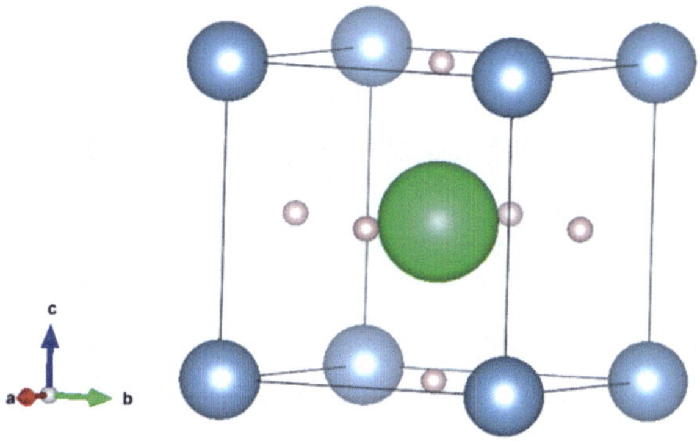

Fig. 29.1 The crystal structure of $SrAlH_3$. Sr, Al, and H atoms are shown as green (big), blue (medium), pink (small) spheres, respectively

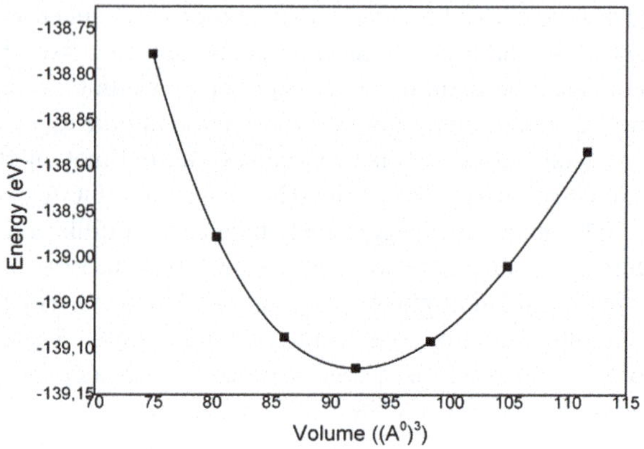

Fig. 29.2 Total energy vs. cell volume for $SrAlH_3$

Table 29.1 Optimized equilibrium structural parameters for $SrAlH_3$ predicted structure

Prototype [23]	Unit cell (Å)	Atomic position	Volume (eV/f.u)	B_0 (GPa)	B'
Pm-3m (Space group N° 221)	a = 4.515	Al (1a) 0,0,0 Sr (1b) 0.5,0.5,0.5 H (3d) 1/2,1/2,0	−139.121	19.47	3.74

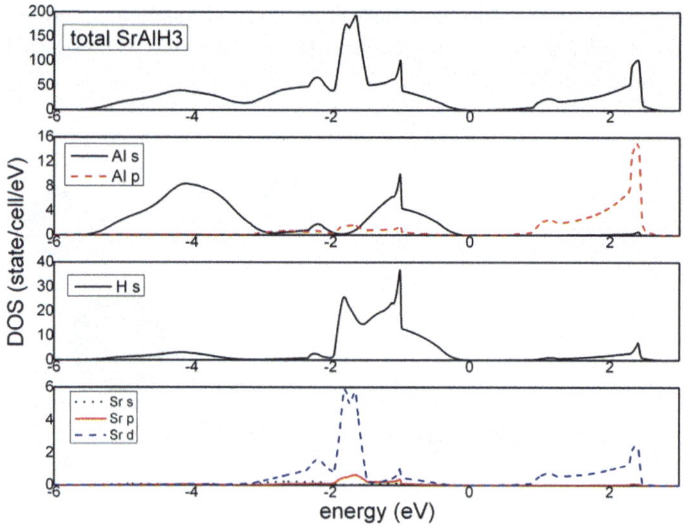

Fig. 29.3 Total and partial DOS for $SrAlH_3$. Fermi level is set to zero

The calculated DOS (Fig. 29.3) indicates that $SrAlH_3$ has a band gap of 0.6 eV. This material can be considered as a semiconductor. It must be noted that the calculated band gaps are equal or smaller than the real band gap, due to the well-known band-gap underestimation in DFT calculations. To the best knowledge of authors, no previous experimental or theoretical research works reported the band gap energy of $SrAlH_3$. From the partial DOS (Fig. 29.3), we observed the existence of two well-separated regions:

1. Valence band (below 0 eV) which is subdivided into two regions: (a) the region from −2 to 0 eV mainly dominated by the Sr d-states, Al s-states and hydrogen atoms; (b) the region below −3 eV is mainly dominated by the H states with a little contribution of Al s-states.
2. Conduction band (below 0.6 eV) dominated by Al p-states and Sr d-states indicating the existence of ionic bond in $SrAlH_3$.

The existence of sp hybridization between Al and H atoms in the valence band indicates that the chemical bonding in $SrAlH_3$ is a mixte of covalent and ionic nature.

Moreover, the charge density of the studied hydrides has been examined to further understand the nature of the chemical bonding, especially that the PDOS don't show a strong correlations between Sr, Al and H in $SrAlH_3$. Figure 29.4 displays the charge density of $SrAlH_3$ compounds. It can be noted that Sr, Al and H in $SrAlH_3$ perovskite structure are not bonded ionically.

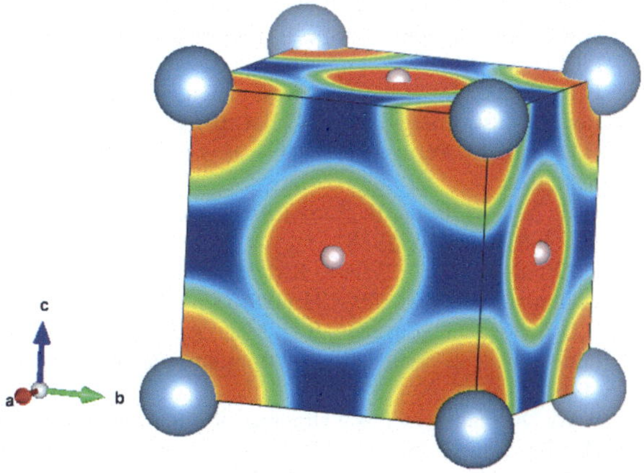

Fig. 29.4 The charge density of $SrAlH_3$ isosurface is at level of 0.01. Sr, Al, and H atoms are shown as green (big), blue (medium) and pink (small) spheres, respectively

29.4 Conclusion

In this work, we have used density functional theory in order to suggest and systematically investigate new type of perovskite materials for solid-state storage of hydrogen. Indeed, we have studied structural, electronic and mechanical properties of $SrAlH_3$. Electronic band structures and their corresponding density of states of compounds are obtained. The results indicate that the compound has semiconductor nature with gap energy equal to 0.6 eV. The bond chemical nature was analyzed using ELF combined with total and projected DOS. We have found that the chemical bonding in $SrAlH_3$ is a mixte of covalent and ionic nature.

References

1. T.Y. Wei, K.L. Lim, Y.S. Tseng, S.L.I. Chan, A review on the characterization of hydrogen storage materials. Renew. Sust. Energy Rev. **79**, 1122–1133 (2017)
2. R. Chaubey, S. Sahu, O.O. James, S. Maity, A review on development of industrial processes and emerging techniques for production of hydrogen from renewable and sustainable sources. Renew. Sust. Energy Rev. **23**, 443–462 (2013)
3. Y. Li, J. Yang, J. Song, Structure models and nano energy system design for proton exchange membrane fuel cells in electric energy vehicles. Renew. Sust. Energy Rev. **67**, 160–172 (2017)
4. Q.X. Wu, Z.F. Pan, L. An, Recent advances in alkali-doped polybenzimidazole membranes for fuel cell applications. Renew. Sust. Energy Rev. **89**, 168–183 (2018)
5. I. Sreedhar, K. Kamani, B.M. Kamani, B. Reddy, A. Venugopall, A Bird's Eye view on process and engineering aspects of hydrogen storage. Renew. Sust. Energy Rev. **91**, 838–860 (2018)
6. H.G. Shiraz, O. Tavakoli, Investigation of graphene-based systems for hydrogen storage. Renew. Sust. Energy Rev. **74**, 104–109 (2017)
7. H.H. Park, M. Pezat, B. Darriet, A new ternary hydride: $CsCaH_3$. Rev. Chim. Min. **23**, 323–328 (1986)
8. H.H. Park, M. Pezat, B. Darriet, Deux nouveaux hydrures: $RbCaH_3$ et Rb_2CaH_4. C. R. Acad. Sci. **306**, 963–966 (1988)
9. R. Schuhmacher, A. Weiss, $KMgH_3$ single crystals by synthesis from the elements. J. Less Common Met. **163**, 179–183 (1990)
10. C.E. Messer, J.C. Eastman, R.G. Mers, A.J. Maeland, Ternary perovskite phases in systems of lithium hydride with barium, strontium, and calcium hydrides. Inorg. Chem. **3**, 776–778 (1964)
11. A.W. Overhauser, Crystal structure of lithium beryllium hydride. Phys. Rev. B **35**, 411–414 (1987)
12. Y. Bouhadda, Y. Boudouma, N. Fenineche, A. Bentabet, Ab initio calculations study of the electronic, optical and thermodynamic properties of $NaMgH_3$, for hydrogen storage. J. Phys. Chem. Solids **71**, 1264–1268 (2010)
13. Y. Bouhadda, N. Fenineche, Y. Boudouma, Hydrogen storage lattice dynamics of orthorhombic $NaMgH_3$. Phys. B **406**, 1000–1003 (2011)
14. Y. Bouhadda, M. Bououdina, N. Fenineche, Y. Boudouma, Elastic properties of perovskite-type hydride $NaMgH_3$ for hydrogen storage. Int. J. Hydrogen Energy **38**, 1484–1489 (2013)

15. Y. Bouhadda, N. Kheloufi, A. Bentabet, Y. Boudouma, N. Fenineche, K. Benyalloul, Thermodynamic functions from lattice dynamic of $KMgH_3$ for hydrogen storage applications. J. Alloy. Compd. **509**, 8994–8998 (2011)
16. X. Gonze, J.M. Beuken, R. Caracas, F. Detraux, M. Fuchs, G.M. Rignanese, L. Sindic, M. Verstraete, G. Zerah, F. Jollet, M. Torrent, A. Roy, M. Mikami, P. Ghosez, J.Y. Raty, D.C. Allan, First principles computation of material properties: the ABINIT software project. Comput. Mater. Sci. **25**, 478–492 (2002)
17. W. Kohn, L.J. Sham, Self-consistent equations including exchange and correlation effects. Phys. Rev. A **140**, 1133–1138 (1965)
18. S. Goedecker, Fast radix 2, 3, 4, and 5 kernels for fast Fourier transformations on computers with overlapping multiply-add instructions. SIAM J. Sci. Comput. **18**, 1605–1611 (1997)
19. M.C. Payne, M.P. Teter, D.C. Allan, T.A. Arias, J.D. Joannopoulos, Iterative minimization techniques for ab initio total-energy calculations: molecular dynamics and conjugate gradients. Rev. Mod. Phys. **64**, 1045–1097 (1992)
20. X. Gonze, Towards a potential-based conjugate gradient algorithm for order-N self-consistent total energy calculations. Phys. Rev. B **54**, 4383–4386 (1996)
21. J.P. Perdew, K. Burke, M. Ernzerhof, Generalized gradient approximation made simple. Phys. Rev. Lett. **77**, 3865–3868 (1996)
22. M. Fuchs, M. Scheffler, Ab initio pseudopotentials for electronic structure calculations of poly-atomic systems using density-functional theory. Comput. Phys. Commun. **119**, 67–98 (1999)
23. J. Daams, R. Gladyshevskii, O. Shcherban, V. Dubenskyy, V. Kuprysyuk, O. Pavlyuk, I. Savysyuk, S. Stoyko, R. Zaremba, *Group III: Condensed Matter Volume 43 Crystal Structures of Inorganic Compounds, Subvolume A, Structure Types, Landolt-Börnstein Numerical Data and Functional Relationships in Science and Technology New Series*, Germany (2013)

Chapter 30
Utilization of Double-Layered Hydroxides for Enhancement of Dissolved Oxygen Reduction in Microbial Fuel Cell: An Approach for the Evaluation of Coulomb Efficiency

Meriem Djellali, Mostefa Kameche, Hakima Kebaili, Abdallah Benhamou, Mustapha Bouhent, and Christophe Innocent

Abstract Following their successful use as novel bioanodes in Microbial Fuel Cells (MFCs) [1], Lamellar Double Hydroxides (LDH) were tested as promising cathodes for reducing electrons in the presence of dissolved oxygen. The biofilm on anode, allowed oxidation of organic matter, yielding electrons and protons. The electrons circulated in external circuit, whilst the protons circulated in internal circuit, crossing a separator between the anolyte and catholyte. The nickel-aluminium LDH material was synthesized and added by deposition on Carbon Felt fibers (CF). It was then tested as cathode in the MFC. The electrochemical performances of the material were characterized by cyclic voltammetry (VC) and electrochemical impedance spectroscopy (EIS). Moreover, the MFC performance was evaluated from its Coulomb Efficiencies (CE). Two CEs were calculated graphically from the integral of the voltage between initial and final times, i.e. the area of the domain delimited by the representative curve, the abscissa and the vertical lines. One CE was calculated over the whole voltage curve and another from the triangular peak resulting from the addition of the substrate. It was revealed that this electrode allowed much oxygen reduction and therefore much electrical energy than that using non-modified cathode with the MFC inoculated with a fruit peeling leachate. In comparison to enzymatic fuel cell [2], it seemed to be relatively

M. Djellali (✉) · M. Kameche · H. Kebaili · M. Bouhent
Laboratory of Physico-Chemistry of Materials, Catalysis and Environment, University of Sciences and Technology of Oran-Mohammed Boudiaf, M'Nouar, BP 1505, Oran, Algeria
e-mail: meriem.djellali@univ-usto.dz

A. Benhamou
Laboratory of Environmental Processes Engineering, University of Sciences and Technology of Oran-Mohammed Boudiaf, M'Nouar, BP 1505, Oran, Algeria

C. Innocent
European Membrane Institute, UMR 5635, University of Montpellier, CC 047, Place Eugene Battalion, 34095 Montpellier Cedex 5, France

A. Belasri and S. A. Beldjilali (eds.), *ICREEC 2019*, Springer Proceedings in Energy,
https://doi.org/10.1007/978-981-15-5444-5_30

lower. However, the biomolecule (enzyme) is more expensive and difficult to handle experimentally. Thus, the low-cost biomaterials (CF) used in the MFC was very promising for practical application in bioenergy production and treatment of fruit wastes.

Keywords Microbial Fuel Cell · Layered double hydroxide · Coulomb Efficiency

30.1 Introduction

Microbial Fuel Cells (MFCs), also known as microbial biofuels, are new electrochemical systems that exploit electrogenic bacteria to produce electrical energy while degrading organic pollutants in wastewater [1, 4]. However, optimization of these innovative biological fuel cells that will ultimately facilitate energy recovery remains one of the factors that seriously compromise the effectiveness of such applications. It is therefore interesting to look for ways to improve their electrical performance.

Nowadays, serious development prospects and a lot of research are focused on improving electron transfer between bacteria and electrodes using various approaches, materials and electrode design. Hence there is a need to develop new and simple strategies to make efficient anodes to increase bacterial load and improve therefore transfer capacity.

Clays have been frequently seen as one of the most promising materials for low-cost and easily utilizable as electrodes, capable to store electrical energy due to their unique properties, such as the presence of metal lamellae, large surface area, controllable particle size and high thermal stability [5]. In particular, anionic clays, namely hydrotalcite (HT) compounds, which comply with environmental standards and are relatively easy to synthesize. HTs have a wide range of versatility that allows their synthesis with different combinations of divalent and trivalent cations for specific applications [6, 7]. Studies on the electrochemical behavior of HDLs with reference to the nature of the cations are very limited, they are dedicated to HDL with transition metals such as Ni and Co [8–11]. Due to their beneficial properties including low cost, good biocompatibility, high catalytic activity, and high chemical stability, these materials could find potential application as anode materials for fuel cells as reported very recently in the literature [12].

Owing to the low efficiency of the MFC, limited by oxygen reduction, these novel materials will used as catalyst to increase the charge transfer to the cathode. This property has been determined by calculating the Coulomb Efficiency (CE) i.e. the ratio between the charge density of the electron oxidized or reduced and density charge that should be involved in the process. Two ways will tested in the present investigation, depending upon the graphical analysis of the area in the evolution curve current density versus time.

30.2 Experimental Techniques

30.2.1 Preparation of LDH

Solution of a 50 ml of divalent and trivalent metal salts (Ni_2^{++} A_3^{+}) of total concert tion (1 M) with a molar ratio MII/MIII = 3 is added at constant flow rate (0.5 mL/min) in a reactor containing previously 50 ml of decarbonized water stirred magnetically. The pH (pH meter METLER DELTA 350) of the solution was kept constant by the simultaneous addition of a sodium hydroxide solution of concentration 1 M [16]. In order to avoid the contamination of the phases by carbonated anions resulting from the CO^2 of the air, the synthesis was carried out under nitrogen flow at 25 °C using a fully automated device equipped with pumps and a pH meter, which allows to regulate the flows of solutions of salts and soda). The final product is recovered after several cycles of washing/centrifugation with deionized water and then dried at 60 °C [6, 17].

30.2.2 Preparation of LDH Modified Electrode

The modified carbon felt electrodes were prepared by depositing a mixture of aqueous HDL (NiAl) suspension on the surface of the electrode by immersion for 24 h. First, a piece of 14.88 cm^2 active surface carbon felt is washed well with 0.1 M hydrochloric acid. Afterwards, it was immersed in a suspension of LDH 0.1 mg/ml and then stirred overnight. Finally, it was dried at room temperature.

30.2.3 Microbial Fuel Cell Design

The MFC microbial fuel cell consisted of two glass cells separated by a cation exchange membrane (Nafion117) and CF electrodes (Fig. 30.1). Half of the cell contained virgin CF in the form of anode and leachate of fruit waste for the formation of an electroactive biofilm. Oxidation of organic matter, particularly sodium acetate, has generated electrons and protons [18]. The electrons are transferred along the outer circuit, while the protons are scattered across the membrane. The other half of the cell contained the NiAl-LDH-modified CF as the cathode and an electrolyte solution (KCl or $NiCl_2$ or $ALCl_3$) that received the electrons from the external circuit. In addition, dissolved oxygen in the catholyte has improved proton reduction to produce water molecules. The circuit of the microbial fuel cell was closed with an external resistance of 1000 Ω. At first, the cell voltage varied due to the naturalization of the electrode in the middle and then became more stable, giving more significant current values. In general, it took a week.

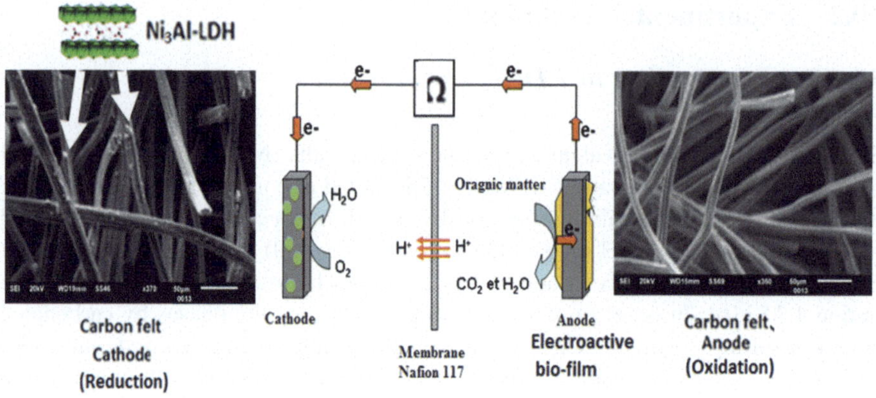

Fig. 30.1 Schematic of a typically employed two-chamber microbial fuel cell highlighting the various electrochemical and electro-microbiological processes

30.2.4 Coulomb Efficiency (CE)

Figure 30.2 displays the current density produced as a function of time. According to Faraday's 1st law, the integral of the current density with respect to time gives the quantity of electricity generated Qexp according to (Eq. 30.2). Graphically, this integral represents the area between the curve representative of the current density and the initial and final times. Theoretically, the amount of electricity available in the Qthsubstrate involved in the oxidation reaction [19]. Its expression is given by Eq. 30.2, according to Faraday's 2nd law. The oxidation reaction of the acetate is given by Eq. 30.3. The ratio between the two densities of charges gives faradic efficiency i.e. Coulomb Efficiency (CE).

Fig. 30.2 MFC faradic yield with CF-NiAl-LDH modified cathode in the presence of $NiCl_2$ electrolyte

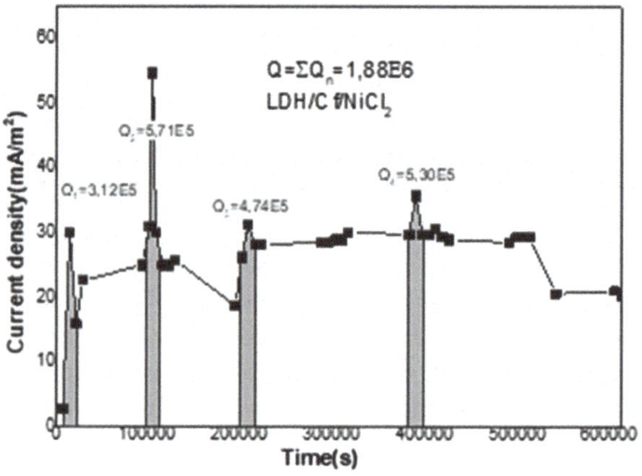

Table 30.1 Electric charge during the MFC functioning

Time (h)	Electric charge 10^5C
4	3.12
29	5.71
108	4.74
174	5.30

$$Q_{exp} = \int_0^t idt \quad (30.1)$$

$$Q_{th} = \frac{nFCV}{M} \quad (30.2)$$

$$CH_3COO^- + 4H_2O \rightarrow 2HCO_3^- + 8H^+ + 9é \quad (30.3)$$

$$EC = \frac{Qexp}{Qth} * 100 \quad (30.4)$$

A: Electric current.
t: MFC operating time (s).
n: number of moles of electrons exchanged during the oxidation of acetate (n = 8).
F: Faraday constant (96500 C mol^{-1}).
C: concentration of substrate consumed (60 mM).
V: volume of the analyte (V = 70 ml)

Experimentally, the value of the CE for each period, is calculated graphically by using Eq. 30.2. Thence, the values of quantity of charge produced by oxidation upon the addition of the same amount of acetate substrate (CH_3COO-), during different periods are given in Table 30.1. It can be noticed that approximately the same amount of electric charge is produced, which attests that the MFC was becoming mature and stable during the whole period. However, CE in the beginning of the functioning of the MFC was a bit less because the cell is not yet mature.

30.3 Conclusion

The NiAl-LDH was used successfully used as a potential material for electron transfer in cathode of MFC. It therefore makes possible to control the electrochemical response and application in the field of energy as an electrode modification material for a biological fuel cell.

The results reveal that the power density of the carbon felt modified with NiAl-LDH, is increased. Besides, the quantities of electron charge produced by oxidation of acetate during different periods, are approximately equal which attested that the MFC is becoming mature and stable. Finally, the performances obtained with this novel biological device recommend its use for the production of small amount of bio-energy and simultaneously preservation of environment from domestic wastes.

References

1. S.M. Tiquia-Arashiro, M. Mormile, Sustainable technologies: bioenergy and biofuel from biowaste and biomass. Environ. Technol. **34**(13-14), 1637–2211 (2013)
2. S. Korkut, M.S. Kilic, S. Uzuncar, B. Hazer, Novel graphene-modified poly (styrene-b-isoprene-b-styrene) enzymatic fuel cell with operation in plant leaves. Anal. Lett. **49**, 2322–2336 (2016)
3. R. Karthikeyan, H.P. Uskaikar, S. Berchmans, Electrochemically prepared manganese oxide as a cathode material for a microbial fuel cell. Anal. Lett. **45**, 1645–1657 (2012)
4. C. Mousty, V. Prévot, Hybrid and biohybrid layered double hydroxides for electrochemical analysis. Anal. Bioanal. Chem. **405**, 3513–3523 (2013)
5. V. Rives, *Layered Double Hydroxides: Present and Future* (Nova, New York, 2001). PII: S0169-1317(02)00112
6. S. Carlino, The intercalation of carboxylic acids into layered double hydroxides: a critical evaluation and review of the different methods. Solid State Ionics **98**, 73–84 (1997)
7. Y. Wang, D. Zhang, W. Peng, L. Liu, M. Li, Electrocatalytic oxidation of methanol at Ni–Al layered double hydroxide film modified electrode in alkaline medium. Electrochim. Acta **56**, 5754–5758 (2011)
8. Z. Gao, J. Wang, Z. Li, W. Yang, B. Wang, M. Hou, Y. He, Q. Liu, T. Mann, P. Yang, M. Zhang, Graphene nanosheet/Ni2+/Al3+ layered double-hydroxide composite as a novel electrode for a supercapacitor. Chem. Mater. **23**(15), 3509–3516 (2011)
9. M. Jitianu, D.C. Gunness, D.E. Aboagye, M. Zaharescu, A. Jitianu, Nanosized Ni–Al layered double hydroxides—structural characterization. Mater. Res. Bull. **48**, 1864–1873 (2013)
10. J. Pérez-Ramírez, G. Mul, F. Kapteijn, J.A. Moulijn, In situ investigation of the thermal decomposition of Co–Al hydrotalcite in different atmospheres. J. Mater. Chem. **11**, 821–830 (2001)
11. A. Zebda, S. Tingry, C. Innocent, S. Cosnier, C. Forano, C. Mousty, Hybrid layered double hydroxides-polypyrrole composites for construction of glucose/O2 biofuel cell. Electrochim. Acta **56**, 10378–10384 (2011)
12. F. Wypych, K.G. Satyanarayana (eds.), *ClaySurfaces: Fundamentals and Applications* (Elsevier, Amsterdam, 2004), pp. 345–373
13. I. Carpani, M. Berrettoni, M. Giorgetti, D. Tonelli, Intercalation of iron (III) hexacyano complex in a Ni, Al hydrotalcite-like compound. J. Phys. Chem. B **110**(14), 7265–7269 (2006)
14. S. Therias, C. Mousty, C. Forano, J.P. Besse, Electrochemical transfer at anionic clay modified electrodes. Case of 2, 2 '-Azinobis (3-ethylbenzothiazoline-6-sulfonate). Langmuir **12**, 4914–4920 (1996)
15. U. Mardiana, C. Innocent, M. Cretin, B. Buchari, S. Gandasasmita, Yeast fuel cell: application for desalination. Mater. Sci. Eng. **107**, 1–13 (2016)

Chapter 31
Electrical Characterization and Modeling of a Renewable Energy System Combining Photovoltaic Panels and Bacterial Fuel Cell

Mohammed Benghernit, Fatima Zohra Zerhouni, and Mostefa Kameche

Abstract Renewable energy is expected to be a promising alternative to current energy sources. In our PV photovoltaic system, the photovoltaic panel is the master part. Its electrical model with corresponding equations are elaborated. We determined the intrinsic descriptive parameters of the PV panel, in order to carry out our simulation, in order to better understand and predict its behavior towards external factors. We superimposed the characteristics we obtained, after our characterization, with those provided by the manufacturer. The results coincide. The RMSE calculation yielded conclusive results. A microbial fuel cell is also adopted using leachate from sheep manure. This bio-fuel replaces the platinum-based mineral catalyst for the traditional fuel cell with a bacteria that forms a biological film on the electrode surface. In this paper, we describe the protocol for developing a graphite carbon based bio-anode. This bio-anode has been characterized with the two electrochemical techniques adopted which are chronoamperometry and cyclic voltammetry. Another party is interested in characterizing and modelling the electrical behavior of the battery. The results obtained are advanced. After an analysis of the energy context, a state of the art of electrochemical components coupling hydrogen and electricity is presented, particularly on electrolysers and regenerative or unitized reversible fuel cells. Finally, this model is used to study the modularity of components, including electrical and thermal imbalances in series or parallel associations of fuel cells or electrolysers. An architecture combining these

M. Benghernit (✉) · F. Z. Zerhouni
Department of Electronics, Faculty of Electrical Engineering,
University of Sciences and Technology of Oran Mohamed Boudiaf USTO-MB,
BP 1505 El M'naouar, 31000 Oran, Algeria
e-mail: benghernit81@gmail.com

M. Kameche
Laboratory of Physical Chemistry of Materials, Catalysis and Environment,
University of Sciences and Technology of Oran Mohamed Boudiaf USTO-MB,
BP 1505 El M'naouar, 31000 Oran, Algeria

A. Belasri and S. A. Beldjilali (eds.), *ICREEC 2019*, Springer Proceedings in Energy,
https://doi.org/10.1007/978-981-15-5444-5_31

elements with a photovoltaic generator to power an electric load or an electrical grid is finally submitted.

Keywords Microbial fuel cell · Photovoltaic panel · Electrolysers · Electrical model

31.1 Introduction

A photovoltaic cell is a device that transforms light energy into electric current. The first photo-pile was developed in the United States in 1954 by BELL laboratory researchers, who discovered that the photosensitivity of silicon could be increased by adding impurities. This is a technique called doping that is used for all semi-conductors. Photovoltaic solar energy transforms a part of solar radiation into electrical energy. The energy conversion can be achieved by means of a photo-voltaic (PV) cell, based on photovoltaic effect.

In general, solar cells are combined either in series or in parallel. They can be encapsulated under glass to obtain a photovoltaic module. A photovoltaic generator (PV) consists of interconnected modules to form a unit yielding high continuous power compatible with conventional electrical devices. The PV modules increase both the voltage and current output. A photovoltaic module is the combination of Ns cells in series and Np cells in parallel.

As renewable energy systems, Microbial fuel cells (MFCs) are bio-electrochemical devices that harvest the power from organic substrates directly into electrical energy using biocatalysts, in general micro-organisms and bacteria. The MFC transforms chemical energy into electricity using oxidation at anode and reduction at cathode. It can be used to treat wastewater and produces low energy high enough to functioning small electrical devices.

31.2 Electric Modeling of Photovoltaic Solar Module

31.2.1 Equivalent of Solar Cell Circuit

The equivalent diagram of the actual photovoltaic cell takes into account the parasitic resistive effects due to manufacture and is shown in Fig. 31.1. This circuit can be used for both an elementary cell and a module or a panel made up of several modules [3–5].

Fig. 31.1 Equivalent circuit of photovoltaic cell [4]

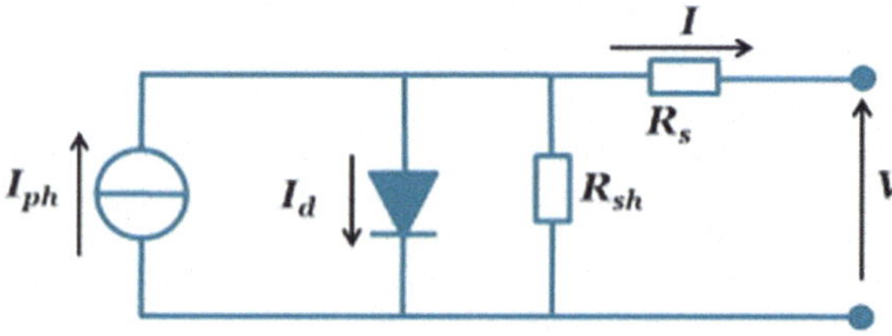

31.2.2 (I, V) and (P, V) Characteristics for Different Illuminations

To check the validity of our model, we give on the Figs. 31.1 and 31.2 the evolution of the two main (I, V) and (P, V) characteristics for different illuminations due to global radiation.

31.2.3 Influence of Different Parameters

Influence of Illumination. In Figs. 31.3 and 31.4, we show the results for the I (V) and P (V) characteristics of the panel at 25 °C obtained for various illuminations.

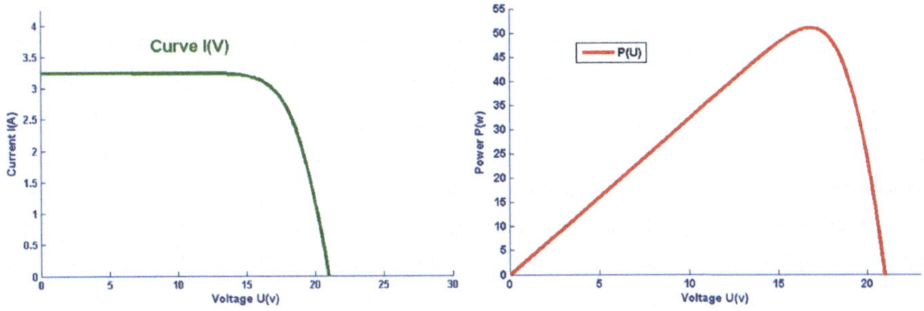

Fig. 31.2 Evolution of (I, V) and (P, V) characteristics using an illumination of 1000 W/m^2 and temperature 25 °C for the ZYTECH module

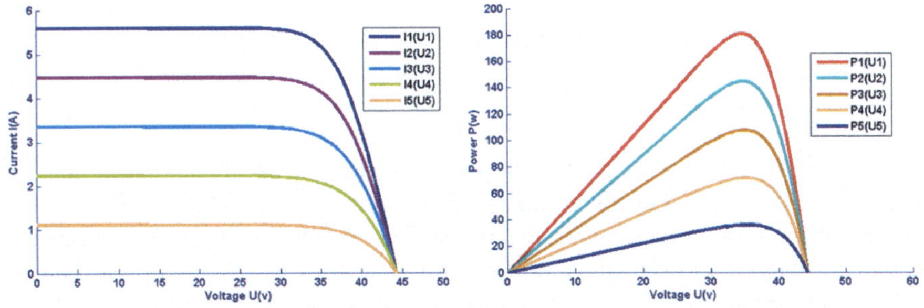

Fig. 31.3 Influence of illumination on I (V), P (V) Characteristics; $T_{ref} = 25\,°C$

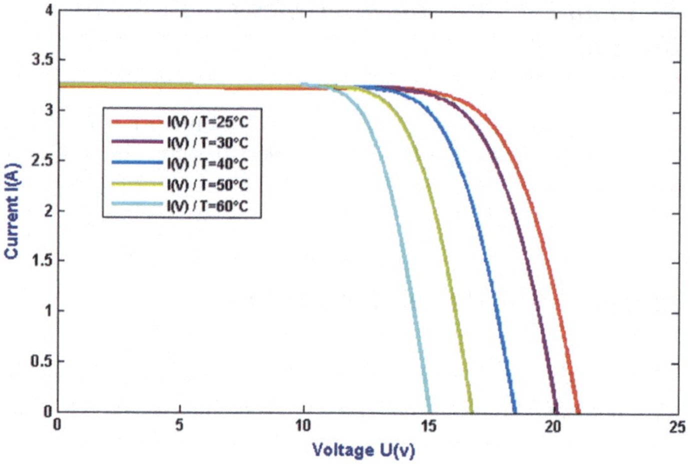

Fig. 31.4 Influence of temperature (T = 25 °C, 30 °C, 40 °C, 50 °C, 60 °C) on Characteristics I(V) to 1000 W/m^2 for the ZYTECH module

Influence of Temperature. Temperature is a very important parameter in the behavior of PV cells. Figure 31.4 describes the behavior of the module with 1000 W/m^2 illumination and at temperatures (25 °C, 30 °C, 40 °C, 50 °C, 60 °C). We notice that the current increases with the increase in temperature. However, the open circuit voltage decreases. This results in a decrease in maximum power.

31.3 Microbial Fuel Cell (MFC): Another Renewable Clean Energy

A microbial fuel cell using the leachate sheep manure was undertaken. This bio-fuel cell replaces the inorganic catalyst based on platinum for traditional fuel cell. In this cell, the bacterium forms a bio-film on the surface of the electrode. In the present work, we describe the development of a graphite carbon-based bio-anode protocol. This bio-anode was characterized with the two electrochemical techniques that are chronoamperometry and cyclic voltammetry. Indeed, the first technique allowed the bacterial bio-film to properly adhere to the surface of the bio-anode. As regards the second technique, showed the redox reactions at both electrodes of the battery cathode and anode. The latter allows oxidation of the fuel (sodium acetate or glucose) in the presence of the biological catalyst (bacterial bio-film). At the cathode, the proton generated by oxidation recombines with oxygen to give water. The discharge of the bio-fuel cell to an external resistor 1000 Ω, yields an electric potential of 300 mV after 10 days and after that it fall gradually. By adding new fuel, the potential increases again making in evidence the oxydation of the sustrate and as a result the depollution of the environment.

31.3.1 Principle of MFC

The operating principle of the MFC offers a new opportunity for sustainable energy production. MFC can be defined as electrochemical devices that use the catalytic action of bacteria organized in bio-films to achieve the oxidation of organic or inorganic compounds by producing current. These bacteria are well qualified because they are able to transfer electrons out of their cells. MFC can be completely microbial when the catalysis of reactions to the two electrodes is carried out by microorganisms.

31.3.2 Design of Two-Compartment MFC

A Microbial Fuel Cell consists of two compartments separated by a cation exchange membrane in general Nafion (Fig. 31.5). The two gel-filled flexible compartments containing 120 ml of electrolyte were equipped with a membrane separator that maintains ionic conductivity between the two compartments and the battery is sealed by a silicone seal. In some tests, the anolyte and catholyte are agitated at 250 rpm thanks to a magnetic bar 1 cm long and 0.4 cm thick. A scheme of the electrochemical assembly of two-compartment MFC is given in Fig. 31.5. The experimental set up of the MFC is depicted in Fig. 31.6.

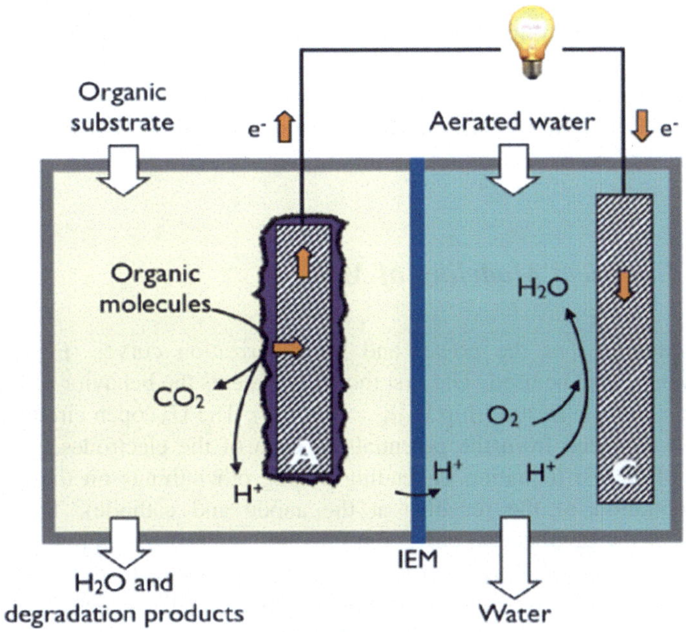

Fig. 31.5 Electrochemical assembly of two-compartment MFC

Fig. 31.6 Experimental of two-compartment MFC

31.4 Modeling of Characteristics of MFC

Microbial fuel cells (MFCs) are part of the environmental energy collection systems [1]. Micro-generators such as these can power low-power circuits such as sensors or stand-alone field equipment. To supply them correctly, the association of several MFC is necessary. A series combination allows an increase of the output voltage while a parallel mounting increases its intensity. However, coupling several batteries leads to structural problems (hydraulic coupling) or electrical problems (non-homogeneity). The use of balancing circuits can minimize these. The modeling of these assemblies makes it possible to define and anticipate the number of batteries as well as the balancing circuits [2].

31.4.1 Electrical Modeling of MFC

Static Modeling. For the power and the polarization curves (Fig. 31.7), two possible models can be used. The first model represents the behavior of MFC in the form of an algebraic relationship $U_{MFC} = f(U_{MFC})$. The U_{oc} open circuit voltage of the MFC is deduced from the potentials of each of the electrodes, given by the equations of Nernst (equation depending among other things on the temperature and concentration of the reagents at the anode and cathode). The analytical expression of the polarization curve can thus be obtained by subtracting the three types of loss described above (activation, ohmic, diffusion) from the U_{co} voltage. On the basis of the above-mentioned modeling used for PV panels, the evolution, polarisation and power curves of the MFC were obtained.

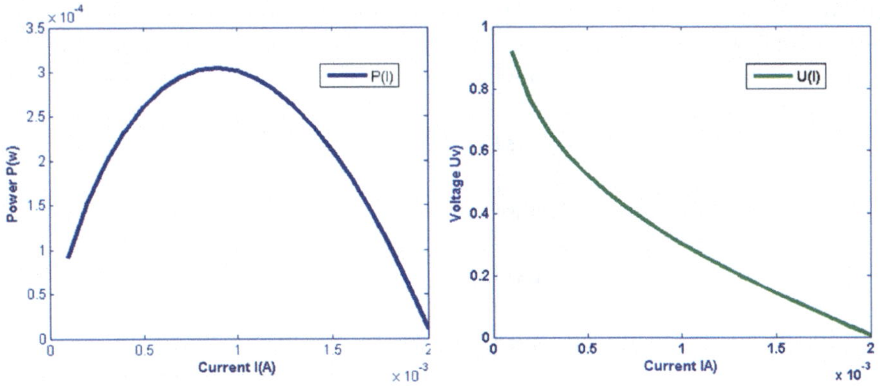

Fig. 31.7 Polarization and power curves of MFC

Dynamic Modeling. Randles electric circuits often used to model the electrical behavior of each electrode [3] (Fig. 31.8). It consists of a electrolyte resistance in series of a parallel CDL/Rct circuit where CDL and Rct are the capacitance of the Double Layer and the charge transfer resistance respectively. In general, the electrode resistor groups together the activation, diffusion and load transfer losses related to the electrode in question. The capacity represents the so-called Double Layer (DL) at the electrode/electrolyte interface. It is correlated to bio-film and its ability to store of charges. It is important because the interface surface is large, i.e. the porosity of the material used for the electrodes is high [4]. Moreover, dynamic component can be added to the model to illustrate the dynamics of distinguishable scatter losses at low frequencies [5]. On the basis of this model, we tabulated values of the MFC voltage, and ploletted them versus time. As shown in the Fig. 31.9, the voltage dropped off during the first hours and then remained almost constant during the rest of the experiment.

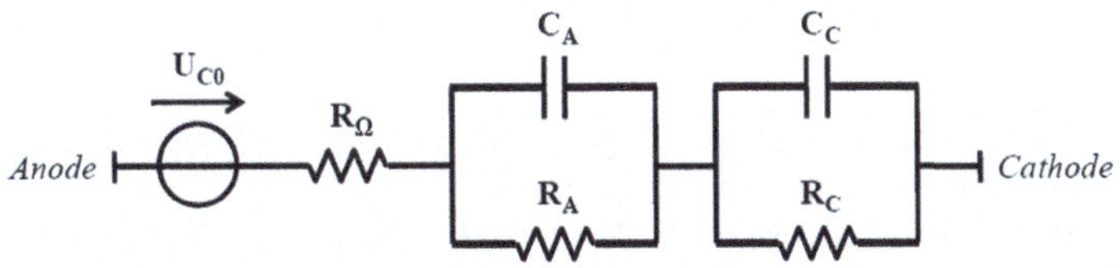

Fig. 31.8 Equivalent Randles circuit for dynamic behavior of MFC

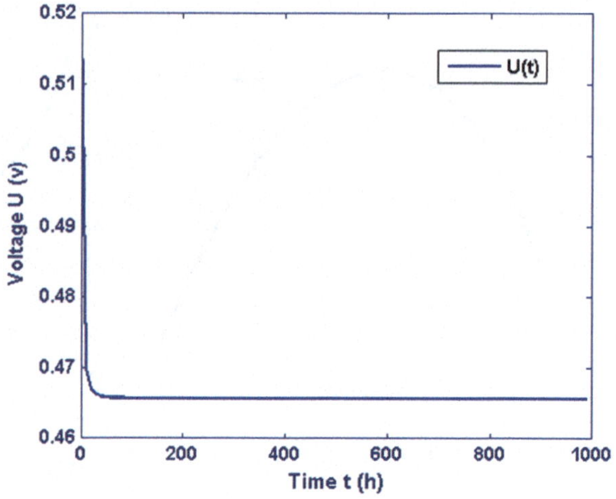

Fig. 31.9 Evolution curve of MFC cell voltage

31.5 Conclusion

The modeling of photovoltaic solar module has been investigated by using the adequate electric equivalent circuit. The current-voltage and power characteristics were obtained for different illuminations and temperatures. The power increased with increasing illumination. Besides, the influence of temperature became pronounced beyond 15 °C. Furthermore, the implementation of the model was also successfully applied to the microbial fuel cell.

References

1. Y. Fan, E. Sharbrough, H. Liu, Quantification of the internal resistance distribution of microbial fuel cells. Environ. Sci. Technol. **42**, 8101–8107 (2008)
2. A.M. Dhirde, N.V. Vale, H. Salehfar, M.D. Mann, T.H. Han, Equivalent electric circuit modeling and performance analysis of a PEM fuel cell stack using impedance spectroscopy. IEEE Trans. Energy Convers. **25**, 778–786 (2010)
3. N. Sekar, R.P. Ramasamy, Electrochemical spectroscopy for microbial fuel cell characterization. J. Microbial Biochem. Technol. S6-004 (2013)
4. V. Agostino, D. Ahmed, A. Sacco, V. Margaria, C. Armato, M. Quaglio, Electrochemical analysis of microbial fuel cells based on enriched biofilm communities from freshwater sediment. Electrochimica Acta **237**, 133–143 (2017)
5. L.P. Fan, J.J. Li, Overviews on internal resistance and its detection of microbial fuel cells. Int. J. Circuits Syst. Signal Process. **10**, 316–320 (2016)

Chapter 32
Ionic Exchange Membrane Used as Separator in Methanol Fuel Cell: An Electrochemical Characterization Using Hydro-Organic Solution

Mhamed Mehouen, Mostefa Kameche, and Christophe Innocent

Abstract In addition to their utilization in membrane separation processes (electrodialysis, dialysis, electro-deionization, etc.), polymer ionic membranes are utilized as a separator in fuel cells (hydrogen, biological and direct methanol fuel cells, etc.) [1]. In the latter, the oxidation of methanol generates electrons that circulate in the external circuit to feed a load (electrical resistance) and the protons migrate through the membrane, and combined with oxygen at cathode to yield droplets of water [2]. In this work, the proton transfer through the membrane was studied with voltamperometry by plotting current-voltage curves. The effect of methanol during proton transfer has also been studied by giving the limiting currents. The transport of proton through the membrane, has been investigated in two media: aqueous solution and hydro-organic solution. Moreover, the stability of applied current through the Nafion membrane, was also studied by using the Hittorf cell for determining the transport numbers of metal ions.

Keywords Fuel cell · Membrane · Characterization

32.1 Introduction

Nowadays, membrane separation processes have been widely developed for applications to industrial utilization [3]. We distinguish various separation processes such as microfiltration, reverse osmosis, dialysis and electrodialysis. In this

M. Mehouen (✉) · M. Kameche
Laboratory of Physico-Chemistry of Materials, Catalysis and Environment, University of Sciences and Technology of Oran, Mohammed-Boudiaf, M'Nouar, 1505 Oran, Algeria
e-mail: mehouenmed@gmail.com

C. Innocent
European Institute of Membranes, University of Montpellier, Montpellier, France

A. Belasri and S. A. Beldjilali (eds.), *ICREEC 2019*, Springer Proceedings in Energy,
https://doi.org/10.1007/978-981-15-5444-5_32

latter, ionic exchange membranes are used, to separate anode and cathode compartments.

An ionic exchange membrane is composed of macromolecular chain more or less cross-linked to form an insoluble three-dimensional network, on which ionizable functional groups are fixed, that confer its specificity [4]. When the functional group is negative, the membrane exchanges cations and it is called Cation Exchange Membrane (CEM). Whilst, it is positive, the membrane exchanges the anions and it is therefore called anion exchange membrane (AEM).

Few data are available on the properties of Ionic Exchange Membranes (IEM), when organic solvent is used (e.g. organic electrosynthesis). The transport of organic molecules is reduced by the increased resistance of the membrane in the presence of solvent. When methanol molecules migrate through the ionic conducting polymer, the performance of the fuel decreases, as the numbers electrons produced by oxidation is lowered [5].

In the present, we study the behavior the cation exchange membrane Nafion conditioned in organic solvent, by using the electrochemical technique (voltamperometry).

32.2 Experimental

Before characterization by using the voltamperometry, samples of the cation exchange membrane (Nafion117) were stored in hydrochloric acid during 24 h. After that, they were equilibrated separately in salt solutions of alkali metal salts LiCl, NaCl and KCl for 24 h. At the end, they were washed up with deionized water. Measurements of current-voltage were done using the two compartment Plexiglas (Fig. 32.1). Two titanium sheets coated with platinum, were utilized to apply a constant current. Whilst, two silver-silver chloride electrodes were placed close to the membrane sides, for measuring the transmembrane potential difference. Two digital multimeters were utilized to control both the current intensity and the voltage.

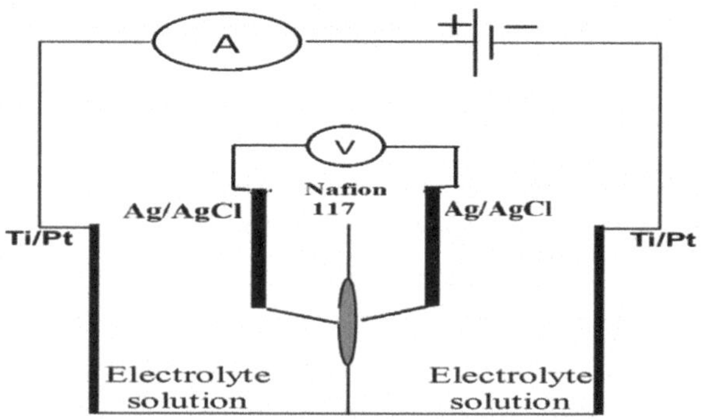

Fig. 32.1 Scheme of electrochemical cell for measurements of current–voltage using Nafion membrane

32.3 Results and Discussion

The current–voltage characteristics of the Nafion membrane are shown in Figs. 32.2, 32.3 and 32.4 for solutions of LiCl, NaCl and KCl. These figures present three regions: the first ohmic region is obtained between the applied current and the voltage drop. When the current is increased, the polarization becomes important. As a result, the concentration in the dilute boundary layer is decreased sharply and the resistance rises up. Consequently, a limiting current is reached, yielding a plateau. An over-limiting current due to water dissociation, can be created by increasing again the applied voltage [6]. The three methods, Cowan Brown, derivative and cross-section tangents can be utilized to determine the limiting current. Among

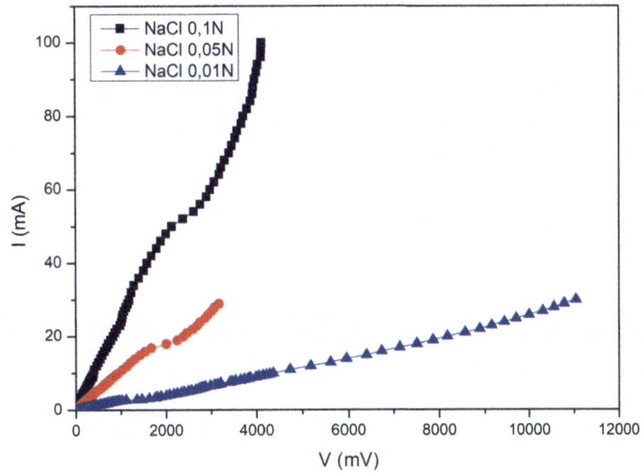

Fig. 32.2 I–V characteristics of Nafion117 membrane conditioned with aqueous solution of NaCl: concentration effect

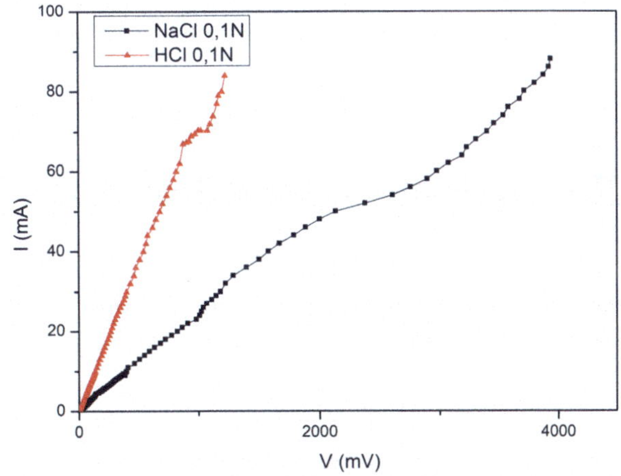

Fig. 32.3 I–V characteristics of Nafion117 membrane conditioned with different aqueous: salt effect

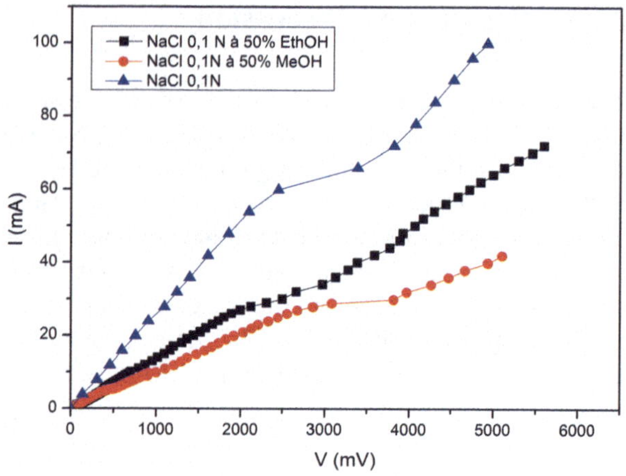

Fig. 32.4 I–V characteristics of Nafion117 membrane conditioned with hydro-organic solutions: solvent effect

Table 32.1 Values of the limiting currents of Nafion117 equilibrated with aqueous solutions of NaCl

Concentration (mol/l)	Limiting current (mA)
0.01	5
0.05	19
0.1	55

these methods, cross-section method was used simply by drawing the two slopes of the Ohmic region and plateau which intersect at the limiting current. The values obtained with the other methods are quite different; in general we choose the lowest. Nevertheless, with respect to Fig. 32.2, the values of the limiting currents of Nafion117 equilibrated with aqueous solutions of NaCl are given in Table 32.1.

Indeed, as shown in this table, when the value of the salt concentration is raised, the limiting current is also raised, making in evidence the pronounced effect of the concentration polarization when the concentration is lower. Moreover, we have noticed the other effects of electrolyte and the presence of the solvent where remarkable differences appear, in particular when the membrane is conditioned in acid solution.

32.4 Application to Direct Methanol Fuel Cell

Owing to the highest the limiting current of the Nafion117 membrane in the proton form, it is commonly used as a separator in direct methanol fuel cell. The oxidation and reduction reactions occur at anode and cathode respectively (Fig. 32.5).

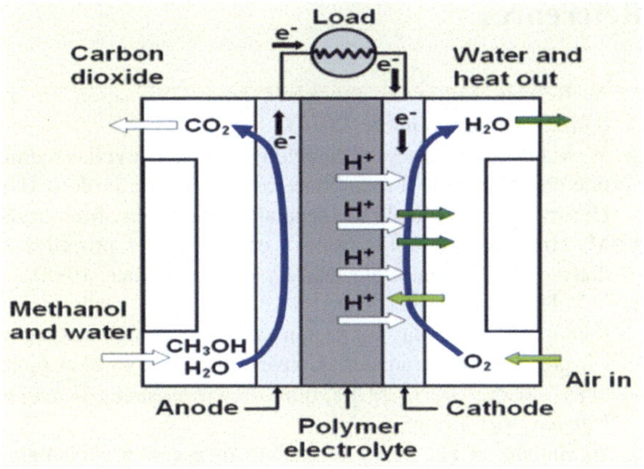

Fig. 32.5 Scheme of direct methanol fuel cell [7]

At the Anode, there is oxidation of methanolate (Eq. 32.1), to yield electrons, protons and carbone dioxide. The electrons travel in the external circuit, while the protons move and diffuse through the Nafion membrane.

$$CH_3OH + H_2O \rightarrow 6\,H^+ + 6\,e^- + CO_2 \qquad (32.1)$$

At the Cathode, there is reduction of protons and oxygen (Eq. 32.2), to yield droplets of water needed to humidify the membrane. However, the operating system of this fuel cell is limited by the transfer of the methanol molecules by electro-osmosis. In other words, instead of being oxidized, the methanol molecules are transferred through the membrane [8], have modified the per sulphonated Nafion membrane with conducting polymer polypyrol, in which its permeability was quite reduced.

$$O_2 + 6\,H^+ + 6\,e^- \rightarrow 6\,H_2O \qquad (32.2)$$

32.5 Conclusion

The I–V characteristics of the Nafion117 were given for various aqueous electrolytic solutions. Three regions were found: ohmic region, plateau current and water dissociation. Besides, owing to the highest limiting current obtained with this membrane in the proton form, it can be therefore used as a solid electrolyte separator in direct methanol fuel cell.

References

1. A. Besson, Etude de polymères pour l'utilisation en membranes de piles à combustible (Université de Grenoble, 2014)
2. W. Mabrouk, Synthèse et caractérisation de nouvelles membranes protoniques: Applications en pile à combustible à membrane échangeuse de protons (Université de Tunis, 2012)
3. H. Strathmann, Membrane separation processes. Membr. Sci. **9**, 121–189 (1981)
4. M. Hichour, Défluorurationdes eaux par des procédés à membranes échangeuses d'ions: dialyse de Donnan et électrodialyse (Montpellier, 1998)
5. N.V. Long et al, Novel Pt and Pd based core-shell catalysts with critical new issues of heat treatment stability and durability for proton exchange membrane fuel cells and direct methanol fuel cells, in Chapter 3, Heat Treatment –Conventional and Novel A pplications, ed by F. Czerwinski (2012). https://doi.org/10.5772/2798. https://www.researchgate.net/publication/232318194_Chapter_3_Nguyen_Viet_Long
6. H. Hamani et al., Study of fouling of a cation exchange membrane with a surfactant using voltamperometry. Sep. Sci. Technol. **46**, 2322–2331 (2011)
7. https://www.researchgate.net/figure/Chemical-reactions-of-direct-methanol-fuel-cell-DMFC_fig1_232318194
8. F. Xu, Chemical modification of perflurosulfonated membrane with pyrrol for fuel cell application: preparation, characterization and methanol transport. Fuel Cells **3**(5), 398–405 (2005)

Materials for Energy

Chapter 33
Boron-Oxygen Metastable Defects Degradation of Carrier Lifetime by Illumination in Cz and Multicrystalline P-Type Silicon Wafers

D. Bouhafs, C. Tahraoui, Y. Kouhlane, N. Khelifati, R. Si-Kaddour, H. Amrouch, and L. Baba Ahmed

Abstract Light induced degradation (LID) of carrier lifetime and electrical performances were investigated in this study on Boron-doped silicon wafer and solar cells. Boron-doped P-type Silicon Czochralski monocrystalline (Cz-Si) and multicrystalline (Mc-Si) wafers are used in this work. Measured Minority carrier lifetime before and after a prolonged illumination show a very fast degradation in the first few minutes and reach a complete degradation after four hours in Si-cm wafers and after 28 h in Cz-Si one. The normalized effective density of the metastable Boron-Oxygen (BO) defects was calculated by means of the difference between the inverse of the measured lifetime before and after illumination and it is proportional to the boron concentration. Analysis of the lifetime minority vs. injection level Δn using Shockley-Read-Hall theory shows that the prolonged illumination generate a deep BO defects Ec-Et = 0.45 eV with a ratio of the capture cross section $k = \sigma_n/\sigma_p = 15$. The obtained results show that the Cz-Si is more sensible to the electrical degradation properties under illumination than the Mc-Si wafers and cells despite the higher boron concentration. The second phenomena related to the BO metastable defects saturation density N^{*}_{tsat} is reached rapidly in the first 4 h compared by the 28 h in the Cz-Si samples. This can be explained by the predominance

D. Bouhafs (✉) · N. Khelifati · R. Si-Kaddour · H. Amrouch
DDCS-Centre de Recherche en Technologie des Semi-conducteurs pour l'Energétique, 2 Bd Frantz Fanon, BP 140 Alger 7-Merveilles, 16038 Algiers, Algeria
e-mail: bouhafsdjoudi@CRTSE.DZ

C. Tahraoui · L. Baba Ahmed
Department of Physics, Saad Dahlab University, Blida, Algeria

Y. Kouhlane
Department of Physics, Bordj Badji Mokhtar University, Annaba, Algeria

A. Belasri and S. A. Beldjilali (eds.), *ICREEC 2019*, Springer Proceedings in Energy,
https://doi.org/10.1007/978-981-15-5444-5_33

of the oxygen content in the creation of the N_t^* concentration than the boron and the high density of the vacancy sites in the Mc-Si which accelerate the oxygen dimers (O_{2i}) diffusivity and their reaction with the substitutional boron atoms.

Keywords Solar cells · Silicon · Lifetime · BO defects · Lifetime degradation

33.1 Introduction

Actually more than 90% of the fabricated solar cells and modules photovoltaic (PV) is based on c-Si both; mono and multicrystalline silicon wafers. Boron-doped P type wafers remain to dominate the PV market with 60% for casted material and 40% for mono [1]. Additional to the standard-BSF solar cells the upcoming of new cell structures such as PERC and bifacial those require the use of Cz-Si grown material to achieve higher efficiencies greater than 20% [2]. Unfortunately, p-type boron-doped silicon known to undergo electrical degradation under the prolonged illumination due to generation of metastable related to the boron-Oxygen (BO) defects and which commonly known as LID phenomena [3, 4]. This effect can severely degrade the effective of minority carrier lifetime in oxygen and boron rich p-type monocrystalline Cz silicon and it is less pronounced in p-type multicrystalline due to the lower oxygen concentration [5]. The most configuration of the BO metastable defects involve a substitutional boron atoms Bs and two oxygen atoms (Dimers) O_{2i} which are activated under illumination [6].

In this work we have investigated the LID phenomena on Mc-Si and Cz-Si wafers and standards-BSF solar cells. Lifetime degradation and electrical solar cell performances relative degradation were measured. Discussions about the predominance on the [b] and [Oi] concentration in the metastable defects generation will be discussed in the light of the obtained results.

33.2 Experiment

Wafers investigated in this work are P-type boron doped 2.7 Ω.cm <100> oriented Cz-Si and 1 Ω.cm Mc-Si. Before Light induced degradation setup, the wafer surface was chemically cleaned with alkaline solution of NaOH:H_2O at 80 °C for 10 min to remove the saw damage layer and piranha-etch for organic contaminants. The final wafer thickness is around 320 ± 20 μm. The minority effective lifetime τ_{eff} vs. excess carrier concentration Δn is measured by the Quasi-Steady-State Photoconductance technique using the WCT-120 from Sinton–Instruments corp. Before each measurement the samples were cleaned in 10% diluted hydrofluoride (HF) acid for 2 min (mn). The QSSPC measurements were implemented in polyethylene bag in which the surfaces of the silicon wafer are passivated with the iodine-ethanol (IE.) solution. For the LID setup, Cz-Si and Mc-Si wafers were exposed to illumination under a halogen lamp with a power of 500 W. With a tuner we can vary the illumination intensity of from 0 to 0.5 suns. In all LID experiment we have used 0.5 suns equivalent to 50 mW/cm^2 and the temperature of the wafers and/ or cells is controlled by means of a fan in order to remain in a range of 40 to 60 °C.

33.3 Results and Discussion

Prior to the illumination degradation experiment, the initial effective life τ_0 at t = 0 of the two Cz-Si and Mc-Si wafers was measured and the results obtained are illustrated in Fig. 33.1 which shows the variation of the τeff as a function of the injection level Δn of the minority charge carriers. The values τ0 were taken at an injection level $\Delta n = 0.1 \times P_0 \approx 0.1 \times Na$, where Na is the doping level of boron. After LID during 28 h, the Cz-Si sample show 84% of effective lifetime degradation from 290 μs to 46 μs as illustrated in Fig. 33.2 and 23% of τ_{eff} degradation the Mc-Si has been observed.

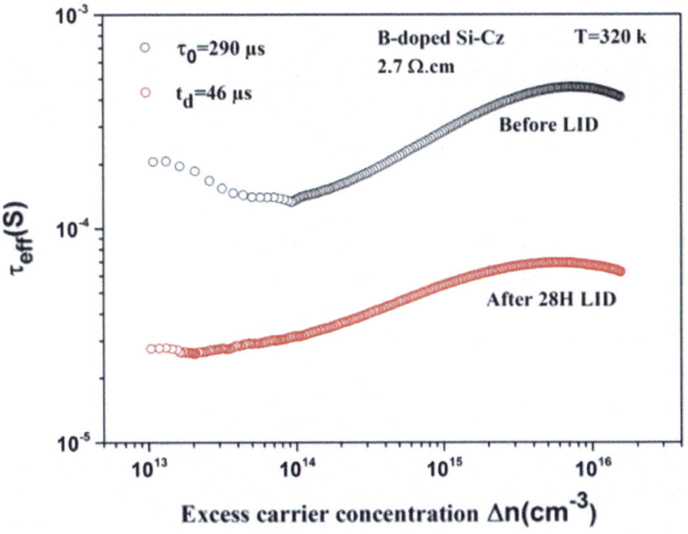

Fig. 33.1 τ_{eff} vs. excess carrier injection level of 1 Ω.cm boron-doped Mc-Si wafer before and after LID during 28 h

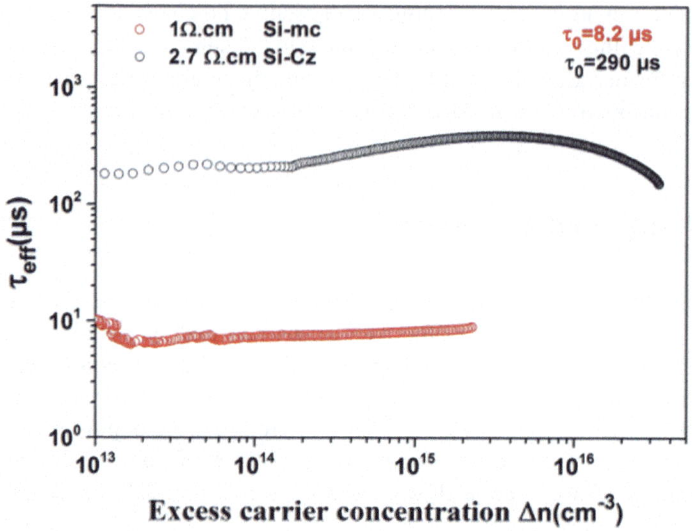

Fig. 33.2 τ_{eff} vs. excess carrier injection level of 2.7 Ω.cm boron-doped Cz-Si wafer before and after LID during 28H

33.3.1 Quantification of Metastable Complexes Related to the BO

Compared to defects related to the transition metal impurities in silicon which we know their capture cross- sections σn and σp allowing to go up to their density in atoms/cm^{-3}, it is practically impossible to evaluate the concentration of the metastable defects related to the BO because the effective electron/hole capture cross-sections [7]. As the lifetime in silicon under illumination is sensitive only to the defects related to the BO, one can define an effective defect concentration N_t^* proportional to the concentration of the metastable defects BO and is function of the inverse of the lifetime [8]:

$$N_t^*(t) = \frac{1}{\tau_t} - \frac{1}{\tau_0} \tag{33.1}$$

$N_t^*(t)$ is the effective defect density generated during the illumination time t. At the complete degradation: $N_t^*(t) = N_{tsat}^* = 1/\tau_d$.

Using the LID degradation setup, the τ_{eff} was measured after each illumination interval which starts with short durations of 20 min to several hours. We have observed that rapid lifetime deterioration occurs in the first minutes and continues slowly after the first hour until almost total degradation τd after 26H is reached for Cz-Si wafers as illustrated in Fig. 33.3. On the other hand, Fig. 33.4 shows that the

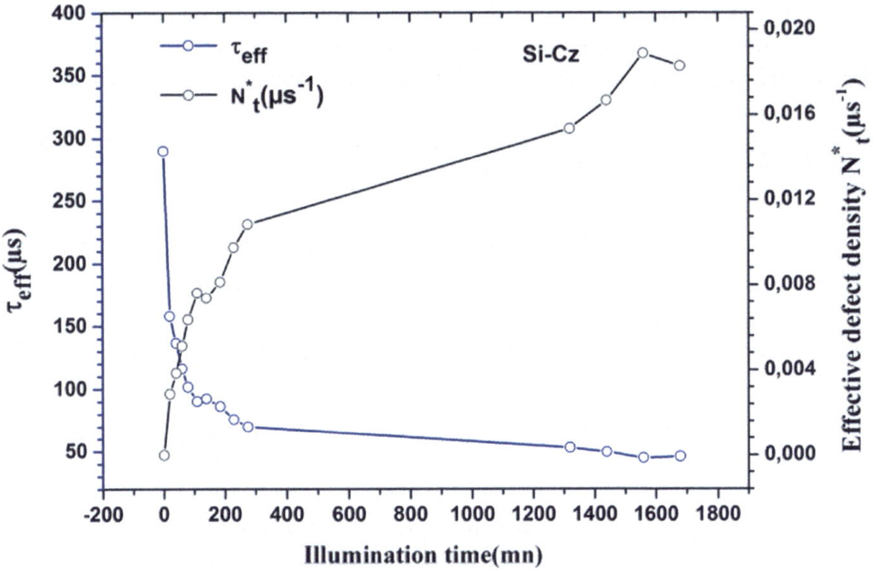

Fig. 33.3 Measured effective carrier lifetime and the effective BO defect density N_t^* as a function of the illumination time for Cz-Si wafer

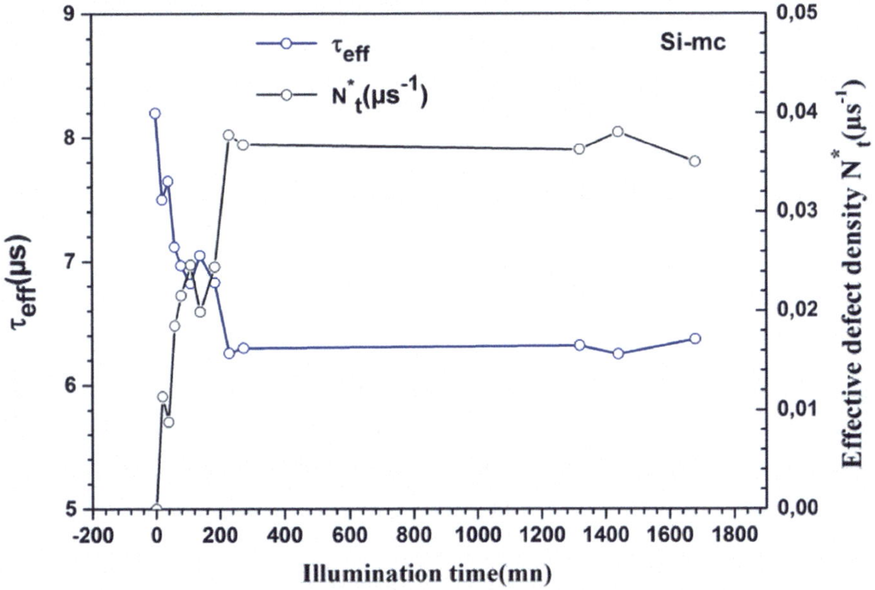

Fig. 33.4 Measured effective carrier lifetime and the effective BO defect density N_t^* as a function of the illumination time for Mc-Si wafer

complete degradation is observed in the first 4 hours for multicrystalline silicon without a significant change during 28H illumination. This can be explained by a higher concentration of vacancy sites (V) in Mc-Si than in Si-Cz due in particular to crystalline defects; favoring rapid diffusivity of the O_{2i} dimmers in the material and the formation of metastable BO related defects B_sO_{2i} bonds with the immobile B_s following the kinetic: $B_s + O_{2i} \rightleftarrows B_sO_{2i}$ [9].

33.3.2 *Shockley-Read Hall Analysis of the Light Induced Due to the BO Defects*

During the LID setup the other defects such as metallic elements and crystalline defects and surface passivation quality are assumed to not affect the effective carrier lifetime. So the τ_{eff} value dominated by the SRH recombination mechanism and the corresponding τ_{SRH} lifetime is the result of the metastable BO defects. The SRH lifetime illustrated on Cz-Si wafers before and after LID degradation in Fig. 33.5 is obtained by the reciprocal of the difference of inverse carrier lifetime at t_0 and t_{sat} corresponding to the complete effective lifetime degradation τ_d:

$$\tau_{SRH} = \left(\frac{1}{\tau_d} - \frac{1}{\tau_0}\right)^{-1} \tag{33.2}$$

With τ_d the effective lifetime after LID degradation and τ_0 the initial value before illumination. In order to modeling the injection dependent of the τ_{SRH} using one defect model we have use the standard SRH equation

$$\tau_{SRH} = \tau_{n0}\left[\frac{(P_0 + P_1 + \Delta n)}{(n_0 + P_0 + \Delta n)} + k\frac{n_0 + n_1 + \Delta n}{(n_0 + P_0 + \Delta n)}\right] \tag{33.3}$$

where $k = \sigma_n/\sigma_p$ is the Capture Cross-Section ratio of electrons and holes and n_0 and p_0 are the electron and holes concentration at the thermal equilibrium. The densities n_1 and p_1 are given by the expressions:

$$n_1 = N_c \exp\left(\frac{E_t - E_c}{k_B T}\right) \tag{33.4}$$

$$p_1 = N_V \exp\left(\frac{E_V - E_t}{k_B T}\right) \tag{33.5}$$

where E_t is the defect energy level in the band-gap and Nc, Nv represent the effectives electronic state densities in the conduction and the valence bands respectively.

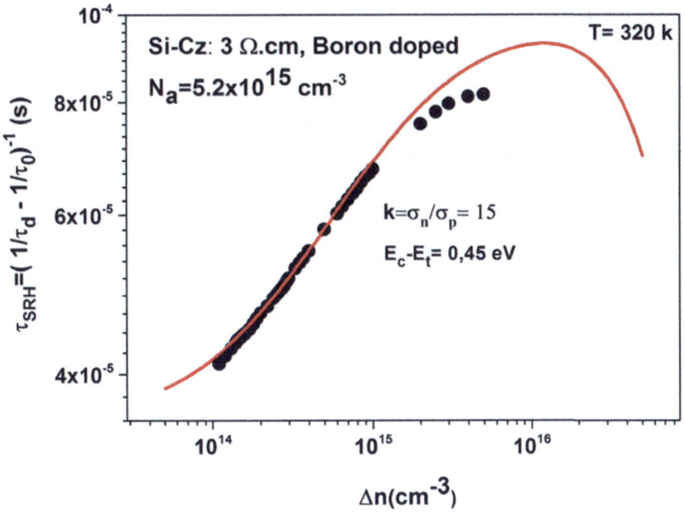

Fig. 33.5 Experimental $\tau_{SRH}(1/\tau d - 1/\tau 0)^{-1}$ as a function if carrier injection level for the Cz-Si sample (filled circles) and the theoretical fit (red curve) using the one defect model SRH lifetime equation

Considering that the surface passivation does not affect the τ_{eff} (Δn) QSSPC measurements, the study of the effective lifetime in the investigated Cz-Si wafer is dominated by the bulk recombination activity. We can observe in Fig. 33.2 that the τ_{eff} increase with the Δn injection level and the shape of the two curves before (dark) and after complete light-Induced Degradation (red) shows that the τ_0 before degradation increases is higher than that measured after LID. This denote than the defect recombination centers are different in the two cases. Also, we can observe the presence of the trap activity at injection level density $\Delta n < 1 \times 10^{14}$ cm^{-3} which is neutralized after LID in the range >2 × 10^{13} cm^{-3}. At high injection level $\Delta n \gg$ Na the τ_{eff} decrease as showed in the Fig. 33.2. Schmidt and Cuevas attributed this rise of shallow defects recombination center activated by the illumination [10] but Rein and Glunz refutes this hypothesis, arguing that under low level injection of about 1016 cm^{-3} the τ_{SRH} can be simulated by the single deep defect model [7]. To highlight the role of the effective BO defects generated during the light prolonged exposure in the Cz-Si samples, the investigation of the quantity $(1/\tau_d - 1/\tau_0)^{-1}$ can give us information about the level defects and should equal to the τ_{SRH} due to the LID degradation. Figure 33.5 illustrate the SRH lifetime (filled square) obtained from the effective lifetime measured QSSPC data using Eq. 2. We observe that the τ_{SRH} increase with increasing carrier concentration and the simulated curves with the one deep model using the Eq. 3 (red curve) was obtained by varying the symmetry factor k to obtain the best fit. The energy level defect (Ec–Et) was also varied between 0.4 and 0.8 eV. The value 0.45 eV with $k = \sigma_n/\sigma_p = 15$ correlate with the experimental τ_{SRH} curve up to 10^{15} cm^{-3}.

33.4 Conclusion

Effective lifetime Light-Induced Degradation of P-type mono and multicrystalline silicon wafer have been investigated. A complete degradation of τ_{eff} is observed in the first 4 hours for Mc-Si samples despite the high boron concentration (1×10^{16} cm^{-3}). On the other hand rapid lifetime deterioration occurs in the first minutes and continues slowly after the first hour until almost total degradation τd after 26H is reached for Cz-Si wafers with a doping level of about 5.2×10^{15} cm^{-3}. This can be explained by higher density of vacancy sites (V) in Mc-Si than in Si-Cz due in particular to crystalline defects concentration which enhancing the rapid diffusivity of the O_{2i} dimmers in the material and the formation of metastable BO related defects B_sO_{2i}. Another observation in the kinetic formation of the BO defects during LID setup suggest that the oxygen concentration dominate the formation of the effective defect concentration N_t^* than the boron in the p-type silicon. This is verified in the Mc-Si samples which contain a high boron concentration and a low oxygen (5×10^{17} cm^{-3}) compared to the Cz-Si samples with a low boron and a high oxygen concentration (8×10^{17} cm^{-3}). To highlight the role of the effective BO defects generated during the light prolonged exposure in the Cz-Si samples, the SRH lifetime $(1/\tau d - 1/\tau 0)^{-1}$ curve have been traced from the measured experimental data of the QSSPC lifetime before and after the complete τ_{eff} degradation dependent on the injection level Δn. We have used the standard SRH equation with one defect model to simulate the experimental τ_{SRH} and the fitting where implemented by varying the $k = \sigma_n/\sigma_p$ and the energy BO defect level Ec-Et. The best fit is obtained with value 0.45 eV and $\sigma_n/\sigma_p = 15$ which correlate with the experimental τ_{SRH} curve up to 10^{15} cm^{-3}.

Acknowledgements This work is realized in the DDCS Laboratory/CRTSE and supported by the National fund of Research-DGRSDT, Higher Education Ministry of Algeria.

References

1. ITRPV 2018, International Technology Roadmap for Photovoltaic, 9th edn., September 2018, http://www.itrpv.net/Reports/Downloads/2018/
2. G. Hahn, S. Wilking, A. Herguth, Solid State Phenom. **242**, 80–89 (2016)
3. N. Nampalli, B. Hallam, C.E. Chan, M.D. Abbott, S. Wenham, 42nd IEEE Photovoltaic Specialist Conference, vol. 5 (2015), pp. 1580–1585
4. A. Herguth, G. Schubert, M. Kaes, G. Hahn, Proceedings of 4th IEEE WCPEC, vol. 1 (2006), pp. 940–943
5. B. Hallam, M.D. Abbott, N. Nampalli, P.G. Hamer, S. Wenham, IEEE J. Photovolt. **69**(1), 92–98 (2016)
6. K. Both, J. Schmidt, Solid State Phenom. **9596**, 223–228 (2004)
7. S. Rein, S.W. Glunz, Appl. Phys. Lett. **82**(7), 1054–1056 (2003)
8. D.C. Walter, B. Lim, K. Both, V.V. Voronkov, R. Flaster, J. Schmidt, Appl. Phys. Lett. **104**, 042111 (2014)
9. B. Lim, K. Both, J. Schmidt, Phys. Stat. Sol. (RRL) **2**(3), 93–95 (2008)
10. J. Schmidt, A. Cuevas, J. Appl. Phys. **86**(6), 3175–3180 (1999)

Chapter 34
Study of the Stoichiometry Effect on the Interaction of Hexagonal HgSe with Electromagnetic Radiation

F. Ghalouci, L. Ghalouci, M. Safer, F. Belkheir, and F. Djali

Abstract In the framework of the Density Functional Theory (DFT) we have exploited the (FP-APW + lo) method to explore the optical properties of HgSe wurtzite unit cell and $Hg_{1-x}Se_x$ (x = 0.37, 0.50) supercells considered in a hexagonal phase and under the stoichiometry constraint. Our results, exploiting the mBJ correction, show that the HgSe wurtzite unit cell and $Hg_{0.50}Se_{0.50}$ hexagonal supercell exhibit same reflectivity behavior, however the latter is slightly less opaque. Moreover, The $Hg_{0.37}Se_{0.63}$ supercell has, optically, a semiconductor character only for electromagnetic radiations with // incidence. Our results show also that beyond 16 eV, the refractive character is insensitive to the direction of incidence for both the HgSe wurtzite unit cell and $Hg_{1-x}Se_x$ (x = 0.37, 0.50) hexagonal supercells. Also, the energy loss by absorption, for our supercells, is more important for ⊥ incident radiations within energies lower than 5 eV. Our photonic study shows that the birefringent character is more pronounced for the $Hg_{0.37}Se_{0.63}$ supercell.

Keywords FP-APW + lo · HgSe · Hexagonal · mBJ

34.1 Introduction

In a context of investigation on mercury chalcogenides, the HgSe compound holds a remarkable place within the scientific community because of its interesting properties in both industrial and medical fields. An overview of the literature shows

F. Ghalouci · L. Ghalouci (✉) · F. Belkheir
Physics Engineering Department, Faculty of Physics,
University of Sciences and Technology of Oran USTO-MB, Oran, Algeria
e-mail: lahouari.ghalouci@univ-usto.dz

M. Safer
University Center of Relizane, Relizane, Algeria

F. Djali
Hassiba Benbouali University, Chlef, Ouled Fares, Algeria

A. Belasri and S. A. Beldjilali (eds.), *ICREEC 2019*, Springer Proceedings in Energy,
https://doi.org/10.1007/978-981-15-5444-5_34

that it is involved in optoelectronic, magnetic and spintronic [1–6] applications, hollow nanofiber synthesis [7] and in the infrared domain development of detectors and lasers [8, 9]. The literature also shows that the physical and optoelectronic properties of HgSe have been invested mainly for a zinc-blende structure [10, 11]. Indeed, the latter is proclaimed stable phase in the ambient conditions by many experimental and theoretical works [10, 12–17]. Under ascending pressure, HgSe exhibits several phase transitions from zinc-blende structure to CsCl, respectively passing through the cinnabar structure (P3221, #154) → Rock-salt → Cmcm [1, 16–18]. Although HgSe in a wurtzite phase has not been observed experimentally, the theoretical investigation of Torres et al. [19] revealed that this structure has a semiconductor character; moreover, it is predicted as a potential 3D topological insulator. Radescu et al. [17] exhibit in their work the near similarity between zinc-blende and wurtzite structures in terms of ground state energy and volume. Biering [18], from a non-relativistic point of view, stated that the cohesion energy in a wurtzite structure is higher than in the zinc-blende one, which makes the former better adapted to the HgSe. The cohesion energy small difference between zinc-blende and wurtzite structures of HgSe was considered also by Yeh et al. [20]. Despite what has been predicted by aforementioned works, as interesting features for the wurtzite HgSe, to our knowledge, the literature suffers, as well, from a lack of information regarding selenium mercury in this hexagonal phase, related perhaps to the fact that it hasn't be seen experimentally. However, it is worth mentioning that in the real world a crystal is not ideal; the periodic arrangement can be disrupted by localized defects that can arise during crystal growth leading to the emergence of new phases. In addition, the control of crystal growth is increasingly tuned by technological progress [21]. That being said, and in the hope of revitalizing further the debate on the wurtzite structure of this compound, the purpose of this work is to shed more light on the optoelectronic properties of the HgSe compound with such configuration, also to study the influence of stoichiometry on these properties. Hence, this introduction will be followed by the presentation of our calculation methodology, we will present and discuss our results and we will conclude with a summary of the main part of this work.

34.2 Computational Details

In the Density Functional Theory framework [22], the full potential augmented plane waves plus local orbitals (FP-APW + lo) method was used [23] as implemented in the WIEN2K code [24]. The generalized gradient approximation (PBE-GGA) [25, 26] was chosen to describe the exchange–correlation functional energy. To separate valence and core states, a cutoff energy of −7.0 Ry was considered. Thus, 3d, 4s and 4p orbitals of Se atoms and 5p, 5d and 6s orbitals of Hg atoms are treated as valence states. Inside muffin-tin spheres, electron wave functions are expanded in terms of spherical harmonics up to lmax = 10, whereas, in the interstitial region, a plane waves (PWs) basis set, with a wave vector cutoff

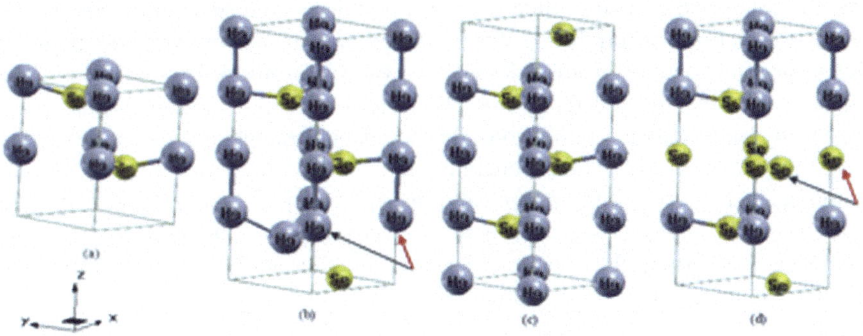

Fig. 34.1 Crystalline configuration of the HgSe compound considered in: **a** wurtzite unit cell structure and **b** $Hg_{0.63}Se_{0.37}$, **c** $Hg_{0.50}Se_{0.50}$, and **d** $Hg_{0.37}Se_{0.63}$ hexagonal supercell structures. Big grey and small yellow balls represent Hg and Se atoms respectively. Arrows indicate substituted atoms (black arrow) and their 1st nearest neighbor (red arrow)

Kmax = 8.5/Rmt, was chosen to describe electron wave functions. Here, Rmt states for the less muffin-tin radii. In our calculations we adopted Rmt(Se) = 2.3 a.u and Rmt(Hg) = 2.35 a.u. The convergence stability, in our calculations, was checked in terms of the plane wave basis set size and the k-points sampling within the irreducible part of the Brillouin Zone (BZ). The k-integration over the BZ is performed using the Monkhorst-Pack mesh [27]. Hence, 1000 and 4000 k-points respectively are chosen in the structural optimization study and optoelectronic properties investigation. In this later, the modified Becke-Johnson (mBJ) correction [28] within GGA was applied due to its accuracy for semiconductor and insulator compounds. The minimization of total ground state energies versus c/a ratios and volumes of our hexagonal structures was carried out by choosing simultaneously the total energy variation ($\leq 10^{-4}$ Ry) and the charge distance ($\leq 10^{-4}$ e^{-}) as convergence criteria during Self-Consistent Field (SCF) iterations.

The internal atomic positions in our structures were fully relaxed through minimization of the quantum-mechanical forces on each atom to be below 1.0 mRy/bohr. We have at first formed and optimized the wurtzite unit cell from which three supercells (see Fig. 34.1) were generated and relaxed.

34.3 Results and Discussion

34.3.1 Optical Properties

We have conducted a study to determine the effect of stoichiometry on the photonic behavior of our hexagonal structures: HgSe wurtzite unit cell and $Hg_{1-x}Se_x$ (x = 0.50 and 0.63) supercells. The structure $Hg_{0.63}Se_{0.37}$ was omitted here because of its semi-metallic character, defined during our electronic investigation (not included here). The number of k-points chosen is 4000. Indeed, this densification aims to reduce the parasitic effect in the different calculated entities curves (the dielectric function, absorption, reflectivity, etc.). Our electromagnetic radiation covers the energy range from ~13 meV to 60 eV (arbitrary choice). This range corresponds to the spectral range from IR to near UV. Our interest in this area is justified by photovoltaic applications. It should be remembered that a hexagonal (uniaxial) crystalline structure generally admits two directions of the incident photon beam, hence the study of optical properties requires an investigation of the // (index zz) and $\perp$ (index xx or yy) incident electromagnetic radiation to the z axis. Our results, concerning the complex dielectric function, reflectivity, absorption and refractive index are illustrated in Figs. 34.2, 34.3, 34.4 and 34.5.

The Real and Imaginary Dielectric Function. Figure 34.2(a1 and b1) shows respectively the real part of the dielectric function of HgSe wurtzite unit cell and $Hg_{0.50}Se_{0.50}$ hexagonal supercell. We can see that both structures do not exhibit a great difference in their real part dielectric function shape, for both // and $\perp$ incident photon direction, except at the energy point ~4 eV where the unit cell present relatively consistent peak intensities. The static permittivity $\varepsilon_1(h\nu = 0)$, denoted ε_0, is the same ($\varepsilon_0 \approx 5$) for both considerations (HgSe wurtzite unit cell and $Hg_{0.50}Se_{0.50}$ supercell). The metallic character of both structures disappears above 12 eV. The supercell $Hg_{0.37}Se_{0.63}$ (Fig. 34.2(c1)) exhibits a different shape of $\varepsilon_1(h\nu)$ compared to the balanced structure $Hg_{0.50}Se_{0.50}$, in addition we can notice an enhancement in the static dielectric function value ($\varepsilon_{0xx} \cong 7.5$, $\varepsilon_{0zz} \approx 52$). This discrepancy disappears beyond 12 eV. The $Hg_{0.37}Se_{0.63}$ exhibits a slight metallic character for $\perp$ incident electromagnetic radiation. The imaginary part representation of the dielectric function for our supercells (see Fig. 34.2(b2 and c2)) shows that the energy loss by absorption is clearly important for $\perp$ incident radiation with energy lower than 5 eV. However, HgSe unit cell (Fig. 34.2(a2)) exhibits larger energy loss peaks compared to our supercells. It is worth noting the disappearance of the loss energy peak around 4 eV for the $Hg_{0.37}Se_{0.63}$ structure for // incident electromagnetic radiation.

The Reflectivity. Figure 34.3(a) shows that the HgSe wurtzite unit cell exhibits a variation in reflectivity up to 17 eV. A maximum of about 54% is felt for $\perp$ incident electromagnetic radiation (at 3.4 eV), followed by another about 50% at 10 eV. This reflectivity behavior is preserved in the $Hg_{0.50}Se_{0.50}$ supercell, as can be seen in Fig. 34.3(b), with the difference that it is relatively slightly less opaque (the first maximum of about 46% (at 3.6 eV) and the second of about 40% (at 9.9 eV), both for $\perp$ incident photon). Figures 34.3(a and b) respectively also show the anisotropic behavior of the reflectivity between // and $\perp$ incident electromagnetic radiation in the HgSe wurtzite unit cell and the $Hg_{0.50}Se_{0.50}$ hexagonal supercell. This anisotropic behavior goes off above ~20 eV. The structure $Hg_{0.37}Se_{0.63}$ (Fig. 34.3(c)) has rather a fair regular reflectivity decay for $\perp$ incident photon ray within energy more than 10 eV, and a fair constant reflectivity for // electromagnetic radiation with energy between 6 eV and 12 eV. Beyond 16 eV the anisotropic character disappears and the material exhibits almost the same reflectivity behavior for both directions of incident photon.

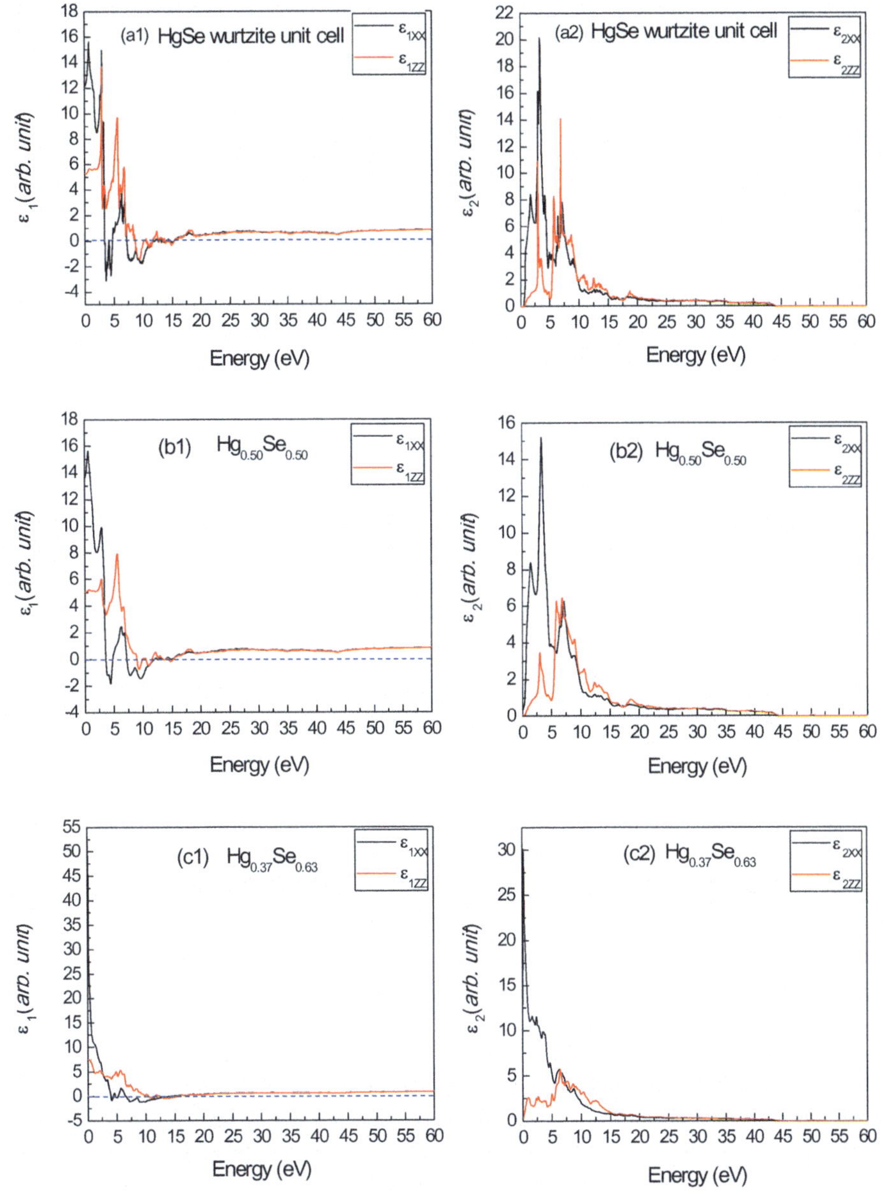

Fig. 34.2 Evolution of the real (ε_1) and imaginary (ε_2) part of the dielectric function according to the incident photons energies for the HgSe compound in: (a1, a2) wurtzite unit cell structure and (b1, b2) $Hg_{0.50}Se_{0.50}$, (c1, c2) $Hg_{0.37}Se_{0.63}$ hexagonal supercells. Subscripts *xx* and *zz* stand respectively for ⊥ and // incident photon direction

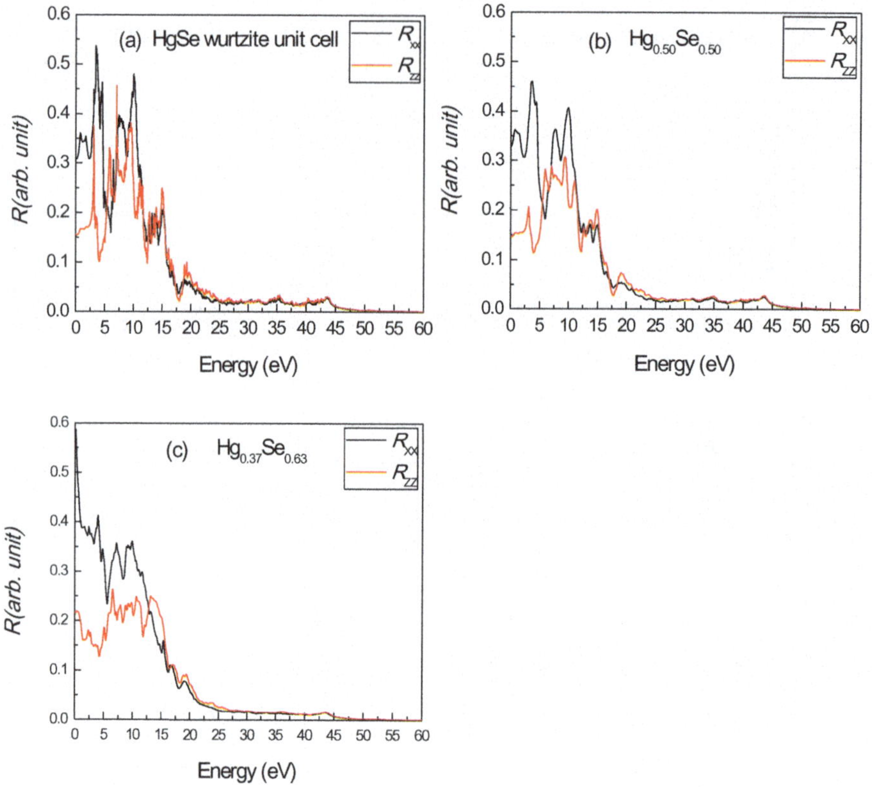

Fig. 34.3 Evolution of the reflectivity (R) according to the incident photons energies for the HgSe compound in: **a** wurtzite unit cell structure and **b** $Hg_{0.50}Se_{0.50}$, **c** $Hg_{0.37}Se_{0.63}$ hexagonal supercell. Subscripts xx and zz stand respectively for ⊥ and // incident photon direction

The Absorption. Our calculations show that the absorption phenomenon for HgSe (unit cell) manifests at 0.44 eV and 0.47 eV respectively for electromagnetic radiations with ⊥ and // incidence (see Fig. 34.4(a)). The absorption curve, with maxima between 6 eV and 12 eV, goes off at 44.24 eV. However, there is another window of low intensity around the energy range [49 eV, 51 eV]. The $Hg_{0.50}Se_{0.50}$ supercell structure (Fig. 34.4(b)) exhibits an absorption rate similar to that of the wurtzite unit cell structure, but with maxima of low intensities. Also, the absorption phenomenon manifests, albeit rarely, around 13.6 meV. It should be noted that our three structures have better absorption for electromagnetic radiation with energy less than 5 eV (see Fig. 34.4(a–c)). The $Hg_{0.37}Se_{0.63}$ supercell structure (Fig. 34.4 (c)) has, optically, a semiconductor character only for electromagnetic radiations with // incidence. The deficit in atoms of Hg causes a decrease of the absorption between 25 eV and 45 eV.

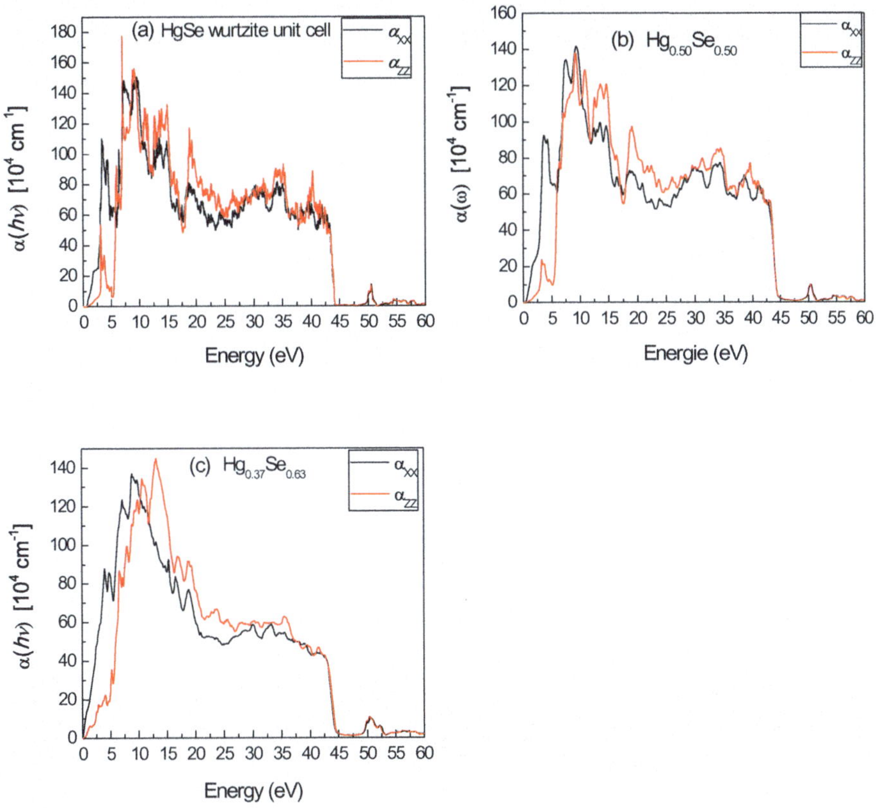

Fig. 34.4 Evolution of the absorption coefficient (α) according to the incident photon energies for the HgSe compound in: **a** wurtzite unit cell structure and **b** $Hg_{0.50}Se_{0.50}$, **c** $Hg_{0.37}Se_{0.63}$ hexagonal supercell. Subscripts *xx* and *zz* stand respectively for ⊥ and // incident photon direction

The Refractive Index. The analysis of Fig. 34.5(a and b) shows that, for photons with ⊥ incidence, the static refractive index $n_{xx}(h\nu = 0)$ goes from 3.48 (for the HgSe wurtzite unit cell) to 3.69 (for the $Hg_{0.50}Se_{0.50}$ supercell), whereas, for radiations with // incidence, we can see a slight decrease in the static refractive index $n_{zz}(h\nu = 0)$ which goes from 2.30 (for the HgSe wurtzite unit cell) to 3.24 (for the $Hg_{0.50}Se_{0.50}$ supercell). Moreover, maxima in the refraction index for the unit cell are around 0.8 eV ($n_{xx} = 3.96$, $n_{zz} = 2.39$), whereas, for the $Hg_{0.50}Se_{0.50}$ supercell, maxima for ⊥ and // photons are clearly at different energy levels ($n_{xx}(0.72\text{ eV}) = 3.75$, $n_{zz}(5.78) = 2.91$). The $Hg_{0.37}Se_{0.63}$ structure (Fig. 34.5(c)) exhibits a remarkable difference between its two static refractive indexes ($n_{xx}(h\nu = 0) = 2.72$ and $n_{zz}(h\nu = 0) = 7.27$). We can also notice the quasi disappearance of the maxima already present in the stoichiometric balanced structure $Hg_{0.50}Se_{0.50}$. Beyond 16 eV, the refractive character is insensitive to the direction of incidence for all structures considered here.

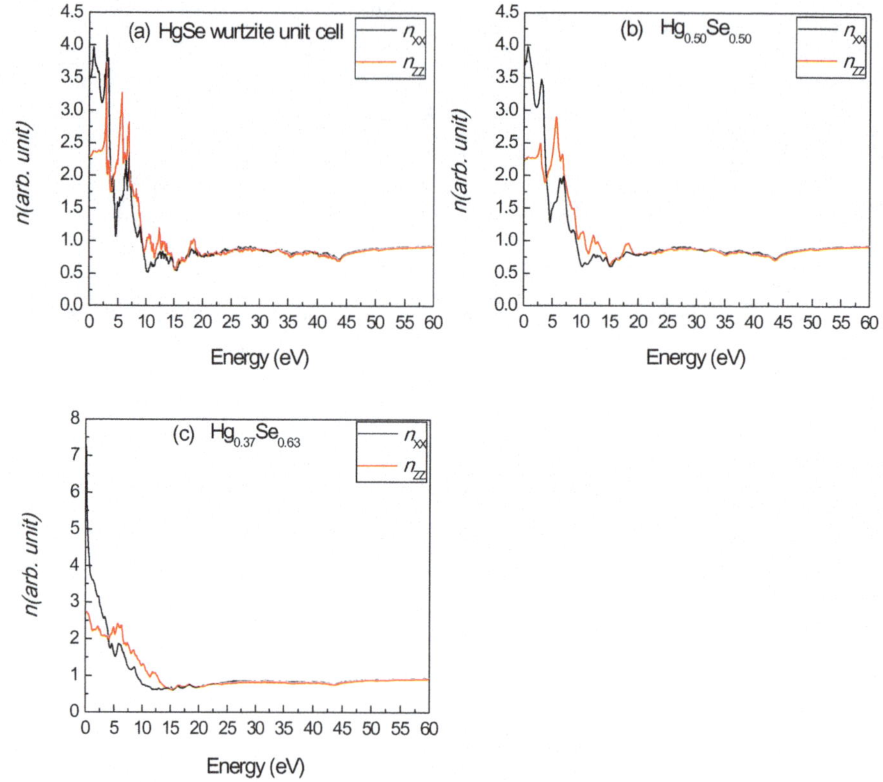

Fig. 34.5 Evolution of the refractive index (n) according to the incident photon energies for the HgSe compound in: **a** wurtzite unit cell structure and **b** $Hg_{0.50}Se_{0.50}$, **c** $Hg_{0.37}Se_{0.63}$ hexagonal supercell. Subscripts *xx* and *zz* stand respectively for ⊥ and // incident photon direction

34.4 Conclusion

Our study of the stoichiometry effect on optical properties of HgSe wurtzite unit cell and $Hg_{1-x}Se_x$ (x = 0.37, 0.50) hexagonal supercells, led to the following results:

The energy loss by absorption, for our supercells, is more important for ⊥ incident radiations within energies lower than 5 eV. The HgSe wurtzite unit cell and $Hg_{0.50}Se_{0.50}$ hexagonal supercell exhibit same reflectivity behavior, however the latter is slightly less opaque. The $Hg_{0.37}Se_{0.63}$ supercell has, optically, a semiconductor character only for electromagnetic radiations with // incidence. Beyond 16 eV, the refractive character is insensitive to the direction of incidence for all structures considered here. Our photonic study shows that the birefringent character is more pronounced for the $Hg_{0.37}Se_{0.63}$.

References

1. I. Düz, I. Erdem, K.S. Ozdemir, V. Kuzucu, DFT study on phase transition behavior and mechanical properties of HgSe polymorphs under high pressure. Curr. Appl. Phys. **18**(4), 424–436 (2018). https://doi.org/10.1016/j.cap.2018.01.013
2. A.T. Lonchakov, S.B. Bobin, V.V. Deryushkin, V.I. Okulov, T.E. Govorkova, V.N. Neverov, Peculiar behavior of magnetoresistance in HgSe single crystal with low electron concentration. Appl. Phys. Lett. **112**(8) (2017). https://doi.org/10.1063/1.5018357
3. J. Choi, J. Yun, K. Cho, S. Kim, Vertical NOR-logic circuits constructed using nanoparticle films on plastic substrates. Jpn. J. Appl. Phys. **53**, 08NE02 (2014). https://doi.org/10.7567/JJAP.53.08NE02
4. T.E. Govorkova, I.V. Zhevstovskikh, A.T. Lonchakov, V.I. Okulova, K.A. Okulova, S.M. Podgornykh, S.B. Bobin, V.V. Deryushkin, L.D. Paranchich, New data and developments pertaining to ideas about the electron system of hybridized states of cobalt impurity atoms in a mercury selenide crystal. Low Temp. Phys. **43**(4), 508 (2017). https://doi.org/10.1063/1.4983997
5. M. Sozanskyi, V. Stadnik, R. Chaykivska, R. Guminilovych, P. Shapoval, I. Yatchyshyn, Synthesis and properties of mercury selenide films deposited by using potassium iodide as complexing agent. Chem. Chem. Technol. **11**(4), 445–448 (2017). https://doi.org/10.23939/chcht11.04.445
6. J. Li, C. He, L. Meng, H. Xiao, C. Tang, X. Wei, J. Kim, N. Kioussis, G.M. Stocks, J. Zhong, Two-dimensional topological insulators with tunable band gaps: single-layer HgTe and HgSe. Sci. Rep. **5**, 14115 (2015). https://doi.org/10.1038/srep14115
7. N.G. Ndifor-Angwafor, M. Nchare, T.T. Donald Raoul, N. Sadeu Christian, A. Gabche, V. Myung, Synthesis of ultra-long hollow mercury selenide (HgSe) chalcogenide nanofibers from Co and Ni sacrificial nanofibers. Chem. Sci. Int. J. **20**(3), 1–8 (2017). https://doi.org/10.9734/CSJI/2017/36678
8. E. Izquierdo, Synthèse et caractérisation d'homostructures et d'hétérostructures de nanoplaquettes de chalcogénures de mercure. Ph.D. thesis in Materials Chemistry, Paris-Sciences-et-Lettres (PSL) University, France, October (2018)
9. B. Martinez, Étude Des Propriétés Optoélectroniques De Nano cristaux colloïdaux À Faible Bande Interdite: Application À La Détection In-Frarouge. Science des matériaux [cond-mat. mtrl-sci] (Sorbonne Université, Français, 2019), https://hal.archives-ouvertes.fr/tel-02270058/document
10. I. Düz, I. Erdem, K.S. Ozdemir, V. Kuzucu, First principles investigations of HgX (X = S, Se and Te). Arch. Mater. Sci. Eng. **79**(1), 5–11 (2016). https://doi.org/10.5604/18972764.1227656
11. J.O. Akinlami, O.O. Odeyemi, Electronic structure and optical properties of HgSe. Semicond. Phys. Quantum Electron. Optoelectron. **21**(3), 288–293 (2018). https://doi.org/10.15407/spqeo21.03.288
12. K. Persson, Materials Data on HgSe (SG:216) by Materials Project (2016). https://doi.org/10.17188/1282139
13. P. Capper, Bulk crystal growth: methods and materials, in *Springer Handbook of Electronic and Photonic Materials*, ed. by S. Kasap, P. Capper (Springer, Cham, 2017), pp. 281–285. https://doi.org/10.1007/978-3-319-48933-9_12
14. N.Z. Boctor, G. Kullerud, Mercury selenide stoichiometry and phase relations in the mercury-selenium system. J. Solid State Chem. **62**(2), 177–183 (1986). https://doi.org/10.1016/0022-4596(86)90229-X
15. W. Paul, T. Suski, *High Pressure Semiconductor Physics I*, vol. 54, 1st edn. (Academic Press, Cambridge, 1998), pp. 228–229. ISBN-13 978-0-12-752162-6, ISBN 0-12-752162-3
16. E.Yu. Tonkov, *High Pressure Phase Transformations: A Handbook*, vol. 2 (Gordon and Breach Science Publishers, Philadelphia, 1992), pp. 481–482. ISBN 288124758X, 9782881247583. ISBN 2881247598, 9782881247590

17. S. Radescu, A. Mujica, J. López-Solano, R.J. Needs, Theoretical study of pressure-driven phase transitions in HgSe and HgTe. Phys. Rev. B **83**(9), 094107 (2011). https://doi.org/10.1103/PhysRevB.83.094107
18. S. Biering, The unusual structure of the mercury chalcogenides: relativistic effects in the solid state. Ph.D. thesis, Massey University, Albany, New Zealand, 10 September 2010
19. D.D. Torres, P. Banerjee, S. Pamidighantam, P.K. Jain, A non-natural wurtzite polymorph of HgSe: a potential 3D topological insulator. Chem. Mater. **29**(15), 6356–6366 (2017). https://doi.org/10.1021/acs.chemmater.7b01674
20. C.-Y. Yeh, Z.W. Lu, S. Froyen, A. Zunger, Zinc-blende–wurtzite polytypism in semiconductors. Phys. Rev. B **46**, 10086 (1992). https://doi.org/10.1103/PhysRevB.46.10086
21. B.M. Hudak, J. Song, H. Sims, M.C. Troparevsky, T.S. Humble, S.T. Pantelides, P.C. Snijders, A.R. Lupini, Directed atom-by-atom assembly of dopants in silicon. ACS Nano **12** (6), 5873–5879 (2018). https://doi.org/10.1021/acsnano.8b02001
22. P. Hohenberg, W. Kohn, Inhomogeneous electron gas. Phys. Rev. **136**(3B), B864–B871 (1964). https://doi.org/10.1103/PhysRev.136.B864
23. S. Cottenier, Density Functional Theory and the Family of (L)APW-Methods: A Step-by-Step Introduction, 2nd edn. (2002–2013), p. 25. http://www.wien2k.at/reguser/textbooks. ISBN 978-90-807215-1-7
24. P. Blaha, K. Schwarz, G.K.H. Madsen, D. Kvasnicka, J. Luitz, WIEN2k, an augmented plane wave plus local orbitals program for calculating crystal properties (Vienna University of Technology, Vienna, 2001). ISBN 3-9501031-1
25. J.P. Perdew, K. Burke, M. Ernzerhof, Generalized gradient approximation made simple. Phys. Rev. Lett. **77**(18), 3865–3868 (1996). https://doi.org/10.1103/PhysRevLett.77.3865
26. J.P. Perdew, K. Burke, M. Ernzerhof: Generalized gradient approximation made simple [Phys. Rev. Lett. **77**, 3865 (1996)]. Phys. Rev. Lett. **78**(7), 1396 (1997). https://doi.org/10.1103/PhysRevLett.78.1396
27. H.J. Monkhorst, J.D. Pack, Special points for Brillouin-zone integrations. Phys. Rev. B **13** (12), 5188–5192 (1976). https://doi.org/10.1103/PhysRevB.13.5188
28. F. Tran, P. Blaha: Accurate band gaps of semiconductors and insulators with a semilocal exchange-correlation potential. Phys. Rev. Lett. **102**(22), 226401–1:4 (2009). https://doi.org/10.1103/PhysRevLett.102.226401

Chapter 35
Analysis of $Cu_2ZnSn(S,Se)_4$ (CZTSSe) Thin Film Solar Cells Performances with CZTSe Stacked Layer by Using SCAPS-1D

Abd Elhalim Benzetta, Mahfoud Abderrezek, and Mohammed Elamine Djeghlal

Abstract In this work, a simulation of the CZTSSe solar cell with Al/ZnO: Al/ZnO (i)/CdS/CZTSSe/Mo structure have been studied using the SCAPS-1D (Solar Cell Capacitance Simulator in one Dimension). The simulation results have been validated with real experimental results. Next, a novel structure is proposed in which a CZTSe layer is stacked in the back side of the solar cell in order to boost its performances. The efficiency of CZTSSe solar cell increases from 12.3% to 15.3% by inserting the CZTSe layer. Finally, we carried out an optimization of different physical parameters (Thickness and doping concentration) of CZTSe stacked layer to determine the optimum values. The proposed structure of CZTSSe solar cells with optimum parameters showed higher functional properties. The maximum value of efficiency achieved was 16.77% with J_{SC} = 36.81mA/cm^2, V_{OC} = 0.657V and FF = 69.19% under AM1.5G illumination at 10 nm thickness and 5E19 cm^{-3} doping concentration.

Keywords BSF · CZTSSe · Solar cell · SCAPS-1D · Thin film

A. E. Benzetta (✉)
Laboratoire Génie des Matériaux, Ecole Militaire Polytechnique,
BP17, 16046 Bordj El-Bahri, Algiers, Algeria
e-mail: halimbenz40@yahoo.fr

M. Abderrezek
Unité de Développement des Equipements Solaires, UDES/Centre de
Développement des Energies Renouvelables, CDER, 42415 Tipaza, Algeria

M. E. Djeghlal
Laboratoire de Sciences et Génie des Matériaux, Ecole Nationalle Polytechnique,
B.P. 182 10, Avenue Hassen Badi, El-Harrach, Algiers, Algeria

A. Belasri and S. A. Beldjilali (eds.), *ICREEC 2019*, Springer Proceedings in Energy,
https://doi.org/10.1007/978-981-15-5444-5_35

35.1 Introduction

Due to the promising performances joined to earth-copious components of thin-film solar cells based on $Cu_2ZnSn(S,Se)_4$ (CZTSSe) [1, 2], they turned into an interesting alternative to other thin-film photovoltaic technologies [3]. They have a useful band gap energies range between 1.0–1.5 eV and high absorption coefficients surpassing 10^4 cm^{-1}. Despite that, the maximum achieved efficiency of CZTSSe solar cells (η of 12.6%) still low in comparison with CdTe (η of 22.1%) and CIGS (η of 22.0%) [4] and it needs to be improved. To date, the real weakness point of CZTSSe solar cells is the open-circuit voltage (V_{OC}) and the reasons for this issue has been a subject of intense discussion [5]. One of the promising methods for upgrading the efficiency of CZTSSe solar cell is by utilizing a heavily doped layer at the back side of the solar cell between the absorber layer and the back contact, it helps to make the back/bottom side acts as a back driving force to push the minority carriers to the space charge region (SCR) [6]. Additionally, it limits recombination process at the back contact of the solar cell, which permit to improve its electrical performances, particularly the Voc. The materials used for the elaboration of this layer must have a lattice parameters near the lattice parameters of the absorber layer and high doping of the order of 1E18 cm^{-3} [7]. In this research paper, we aim to boost the CZTSSe solar cells efficiency, the proposed method consists of adding a stacked CZTSe layer between the absorber CZTSSe layer and the Mo back contact. The main objective is to improve the electrical parameters of CZTSSe solar cell, V_{OC} and J_{SC}, this improvement contributes in the efficiency enhancement of the solar cell. At the beginning, the one-dimensional simulator SCAPS-1D is utilized to numerically investigate the performances of the CZTSSe solar cells configuration without CZTSe stacked layer and compared with previously reported experimental data [8]. Next, we analyze the behavior of the new proposed configuration of CZTSSe solar cell with CZTSe stacked layer. Finally, an optimization of the proposed CZTSe stacked layer physical properties (Thickness and doping concentration) is carried out to determine the optimum parameters.

35.2 Methodology

In this study, we consider the structure of CZTSSe thin film for simulation which were made by Yang et al [8], this cell was prepared with several ratios of Se/SeS_2, an electrical efficiency of 12.3% was achieved with an energy gap of 1.097 eV corresponding to 38% of Se/SeS_2 ratio [8].

The software SCAPS-1D is our tool for the simulation of CZTSSe solar cell structures, SCAPS-1D requires numerous physical and optical parameters of semiconductor material [9] which includes; band gap (Eg), electron affinity (χ), dielectric permittivity (ε), conduction band density of states (NC), valence band density of states (NV), electron thermal velocity (Vthn), hole thermal velocity

Table 35.1 Parameter set for the simulation of the CZTSSe solar cell

	ZnO:Al	ZnO-i	CdS	CZTSSe	CZTSe
Thickness (nm)	300	50	50	1800	50
Electron affinity (eV)	4,4	4,4	4,2	4,1	4.35
Band gap (eV)	3,3	3,3	2,4	1,096	0.95
Dielectric permittivity (relative)	9	9	10	13,6	13.6
Effective conduction band density (cm^{-3})	2,2E18	2,2E18	2,2E18	2,2E18	2,2E18
Effective valence band density (cm^{-3})	1,8E19	1,8E19	1,8E19	1,8E19	1,8E19
Electron mobility (cm^2/Vs)	1E2	1E2	1E2	1E2	1E2
Hole mobility (cm^2/Vs)	2,5E1	2,5E1	2,5E1	2,5E1	2,5E1
Electron thermal velocity (cm/s)	1E7	1E7	1E7	1E7	1E7
Hole thermal velocity (cm/s)	1E7	1E7	1E7	1E7	1E7
Doping concentration of acceptors (cm^{-3})	0	1E18	0	1E15	1E18
Doping concentration of donors (cm^{-3})	1E18	1E18	1E17	0	0

(Vthp), electron mobility (μn), hole mobility (μp), donor density (NA), acceptor density (ND) and Absorption coefficient (α). The physical parameters used in this work for different material layers were all cited literatures [10, 11], which are summarized in Table 35.1. The simulation of CZTS solar cell is done under a global solar irradiance AM 1.5G (1000 W/m^2) and an operating temperature of 300°K (Fig. 35.1).

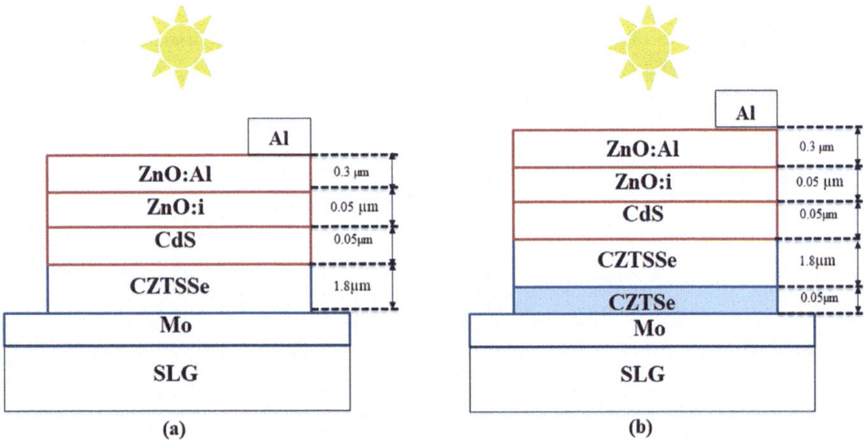

Fig. 35.1 Schematic diagram of the studied CZTSSe structures: **(a)** experimental structure [8] **(b)** the proposed structure

35.3 Results and Discussion

In this work, the effect of CZTSe as BSF layers has been studied, by using the parameters of different solar cell layers presented in Table 35.1, the CZTSSe solar cell without CZTSe stacked layer is simulated, an efficiency of 12,4% is achieved (VOC = 0.532 V, JSC = 33.75 mA/cm^2, FF = 68.90%). These results are close in values to the experimental results in reference [8]. The comparison between experimental and simulation results of J-V curves and quantum efficiency of CZTSSe solar cell are shown in Fig. 35.2. A good agreement is noticed between simulation and experimental results curves. Table 35.2 summarized the obtained electrical parameters of simulation results in comparison with experimental data of reference [8].

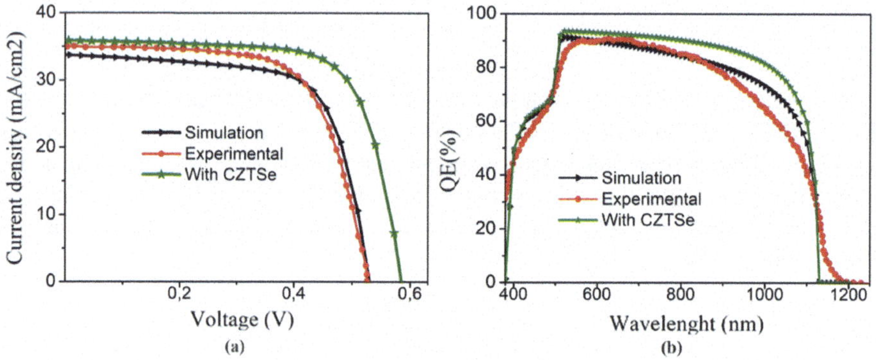

Fig. 35.2 Performances of CZTSSe solar cells (Experimental (red) and simulation (black)) and with CZTSe stacked layer (green): **(a)** J-V characteristic, **(b)** QE curves

Table 35.2 Functional parameters of the CZTSSe solar cells with and without BSF layer

Cell		V_{OC} (Volt)	J_{SC} (mA/cm^2)	FF (%)	η (%)
With BSF	Simulation	0.59	36.30	71.43	15.23
Without BSF	Experimental	0.520	34.64	67.20	12.10
	Simulation	0.532	33.75	68.90	12.40

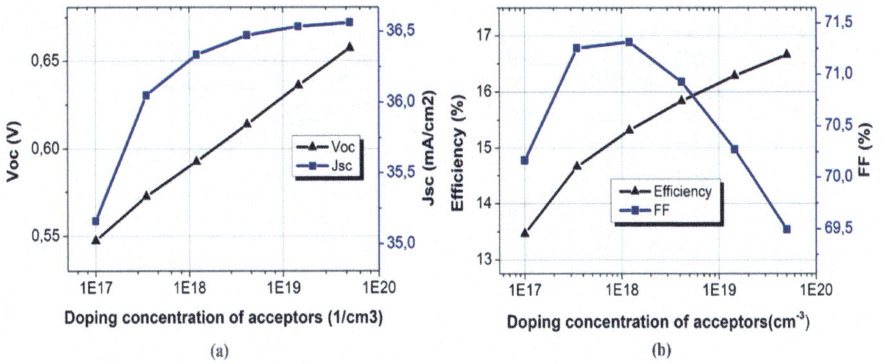

Fig. 35.3 Variation of CZTSSe solar cell functional parameters (V_{oc}, J_{sc}, η and FF) as function of BSF layer doping concentration

35.4 Effect of CZTSe BSF Layer Doping Concentration

In this part, the numerical analysis has been carried out to investigate the influence of CZTSe layer doping concentration on the functional properties of CZTS solar cell. The doping concentration is varied from 1E17 to 5E19 cm^{-3} in CZTSe stacked layer of 50 nm thickness as exhibited in Fig. 35.3. It is observed that these parameters are strongly dependent on BSF layer doping concentration. An increased trend of J_{sc}, V_{oc}, η and FF of the solar cell is noticed as CZTSe doping concentration increases. Figure 35.3a shows that J_{sc} and V_{oc} rise gradually with the increase of doping concentration. From Fig. 35.3b we noted that η and FF increases gradually whereas FF decreases after 1E18 cm^{-3} with maximum efficiency achieved of 16.66% at 5E19 cm^{-3} doping concentration.

35.5 Effect of CZTSe BSF Layer Thickness

The effect of CZTSe layer thickness on the electrical performances of CZTSSe solar cell is studied in this part. The CZTSe stacked layer thickness is varied from 10 to 100 nm with a fixed doping concentration at 5E19 cm^{-3}. Figure 35.4 shows the variation of Jsc, Voc, η and FF in respect to the CZTSe thickness. We notice that these parameters are fairly dependent on CZTSe layer thickness in comparison with doping concentration. Figure 35.4 exhibits a decreased trend of Jsc, η and FF with the increase in CZTSe thickness, the maximum obtained efficiency is 16.77% at 10 nm thickness (V_{OC} = 0.657 V, JSC = 36.81 mA/cm^2, FF = 69.19%). Whereas for Voc, it slightly increase until the maximum value of 0.658 V at 100 nm thickness. The Fill Factor is slightly affected to CZTSe thickness variation, it varies in the range of 69.42% and 69.56%.

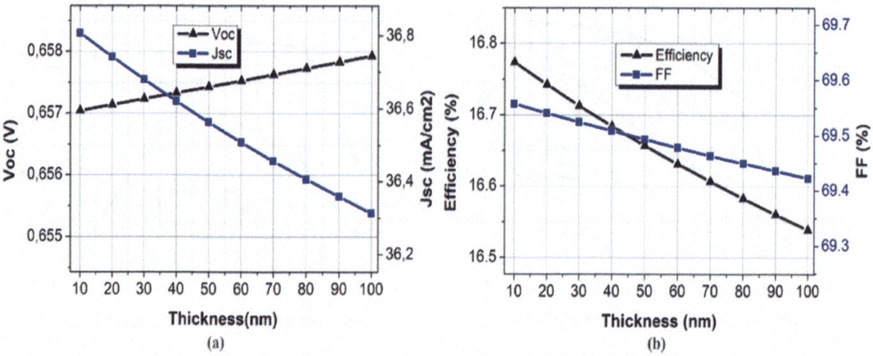

Fig. 35.4 Variation of CZTSSe solar cell functional parameters (Voc, Jsc, η and FF) as function of BSF thickness

35.6 Conclusion

In this paper we numerically investigate CZTSSe solar cell performances by inclusion of CZTSe layer. At the beginning; our simulated model without CZTSe stacked layer was validated with published experimental data in the aim to improve the accuracy of the model. Next, we investigate the CZTSSe solar cell performances with CZTSe stacked layer, by adding this layer, the efficiency of CZTSSe solar cell is increased from 12.3% to 15.23% ($J_{SC} = 36.06$ mA/cm^2, $V_{OC} = 0.59$ V, FF = 71.29%) with an increasing of 25.86% in comparison with experimental results. Finally, an optimizing of doping concentration and thickness of the CZTSe layer is carried out. A high functional parameters achieved, an efficiency of 16.77% ($V_{OC} = 0.69$ V, JSC = 36.31 mA/cm^2) with a rising of 38.60% in comparison with experimental results at doping concentration value of 5E19 cm^{-3} and 10 nm thickness. The quantum efficiency increased in the large wavelength (range of 650–1100 nm). The proposed model with CZTSe stacked layer contributes in the efficiency boosting of CZTSSe solar cell compared with the work reported so far.

Acknowledgements The authors wish to thank Dr. M. Bin the burgelman's group of Electronics and Information Systems (ELIS). University of Gent for the SCAPS-1D program tool.

References

1. J. Li, D. Wang, X. Li, Y. Zeng, Y. Zhang, Cation substitution in earth-abundant kesterite photovoltaic materials. J. Adv. Sci. News **5**(4), 1700744 (2018)
2. S. Mohammadnejad, A. Baghban, CZTSSe solar cell efficiency improvement using a new band-gap grading model in absorber layer. Appl. Phys. A **123**, 758 (2017)
3. M. Asaduzzaman, A.N. Bahar, M.M. Masum, M.M. Hasan, Cadmium free high efficiency $Cu_2ZnSn(S,Se)_4$ solar cell with $Zn_{1-x}SnxOy$ buffer layer. J. Alex. Eng. **56**(2), 225–229 (2017)

4. M.A. Green, Y. Hishikawa, E.D. Dunlop, D.H. Levi, J. Hohl-Ebinger, A.W.Y. Ho-Baillie, Solar cell efficiency tables (version 51). Prog. Photovolt. Res. Appl. **26**(1), 3–12 (2018)
5. T. Gershon et al., Unconventional kesterites: the quest to reduce band tailing in CZTSSe. Curr. Opin. Green Sustain. Chem. **4**, 29–36 (2017)
6. C. Yan, K. Sun, F. Liu, J. Huang, F. Zhou, X. Hao, Boost Voc of pure sulfide kesterite solar cell via a double CZTS layer stacks. Sol. Energy Mater. Sol. Cells **160**, 7–11 (2017)
7. M.K. Omrani, M. Minbashi, N. Memarian, D.H. Kim, Improve the performance of CZTSSe solar cells by applying a SnS BSF layer. J. Solid State Electron. **141**, 50–57 (2017)
8. K.-J. Yang et al., A band-gap-graded CZTSSe solar cell with 12.3% efficiency. J. Mater. Chem. A **4**(26), 10151–10158 (2016)
9. C. Frisk, Y. Ren, S. Li, C. Platzer-Björkman, CZTS solar cell device simulations with varying absorber thickness, in *IEEE 42nd Photovoltaic Specialist Conference (PVSC)*, New Orleans, LA, USA (2015), pp. 1–3
10. O.K.K. Simya, A. Mahaboobbatcha, K. Balachander, A comparative study on the performance of kesterite based thin film solar cells using SCAPS simulation program. J. Superlattices Microstruct. **82**, 248–261 (2015)
11. D. Cozza et al., Modeling the back contact of $Cu_2ZnSnSe_4$ solar cells. IEEE J. Photovolt. **2576678**(1), 1–6 (2016)

Chapter 36
Hydrostatic Pressure Effect on Mechanical Stability and Optoelectronic Properties of $MgGeAs_2$-Chalchopyrite

H. Bouafia, B. Sahli, B. Djebour, S. Hiadsi, and B. Abidri

Abstract The technology of photovoltaic energy conversion is essentially based on semiconductors that have different physical properties, which consequently result in photovoltaic panels of different yield and lifespan but also different costs. Among these semiconductors, those belonging to the chalcopyrite family have attracted the attention of researchers around the world given their very varied physical behaviours and their syntheses, which are not too complicated. In this work, we are interested in $MgGeAs_2$-chalcopyrite. This compound has been studied in detail by several authors except that several physical properties remain unknown including some of its elastic properties and the effect of hydrostatic pressure on its mechanical stability and its narrow band-gap. The study has been carried out with FP-LAPW method as implemented in WIEN2k code. The found results for the structural properties are in good agreement with the previous results. Structural parameters allow the determination of the elastic constants under different pressures. The mechanical stability of this compound has been studied. The value of band-gap energy of this compound has been obtained by several methods including TB-mBJ and GLLB-SC. Optical properties are also the subject of our work, from which the dielectric function and the absorption coefficient of this compound have been analyzed.

Keywords $MgGeAs_2$ · FP-LAPW · Hydrostatic pressure effect · Mechanical stability · Optoelectronic properties

---

H. Bouafia (✉) · B. Sahli · B. Djebour
Physical Engineering Laboratory, University of Tiaret, Tiaret, Algeria
e-mail: bouafia.hamza@univ-tiaret.dz

S. Hiadsi
Laboratoire de Microscope Electronique et Sciences des Matériaux, Département de Génie Physique, Université des Sciences et de la Technologie Mohamed Boudiaf, BP1505 El M'naouar, Oran, Algeria

B. Abidri
Laboratoire des Matériaux Magnétiques, Université Djillali Liabés, 22000 Sidi Bel-Abbes, Algeria

A. Belasri and S. A. Beldjilali (eds.), *ICREEC 2019*, Springer Proceedings in Energy,
https://doi.org/10.1007/978-981-15-5444-5_36

36.1 Introduction

Actually, first-principles methods have become one of the main axes of scientific research thanks to their great contribution in terms of scientific publications and new results. These methods have considerably contributed in the reduction of the research costs as they have proposed a large number of compounds, which are not yet synthesized by the demonstration of their very important and very varied physical properties [1]. Several studies have been performed on the physical properties of $MgGeAs_2$ showing that it is mechanically and dynamically stable [2–4]. To our knowledge, there is no study that has shown that this compound has been synthesized and all the found studies are only predictions. According to these studies, $MgGeAs_2$ has very fascinating physical properties such as its semiconductor behavior [5, 6] and its mechanical stability. Despite the number of studies done on the physical properties of this compound, several information about some of its properties are still unknown. Therefore, in this work, we will, for the first time, study its elastic and optoelectronic behaviors under the effect of the hydrostatic pressure by highlighting its mechanical stability and the variation of its band-gap energy under the effect of an interval of pressures less than or equal to 10 GPa.

36.2 Results and Discussions

36.2.1 Structural Properties

Several previous studies have determined the structural properties of $MgGeAs_2$ [2, 3]. In most of these works, these properties were determined by semi-local functionals «LDA/GGA». It is known that these functionals do not take into account the non-local electron-electron correlations [7], which are non-negligible in layered structures such as chalcopyrite. In this work, using WIEN2k code [8], in addition to GGA-PBE and GGA-PBEsol functionals [9, 10], we have also used optPBE-vdW functional [7, 11] which takes into account these kind of interactions. This study is based on the analysis of the total energy variation of $MgGeAs_2$ unit-cell as a function of its volume by Murnaghan equation [12]. This analysis has been performed after a relaxation of the internal parameter «xAs» of As-atom and the optimization of c/a ratio. The obtained results for the lattice parameters, Bulk modulus and its pressure derivative, as well as the cohesive energy are shown in Table 36.1. From the obtained results, it is possible to clearly notice the very good agreement of our results with those obtained with the same functionals for the lattice parameters, the internal parameter and Bulk modulus. We note the absence of comparison results for the cohesive energy. On the other hand, we know that optPBE-vdW is based on GGA-PBE functional, from which we note that the lattice parameter «a0» obtained by optPBE-vdW is very close to that obtained by

Table 36.1 Calculated lattice parameters «a_0 and c_0» (Å), c_0/a_0 ratio, the internal parameter of As-atom «x_{As}», Bulk modulus «B_0» (GPa) and its pressure derivative «B'» and the cohesive energy «E_{coh}» (eV/cell) in comparison with other theoretical results

$MgGeAs_2$		a_0 (Å)	c_0 (Å)	c_0/a_0	x_{As}	B_0	B'	E_{coh}
Present work	GGA-PBEsol	5.971	11.058	1.852	0.275	58.557	4.436	−26.853
	GGA-PBE	6.032	11.232	1.861	0.273	53.065	4.506	−24.173
	optPBE-vdW	6.034	11.313	1.874	0.270	51.646	4.535	−125.72
Other theoretical results		6.015[a]	11.284[a]		0.273[a]	58.25[b]		
		5.990[b]	11.029[b]					
		5.841[c]	11.086[c]		0.276[c]			
		5.983[d]	11.105[d]		0.275[d]		–	–
		5.841[e]			0.276[e]			
		6.009[f]	11.270[f]	1.876[f]	0.273[f]			
Experiment		–	–	–	–	–	–	–

[a]Ref. [4], [b]Ref. [13], [c]Ref. [5], [d]Ref. [2], [e]Ref. [6], [f]Ref. [14]

GGA-PBE, while «c0» that are obtained by these two functionals are different. This indicates that the non-local electron-electron correlations are very weak along x and y directions whereas they are non-negligible in the z direction, which confirms the importance of this functional (optPBE-vdW) in this study. The negative sign of the cohesive energy confirms the energy stability of this compound.

36.2.2 Elastic Properties and Mechanical Stability

The elastic constants of $MgGeAs_2$ for pressures lower or equal to 10 GPa have been calculated by the theoretical model implemented in IRelast package [15, 16]. The criteria for the mechanical stability of a solid compound under pressure have been revised because the mechanical stability does not only have to depend on the elastic constants but also on the hydrostatic pressure. For a chalcopyrite structure, the generalized stability criteria Mi are given by the following expressions [1]:

$$M_{1,2,3} = C_{ii} - P > 0 \text{ (with: } i = 1, 4, 6) \tag{36.1}$$

$$M_4 = C_{11} - C_{12} - 2P > 0 \tag{36.2}$$

$$M_5 = (C_{33} - P)(C_{11} + C_{12}) - 2(C_{13} + P)^2 > 0 \tag{36.3}$$

The different results obtained for the generalized stability criteria M_i are shown in Table 36.2 from which it is noted that all the obtained values are positive, which shows that $MgGeAs_2$ remains mechanically stable under a pressure of less than or equal to 10 GPa. On the other hand, it is noted that the values of M_4 and M_5 decrease as a function of pressure unlike other M_i values, which shows that the

Table 36.2 Calculated elastic constants «C_{ij}» (GPa) and the generalized stability criteria «M_i» (GPa) for different pressures in comparison with other theoretical results for $MgGeAs_2$

$MgGeAs_2$	Pressure	C_{11}	C_{12}	C_{13}	C_{33}	C_{44}	C_{66}	M_1	M_2	M_3	M_4	M_5
Present work	0	84.134	36.267	40.466	64.153	53.670	22.125	84.134	53.670	22.125	47.867	4449.140
Other theoretical results		83.834[a]	42.475[a]	46.107[a]	66.439[a]	29.789[a]	34.701[a]	–	–	–	–	–
		80.127[b]	53.142[b]	48.801[b]	71.367[b]	34.851[b]	29.964[b]					
Our results	2.5	93.464	45.944	50.579	72.407	62.799	29.554	90.964	60.299	27.054	42.519	4110.875
	5	102.640	55.212	60.910	81.467	71.391	37.314	97.640	66.391	32.314	37.428	3382.270
	7.5	111.216	64.252	70.979	89.863	79.089	44.505	103.716	71.589	37.005	31.963	2134.075
	10	119.039	72.827	79.421	95.462	86.626	51.223	109.039	76.626	41.223	26.212	405.083

[a]Ref. [2], [b]Ref. [3]

mechanical stability of this compound decreases with the increase in pressure. To determine the minimum transition pressure for which this compound loses its mechanical stability, the variations of M_4 and M_5 as a function of pressure have been polynomially fitted (Fig. 36.1), from which it was found that the critical pressures for the two variations are respectively 10.515 GPa and 20.781 GPa indicating that this compound loses its mechanical stability for a pressure of 10.515 GPa. The results of the elastic constants obtained for zero pressure are close to the previous found values (Table 36.2).

36.2.3 Electronic Properties

Semilocal functionals largely underestimate the band-gap energy, so several methods have been proposed in order to overcome this problem [17, 18]. In this work, to ensure adequate accuracy of our results, the band-gap energy of MgGeAs2 has been estimated by TB-mBJ [19] and GLLB-SC [20] potentials. In Fig. 36.2, we show the variations of the band-gap energy of MgGeAs2 as a function of pressure that are obtained by the different functionals and potentials. We can notice that for a given pressure, the band-gap values obtained by GGA-PBE, GGA-PBEsol and optPBE-vdW are far from those obtained by TB-mBJ and GLLB-SC which is obvious because the semilocal functionals underestimate the band-gap energy. These variations show the good accuracy of TB-mBJ and GLLB-SC potentials. We also note that for all the variations, the band-gap energy increases almost linearly (or polynomially) with increasing pressure but they begin to decrease for a pressure of 5 GPa for the variations which are obtained by GGA-PBEsol, TB-mBJ and

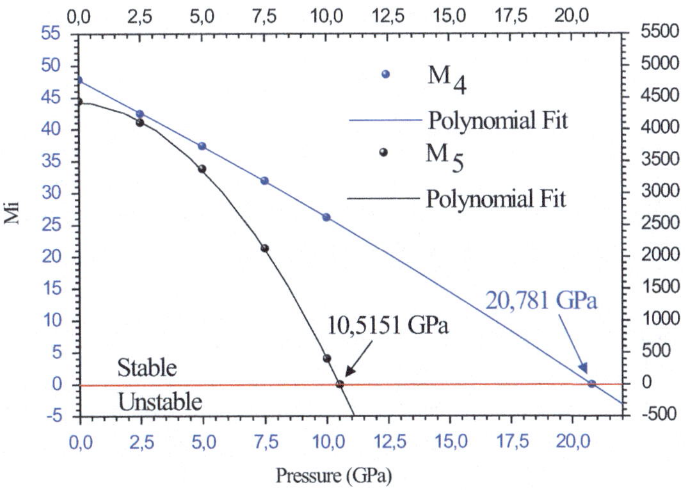

Fig. 36.1 Variation of M_4 and M_5 generalized stability criteria as function of hydrostatic pressure

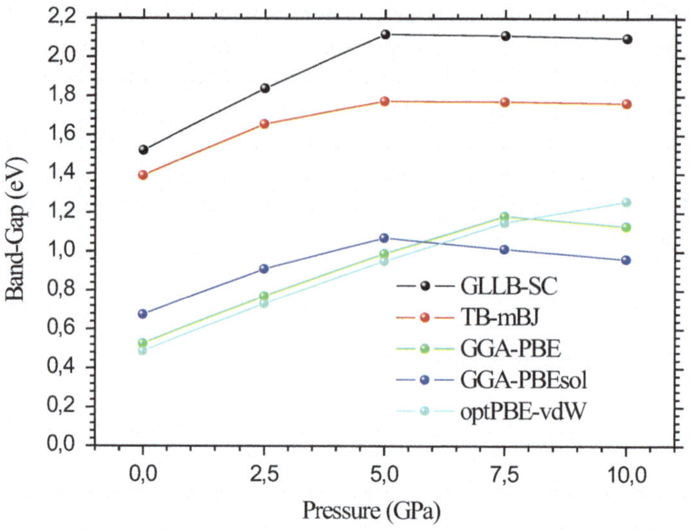

Fig. 36.2 Variations of the band-gap energy as function of hydrostatic pressure

GLLB-SC and for a pressure of 7.5 GPa for that obtained by GGA-PBE while that obtained by optPBE-vdW showed an almost linear (or polynomial) behavior. For zero pressure, the obtained values of band-gap energy are close to the previous values obtained by the same methods (see Table 36.3).

36.2.4 Optical Properties

The electronic transitions between the valence band top and the conduction band bottom can be described by the dielectric function. Its variations (//and ⊥) that are obtained by TB-mBJ are shown in Fig. 36.3(a). It is noted that there is a same shift order that is obtained in the electronic part for zero pressure because these properties are closely related to the band-gaps. It is also noted that the curve of variation for the parallel direction (//) to that of Z is different from that perpendicular (⊥) because of the birefringent nature of $MgGeAs_2$.

Table 36.3 Band-gap energy obtained by the different used functionals and potentials in comparison with other theoretical results

$MgGeAs_2$	GLLB-SC	mBJ	GGA-PBE	GGA-PBEsol	optPBE-vdW
Our results	1.5222	1.3943	0.5313	0.6799	0.4898
Other theoretical results	$0.696^{a(GGA-AM05)}$, $0.624^{b(WC-GGA)}$, $1.620^{b(mBJ)}$, $0.57^{c(GGA)}$, $1.29^{c(HSE06)}$				
	1.920^{d}, 1.6^{e}, $1.39^{f(mBJ)}$, $1.27^{f(GW)}$, 1.6^{g}, 2.1807^{h}				

[a]Ref. [2], [b]Ref. [3], [c]Ref. [4], [d]Ref. [5], [e]Ref. [6], [f]Ref. [14], [g]Ref. [21], [h]Ref. [22]

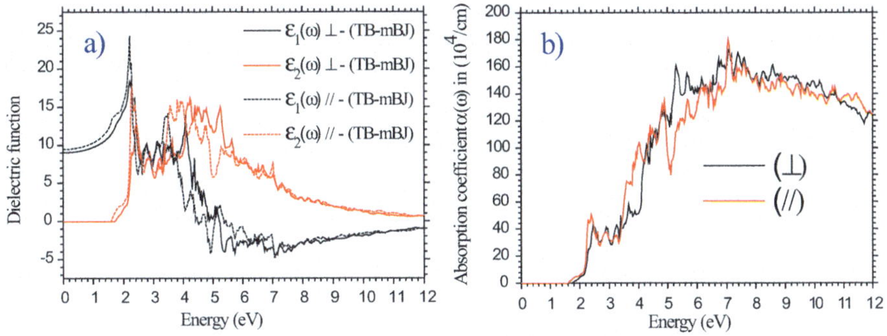

Fig. 36.3 Variations of parallel and perpendicular parts of: **a** the dielectric function and **b** the absorption coefficient as a function of energy

The variations of absorption coefficient of this compound (//and ⊥) are shown in Fig. 36.3(b). We notice that for the two optical directions, it begins for energy equal to the value of that of the band-gap, which confirms the direct nature of the latter. On the other hand, the maximum absorption value for the two optical directions is close to 7 eV, which shows that $MgGeAs_2$ has a large absorption zone.

36.3 Conclusion

The present work is a study of the effect of pressure on the mechanical stability and optoelectronic properties of $MgGeAs_2$. The structural results are in agreement with the previous results. The variations of the generalized stability criteria have shown that $MgGeAs_2$ is mechanically stable for any pressure lower than 10.515 GPa. The variations of band-gap energy as a function of pressure have been analyzed. Optical transitions and the optical absorption in both optical directions have shown that $MgGeAs_2$ has a large absorption zone and is birefringent.

References

1. D. Kerroum et al., Optik **139**, 315–327 (2017)
2. B. Kocak, Y.O. Ciftci, Mater. Res. Bull. **77**, 300–306 (2016)
3. S. Cheddadi, K. Boubendira, H. Meradji et al., Pramana J. Phys. **89** (2017)
4. J. Xiao et al., Mater. Res. Express **5**, 035907 (2018)
5. F. Boukabrine et al., Opt. Mater. **54**, 200–206 (2016)
6. J.E. Jaffe, A. Zunger, Phys. Rev. B **29**, 1882 (1984)
7. J. Klimeš, D.R. Bowler, A. Michaelides, Phys. Rev. B **83**, 195131 (2011)

8. P. Blaha, K. Schwarz, G.K.H. Madsen, D. Kvasnicka, J. Luitz, R. Laskowski, F. Tran, L.D. Marks, WIEN2k, an augmented plane wave + local orbitals program for calculating crystal properties (Karlheinz Schwarz, Techn. Universität Wien, Austria, 2018). ISBN 3-9501031-1-2
9. J.P. Perdew, K. Burke, M. Ernzerhof, Phys. Rev. Lett. **77**, 3865 (1996)
10. J.P. Perdew, A. Ruzsinszky, G.I. Csonka, O.A. Vydrov, G.E. Scuseria, L.A. Constantin et al., Phys. Rev. Lett. **100**, 136406 (2008)
11. F. Tran, J. Stelzl, D. Koller, T. Ruh, P. Blaha, Phys. Rev. B **96**, 054103 (2017)
12. F.D. Murnaghan, Proc. Natl. Acad. Sci. U.S.A. **30**, 244 (1944)
13. S. Cheddadi et al., Phys. B: Condens. Matter **530**, 24–29 (2018)
14. V.L. Shaposhnikov, A.V. Krivosheeva, V.E. Borisenko, Phys. Rev. B **85**, 205201 (2012)
15. A.H. Reshak, M. Jamal, Int. J. Electrochem. Sci. **8**, 12252–12263 (2013)
16. M. Jamal, M. Bilal, I. Ahmad, S. Jalali-Asadabadi, J. Alloys Compd. **735**, 569–579 (2018)
17. E. Engel, S.H. Vosko, Phys. Rev. B **47**, 13164 (1993)
18. P. Dufek, P. Blaha, K. Schwarz, Phys. Rev. B **50**, 7279 (1994)
19. F. Tran, P. Blaha, Phys. Rev. Lett. **102**, 226401(1)–226401(4) (2009)
20. F. Tran, S. Ehsan, P. Blaha, Phys. Rev. Mater. **2**, 023802 (2018)
21. Z. Zhaochun et al., Mater. Sci. Eng. B **54**, 149–152 (1998)
22. C. Suh, K. Rajan, Appl. Surf. Sci. **223**, 148–158 (2004)

Chapter 37
Study of Carbothermal Reduction of Silica, Alumina and Titania Under Argon Gas

Amina Chahtou, Abderrahmene Boucetta, Rabie Benioub, Asmaa Boualem, Saad Hamzaoui, and Kenji Itaka

Abstract Carbothermal reduction of light metals and semiconductor such as Al, Ti and Si under argon gas not only can reduce the reaction temperatures but can also minimize the formation of carbides and oxy-carbides intermediate compounds. The calculation of the gas phase diagram of Al-O-C, Ti-O-C and Si-O-C system suggests the possibility of the enhancement of the Al, Ti and Si product yield by the increase of the ratio of the partial pressure Al_2O/CO, TiO/CO and SiO/CO. Thermodynamic considerations associated with this process are discussed along with the experimental results obtained over a wide range of parameters under argon gas using an induction heating furnace (IH) in a high temperature reactor. Powder mixtures were prepared of stoichiometric mixtures using powders either of alumina or silica or Titania with carbon. Experimental results obtained in induction furnace setup such as the pure element's yields as a function of the CO partial pressure, P_{CO}, temperatures range in the reaction zone of 1400–1800 °C and different deposition temperatures are presented. In this research work, we calculated the thermodynamic phase diagram with the partial pressure ratio of Al_2O/CO, TiO/CO and SiO/CO. This diagrams suggests the possibility of enhancement of Al, Ti and Si yield by the increase of P(Al_2O, TiO and SiO)/P(CO). At the optimal condition, the Al and Si yield was improved comparably to Ti which still under research.

A. Chahtou (✉) · K. Itaka
Institute of Regional Innovation (IRI), Hirosaki University,
2-1-3, Matsubara, Aomori 030-0813, Japan
e-mail: a.chahtou@cder.dz

A. Chahtou · A. Boualem · S. Hamzaoui
Université des Sciences et de la Technologie d'Oran Mohamed Boudiaf, USTO-MB,
BP 1505 El M'naouer, 31000 Oran, Algeria

A. Chahtou
Centre de Développement des Energies Renouvelables, CDER,
BP. 62 Route de l'Observatoire Bouzareah, 16340 Alger, Algérie

A. Boucetta
Graduate School of Engineering, Nagoya University, Nagoya 464-8603, Japan

R. Benioub
SEAVAC, INC, 1-12-6 Kuise-Minami-Shinmachi, Amagasaki-shi, Hyogo 660-0822, Japan

A. Belasri and S. A. Beldjilali (eds.), *ICREEC 2019*, Springer Proceedings in Energy,
https://doi.org/10.1007/978-981-15-5444-5_37

Keywords Carbothermal reduction · Silica · Alumina · Titania · High temperature

37.1 Introduction

Carbothermal reduction of stable metal oxides such as SiO2, TiO2 and Al2O3 requires high temperatures [1, 3]. At these temperatures, reduction occurs from molten slag, in which activity of an oxide can below, while carbon monoxide pressure at the reaction interface has to be above 1 atm to provide conditions for argon gas.

Thus, the reduction of silicon oxides from mixture with carbon in the induction furnace occurs at about 1400 °C [4], the temperature in aluminum production is above 1750 °C [5]. Carbothermal reduction of Titania to titanium is implemented at 1300–1800 °C [6]. High temperature for carbothermal reduction of alumina is a major obstacle in development of carbothermal technology for aluminum production. However, the carbothermal reduction of Al_2O_3 and TiO_2 is similar to the carbothermal reduction of silica (SiO_2) concerning the appearance of sub-oxide gas phase. In the case of silicon (Si), the thermodynamic gaseous phase diagram was utilized to understand the complicated reaction through the SiO gas phase and silicon carbide (SiC) solid phase [7].

This paper presents results obtained in carbothermal reduction of silica, alumina and Titania oxides under argon gas. Therefore, the carbothermal reduction reaction of mixtures of SiO_2, Al_2O_3 and TiO_2 with carbon was studied using induction furnace but with special regard to the formation of CO, SiO, AlO. We performed the optimization of additives under carbothermal reduction of SiO_2 and Al_2O_3. At the optimal condition, the Si and Al yield was improved.

37.2 Experimental Procedure

The carbothermic reduction process was carried out by a high frequency induction furnace of 30 kW power (Toei Scientific Industrial Co. Ltd.). The silica powder (diameter 20–100 μm, Taiheiyo Cement Corporation, Japan), the Al_2O_3 powder (diameter 1 μm), the Titania (diameter 21 μm Kojundo Chemical Laboratory, Ltd.), and glassy carbon powder (diameter 20 μm Tokai Carbon, Ltd.) with a stoichiometric ratio of (SiO_2:C = 1:3, Al_2O_3:C = 1:3, TiO_2:C = 1:1) in molar ratio) were mixed respectively as a base raw material. The mixtures were loaded to a high purity graphite crucible with an inner diameter of 40 mm and height of 70 mm covered with carbon felt and surrounded by quartz tube as an insulator and was installed in the center of the induction furnace (see Fig. 37.1).

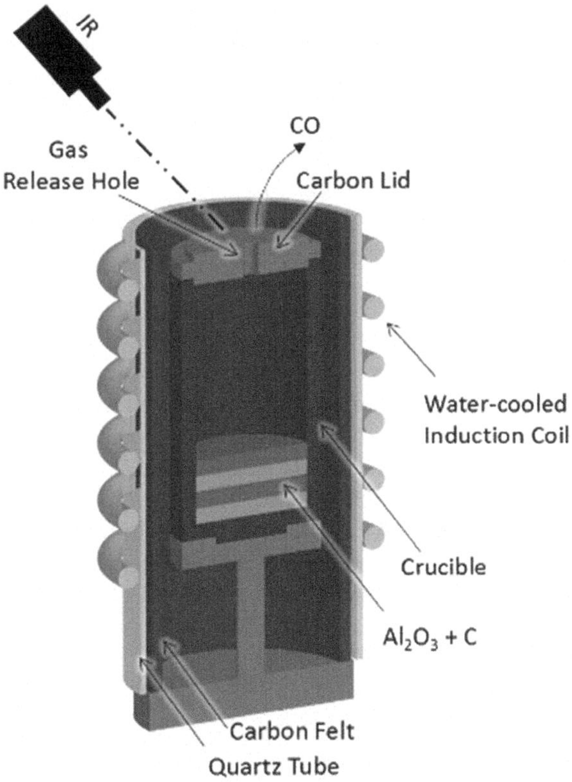

Fig. 37.1 Schematic figure of the experimental apparatus. The carbon crucible was surrounded by thermal insulator (carbon felt) and quartz tube as electrical insulation

37.3 The Reaction Sequence

The reaction mechanism of the carbothermal reduction with argon gas is discussed in a large number of publications, but the description of the reaction sequence is often contradictory. Thus, it has to be suspected that the processes occurring during the reaction have not yet been fully explain.

In silicon case, the SiO_2 react with Carbon to generate SiO and CO gases via reaction (37.1). And than SiO gas react with Carbon to generate SiC (Eq. 37.2). Finally, SiO gas will react with SiC to generate Si metal via reaction in Eq. (37.3).

$$SiO_2 + C = SiO \uparrow + CO \tag{37.1}$$

$$SiO + 2C = SiC + CO \tag{37.2}$$

$$SiO + SiC = 2Si + CO \tag{37.3}$$

Based on the temperature difference, the furnace can be divided into two zones: the low-temperature zone (outer zone) and the high-temperature zone (inner zone).

In the inner zone, molten alumina and aluminum carbide react with each other and form Al_2O gas and CO gas according to reactions 4 and 6, for reactions 5 and 7 the Al_4C_3 and Al form, which is the concern of this study.

$$Al_2O_3(s) + 2C(s) = Al_2O(g) + 2CO(g) \tag{37.4}$$

$$2Al_2O(g) + 5C(s) = Al_4C_3(s) + 2CO(g) \tag{37.5}$$

$$5Al_2O_3(s) + 2Al_4C_3(s) = 9Al_2O(g) + 6CO(g) \tag{37.6}$$

$$3Al_2O(g) + Al_4C_3(s) = 10Al(s) + 3CO(g) \tag{37.7}$$

At first, Al_2O_3 reacts with carbon to generate Al_2O and CO gasses via reaction in Eq. (37.4). And then, Al_2O gas reacts with carbon to generate Al_4C_3 via reaction in Eq. (37.5). On the other hand, Al_2O_3 reacts with Al_4C_3 to generate Al_2O and CO gasses via reaction in Eq. (37.6). Finally, Al_4C_3 will react with Al_2O gas to generate Al metal via reaction in Eq. (37.7).

In the study of carbothermic reduction of titanium, the titanium dioxide react with carbon to generate Ti_2O_3 and CO gas via reaction (37.8). The Ti_2O_3 react with carbon in Eq. (37.9) to generate TiO and CO gas. Finally the TiO react with carbon the generate TiC via Eq. (37.10).

$$2TiO_2 + C = Ti_2O_3 + CO \tag{37.8}$$

$$Ti_2O_3 + C = 2TiO + CO \tag{37.9}$$

$$TiO + 2C = TiC + CO \tag{37.10}$$

37.4 Chamber Gas Analysis

Figure 37.2 shows the temporal change of the relative mass peaks (m/Z = 28) in the case of Al_2O_3+C reduction. The peak with m/Z = 28 represents residual N2 gas and CO gas evolution during heating which mainly occurred due to the carbothermic reduction of alumina. This increase of CO gas was mainly generated from carbothermal reduction because residual N_2 gas is regarded as constant background.

The gas losses of CO and Al_2O were estimated in Eqs. (37.11) and (37.12):

$$M(CO)mol = Pfurnace * k * Vfurnace/R * Tfurnace \tag{11}$$

$$M(Al2O) = (Wloss - M(CO)mol * 28)/70 \tag{12}$$

where R is gas constant, Pfurnace, Vfurnace, and Tfurnace are the pressure (Pa), volume (m^3) and temperature (K) of the total gas in the furnace. k is the ratio of the mass peak intensity between CO and Ar in quadrupole mass spectroscopy at the end

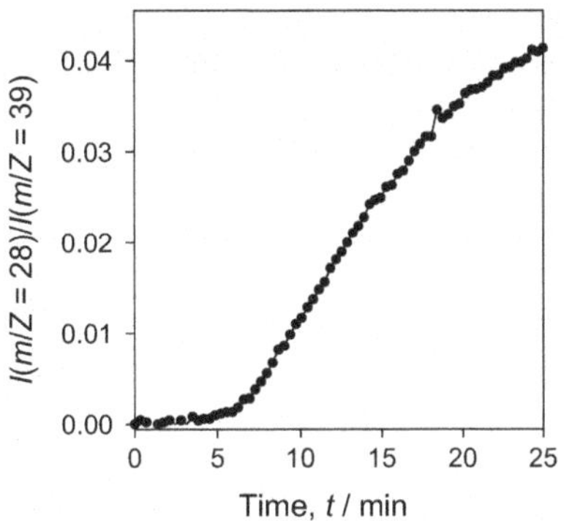

Fig. 37.2 Real-time evolution of the released CO gas during the reduction process product analysis

of the reaction as shown in Fig. 37.2. The increase of the peak intensity of m/ Z = 28 was used to estimate the factor k, which will be used for the estimation of SiO, A12O and TiO gas release during the reduction.

37.5 Quantification of Products

37.5.1 X-Ray Diffraction

Figure 37.3 show the x-ray diffraction patterns of SiO_2 and carbon after heating. SiC and SiO_2 appeared due to incomplete reaction.

Figure 37.4 shows the x-ray powder diffraction patterns for the milled products. The peaks related to Al_2O_3 and C represent the residual raw material due to incomplete reduction caused by gasses partial pressure inside the crucible or in case of only remains Al_2O_3 due to the exhaustion of carbon raw material. The carbon peaks can be explained by excess carbon powder or the graphite crucible.

Figure 37.5 shows the x-ray powder diffraction for the product. The titanium carbide peaks appeared because incomplete reduction of TiO and TiC.

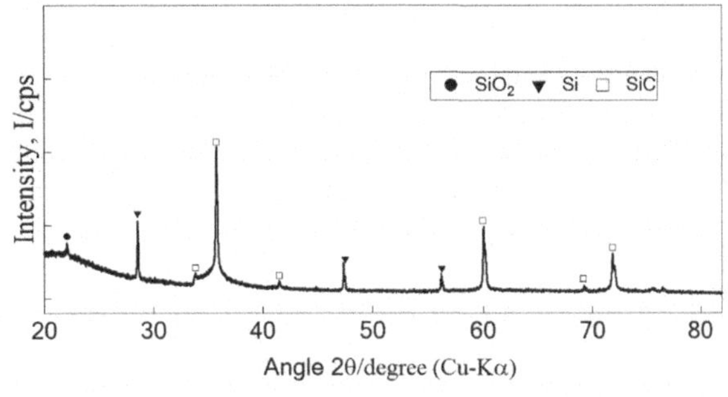

Fig. 37.3 X-ray diffraction patterns of obtained product SiO_2+C under heating temperature of 1500 °C

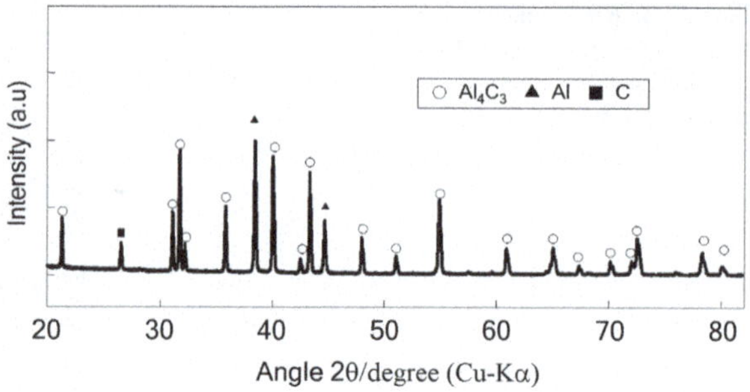

Fig. 37.4 X-ray diffraction patterns of obtained product Al_2O_3+C under heating of 1800 °C

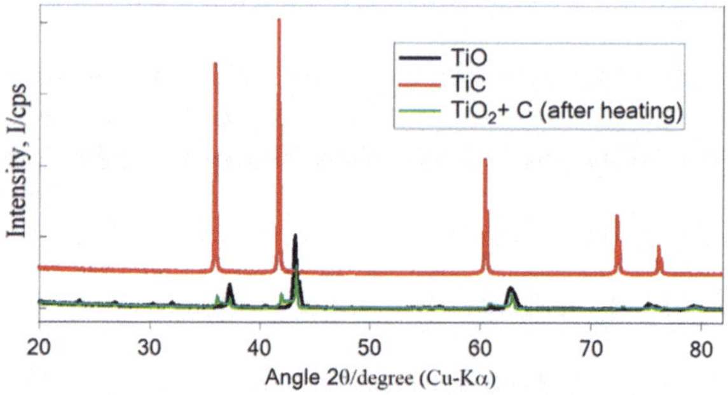

Fig. 37.5 X-ray diffraction patterns of obtained product TiO_2+C under heating of 1800 °C

Table 37.1 Experimental story of SiO_2, Al_2O_3 and TiO_2 carbothermal reduction profiles corresponding to each methods sample

Samples	Yield % (product + carbon)	Granulation (PVA)	Al_4C_3 additive
SiO_2+C	10	33.12	–
Al_2O_3+C	10	–	16.02
TiO_2+C	0	–	–

The yield obtain of silica and carbon powder mixture under argon gas, was 10%, Boucetta et al. [7] have optimize this yield of based on granulation using PVA, with 33.12% (see Table 37.1).

Chahtou et al. [2, 5] was elaborate the carbothermal reduction of alumina. Therefore, the yield obtain on temperature of 1750 °C is 10%. To increase this yield, the Al_4C_3 was added as additive, the yield obtain is 16.02% in 0.05 mol ratio. In titanium case, there no peaks in XRD analysis after heating (see Fig. 37.2). The titanium oxide was obtain in our case (see Table 37.1).

37.6 Conclusion

The experimental verification of heating reduction process showed that silicon and aluminum production can be formed by carbothermal reduction of silica and alumina containing starting materials which is SiO_2+C and Al_2O_3+C.

By contrast, the reduction of SiO_2 and Al_2O_3 by C in the presence of argon gas not only would sharply increase the yield of Si and Al, but also produce by-product which is SiC and Al_4C_3 respectively.

The carbothermal reduction of TiO_2 and carbon, was strongly affected by temperature and argon gas. The formation of titanium oxy-carbide started at 1300 °C. The titanium not affected in this case.

Acknowledgements This research was supported by JST-JICA, SATREPS. A. Chahtou, R. Benioub and A. Boucetta gratefully acknowledge the scholarship from Ministry of Education, Culture, Sport, Science, and Technology (MEXT) of Japan.

References

1. K. Itaka, T. Ogasawara, A. Boucetta, R. Benioub, M. Sumiya, T. Hashimoto, H. Koinuma, Y. Furuya, J. Phys. Conf. Ser. **596**(1), 012015 (2015)
2. A. Chahtou, R. Benioub, K. Itaka, Int. J. Biosen. Bioelectron. **4**(2), 67–72 (2018)
3. G.-G. Lee, B.-K. Kim, Mater. Trans. **44**(10), 2145–2150 (2003)
4. R. Benioub, A. Boucetta, A. Chahtou, S.M. Heddadj, M. Adnane, Y. Furuya, K. Itaka, Mater. Trans. **57**, 1930–1935 (2016)
5. A. Chahtou, R. Benioub, A. Boucetta, L. Zeng, H. Kobatake, K. Itaka, J. New Technol. Mater. **08**(02), 77–82 (2018)
6. L.M. Berger, W. Gruner, Int. J. Refract. Metals Hard Mater. **20**, 235–251 (2002)
7. A. Boucetta, R. Benioub, A. Chahtou, S.M. Heddadj, T. Ogasawara, Y. Furuya, S. Hamzaoui, K. Itaka, J. Mater. Trans. **57**, 1936–1944 (2016)

Chapter 38
Simulation of a Silicon Based Solar Cell Using TCAD-Silvaco Tools

A. Ghazli, A. Aissat, and J. P. Vilcot

Abstract The most well-known solar cells are made of semiconductors, mainly based on crystalline silicon (mono- or poly-crystalline). It consists in converting solar radiation into electricity. Generally, the solar cell device that can carry out this function is essentially a single PN junction with large surface. The most of solar cells on the market today are based on silicon crystal. Currently, multiple researches are in progress in order to realize cells with multi junctions, tandem, by connecting to the silicon cell, another cell based on a material with wide gap energy in order to obtain a better efficiency. In this paper, we studied the efficiency of a silicon solar cell by using TCAD—Silvaco tools. The silicon solar cell structure was defined using Athena 2D process simulator that permit to create the structure in order to study it and use it in predictive simulation. On the other hand, the electrical simulation was performed using Atlas simulator. After simulation, the final output parameters are Voc = 0.61 V, Jsc = 23.267 mA.cm^{-2}, FF = 0.795 and 11.30% of efficiency.

Keywords Silicon based solar cell · PN junction · TCAD-Silvaco

38.1 Introduction

The light we receive from the sun provides a source of energy that is available all year round and can be used without damage for the environment or for human. The manufacturing of solar cells have underwent, until today, many development steps from generation to another one. Before any photovoltaic project, it is very interesting to start by silicon solar cell that belongs to the first generation solar cells.

A. Ghazli (✉) · A. Aissat
Saad Dahlab University of BLIDA, BP 270 Route de Soumâa, 09000 Blida, Algeria
e-mail: aghazli@univ-blida.dz

A. Aissat · J. P. Vilcot
University of Sciences and Technologies of Lille 1, 60069 Avenue Poincare, 59652 Villeneuve of Ascq, France

A. Belasri and S. A. Beldjilali (eds.), *ICREEC 2019*, Springer Proceedings in Energy,
https://doi.org/10.1007/978-981-15-5444-5_38

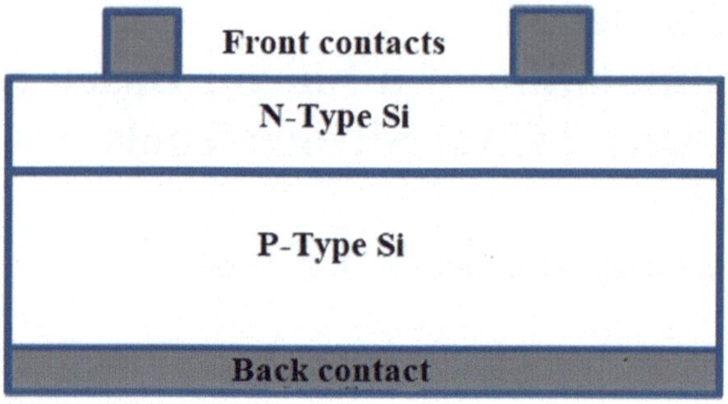

Fig. 38.1 The based structure of silicon solar cell

In this generation, the device is built on Si wafers [1]. The solar cell is PN junction diode [2], which has the role to convert the sunlight into electricity [3] (Fig. 38.1).

Silicon absorbs the photons incident and generates at least one electron-hole pairs for each photon [3]. The electric field generated in the PN junction will prevent the diffusion of the electrons to the side of P-type Si and the holes to the side of N-type Si. Silicon absorbs only photons with higher energy than its own band gap, 1.12 eV. However, only photons with energy higher than the bandgap energy of Silicon will be used and the rest will be a source of heating for the solar cell device. [3]. In general, silicon solar cell is characterized by the concentration of the dopants, acceptor N_A, of the substrate from 10^{16} atm.cm^{-3} to 10^{17} atm.cm^{-3} and the concentration of dopants, donor N_D, of the emitter from 10^{18}–10^{19} [3]. Since silicon has, a small absorption coefficient compared with the other solar materials and that because it is a semiconductor with indirect band gap. Therefore, Silicon need to be thicker so it can absorb photons with longer wavelength [3].

38.2 Materials and Method

The optimization of the structure that define the silicon cell is created by simulation using TCAD-Silvaco device simulator. Technology Computer Aided Design (TCAD) refers to the use of computer simulation to study, develop and optimize devices and semiconductor processing technologies. Silvaco-ATLAS software predicts the electrical characteristics of physical structures [4]. Athena is 2D process simulation software that provides means to create structures to be studied and it is used for application including deposition, implantation, etching… etc. [5].

To create the structure, we used a p-Type crystalline silicon wafer with orientation <100>, 100 μm of thickness and 100 μm of width. The emitter is formed by adding phosphorous atoms into Silicon wafer using the implantation method. To define front contacts, we used Aluminum material with 5 μm of thickness [3]. In order to optimize the doping concentration of our structure, we vary the doping concentration (N_A) of the substrate from 10^{15} atm.cm^{-3} to 10^{17} atm.cm^{-3} and for each N_A, we vary N_D of the emitter, from 10^{16} atm.cm^{-3} to 10^{19} atm.cm^{-3} where

N_D is the concentration of the donor dopants and N_A is the concentration of the acceptor dopants. The output parameters of the cell investigated above: the open circuit voltage (Voc), the short circuit voltage (Jsc), the Fill Factor (FF) and the efficiency (η) are extracted from the J-V curve of the cell, measured using Atlas device simulator and under illumination. The spectrum AM1.5G is used and the beam is performed with incident angle 90 °C.

38.3 Results and Discussion

38.3.1 Doping and Electric Characteristics

As described above, we study the physical structure defined in the Fig. 38.2 using Athena in order to optimize its doping characteristic. The simulation result predicted under illumination is giving in Fig. 38.3. As shown, Voc increase with the doping concentration N_A of the substrate and that because the doping concentration of C-Si absorber change the cell electrical resistivity. However, the courant Isc decrease and that because of the recombination losses essentially Auger and shockly-read-hall (SRH) [6] and the optic losses caused by the series resistance, reflection at the surface and the shadow losses [7]. It is evident that the doping concentration of C-Si absorber mainly affects the open circuit voltage (Voc).

Therefore, the optimum doping concentration for the structure defined above are $N_A = 10^{17}$ atm.cm^{-3}, $N_D = 10^{18}$ atm.cm^{-3} and the electrical characteristic of the cell is giving in the Fig. 38.4.

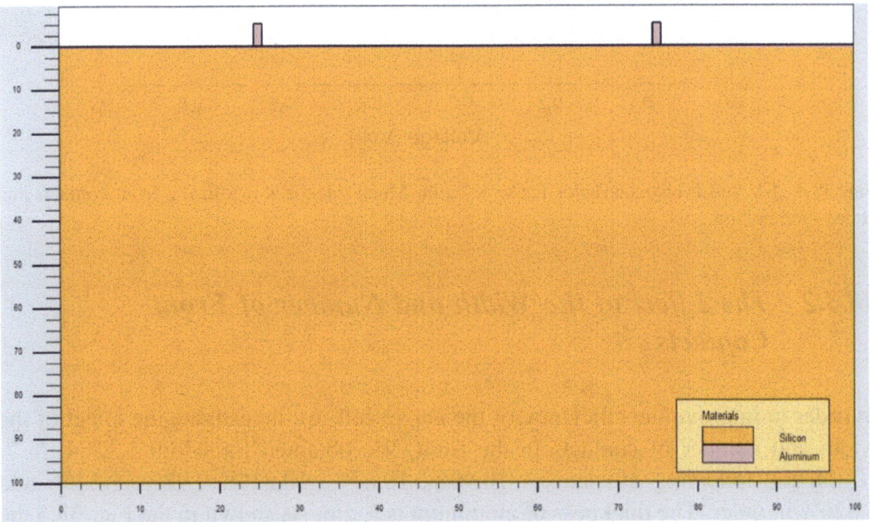

Fig. 38.2 Silicon solar cell structure and its doping distribution defined using Athena

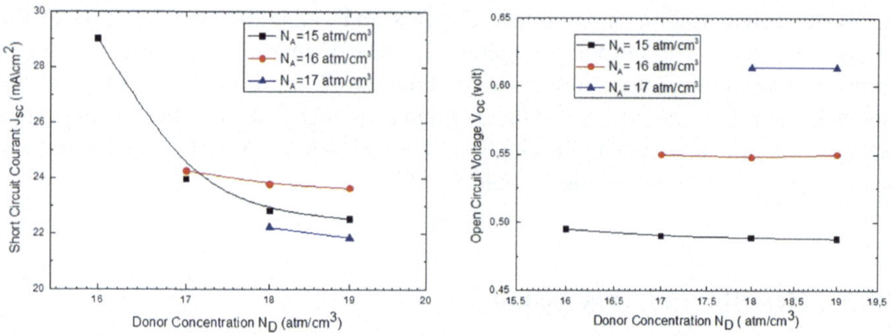

Fig. 38.3 The effect of doping concentration on the output parameters Voc and Jsc of the cell

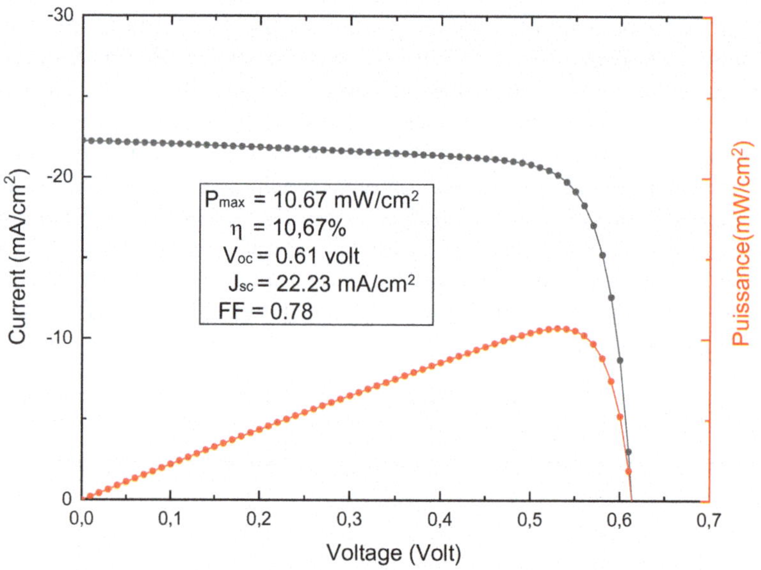

Fig. 38.4 J-V and P characteristics for w = 2 μm, Metal thickness = 5 μm, 2 front contacts and under illumination

38.3.2 The Effect of the Width and Number of Front Contacts

In order to improve the efficiency of the above cell, we investigate the effect of the width and number of contacts in the front. We obtained for width 1, 2, 4, 5, 7, 10 μm the following efficiencies 10.90%, 10.67%, 10.21%, 9.98%, 9.52%, and 8.826% in order. The thickness of aluminum is 5 μm. As shown in the Fig. 38.5 the simulation results show a better efficiency for a width of 1 μm.

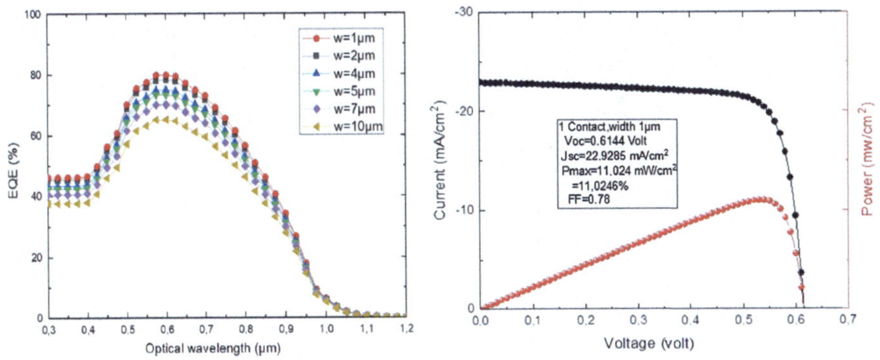

Fig. 38.5 The effect of the front contact width on the cell performance

It is evident through the spectral response that there is more optic losses with increasing the surface shadow because the metal will prevent the light to pass to C-Si absorber since it has a large reflection index [7]. Moreover, to improve more the efficiency, we found that by decreasing the number of front contacts to one, the simulation result gave 11.0246% of efficiency. This situation can be explain by the fact that with the decreasing the number of contacts, we are decreasing the series resistance that need to be low as possible as it can [7]. The open circuit voltage, short circuit current and fill factor are 0.6144 V, 22.9285 $mA.cm^{-2}$ and 0.7825 in order.

38.3.3 Surface Texturing and ARC Layer

Crystalline silicon material has a high refractive index. The reflection is minimized by using an anti-reflective coating (ARC) at the surface and modifying the geometry of Silicon surface (surface texturing) [7]. In Atlas simulator, the reflectivity is ignored by default (R = 0) [8]. If we take it in consideration, the efficiency drop down to 6.95% and the current Jsc to 14.67 $mA.cm^{-2}$. In order to increase this efficiency, we investigate the best antireflective coating for the structure defined in this paper and look for better texturing which give a better result of efficiency. Texturing the surface of silicon is done using Athena process simulator.

The Antireflective Coating: As shown in Fig. 38.6, there is a clear improvement in the efficiency of the cell. The simulation results show that using nitride layer or nitride with oxide double layers coating give a better result than that of oxide. As shown, the reflectivity decrease significantly and can react 10% to 0% for the visible spectrum and therefore the absorption will increase automatically. The antireflective layer is a very important part in the silicon solar cell.

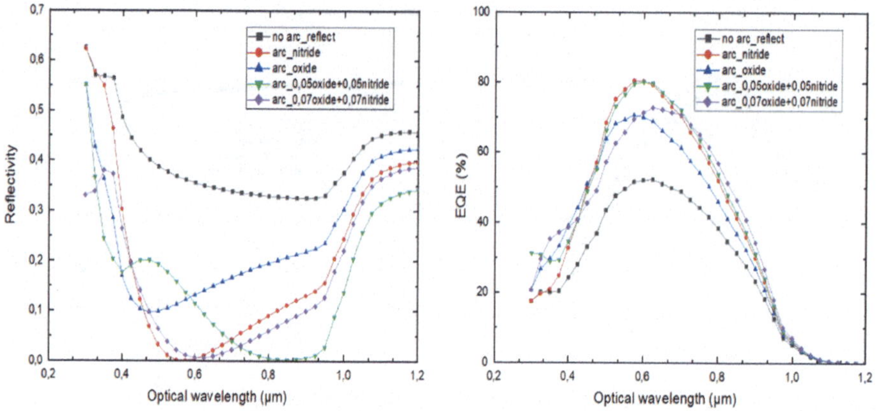

Fig. 38.6 Simulation result for the silicon cell with antireflective coating

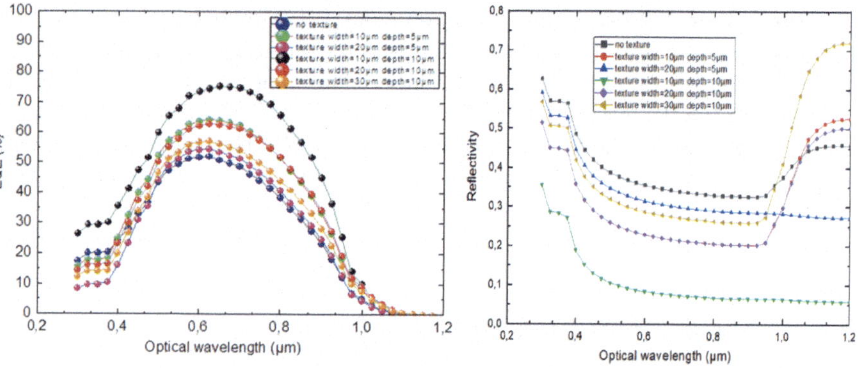

Fig. 38.7 Simulation result for the silicon cell with textured surface

Surface Texturing: The texturing process helps to minimize the number of reflection rays at the surface by light trapping. This process will improve the performance especially, when using a material semiconductor (absorber) with relatively low absorption coefficient [9].

As shown in the Fig. 38.7, the simulation results show a significant improvement in the efficiency by decreasing the reflectivity and increasing the absorption. The best result obtained, 11.2987% of efficiency, is corresponding to the texture with 10 µm of width and 10 µm of depth.

38.4 Conclusion

In this paper, we studied a silicon solar cell by using TCAD-Silvaco tools. The physical structure which define the cell to be optimized was created using Athena process simulator. On the other hand, we used Atlas simulator to calculate the output parameters of the cell. Since silicon has a high reflective index which affect negatively the cell performance. Therefore, a surface texturing with an anti-reflective coating is used to minimize the reflection on the surface. The final structure gave the output electrical parameters: Voc = 0.61 V, Isc = 23.267 mA.cm^{-2}, FF = 0.795 and 11.30% of efficiency. This efficiency need to be improved more especially the current. Therefore, the thickness of silicon need to be increased because silicon is known by his small absorption coefficient, To absorb photons with longer wavelength, Silicon need to have a thickness in range 200 μm–300 μm [3]. On other hand, the recombination in the back wafer need to be decreased and that by using the back selective field [10].

References

1. S. Sharma, Solar cells: in research and applications—a review. Mater. Sci. Appl. **6**, 1145–1155 (2015)
2. E.M.G. Rodrigues, Simulation of a solar cell considering single-diode equivalent circuit mode. RE&PQJ **1**(9), 13–15 (2011)
3. A. Zekry, A. Shaker, Solar cells and arrays: principles, analysis and design, in *Advances in Renewable Energies and Power Technologies*, 1st edn., chap. 1 (Elsevier Science, 2018)
4. SILVACO International Inc., ATLAS User's Manual: Device Simulation Software. Copyright 2006
5. SILVACO International Inc., ATHENA User's Manual: 2D Process Simulation Software. Copyright 2000
6. M. Mohammedi, Etude et simulation de cellules photovoltaïques à contacts arrières interdigités sous environnement Silvaco. Thesis, Faculty of Technology, Algéria, 2012
7. PVEDUCATION Homepage, https://www.pveducation.org/pvcdrom/design-of-silicon-cells. Accessed 18 Oct 2019
8. SILVACO International Inc., Atlas User's Manual: Device Simulation Software. Copyright 1998
9. H. Bashiri, M.A. Karami, Crystalline silicon solar cell engineering to improve fill factor, open circuit voltage, short circuit courant, overall cell efficiency. Modares J. Electr. Eng. **14**(4), 6–14 (2014)
10. R.A.Z. Razera, Passivation analysis of the emitter and selective back surface field of silicon solar cells (IEEE, 2017)

Chapter 39
Synthesis and Characterization of Tio_2 Thin Films for Photovoltaic and Optoelectronic Applications

Houssem Eddine Doghmane, Tahar Touam, Azeddine Chelouche, Fatiha Challali, and Djamel Djouadi

Abstract In this paper, TiO_2 thin films were first deposited by RF magnetron sputtering on glass substrates at different temperatures. Then, the obtained samples were characterized by several techniques: X-ray diffraction, scanning electron microscopy, atomic force microscopy and ultraviolet-visible spectroscopy. Hence, after a brief description of the preparation conditions we investigate the effects of substrate temperature, T_s, (RT, 200, 300 and 400 °C) on structural, morphological and optical properties of all prepared TiO_2 thin films. It was found that increasing T_s leads to several interesting phenomena: (i) more uniform distribution of densely packed well-defined grains (ii) the root mean square roughness increases from 3.88 to 9.56 nm, (iii) improvement of the crystalline structure, (iv) the crystallite size increases from 13.4, to 22.3 nm, (v) the films are highly transparent in the visible region and (vi) the direct optical band gap value of the films is found to decrease from 3.64 to 3.46 eV. These dependences, in agreement with literature, are of great importance in TiO_2 potential material applications in optoelectronic and photovoltaic fields.

Keywords TiO_2 · RF magnetron sputtering · Photovoltaics

H. E. Doghmane (✉)
Laboratoire des Semi-conducteurs, Université Badji Mokhtar-Annaba, 23000 Annaba, Algeria
e-mail: h.doghmane@yahoo.fr

T. Touam
Unité de Recherche en Optique et Photonique CDTA, Université de Sétif 01, 19000 Sétif, Algeria

A. Chelouche · D. Djouadi
Laboratoire de Génie de l'Environnement, Université de Bejaia, 06000 Bejaia, Algeria

F. Challali
Laboratoire des Sciences des Procédés et des Matériaux, Université Paris 13, Sorbonne Paris Cité, 99-Jean-Baptiste Clément Avenue, 93430 Villetaneuse, France

A. Belasri and S. A. Beldjilali (eds.), *ICREEC 2019*, Springer Proceedings in Energy,
https://doi.org/10.1007/978-981-15-5444-5_39

39.1 Introduction

Titanium oxide TiO_2 thin films have attracted tremendous research activities in recent years. They have emerged, among transparent conducting oxides, as the most promising materials due to their remarkable and multifunctional properties. With such properties, TiO_2 has been considered for various applications involving energy production through photon absorption and photo-electron generation such as solar cells [1], optoelectronics [2] and photovoltaics [3]. Moreover, TiO_2 films are important as optical films owing to their high refractive index, high bandgap and transparency on top of wide spectral range, adjustable electrical conductivity and ecological non-toxicity [4]. TiO_2 can be deposited through various methods [5] such as chemical vapor deposition, pulsed laser deposition, electrophoretic deposition, magnetron sputtering and sol-gel dip-coating. Among these techniques, RF magnetron sputtering process [6] has proven to be an efficient method for the deposition at room temperature (RT) of high quality thin films of simple as well as complex oxides beside many advantages, like very good adhesion, fast deposition rates and good control of the film structure.

In this context, we deposited TiO_2 thin films by RF magnetron sputtering on glass substrates at different substrate temperatures, T_s, (RT, 200, 300 and 400 °C). Then, we investigate the structural, morphological and optical properties of the prepared TiO_2 thin films. To do so, we used several characterization techniques: X-ray diffraction (XRD), scanning electron microscopy (SEM), atomic force microscopy (AFM) and ultraviolet-visible (UV-Vis) spectroscopy.

39.2 Experimental Procedure

39.2.1 Preparation Conditions

Several TiO_2 thin films were first prepared on glass substrates, at different temperatures (RT, 200 °C, 300 °C and 400 °C), in a magnetron sputtering system. The following preparation conditions and parameters were chosen in order to offer better control, reproducibility and ease of film fabrication that can be scaled up for industrial applications. A high purity TiO_2 (99.99%) target, situated at 150 mm from the substrate, was used. The sputtering was performed at a pressure of 2×10^{-3} mbar and a flow rate of pure Argon of 30 cm^3. An RF power of 200 W was applied for 60 min deposition.

39.2.2 Characterization Techniques

The crystalline, morphological, and optical properties of the prepared TiO_2 thin films were investigated by various characterization techniques: XRD, SEM, AFM and UV-Vis spectroscopy, as follows:

- To investigate crystalline properties, TiO_2 films were analyzed by X-ray diffraction (XRD) using a PanAnalytical type diffractometer whose X-rays are produced from CuKα radiation source (λ = 1.54 Å).
- To examine the surface morphology of the films, we used a Raith Pioneer SEM system with a real magnification of, 50,000x. Whereas, the surface topography analysis is carried out by atomic force microscopy using a Nanosurf Flex-AFM system. AFM images have been scanned in contact mode over an area of 2 μm × 2 μm, and Gwyddion software was employed for image processing and surface roughness calculations [7].
- The optical transmittance of the TiO_2 films was measured at the room temperature using a Safas UVmc^2 ultraviolet–visible (UV–Vis) spectrophotometer. From the transmission data, the optical bandgap energy is extracted.

39.3 Results and Discussion

39.3.1 Morphology and Topography Analysis

In order to put into evidence the effects of the substrate temperature on the morphology and topography of deposited films, we show in Fig. 39.1 (i) SEM micrographs and (ii) AFM topography images of TiO_2 thin films grown at different

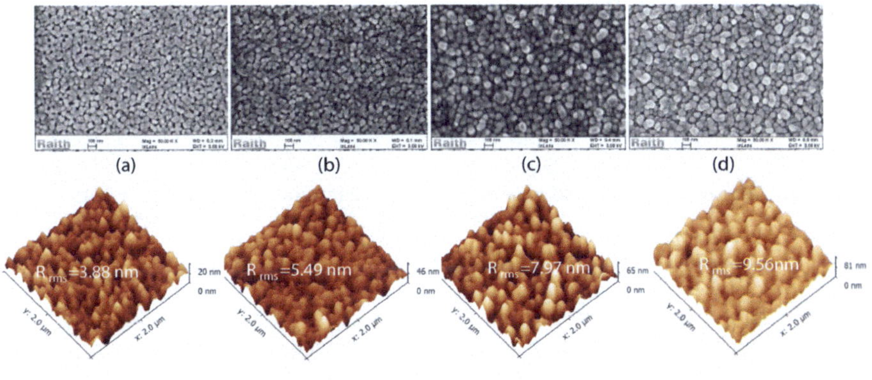

Fig. 39.1 SEM micrographs and AFM topography images of TiO_2 thin films grown at different substrate temperatures: (a) RT, (b) 200 °C, (c) 300 °C and (d) 400 °C

substrate temperatures: (a) RT, (b) 200 °C, (c) 300 °C and (d) 400 °C. Several interesting observations can be easily made, as follows:

all the films consist of spherical grains homogeneously distributed throughout the surface of the film.

The morphology of the films appears to be dependent on T_s.

As T_s increases up to 400 °C, the surface of the films presented more uniform distribution of densely packed well-defined grains and the grain size becomes larger, which confirms that the crystallinity of the films is significantly improved.

Surface roughness, is clearly put into evidence by the 3D AFM images; it is very smooth for RT sample and becomes rough for higher T_s.

From the data analysis, the root mean square roughness (R_{rms}) values of the TiO_2 films deposited at RT, 200, 300 and 400 °C are found to be 3.88, 5.49, 7.97 and 9.56 nm, respectively (Table 39.1). This behavior is in agreement with literature [8].

39.4 Microstructural Investigation

The effects of substrate temperatures on the microstructure of TiO2 thin films are illustrated, by XRD patterns of the TiO2 thin films, in Fig. 39.2. Some observations, in agreement with literature [9, 10] can be made. RT films are crystalline with an anatase structure with preferred orientation along the (101) direction: as Ts increases the (101) anatase peak is enhanced, indicating the improvement of the crystalline structure. Moreover, new diffraction peaks belonging to anatase (004), (200) and (105) planes are also detected.

To enrich this investigation, we used Scherer formula [11], to deduce average crystallite size from the full-width at half-maximum (FWHM) of the most intense peak of anatase structure corresponding to (101). The obtained data are summarized in Table 39.1. It is clear that the crystallite size increases from 13.4 nm, to 22.3 nm when Ts increases from RT to 400 °C indicating that films crystallinity is improved for higher Ts, in agreement with literature [12]. In fact, during the sputtering process, the substrate heating can lead to a sufficient energy to produce TiO2 films with a high crystallinity and a low defect density.

Table 39.1 Measured and deduced parameters of TiO_2 thin films prepared at different T_S

Samples	Substrate temperature (°C)	R_{rms} (nm)	Crystallite size (nm)	Average transmittance %	Optical gap (eV)
S_1	25	3.88	13.4	84	3.64
S_2	200	5.49	15.6	79	3.54
S_3	300	7.97	19.8	76	3.51
S_4	400	9.56	22.3	74	3.46

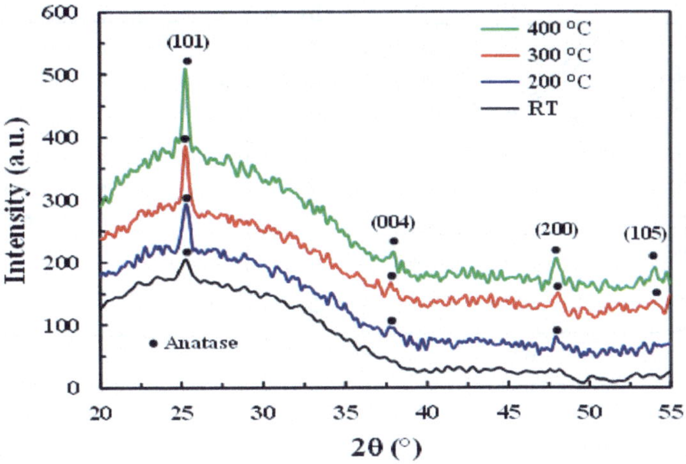

Fig. 39.2 XRD patterns of TiO_2 thin films deposited at different substrate temperatures

39.4.1 Transmittance and Optical Band Gap

Optical properties of thin films are of great importance in several potential fields, in particular photovoltaic and optoelectronic applications. Therefore, all prepared TiO2 thin films were studied in the 300-900 nm wavelength range in order to obtain optical constants such as transmission (T) and optical band gap (Eg). The obtained results are shown in Fig. 39.3a for the optical transmittance spectra of TiO_2 thin films grown at different substrate temperatures and measured with respect to air. It is clear that the UV-visible spectroscopy analysis put into evidence that the film deposited at RT is highly transparent in the visible region with average transmittance greater than 84%. Higher substrate temperatures result in a slight decrease of transmittance. Moreover, with increasing substrate temperature, despite the good optical transparency kept by the films, they show a slow decrease which reaches 74% for the sample with the highest Ts = 400 °C (see Table 39.1). The high transparency is an indicator of surface smoothness and homogeneity. The decrease in optical transmittance with substrate temperature indicates a slightly higher light absorbability of the films [13].

In order to calculate the direct optical band gap of the TiO_2 films, we use the method based on the first derivative of transmittance (T), dT/dE, as a function of energy (E) [14]. The obtained data (in Table 39.1) showed that the direct optical band gap value of the films is found to decrease from 3.64 to 3.46 eV with increasing the substrate temperatures from RT to 400 °C, respectively. This behavior is better illustrated in Fig. 39.3b in terms of Eg versus Ts. The importance of this curve lies in its applicability to the determination of TiO_2 energy gap band by just knowing the substrate temperature. The knowledge of the exact Eg is of great importance in potential material applications in optoelectronic and photovoltaic fields.

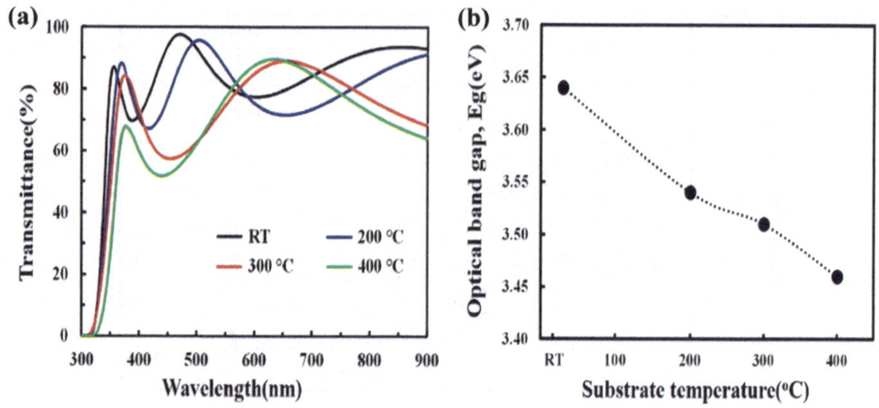

Fig. 39.3 (a) Transmittance spectra of TiO_2 thin films deposited at different substrate temperatures and (b) optical band gap variation with substrate temperature

39.5 Conclusions

Several TiO_2 thin films were first prepared by RF magnetron sputtering technique at different substrate temperatures (RT, 200, 300 and 400 °C) and then characterized by various techniques: XRD, SEM, AFM and UV-Vis spectroscopy. It was found that structural, morphological and optical properties of the prepared TiO_2 thin films depend on T_s. TiO_2 thin films are polycrystalline having anatase phase only with a preferential orientation of (101), they presented more uniform distribution of densely packed well-defined grains and the grain size becomes larger with increasing substrate temperature. The R_{rms} increases, with subtrate temperature, from 3.88 to 9.56 nm. The (101) anatase peak is increased indicating the improvement of the crystalline structure. The crystallite size increases from 13.4, to 22.3 nm. Furthermore, the films are found to be highly transparent in the visible region and the direct optical band gap value decreases from 3.64 to 3.46 eV. These dependences, in agreement with literature, are of great importance in potential application of TiO_2 potential material applications in optoelectronic and photovoltaic fields.

References

1. K. Al-Attafi, A. Nattestad, Q. Wu, Y. Ide, Y. Yamauchi, S.-X. Dou, J.-H. Kim, The effect of amorphous TiO_2 in P25 on dye-sensitized solar cell performance. Chem. Commun. **54**(4), 381–384 (2018)
2. Y. Zhu, K. Deng, H. Sun, B. Gu, H. Lu, F. Cao, J. Xiong, L. Li, TiO_2 phase junction electron transport layer boosts efficiency of planar perovskite solar cells. Adv. Sci. **5**(3), 1700614 (2018)

3. Y. Bai, I. Mora-Seró, F. De Angelis, J. Bisquert, P. Wang, Titanium dioxide nanomaterials for photovoltaic applications. Chem. Rev. **114**(19), 10095–10130 (2014)
4. V.E. Vrakatseli, E. Amanatides, D. Mataras, Comparative study of RF reactive magnetron sputtering and sol-gel deposition of UV induced superhydrophilic TiOx thin films. J. Phys. Conf. Ser. **700**, 012039 (2016)
5. M.K. Patil, S. Shaikh, I. Ganesh, Recent advances on TiO_2 thin film based photocatalytic applications (a review). Curr. Nanosci. **11**, 271–285 (2015)
6. I. Hadjoub, T. Touam, A. Chelouche, M. Atoui, J. Solard, M. Chakaroun, A. Fischer, A. Boudrioua, L.-H. Peng, Post-deposition annealing effect on RF-sputtered TiO_2 thin film properties for photonic applications. Appl. Phys. A **122**, 78–80 (2016)
7. D. Nečas, P. Klapetek, Gwyddion: open-source software for SPM data analysis. Centr. Eur. J. Phys. **10**(1), 181–188 (2012)
8. G. Balakrishnan, S. Manavalan, R. Venkatesh Babu, J.I. Song, Effect of substrate temperature on microstructure and properties of nanocrystalline titania thin films prepared by pulsed laser deposition. Nanosyst. Phys. Chem. Math. **7**(4), 621–623 (2016)
9. A. Karuppasamy, A. Subrahmanyam, Studies on the room temperature growth of nanoanatase phase TiO_2 thin films by pulsed DC magnetron with oxygen as sputter gas. J. Appl. Phys. **101**, 064318 (2007)
10. R. Ananthakumar, B. Subramanian, S. Yugeswaran, M. Jayachandran, Effect of substrate temperature on structural, morphological and optical properties of crystalline titanium dioxide films prepared by DC reactive magnetron sputtering. J. Mater. Sci. Mater. Electron. **23**(10), 1898–1904 (2012)
11. J.I. Langford, A.J.C. Wilson, Scherrer after sixty years: a survey and some new results in the determination of crystallite size. J. Appl. Crystallogr. **11**, 102–113 (1978)
12. A. Kennedy, K. Viswanathan, K. Pradeev Raj, Study of the influence of substrate temperature on structural, optical, and electrical properties of Zn-doped $MnIn_2S_4$ thin films prepared by chemical spray pyrolysis. Phys. Lett. A **380**(36), 2842–2848 (2016)
13. Y. Zhang, X. Ma, P. Chen, D.J. Yang, Effect of the substrate temperature on the crystallization of TiO_2 films prepared by DC reactive magnetron sputtering. J. Cryst. Growth **300**(2), 551–554 (2007)
14. T. Touam, M. Atoui, I. Hadjoub, A. Chelouche, B. Boudine, A. Fischer, A. Boudrioua, A. Doghmane, Effects of dip-coating speed and annealing temperature on structural, morphological and optical properties of sol-gel nano-structured TiO_2 thin films. Eur. Phys. J. Appl. Phys. **67**(3), 30302 (2014)

Chapter 40
Dielectric and Impedance Measurements Within 500 KHz–1 MHz Range of Organic Diode for Renewable Energy

Mostefa Benhaliliba, Tarik Asar, and Suleyman Özçelik

Abstract The Ag/MgPc/GaAs/Au-Ge organic heterojunction diode is fabricated by low cost spin coating technique. Dielectric and impedance measurements within 500 kHz–1 MHz range of organic heterojunction diode are investigated at room temperature for renewable energy applications. A broaden peak and decay of capacitance to negative values are recorded for 500, 700 kHz and 1 MHz applied frequencies within −2, +5 V biasing voltage range. The negative capacitance (NC) of our organic device is observed inside the forward bias voltage range. Increase of conductance with decline in frequency within forward bias voltage is recorded inside the same voltage range. Ac conductivity is frequency dependent in particular within forward voltage range and exceeds 3.5 μS/m at lower frequency. Dielectric constant of organic diode follows similar variation as that of C-V for measured frequencies. Dimensionless parameter M″ versus M′ for applied frequencies is represented in Cole-Cole diagram. It is recorded a highest value of 0.6 for 1 MHz.

Keywords Organic diode · Dielectric · Impedance study · Conductance-voltage characteristics · Cole-Cole diagram

---

M. Benhaliliba (✉)
Film Device Fabrication-Characterization and Application FDFCA Research Group, USTOMB, 31130 Oran, Algeria
e-mail: mbenhaliliba@gmail.com

M. Benhaliliba
Physics Faculty, USTOMB University, P.O. Box 1505 Mnaouer, 31130 Oran, Algeria

T. Asar · S. Özçelik
Physics Department, Faculty of Sciences, Gazi University, 06500 Ankara, Turkey

T. Asar · S. Özçelik
Photonics Application and Research Center, Gazi University, 06500 Ankara, Turkey

A. Belasri and S. A. Beldjilali (eds.), *ICREEC 2019*, Springer Proceedings in Energy,
https://doi.org/10.1007/978-981-15-5444-5_40

40.1 Introduction

Recently many researchers are interested to organic materials and their use in device fabrication. It is worthy to produce organic heterostructures and Schottky diode with organic molecules like pentacene [1], P3Ht [2], MEH-PPV [3], because they are facile to deposit in thin layer, non-toxic and owing semiconducting properties. MgPc as a semiconductor material presents a photoabsorption coefficient of 2×10^5 cm^{-1} and an energy gap from 1 to 2.5 eV as mentioned in literature [4]. The trap content in semiconductors was measured by several methods including PICTS, DLTS, laser DLTS and capacitance-frequency methods [5]. Throughout this research, the investigation is based on capacitance frequency method. Throughout this paper, the electrical and dielectric parameters of the organic fabricated heterostructure have been entirely characterized.

40.2 Experimental Procedure

The organic heterostructure based on MgPc/GaAs contact is made using spin-coating route. Organic layer has grown on GaAs substrate by spincoat G3P-8 equipment at 2000 rpm for 1 min. Front Ag contact is fabricated by thermal evaporation process under 10^{-6} Torr. The n-GaAs is used as substrate having (111) crystal direction, 1.7×10^{-3}–2.37×10^{-3} Ω cm of resistivity and 350 μm of thickness. Firstly it is cleaned according to the protocol mentioned before [4]. A quantity of 0.02 g of MgPc is dissolved in 25 mL of chloroform; so the obtained solution is poured on a suitable substrate and the film produced by the process cited above. The organic structure contacts are characterized by HP 4192 A LF impedance analyzer.

40.3 Results and Discussion

Figure 40.1 displays the capacitance of MgPc/GaAs for the 500 kHz–1 MHz frequency range. Inset of the Fig. 40.1 indicates the configure of the Ag/MgPc/GaAS/Au-Ge heterostructure. Inside the reverse biasing voltage the capacitance slightly increases till a broaden peak of 5.93 nF within 0.5–0.6 V then it decreases drastically and rapidly to negative values for 700 kHz and 1 MHz but slowly for 500 kHz. Similar capacitance decay is recorded for Al/(CdSe-PVA)/p-Si/Al (MPS) structure as reported by Büyükbaş et al. [6]. As shown in Fig. 40.1, while C decreases with increasing V in accumulation region and then goes to negative values in forward voltage. At 700 kHz, the capacitance of such organic heterostructure attains the lowest value of −30 nF. So capacitance becomes negative in the accumulation region (a) and G/ω attains a maximum in the same region. Namely, the negative value of C-V in the accumulation region relates to the maximum value of G/ω due to inductive behavior of the device as shown in Fig. 40.2 for 1 MHz–500 kHz applied frequency range. It is observed that NC means that arise in the voltage creates a decline in the number of charges on the electrodes [5]. Besides, the inductive behavior results from the relaxation-like nature of the material due to injection of holes those recombine easily with the electron free carriers of the dipole near the junction and hence lead to a decrease in the charge dipole as mentioned by Jones [5]. G/ω increases with increasing V at the accumulation region for 1 MHz–500 kHz applied frequency range as shown in Fig. 40.2. It is observed in accumulation region that lower capacitance coincide with the higher value of conductance as plotted in Figs. 40.1 and 40.2. A negative capacitance (NC) is observed in forward voltage for the applied frequency 500 kHz–1 MHz range as depicted in Fig. 40.1. It becomes negative at 2.6, 1.65 and 1.5 V respectively for 500, 700 kHz and 1 MHz where the following −1000, −118, −215 pF values are respectively recorded. This behavior of negative capacitance occurs in accumulation region and it is due to traps as reported previously [7]. The NC is observed in the accumulation region to be more important because it involves that an increment in bias voltage produces a decline in the number of charges on the electrodes as mentioned by Jones et al. [5]. The (NC) phenomenon takes place also in the silicon device and affects the value at low negative bias to produce the peak, which is characteristic of diodes made from relaxation-like material as Jones et al. reported [5]. These curves are temperature and frequency dependent which is caused by the presence of deep traps, but it is mentioned that the low frequency/high temperature limit where the traps respond completely to the applied measurement Ac signal. The Fig. 40.3 shows the conductance-voltage measurements in dark and room temperature conditions of MgPc/GaAs organic heterostructure device. The conductance G/ω-V, within 500 kHz–1 MHz, is increasing when frequency decreases in the forward bias voltage range. It is indicated that the maximum of conductance (Gmax) and capacitance of oxide layer Cox decrease with a rise of frequency. As mentioned elsewhere that the Cr/p-Si SBD with Polypyrrole (PPy) interfacial layer exhibited NC for low frequencies [7]. Ac conductivity in terms of angle loss and permittivity function is expressed as [4];

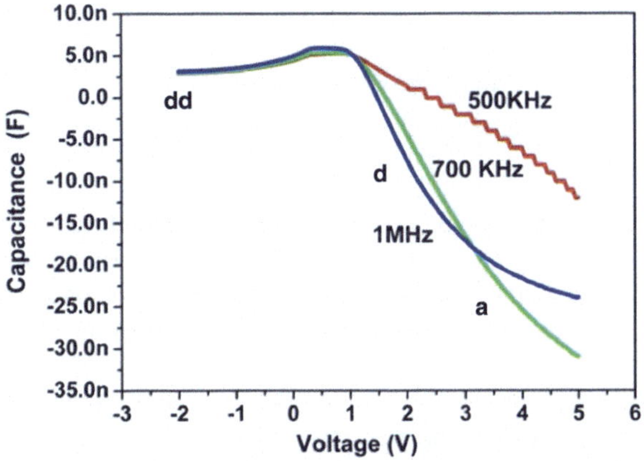

Fig. 40.1 C-V plots of Ag/MgPc/GaAs/Au-Ge organic heterojunction for 500 kHz–1 MHz frequency range. Accumulation (a), depletion (d) and deep depletion (dd) regions are marked. Inset displays the Ag/MgPc/GaAs/Au-Ge heterojunction scheme

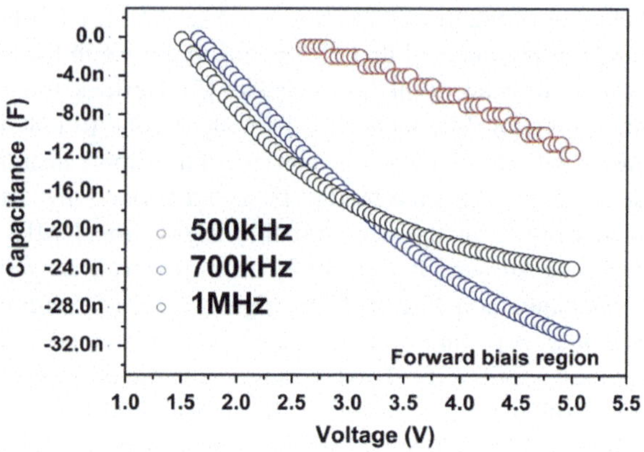

Fig. 40.2 Negative capacitance profile of Ag/MgPc/GaAs/Au-Ge organic heterojunction for 500 kHz–1 MHz frequency range within forward bias voltage region

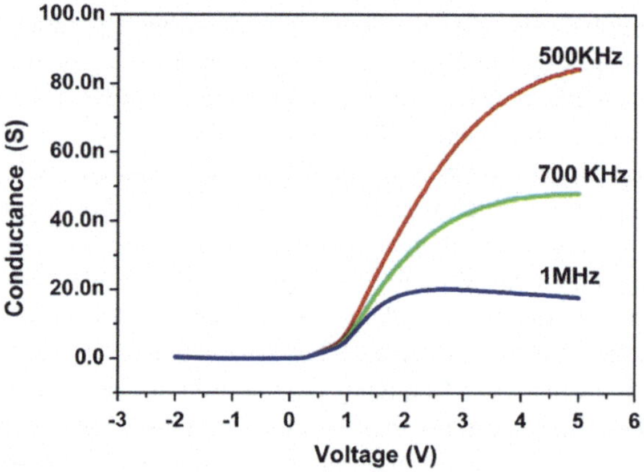

Fig. 40.3 G-V plots of Ag/MgPc/GaAs/Au-Ge organic heterojunction for 500 kHz–1 MHz frequency range

$$\sigma' = \sigma_{AC} = \omega \varepsilon_0 \varepsilon' \tan \delta \tag{1}$$

It is worthy to know the conduction mechanism in such organic heterostructure device, so we calculate the exponents s1 and s2 using following expression and the lnσ plots versus lnω (not shown here) [4];

$$\sigma = \sigma_0 + A\omega^{s1} + B\omega^{s2} \tag{2}$$

The Fig. 40.3 depicts the G-V characteristics plots Ag/MgPc/GaAs/Au-Ge organic heterojunction for 500 kHz–1 MHz. G values are so weak then it is observed that an increase of conductance is recorded within forward bias range with a decrease of frequency. The highest value exceeds 85 nS at 5 V for 500 kHz. From the linear distinct region, the values between 0 and 1 corresponding to low or intermediate frequency region is considered to be a result of the interaction among charge carriers and trap states.

No change in Ac conductivity is observed within reverse voltage region till 0.3 V, then a strong growth is recorded from 15 nS/m to 3.75 μS/m at 500 kHz then value becomes gradually lesser to 3 and 1.6 μS/m at 5 V as shown in Fig. 40.4. Consequently, Ac conductivity σ_{Ac} of MgPc/GaAs organic device is strongly frequency-dependent. Many works of Ac conductivity which describe such behaviors and dielectric response to Ac signal are reported [8] (Fig. 40.4).

The Fig. 40.5 displays the dielectric constant of Ag/MgPc/GaAs/Au-Ge organic heterojunction for 500 kHz–1 MHz frequency range. A slight increase till a prolonged peak as indicated by arrow in Fig. 40.5 is recorded. Hence, a rapid decay of dielectric function within the forward voltage range occurs till negative values of −5 at 5 V. The Cole-Cole graphs displays the imaginary M″ and real M′ parameters of impedances as shown in Fig. 40.6 at discrete excitation frequencies within 500 kHz–1 MHz.

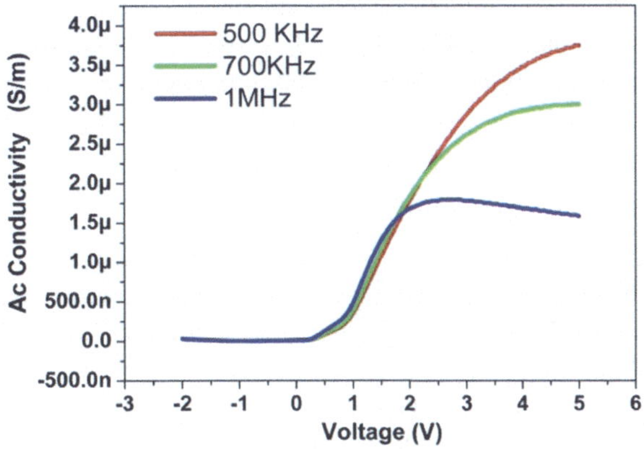

Fig. 40.4 Ac conductivity plots against bias voltage of Ag/MgPc/GaAs/Au-Ge organic heterojunction for 500 kHz–1 MHz frequency range

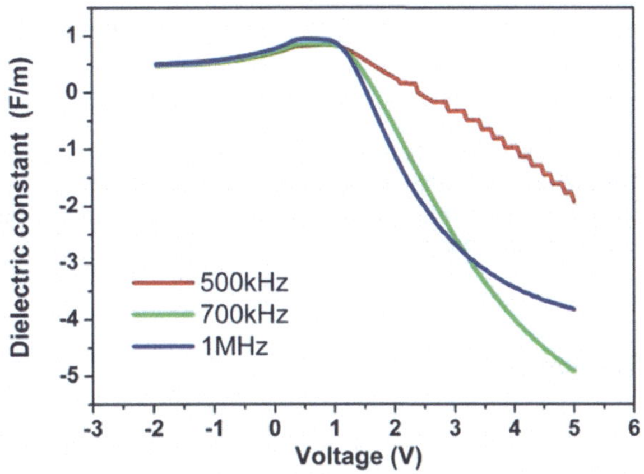

Fig. 40.5 Dielectric function vs. voltage of Ag/MgPc/GaAs/Au-Ge organic heterojunction for 500 kHz–1 MHz frequency range

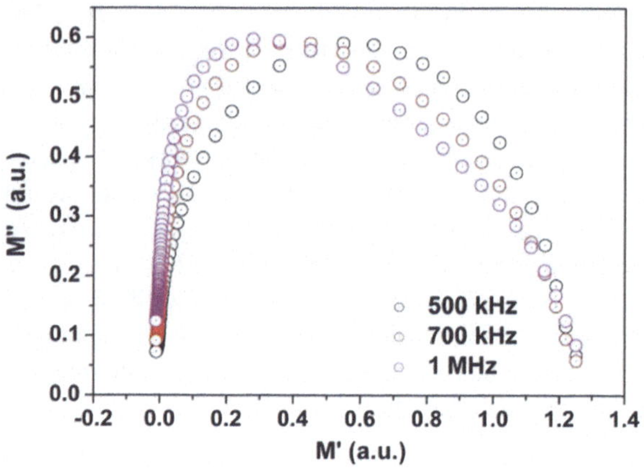

Fig. 40.6 Cole-Cole plots of Ag/MgPc/GaAs/Au-Ge organic heterojunction for 500 kHz–1 MHz frequency range

The semicircle tends to be asymmetric, having a width of 1.3, when frequency is higher than 500 kHz. The radii of semicircle are roughly the same, M″ reaches the value of 0.6 for 1 MHz and slightly decreases for the lower frequencies.

40.4 Conclusion

Throughout this research the capacitance, conductance and Cole-Cole characteristics of MgPc/GaAs organic heterostructure for higher frequencies (500 kHz–1 MHz) are investigated. Ac signal conductivity rises with voltage value but decreases with increase in frequency from 15 nS/m to 3.75 μS/m. Dielectric function versus voltage of Ag/MgPc/GaAs/Au-Ge organic heterojunction for 500 kHz–1 MHz frequency range obeys to the same profile as that of C-V and negative values are recorded within forward bias voltage range. Cole-Cole diagram presents semicircles indicating an asymmetric profile and high value of M″ is 0.6.

Acknowledgements The work is inserted in PRFU 2018 PROJECT under contract B00L02UN31 0220180011. http://www.prfu-mesrs.dz/. https://www.mendeley.com/impact/mostefa-benhaliliba, https://orcid.org/0000-0001-6507-3663. This work is also supported by the Ministry of Development of Turkey under project number: 2016K121220. We dedicate this work to every researcher in the world, especially in biology and virology, to all doctors and all those working in nursing and pharmacy.

References

1. M. Bok, J.-H. Jeong, E. Lim, Mater. Chem. Phys. **227**(1), 250–254 (2019)
2. R.A.M. Gamboa, O.A. Jaramillo-Quintero, Y.A.A. Altamirano, M.O. Concha-Guzmán, M.E. Rincón, J. Colloid Interface Sci. **535**, 400–407 (2019)

3. D. Hewidy, A.-S. Gadallah, G. AbdelFattah, Phys. B Condens. Matter **507**, 46–50 (2017)
4. M. Benhaliliba, T. Asar, I. Missoum, Y.S. Ocak, S. Özçelik, C.E. Benouis, A. Arrar, Ac conductivity and impedance spectroscopy study and dielectric response of MgPc/GaAs organic heterojunction for solar energy application. Phys. B Phys. Condens. Matter (2019). https://doi.org/10.1016/j.physb.2019.411782
5. B.K. Jones, J. Santanat, M. McPherson, Solid State Commun. **107**(2), 47–50 (1998)
6. A. Büyükbaş-Uluşan, A. Tataroğlu, Y. Azizian-Kalandaragh, M. Koşal, J. Mater. Sci.: Mater. Electron. **30**(10), 9572–9581 (2019). https://doi.org/10.1007/s10854-019-01291
7. Ç. Bilkan, A. Gümüş, Ş. Altındal, Mater. Sci. Semicond. Process. **39**, 484–491 (2015)
8. A.A.M. Farag, A. Ashery, M.A. Salem, Superlattices Microstruct. **135**, 106277 (2019)

Chapter 41
Classification of the Categories of Amorphous Hydrogenated Silicon Oxynitride Films Using Infrared Spectroscopy

Mohammed Boulesbaa and Ahmed Bouchekhlal

Abstract In this paper, a contribution to the classification of the various categories of amorphous hydrogenated silicon oxynitride films using infrared spectroscopy has been carried out. A wide variety of these films were prepared by plasma enhanced chemical vapor deposition technique. As the gas ratio of $N_2O/(N_2O + NH_3)$ flow rates increases from 0 to 0.97, four categories of amorphous films were identified by infrared spectroscopy namely: the hydrogenated silicon nitride, the nitrogen-rich silicon oxynitride, the oxygen-rich silicon oxynitride and the highly oxygen-rich silicon oxynitride films. Finally, the concentration of Si-O and Si-N bonds present in the hydrogenated silicon oxynitride films were successfully correlated with the gas ratio $R = N_2O/(N_2O + NH_3)$, in the context to optimize of their vibrational properties for different applications.

Keywords Silicon oxynitride · Infrared spectroscopy · Gaussian deconvolution

41.1 Introduction

In recent years, increasing attention has been paid to silicon based dielectrics such as; silicon oxide, silicon nitride and silicon oxynitride as potential materials for the development of solar cells application [1] and optoelectronics devices [2].

M. Boulesbaa (✉)
Laboratoire de Rayonnement et Plasmas et Physique des Surfaces (L.R.P.P.S), Université Kasdi Merbah Ouargla, 30000 Ouargla, Algeria
e-mail: boulesbaa.mohammed@univ-ouargla.dz

M. Boulesbaa
Département d'Electronique et des Télécommunications, Université Kasdi Merbah Ouargla, 30000 Ouargla, Algeria

A. Bouchekhlal
Laboratoire de Microsystème et Instrumentation, Université de Constantine 1, Constantine, Algeria

A. Belasri and S. A. Beldjilali (eds.), *ICREEC 2019*, Springer Proceedings in Energy,
https://doi.org/10.1007/978-981-15-5444-5_41

SiO_xN_y:H material has an important optical properties such as transparency that extends from the visible to near infrared and refractive index adjusted to a wide range (n = 1.45–2.0). A variety of deposition techniques has been employed to steer the nitrogen and oxygen concentration in the films composition. A wide variety of these films were prepared by plasma enhanced chemical vapor deposition (PECVD) technique. Infrared absorption spectroscopy (FTIR) is a powerful tool for providing information in terms of identifying and quantifying the various chemicals bonds incorporated in SiO_xN_y:H films such as Si-N, Si-O, N-H and Si-H chemical bonds. In this work, a contribution to the classification of the various categories of amorphous hydrogenated silicon oxynitride films using infrared spectroscopy has been carried out.

41.2 Materials and Methods

To complete the previously reported works [3] by an infrared study, we used the same SiO_xN_y:H thin films. They are obtained by a mixture of silane (SiH_4) diluted in the nitrogen (N_2), ammonia (NH_3) and the nitrogen protoxide (N_2O) using PECVD technique. The films were deposited on single crystalline silicon substrates. In this study, the R ($R = N_2O/(N_2O + NH_3)$) gas ratio values are varied between 0 and 0.97 by changing N_2O flow rate from 0 to 500 standard cubic centimeter per minute (sccm), while the flow rates of SiH_4 and NH_3 kept constant at 35 and 15 sccm, respectively.

The physicochemical analysis of amorphous silicon oxynitride thin films elaborated by PECVD technique has been carried out by infrared spectrometry using Nicolet Avatar 360 spectrometer. The infrared spectra were acquired in absorption mode with a wave number ranging from 400 to 4000 cm^{-1}. To accurate identification and quantification of Si-O and Si-N bond concentrations, the Gaussian deconvolution has been used for decomposing the broad band into several separate components versus the gas ratio R.

In order to quantify the Si-O and Si-N bond concentrations, we used the following equation [4]:

$$N_{Si-X} = \frac{K_{Si-X}}{\omega_0} \int_{\omega} \alpha(\omega) d\omega \tag{41.1}$$

where $N_{Si\text{-}X}$ represents the absolute concentration of the Si-X, $\alpha(\omega)$ the absorption coefficient at the wave number ω and ω_0 corresponding to the maximum absorption. $K_{Si\text{-}X}$ is a called calibration factor linked to the bond oscillator force. The values of the proportionality coefficient used to calculate of the bond concentrations are: $K_{Si\text{-}O} = 1.48 \times 10^{19} cm^{-2}$ [5] and $K_{Si-N} = 6.30 \times 10^{18} cm^{-2}$ [6].

41.3 Results and Discussion

Figure 41.1 shows the FTIR spectra of amorphous SiO_xN_y:H thin films deposited by PECVD technique at different gas ratios R. This figure illustrates a broad predominant absorption band localized between about 650 cm^{-1} and about 1350 cm^{-1}. In addition, these spectra illustrate the appearance of peaks series. Two peaks with low intensities were located at 3350 and 2190 cm^{-1} corresponding to N-H stretching vibration mode and Si-H stretching vibration mode, respectively [7–9]. The peak localized at around 465 cm^{-1} is attributed to the Si-O rocking vibration mode [10] in SiO_xN_y:H thin films deposited with a R equal or higher than 0.83. On the other hand, this same peak is attributed to the Si-N breathing vibration mode (475 cm^{-1}) [11] in thin films elaborated with R lower than or equal to 0.77. We also observe that the main peak or predominant band moves from the Si-N stretching mode situated at around 846 cm^{-1} to the Si-O symmetric stretching mode situated at around 1025 cm^{-1} when the R change from the 0 to 0.97. The shoulder at the left or right side of broad band is also detected. On the other hand, a remarkable shift of the main peak to great wave number has been observed in the infrared spectra. These complex phenomena can be explained by the Gaussian deconvolution of this band.

The Gaussian deconvolution of the broad band of infrared spectrum of the film deposited with a ratio R = 0.70 allows us to obtain a three Gaussian components (see Fig. 41.2). The component situated at 836 cm^{-1} is attributed to the Si-N stretching mode [5, 11, 12]. The component localized at 946 cm^{-1} is attributed to

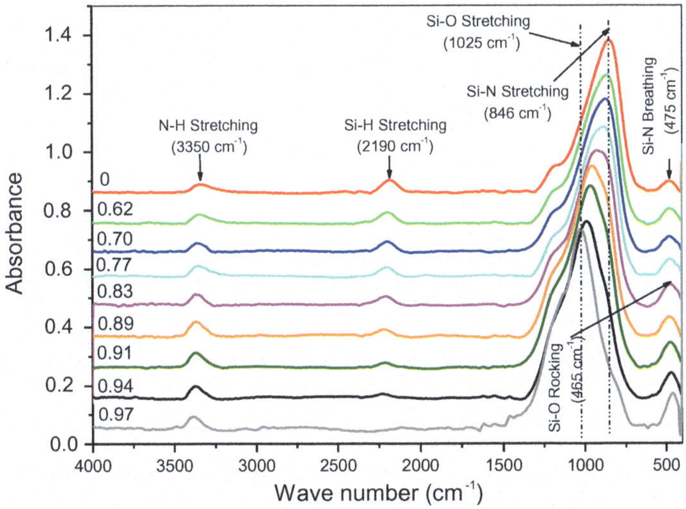

Fig. 41.1 Infrared absorption spectra of amorphous hydrogenated silicon oxynitride films with various R ratios showed above each spectrum

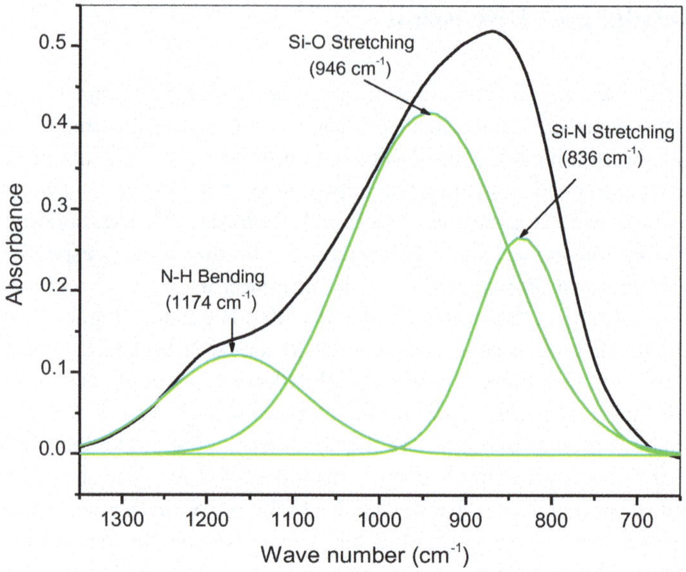

Fig. 41.2 Example the Gaussian deconvolution of the broad band of infrared spectrum of the thin film deposited at R = 0.70

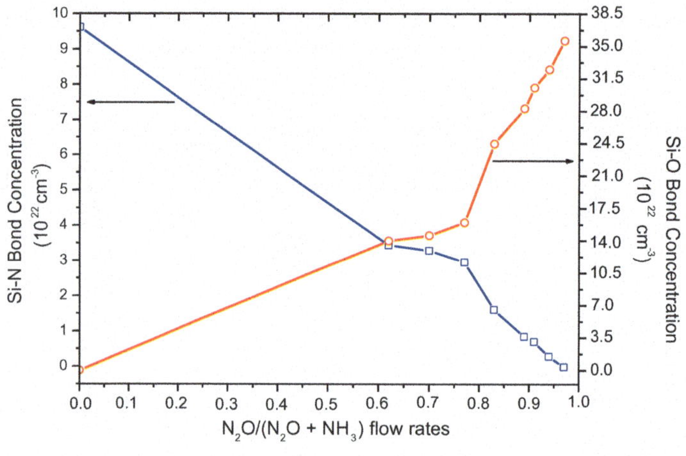

Fig. 41.3 Si-O and Si-N chemical bonds concentration in the amorphous hydrogenated silicon oxynitride films plotted as a function of R ratio

Si–O stretching mode [5, 6, 13]. Finally, the component situated at 1174 cm^{-1} is attributed to the bending of the N-H mode [14].

Figure 41.3 presents the variation of the Si-O and Si-N bonds concentration in the SiO_xN_y:H films as a function of R ratio. These results can be explained by the progressive incorporation of oxygen atoms in the films with increasing of R ratio. This progressive incorporation of the oxygen atoms led to form Si-O bonds instead of Si-N ones because of the preferably linked of the silicon atoms with oxygen in compared with nitrogen.

From the identification of the different absorption bands (Si-N, Si-O, Si-H and N-H) characteristic of the amorphous SiO_xN_y:H films, the elaborated samples can be mainly classified into four categories with respect to the R values:

- Category 1 (R = 0): This category is characterized by the presence of a series of Si-N bonds in the infrared spectra. Also, this film is characterized by a higher content of hydrogen incorporated in the form of Si-H and N-H bonds; a small amount of excess silicon; a fine layer of native oxide. This category of film can be called "hydrogenated silicon nitride" and the film composition can be written as "SiN_X:H".
- Category 2 ($0.62 \leq R \leq 0.77$): It is similar to the previous category (1) with lower incorporation of oxygen in films in the form of Si-O bond. In this case, the peak associated to Si-N vibration mode is predominant in the infrared spectra. This category of films can be called "nitrogen-rich silicon oxynitride" and the film composition can be written as "nitrogen-rich SiO_xN_y:H".
- Category 3 ($0.83 \leq R \leq 0.94$): It is characterized by higher incorporation of oxygen atoms in the films in the form of many vibration types of Si-O bonds. In this case, the peak associated with Si-O vibration is predominant in infrared spectra. These films are characterized also by a lower incorporation of nitrogen and hydrogen atoms in the films with the disappearance of the N-H bending vibration. Therefore, the film composition can be written as "oxygen-rich SiO_xN_y:H".

Category 4 (R = 0.97): It is a transition zone towards silicon oxide (SiO_2) characterized by a total disappearance of the Si-N and Si-H bonds in the films. Thus, deposit composition can be written as "highly oxygen-rich SiO_xN_y:H film".

From the calculation of the Si-N and Si-O bonds concentrations, we presented the evolution of the concentration of $C_{Si\text{-}O}/(C_{Si\text{-}O}+C_{Si\text{-}N})$ ratio, (concentration of Si-O/(concentration of Si-O + concentration of Si-N)), as a function of R ratio in the Fig. 41.4. This figure shows that the composition of our films evolved almost linearly from SiN_x:H to the highly oxygen-rich SiO_xN_y:H film one. If we consider that the silicon atoms are linked with oxygen and nitrogen atoms without any preference or priority, all of the points in the bond concentration ratio would be on the dashed line drawn in Fig. 41.4. So, the value of $C_{Si\text{-}O}/(C_{Si\text{-}O}+ C_{Si\text{-}N})$ ratio is the same as $R = N_2O/(N_2O+NH_3)$ for each SiO_xN_y:H thin film. We also observe that every point the $C_{Si\text{-}O}/(C_{Si\text{-}O}+ C_{Si\text{-}N})$ concentration ratio is situated above the dashed line, except the ratio corresponding to the hydrogenated silicon nitride film (R = 0). This suggests that silicon atoms preferentially establish chemical bonds with oxygen atoms rather than nitrogen atoms.

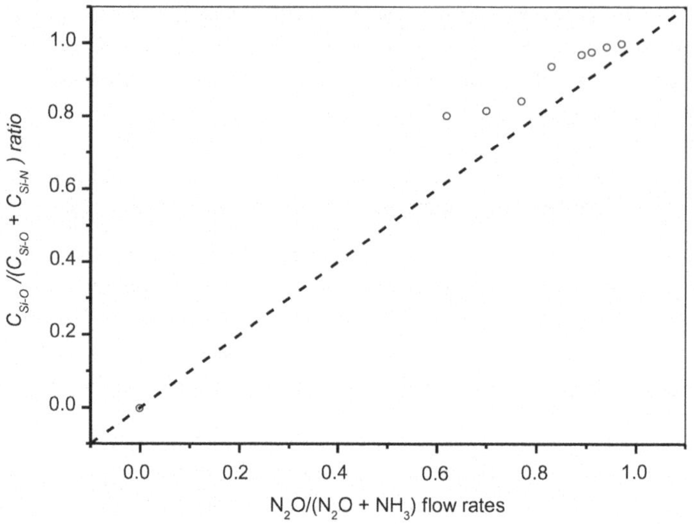

Fig. 41.4 Proportion of Si-N and Si-O densities as a function of R ratio (open circle share). The dashed line represents the ideal case where oxygen and nitrogen atoms are bonded to silicon atom without any preference or priority order

41.4 Conclusion

Hydrogenated amorphous silicon oxynitride films were deposited by PECVD technique from the following reactive gases: NH_3, SiH_4 and N_2O. Four types of amorphous thin films were identified in this study namely; the hydrogenated silicon nitride (R = 0), the nitrogen-rich SiO_xN_y:H ($0.62 \leq R \leq 0.77$), the oxygen-rich SiO_xN_y:H ($0.83 \leq R \leq 0.94$) and the highly oxygen-rich SiO_xN_y:H film (R = 0.97). We found that the oxygen atoms were gradually substituting the nitrogen ones to form Si-O bonds in detriment of the Si-N ones as the gas ratio R increases. This difference in chemical composition caused a shift of the absorption maximum of the broad bands in the infrared spectra to higher wave numbers due to the more electronegativity of oxygen atoms compared to the nitrogen ones. This study also showed that the silicon atoms are preferentially bonded to oxygen compared with nitrogen. This work displays a good correlation between the chemical composition (Si-O, Si-N bonds densities) and the deposition conditions.

References

1. N. Balaji, H.T.T. Nguyen, C. Park, M. Ju, J. Raja, S. Chatterjee, R. Jeyakumar, J. Yi, Electrical and optical characterization of SiO_xN_y and SiO_2 dielectric layers and rear surface passivation by using SiO_2/SiO_xN_y stack layers with screen printed local Al-BSF for c-Si solar cells. Curr. Appl. Phys. **18**, 107–113 (2018)

2. M. Boulesbaa, A. Bouchekhlal, F. Chebbara, B. Lakehal, L. Lakhdar, Modeling of birefringence properties in SiO_xN_y film rib optical waveguides, in *2018 International Conference on Communications and Electrical Engineering (ICCEE)* (2018)
3. R. Mahamdi, M. Boulesbaa, L. Saci, F. Mansour, C. Molliet, M. Collet, P. Temple-Boyer, Ellipsometric and Rutherford back scattering spectroscopy studies of SiO_xN_y films elaborated by plasma-enhanced chemical vapor deposition. J. Nanosci. Nanotechnol. **11**, 9118–9122 (2011)
4. M.H. Brodsky, M. Cardona, J.J. Cuomo, Infrared and Raman spectra of the silicon-hydrogen bonds in amorphous silicon prepared by glow discharge and sputtering. Phys. Rev. B **16**, 3556–3571 (1977)
5. L.N. He, D.M. Wang, S. Hasegawa, A study of plasma-deposited amorphous SiOx:H ($0 \leq x \leq 2.0$) films using infrared spectroscopy. J. Non-Cryst. Solids **261**, 67–71 (2000)
6. F. Rebib, E. Tomasella, M. Dubois, J. Cellier, T. Sauvage, M. Jacquet, SiO_xN_y thin films deposited by reactive sputtering: process study and structural characterization. Thin Solid Films **515**, 3480–3487 (2007)
7. F. Ay, A. Aydinli, Comparative investigation of hydrogen bonding in silicon based PECVD grown dielectrics for optical waveguides. Opt. Mater. **26**, 33–46 (2004)
8. J. Dupuis, E. Fourmond, D. Ballutaud, N. Bererd, M. Lemiti, Optical and structural properties of silicon oxynitride deposited by plasma enhanced chemical vapor deposition. Thin Solid Films **519**, 1325–1333 (2010)
9. B. Hallam, B. Tjahjono, S. Wenham, Effect of PECVD silicon oxynitride film composition on the surface passivation of silicon wafers. Sol. Energy Mater. Sol. Cells **96**, 173–179 (2012)
10. K.F. Albertin, I. Pereyra, Improved effective charge density in MOS capacitors with PECVD SiO_xN_y dielectric layer obtained at low RF power. J. Non-Cryst. Solids **354**, 2646–2651 (2008)
11. A. Benami, G. Santana, A. Ortiz, A. Ponce, D. Romeu, J. Aguilar-Hernandez, G. Contreras-Puente, J.C. Alonso, Strong white and blue photoluminescence from silicon nanocrystals in SiN_x grown by remote PECVD using $SiCl_4/NH_3$. Nanotechnology **18**, 1–5 (2007)
12. Y. Ogawa, K. Ohdaira, T. Oyaidu, H. Matsumura, Protection of organic light-emitting diodes over 50000 hours by Cat-CVD SiN_x/SiO_xN_y stacked thin films. Thin Solid Films **516**, 611–614 (2008)
13. Y. Liu, I.K. Lin, X. Zhang, Mechanical properties of sputtered silicon oxynitride films by nanoindentation. Mater. Sci. Eng. A **489**, 294–301 (2008)
14. M. Boulesbaa, Optical and physicochemical properties of hydrogenated silicon nitride thin films: effect of the thermal annealing. Spectrosc. Lett. **50**, 1–5 (2017)

Chapter 42
Structural and Optical Parameters of SnO_2 Nanoparticle Deposited by Spray Pyrolysis

Y. Bakha, H. Khales, A. Tab, A. Smatti, and S. Hamzaoui

Abstract SnO_2 nanoparticle thin films are obtained using simple spray pyrolysis chemical process with 0.1 molar concentrations on glass substrate at: T = 300, 325 and 350 °C respectively. The deposited films were investigated using the appropriate method of characterization. The X-ray diffraction revolved that the deposited SnO_2 nanoparticle thin films have tetragonal rutile structure. The scanning microscopy (SEM) and Atomic force microscopy (AFM) indicate that films are homogenous and uniform in all direction with small grain size (10–14) nm. The UV-visible spectrophotometry showed that the transmittance of the films is above 85% in the visible and band gap Eg around 4 eV.

Keywords SnO_2 · Nanoparticle · Spray pyrolysis

42.1 Introduction

Tin dioxide films SnO_2 is one of most important materials with direct band gap around (3.8–4.3) and indirect gap (2.7–3.1) with transparency in the visible region about 80% [1]. Many researches were interested on tin dioxide for his various and interesting properties in chemical, physical and structural parameters, and high optical transmittance in the visible, It is popular materials used in large applications such as transparent conductive film and window materials in solar cell applications [2], gas sensing [3–5], optoelectronic materials [6], and in several works in anode material for lithium ion batteries [7, 8]. Actually many researches are focused on

Y. Bakha (✉) · H. Khales · A. Tab · A. Smatti
CDTA Centre for Development of Advanced Technologies,
Cité 20 Août 1956, Baba Hassen, BP. 17, 16303 Algiers, Algeria
e-mail: ybakha@cdta.dz

A. Tab · S. Hamzaoui
Laboratoire de Microscope Electronique & Sciences des Matériaux,
Université des Sciences et de la Technologie d'Oran Mohamed Boudiaf,
El-Mnaouar, BP 1505, 31000 Oran, Algeria

A. Belasri and S. A. Beldjilali (eds.), *ICREEC 2019*, Springer Proceedings in Energy,
https://doi.org/10.1007/978-981-15-5444-5_42

nanoparticle properties of SnO_2 thin films deposited by different process. The quality and film application is related to the process. We can divide them into two parts: physical deposition such as sputtering [9], laser ablation [10], thermal physical vapor deposition [11, 12]; and the chemical process as; spray pyrolysis [13], sol gel [7, 14] and spin coating [15]. In our purpose nanoparticle of SnO_2 were obtained using spray pyrolysis method. This chemical spray process is advantageous for his simplicity, low cost and process yield. We demonstrate that we can obtain nanoparticle materials in the range of 10 nm by using this simple technique. The properties of our films are considered through the characterization by scanning microscopy (SEM) and Atomic force microscopy (AFM), X-ray diffraction (XRD) and transmittance.

42.2 Experimental Details

Deposition of SnO_2 nanoparticle thin films was performed using spray pyrolysis process in ambient atmosphere, with 0.1 M concentrations at: T = 300, 325 and 350 °C, using a starting solution of tin chloride ($SnCl_2$, $2H_2O$) diluted in methanol 99.99% as solvent (CH_3OH). The distance between the substrate and the spray gun nozzle was fixed at 33 cm. A schematic drawn of the spray pyrolysis apparatus for the synthesis of the SnO_2 nanoparticle thin films is illustrated in Fig. 42.1.

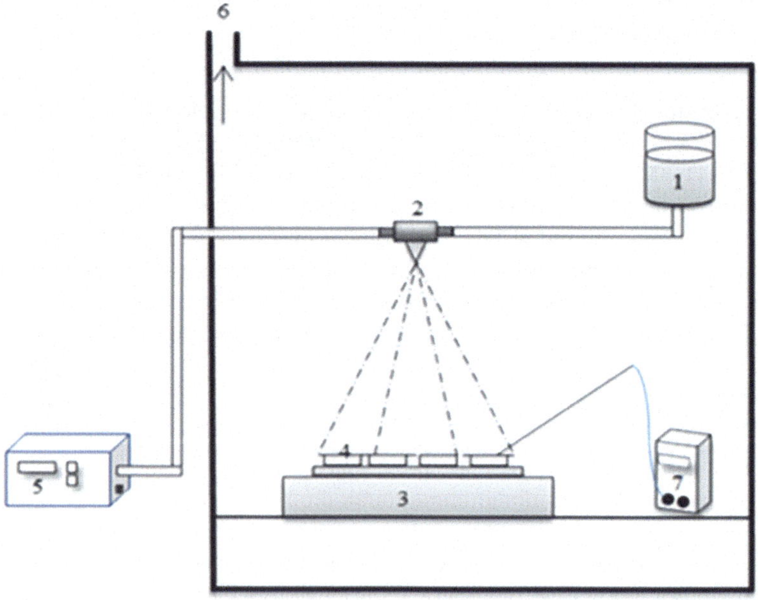

Fig. 42.1 Schematic draws of chemical spray pyrolysis process. 1—Solution, 2—Atomizer, 3—Heater system, 4—Substrate, 5—Air compressor, 6—Exhaust and 7—Thermocouple

42.3 Results and Discussion

In this paper, we discuss the structural and optical properties of SnO_2 nanoparticle thin films deposited at several temperatures in the range of 300–350 °C for 0.1 M concentrations. The surface morphologies of films were examined by scanning microscopy (SEM) (JEOL JSM-6610LA) and Atomic force microscopy JEOL JSPM 5200 AFM. The investigation of structural properties of the SnO2 thin films was performed through X-ray diffraction (XRD) D8 Advance Bruker, using the Cu Kα radiation (λ = 0.15406 Å). Measurements of transmittance were performed in the range of 300 to 1200 nm UV-Visible spectrophotometer (Perkin Elmer lambda 20).

42.3.1 Scanning Electron Microscopy (SEM)

Figure 42.2 shows the scanning electron microscope (SEM) images of SnO_2 thin films deposited at (a) T = 300 °C, (b) T = 325 °C, (c) T = 350 °C. The surface of the thin films displays a homogenous appearance in all scanned areas.

42.3.2 Atomic Force Microscopy (AFM)

The surface morphologies of the SnO_2 thin films were investigated by using atomic force microscopy (AFM), as illustrated in Fig. 42.3. The atomic force microscopy surface 3D images showed the surface roughness variations of SnO_2 thin films with 0.1 M as function of temperature of deposition, the grains can be seen almost all over the substrate.

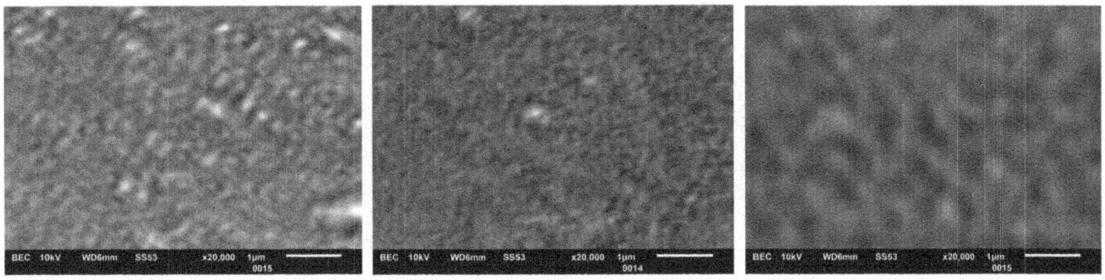

(a) **(b)** **(c)**

Fig. 42.2 SEM images of SnO_2 thin film deposited at: **a** T = 300 °C, **b** T = 325 °C and **c** T = 350 °C

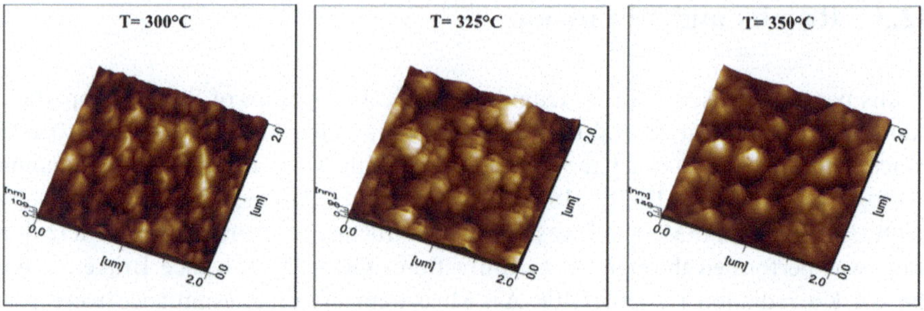

Fig. 42.3 AFM surface 3D images of SnO_2 (0.1 M) at: T = 300 °C, T = 325 °C and T = 350 °C

42.3.3 X-ray Diffraction Characterization

The X-ray diffraction analysis of the SnO_2 sample was used in order to identify the crystal structure and to estimate average grain size. The X-ray diffraction pattern of SnO_2 sample deposited by spray pyrolysis technique is as illustrated in Fig. 42.4.

The crystallite size of the thin films was determined using the Scherer's formula [16]:

$$D = \frac{K\lambda}{\beta \, cos\theta_{(hkl)}} \quad (1)$$

where K = 0.9, D is the crystallite size, λ = 0.15406 nm the mean wavelength of Cu Kα radiation and β the full-width at half maximum (FWHM) of Bragg peak observed at Bragg angle θ (rad). The average grain sizes in different orientations were determined from the diffraction peak width.

This investigation can clearly reveals formation of SnO_2 nanoparticles with polycrystalline phase demonstrating rutile structure comparing with JCPDS Card

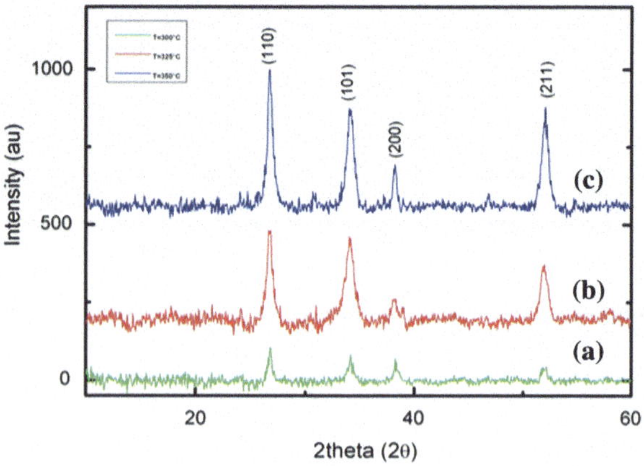

Fig. 42.4 XRD patterns of SnO_2/0.1 M thin films deposited at: **a** T = 300 °C, **b** T = 325 °C and **c** T = 350 °C

No: 41-1445, a = 4.743 Å. C = 3.1859 Å. The crystal planes (110) is most intense peak which is observed for all samples and other peaks assigned as (101), (211) orientations are also observed in XRD indicating of SnO_2 thin films. The highest density of the atoms can explain the highest intensity of the peak (110) in all planes of the rutile crystal structure; therefore, the surface energy of the (110) plane is the lowest.

The value of lattice constant of the deposited sample is determinate from different (hkl) planes using the formula [17]:

$$\frac{1}{d^2} = \left(\frac{h^2 + k^2}{a^2}\right) + \frac{l^2}{c^2} \tag{2}$$

where d is inter-planer distance and (hkl) are Miller indices. The calculated average values of lattice constant were a = 4.7020 Å, c = 3.1634 Å. No impurity phase was observed. The average crystallite size (D) was determined from the XRD spectra using Scherer's formula was (10 to 14) nm.

42.3.4 Optical Transmission

The transmittance spectra of SnO_2 nanoparticle thin films on glass substrate are shown in Fig. 42.5a. It is found that the transmittance is above 85%. The optical band gap with direct transition Eg was obtained by extrapolating the linear portion of the plot (αhν)2 versus (hν) to α = 0 according to the following equation [18]:

$$\alpha = A(h\nu - Eg)^{1/2} \tag{3}$$

The optical band gap Eg is illustrated in Fig. 42.5b. It rises to 3.80 eV, this value is higher than the value of Eg = 3.57 eV reported for single crystal SnO_2 [19].

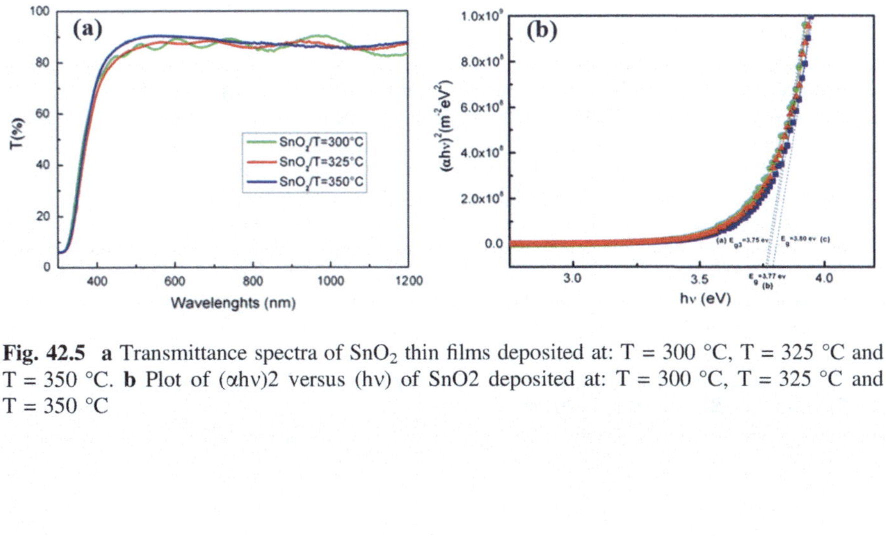

Fig. 42.5 **a** Transmittance spectra of SnO_2 thin films deposited at: T = 300 °C, T = 325 °C and T = 350 °C. **b** Plot of (αhν)2 versus (hν) of SnO2 deposited at: T = 300 °C, T = 325 °C and T = 350 °C

42.4 Conclusion

SnO2 nanoparticle thin films have been prepared using spray pyrolysis chemical deposition technique. In this method the most important deposition parameters are precursors, the solution concentration and the temperature. After having examined all the results obtained by the different techniques of characterization we have found that a high quality of thin films was obtained. The AFM image revolves a homogenous surface in all samples with small grain size which confirmed by X-Ray diffraction, the grain sizes are above (10–14) nm. A high optical transparency with band gap around 4 eV has been obtained by the transmittance characterization. These interesting properties of the film material are useful for both solar cell and gas sensing applications.

References

1. Z.M. Jarzebski, J.P. Marton, Physical properties of SnO_2 materials I. Preparation and defect structure. J. Electrochem. Soc. **123**(7), 199C–205C (1976)
2. H.J. Snaith, C. Ducati, SnO_2-based dye-sensitized hybrid solar cells exhibiting near unity absorbed photon-to-electron conversion efficiency. Nano Lett. **10**(4), 1259–1265 (2010)
3. A. Yang, X. Tao, R. Wang, S. Lee, C. Surya, Room temperature gas sensing properties of SnO_2/multiwall-carbon-nanotube composite nanofibers. Appl. Phys. Lett. **91**(13), 133110 (2007)
4. S. Lee, S. Kim, B. Hwang, S. Jung, D. Ragupathy, I. Son, J. Kim, Improvement of H_2S sensing properties of SnO_2-based thick film gas sensors promoted with MoO_3 and NiO. Sensors **13**(3), 3889–3901 (2013)
5. A. Debataraja, D.W. Zulhendri, B. Yuliarto, B. Sunendar, Investigation of nanostructured SnO_2 synthesized with polyol technique for CO gas sensor applications. Procedia Eng. **170**, 60–64 (2017)
6. M. Diantoro, Kholid, A.A. Mustikasari, Yudiyanto, The influence of SnO2 nanoparticles on electrical conductivity, and transmittance of PANI-SnO2 films. IOP Conf. Ser. Mater. Sci. Eng **367**, 012034 (2018)
7. H. Kose, A.O. Aydin, H. Akbulut, Sol-gel synthesis of nanostructured $Sn0_2$ thin film anodes for li-ion batteries. Acta Phys. Pol. Ser. Gen. Phys. **121**(1), 227 (2012)
8. X. Fan, J. Shao, X. Xiao, X. Wang, S. Li, H. Ge, C. Wang, In situ synthesis of SnO_2 nanoparticles encapsulated in micro/mesoporous carbon foam as a high-performance anode material for lithium ion batteries. J. Mater. Chem. A **2**(43), 18367–18374 (2014)
9. Z. Banyamin, P. Kelly, G. West, J. Boardman, Electrical and optical properties of fluorine doped tin oxide thin films prepared by magnetron sputtering. Coatings **4**(4), 732–746 (2014)
10. H. Kim, R.C.Y. Auyeung, A. Piqué, Transparent conducting F-doped SnO_2 thin films grown by pulsed laser deposition. Thin Solid Films **516**(15), 5052–5056 (2008)
11. S.H. Luo, Q. Wan, W.L. Liu, M. Zhang, Z.T. Song, C.L. Lin, P.K. Chu, Photoluminescence properties of SnO_2 nanowhiskers grown by thermal evaporation. Prog. Solid State Chem. **33** (2–4), 287–292 (2005)
12. S.J. Ikhmayies, Properties of amorphous SnO_2 thin films prepared by thermal evaporation. Int. J. Mater. Chem. **2**(4), 173–177 (2012)
13. C. Nassiri, A. Hadri, F.Z. Chafi, A. El Hat, N. Hassanain, M. Rouchdi, A. Mzerd, Structural, optical and electrical properties of Fe doped SnO_2 prepared by spray pyrolysis. J. Mater. Environ. Sci. **8**(2), 420–425 (2017)

14. M. Aziz, S.S. Abbas, W.R.W. Baharom, Size-controlled synthesis of SnO_2 nanoparticles by sol–gel method. Mater. Lett. **91**, 31–34 (2013)
15. W. Izydorczyk, K. Waczyński, J. Izydorczyk, P. Karasiński, J. Mazurkiewicz, M. Magnuski, W. Filipowski, Electrical and optical properties of spin-coated SnO_2 nanofilms. Mater. Sci. Pol. **32**(4), 729–736 (2014)
16. G. Scherrer, Nachr. **2**, 98 (1918)
17. A. Abdelkrim, S. Rahmane, O. Abdelouahab, A. Hafida, K. Nabila, Optoelectronic properties of SnO_2 thin films sprayed at different deposition times. Chin. Phys. B **25**(4), 046801 (2016)
18. A. Koçyiğit, D. Tatar, A. Battal, M. Ertuğrul, B. Düzgün, Highly efficient optoelectronic properties of doubly doped SnO_2 thin film deposited by spin coating technique. J. Ovonic Res. **8**(6), 171–178 (2012)
19. R. Summitt, J.A. Marley, N.F. Borrelli, The ultraviolet absorption edge of stannic oxide (SnO_2). J. Phys. Chem. Solids **25**(12), 1465–1469 (1964)

Chapter 43
N Type Microcrystalline Silicon Oxide Layer Effect in P-I-N Ultra-Thin Film Solar Cell

Wafa Hadj Kouider, Abbas Belfar, Mohammed Belmekki, and Hocine Ait Kaci

Abstract Light trapping plays an important role in improving the power conversion efficiency of thin-film hydrogenated amorphous silicon solar cells. Therefore, doped microcrystalline silicon oxide has received much attention due to its versatile applicability in photovoltaic technologies. It is a mixed phase material of an oxygen rich amorphous silicon oxide phase, which supplies low refractive index, a wide band gap and high optical transparency and the doped microcrystalline silicon phase, which guaranties agreeable electrical conductivity. In order to evaluate the microcrystalline silicon oxide based single junction p-i-n solar cell efficiency, a simulation study is executed using AMPS-1D (Analysis of Microelectronic and Photonic Structures) simulator.

The simulation results show that hydrogenated microcrystalline silicon oxide n-layer has a significant effect on the cell performance and that amorphous silicon a-Si:H solar cells using hydrogenated microocrystalline silicon oxide instead of the conventional hydrogenated amorphous silicon n-layer show good electrical performance, which is apparent as an increase of efficiency from 6.513% to 8.736%.

Keywords Ultra-thin film · AMPS-1D · n-layer · nc-SiOx:H · Solar cell

43.1 Introduction

Semiconductors have become the preferred material for processing, transfer and the storage of information [1]. Hydrogenated amorphous silicon (a-Si:H) based thin-film solar cells are very promising for large-area and low cost photovoltaic

W. Hadj Kouider (✉) · A. Belfar · M. Belmekki · H. Ait Kaci
Laboratory of Plasma Physics, Conductor Materials and their Applications,
Faculty of Physics, University of Sciences and Technology of Oran Mohamed Boudiaf,
USTO-MB, BP1505 Oran, Algeria
e-mail: wafa.hadjkouider@univ-usto.dz

A. Belfar
e-mail: abbas.belfar@univ-usto.dz

A. Belasri and S. A. Beldjilali (eds.), *ICREEC 2019*, Springer Proceedings in Energy,
https://doi.org/10.1007/978-981-15-5444-5_43

application [2], unfortunately, they suffer appreciable light-induced degradation for thickness greater than 200 nm [3]. In recent years, there has been an increasing interest in developing ultrathin film solar cells due to their considerable savings in material and processing time [4]. But what determines the ultrathin hydrogenated amorphous silicon solar cell performance, is the carrier transport [5].

The low value of JSC is the major limitation of energy conversion efficiency of ultra-thin a-Si:H solar cells because the thin i-layer does not absorb enough sun light to generate photocurrent [6]. That'why materials such as nc-SiOx:H with wide band gap, low refractive index and high conductivity are recommended. The hydrogenated nanocrystalline silicon oxide layers have been widely used in thin-film solar cells, due to the low refractive index and the wide band gap [7].

In a study conducted by Fang et al. [6] on the impact of using n-type hydrogenated nanocrystalline silicon oxide as a back reflector, it has been revealed that the cell performance has improved significantly. The fundamental goal of all these studies of solar cells is to improve the power-conversion efficiency (PCE). In addition to various parameters: the short-circuit current density (JSC), open-circuit voltage (VOC), and fill factor (FF).

Motivated by the desire to elevate the performance of a-Si: H-based solar cells further, we investigated the effect of n-type hydrogenated nanocrystalline silicon oxide (n-nc-SiOx:H) layers on the electrical performance of a-Si:H single-junction solar cells based on the material research of n-type nc-SiOx:H films.

43.2 Methodology

43.2.1 Simulation Model

We have used AMPS-1D (One Dimensional Device simulation program for the Analysis of Microelectronic and Photonic Structures), which is designed for crystalline and amorphous materials to analyze the transport behavior of semiconductor electronic and optoelectronic device structures. It resolves the first-principles continuity (43.2) and (43.3) and Poisson's (43.1) equations using finite differences and Newton-Raphson numeric methods [8].

Poisson's equation:

$$\frac{d}{dx}\left(-\varepsilon(x)\frac{d\Psi'}{dx}\right) = q.\left[p(x) - n(x) + N_D^+(x) - N_A^-(x) + p_t(x) - n_t(x)\right] \quad (43.1)$$

Continuity equations:
For electrons:

$$\frac{1}{q}\left(\frac{dJ_n}{dx}\right) = -G_{op}(x) + R(x) \tag{43.2}$$

For holes:

$$\frac{1}{q}\left(\frac{dJ_p}{dx}\right) = G_{op}(x) - R(x) \tag{43.3}$$

43.2.2 Device Structure and Input Parameters

In order to determine whether using hydrogenated microcrystalline silicon oxide as n-layer has an effect on a-Si:H solar cell. We have considered a device based on hydrogenated amorphous silicon (a-Si:H), already carried out experimentally.

Two single junction solar cells are designed by varying n-layer material (a) n-a-Si:H (b) n-nc-SiOx:H. The structures of single junction simulated solar cells are shown in Fig. 43.1. Table 43.1 resumes the different simulation parameters based on initial values from our previous works.

In order to explore the effects of the n-layer material on the performance of ultra-thin amorphous silicon solar cells, the p and i layers physical properties in both simulated structures are kept constant.

The optical band gap of the conventional n-a-Si:H and n-nc-SiOx:H layers have been taken from Fang [6].

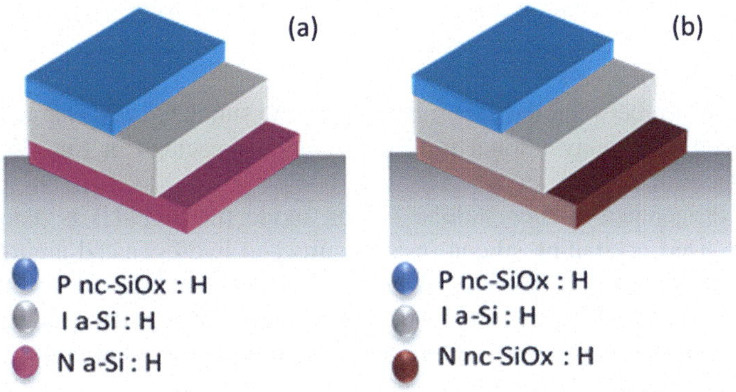

Fig. 43.1 Schematic diagram of single junction solar cell structures with two different n-layers. **a** n-a-Si:H et **b** n-nc-SiOx:H

Table 43.1 Layer input parameters

Parameters	Pnc-SiOx:H	I a-Si:H	N a-Si:H	N nc-SiOx:H
L (nm)	10	70	25	25
ε_r	11.9	11.9	11.9	11.9
χ (eV)	3.76	4	4	4
Eg (eV)	2.1	1.74	1.82	2.31
Nc (cm^{-3})	10^{20}	$2.5 \cdot 10^{20}$	$2.5 \cdot 10^{20}$	$2.5 \cdot 10^{20}$
Nv (cm^{-3})	10^{20}	$2.5 \cdot 10^{20}$	$2.5 \cdot 10^{20}$	$2.5 \cdot 10^{20}$
N_A (cm^{-3})	10^{18}	0	0	0
N_D (cm^{-3})	0	0	10^{18}	10^{16}
μ_{e-} ($cm^2V^{-1}s^{-1}$)	5	20	20	400
μ_{h-} ($cm^2V^{-1}s^{-1}$)	0.5	4	4	20
G_{DO}/G_{AO} (cm^{-3} eV^{-1})	$10^{22}/10^{22}$	$4 \cdot 10^{21}/4 \cdot 10^{21}$	$4 \cdot 10^{21}/4 \cdot 10^{21}$	$10^{20}/10^{20}$
E_D/E_A (eV)	0.06/0.03	0.04/0.03	0.05/0.03	0.05/0.03
N_{DG} (cm^{-3})	10^{18}	$2 \cdot 10^{15}$	$5 \cdot 10^{18}$	10^{15}
N_{AG} (cm^{-3})	10^{18}	$2 \cdot 10^{15}$	$5 \cdot 10^{18}$	10^{15}
E_{DG}/E_{AG} (eV)	1.25/1.15	1.08/0.98	1.05/0.98	1.15/0.98
σ_{de} (cm^2)(tails)	$4 \cdot 10^{-16}$	10^{-15}	10^{-15}	10^{-15}
σ_{dh} (cm^2)(tails)	$2 \cdot 10^{-17}$	10^{-16}	10^{-17}	10^{-17}
σ_{ae} (cm^2)(tails)	$2 \cdot 10^{-16}$	10^{-16}	10^{-17}	10^{-17}
σ_{ah} (cm^2)(tails)	$4 \cdot 10^{-15}$	10^{-15}	10^{-15}	10^{-15}
σ_{de} (cm^2)(gauss)	10^{-14}	10^{-15}	10^{-15}	10^{-15}
σ_{dh} (cm^2)(gauss)	10^{-16}	10^{-16}	10^{-16}	10^{-16}
σ_{ae} (cm^2)(gauss)	10^{-16}	10^{-16}	10^{-16}	10^{-16}
σ_{ah} (cm^2)(gauss)	10^{-14}	10^{-15}	10^{-15}	10^{-15}

43.3 Results and Discussion

This section analyzes the obtained results. The effect of n-layer type on the open circuit voltage (VOC), short circuit current (JSC), fill factor (FF) and efficiency (η) of a-Si:H p-i-n solar cell is described.

A comparison between the experimental and the simulated J-V characteristics of the solar cell with conventional n-a-Si:H layer is presented in Fig. 43.2. Under AM1.5 spectrum.

The hydrogenated nanocrystalline silicon oxide (nc-SiOx:H) is a mixture of nanometer sized crystalline silicon (c-Si) grains and hydrogenated amorphous silicon oxide (a-SiOx:H) tissues. This material has a wide band gap. That is the reason of using it in doped layers to reduce the parasitic absorption [6]. The doped nc-SiOx:H layers were used to replace the conventionally used non-alloyed silicon layers (μc-Si:H or a-Si:H) [9].

Table 43.2 compares the performance parameters of experimental [6] and simulated solar cells with n-nc-SiOx:H layer.

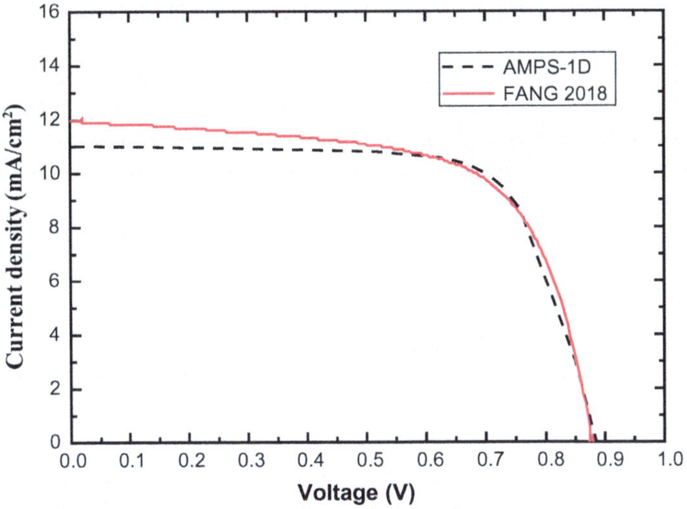

Fig. 43.2 Experimental (solid line) and simulated (dashed line) J-V characteristics of a single junction solar cell with a-Si:H n-layer

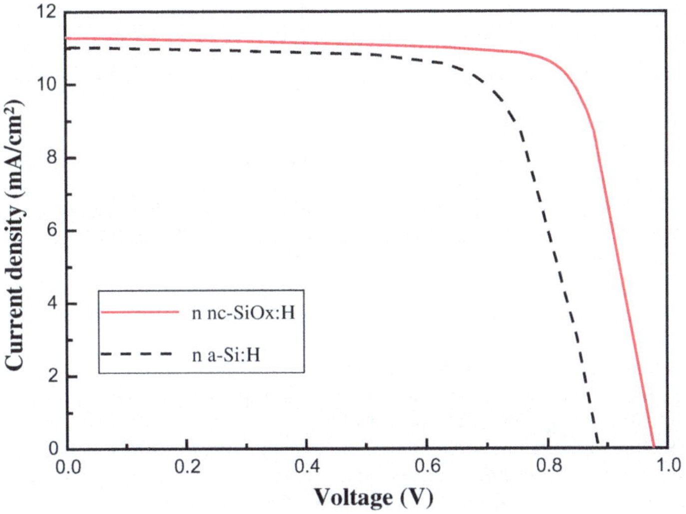

Fig. 43.3 Comparison of simulation results of the two simulated structures (dashed line) for a-Si:H n-layer and (solid line) for nc-SiOx:H n-layer

Table 43.2 Comparison between the performance parameters of experimental [6] and simulated solar cells with n-nc-SiOx:H layer

	V_{OC}	J_{SC}	FF	E_{FF}
FANG	915	13.83	69.5	8.79
AMPS-1D	980	11.35	78.5	8.73

The results obtained from the simulations of n-a-Si:H and n-nc-SiOx:H structures can be compared in Fig. 43.3.

As can be seen in Fig. 43.3, the open circuit voltage and the short circuit current density increase when using nanocrystalline silicon oxide in the n-layer.

The major limitation of energy conversion efficiency of ultra-thin a-Si:H solar cells is due to the low J_{SC} because the absorption in the thin absorber layer (i-layer) is not enough to generate a significant photo-current [6].

43.4 Conclusion

This study suggests the application of nanocrystalline silicon oxide as n-layer. For this, we used the device simulator AMPS-1D to simulate the performances of ultra-thina-Si:H solar cells. As a result, the simulated solar cells with n-nc-SiOx:H layer showed an improvement in the values of VOC and JSC which causes an increase in the conversion efficiency value. The reduction of the refractive index for nc-SiOx:H n-layer, which is in contact with the back reflector, can also lead to a reduction of parasitic absorption and therefore a gain in the current can be obtained.

References

1. R. Könenkamp, Photoelectric properties and applications of low-mobility semiconductors (2000)
2. Y.H. Chen, J.C. Liu, Y.R. Chen, J.W. Lin, C.H. Chen, W.H. Lu, C.N. Li, Enhancing performance of amorphous SiGe single junction solar cells by post-deposition thermal annealing. Thin Solid Films **529**, 7–9 (2013)
3. P. Chatterjee, P.R.I. Cabarrocas, Influence of N-type μ c-SiO x :H intermediate reflector and top cell material properties on the electrical performance of "micromorph" tandem solar cells. AIP Adv. **8** (2018)
4. H.-L. Chen, A. Cattoni, R. De Lépinau, A.W. Walker, O. Höhn, D. Lackner, G. Siefer, M. Faustini, N. Vandamme, J. Goffard, B. Behaghel, C. Dupuis, N. Bardou, F. Dimroth, S. Collin, A 19.9%-efficient ultrathin solar cell based on a 205-nm-thick GaAs absorber and a silver nanostructured back mirror. Nat. Energy (2019)
5. S. Nunomura, I. Sakata, K. Matsubara, Electronic properties of ultrathin hydrogenated amorphous silicon. Appl. Phys. Express. **10** (2017)
6. J. Fang, B. Yan, T. Li, C. Wei, Q. Huang, X. Chen, G. Wang, G. Hou, Y. Zhao, X. Zhang, Improvement in ultra-thin hydrogenated amorphous silicon solar cells with nanocrystalline silicon oxide. Sol. Energy Mater. Sol. Cells **176**, 167–173 (2018)
7. J. Fang, Z. Chen, G. Hou, F. Wang, X. Chen, C. Wei, G. Wang, J. Sun, D. Zhang, Y. Zhao, X. Zhang, High-quality hydrogenated intrinsic amorphous silicon oxide layers treated by H2 plasma used as the p/i buffer layers in hydrogenated amorphous silicon solar cells. Sol. Energy Mater. Sol. Cells **136**, 172–176 (2015)
8. J. Arch, J. Cuiffi, J. Hou, W. Howland, P. Mcelheny, A. Moquin, M. Rogosky, T. Tran, H. Zhu, F. Rubinelli, AMPS-1D Analysis of Microelectronic and Photonic Structures
9. A. Lambertz, Development of Doped Microcrystalline Silicon Oxide and its Application to Thin-Film Silicon Solar Cells Ontwikkeling van gedoteerd microkristallijn siliciumoxide en de toepassing in dunne film zonnecellen Proefschrift Andreas Lambertz (2015)

Chapter 44
Ab Initio Calculations of Structural, Mechanic, Electronic and Optical Properties of $Ag_2BaSn(Se, S)_4$ in Kesterite Structure

Rabia Rahmani, Bouhalouane Amrani, Kouider Driss Khodja, and Hayat Sediki

Abstract The *ab intio* method within the recently modified Becke-Johnson potential explore that the $Ag_2BaSn(Se, S)_4$ compounds are narrow band gap semiconductors. The lattice parameters, bulk modulus and its derivative are predicted. The calculated elastic constants satisfy Pugh's criteria. The mBJ potential was selected for further explanation of optical properties of these compounds. The two non-zero dielectric tensor components show isotropy behavior between the perpendicular and parallel components. Based on the results our finding that the $Ag_2BaSn(Se, S)_4$ compounds are efficient materials for energy conversion.

Keywords Kesterite · DFT · TB-mBJ · Electronic band-structure · $Ag_2BaSn(Se, S)_4$ · Photovoltaic (PV) cells

R. Rahmani (✉) · B. Amrani · K. D. Khodja
Laboratoire de Théorie et Simulation des Matériaux, Université Oran1 Ahmed Ben Bella, Oran, Algeria
e-mail: rah.physique@gmail.com

R. Rahmani
Faculté de Physique, Département de Génie Physique, Université des Sciences et de la Technologie d'Oran-Mohamed-Boudiaf, Oran, Algeria

H. Sediki
Laboratoire de Chimie Physique Macromoléculaire, Université d'Oran1 Ahmed Ben Bella, Es-Senia, Oran, Algeria

H. Sediki
Faculté de Physique, Département de Technologie des Matériaux, Université des Sciences et de la Technologie d'Oran- Mohamed- Boudiaf, Oran, Algeria

A. Belasri and S. A. Beldjilali (eds.), *ICREEC 2019*, Springer Proceedings in Energy,
https://doi.org/10.1007/978-981-15-5444-5_44

44.1 Introduction

The conversion efficiencies of typical energy of monocrystalline, polycrystalline and amorphous silicon-based photovoltaic cells are 25, 20 and 10% respectively [1]. The high manufacturing costs still prevent the massive use of these devices worldwide. The band gap is a fundamental property determining efficiency, with Eg ≈1.4−1.6 eV being the most efficient solar absorbers [2–8]. Some of quaternary semiconductors had been synthesized, (I2-II-IV-VI 4 with I = Cu, Ag, II = Zn, Cd, IV = Si, Ge, Sn, Pb, and VI = S, Se) [9]. These quaternary semiconductors have different structures types, including kesterite (KS), stannite (ST), wurtzitekesterite, and wurtzite-stannite, among others [10–14]. Previous studies show that the photoelectrochemical (PEC) performances of Ag based semiconductors are better than those of Cu based semiconductors [15–18].

A description of synthesized, properties and the X-ray investigations of orthorhombic Ag_2BaSnS_4 is given. It has an unit cell: a = 7.12 Å, b = 8.11 Å, c = 6.85 Å [19, 20]. The X-ray investigation of the system Cu_2 –xAgxBaSnS$_4$ ($0 \leq x \leq 2$) yields a homogenity range of ($1.7 < x \leq 2$) [21].

In this work, a series of I2-II-IV-VI4 semiconductors are studied through the first-principles calculations. The results show that although the substitution of VI by S and Se can lead to stable quaternary compound semiconductors.

To our knowledge, there are no experimental results on the physical properties of our materials. We hope that this preliminary work could serve as a reference for further experimental studies to further promote the photovoltaic applications of these studied materials.

44.2 Calculation Methods

The kesterite $Ag_2BaSn(S, Se)_4$ compound crystallizes in tetragonal kesterite structure with space group I-4 (N°82). The unit cell of a crystal contains 8atoms, where the atoms Ag occupy the 2a (0, 0, 0, 0) and 2c (0.1/2, 1/4) positions, one Ba and one Sn atoms are located at the 2d (1/2, 0, 1/4) and 2b (1/2, 1/2, 1/2, 02) positions while S (Se) atoms are located at 8 g (0.7560, 0.7566, 0.8722) (see Fig. 44.1).

In the present work, we performed calculations using two methods, full-potential linear Augmented-plane-wave method (FP-LAPW) [22, 23] implemented in Wien2k and the plane-wave pseudopotential (PP-PW) implemented in the Cambridge Serial Total Energy Package (CASTEP) code [24]. In both cases, the exchange correlation energy is described by the generalized gradient approximation of Perdew et al. (GGA-PBE) [25].

The states Ag ($5s^1$, $4d^{10}$), Ba ($6s^2$), Sn ($5s^2$ $5p^2$), Se ($4s^2$ $4p^4$), S ($3s^2$ $3p^4$) were treated as valence electrons. The Brillouin zone was sampled by a 52 ($7 \times 7 \times 7$) uniform k-point mesh according to the Monkhorst-Pack scheme grids [26].

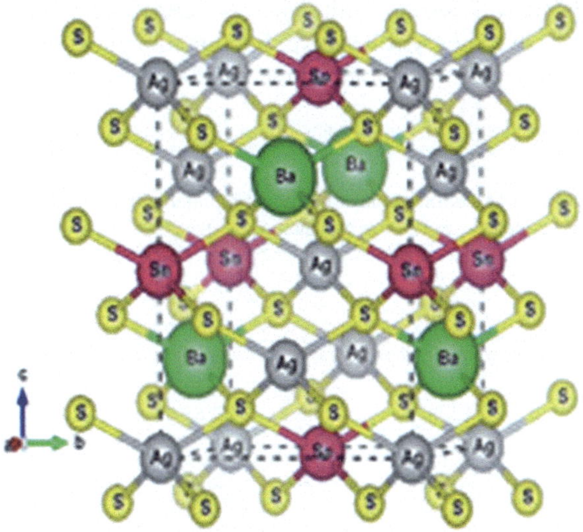

Fig. 44.1 The unit cell of Ag_2BaSnS_4

The enhanced and linearized plane wave method with a total potential FP-LAPW [22, 23] that was developed by Blaha, Schwarz and their collaborators [4, 5] and taking the cutoff energy equal to lmax = 10 and Rmt × Kmax = 8. The iterative process is repeated until the calculated total energy error is less than 0.0001 Ryd. The values of the muffin-tin rays are 2.30, 2.31, 2.38 and 2.19 au for the Ag, Ba, Sn, and S (Se) atoms respectively. Both the muffin-tin radius and the number of *k*-points are varied to ensure convergence. The total energies were calculated as a function of volume and the obtained data are fitted according to the Murnaghan equation of state [27].

The calculations with the CASTEP code [24] are done using a pseudo-potential ultra-soft for a good calculation with plane wave cutoff energy of 400 eV. Convergence is obtained with ultrafine quality which is equal to 5×10^{-7} eV/atom. The Convergence is reached with an accuracy of 10^{-6} eV/atom for the total energy.

44.3 Results and Discussions

44.3.1 Structural Properties

We performed the structural optimization by minimizing the total energy with respect to the crystal volume and fitting by the Murnaghan's equation of state [27] . The numerical results obtained are presented in Table 44.1.

Our values of the lattice parameters and bulk modulus B and their derivative B′ obtained by the two different methods (total potential and pseudo potential) are practically the same. Unfortunately, there are no available experimental or theoretical results until now. Nevertheless, they are stable compounds as the cohesive energy is negative. Importantly, interatomic distances, calculated between such average positions on the band structure (See Table 44.2).

Table 44.1 Lattice constant a, c/a ratio, internal parameters U, bulk modulus B, first derivative B′, and cohesive energy for $Ag_2BaSn(S, Se)_4$

	a (Å)	$\delta = c/a$	U_X U_Y U_Z	B (GPa)	B′	E_{coh} (eV/cell)	*Method*
Ag_2BaSnS_4	6.2202[a]	2.002[a]	0.2442329[a] 0.2433032[a] 0.1276891[a]	36.5340[a]	3.9888[a]	−1.9801[a]	FP-LAPW
	6.2180[b]	2.000[b]		37.8284[b]			PP-PW
$Ag_2BaSnSe_4$	6.4573[a]	1.9726[a]	0.2285458[a] 0.2957334[a] 0.1081571[a]	36.4973[a]	3.2099[a]	−1.8874[a]	FP-LAPW
	6.404[b]	1.9708[b]		36.572[b]			PP-PW

[a]Our work (FP-LAPW). [b]Our work PP-PW

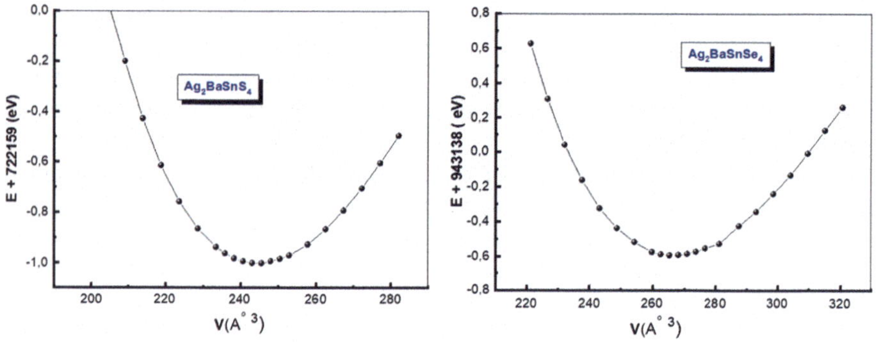

Fig. 44.2 Total energy as a function of volume for Ag_2BaSnX_4

Table 44.2 Interatomic distances calculated in units of (Å) for Ag_2BaSnX_4 compounds

Distance (Å)	Ag_1-Ag_2	Ag_1-Sn	Ag_1-X	Sn-X	X-X	Ag_2-X	X-Ba
Ag_2BaSnS_4	4.40065	4.39840	2.66963	2.75848	4.28877	2.67998	2.67909
$Ag_2BaSnSe_4$	4.53492	4.56667	2.77901	2.59047	4.14936	2.67237	3.15974

44.3.2 Elastic Properties

To the best of our knowledge, the elastic properties for Ag_2BaSnX_4 have never been reported so far. For a tetragonal structure type I, the number of elastic stiffness constants is reduced by symmetry to six independent constants such as C_{11}, C_{12}, C_{13}, C_{33}, C_{44} et C_{66}. Whereas the Type II has seven such as C_{11}, C_{12}, C_{13}, C_{33}, C_{44}, C_{66} and C_{16} with $C_{16} = -C_{26}$. Using the Mehl model [28], the elastic constants for $Ag_2BaSn(S, Se)_4$ are grouped in Table 44.3.

For tetragonal crystals, the mechanical stability requires the elastic constants satisfying the well-known Born stability criteria [29–31],

$$\mathbf{C_{11} > 0,\ C_{33} > 0,\ C_{44} > 0,\ C_{66} > 0,\ C_{11} - C_{12} > 0,}$$
$$\mathbf{C_{11} + C_{33} - 2C_{13} > 0,\ [2(C_{11} + C_{12}) + C_{33} + 4\ C_{13}] > 0}$$

The above criteria are satisfied indicating that the Ag_2BaSnS_4 and $Ag_2BaSnSe_4$ are mechanically stable. To our knowledge, there are no experimental data or theoretical results available.

44.3.3 Electronic Properties

In order to understand the origin of the gap states and the role of chemical composition, we have presented in Fig. 44.3, the total and partial state densities for the two compounds respectively. The electronic properties were calculated within the (mBJ-GGA) scheme. Fermi level is taken as the zero of energy.

For Ag_2BaSnS_4, the lower part *of the valence band (around −6 eV)* is attributed to *Ba-s, S-s* and *Sn-s,* while the upper part (−4 to −0 *eV)* is mostly formed *by Ag-d S-p,* giving rise to the *p-d hybridization.*

Figure 44.3 illustrates the calculated band structures using the modified Becke and Johnson mBJ-GGA exchange approach. It is easily observed from Fig. 44.3 that the minimum of the conduction band and the maximum of the valence band for Ag_2BaSnX_4 lie in the same K-point (Γ), indicating that the both compounds are a direct-gap semiconductor. The different values of gap energy are given in Table 44.4 and compared with the experimental value of silicon [32] (Fig. 44.4).

44.3.4 Optical Properties

The complex dielectric function ε(ω) which describes the response of a material subjected to the effect of external excitation (radiation) is written in the form

$$\varepsilon(\omega) = \varepsilon_1(\omega) + \varepsilon_2(\omega) \tag{44.1}$$

Table 44.3 Calculated elastic constants Cij (in GPa) for $Ag_2BaSn(S, Se)_4$

Cij (GPa)	C_{11}	C_{12}	C_{13}	C_{33}	C_{44}	C_{66}	C_{16}	C_{26}
Ag_2BaSnS_4	41.507	23.455	29.256	41.961	13.785	5.484	1.412	−1.412
$Ag_2BaSnSe_4$	40.514	29.415	32.414	46.421	10.109	11.109	1.456	−1.456

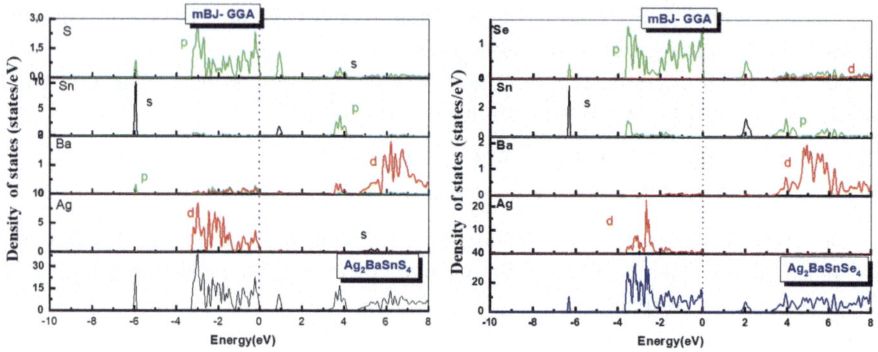

Fig. 44.3 The total and partial DOS of Ag_2BaSnX_4 within the GGA-TB-mBJ approximation

Table 44.4 Calculated values of direct band (Γ-Γ) gap (in unit eV) within different approximations compared with the silicon

	E_g (eV)	Ag_2BaSnS_4	$Ag_2BaSnSe_4$	*Si* [32]
	GGA	0.209	0.406	
E_g (eV)	EV-GGA	0.209	0.406	
Nos calculs	LDA	0.131	0.231	
	PBESOL	0.199	0.330	
	PP-PW (GGA)	0.24	0.594	
	mBJ-GGA	0.918	1.273	
Exp				1.17

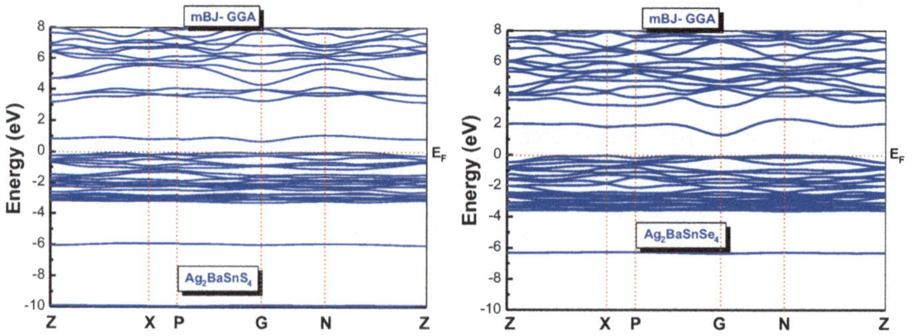

Fig. 44.4 The band structure of Ag_2BaSnX_4 within the GGA-TB- mBJ approximation

The calculated optical properties at the equilibrium lattice constant, using the mBJ-GGA approximation, are presented in Fig. 44.5. *From this figure it is clear to see that* the two non-zero dielectric tensor components show isotropy behavior between the perpendicular and parallel components. Our analysis of $\varepsilon_2(\omega)$ curve shows that the threshold energy (first critical point) of the dielectric function occurs at 0.918 eV for Ag_2BaSnS_4 and at 1.273 eV for $Ag_2BaSnSe_4$. This point is Γ_V–Γ_C splitting which gives the threshold for direct optical transitions between the highest valence and the lowest conduction band. This is known as the fundamental absorption edge.

The real part $\varepsilon_1(\omega)$ of dielectric function can be evaluated from the imaginary part $\varepsilon_2(\omega)$ by using Kramer–Kronig relationship. The static dielectric constant $\varepsilon_1(0)$ is given by the low energy limit of $\varepsilon_1(\omega)$. Note that we do not include phonon contributions to the dielectric screening (Table 44.5).

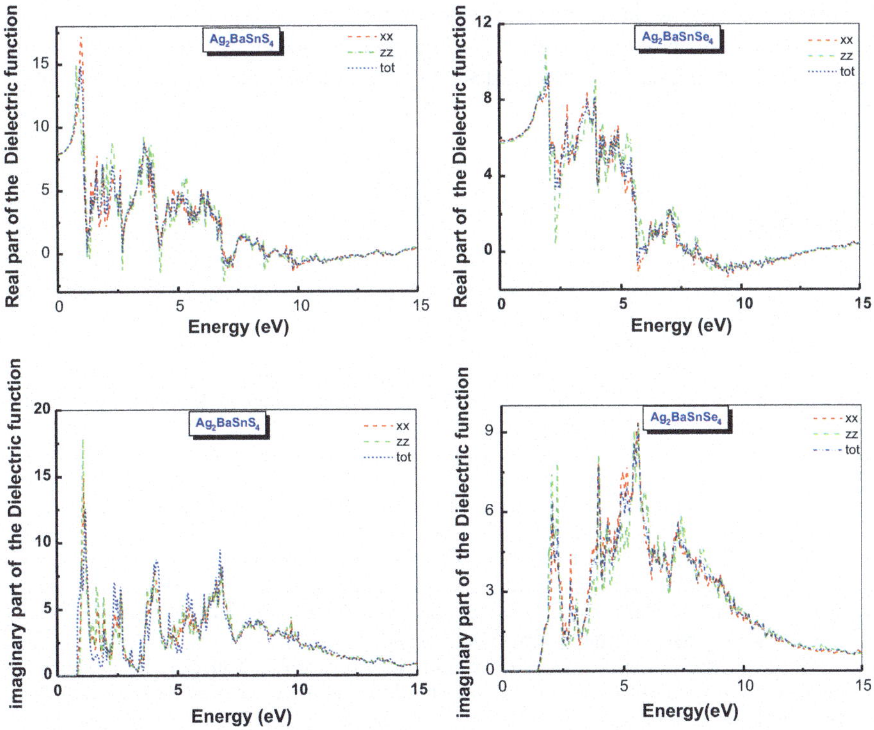

Fig. 44.5 Real part $\varepsilon_1(\omega)$ and imaginary part $\varepsilon_2(\omega)$ of the dielectric function for Ag_2BaSnX_4 within the GGA-TB-mBJ approximation

Table 44.5 Calculated theoretical values of static constants $\varepsilon_{//}(0)$, $\varepsilon_{\perp}(0)$, $\varepsilon(0)$, $\Delta\varepsilon(0)$ and $\Delta n(0)$

	$\varepsilon_{1//}(0)$	$\varepsilon_{1\perp}(0)$	$\varepsilon_1(0)$	$\Delta\varepsilon_1(0)$	$\Delta n(0)$
Ag_2BaSnS_4	7.95823	7.85944	7.9253	0.016253	0.01756
$Ag_2BaSnSe_4$	5.86774	5.72091	5.81879	0.01852	0.03049

44.4 Conclusions

Using PP-PW method and FP-LAPW, we have studied the structural, elastic, electronic and optical properties of Ag_2BaSnX_4. We have calculated the structural parameters at equilibrium. The results of formation energy and elastic constants confirm that both compounds are mechanically stable.

The band structures show that these two compounds have a direct gap in Γ direction. In addition, with the Beck-Johnson modified potential (mBJ) approach, the values of the calculated gap energies can be improved.

The optical properties obtained confirm the electronic results. Based on the results our finding that the $Ag_2BaSn(Se, S)_4$ compounds are efficient materials for energy conversion.

References

1. M.A. Green, K. Emery, Y. Hishikawa, W. Warta, Solar cell efficiency tables (version 34). Prog. Photovolt. Res. Appl. **17**, 320–326 (2009)
2. S. Ruhle, Sol. Energy **130**, 139 (2016)
3. K. Ito, T. Nakazawa, Jpn. J. Appl. Phys., Part 1 **27**, 2094 (1988)
4. H. Katagiri, K. Saitoh, T. Washio, H. Shinohara, T. Kurumadani, S. Miyajima, Sol. Energy Mater. Sol. Cells **65**, 141 (2001)
5. J.-S. Seol, S.-Y. Lee, J.-C. Lee, H.-D. Nam, K.-H. Kim, Sol. Energy Mater. Sol. Cells **75**, 155 (2003)
6. T. Tanaka, T. Nagatomo, D. Kawasaki, M. Nishio, Q. Guo, A. Wakahara, A. Yoshida, H. Ogawa, J. Phys. Chem. Solids **66**, 1978 (2005)
7. K. Moriya, J. Watabe, K. Tanaka, H. Uchiki, Phys. Status Solidi C **3**, 2848 (2006)
8. J.J. Scragg, P.J. Dale, L.M. Peter, Electrochem. Commun. **10**, 639 (2008)
9. G.H.M. Bohm, A. MacKinnon, O. Madelung, A. Scharmann, E.-G. Scharmer, *Physics of Ternary Compounds* (Springer, New York, 1985)
10. T. Gurel, C. Sevik, T. Cagın, Characterization of vibrational and mechanical properties of quaternary compounds Cu_2ZnSnS_4 and $Cu_2ZnSnSe_4$ in kesterite and stannite structures. Phys. Rev. B **84**, 205201 (2011)
11. H.-R. Liu, S. Chen, Y.-T. Zhai, H.J. Xiang, X.G. Gong, S.-H. Wei, First-principles study on the effective masses of zinc-blend-derived Cu_2Zn-IV-VI_4 (IV = Sn, Ge, Si and VI = S, Se). J. Appl. Phys. **112**, 093717 (2012)
12. M.D. Regulacio, M. Han, Multinary I-III-VI_2 and I_2-II-IV-VI_4 semiconductor nanostructures for photocatalytic applications. Acc. Chem. Res. **49**, 511–519 (2016)
13. C. Wang, S. Chen, J. Yang, L. Lang, H. Xiang, X. Gong, A. Walsh, S. Wei, Design of I_2–II–IV–VI_4 semiconductors through element substitution: the thermodynamic stability limit and chemical trend. Chem. Mater. **26**, 3411–3417 (2014)

14. F. Ozel, A. Sarılmaz, B. Istanbullu, A. Aljabour, M. Kus¸ S. Sonmezoglu, Penternary chalcogenides nanocrystals as catalytic materials for efficient counter electrodes in dye-synthesized solar cells. Sci. Rep. **6**, 29207 (2016)
15. K.W. Cheng, C.H. Yeh, Int. J. Hydrog. Energy **37**, 13638 (2012)
16. C.C. Chang, C.J. Liang, K.W. Cheng, Sol. Energy Mater. Sol. Cells **93**, 1427 (2009)
17. K.W. Cheng, W.C. Lee, M.S. Fan, Electrochim. Acta **87**, 53 (2013)
18. K.W. Cheng, B.H. Liu, Sol. Energy Mater. Sol. Cells **95**, 1859 (2011)
19. H. Chen, P.F. Liu, B.X. Li, H. Lin, L.M. Wu, X.T. Wu, Experimental and theoretical studies on the NLO properties of two quaternary non-centrosymmetric chalcogenides: $BaAg_2GeS_4$ and $BaAg_2SnS_4$. Dalton Trans. **47**, 429–437 (2018)
20. L.Y. Nian, J.B. Huang, K. Wu, Z. Su, Z.H. Yang, S.L. Pan, $BaCu_2MIVQ_4$ (MIV = Si, Ge, and Sn; Q = S, Se): synthesis, crystal structures, optical performances and theoretical calculations. RSC Adv. **7**, 29378–29385 (2017)
21. L. Chr, O. Teske, Z. Vetter, Anorg. Allg. Chem. **426**, 281–287 (1976)
22. P. Blaha, K. Schwarz, G.K.H. Madsen et al., *WIEN2k (Schwarz K')* (Techn. Universität Wien, Austria, 2001)
23. O.K. Andersen, Phys. Rev. B: Condens. Matter **12**(8), 3060–3083 (1975)
24. M.D. Segall, P.J.D. Lindan, M.J. Probert, C.J. Pickard, P.J. Hasnip, S.J. Clark, M.C. Payne, First-principles simulation: ideas, illustrations and the CASTEP code. J. Phys.: Condens. Matter **14**(11), 2717–2744 (2002)
25. J.P. Perdew, Y. Wang, Phys. Rev. B **45**, 13244 (1992)
26. J.P. Perdew, K. Burke, M. Ernzerhof, Phys. Rev. Lett. **77**, 3865 (1996)
27. F.D. Murnaghan, Natl. Acad. Sci. U.S.A. **30**, 5390 (1994)
28. M.J. Mehl, J.E. Osburn, D.A. Papaconstantopoulos, B.M. Klein, Phys. Rev. B **41**, 10311 (1990)
29. M. Born, Proc. Cambridge Philos. Soc. **36**, 160 (1940)
30. M. Born, K. Huang, *Dynamical Theory of Crystal Lattices* (Oxford University Press, London and New York, 1954)
31. J. Wang, S. Yip, S.R. Phillpot, D. Wolf, Phys. Rev. Lett. **71**, 4182 (1993)
32. A. Chovet, P. Masson, "Physique des semi-conducteurs", Cours de l'école polytechnique universitaire de Marseille (2004)

Chapter 45
Modelling the Dielectric Response of $BaTiO_3$ Doped with Conductive Inclusions Using Random RC Networks

Ahmed Benyahia and Rachid Bouamrane

Abstract We investigate the dielectric properties of random resistor-capacitor networks (RRCN) modelling barium titanate ($BaTiO_3$) ceramics doped with different rare earth additives. The host matrix of $BaTiO_3$ is regarded as a dielectric RRCN, where the effect of conductive inclusions is modelled by a random incorporation of resistors having small resistance. In order to model the effects of blending silver, nickel, etc., into barium titanate on the effective dielectric properties of the composite, Frank and Lobb algorithm is used to calculate the effective conductivity, resistivity and permittivity of the network at different proportions of conductive inclusions. This network approach indicates similar properties trends to that observed experimentally, a qualitative agreement with the Maxwell equation is obtained at low proportions of conductive inclusions.

Keywords Composite materials · Dielectric properties · Random RC networks

45.1 Introduction

Heterogeneous materials which comprise conductive and dielectric phases are present on a large variety of materials, including ceramics, polymers, and composites [1]. Barium titanate (BaTiO3, BT) as an electro-active ceramic, is widely investigated and used in electronic industry because of its high dielectric constant [2] and ferroelectric properties [3, 4]. Often, it is assumed that the dielectric properties can be improved significantly by dispersion of conducting particles in a dielectric matrix [5, 6]. Feng Li et al. [7] stated that the relative permittivity of ceramic-metal composites can be predicted using the Maxwell's equation $\varepsilon(p) = \varepsilon_m(1 + 2p)/(1 - p)$, where ε_m is the relative permittivity of the matrix and p is the metal volume's fraction [8]. According to this model, the increase in dielectric

A. Benyahia (✉) · R. Bouamrane
LEPM, Faculté de Physique, USTO-MB, BP 1505 El M'Naouer, 31000 Oran, Algeria
e-mail: ahmed.benyahia91@univ-usto.dz

A. Belasri and S. A. Beldjilali (eds.), *ICREEC 2019*, Springer Proceedings in Energy,
https://doi.org/10.1007/978-981-15-5444-5_45

permittivity can be explained by the increase of the effective electric field in the dielectric phase, because the distance of the electrodes is shortened by conductive particles [9].

Several papers have studied the dielectric properties of BaTiO3-Ag composites [10–12], where the observed increase in permittivity has been suggested to be due to the silver particles acting as 'internal' electrodes within the barium titanate and increasing the local electric field. Kwan et al. proposed the modified Maxwell equation, $\varepsilon_n = \varepsilon_m\left[3 - 2(1 - p_n)^{1/n}\right]^n/(1 - p_n)$, for a non-continuous metallic phase to fit this process. Where, ε_n is the effective dielectric constant of a composite sphere having a volume fraction, p_n, of conducting filler. When the number of conducting particles, n, is sufficiently large, ε_n approaches the limit: $\varepsilon(p) = \varepsilon_m/(1 - p)^3$ [13].

In some cases of multilayer ceramic capacitors (MLCC), precious metals like silver are replaced with base metals as nickel to reduce the production costs [14, 15]. By using electrical measurements with respect to the nature of the material under study, separation of electroactive components effects such as grain boundary and grain is possible [16]. As a result, relaxation processes within the electronic ceramics can be modelled by an impedance model envisaged as a series of relaxators, i.e. grouped elements of parallel RC circuits, each corresponds to a different process happening at the electrode interfaces, at the grain boundary or within the grains themselves. Therefore, microstructure properties of BT based materials, expressed in grain boundary contacts, are of fundamental importance for electric properties of this material [18].

More often than not, for electronic materials design, the microstructure of this kind of ceramics can be depicted by a brick wall model [18]. This model demonstrates the polarization and the conduction contributions to the system impedance, which can be described by an equivalent electrical circuit comprising three RC branches. The RC elements of the equivalent circuit may be ascribed to the permittivity, conductivity, and geometrical extensions of the grain bulk, grain boundary regions, and electrical interfaces. Therefore, it is necessary to have a generalized model to provide a realistic representation of the electrical properties [17]. This is important for modern devices applications and presents a challenge for simulation. Due to the limited solid solubility of these metals in the host matrix, no significant solid-solutions or reaction phases are expected to form. Then, we propose that changes in dielectric properties of these composites can be modeled using RRCN, which are widely used in modeling conductor-dielectric composites [1, 19, 21, 22].

In this paper, the addition effect of conductive inclusions into the dielectric matrix will be modeled by the random incorporation of conducting bonds having small resistance r in a dielectric RRCN (DRRCN) as in [23], the difference is that we will use 2D networks rather than 3D. In addition, we will compare numerical results to those predicted by the analytical formulae of Maxwell equation, the modified Maxwell equation and the percolation law, this is achieved using the Franck and Lobb algorithm implemented in FORTRAN code [24].

45.2 RC Model

The RC model of a conductor-dielectric mixture is defined as follows: the occupied connections of a percolation network are considered as pure resistors, having resistance R and distributed randomly with a proportion of $1-p$, while the empty links which represent the polarizability of the dielectric medium, are considered as perfect capacitors having capacitance C and distributed randomly with a proportion p. Hence, the conductance of the two links' types at the frequency $f = 2\pi\omega$ are $\sigma_1 = 1/R$ and $\sigma_2 = i\omega C$, where $\omega_0 = 1/RC$ is the microscopic frequency scale of the model. In the static or DC (i.e. zero-frequency) limit, the capacitors conductivity becomes null which makes the RC model becomes the conductor-insulator model [19, 20]. The electrical response of these networks exhibits a strong similarity to that of many materials, which exhibits the anomalous power-law frequency dependence of permittivity or AC conductivity at intermediate frequencies, where $\sigma \propto \omega^n$ and $\varepsilon \propto \omega^{n-1}$ with $n \approx p$ [25–27]. These network characteristics are similar to the familiar dielectric response of Cole-Davidson model, for which the permittivity is given by $\varepsilon = \propto 1/(1+i\omega\tau)^{\beta}$. Three domains of frequencies can be distinguished:

- at low frequencies, the resistors conduct AC currents well, while capacitors act as open circuits. If a resistive percolation path is present through the network, AC currents will flow preferably through it;
- at high frequencies, the capacitors conduct AC currents well and act as short circuits because of their high conductance. If a capacitive percolation path is present through the network, AC currents will flow preferably through it;
- at intermediate frequencies, resistors' and capacitors' conductance are equivalents, which makes the presence of a resistive or capacitive percolation path having a little influence on the AC currents flow through the network [28].

45.3 Frank and Lobb Algorithm

The effective properties of such random electrical networks are calculated using a very efficient and fast numerical method known as Frank and Lobb algorithm. This algorithm consists of a reduction scheme that transforms the network into one element using the so-called triangle-star, parallel and simple series transformations [24].

In this work, we use this method to estimate the proportion effect on the dielectric properties of such networks. We consider a square network with the size of $l \times l$ nodes and hence a total of $2l^2$ components; the relative permittivity $\varepsilon = \varepsilon' - i\varepsilon''$ is obtained from the equivalent impedance $Z(\omega)$ of this network by the relation $\varepsilon = 1/i\omega\varepsilon_0 Z(\omega)$ [21]. Comparing to other approaches, the reduction of the network to a single link is in principle less expensive in calculation. It was found that Frank and Lobb algorithm has a consistent resolution time $t \propto N^{1.5}$ [25].

45.4 Results and Discussion

We model the barium titanate ceramics doped with a conductive additive using the network approach in modelling conductor-dielectric composites. The barium titanate is modelled as a DRRCN by taking networks of 70% capacitors and 30% resistors to ensure the dielectric character of the network, which represents the host matrix of BT. The capacitors are 1 nF, while the resistors are 10 $k\Omega$, both types of elements are distributed randomly in a square node lattice using a uniform distribution.

The addition effect of conductive particles is modelled by adding a second type of resistors of 1 μΩ, representing the high conductivity of Ag, Ni, etc. inclusions, the calculations are performed for network sizes of 64 and 512 at 1 kHz. The results are averaged over 1000 samples and shown below for the permittivity, resistivity and conductivity (Fig. 45.1).

The results show that the network size is enough and there are no finite size effects, this can be seen clearly from the superposition graphs. Addition of conductive inclusions in the network, makes the permittivity and the dielectric loss increase together, with guarding the permittivity magnitude greater than the dielectric loss. In addition, the conductivity increases, while the resistivity decreases.

The relative permittivity can also be estimated by using a percolation law [29] as $\varepsilon(p) = \varepsilon_m |(p_c - p)/p_c|^{-q}$, where p_c is the percolation threshold and q is a constant. The percolation theory predicts a dramatic increase in relative permittivity as the volume fraction is approaching its percolation threshold. Data are compared to the Maxwell equation, Modified Maxwell equation and to the power law of percolation theory (Fig. 45.2).

The results show a good agreement with the power law of percolation theory, but with high values for p_c, which theoretically equals 0.16. In our results, $p_c = 0.75$ for ε', while $q = 1.88$. Authors of [30] have reported values of $p_c = 0.37$ and $q = 0.6$ for Ni Nano-particles addition to barium titanate ceramic matrix, where it should be noted that the value of the percolation threshold is abnormally high according to the percolation theory of two-phase random composites.

The differences of percolation threshold can be attributed firstly to the network dimension and secondly to the fact that these networks contain more than two phases (three), which are dielectric bonds of capacitance C and two types of resistors R and r. The results show a qualitative agreement with the Maxwell equation for low proportions, and they intersect at a proportion 0.16 of r content. Also, they indicate similar permittivity and resistivity trends to that observed experimentally in [23].

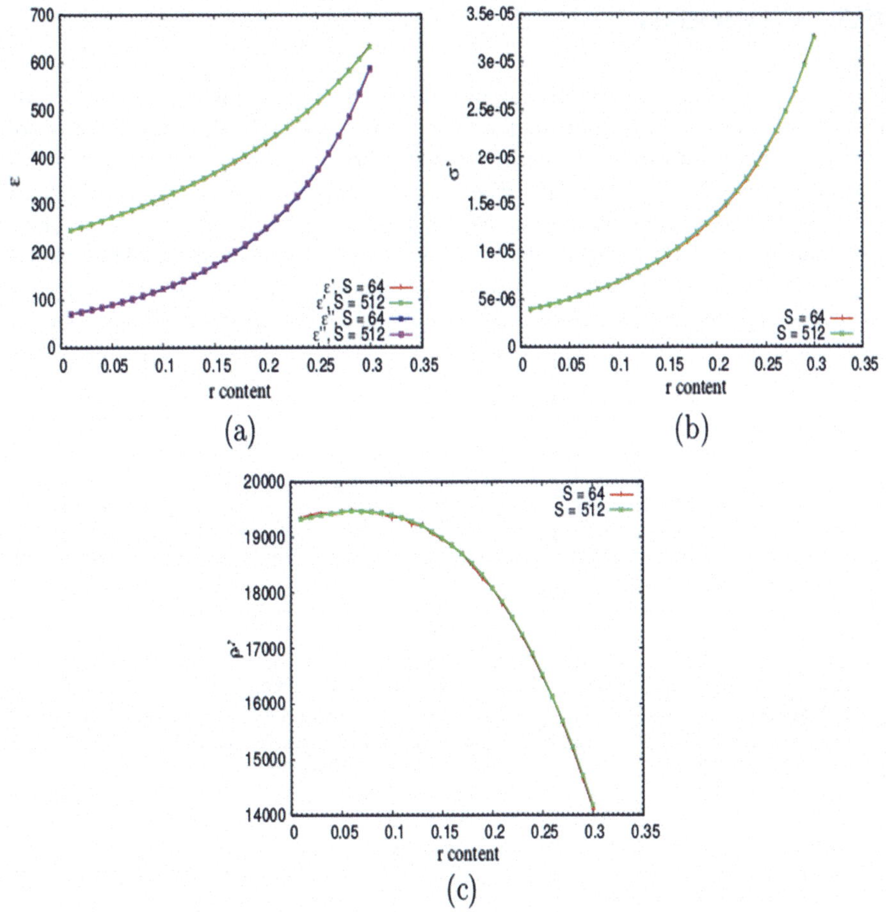

Fig. 45.1 The dielectric response of $l \times l$ square node RRCN for 1000 different samples, showing the effect of r content, which represent the conductive particles content in BT matrix for $l = 64$ and $l = 512$: **a** Permittivity. **b** Conductivity. **c** Resistivity

Fig. 45.2 Addition effect of conductive inclusions on the DRRCN permittivity

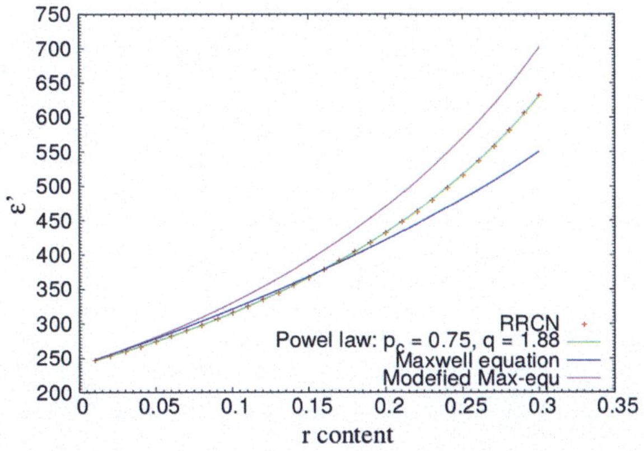

45.5 Conclusions

This work was aiming to establish an impedance model using RRCN. In this study, after giving an overview on the dielectric response of RRCN, the addition effect of conductive inclusions into BT matrix is modelled by random incorporation of small resistors into a DRRCN. The equivalent impedance of these networks is calculated using Frank and Lobb algorithm. The permittivity results show a qualitative agreement with the Maxwell equation for low proportions of r resistors. High values of percolation threshold and power-law exponent have been obtained. One raison for these abnormally high values, can be attributed to the network dimension.

References

1. S. Panteny, R. Stevens, C.R. Bowen, The frequency dependent permittivity and AC conductivity of random electrical networks. Ferroelectrics **319**, 199–208 (2005)
2. D. Hennings, Barium titanate based ceramic materials for dielectric use. Int. J. High Technol. Ceram. **3**(2), 91–111 (1987)
3. A. Rae, M. Chu, V. Ganine, Barium titanate-past, present and future. Ceram. Trans. **100**, 1–12 (1999)
4. T. Li, L.T. Li, Y. Kou, Z.L. Gui, Stable temperature dependence of dielectric properties in $BaTiO_3$-Nb_2O_5-Co_3O_4-Gd_2O_3 system. J. Mater. Sci. Lett. **19**(11), 995–997 (2000)
5. H.J. Hwang, K. Watari, M. Sando, M. Toriyama, K. Niihara, Low-temperature sintering and high-strength Pb(Zr, Ti)O_3-matrix composites incorporating silver particles. J. Am. Ceram. Soc. **80**, 791 (1997)
6. H.J. Hwang, T. Nagai, T. Ohji, M. Sando, M. Toriyama, K. Niihara, Curie temperature anomaly in lead zirconate titanate/silver composites. J. Am. Ceram. Soc. **81**, 709 (1998)
7. J.-F. Li, K. Takagi, N. Terakubo, R. Watanabe, Electrical and mechanical properties of piezoelectric ceramic/metal composites in the Pb(Zr, Ti)O_3/Pt system. Appl. Phys. Lett. **79**, 2441 (2001)
8. P. Chỳlek, V. Srivastava, Effective dielectric constant of a metal-dielectric composite. Phys. Rev. B **30**, 1008 (1984)
9. N. Halder, A.D. Sharma, S.K. Khan, A. Sen, H.S. Maiti, Effect of silver addition on the dielectric properties of barium titanate based low temperature processed capacitors. Mater. Res. Bull. **34**, 545 (1999)
10. C.Y. Chen, W.H. Tuan, Mechanical and dielectric properties of $BaTiO_3$/Ag composites. J. Mater. Sci. Lett. **18**, 353 (1999)
11. P. Ren, H. Fan, X. Wang, J. Shi, Effects of silver addition on microstructure and electrical properties of barium titanate ceramics. J. Alloy. Compd. **509**, 6423 (2011)
12. S. Panteny, C.R. Bowen, R. Stevens, Characterisation of barium titanate-silver composites part II: electrical properties. J. Mater. Sci. **41**, 3845 (2006)
13. S.H. Kwan, F.G. Shin, W.L. Tsui, Dielectric constant of silver-thermosetting polyester composites. J. Mater. Sci. **19**, 4093 (1984)
14. C. Pecharromán, F. Esteban-Betegón, J.F. Bartolomé, S. López-Esteban, J.S. Moya, New percolative $BaTiO_3$-Ni composites with a high and frequency-independent dielectric constant (εr = 80000). Adv. Mater. **13**, 154 (2001)
15. H. Emoto, J. Hojo, Sintering and dielectric properties of $BaTiO_3$-Ni composite ceramics. J. Ceram. Soc. Jpn. **100**, 555 (1992)

16. N. Hirose, A. West, Impedance spectroscopy of undoped $BaTiO_3$ ceramics. J. Am. Ceram. Soc. **79**, 1633 (1996)
17. M. Vollman, R. Waser, Grain boundary defect chemistry of acceptor-doped titanates: space charge layer width. J. Am. Ceram. Soc. **77**, 235 (1994)
18. V. Mitic, V.B. Pavlovic, Lj. Kocic, V. Paunovic, D. Mancic, Application of the intergranular impedance model in correlating microstructure and electrical properties of doped $BaTiO_3$. Sci. Sinter. **41**, 247 (2009)
19. J.P. Clerc, G. Giraud, J.M. Laugier, J.M. Luck, The electrical conductivity of binary disordered systems, percolation clusters, fractals and related models. Adv. Phys. **39**, 191 (1990)
20. R.K. Chakrabarty, K.K. Bardhan, A. Basu, Measurement of AC conductance, and minima in loss tangent, of random conductor-insulator mixtures. J. Phys.: Condens. Matter **5**, 2377 (1993)
21. R. Bouamrane, D.P. Almond, The 'emergent scaling' phenomenon and the dielectric properties of random resistor-capacitor networks. J. Phys.: Condens. Matter **15**, 4089 (2003)
22. C.R. Bowen, D.P. Almond, Modelling the 'universal' dielectric response in heterogeneous materials using microstructural electrical networks. Mater. Sci. Technol. **22**, 719 (2006)
23. S. Panteny, R. Stevens, C.R. Bowen, Characterisation and modelling of barium titanate-silver composites. Integr. Ferroelectr. **63**, 131 (2004)
24. D.J. Franck, C.J. Lobb, Highly efficient algorithm for percolative transport studies in two dimensions. Phys. Rev. B **37**, 302 (1988)
25. M. Aouaichia, N. McCullen, C.R. Bowen, D.P. Almond, C. Budd, R. Bouamrane, Understanding the anomalous frequency responses of composite materials using very large random. Eur. Phys. J. B **90**, 39 (2017)
26. N.J. McCullen, D.P. Almond, C.J. Budd, G.W. Hunt, The robustness of the emergent scaling property of random RC network models of complex materials. J. Phys. D: Appl. Phys. **42**, 064001 (2009)
27. C.R. Bowen, A.C.E. Dent, D.P. Almond, T.P. Comyn, Modelling power law dependencies of frequency dependent AC conductivity and permittivity of conductor-relaxor composites. Ferroelectrics **370**, 166 (2008)
28. D.P. Almond, B. Vainas, The dielectric properties of random R-C networks as an explanation of the 'universal' power law dielectric response of solids. J. Phys. Condens. Matter **11**, 9081 (1999)
29. A.L. Efros, B.I. Shklovskii, Critical behavior of conductivity and dielectric constant near the metal-non-metal transition threshold. Phys. Status Solid. B **76**, 475 (1976)
30. R. Chen, X. Wang, H. Wen, L. Li, Z. Gui, Enhancement of dielectric properties by additions of Ni nano-particles to a X7R-type barium titanate ceramic matrix. Ceram. Int. **30**, 1271 (2004)

Chapter 46
Numerical Study of P3HT: Graphene Organic Solar Cell

Chahrazed Dridi and Naima Touafek

Abstract Organic cells based on conjugated polymers such as P3HT have attracted many research interests because of their advantages and potential for continuous development. A computational study on the performances in bulk heterojunction (BHJ) organic solar cells of P3HT: graphene was done. We have used the Solar Cell Capacitance Simulator (SCAPS) intended for solar cells; we have choose the effective medium model (EMM) to simulate the active layer lightly doped P. It is assumed that the two materials are very well interpenetrated in the active layer with a distance between all points of the blend and a donor/acceptor interface being as inferior to the diffusion length of the excitons. The appropriate parameters were introduced in platform of (SCAPS) simulator. (J-V) chahracteristics, the quantum efficiency and performance parameters (Voc, Jsc, FF and η) were simulated. The organic solar cell performance parameters (Voc, Jsc, FF and η) are affected by the weight and electron affinity of Graphene in the blend P3HT:graphene.

Keywords Graphene · P3HT · Organic solar cell · SCAPS

46.1 Introduction

Since their creation in 1954, the photovoltaic world has constantly evolved mainly inorganic semi-conductor cells [1]. Photovoltaic cells (PV) open up the possibility of producing electricity directly from solar radiation without moving parts, no heat production and no risk of atmospheric pollution.

C. Dridi (✉)
Laboratoire de Physique Mathématique et Subatomique LPMS, Mentouri Brothers, Constantine 1 University, 2500 Constantine, Algeria
e-mail: Chazed2011@gmail.com

N. Touafek
Higher National School of Biotechnology "Toufik Khaznadar" (ENSB), 2500 Constantine, Algeria

A. Belasri and S. A. Beldjilali (eds.), *ICREEC 2019*, Springer Proceedings in Energy,
https://doi.org/10.1007/978-981-15-5444-5_46

However, despite the subsidization of the solar energy sector, prices of power station or solar panels remain expensive compared to other combustion techniques [2], then another approach to reduce the cost of manufacturing solar cells has emerged is the use of carbon derivatives that are available and require little treatment.

These organic materials are increasingly required to replace conventional materials in functions such as light emission they are especially those based on polymer. The Poly (3-hexylthiophene) (P3HT) is one of the most studied conjugated polymers that is suitable for organic solar cells. It has a high electrical conductivity and P3HT is a donor material and a p-type semiconductor [3], it has a high hole mobility $\sim$0.2 cm^2/Vs [4] but remains lower than that of inorganic materials. P3HT has a low photocurrent due to very low electron mobility it must be mixed to a material with high mobility to increase the electrons mobility.

In this work, we chose graphene, when it is blended with the polymer P3HT, offers a large interfacial surface to the p3HT molecules. Thanks to its dimensional structure 2D. As it has a high electron affinity ($\sim$4.5 eV) [9] compared to that of P3HT (LUMO $\sim$2.8 eV) [10], it can be considered as an excellent acceptor so it can improve the dissociation of excitons.

By its remarkably high carrier mobility (1e4 cm^2/Vs), graphene contributes to the improvement of electron transport through the active layer, it is highly transparent and its electrical conductivity is very high [4].

We have studied numerically the effect of the presence of graphene, namely the electronic affinity and the ratio of the weight in the organic cell P3HT: Graphene. We have use different weight content of graphene in the P3HT: Graphene (0wt%, 0.1wt%, 0.5wt%, 1wt%, 3wt%, 5wt% and 10wt%) [4].

To simulate this structure, we have used the effective medium based on the theory of meta-materials [6, 7, 8], which amounts to consider the nanostructure as a single semiconductor layer. In this effective medium, almost all properties are related to carriers of charges either n or p. The appropriate parameters have been introduced in the SCAPS (a Solar Cell Capacitance Simulator) simulator.

46.2 Theoretical Model

The most general strategy today in organic cells, is the heterojunction cell, in which electron donors (P3HT) and electron acceptors (Graphene) have mixed to form a blend layer.

In organic materials, the process of photo generation of charge carriers is generally, attributed to the dissociation of molecules in the excited state in the presence of an internal electric field generated by an interface, an impurity or defects [5]. The Light absorption leads to the creation of electron-hole pairs strongly bound by electrostatic attraction.

The latter, called excitons diffuse by jump from one molecule to another, towards a contact, an interface or an impurity; they dissociate if they are at a distance less than the diffusion distance of the excitons; transport and collection of charges at the electrodes are controlled by the mobility of the charge carriers in the organic layer. The equations that govern the studied model are based on solving the basic semiconductors equations: Poisson equation and continuity equations for electrons and holes:

$$\frac{d^2\varphi}{dx^2} = -\frac{q}{\varepsilon\,\varepsilon r}\left(p - n - NA^- - ND^+ + \frac{\rho t}{q}\right) \tag{46.1}$$

$$\frac{1}{q}\frac{dJn}{dx} = (G - Rn) \tag{46.2}$$

$$\frac{-1}{q}\frac{dJp}{dx} = (G - Rp) \tag{46.3}$$

where φ describes electrical potential, q is the unit charge, $\varepsilon_0, \varepsilon_r$ is the dielectric constant of vacuum and the semiconductor respectively; n and p are electron and hole density, respectively; N_D^+ *and* N_A^- the density of ionized donors and acceptors, ρ_t the charge density of defects, J_n and J_p the electron and hole current density; G is the generation rate and Rn(p) is the recombination loss.

The next equations define the transport charge carriers' model: drift and diffusion

$$Jn = -qDn\frac{dn}{dx} + q\mu n\frac{d\varphi}{dx} \tag{46.4}$$

$$Jp = -qDp\frac{dn}{dx} - q\mu p\frac{d\varphi}{dx} \tag{46.5}$$

To which must be added the exciton conservation equation:

$$\frac{Dx}{dt} = Gph \tag{46.6}$$

Where $\mu_{n(p)}$ is the electron/hole mobility and Dn(p) is the diffusion coefficient.

46.3 Methodology

The effective medium model is characterized by its HOMO which is that of the P type material (P3HT) and it's LUMO that of the N type material (graphene).

This hypothesis amounts to considering an active layer where the two materials are very well interpenetrated, that is to say that the distance between all points of the

Table 46.1 Parameters used in simulation

Parameter	Symbol	Value	Unit
Thickness	D	100	[nm]
Band gap	Eg	2.065	[ev]
Electron affinity	χ	4.5	[ev]
Effective density of states	Nc, Nv	1E18	[cm^{-3}]
Electron mobility	μn	1E4	[cm^2/Vs]
Hole mobility	μp	2E − 1	[cm^2/Vs]
Dielectric constant	ε	6.3	

blend and a donor/ acceptor interface is less than the diffusion length of the excitons.

A single, lightly doped semiconductor layer P represents the nanostructure p-n. To create a driving force for the separation of the electron hole pairs contacts are considered selective, that is to say, a contact accepts only the electrons and the other accepts only the holes [11].

Parameters necessary for the simulation of the active layer are extracted from the literature such as the permittivity, the band gap as well as the absorption coefficient of the P3HT: graphene blend [4]. Subsequently are introduced on the platform of the simulator capacity of solar cells in one dimension (SCAPS-1D) and are given by Table 46.1.

46.4 Results and Discussion

According to the results of the simulation presented in Fig. 46.1a, it can be seen that the graphene weight content of 1wt% shows the greatest current density value of all the concentrations; This proves that only a small amount of graphene is needed to increase the current.

In Fig. 46.1b. The max efficiency (η), open circuit voltage (V_{oc}), short circuit current (Jsc) and fill factor (FF) are obtained for the concentration of 1% by weight of graphene.

In Fig. 46.2 (a, b), the absorption coefficient corresponding to 1wt% of graphene content in the p3ht: graphene blend was used [4]. By varying graphene electronic affinity, it is found that the current density (Jtot) and short circuit current (Jsc) remains unchanged, which implies that it directly relates to the graphene weight (wt%) present in The mixture. Whereas the open circuit voltage (V_{oc}), the efficiency (η) and the fill factor (FF) decrease with increasing graphene electronic affinity. Graphene has a remarkable influence on the characteristics of the solar cell. Therefore reducing graphene affinity leads to the increase all performance parameters specially the open circuit voltage and efficiency.

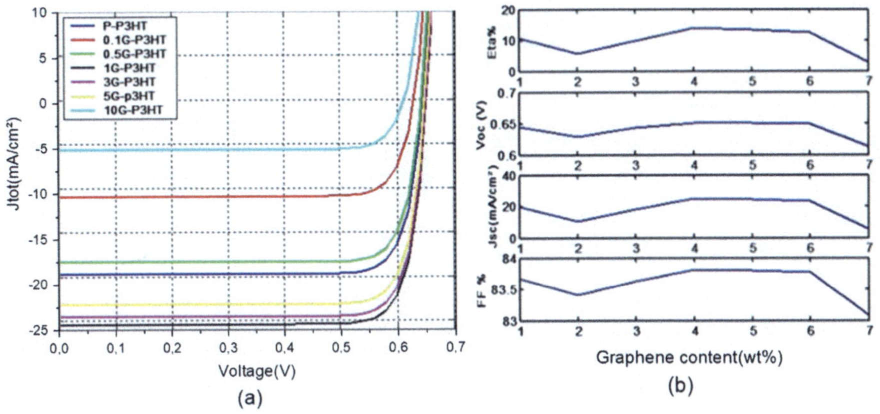

Fig. 46.1 For every graphene weight content: **a** Current density (Jtot), **b** parameters performance (η), (V_{oc}), (Jsc), (FF)

The best quantum efficiency (QE) was obtained for (1wt%) of graphene Fig. 46.3 (a); by varying the affinity of the graphene electrons, the quantum efficiency remains constant for all the values given in Fig. 46.3 (b). QE is influenced by the graphene content in the active layer.

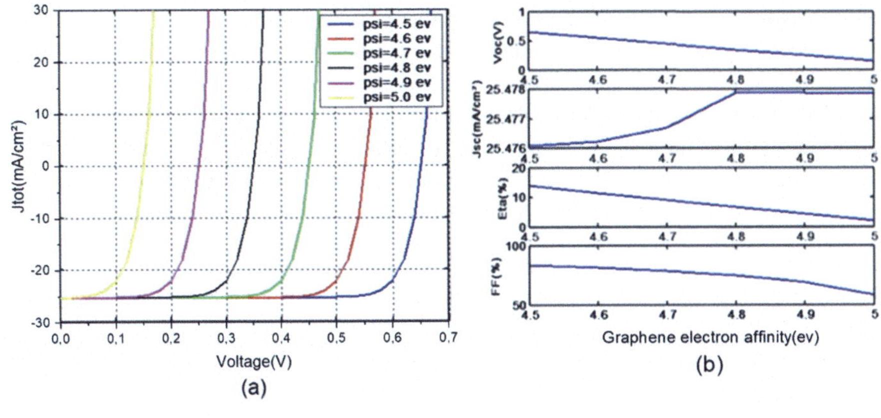

Fig. 46.2 For every electronics affinity: **a** Current density (Jtot), **b** parameters performance (η), (V_{oc}), (Jsc), (FF)

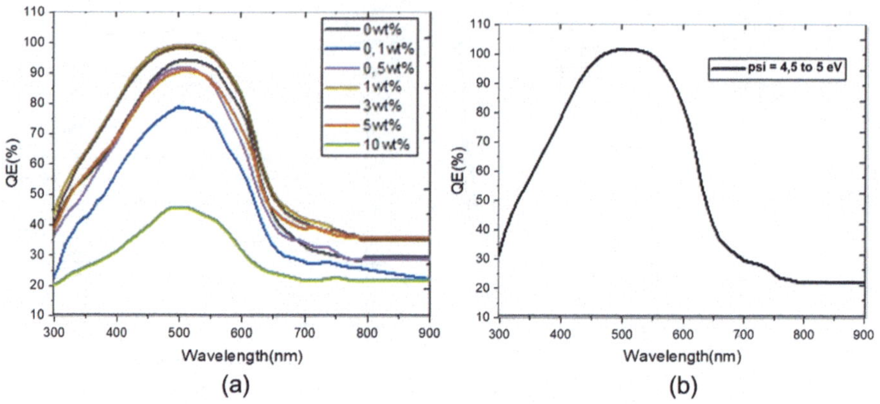

Fig. 46.3 Variation of quantum efficiency: **a** weight graphene, **b** graphene electron affinity

46.5 Conclusion

The bulk heterojunction organic solar cell architecture Ag/P3HT: Graphene/Al was simulated by SCAPS software for the analysis of photovoltaic cells.

The influence of the presence of graphene with different weights in the mixture p3ht: graphene; on the electrical parameters of the cell was the main objective of this work. The weight of (1wt%) of graphene in the mixture constituting the active layer revealed good performance of the p3ht: graphene cell.

The second goal in this study was to improve the cell's performance by lowering the electronic affinity of the graphene, the open circuit voltage, the efficiency and the fill factor increased considerably with the decrease in affinity while the current density remained almost constant. This simulation work will therefore provide useful information on the active P3HT: Graphene solar cell layer in order to reasonably choose the graphene (acceptor) parameters and achieve good performance.

Acknowledgements The authors would like to thank Marc Burgelman and his team at Department of Electronics and Information Systems (ELIS) of the University of Gent, Belgium for the access of simulator SCAPS.

References

1. A. Goetzberger, C. Hebling, H.-W. Schock, Photovoltaic materials, history, status and outlook. Mater. Sci. Eng. **R40**, 1 (2003)
2. A.M. Bagher, Introduction to organic solar cells. Sustain. Energy **2**(3), 85–90 (2014)
3. J. Ye., X. Li., Z. Jianjun, M. Xuelan, L. Qian, Efficient and stable perovskite solar cells based on functional graphene-modified P3HT hole transporting layer. RSC Adv. **6**, 36356–36361 (2016)

4. V. Saini, O. Abdulrazzaq, S. Bourdo, E. Dervishi, A. Petre, V.G. Bairi, T. Mustafa, L. Schnackenberg, T. Viswanathan, A.S. Biris, Structural and optoelectronic properties of P3HT-graphene composites prepared by in situ oxidative polymerization. J. Appl. Phys. **112** (5), 54327–54329 (2012)
5. M. Oukachmih, *Les cellules photovoltaïques à base de matériaux organiques Discotiques* (Thèse de Doctorat à l'Université Paul Sabatier, Toulouse, 2003)
6. B. Minnaert, M. Burgelman, Modelling MEH-PPV: PCBM (1:4) bulk heterojunction solar cells. Numos, 327–339 (2007)
7. C. Brabec, V. Dyakonov, J. Parisi, N. Sariciftic, Efficient Polyfluorene Based Solar Cells (Springer, Berlin 2003), pp. 183–185
8. B. Minnaert, C. Grasso, M. Burgelman, An effective medium model versus a network model for nanostructured solar cells. C.R. Chimie **9**, 735–741 (2006)
9. N.S.M. Shariff, P.S.M. Saad, M.R. Mahmood, Capacitance Voltage of P3HT: Graphene Nanocomposites Based Bulk-Heterojunction Organic Solar Cells
10. U. Zhokhavets, T. Gobsch, G. Erb, M. Al-Ibrahim, O. Ambacher, Chem. Phys. Lett. **418**(4), 47–350 (2006)
11. C. Bendina, Etude et Simulation des Proprietes Electriques et Optiques des Cellules Photovoltaiques organiques multicouches. thése Doctorat, Université Abdelhamid ibn Badis Mostaganem, Algerie, Fev. (2019)

Chapter 47
Structural and Electronic Properties for Chalcopyrite Semiconducting Materials Based on CuXS2 (X = In, Ga and Al): AB-Initio Computational Study

F. Belarbi, M. Adnane, and F. Boutaiba

Abstract A theoretical study of structural and electronic properties of chalcopyrite semiconducting materials is presented using the full-potential linearized augmented plane wave method implemented in Wien2k computational package. In this approach, the Generalized Gradient approximation was used for exchange correlation potentials. Our interest in these semiconductor materials is motivated by their promising roles in the manufacture of thin-film solar cells because of their interesting structural and electronic properties.

Results are given for lattice constant, bulk modulus and its pressure derivative for chalcopyrite materials specifically based on copper indium sulfur (CIS), copper gallium sulfur (CGS) and copper aluminum sulphur (CAS). Band structure are also given. The results are compared with previous calculations and with experimental measurements.

Keywords Solar cell · Chalcopyrite thin film · Wien2k

47.1 Introduction

Recently A I B III C VI chalcopyrite semiconductors revealed a lot of interest in the materials community and been extensively studied for photovoltaic solar cell and optoelectronic due to the large absorption coefficient (>105 cm-1) and suitable bond gap for these application [1–5]. For example, the polarized absorption spectra of $CuGaS_2$, $CuInS_2$ were reported in Ref. [6]. We know that the Understanding of material physics requires the fundamental knowledge of its various structural and

F. Belarbi (✉) · M. Adnane · F. Boutaiba
University of Sciences and Technology of Oran Mohamed Boudiaf USTO-MB, Oran, Algeria
e-mail: Faten.belarbi@usto-univ.dz; belarbifaten84@gmail.com

A. Belasri and S. A. Beldjilali (eds.), *ICREEC 2019*, Springer Proceedings in Energy,
https://doi.org/10.1007/978-981-15-5444-5_47

optoelectronic properties, and to achieve a better understanding of these proprieties of the $CuXS_2$ chalcopyrite compounds, we have used theoretical methods more solicited called ab initio methods which allow to describe the physico-chemical properties in the ground state. Ab initio simulations based on density functional theory (DFT), which only use atomic constants as input parameters for the resolution of Schrödinger's equation, these methods have become today a tool for basis for the study of different properties of molecules and materials. Among the ab initio methods that compete with the world of digital physics, the FP-LAPW method (Full Potential-Linearized Augmented Plane Wave) which is proven in the precision of calculation.

In this paper we present a combined ab initio study of the electronic, structural, properties of CXS2, using theWIEN2K code [7] within the density functional formalism (DFT) with the TB-mBJ potential [9–11] and also Perdew–Burk–Ezrenhof (PBE) potential [13] are reported.

47.2 Computational Details

The calculations related to $CuXS_2$ have been performed using the full potential linear augmented plane wave (FP-LAPW) method as implemented in Wien2K code. This method is based on density functional theory (DFT) which is a universal quantum mechanical approach for many body problems in this method the unit cell volume is partitioned into the non-overlapping Muffin-Tin spheres surrounding every atom and the remaining interstitial region. We have taken the product of minimum of the Muffin Tin radius of atoms (Rmt) and maximum value of the wave vector (K max), Rmt * Kmax 7.0 and allowed the expansion of the valence wave functions inside the Muffin Tin sphere into spherical harmonics up to l max 10 whereas the Fourier expansion of the charge density was up to Gmax = 12 (a.u.) while in the interstitial region plane waves are used. The Muffin Tin radius are chosen in such a way that they do not overlap each other, and core charge do not leak out of the spheres. We have employed the generalized gradient approximation (GGA PBE) due to Perdew, Burke and Ernzerhof [8] to obtain exchange correlation potential and Tran–Blaha modified Beck and Johnson (TB-mBJ) to extract optoelectronic properties. The lattice constants for CuXS2 are taken to be a = b = 5.305 Å, c = 10.610 Å for CuInS2 and a = b = 5.097 Å, c = 10.194 for $CuGaS_2$ Å and a = b = 5.029 Å, c = 10.058 for CuAlS2 Å. Total 400k points are taken to perform present computations.

47.3 Results and Discussion

47.3.1 Structural Properties

The most important step in an ab initio calculation is the determination of structures properties of the studied material. Knowledge of this information allows us to access subsequently to other physical properties (electronic, optical,…) The CuX_2 (X = In, AL and Ga) materials crystallize in the tetragonal structure of Chalcopyrite, with space group I-42d with four formula units in a unit cell, (ranked 122 in the International Table of crystallography). We performed a self-consistent calculation of the total energy for several values of the network parameters a and c taken in the vicinity of the experimental value. To determine the structural properties of the static equilibrium such as the lattice parameter a0, the compressibility module B0 and its derivative relative to at the pressure B′ of the chalcopyrite structure, we must optimize the following independent parameters: the volume V of the mesh by calculating the total energy E_{tot} for different volume values and then adjust by the Murnaghan state Eq. (1944) [8].

$$\mathrm{E(v)} = \mathrm{E_0} + \frac{\mathrm{B}}{\mathrm{B'(B'-1)}}\left[V\left(\frac{V0}{V}\right)^{B'} - V0\right] + \frac{B}{B'}(V - V0) \tag{47.1}$$

Where V_0 is the static equilibrium volume of the primitive mesh, E_0 the total energy per primitive mesh of the equilibrium state.

B: the compressibility modulus is determined by the following equation:

$$B = V\frac{\partial^2 E}{\partial V^2} \tag{47.2}$$

B′: the derivative of the compressibility module:

$$\mathrm{B}' = \frac{\partial B}{\partial P} \tag{47.3}$$

In Table (47.1) we have collected all the equilibrium quantities such as the lattice parameter a, the compression modulus B and derivative B′, calculated ab initio using GGA. PBE for structure chalcopyrite. We have also included in the Table (47.1) the experimental values and other theoretical calculations to facilitate the comparison.

Table 47.1 Equilibrium lattice parameters, Bulk modulus, and Band gap with available previous reported value for CuXY2 (X = In, Ga, Al)

Compounds	a(Å)	c(Å)	V_0	B	B′
CuInS2 (CIS)	5.305, 5.505[a],	10.61	1195.2690	67.67	5.045
CuAlS2(CAS)	5.029, 5.332[b],	10.258	1017.6608	82.51	5.163
CuGaS2(CGS)	5.097, 5.37[c],	10.194	1017.6608	71.92	4.793

[a] Reference [13], [b] Ref. [4], [c] Ref [15]

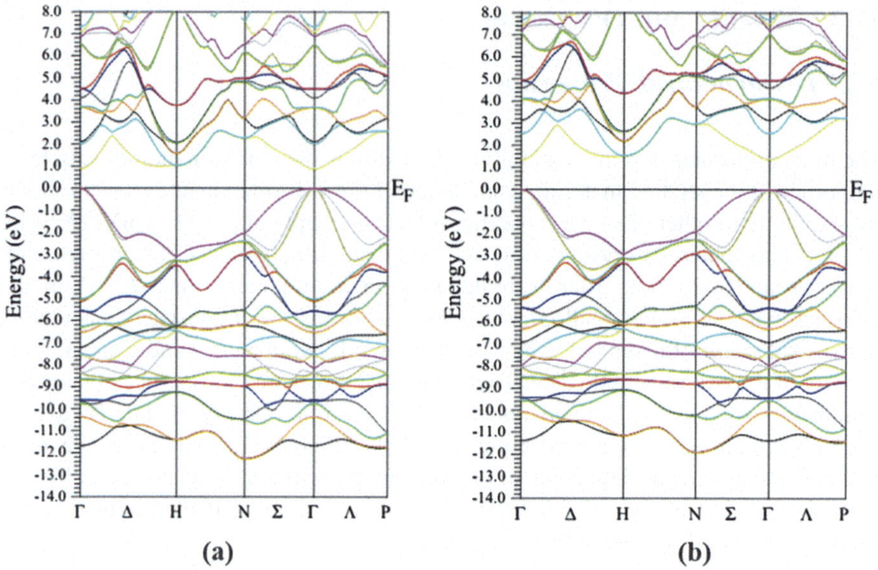

Fig. 47.1 The electronic band structure of $CuInS_2$ calculated with **a** the GGA–PBE **b** the mbJ-GGA

47.4 Electronic Properties

We are interested in the calculations of energy band structures, relating to the $CuInS_2$ chalcopyrite ternary compounds, $CuGaS_2$. and $CuAlS_2$ using the optimized. parameters of the mesh. et presented the dispersion equation E = f (k). Calculations have were performed using the GGA-PBE and mBJ-GGa approximations. This result is adopted to improve energy gaps compared to those of experience. We notice that our materials are semiconductors materials with Direct gap at point Γ (Fig. (47.1), (47.2), (47.3)).

The values of the gap of energy that we found with the approach of GGA - PBE and TB mBJ in our calculation, is given in Table (47.2), compare with other experimental results.

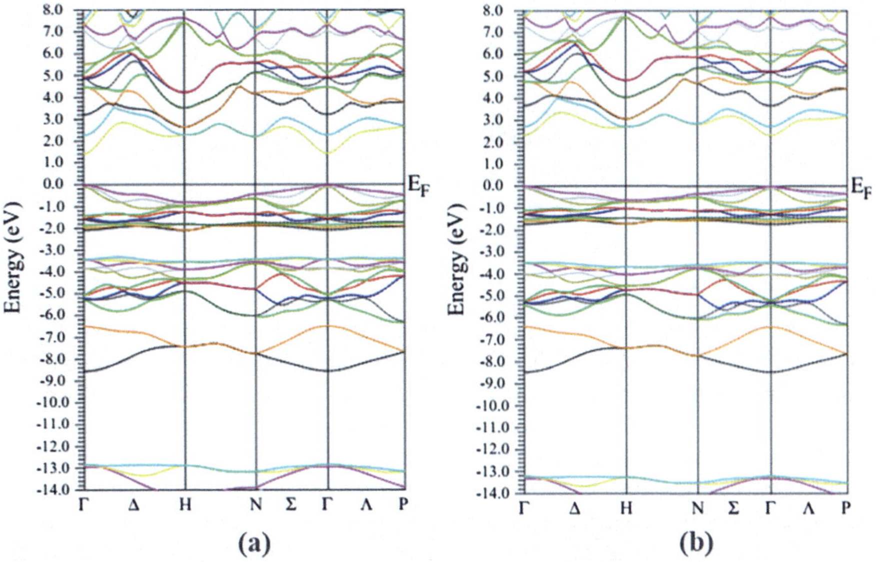

Fig. 47.2 The electronic band structure of the $CuGaS_2$ calculated with **a** the GGA-PBE **b** the mbJ-GGA

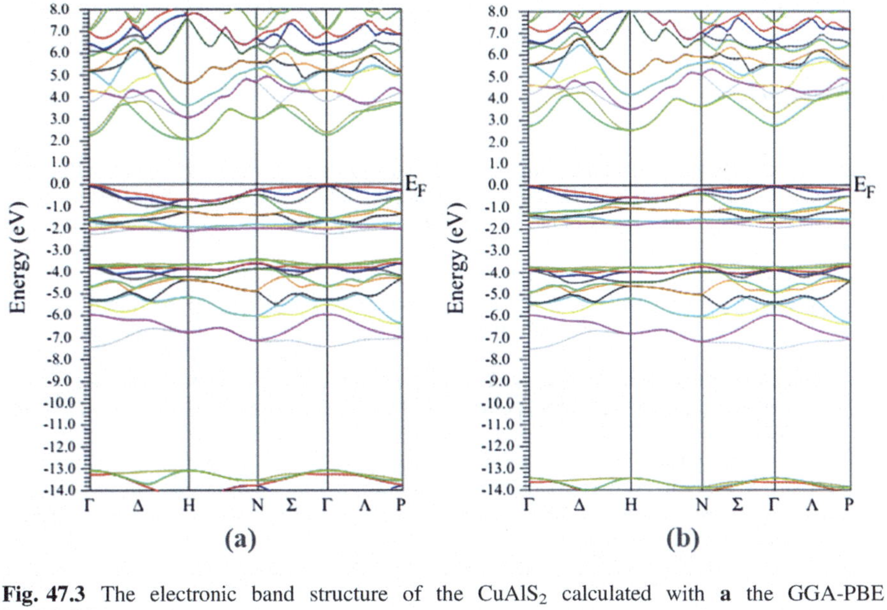

Fig. 47.3 The electronic band structure of the $CuAlS_2$ calculated with **a** the GGA-PBE **b** mbJ-GGA

Table 47.2 The values of gap energy for $CuXS_2$ (x = In, Al and Ga) materials

	CuInS2	CuGaS2	CuAlS2
Eg(eV)calculated (GGA-PBE)	0.8	1.7	2.1
Eg(eV)calculated (mbj-GGA)	1.35	2.30	2.9
Eg(ev)experimental	1.53[a]	2.43[b]	3.49[c]

[a] Reference [8], [b] Ref. [9], [c] Ref [14]

47.5 Conclusion

This work is a contribution to the study of structural properties (lattice parameter, equilibrium positions, compressibility module and its derivative), the electronic properties (band structures, electronic density of states) of compounds $CuXS_2$ (X = In, Ga and Al) in the chalcopyrite phase.

Our results concerning the structural properties of steady state values obtained are not far from the experimental values.

The total energy calculations were performed using the GGA-PBE et TB-mBJ exchange potential as implemented in WIEN2K (FP-LAPW) method to investigate the electronic and optical properties of $CuXS_2$ (X = In, Ga, Al) compounds are in good agreement with available experimental results and other reported theoretical results.

References

1. J.E. Jaffe, A. Zunger, Electronic structure of the ternary chalcopyrite semiconductors CuAlS2, CuGaS2, CuInS2, CuAlSe2, CuGaSe2, and CuInSe2. Phys. Rev. B **28**, 5822 (1983)
2. J.E. Jaffe, A. Zunger, Theory of the band-gap anomaly in ABC2 chalcopyrite semiconductors. Phys. Rev. B **29**, 1882 (1984)
3. R. Ahuja, S. Auluck, O. Eriksson, J.M. Wills, B. Johansson, Calculated optical properties of a solar energy material: CuGaS2 Sol. Energy Mater. Sol. Cells **53**, 357–366 (1998)
4. J. Lazewski, P.T. Jochym, K. Parlinski, Band structure, born effective charges, and lattice dynamics of CuInS2 from ab initio calculations. J. Chem. Phys. **117**, 2726 (2002)
5. S. Laksari, A. Chahed, N. Abbouni, O. Benhelal, B. Abbar, First-principles calculations of the structural, electronic and optical properties of CuGaS2 and AgGaS2. Comput. Mater. Sci. **38**, 223–230 (2006)
6. I.-H. Choi, S.-D. Han, S.-H. Eom, W.-H. Lee, H.C. Lee, J. Kor. Phys. Soc. **29**, 377 (1996)
7. P. Blaha, K. Schwarz, G.K.H. Madsen, D. Kvasnicka, J. Luitz, *WIEN2K an augmened plane wave local orbitals program for calculating crystal properties, Karlheinz Schwarz, Techn* (University at Wien, Austria, 2001)
8. M.I. Alonso, K. Wakita, J. Pascual, M. Garriga, N. Yamamoto, Optical functions and electronic structure of CuInSe2, CuGaSe2, CuInS2, and CuGaS2. Phys. Rev. B **63**, 075203 (2001)
9. O. Madelung, Semiconductors: Data Handbook, vol 3 (University of Marburg, Marburg 2004)

10. W.B. Pearson, *Handbook of Lattice Spacing and Structures of Metals and Alloys*, vol. 2 (Pergamon Press, Oxford, 1967)
11. A. Ghosh, R. Thangavel, Rajagopalan M, (2014) Electronic and optical modeling of solar cell compound CuXY2 (X = In, Ga, Al; Y = S, Se, Te): first-principles study via Tran–Blaha-modified Becke-Johnson exchange potential approach. J. Mater. Sci. **50**, 1710–1717 (2015)
12. K.M.A. Hussain, J. Podder, D.K. Saha, M. Ichimura, Structural, electrical and optical characterization of CuInS2 thin flims deposited by spray pyrolysis. Indian J. Pure Appl. Phys. **50**, 117–122 (2012)
13. S. Chen, X.G. Gong, S.-H. Wei, Band-structure anomalies of the chalcopyrite semiconductors CuGaX2 versus AgGaX2(X = S and Se) and their alloys. Phys. Rev. B **75**, 205209 (2007)
14. L. Zihan, F. Xiaofang, M. Qiang Ma, X. Gaoa, X. Su, CuInS2 quantum dots@silica near-infrared fluorescent nanoprobe for cell imaging. New J. Chem. **38**, 90–96 (2014)

Chapter 48
Comparative Analysis of Two Methods of Modeling a Mono-Crystalline Silicon PV Module

Fatma Zohra Kessaissia, Abdallah Zegaoui, Aicha Aissa Bokhtache, Asma Toualbia, and Hadj Allouache

Abstract The main purpose of this work is to analyze the comparison between the current voltage characteristic modeling of a mono-crystalline photovoltaic panel using classical and Design of Experiments methods. These methods based on mathematical models have been solved using numerical method, therefore, the main difference between them is in the process of the response output calculation that the classical one need several input parameters, which is obtain analytically. By cons, the design of experiments method is a fast tool to calculate the response, it consider only input and output parameters, discuss the obtained models and the associated calculation methods that we have evaluated via Matlab and Minitab development soft wares. In this contribution, we present the important steps to model the studied system, and we consider experiments concept allows finding the best possible model that both represents the experimental data and enables accurate predictions of the responses according to input variation factors in a strict definite study domain.

Keywords Design of experiments · Mathematical model · Study domain

48.1 Introduction

Photovoltaic module (PV) is one of the most important devices, subject of numerous research projects, which represents several papers that divided into three main areas: modeling, characterization and optimization. In that, we have chosen to analyze the comparison between the analytic and the practical methods. Modeling a photovoltaic module is the first step of the evaluation of the efficiency of photovoltaic energy production systems, which provide a mathematical model of the

F. Z. Kessaissia (✉) · A. Zegaoui · A. Aissa Bokhtache · A. Toualbia · H. Allouache
Hassiba Benbouali University, BP C78, Ouled Fares, Chlef, Algeria
e-mail: f.kessaissia@univ-chlef.dz

A. Zegaoui
University of Lorraine, EA-4423, 2 rue E. Belin, 57070 Metz, France

A. Belasri and S. A. Beldjilali (eds.), *ICREEC 2019*, Springer Proceedings in Energy,
https://doi.org/10.1007/978-981-15-5444-5_48

studied system. In this paper, we treat two kinds of modeling a PV module, one can used analytical method, or experimental, all these methods need experimental data, but differ in the resolution of those mathematical models.

Analytical or classical method [1–3] is the most used method in differ literature, it given by an implicit model that reproduce the behavior of the PV module taking into account the temperature and the illumination variations. It allows presenting the I-V characteristics of a PV module depending on a set of parameters, these parameters are function of the operating condition of illumination of PV cells and temperature, which are analytically calculated from some experimental measured variables. The resolution of the mathematical model requires at first to extract the parameters characterizing the model, then calculated the correspond responses using numerical method to solve the equation model.

By cons, the design of experiments method (DoE) [4, 5] is a practical and fast way to evaluate the current voltage characteristic, it gives more information on the system in few time and with some experience data. However, the DoE appears as an alternative method for evaluating the significant factors, correlation between several factors and their influence on the response of the studied system.

48.2 Method and Modeling

The main objective of present contribution is to find a mathematical model presenting the maximum power P_m output response under different input parameters variations of illumination E and temperature T, this mathematical model need a set of experiments.

Classical method based on the five physical and electrical parameters of the module, which are the photo-current I_{ph}, the saturation current I_s, the series resistance R_s, the shunt resistance R_{sh}, and the ideality factor m.

The characteristic equation related to this model is given by the formula [6, 7]:

$$I = I_{ph} - I_s\left[exp\left(\frac{V + R_s I}{mN_s V_t}\right) - 1\right] - \frac{V + R_s I}{R_{sh}} \tag{48.1}$$

Within:

$$V_t = \frac{KT_c}{q} \tag{48.2}$$

Thanks to the I-V curve of a PV module that is obtained experimentally, for a given illumination and temperature, we will analytically determine the values of the following five parameters: I_{ph}, I_s, R_s, R_{sh} and m, using any points that are well defined of this curve.

The design of experiments method divides into three important parts:

- Construction of the experimental design by presenting the purpose of the study, then consider the correspondent factorial design, factors, domain variation of factors and the response.
- Realization of the experiments mentioned at the experimental design in the first step.
- Interpretation of the obtaining results.

However, the DoE method considers the studied system as a black box, which taking into account only input parameters and outputs of the studied system, and in absence of any specific information about the links between these inputs and outputs, one can model all systems by a general polynomial relation:

$$y = f(x) \tag{48.3}$$

With four trials for the experimental design, the method will produce the first order mathematical model for the considered response. For these four experiments, the mathematical model can be written, as indicated in Eq. 48.4:

$$y = a_0 + a_1x_1 + a_2x_2 + a_{12}x_1x_2 \tag{48.4}$$

Where, in our case y is the measured response, $\times_1$ and $\times_2$ are the illumination and the temperature factors respectively, a_0 is the coefficient representing the central value and a_1, a_2 and a_{12} are the coefficients associated to the respective contributions and interaction between them of the factors $\times_1$, $\times_2$.

Therefore, these experiments measurement consist to define a study domain with intersection of the factors variations domain constitute by the way a full factorial design named 2^2, when each point of intersection present an experiment. However, in order to obtain the same unit of factor variations, we use the coded units (CU) and each factor variation domain is defined by two levels; the low level designed by '−1' and the high level designed by '+1'. Figure 48.1 presents the study domain of the defining experimental design limited by the intersection of input factors variations domain, given by the way an experimental points, each point of the study domain define an experiment under the variations of illumination and temperature.

In order to obtain the same unit, after acquisition of experimental data, firstly one can translated the input factors values into the coded units using the Eq. 48.5 mentioned by the formula:

$$x = \frac{A - A_0}{Step} \tag{48.5}$$

Where A_0 is the central value in the coded units and *Step* is the half of the difference between the upper and the lower levels, both expressed in the original unit A of the factor.

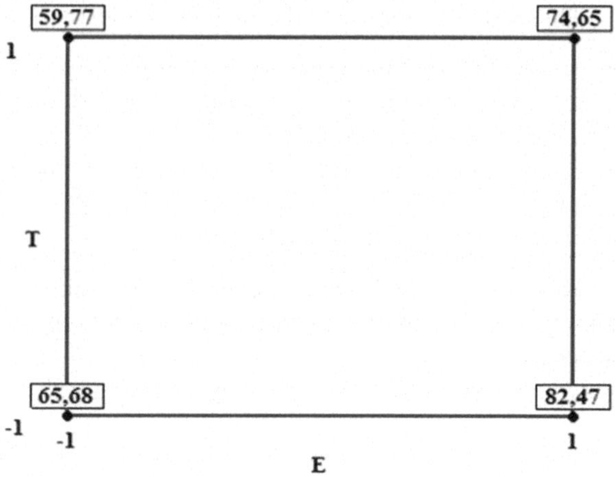

Fig. 48.1 Factorial design of the studied response

48.3 Experimental Setup

The experienced mono-crystalline module is the BP Saturn with a maximum power of P_m = 85 W realized at voltage of V_{mp} = 29,2 V and a current I_{mp} = 2,3 A. Its open circuit voltage is V_{oc} = 35,9 V and its short circuit current is I_{sc} = 3,0 A. This PV module is built within 60 cells in series and 1 in parallel. These values are extracted from the datasheet of the panels.

Table 48.1, gives the experimental design measurement and the observed response: factors: Illumination E, Temperature T and response: maximum Power P_m respectively.

The deduced upper and lower limits of each factor and them coded unit's values are indicated in Table 48.2. We can notice that irradiation levels and temperatures were recorded during the same indoor experiments.

Table 48.1 The experimental design measurement and the observed response

Factors		Response
E (W/m^2)	T (°C)	P_m (W)
800	25	65,68
800	45	59,77
1000	25	82,47
1000	45	74,65

Table 48.2 The factors variations domain

Factors/Levels	Low level "−1"	High level "+1"
Illumination E (W/m^2)	800	1000
Temperature T (°C)	25	45

48.4 Results and Discussion

Figures 48.2 (a) and (b) present experimental and calculated I-V curves obtained through the analytical method applied to mono-crystalline silicon module for different irradiation levels due to different temperatures of 25 °C and 45 °C.

The implicit model, we note that the calculated I-V characteristics are with a strong agreement with experimental characteristics.

Numerical methods allow obtaining the DoE model coefficients. By else, the substitution of these coefficients in the postulated model gives the calculated values of the responses. After these steps of statistical calculation, we obtain the postulated model presenting the behavior of the mono-crystalline module.

In Fig. 48.3 (a), we can observe and validate that the effect of the irradiation factor is more significant than the effect of the temperature factor. The centered value corresponding to the constant coefficient is defined at the middle of the effect plot slope, which considered as the reference point, on the right of the centered value, the maximum power follows the irradiation increase and on its left side, the maximum power follows the temperature decrease. The effect of the temperature factor acts on the opposite direction of those of the irradiation resulting in a negative slope showing that the variation of the maximum power increases with the irradiation and decreases with the temperature. Thanks to the DoE concept, we analyze the interaction effects between irradiation and temperature of the maximum power response. The interaction is even stronger as slopes are different [5]. In Fig. 48.3 (b), we observe the effect of irradiation at different levels of temperature: low level (black straight line) and high level (red dashed line) respectively. The interaction measures the variation of the effect factor 1 when the factor 2 changes from the low level to the high level.

One can note that the two straight lines are almost the same, which means a small interaction between irradiation and temperature.

In Fig. 48.4 (a) and (b), the surface response and the contour curve responses analysis are displayed, which show the outline of the contours of the response

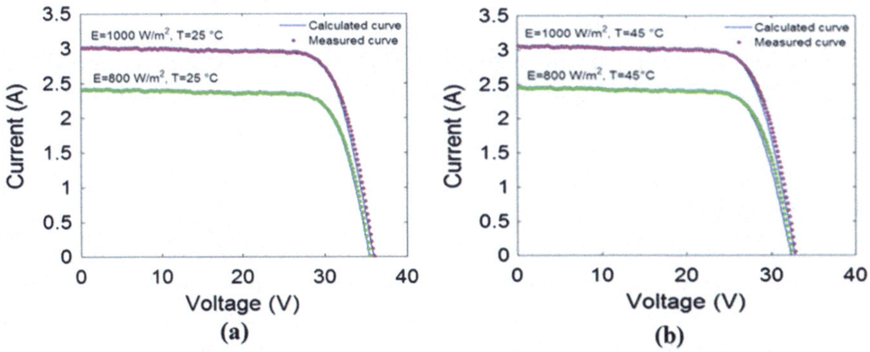

Fig. 48.2 Current voltage characteristic with variation of irradiation **a** At 25 °C, **b** At 45 °C

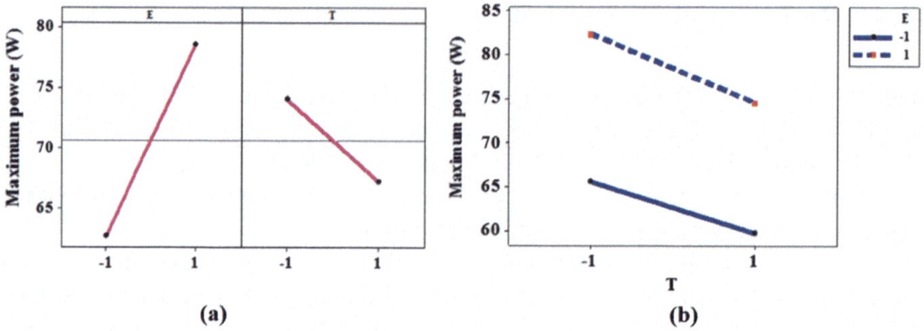

Fig. 48.3 Design of experiments **a** Effects factors on maximum power. **b** Interaction between factors

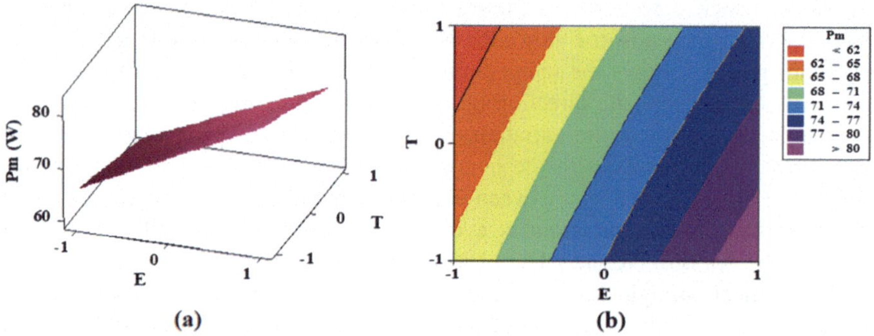

Fig. 48.4 Illustration of the maximum power response: **a** surface response, **b** contour curve

surface performed under Minitab software. As shown in these figures, we see that the evolution of the maximum power, P_m response for the mono-crystalline module follows the same direction as the main effect of illumination but varies in the opposite direction with the main effect of temperature. All this simulation results show that the interactive effect of the two factors influences P_m response in the same direction. This shows that the increase of the irradiation compensates by a dominant effect the decrease of the P_m response caused by the rise of the temperature.

48.5 Conclusion

The present paper deduce for comparative analysis between two modeling methods of a mono-crystalline silicon PV module. First, the parameters of analytical method involved in the physical model are introduced by the experimental data of the manufacturer and modeling lead to the selection of the appropriate mathematical model. Therefore, the Design of Experiments method based on statistical and algebraic calculation. In the current contribution, the DoE method was developed for modeling the behavior of photovoltaic module under indoor illumination and temperature changes. With the DoE method, we can obtain more information on the studied system in few times and using some experiments, contrary to analytical method that need several parameters calculation, which take more time in order to have the best values that checks a good result. Additionally, with the knowledge of the actual response of a PV module, we have analyzed this behavior obtained by simulation. We have shown that the response surface method analysis proved to be reliable for the modelling and studying the main and interactive effects of both factors.

References

1. M. Chegaar, Z. Ouennoughi, F. Guechi, H. Langueu, Determination of solar cells parameters under illuminated conditions. Journal of electron devices **2**, 17–21 (2003)
2. I. Santiago, D. Trillo-Montero, I.M. Moreno-Garcia, V. Pallarés-López, J.J. Luna-Rodríguez, Modeling of photovoltaic cell temperature losses: a review and a practice case in South Spain. Renew. Sustain. Energy Rev. **90**, 70–89 (2018)
3. F.Z. Kessaissia, A. Zegaoui, A. Hadj Arab, L. Loukarfi, M. Aillerie, Comparison of two PV modules technologies using analytical and experimental methods. Energy Procedia **74**, 389–397 (2015)
4. J.P. Charles, F. Hannane, M. El Mossaoui, A. Zegaoui, T.V. Nguyen, P. Petit, M. Aillerie, Faulty PV panel identification using the Design of experiments (DoE). Elect. Power Energy Syst. **57**, 31–38 (2014)
5. J. Goupy, L. Ceighton, *Introduction to design of experiments with JMP Examples*, 3rd edn. (SAS, USA, 2007)
6. M.A. Hasan, S.K. Parida, An overview of solar photovoltaic panel modeling based on analytical and experimental viewpoint. Renew. Sustain. Energy Rev. **60**, 75–83 (2016)
7. A. Yahya-Khotbehsara, A. Shahhoseini, A fast modeling of the double-diode model for PV modules using combined analytical and numerical approach. Sol. Energy **162**, 403–409 (2018)

Chapter 49
Nano Grain of Zinc Oxide Assembled in Hex Nut Deposited by Sol Gel Coating Method

A. Tab, Y. Bakha, A. Abderrahmane, S. Hamzaoui, and M. Zerdali

Abstract Zinc oxide is known for various applications because of its physico-chemical, optoelectronic and piezoelectric properties which are strongly related to the size and morphology of the material. In the present work, we deposited ZnO nanofilm by sol-gel spin coating on glass substrates. X-ray grazing incidence analysis shows a wurtzite structure. The thickness measured by profilometry is about 200 nm and a strong photoluminescence in the visible range has been observed. The atomic force microscopic observation shows a regular arrangement of ZnO grains in the form of hexagonal nut, this morphology improves the roughness of the layer which increases its diffusion of light which is highly desirable for use in the field of solar cells.

Keywords ZnO · Nanofilm · Hex nut · Sol-gel · Spin coating

49.1 Introduction

The study of the properties of thin films of zinc oxide reveals to us every day new morphological aspects: new physicochemical properties that open new perspectives for the use of this material, whether in the field of photovoltaics [1–3], in the field of gas sensors [4–8], in the field of laser diode [9–11], in the field of SAW [12–14], as a photocatalytic and antibacterial [15]. This is an inexpensive material that can be

A. Tab (✉)
PTM Centre de Développement des Technologies Avancées,
cité du 20 aout 1956 Baba Hassen, Algiers, Algeria
e-mail: atab@cdta.dz; Tab.abdelkader@gmail.com

A. Tab · A. Abderrahmane · S. Hamzaoui · M. Zerdali
Laboratoire de Microscope Electronique & Sciences des Matériaux, Université des Sciences et de la Technologie d'Oran Mohamed Boudiaf El-Mnaouar, BP 1505, 31000 Oran, Algeria

Y. Bakha
DMN Centre de Développement des Technologies Avancées,
cité du 20 aout 1956 Baba Hassen, Algiers, Algeria

A. Belasri and S. A. Beldjilali (eds.), *ICREEC 2019*, Springer Proceedings in Energy,
https://doi.org/10.1007/978-981-15-5444-5_49

deposited by several techniques, including that of soft chemistry. In particular the sol gel method, which is a result of hydrolysis reaction followed by a condensation one, that causes the formation of very different grains sizes. Because the condensation is done at random way in space and in time, that's why the idea to make a separation of its agglomerates of molecules by centrifugation of the solution before deposition so as to have a uniform grain size deposit. This study can be used to compare the properties of this film, with others [16–18] and with a theoretical model.

49.2 Experimental Details

We prepared the solution with 0.3 M concentration by dissolving zinc acetate (99.00 wt% $ZnC4H6O4 \cdot 2H2O$, Analytical Reagent) in ethanol (C2H6O 96.00 wt%, Biochem Chemopharma). The solution obtained is subjected to the effect of heating and stirring at a temperature of 60 °C for 30 min, then it is centrifuged for 4 steps of 5 min each. By increasing rotation speed of 1000 rpm until reaching a rotation speed of 4000 rpm, then a drop chosen in the region of low density of the solution, is deposited on a glass substrate, which is centrifuged by a HOLMARC Model HO-TH-05 spin coater with a rotation speed of 1000 rpm for 30 s with an acceleration of 300 rps, then the sample is annealed under ambient atmosphere for 3 h at 450 °C.

49.3 Results and Discussion

This work is focused on the morphology of ZnO thin films, firstly we studied the structural properties by X-ray diffraction grazing incidence by a PANANALYTICAL diffractometer and then we measured the optical transmission using a Spectrophotometer Perkinelmer Lambda20, we used a Bruker profilometer for thickness measurement and a Horiba iHR 550 for the measurement of photoluminescence, and lastly the study of the surface morphology was performed using atomic force microscopy AFM JEOL 5200.

49.3.1 X-ray Diffraction Characterization

The grazing incidence x-ray diffraction spectrum shown in Fig. 49.1 has been identified using the High Score software with the ASTM data sheet referenced by the code: 01-078-4604, which is that of a Wurtzite structure of the ZnO, with the mesh parameters a = b = 3.209 Å and c = 5.177 Å

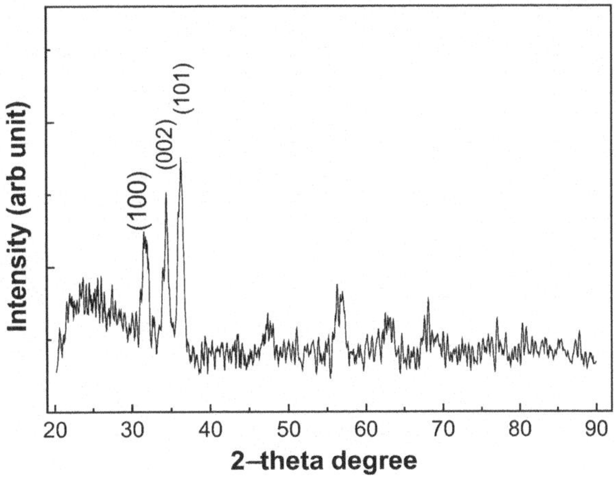

Fig. 49.1 X-ray diffraction spectrum of ZnO nanofilm deposited on a glass substrate

49.3.2 Optical Transmission

Figure 49.2 reveals a strong transmission in the visible region with an attenuation of about 50% in the region 400–300 nm, and for the spectrum in the UV range, it was explained by a strong photoluminescence in the visible range [19] and is confirmed by the photoluminescence characterization.

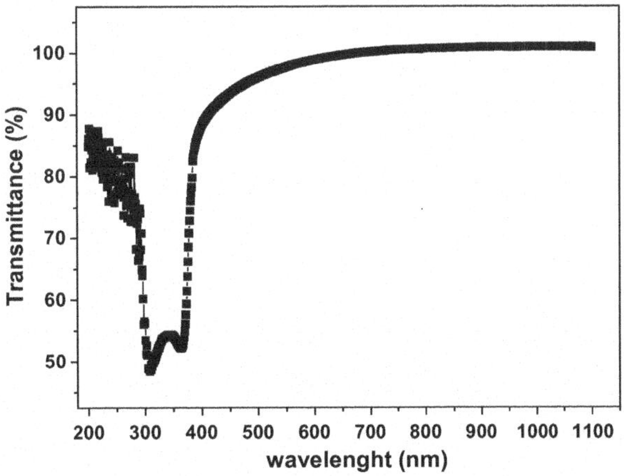

Fig. 49.2 Optical transmission of ZnO nanofilm deposited on a glass substrate

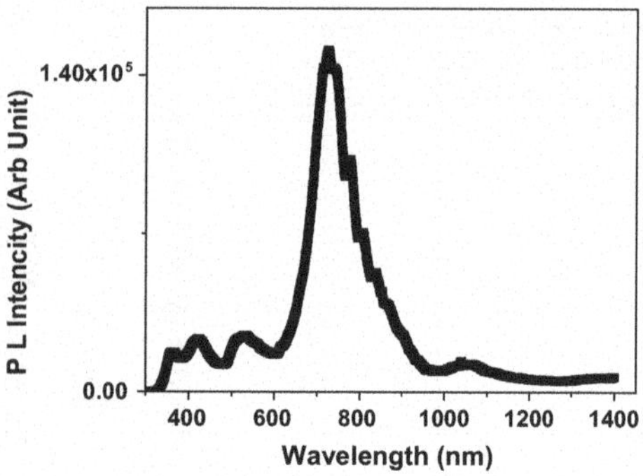

Fig. 49.3 Spectrum of PL of ZnO nanofilm deposited on a glass substrate

49.3.3 Photoluminescence Characterization

The spectrum represented in Fig. 49.3 of the photoluminescence measurement made with a Horiba iHR550 and with 325 nm wavelength laser excitation, shows a strong luminescence in the visible range.

49.3.4 Thickness and Band Gap

The measurement of the thickness carried out using a Bruker profilometer gives us a value of 202 nm and the determination of the Band Gap energy is given by extrapolation of the linear part of the plot (αhν) 2 as a function of (hν) as illustrated in Fig. 49.4, and gives the value of 3.2 eV.

49.3.5 Atomic Force Microscopy (AFM)

The study of the morphology of the layer was made in taping mode using a Jeol 5200 atomic force microscope, the images clearly show a stack of hollow hexagonal shapes having the shape of hexagonal nuts of similar size as shown in Fig. 49.5.

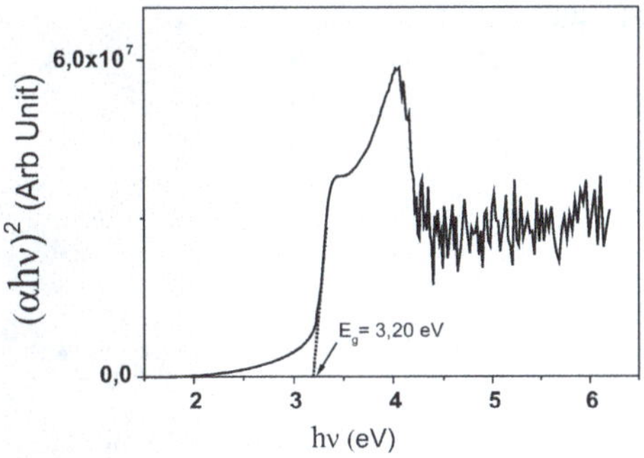

Fig. 49.4 Determination of the Band Gap energy of ZnO nanofilm deposited on a glass

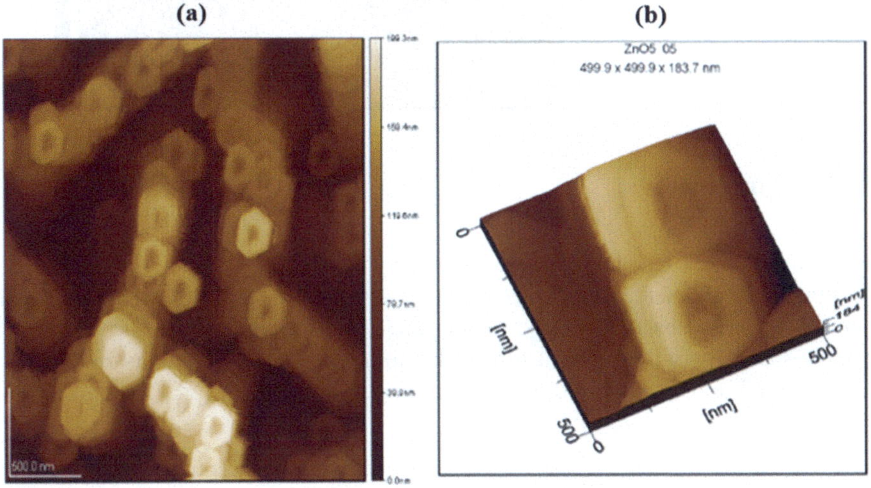

Fig. 49.5 AFM morphology of ZnO nanofilm deposited on a glass **a** 2D and **b** 3D

The image of Fig. 49.6 on a larger scale reveals that the edges of the hexagonal nut are formed of small similar grains.

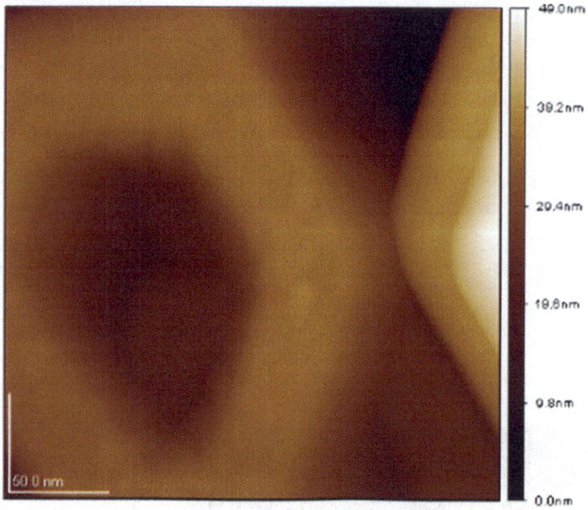

Fig. 49.6 AFM morphology of ZnO nanofilm deposited on a glass

49.4 Conclusion

Nanofilms of zinc oxide have been deposited by the solgel spin coating method, the grazing XRD characterization has revealed a wurtzite structure, a high optical transmission was observed in the visible range, the morphology of the nanofilms reveal a stack of similar hex-nut formed by an arrangement of similar-sized grains was observed under an atomic force microscope.

References

1. K. Iwata, T. Sakemi, A. Yamada, P. Fons, K. Awai, T. Yamamoto et al., Improvement of ZnO TCO film growth for photovoltaic devices by reactive plasma deposition (RPD). Thin Solid Films **480**, 199–203 (2005)
2. S. Faÿ, J. Steinhauser, S. Nicolay, C. Ballif, Polycrystalline ZnO: B grown by LPCVD as TCO for thin film silicon solar cells. Thin Solid Films **518**, 2961–2966 (2010)
3. S. Faÿ, J. Steinhauser, N. Oliveira, E. Vallat-Sauvain, C. Ballif, Opto-electronic properties of rough LP-CVD ZnO: B for use as TCO in thin-film silicon solar cells. Thin Solid Films **515**, 8558–8561 (2007)
4. J. Xu, Q. Pan, Z. Tian, Grain size control and gas sensing properties of ZnO gas sensor. Sens. Actuators B Chem. **66**, 277–279 (2000)
5. M. Wagh, G. Jain, D. Patil, S. Patil, L. Patil, Modified zinc oxide thick film resistors as NH3 gas sensor. Sens. Actuators B Chem. **115**, 128–133 (2006)
6. N. Hongsith, E. Wongrat, T. Kerdcharoen, S. Choopun, Sensor response formula for sensor based on ZnO nanostructures. Sens. Actuators B Chem. **144**, 67–72 (2010)
7. N. Al-Hardan, M. Abdullah, A.A. Aziz, Sensing mechanism of hydrogen gas sensor based on RF-sputtered ZnO thin films. Int. J. Hydrog. Energy **35**, 4428–4434 (2010)
8. Y. Bakha, H. Khales, Zinc oxide thin film prepared by micro-dropping chemical method on QCM for gas sensing

9. X.-L. Guo, J.-H. Choi, H. Tabata, T. Kawai, Fabrication and optoelectronic properties of a transparent ZnO homostructural light-emitting diode. Jpn. J. Appl. Phys. **40**, L177 (2001)
10. D.C. Kim, W.S. Han, B.H. Kong, H.K. Cho, C.H. Hong, Fabrication of the hybrid ZnO LED structure grown on p-type GaN by metal organic chemical vapor deposition. Phys. B Condens. Matter **401**, 386–390 (2007)
11. J.H. Lim, C.K. Kang, K.K. Kim, I.K. Park, D.K. Hwang, S.J. Park, UV electroluminescence emission from ZnO light-emitting diodes grown by high-temperature radiofrequency sputtering. Adv. Mater. **18**, 2720–2724 (2006)
12. N. Emanetoglu, C. Gorla, Y. Liu, S. Liang, Y. Lu, Epitaxial ZnO piezoelectric thin films for saw filters. Mater. Sci. Semicond. Process. **2**, 247–252 (1999)
13. J.-B. Lee, M.-H. Lee, C.-K. Park, J.-S. Park, Effects of lattice mismatches in ZnO/substrate structures on the orientations of ZnO films and characteristics of SAW devices. Thin Solid Films **447**, 296–301 (2004)
14. X. Du, Y. Fu, S. Tan, J. Luo, A. Flewitt, S. Maeng, et al., ZnO film for application in surface acoustic wave device. J. Phys. Conf. Ser. **76**, 012035 (2007)
15. K. Qi, B. Cheng, J. Yu, W. Ho, Review on the improvement of the photocatalytic and antibacterial activities of ZnO. J. Alloy. Compd. **727**, 792–820 (2017)
16. J. Zhang, L. Sun, J. Yin, H. Su, C. Liao, C. Yan, Control of ZnO morphology via a simple solution route. Chem. Mater. **14**, 4172–4177 (2002)
17. Z.R. Tian, J.A. Voigt, J. Liu, B. Mckenzie, M.J. Mcdermott, M.A. Rodriguez et al., Complex and oriented ZnO nanostructures. Nat. Mater. **2**, 821 (2003)
18. C.-H. Lu, C.-H. Yeh, Influence of hydrothermal conditions on the morphology and particle size of zinc oxide powder. Ceram. Int. **26**, 351–357 (2000)
19. A. Tab, A. Abderrahmane, Y. Bakha, S. Hamzaoui, M. Zerdali, Investigation on the optical transmission in UV range of sol-gel spin coated zinc oxide nanofilms deposited on glass. Optik **194**, 163073 (2019)

Chapter 50
Optimization of Ultra-Thin CIGS Based Solar Cells by Adding New Absorber Layers: InGaAs and AlGaAs

F. Merad, A. Guen-Bouazza, A.-A. Kanoun, and A. E. Merad

Abstract In this paper we presented a numerical simulations using SCAPS-1D device model of chalcopyrite (CIG(S,Se)) solar cells. In order to increase the efficiencies of CIGS device, we simulated an alternative ZnO/CdS/CIGS/AlGaAs and ZnO/CdS/CIGS/InGaAs structure. There is an interesting enhancement of the efficiency of CIGS/AlGaAs or /InGaAs and compared to the conventional CIGS solar cells. Our simulation results show the possibility for the present solar cells give conversion efficiency of 28% for CIGS/AlGaAs and 26% for CIGS/InGaAs respectively. The present results showed that the addition of AlGaAs or InGaAs layer thin film CIGS solar cells structure has performance parameters according of Band Gap of CIGS.

Keywords Solar cells · CIGS · Numerical simulations · SCAPS

50.1 Introduction

Midst the evolution of the second generation thin films solar cells, chalcopyrite semiconductors have emerged as promising nontoxic, low-cost and high efficiency materials for thin film solar cell applications [1–3]. Different chalcopyrite semiconductor materials such as CIGS has been used for thin film photovoltaic (PV) solar cells [3]. Recently, CIGS thin-film photovoltaic cell technologies, which act as the absorber layer, achieved a record solar efficiency of 22.6%, establishing a new world record for the thin-film devices [4] and outperforming polycrystalline silicon cells by 1.3% points [2]. It shows that the CIGS based photovoltaic cells

F. Merad (✉) · A. Guen-Bouazza
Unit of Research Materials and Renewable Energies, Department of Electronics, Faculty of Technology, University of Tlemcen, B.P. 119, 13000 Tlemcen, Algeria
e-mail: meradfaiza13@gmail.com; meradfaiza@yahoo.com

A.-A. Kanoun · A. E. Merad
Team: Solid State Physics, Laboratory of Theoretical Physics, Department of Physics, Faculty of Science, University of Tlemcen, B.P. 119, 13000 Tlemcen, Algeria

A. Belasri and S. A. Beldjilali (eds.), *ICREEC 2019*, Springer Proceedings in Energy,
https://doi.org/10.1007/978-981-15-5444-5_50

have become a real alternative to the Si-based technologies that still dominate the market [2, 5]. CIGS based on chalcogenide zinc-blende ($CuIn1\text{-}xGaxSe_2$) alloys, which is obtained by alloying $CuInSe_2$ (CIS) with $CuGaSe_2$ (CGS), have a direct band gap varying from 1.04 to 1.68 eV [6] by adding the gallium content.

In the present work, we used SCAPS-1D (Solar Cell Capacitance Simulator – 1D) simulation model [7, 8] to investigate CIGS solar cell performances by adding new absorber layer with various band gaps.

50.2 Computational Model

The considered solar cell structures simulations were performed using SCAPS-1D [7, 8] thin-film simulation software, under AM 1.5 solar spectrum at 1000 mW/cm^2 for J–V characteristics. CIGS solar cells are built on substrate of glass. Here, CIGS act as the absorber layer and molybdenum is used as back contact between glass substrate and the absorber layer. The device structure is completed by adding CdS as buffer layer and ZnO as transparent conducting electrode layer.

The following equation shows the current-voltage (J-V) characteristic of the solar cell [9]:

$$J = J_{ph} - J_0\left(e^{\frac{V}{nU_t}} - 1\right) \quad (50.1)$$

where J0 is the saturation current density, n is the diode ideality factor and Jph is the photo-current density.

From the J-V characteristic and by taking P_i as a solar incident power, we can easily deduce the short-circuit density J_{sc}, the open circuit voltage V_{co} and the maximum power P_m given respectively by:

$$J_{sc} = J_{ph} \quad (50.2)$$

$$V_{co} = n\,U_t ln\left(\frac{J_{ph}}{J_0}\right) \quad (50.3)$$

$$P_m = J_{max}\,V_{max} \quad (50.4)$$

$$FF = \frac{P_m}{J_{sc}V_{co}} \quad (50.5)$$

$$\eta = \frac{P_m}{P_i} \quad (50.6)$$

We employed the computational modeling as implemented through software SCAPS [10], the CIGS thin film solar cell output characteristic were numerically modeled. To proceed with the simulations, the material parameters used as the inputs file were selected based on the reported literature values or limited to reasonable ranges. The SCAPS simulator requires the input parameters of materials of each layers in structures of solar cells.

50.3 Results and Discussion

To enhance the performance of chalcopyrite solar cells, we investigated the possibility to add a new layer of InGaAS or AlGaAs in CIGS absorber layer. In this simulation, InGaAs or AlGaAs have been added at back contact making interfaces from CIGS/InGaAs or AlGaAs that eliminated any possible recombination loss at the back contact of CIGS solar cells (see Fig. 50.1). Table 50.1 provides the main parameters of considered materials. Figure 50.2 shows the J-V curves of CIGS solar cells with a new layer in order to compare the new layer adding effects. This result reflects the importance of adding layer.

The performance of solar cell relies on to the response of solar spectrum. This is influenced by the optical properties of the absorber layer. The optical properties of absorber layer is an important parameter and plays a critical role in optimizing the performance of the solar cell.

Hence, in Fig. 50.3 we presented the quantum efficiency curves of CIGS solar cells with respect to wavelength. The obtained simulations reveal that the CIGS solar cells with and without new layer do not exhibit any significant difference in the short wavelength region (less than 400 nm). However, for long wavelengths (larger than 500 nm), it can see there is an increase of efficiency. Thus, incorporating the

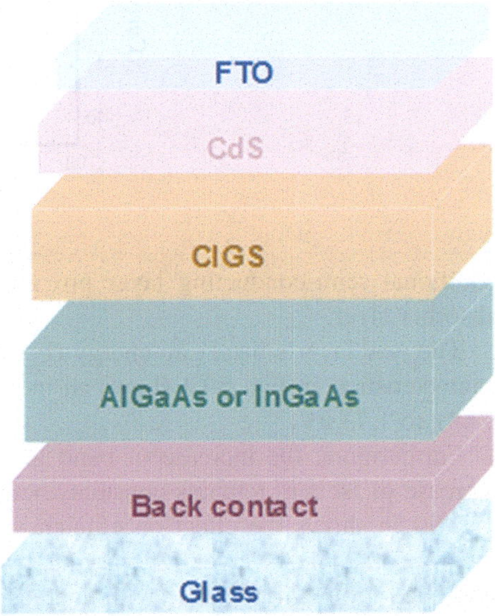

Fig. 50.1 CIGS Solar cell configuration with AlGaAs or InGaAs layer

Table 50.1 Physical parameters of material

Layer properties	ZnO	n-Cds	p-CIGS	InGaAs	AlGaAs
E_g (eV)	3.40	2.40	1.16	0.634–1.425	1.425–1.98
χ (eV)	4.55	4.50	4.80	4.11–4.6095	3.575–4.07
ε_r	10	10	13.60	12.90	12.9
μ_n (cm^2/Vs)	50	100	100	8500	8500
μ_p (cm^2/Vs)	20	25	25	400	400
N_c (cm^{-3})	4×10^{18}	2.2×10^{18}	2.2×10^{18}	1×10^{18}	1×10^{18}
N_v (cm^{-3})	9×10^{18}	1.8×10^{19}	1.8×10^{19}	1×10^{19}	1×10^{19}
N_D (cm^{-3})	10^{17}	2.0×10^{87}	–	–	–
N_A (cm^{-3})	–	–	8.0×10^{16}	1×10^{18}	1×10^{18}
V_e (cm/s)	10^7	10^7	10^7	10^7	10^7
V_h (cm/s)	10^7	10^7	10^7	10^7	10^7
Defect density	10^{14}	10^{14}	10^{14}	10^{14}	10^{14}
Thickness (μm)	0.05	0.05	1	1	1

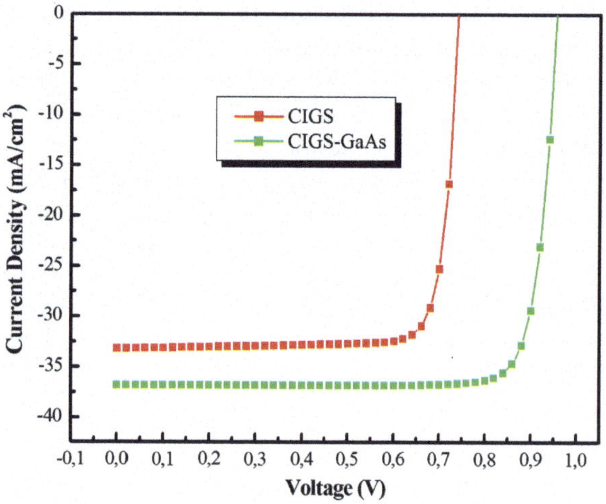

Fig. 50.2 J-V curves of CIGS solar cells

additional semi-conducting layer provides much more efficient than only CIGS absorber layer.

The new layer's band gap energy (E_g) is varied from 0.634 to 1.98 eV, For J-V characteristics, in Fig. 50.4 shows an increasing of V_{oc} when the E_g increase from 0.681 to 1.73 eV.

Furthermore, the increase of band gap energy from 0.6 to 1.2 eV leads to an increase of Jsc and remains constants between 1.5 and 1.9 eV (see Fig. 50.5a). In Fig. 50.5b shows the effect on efficiency and fill factor of CIGS solar cells with InGaAs or AlGaAs layer for varying band gap energy. The FF is significantly improved from 74 to 85% with increasing the band gap energy, and remains

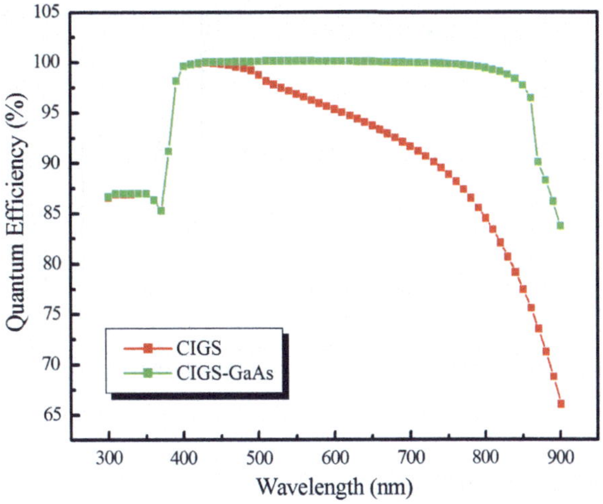

Fig. 50.3 Quantum efficiency curves of CIGS solar cells

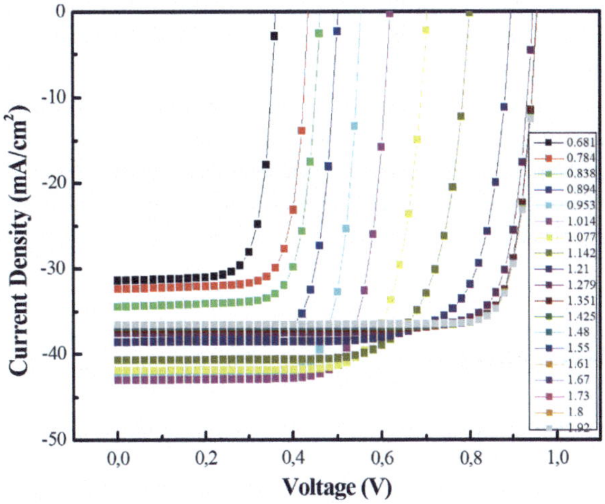

Fig. 50.4 J-V curves of CIGS solar cells with InGaAs or AlGaAs layers

constants when the band gap energy over 1.5 eV. On the other hand, efficiency increase while the band gap energy increase leading to a good band alignment between absorber and back contact.

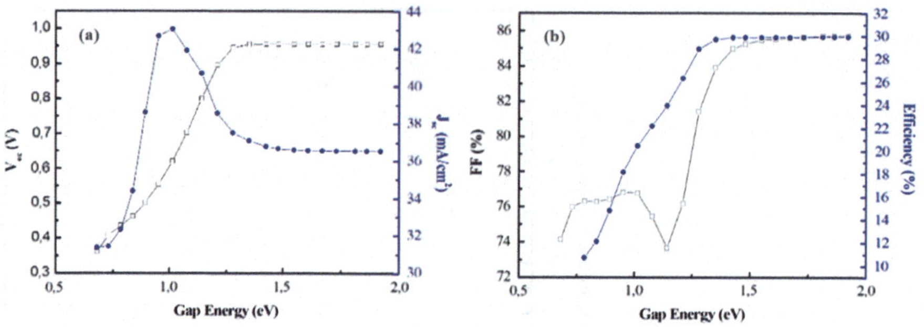

Fig. 50.5 CIGS solar cells with InGaAs or AlGaAs layers parameters: **a** Voc and Jsc, **b** FF and efficiency

50.4 Conclusion

In the present study, we investigated the performance of solar cell device based on CIGS by using SCAPS. The modified structure of CIGS solar cells has been investigated by adding a new semiconducting layer based on InGaAs or AlGaAs. This addition has improved the carrier collection that has led to solar cell performance enhancement, since the quantum efficiency has increased. Therefore, the adding layer specially based on InGaAs or AlGaAs layer, has enhanced the efficiency of the solar cell device and made it harvesting solar energy with for CIGS efficiency.

References

1. K.L. Chopra, P.D. Paulson, V. Dutta, Thin-film solar cells: an overview. Prog. Photovolt. Res. Appl. **12**, 69–92 (2004)
2. National Renewable Energy Laboratory, Best research cell efficiencies (2017), http://www.nrel.gov/ncpv/images/efficiency_chart.jpg
3. A. Chirilă, P. Reinhard, F. Pianezzi, P. Bloesch, A.R. Uhl, C. Fella, L. Kranz, D. Keller, C. Gretener, H. Hagendorfer, D. Jaeger, R. Erni, S. Nishiwaki, S. Buecheler, A.N. Tiwari, Potassium-induced surface modification of Cu(In,Ga)Se2 thin films for high-efficiency solar cells. Nat. Mater. **12**, 1107 (2013)
4. P. Jackson, R. Wuerz, D. Hariskos, E. Lotter, W. Witte, M. Powalla, Effects of heavy alkali elements in Cu(In,Ga)Se_2 solar cells with efficiencies up to 22.6%. Phys. Status Solidi RRL Rapid Res. Lett. **10**(8), 583–586 (2016)
5. S. Siebentritt, M. Igalson, C. Persson, S. Lany, The electronic structure of chalcopyrites—bands, point defects and grain boundaries. Prog. Photovolt. Res. Appl. **18**, 390–410 (2010)
6. T. Tinoco, C. Rincón, M. Quintero, G. Sánchez Pérez, Phase diagram and optical energy gaps for CuInyGa1−ySe2 alloys. Phys. Status Solidi (a) **124**, 427 (1991)
7. M. Burgelman, P. Nollet, S. Degrave, Modelling polycrystalline semiconductor solar cells. Thin Solid Films **361**, 527–532 (2000)

8. M. Burgelman, K. Decock, S. Khelifi, A. Abass, Advanced electrical simulation of thin film solar cells. Thin Solid Films **535**, 296–301 (2013)
9. B. Vermang, J.T. Wätjen, V. Fjällström, F. Rostvall, M. Edoff, R. Kotipalli, F. Henry, D. Flandre, Employing Si solar cell technology to increase efficiency of ultra-thin Cu(In,Ga)Se2 solar cells. Prog. Photovolt. Res. Appl. **22**, 1023–1029 (2014)
10. H. Heriche, Z. Rouabah, N. Bouarissa, High-efficiency CIGS solar cells with optimization of layers thickness and doping. Opt. Int. J. Light. Electron Opt. **127**, 11751–11757 (2016)

Chapter 51
Simulation and Analysis of Perovskite Solar Cell Based on Germanium

A. Lachgueur and K. Rahmoun

Abstract The search for low-cost materials ton turn solar energy into electricity is a challenge for researchers. Lead-free perovskite solar cells have been of interest for research because of the toxicity of this material. The perovskite $CH_3NH_3PbI_3$ solar cells have a conversion efficiency of about 22%. In this work lead is replaced by germanium. The solar cell analysis is performed using a digital tool «Solar Cell Capacitance Simulator (SCAPS)» Solar cell capacitance simulator was developed by Marc Burgelman at the University of Gent. SCAPS is used to analyses the micro and poly crystalline and photonic structure. SCAPS measure various parameters of a solar cell like open circuit voltage, Fill Factor, Short circuit current density, PCE. SCAPS and runs using the governing equations which includes continuity equations, Poisson equation, of electron and hole. The solar cell structures is based on the mixed perovskite compound as an absorbent layer and Pedot:Pss is used as HTM; PCBM is used as ETM. The contact materials are FTO coated glass (Front contact) and Al (back contact) is used. The total cell structure Al/Pedot:Pss/$CH_3NH_3GeI_3$-PCBM/FTO. We study the influence of thickness of absorber layer on solar cell performance, efficiency (η) open circuit voltage (VOC) and short-circuit density (Jsc) and fil factor (FF).

Keywords Perovskite solar cell · $CH_3NH_3GeI_3$ · Simulation

A. Lachgueur (✉) · K. Rahmoun
Unité de Recherche Materiaux et Energies Renouvelables (URMER), Tlemcen, Algeria
e-mail: lgr_abd@yahoo.fr

A. Lachgueur · K. Rahmoun
Département de Physique, Faculté des Sciences, Université Abou Baker Belkaid, BP119, 13000 Tlemcen, Algeria

A. Belasri and S. A. Beldjilali (eds.), *ICREEC 2019*, Springer Proceedings in Energy,
https://doi.org/10.1007/978-981-15-5444-5_51

51.1 Introduction

The huge economic growth in the world requires a large amount of energy that is primarily from fossil fuels that has a detrimental effect on our environment for that the researchers redirect towards research and development of renewable energy technologies. From past seven years, the perovskite solar cell efficiency has been increasing enormously from 3.8% (2009) to to 25.2% in 2019 [1–5].

Excellent opto-electronic properties of perovskite materials such as long diffusion length for carrier transport and broad absorption spectrum that covers the maximum visible region are attractive for developing devices [2, 5, 6].

SCAPS analyses the physics of the model and it explains the recombination profiles, electric field distribution, carrier transport mechanism and individual current densities.

The continuity equations of electrons and holes are:

$$\frac{dj_n}{dx} = G - R \tag{1}$$

$$\frac{dj_p}{dx} = G - R \tag{2}$$

where, j_n is the Electron Current Density, j_p the Holle Current Density, G the Recombination Rate and R the Generation Rate. The Poisson equation is given by

$$\frac{d^2\,\phi(x)}{d^2\;(x)} = \frac{e}{\in 0\;\in \mathrm{r}}\left(p(x) - n(x) + \mathrm{N_D} - \mathrm{N_A} + \rho_\mathrm{p} - \rho_\mathrm{n}\right) \tag{3}$$

where,

$\Phi(\mathrm{x})$ is the Electrostatic potential,

e =	Electric Charge
$\in_\mathrm{r}$ =	Relative Permittivity
p and n	are hole and electron concentration
$\mathrm{N_D}$ =	Charged impurities of donor
$\mathrm{N_A}$ =	Charged impurities of acceptor

ρ_p and ρ_n are holes and electron distribution

The Drift and Diffusion Equations are:

$$Jn = \mathrm{D}n\frac{dn}{dx} + \mu_n\, n\, \frac{d\Phi}{dx} \tag{4}$$

$$Jp = \mathrm{D}p\frac{dn}{dx} + \mu_\mathrm{p}\, p\, \frac{d\Phi}{dx} \tag{5}$$

Jn And Jp Electron and hole current density
Dn = Electron diffusion coefficient
μ and μp are electron and hole mobility

Semiconductor equations in 1-Dimension under steady state condition in Fig. 51.1 explains the simulation process using SCAPS.

51.2 Cell structure and simulation parameter

A typical $CH_3NH_3GeI_3$ based solar cell structure consists of an absorber layer and at the top p-type (PEDOT: PSS) and n-type (PCBM) is arranged at the bottom side. The cell schematically is as shown in Fig. 51.2 [1].

Note that the main material parameters are carefully selected from those reported in experimental data and other theoretical results [1–4, 6, 8, 9].

Table 51.1 recapitulates all the primary parameters used in the simulation.

Pre-factor Aa is 10^5 to obtain the absorption coefficient (α) curve calculated by $\alpha(\lambda) = Aa(h\nu\text{-}Eg)^{1/2}$ [7].

Our study focuses on the effects of the thickness of the perovskite absorber layer $CH_3NH_3GeI_3$. The initial parameters and defect are shown in Table 51.1.

51.3 Results and Discussion

51.3.1 Influence of Absorber Layer Thickness on Solar Cell Characteristics

To absorb the maximum number of photons and to generate electron-hole pair, absorber layer should set for optimum thickness [1]. The thickness of perovskite layer has been varied from 0.2 μm to 0.9 μm, to optimize the thickness and to get the high efficiency. We calculate the photovoltaic parameters for thickness values

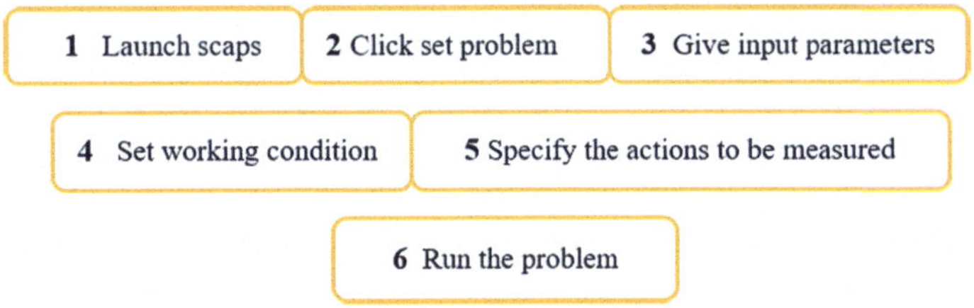

Fig. 51.1 SCAPS working procedure [1]

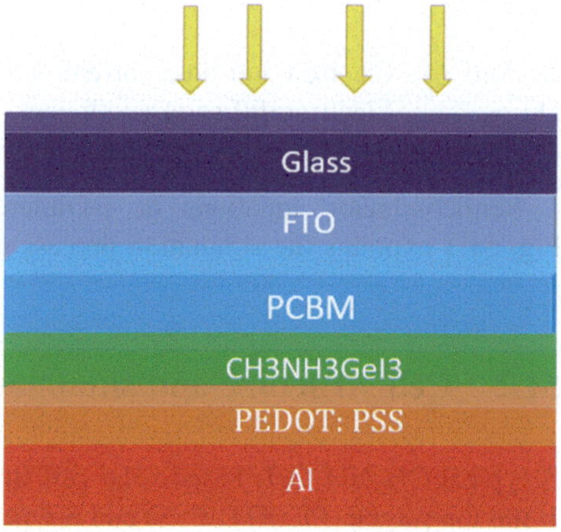

Fig. 51.2 Schematic representation device architecture (Glass/FTO/PCBM/$CH_3NH_3GeI_3$/PEDOT: PSS/Al)

Table 51.1 The parameters set for $CH_3NH_3GeI_3$ based solar cell at 300K and at A.M. 1.5G and solar spectrum at 100 mW/cm^2

Parameters	SnO_2:F	PCBM	$CH_3NH_3GeI_3$	PEDOT: PSS
Thickness (μm)	0.45	0.4	0.4 (variable)	0.4
Bandgap, Eg (eV)	3.5	1.3	1.38	2.2
Electron affinity, χ (eV)	4	3.9	3.9	2.9
Relative permittivity, ε_r	9	4	10	3
Densities of states NC (1/cm^3)	2.2×10^{17}	1×10^{21}	1×10^{16}	2.2×10^{15}
Densities of states NV (1/cm^3)	2.2×10^{16}	2×10^{20}	1×10^{15}	1.8×10^{18}
Electron thermal velocity (cm/s)	1.0×10^{7}	1.0×10^{7}	1.0×10^{7}	1.0×10^{7}
Hole thermal velocity(cm/s)	1.0×10^{7}	1.0×10^{7}	1.0×10^{7}	1.0×10^{7}
Electron mobility (cm^2/Vs)	20	10^{-2}	162×10^{3}	1.0×10^{-2}
Hole mobility (cm^2/Vs)	10	10^{-2}	101×10^{3}	2.0×10^{-1}
Donor densities, ND (cm^{-3})	1.0×10^{15}	1.0×10^{15}	1.0×10^{10}	1×10^{16}
Accepter densities, NA (cm^{-3})			1×10^{14}	1×10^{15}
Capture cross section electrons (cm^2)		1×10^{-12}	1×10^{-14}	1×10^{-17}
Capture cross section holes (cm^2)		1×10^{-15}	1×10^{-15}	1×10^{-12}
Total density (1/cm^3)		1×10^{15}	1×10^{15}	1×10^{15}

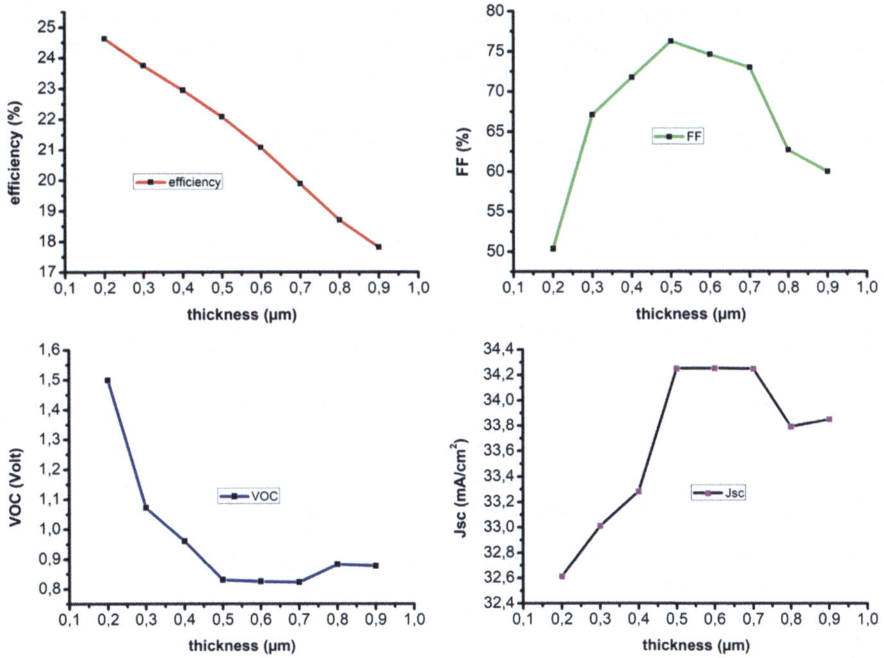

Fig. 51.3 Variation of PV parameters by varying the thickness

between 0.2 μm to 0.9 μm, and keeping the thickness of the other layers constant. The Fig. 51.3 shown the effect of absorber thickness on the characteristic of the perovskite solar cell based on germanium.

The thickness of absorber layer has been varied from 0.2 μm to 0.9 μm. The results show the efficiency decreases from 24.63% to 17.82% and the form factor increases from 50.4% to 76.3% for the value of thickness of the absorbent layer between 0.2 μm and 0.5 μm is to reach the maximum for the value 0.5 μm, then decreases and the current density increases for the value for 0.5 μm the voltage voc decreases slowly by 1.49 volts at 0.87 volts.

51.4 Conclusion

In this paper we have studied the effect of absorber thickness of solar cells having the architecture Glass/FTO/PCBM/$CH_3NH_3GeI_3$/PEDOT:PSS/Al is designed and analyzed using Solar cell capacitance simulator SCAPS-1D on the performance of the solar cell. The thickness of perovskite layer has been varied to optimize and to get the high efficiency.

We obtained a high efficiency 24.63% for a thickness of the absorber layer $CH_3NH_3GeI_3$ of 0.2 μm. We have suggested this architecture for researches to study and improve their characteristics.

Acknowledgements The authors would like to acknowledge Dr. Marc Burgelman (University of Gent) for providing SCAPS software.

References

1. U. Mandadapu, S.V. Vedanayakam, K. Thyagarajan, Simulation and Analysis of Lead based Perovskite Solar Cell using SCAPS-1D. Indian J. Sci. Technol. **2**(5), 99–110 (2016)
2. D. Hui-Jing, W.-C. Wang, J.-Z. Zhu, Device simulation of lead-free $CH_3NH_3SnI_3$ perovskite solar cells with high efficiency. Chin. Phys. B **25**(10), 108802–108810 (2016)
3. U. Mandadapu, S.V. Vedanayakam, K. Thyagarajan, Numerical simulation of $Ch_3Nh_3PbI_3$-XClx perovskite solar cell using SCAPS-1D, in International Journal of Engineering Science Invention. International Conference on Advances in Functional Materials (ICAFM-2017), pp. 40–45
4. A.-A. Kanoun, M.B. Kanoun, A.E. Merad, S. Goumri-Said, Toward development of high-performance perovskite solar cells based on $CH_3NH_3GeI_3$ using computational approach. Solar Energy **182**, 237–244 (2019)
5. nrel.gov. 25 Oct 2019. https://www.nrel.gov/pv/assets/pdfs/best-research-cell-efficiencies.20190923.pdf
6. Y.-Q. Zhao, B. Liu, Yu. Zhuo-Liang, J. Ma, Q. Wan, P.-b. He, M.-Q. Cai, Strong ferroelectric polarization of $CH_3NH_3GeI_3$ with high-absorption and mobility transport anisotropy: theoretical study. J. Mater. Chem. C **5**, 5356–5364 (2017)
7. G. Haidari, Comparative 1D optoelectrical simulation of the perovskite solar cell. AIP Adv. **9**, 085028–085034 (2019)
8. S. Nagane, D. Ghosh, R.L.Z. Hoye, B. Zhao, S. Ahmad, A.B. Walker, M.S. Islam, S. Ogale, A. Sadhanala, Lead-Free Perovskite Semiconductors Based on Germanium-Tin Solid Solutions: Structural and Optoelectronic Properties. J. Phys. Chem. C **122**, 5940–5947 (2018)
9. M. Yue, S. Jie, P. Zhao, Z. Lin, J. Zhang, J. Chang, Y. Hao, Optimizing the performance of $CsPbI_3$-based perovskite solar cells via doping a ZnO electron transport layer coupled with interface engineering. Chin. Phys. B **25**(10), 108802–108808 (2016)

Energy Conversion

Chapter 52
Carbon Dioxide Capture in Fluidized Beds of Nanosilica/$Ca(OH)_2$

H. Moreno, F. Pontiga, and J. M. Valverde

Abstract The use of calcium hydroxide as sorbent of CO_2 in low concentration streams (1% CO_2 vol.) has been experimentally investigated. For that purpose, a CO_2/N_2 dry gas mixture, at ambient temperature and atmospheric pressure, was flowed through a bed of $Ca(OH)_2$, and the concentration of CO_2 in the effluent gas was monitored. In order to improve the fluidizability, both hydrophilic and hydrophobic nanosilica particles were added to the sorbent, and its effect on the absorption of CO_2 was investigated. The results showed that the addition of nanosilica increases the capture capacity of the sorbent (compared to the raw material), and the best results were obtained when using hydrophilic nanosilica. Additionally, to elucidate how the CO_2 capture process is affected by relative humidity during the storage of $Ca(OH)_2$, the sorbent was kept in a controlled CO_2-free atmosphere with constant humidity prior to the experiments. The results showed that samples stored in an atmosphere with high relative humidity exhibit a significantly higher CO_2 absorption capacity.

Keywords CO_2 capture · Ca-based sorbents · Fluidized beds

52.1 Introduction

Anthropogenic emission of carbon dioxide to the atmosphere is contributing to the global climate change, and the largest fraction of CO_2 emissions is originated by fuel combustion at power plants. Indeed, according to the International Energy Agency [1], the generation of electricity and heat accounted for 42% of global CO_2

H. Moreno · F. Pontiga (✉)
Departamento de Física Aplicada II, E.T.S. Ingeniería de Edificación,
Universidad de Sevilla, Av. Reina Mercedes 4a, 41012 Seville, Spain
e-mail: pontiga@us.es

J. M. Valverde
Departamento de Electrónica y Electromagnetismo, Facultad de Física,
Universidad de Sevilla, Av. Reina Mercedes s/n, 41012 Seville, Spain

A. Belasri and S. A. Beldjilali (eds.), *ICREEC 2019*, Springer Proceedings in Energy,
https://doi.org/10.1007/978-981-15-5444-5_52

emissions in 2015. In order to mitigate this problem, carbon capture and storage (CCS) is being considered as a feasible method for short-term reduction of CO_2 generated in power plants [2]. However, the limited efficiencies of capture processes of flue gases still leaves a residual CO_2 that eventually is released to the atmosphere. On the other hand, CCS technologies cannot be applied to mitigate CO_2 emissions from other sectors, like transportation, that also accounts for a significant fraction of global CO_2 emissions ($\sim$24%). Therefore, direct CO_2 capture from the atmosphere is additionally being studied as a method to compensate for these dilute and distributed emissions [3].

In general, alkaline metal oxides and alkaline earth metal oxides have been reported as adsorbents for CO_2 [4], but Ca-based oxides has the additional advantage of being the most abundant among alkaline earth metal oxides in nature. The present work is focused on the use of calcium hydroxide, $Ca(OH)_2$. The absorption of CO_2 occurs in two phases. In a first stage, CO_2 is chemisorbed on the surface of the sorbent, which leads to the production of $CaCO_3$. This phase is known as "fast carbonation". Once the reaction sites on the sorbent begin to saturate, a much slower absorption phase starts, in which CO_2 must diffuse through the layer of $CaCO_3$ [5].

The effectiveness of the absorption process is affected by many factors and, among them, the extent of the contact surface of the sorbent exposed to the gas plays a determinant role. Calcium hydroxide is usually available as a fine powder, which, in principle, should offer a theoretically high surface area for CO_2 absorption. However, due to the effect of van de Waals and capillary forces acting between particles, $Ca(OH)_2$ powder exhibits an elevated cohesiveness and behaves as a Geldart C powder [6]. Thus, it is hardly fluidizable and stable gas channels easily develop inside the sample or at the walls of the fluidization reactor. As shown in Fig. 52.1, scanning electron microscope pictures reveals that $Ca(OH)_2$ tends to form aggregates of size of several tens of microns. All this together limits the efficiency of the gas–solid contact in the fluidized bed, and therefore the CO_2 capture capacity.

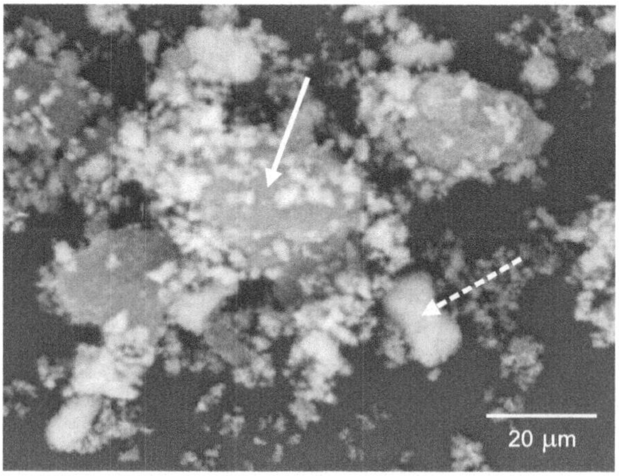

Fig. 52.1 Scanning electron microscope picture of a nano-silica/$Ca(OH)_2$ composite obtained by dry mixing. Nanosilica agglomerates appears as grey granules which are coated by white $Ca(OH)_2$ particles and small aggregates. The solid arrow points at a coated nano-silica agglomerate and the dashed arrow points at a $Ca(OH)_2$ aggregate

In order to improve the fluidization of $Ca(OH)_2$, nanostructured fumed silica powders can be used as additives [7]. Silica particles form highly porous agglomerates of size of tens of microns which can be fluidized uniformly with full suppression of gas bubbles and channels. When $Ca(OH)_2$ is mixed with nanosilicas, silica agglomerates become coated by a monolayer of $Ca(OH)_2$ particles and act as carriers of the $Ca(OH)_2$ fine particles (see Fig. 52.1). In the present work, two different kinds of nanosilicas have been used: Aerosil-R974 and Aerosil-300, both supplied by Evonik [8]. Aerosil-300 is a hydrophilic fumed silica with a loss on drying of less than 1.5 wt% (2 h at 105 °C). In contrast, Aerosil-R974 is a hydrophobic fume silica produced by treating hydrophilic fumed silica with dimethyldichlorosilane, in order to render it hydrophobic, and has a loss on drying below 0.5 wt% (2 h at 105 °C).

52.2 Experimental Setup

Figure 52.2 shows a schematic picture of the experimental setup used in the experiments. Carbon dioxide capture experiments were carried out with a mixture of 1% CO_2 + 99% N_2, at a total flow rate of 0.1 L/min. The mixture of gases was prepared using two mass flow controllers (Alicat) fed with high purity CO_2 (99.995%) and N_2 (99.999%). The fluidized bed reactor consisted of a glass tube with inner diameter of about 1 in., in which a composite sample of 3 g of $Ca(OH)_2$ and 0.45 g of either hydrophobic or hydrophilic nanosilica was deposited above a porous gas distributor. Finally, the whole reactor was placed on a vibration exciter, which helped to achieve a uniform fluidization of the sample.

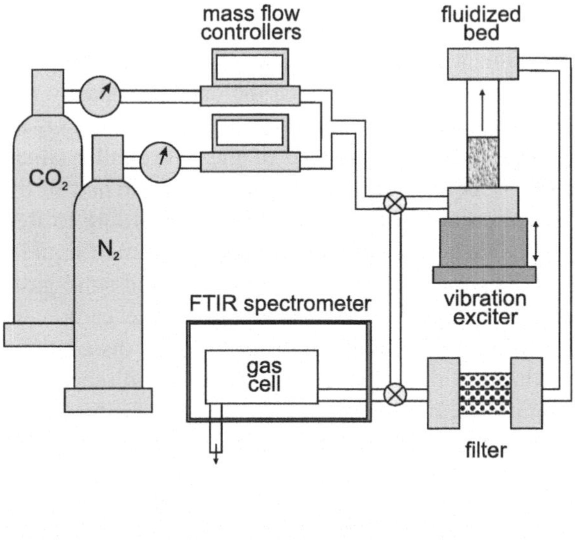

Fig. 52.2 Schematic representation of the experimental set-up used in the experiments

The effluent gas of the fluidized bed reactor was analyzed using infrared spectrophotometry. For this purpose, the gas flow was sent to a gas cell within the sample compartment of a FTIR spectrophotometer (Bruker Vertex 70), and infrared spectra were recorder at regular intervals. Carbon dioxide and vapor water concentration in the effluent gas were obtained from the absorption bands 2390–2280 cm^{-1} and 2000–1250 cm^{-1}, respectively. In order to minimized the interference of ambient CO_2 and H_2O, the optics and the sample compartments of the spectrophotometer were continuously purged with a constant flow of CO_2-free dry air at less than −73 °C dew point.

52.3 Results and Discussion

Figure 52.3 shows the evolution in time of CO_2 volume concentration at the exit of the fluidized bed reactor using three different types of beds: raw calcium hydroxide, a composite of calcium hydroxide and hydrophobic silica, and a composite of calcium hydroxide and hydrophilic silica. Clearly, carbon dioxide is only detected in the effluent gas after a certain time interval, which is usually known as CO_2 breakthrough time (BT). Therefore, before the breakthrough time, CO_2 is completely removed from the gas stream. The shortest BT was found when a bed of raw $Ca(OH)_2$ was used (BT $\approx$ 550 s). In contrast, the composite sample of $Ca(OH)_2$ and hydrophilic silica exhibited the longest BT (BT $\approx$ 1090 s), almost doubling that observed using raw $Ca(OH)_2$. The BT for the mixture of $Ca(OH)_2$ and hydrophobic silica showed an intermediate value, BT $\approx$ 780 s. As mentioned in the Introduction, the CO_2 capture capacity of $Ca(OH)_2$ is strongly influenced by the extent of the surface of the sorbent exposed to the gas. This fact explains the larger breakthrough times observed in beds of $Ca(OH)_2$ + nanosilica, since the addition of nanosilica improves fluidization and, thus, the contact between CO_2 and $Ca(OH)_2$. Once the BT is exceeded, the capture of CO_2 is no longer complete and it appears in the effluent gas. For sufficiently long times, the concentration of CO_2 approaches progressively 1% vol, since the capture capacity of the bed has reached its limit. The longer the BT, the faster the increase of CO_2 after the BT.

The better performance of the hydrophilic silica with respect to hydrophobic silica in composite mixtures must be searched in the role played by water physisorbed on the surface of $Ca(OH)_2$. According to Beruto and Botter [9], if calcium hydroxide is covered with an adsorbed film of liquid water, the reaction of $Ca(OH)_2$ with CO_2 can be transformed into a liquid-solid-gas process, which proceeds more efficiently than a solid-gas reaction. The carbonation of CO_2 would then occur according to the following steps. Firstly, dissolution of CO_2 in the adsorbed water produces carbonic acid, H_2CO_3, which dissociates in CO_3^- and H^+. In a similar manner, dissolution of $Ca(OH)_2$ leads to the formation of Ca^{2+} and OH^- ions. Then, calcium carbonate is formed in the reaction of CO_3^- with Ca^{2+}, and additional water is produced in the reaction of H^+ with OH^-, which facilitates further carbonation of

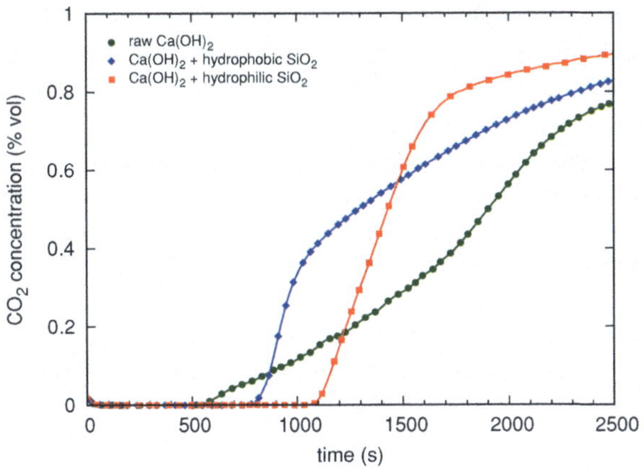

Fig. 52.3 Temporal evolution of CO_2 concentration in a gas flow of 0.1 L/min of dry CO_2/N_2 gas mixture (1 vol% CO_2) after passing through a bed of raw $Ca(OH)_2$ (●), a composite bed of hydrophobic nanosilica and $Ca(OH)_2$ (♦), and a composite bed of hydrophilic nanosilica and $Ca(OH)_2$ (■)

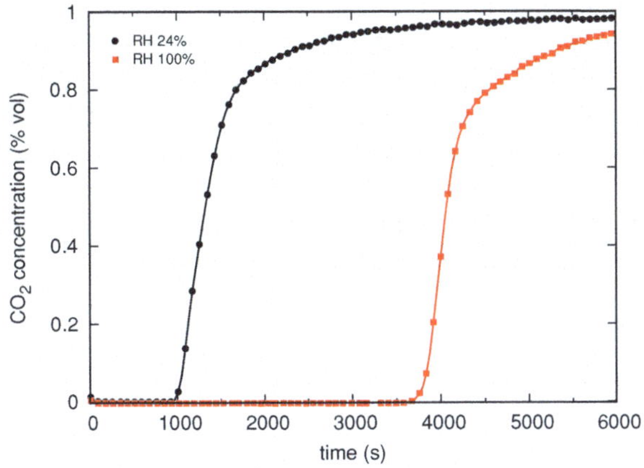

Fig. 52.4 Temporal evolution of CO_2 concentration in a gas flow of 0.1 L/min of dry CO_2/N_2 gas mixture (1 vol% CO_2) after passing through a composite bed of hydrophobic nanosilica and $Ca(OH)_2$. Prior to experiments, $Ca(OH)_2$ was stored at a relative humidity of 24% (●) or 100% (■)

CO_2. Therefore, in our study, the use of hydrophilic silica in the composite beds of $Ca(OH)_2$ + nanosilica helps to stabilize the adsorbed water on $Ca(OH)_2$, which explains the longer BT observed in the experiments.

It is known that carbonation of CO_2 by $Ca(OH)_2$ is strongly affected by the ambient relative humidity [10]. Thus, in order to investigate how the capture capacity of $Ca(OH)_2$/nanosilica beds is influenced by this factor, samples of Ca $(OH)_2$ were stored inside environmental chambers, free of CO_2, where the relative humidity was kept constant with the help of supersaturated salt solutions. Each sample was kept inside the environmental chamber for six days, in order to reach a full equilibrium with the ambient humidity. Then, to test the most unfavorable conditions, composite beds of hydrophobic silica and $Ca(OH)_2$ were prepared, and the CO_2 capture experiments were executed as usual. The results of these measurements are presented in Fig. 52.4, where the performance of two composite nanosilica/$Ca(OH)_2$ beds are compared. In the first sample, $Ca(OH)_2$ was stored at low relative humidity ($\sim 24\%$), while the second was exposed to a very high humidity level ($\sim 100\%$). As it can be readily seen in the figure, the BT corresponding to the second sample is almost three times longer than that found in the first sample.

52.4 Conclusions

This study has shown that the CO_2 capture capacity of a bed of $Ca(OH)_2$ can be significantly extended with the addition of nanosilica (particularly hydrophilic nanosilica), which helps to improve the fluidization of $Ca(OH)_2$ and, thus, to increase the gas-solid contact. Additionally, the storage of $Ca(OH)_2$ in a high humidity environment contributes to the formation of a adsorbed film of liquid water around $Ca(OH)_2$ particles, which increases the efficiency of CO_2 carbonation process. In the most favorable conditions, the breakthrough time was increased by a factor eight with respect to the case of using raw $Ca(OH)_2$.

Acknowledgements This research has been funded by the following organizations: *Ministerio de Economía, Industria y Competitividad, Agencia Estatal de Investigación* and *Fondo Europeo de Desarrollo Regional*, under contract no. CTQ2017-83602-C2-2-R.

References

1. International Energy Agency (IEA), CO_2 emissions from fuel combustion (2017)
2. J. Blamey, E.J. Anthony, J. Wang, P.S. Fennell, The calcium looping cycle for large-scale CO_2 capture. Prog. Energy Combust. Sci. **36**, 260–279 (2010)
3. D.W. Keith, G. Holmes, D.S. Angelo, K. Heidel, A process for capturing CO_2 from the atmosphere. Joule **2**, 1573–1594 (2018)
4. S. Choi, J.H. Drese, C.W. Jones, Adsorbent materials for carbon dioxide capture from large anthropogenic point sources. Chemsuschem **2**, 796–854 (2009)
5. B. Arias, J.C. Abanades, G.S. Grasa, An analysis of the effect of carbonation conditions on CaO deactivation curves. Chem. Eng. J. **167**, 255–261 (2011)
6. D. Geldart, Types of gas fluidization. Powder Technol. **7**, 285–292 (1973)

7. J.M. Valverde, F. Pontiga, C. Soria-Hoyo, M.A.S. Quintanilla, H. Moreno, F.J. Duran, M. J. Espin, Improving the gas–solids contact efficiency in a fluidized bed of CO_2 adsorbent fine particles. Phys. Chem. Chem. Phys. **13**, 14906–14909 (2011)
8. Aerosil Fumed Silica, Technical Overview. Evonik Industries (2015)
9. D.T. Beruto, R. Botter, Liquid-like H_2O adsorption layers to catalyze the $Ca(OH)_2/CO_2$ solid–gas reaction and to form a non-protective solid product layer at 20 °C. J. Eur. Ceram. Soc. **20**, 497–503 (2000)
10. M. Samari, F. Ridha, V. Manovic, A. Macchi, E.J. Anthony, Direct capture of carbon dioxide from air via lime-based sorbents. Mitig. Adapt. Strat. Glob. Chang. (2019)

Chapter 53
Laser Additive Manufacturing

El-Hachemi Amara, Karim Kheloufi, Toufik Tamsaout, Farida Hamadi, Samia Aggoune, Kada Bougherara, and Kamel Bourai

Abstract Metal particles arranged in a powder bed or injected through a nozzle can be used in manufacturing 3D parts by selective laser irradiation or directly deposited on a base material. To obtain complex manufactured parts suitable for a use in specific application in order to improve product performance for industry, medicine, or any other advanced technology, there is still issues to be addressed in relation with the fundamental mechanisms occurring during laser-matter interaction. This can be established through investigations including mainly modeling and simulation approaches. Some results of modeling and experimental approaches are presented for the LMD (Laser Metal Deposition) and the SLM (Selective Laser Melting) processes. The experimental part uses powder jet, it was performed in collaboration with the CSIR-National Laser Center of South Africa in the framework of the ALC projects program, whereas preliminary results for powder bed use in laser additive manufacturing are given for the selective laser sintering/melting technique (SLS/SLM).

Keywords Laser deposition · Selective laser melting · Metals

53.1 Introduction

Industry 4.0, first mentioned in 2011, is expected to fundamentally change the production methods. It involves technologies that are gathered to ensure automated and remote controlled manufacturing of products. In relation to sustainability issues linked to the preservation of the environment additive manufacturing can play a major role since part design optimization and the fabrication process allow to obtain high quality and specific parts by using only the necessary amount of raw material.

E.-H. Amara (✉) · K. Kheloufi · T. Tamsaout · F. Hamadi · S. Aggoune · K. Bougherara · K. Bourai
Laser Material Processing Team, Centre de Développement des Technologies Avancées CDTA, P.O. Box 17 Baba-Hassen, 16303 Algiers, Algeria
e-mail: amara@cdta.dz

A. Belasri and S. A. Beldjilali (eds.), *ICREEC 2019*, Springer Proceedings in Energy,
https://doi.org/10.1007/978-981-15-5444-5_53

Designs prepared by computers are use in additive manufacturing to produce complex parts from 3D models. The optimal use in the industry of this manufacturing process allows the reduction of production times and to obtain parts impossible to achieve by conventional methods. These new techniques are so fast and so well developed that they announce a new conception of the industry in the world. During these last years, in aeronautics industry, Pratt & Whitney [1] and Dassault [2, 3] have adopted additive manufacturing for the design of engine parts. More recently, in 2017 it has been shown that the mechanical properties of metal parts in the automotive industry, obtained by laser additive manufacturing, are excellent [4]. In another area, tailor-made mandibular implants were made of titanium alloy by the selective laser melting process [5]. Usually, in most conventional processes the manufacture of parts is based on the removal of material by using machines such lathes, whereas in additive manufacturing, a 3D process allows to make parts by adding material, where a 3D model is transformed into a physical object, by assembling successive layers. This process makes it possible to manufacture parts that would be very complex to manufacture by conventional means and makes it possible to optimize the production times. The parts are made in one piece, without the need for assembly and with the possibility of providing reinforcements of material on the most fragile or most stressed lines.

In our contribution, we present some results of the modeling approach that has been developed in CDTA. For the LMD process, the experimental part uses powder jet, it was performed in collaboration with the CSIR-National Laser Center of South Africa in the framework of the ALC projects program, whereas preliminary results for powder bed use in laser additive manufacturing are given for the selective laser sintering/melting technique (SLS/SLM). Through the developed work, we give an overview of the challenges to be met to improve our understanding of the physical phenomena that take place during the process.

53.2 Laser Additive Manufacturing (LAM) Process

The central element of the additive manufacturing is a laser beam. It is used for melting the metallic powder, which solidifies to form a high quality piece, or for heating a bed of powder made of agglomeration of the treated products (powders), to give them cohesion and rigidity sufficient without resorting to a merger. In case of powder bed, the particles must be spherical to increase their density, to enhance the packed density and compactness of the powder bed, and to improve the flowability of the powder. Whereas powder particles dimensions used in laser additive manufacturing lies between a human hair thickness and an icing sugar particle.

The main processes that are used for metallic additive manufacturing are the SLM (Selective Laser Melting), SLS (Selective Laser Sintering), and Laser Metal Deposition (LMD). The application of rapid prototyping processes concern the medical and dental fields, cars, aeronautic, architecture, engineering, etc.

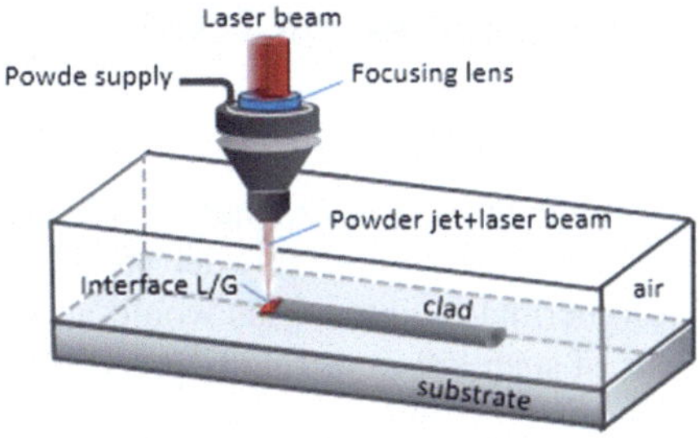

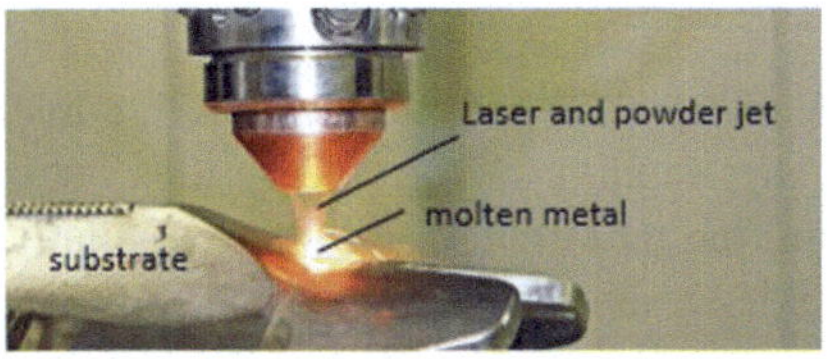

Fig. 53.1 Representation of LMD process

The principle of laser metal deposition (LMD) is to bring by laser irradiation, the feeding material (powder, wire) to its liquid state while it drops on the fused substrate layer, to obtain a metallurgical bound or clad. The powder can be injected laterally or coaxially, Fig. 53.1, in relation with the laser beam direction, and this technique allows to build in very complex parts.

The other techniques using a laser beam in additive manufacturing are the SLM (Selective Laser Melting) and SLS (Selective Laser Sintering) processes. They consist, as shown on Fig. 53.2, in using a laser beam for shinning selectively a surface of a deposited powder bed which after being processed as a designed pattern representing a section following the depth of the part to be realized, another layer is deposited and then processed in same way. And so on by a treatment layer by layer, a more or less complex 3D object or part is obtained. It is important to note that powder that has not been irradiated by the laser beam can be recycled and reused.

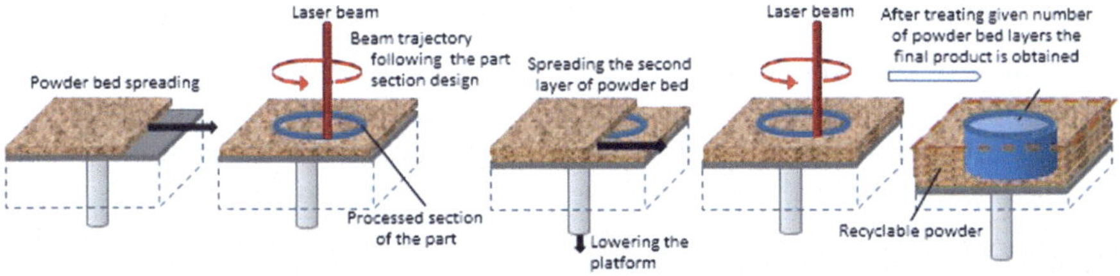

Fig. 53.2 Sequences illustrating the principle of SLS and SLM processes to make a hollow cylinder

53.3 Experimental and Modeling Investigations on LAM Process

A widespread adoption of additive manufacturing technologies depends on a better understanding of the interactions between the laser, the metallic powder particles, and the base metal, occurring at relatively high thermal regimes. This would allow to determine the possibilities and the limitations of such laser machining process. In most of laser material transformation processes, the absorption phenomenon is one of the most important, and the more desired since the treated material can undergo heating, melting, vaporization, till plasma formation. The processing parameters, together with the involved physical phenomena that interact in laser additive manufacturing, make the task more complicated in order to determine the relationships between the desired properties of the product and the suitable control parameters [4]. Optimization of such parameters is thus necessary in order to get the both the wanted accuracy of the part dimensions and its integrity. This allow to manufacture strengthful metallic parts and to reduce defects such as porosities and cracks, and those due to a high residual stress, and the dilution. Numerical modelling constitute one of the most efficient approaches to optimize and to evaluate the effect of the processing parameters, in then to select those offering the better operating ones. An advantage of modeling/simulation is that one can obtain information on quantities that is impossible to measure from experiments, this includes gradients due to temperature variations, the flows in the molten pool, and the rates of heating then cooling.

53.3.1 Laser Metal Deposition

The features of the built part are influenced by the used processing parameters and the operating conditions. These include the laser beam power and its spatial distributions, its defocusing distance, the flow of shielding gas, the powder feeding rate, the particle size and shape, and the thermodynamic properties of the treated material. The main physical mechanisms present during laser deposition include the melting/solidification of the treated material, the free surface dynamics corresponding to the flows occurring at the molten metal and air interface, and obviously the physics related to laser-matter interaction itself.

On Fig. 53.3a we present the simulation result which shows the construction of filled cylinder and the associated temperature field. The corresponding experimental work is shown on Fig. 53.3b, the research on the LMD process [6] was performed in collaboration with the Laser Material Processing Team of the CSIR-National Laser Centre of South Africa. The sequences given on Fig. 53.3a correspond to

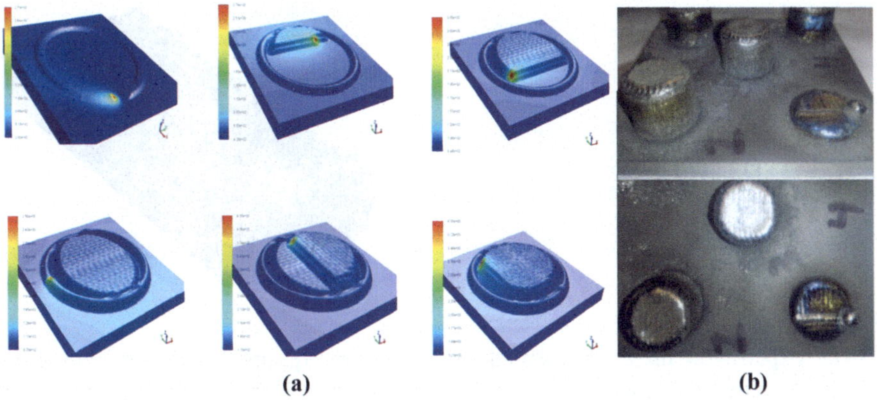

Fig. 53.3 Representation of temperature distribution and the calculated built part for a filled cylinder (Modeling **a** and experiments **b** were performed in CDTA, and CSIR-NLC respectively)

operating parameters related to a powder feed rate which was taken equal to 2 g/min, whereas the laser beam scan speed was equal to 10.58 mm/s. We can observe through the isocolors chart that the deposition occurs for a maximal temperature of T = 2750K (2476.84 °C) during the first pass, whereas it was about 4000K (3726.85 °C) during the second one.

53.3.2 *Selective Laser Sintering/Melting (SLS/SLM) Processes*

With local means and expertise acquired in the field of laser technology and laser process modeling, experimental and theoretical approaches are implemented to seek the possibility of manufacturing 3D metal parts from a bed of powder, by selective laser melting (SLM) or selective laser sintering (SLS) technique

Fig. 53.4 The computational field

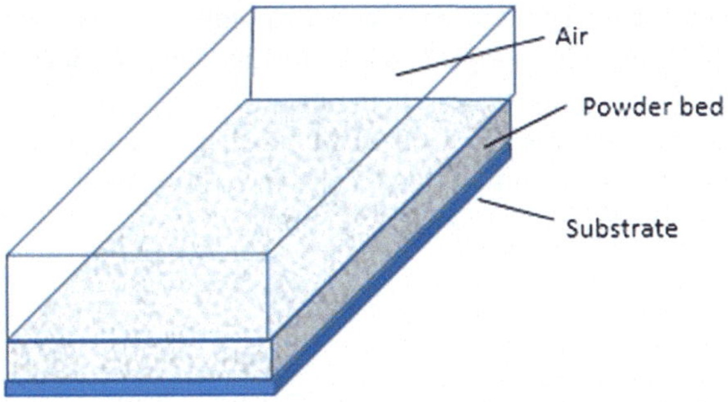

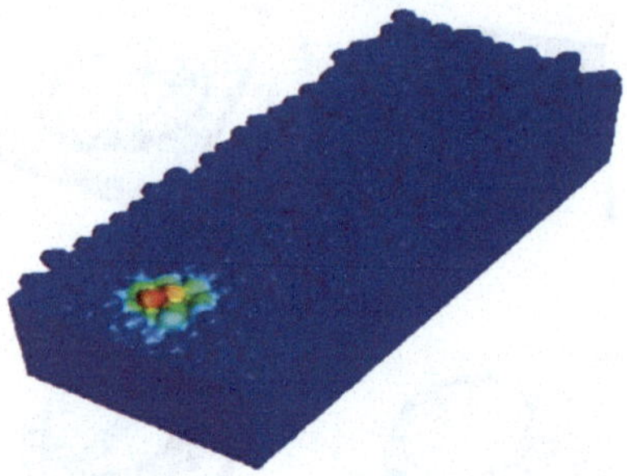

Fig. 53.5 Laser impart on small area of the powder bed

The experiment uses a homemade laser system equipped with galvanometric mirrors to direct the laser beam and thus obtain the selective irradiation of given zone on the powder bed. The experimental setup is composed of a fiber laser delivering an energy up to 1 mj per pulse duration at FWHM of 160 ns, corresponding to an average power of 20 W. The setup for powder bed spreading and its mechanical components is still at design stage. However, concerning the modelling our team succeeded in generating a powder bed numerically, which as shown on Fig. 53.4 must take into account the substrate and the environing air. The powder bed is numerically generated following random particles diameters.

Due to the presence of air as gas and powder particles as solid, we deal with a multiphase problem. The volume of fluid (VOF) method implemented by Ansys/Fluent code is used for tracking the powder/air interface evolution. The selectively local deposited laser energy and the fusion/solidification phenomena are introduced in the computing process in the energy equation through user-defined functions (UDF) procedures, and enthalpy approach respectively. A typical simulation showing laser impart on the powder bed is given on Fig. 53.5.

53.4 Conclusion

We presented a brief overview of laser additive manufacturing, and the main techniques used. Their impact on sustainable development and the environment has been demonstrated from several points of view, and many known manufacturers have already adopted these techniques in their parts manufacturing process. Concerning the research in this field, we give a brief overview of the research carried out at CDTA on SLM processes, and as part of a project of the African Laser Center on the process LMD, in collaboration with the CSIR-NLC. Work on the SLS/SLM processes is ongoing.

References

1. https://www.3dnatives.com/pratt-whitney/
2. https://www.safran-group.com/fr/media/20150618_dassault-systemes-et-safran-annoncent-un-partenariat-strategique-dans-le-domaine-de-la-fabrication-additive
3. https://www.3ds.com/fr/communiques-de-presse/detail-des-communiques-de-presse/dassault-systemes-et-airbus-apworks-sassocient-pour-accelerer-lutilisation-de-la-fabricatio/
4. http://www.themavision.fr/jcms/rw_479843/fabrication-additive-metal-l-automobile-en-phase-de-jauger-son-potentiel
5. M. Regniere, B. Desplanques, S. Saunier, C. Reynaud, P. Bertrand, Ti6Al4V alloy selective laser melted: microstructure evolution through post-heating treatment, in *European Congress and Exhibition on Powder Metallurgy. European PM Conference Proceedings* (The European Powder Metallurgy Association, Shrewsbury, UK 2015), pp. 1–6
6. T. Tamsaout, K. Kheloufi, E.H. Amara, N. Arthur, S. Pityana, CFD model of laser additive manufacturing process of cylinders. S. Afr. J. Ind. Eng. **28**(3), 178–187 (2017)

Chapter 54
Carbon Dioxide Dissociation Using Pulsed DBD with Different Kinds of Dielectric Barriers

F. Pontiga, M. Guemou, H. Moreno, A. Fernández-Rueda, and K. Yanallah

Abstract Carbon dioxide dissociation using dielectric barrier discharge has been experimentally investigated. The electrical discharge was stimulated using high voltage pulses of nanosecond duration, with a repetition rate in the range of 100–1000 Hz. The reactor consisted of two stainless steel plane circular electrodes covered with either fused silica glasses or polytetrafluoroethylene (PTFE) films. Experiments were carried in pure CO_2, and the concentration of carbon monoxide and ozone in the effluent gas was determined using UV/VIS and FTIR spectrophotometry. The results have shown that using silica as dielectric layers results in a higher generation efficiency for both CO and O_3.

Keywords CO_2 dissociation · Dielectric barrier discharge (DBD)

54.1 Introduction

The overwhelming evidence that carbon dioxide emissions are responsible to a large extent of global warming has boosted the research on the application of plasmas to decompose carbon dioxide and to generate valuable chemicals, which may facilitate the production of new fuels [1]. However, the CO_2 molecule is very stable, and breaking the bond OC = O demands a great energy. Thus, conventional techniques require very high temperatures (1600–2000K) for the process to be

F. Pontiga (✉) · H. Moreno · A. Fernández-Rueda
Dpt. Física Aplicada II, E. T. S. Ingeniería de Edificación, Universidad de Sevilla, Av. Reina Mercedes 4a, 41012 Seville, Spain
e-mail: pontiga@us.es

M. Guemou
Dpt. des Sciences et de la Technologie, Faculté des Sciences Appliquées, University of Tiaret, Tiaret, Algeria

K. Yanallah
Dpt. de Physique, Faculté des Sciences de la Matière, Laboratoire de Génie Electrique et des Plasmas, University of Tiaret, Tiaret, Algeria

A. Belasri and S. A. Beldjilali (eds.), *ICREEC 2019*, Springer Proceedings in Energy,
https://doi.org/10.1007/978-981-15-5444-5_54

thermodynamically favorable [2]. Non-thermal plasmas can help to solve this problem since, in these kind of plasmas, the electron temperature is high enough (1–20 eV) to decompose the constituent molecules of the gas [2, 3], while heavy species temperature is close to room temperature. Among others, corona discharge, dielectric barrier discharge (DBD) and microwave discharge are examples of low temperature plasma sources.

Microwave discharge is particularly efficient in dissociating CO_2 molecules by means step-by-step vibrational excitation [4]. However, since microwave discharge must usually be run at reduced pressures, the net yield of products is limited. In contrast, corona and dielectric barrier discharges can be operated at atmospheric pressure, which implies greater experimental simplicity.

Previous studies of CO_2 dissociation using DBD have almost exclusively used AC stimulation in the range of 50 Hz to tens of kHz [5]. In contrast, in corona discharge, high voltage pulses of nanosecond duration have been applied with great success to CO_2 dissociation [6]. Therefore, in this study, this kind of stimulation will be used in DBD to investigate its efficiency for dissociating carbon dioxide and generating other chemicals, like carbon monoxide and ozone. The role played by the dielectric layers used in the reactor will be investigated. Therefore, two different types of dielectrics will be used in the experiments.

54.2 Experimental Setup

Figure 54.1 shows a schematic representation of the experimental setup used in the experiments. The dielectric barrier discharge reactor consisted of two stainless steel plane circular electrodes, 20 mm in diameter, covered with either fused silica glasses (electrical permittivity: $\varepsilon_r \approx 3.8$) or polytetrafluoroethylene films (PTFE, $\varepsilon_r \approx 2$) as dielectrics. The gap between the two dielectric layers was 2 mm. Excluding the electrodes, all remaining components of the DBD reactor were also made of PTFE. The DBD reactor was fed with pure CO_2 (gas purity > 99.995%) and the gas flow rate was kept constant with the help of a mass flow controller (Alicat).

A high voltage nanosecond pulse generator (NPG18-3500N, Megaimpulse), based on drift step recovery diodes (DSRD), was used as power supply for the DBD reactor. This generator produces high voltage pulses with rise times of less than 4 ns, and duration of about 20 ns. The rate of repetition of pulses was controlled externally, with an arbitrary waveform generator (33521A, Agilent), which supplied the trigger signal to the pulse generator. The amplitude of the high voltage pulse was fixed at −27 kV, approximately. However, due to the highly nonlinear nature of the barrier discharge, some reflection of energy inevitably occurs at the reactor. Thus, the first pulse is followed by a subsequent train of pulses of lower amplitude.

The voltage signal provided by the pulse generator was measured with a wide-band high voltage probe with 1000X attenuation (P6015A, Tektronix). A fast current transformer (Bergoz), with a sensitivity of 0.5 V/A, was used for the measurement of the current intensity flowing thought the DBD reactor. Both signals were recorded by a digital oscilloscope with 2.5 GHz bandwidth (DPO7254, Tektronix).

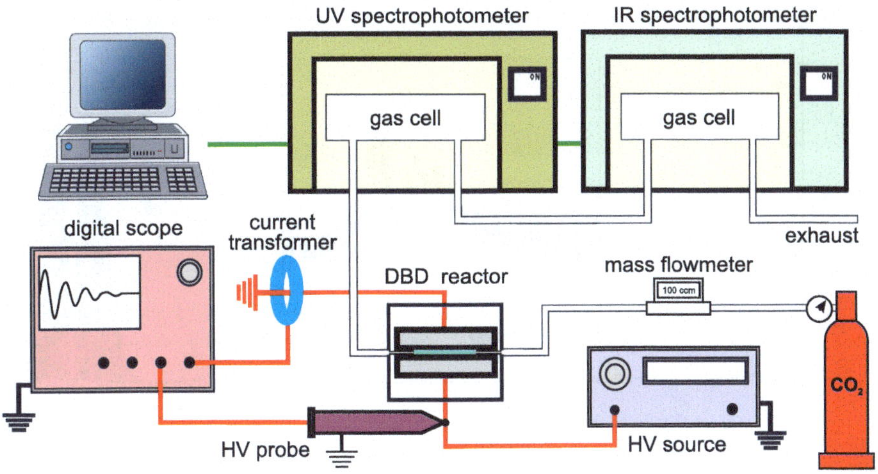

Fig. 54.1 Schematic representation of the experimental set-up used in the experiments

The effluent gas from the DBD reactor was sent to a gas cell within the sample compartment of a UV/VIS spectrometer (Evolution 300, Thermo) and then to another within of a FTIR spectrometer (Vertex 70, Bruker), so as to perform the spectrophotometric analysis of its composition. Spectra were recorded at regular intervals, and the concentration of ozone was determined both from UV spectra (190–320 nm) and IR spectra (1000–1075 cm^{-1}). Regarding carbon monoxide, it was determined from IR spectra in the wavenumber range 2050–2250 cm^{-1} (see Fig. 54.2).

54.3 Results and Discussion

The averaged electrical energy $\langle E \rangle$ delivered to the gas by a train of pulses (i.e., the first high voltage pulse and the subsequent pulses of lower amplitude) was computed from the signals recorded by the oscilloscope as

$$\langle E \rangle \approx \int_0^T \langle V(t) \rangle \langle I(t - t_0) \rangle dt, \tag{54.1}$$

where V is the voltage measured by the high voltage probe, I is the current intensity obtained with the help of the fast current transformer, t_0 is the delay between both signals due to the different cable length of probes, and $\langle\ \rangle$ represents the averaged value over a large number of recorded samples (≥ 400), in order to reduce random fluctuations. The time integration was carried out over a sufficiently long interval, $T \approx 2$ μs, so that the voltage amplitude was negligible at the end of that interval.

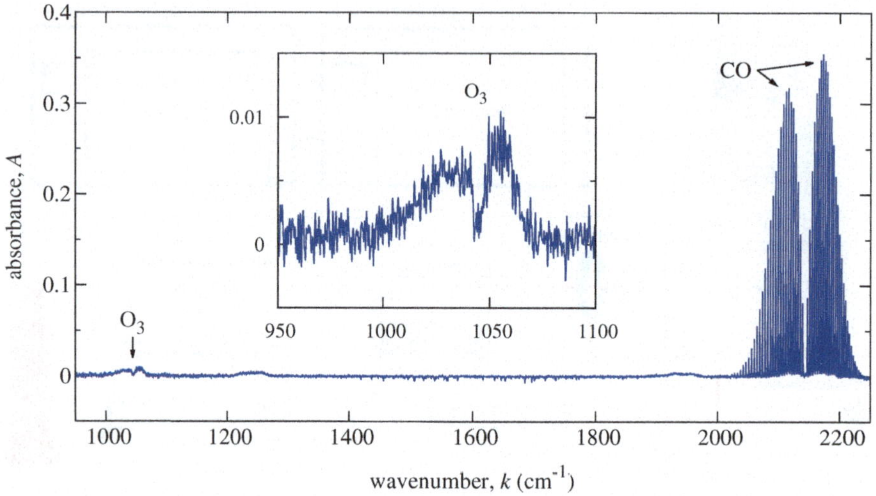

Fig. 54.2 Infrared spectrum of the effluent gas escaping from the DBD reactor when metallic electrodes are covered with silica glasses 1.05 mm thick. The repetition rate of pulses was 750 Hz and the gas flow rate was 100 cm^3/min. The inset shows the absorption band corresponding to ozone

The results of this evaluation showed that the averaged electrical energy was $\langle E \rangle \approx 3.65$ mJ when metallic electrodes were covered with silica glasses of 1.05 mm in thickness, and $\langle E \rangle \approx 3.03$ mJ when dielectrics of PTFE (thickness 1 mm) were used instead of silica glasses. In addition to the different permittivity of dielectrics, many other factors may influence the higher averaged energy delivered with pulses when fuse silica is used as dielectric, like the surface roughness, the thermal conductivity, etc. [7]. The total averaged power consumed in the reactor can then be obtained as

$$\langle P \rangle = f \times \langle E \rangle, \tag{54.2}$$

where f is the repetition rate frequency of the trains of pulses, which was varied from 100 Hz to 1 kHz.

The most important reaction leading to CO_2 splitting in DBD is electron impact dissociation, which proceeds as [8]

$$e^- + CO_2 \rightarrow e^- + CO + O. \tag{54.3}$$

Subsequently, recombination of oxygen atoms produces molecular oxygen and, eventually, reaction of oxygen atoms with oxygen molecules leads to the formation of ozone,

$$O + O + CO_2 \rightarrow O_2 + CO_2, \quad (54.4)$$

$$O_2 + O + CO_2 \rightarrow O_3 + CO_2. \quad (54.5)$$

Of course, this is a simplistic description, because many other reactions are involved in the process of CO_2 dissociation and O_3 generation [6]. Since the rate constant of (54.3) increases with the electron energy, raising the power supplied to the DBD reactor results in an augmentation of carbon monoxide and ozone. Moreover, as stated before, the energy of pulses is higher when fused silica is used as dielectric. Consequently, the concentrations of CO and O_3 are correspondingly higher in that case.

However, from an industrial point of view, it is very convenient to measure the dissociation efficiency of CO_2, and the corresponding generation efficiencies of CO and O_3. These last efficiencies, expressed as grams of product per kilowatt-hour of energy, are shown in Figs. 54.3 and 54.4, respectively, as a function of the pulse repetition rate. As it can be readily seen, the production efficiency of carbon monoxide is higher when metallic electrodes are covered with fused silica dielectrics, except for low values of the pulse repetition frequency (≤ 150 Hz). Moreover, the production efficiency reaches a maximum at 275 Hz, both using fused silica (~ 21 g/kWh) and PTFE (~ 16 g/kWh) as dielectric layers. Above that frequency, the production efficiencies using fused silica declines faster than using PTFE, and they finally have similar values at the highest frequencies.

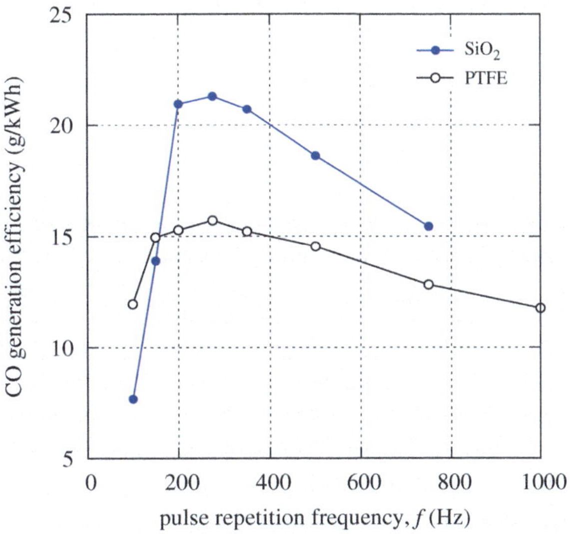

Fig. 54.3 Generation efficiency of carbon monoxide as a function of the pulse repetition rate when both metallic electrodes are covered with either fused silica glasses (●) or polytetrafluoroethylene (PTFE) films (○). The DBD reactor was fed with pure CO_2 at a constant flow rate of 100 cm^3/min

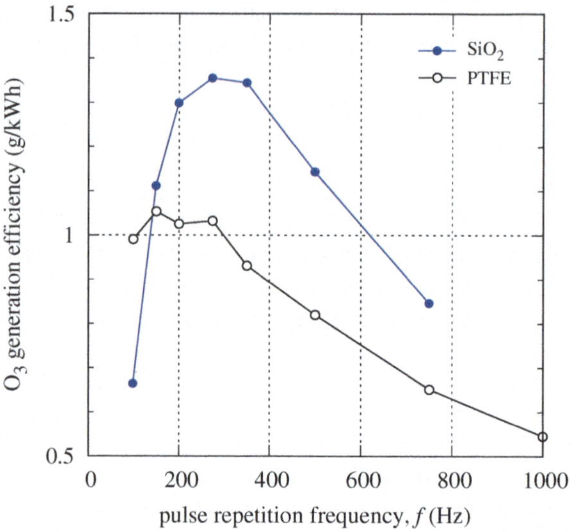

Fig. 54.4 Generation efficiency of ozone as a function of the pulse repetition rate when both metallic electrodes are covered with either fused silica glasses (●) or polytetrafluoroethylene (PTFE) films (○). The DBD reactor was fed with pure CO_2 at a constant flow rate of 100 cm^3/min

Regarding the production efficiency of ozone, it shows a similar trend to carbon monoxide, although the existence of a maximum at a definite frequency is less marked when PTFE is used as dielectric layer. In that case, the production efficiency remains high in the interval 100 Hz $\leq$ f $\leq$ 275 Hz and then declines. However, the generation efficiencies values found for O_3 are more than one order of magnitude lower that those found for CO. This is an expected result, since ozone is formed at the expense of O_2 which involves the intermediate reaction (54.4).

54.4 Conclusions

This study has investigated the dissociation of CO_2 using DBD stimulated by high voltage pulses of nanosecond duration. Special attention has been paid to the generation efficiency of CO and O_3, which are the principal products of CO_2 dissociation, and to the effect of dielectrics used in DBD reactor. The highest efficiencies have been found using fused silica as dielectric, and operating the DBD reactor at a pulse repetition rate of about 275 Hz. Increasing the frequency of pulses further leads to a reduction of the overall efficiency and, at very high frequencies ($\sim$1 kHz), the yields of CO and O_3 depend less on the dielectric (fused silica or PTFE) that is being used as barrier.

Acknowledgements This research has been funded by the following organizations: *Ministerio de Ciencia, Innovación y Universidades*, *Agencia Estatal de Investigación* and *Fondo Europeo de Desarrollo Regional*, under contract no. PGC2018-099217-B-I00.

References

1. R. Snoeckx, A. Bogaerts, Plasma technology – a novel solution for CO_2 conversion? Chem. Soc. Rev. **46**, 5805–5863 (2017)
2. B. Ashford, Tu, X, Non-thermal plasma technology for the conversion of CO_2. Curr. Opin. Green Sustain. Chem. **3**, 45–49 (2017)
3. H. Kim, Nonthermal plasma processing for air-pollution control: a historical review, current issues, and future prospects. Plasma Process. Polym. **1**, 91–110 (2004)
4. G.J. van Rooij, D.C. van den Bekerom, N. den Harder, T. Minea, G. Berden, W.A. Bongers, R. Engeln, M.F. Graswinckel, E. Zoethout, M.C. van de Sanden, Taming microwave plasma to beat thermodynamics in CO_2 dissociation. Faraday Discuss. **183**, 233–248 (2015)
5. R. Aerts, W. Somers, A. Bogaerts, Carbon dioxide splitting in a dielectric barrier discharge plasma: a combined experimental and computational study. ChemSusChem **8**(4), 702–716 (2015)
6. M.S. Moss, K. Yanallah, R.W.K. Allen, F. Pontiga, An investigation of CO splitting using nanosecond pulsed corona discharge: effect of argon addition on CO conversion and energy efficiency. Plasma Sources Sci. Technol. **26**(3), 035009 (2017)
7. A. Ozkan, T. Dufour, A. Bogaerts, F. Reniers, How do the barrier thickness and dielectric material influence the filamentary mode and CO conversion in a flowing DBD? Plasma Sources Sci. Technol. **25**(4), 045016 (2016)
8. A. Fridman, *Plasma Chemistry* (Cambridge University Press, New York, 2008)

Chapter 55
Depth Profiling of Solar Cells Using Laser-Induced Breakdown Spectroscopy (LIBS)

Fatima Zohra Hamdani, Sid Ahmed Beldjilali, Mohamed Amine Benelmouaz, Sabrina Messaoud Aberkane, Kenza Yahiaoui, and Ahmed Belasri

Abstract The quality control of the solar cells in the industrial chain is necessary to detect the responsible defects on the efficiency of the solar panels. In this context, we analyzed a polycrystalline silicon solar cell by Laser Induced Breakdown Spectroscopy (LIBS). The target was irradiated by a Nd: YAG pulsed laser at the fundamental wavelength $\lambda = 1064$ nm with an energy of 20 mJ. Based on the color of the used solar cell, two zones were detected: a blue zone representing the silicon wafer and a white zone representing the contact metal ribbons. Seven elements were detected in the white zone and only four in the blue zone. The distribution of the different elements in the sample is like a depth function called "Depth profiling" which can give us useful information about the constituent layers of this sample. In this work, the depth profiling of the various detected elements versus the number of laser shots was examined to understand the distribution of the elements in the volume of the solar cell.

Keywords Laser-Induced Breakdown Spectroscopy (LIBS) · Solar cells · Polycrystalline silicon · Depth profiling

F. Z. Hamdani (✉) · S. A. Beldjilali · M. A. Benelmouaz · A. Belasri
LPPMCA, Université des Sciences et de la Technologie d'Oran Mohamed Boudiaf USTO-MB, BP 1505 El M'naouar, 31000 Oran, Algeria
e-mail: Fatima.hamdani26@gmail.com

S. Messaoud Aberkane · K. Yahiaoui
CDTA, Centre de Développement des Technologies Avancées, Cité 20 aout 1956, B.P 17, Baba Hassen, Algiers, Algeria

A. Belasri and S. A. Beldjilali (eds.), *ICREEC 2019*, Springer Proceedings in Energy,
https://doi.org/10.1007/978-981-15-5444-5_55

55.1 Introduction

Silicon is the main component used in the fabrication of solar cells, giving it the first place in the field of renewable energies [1, 2]. The detection and qualification of impurities of photovoltaic (PV) solar cells based on polycrystalline silicon are essential steps in the production of PV systems.

Several analytical advanced methods were developed in recent years and the most recent is called "Laser-Induced Breakdown Spectroscopy" (LIBS). This new technique allows us to obtain a qualitative and quantitative analysis [3, 4], of the chemical elemental composition of the different samples (solid, liquid or gases) [5, 6]. Analysis of the emission spectra emitted by the plasma plume allows us to measure the mass fractions of all elements present in the sample without calibration [7, 8].

The principle of the LIBS technique is to focus an impulsive laser beam on the sample to be analyzed to create a very hot micro plasma from a small amount of ejected matter. The spectral analysis of the plasma light determines the nature and concentration of the different chemical elements in the sample.

This analytical technique offers several compelling benefits compared to different elemental analysis techniques. These include: a sample preparation-free activity expertise [9], extremely quick activity time, typically a couple of seconds, for one spot analysis, versatile sampling protocols that embrace quick formation of the sample surface and depth identification, thin-sample analysis while not the fear of the substrate interference.

To get useful information on the constituent layers of the solar cell [10, 11], we have investigated a depth profiling of all elements detected in the sample by measuring their mass fraction versus the depth.

55.2 Experimental Setup

A typical LIBS configuration was used during these measurements to carry out qualitative and quantitative studies of photovoltaic cells based on poly-crystalline silicon. In this experiments, we have used a Nd-YAG laser source operating at its fundamental wavelength $\lambda = 1064$ nm. A 20 mJ of laser energy is focused onto the sample surfaces in air at atmospheric pressure. The plasma emission light is collected via an optical fiber of 400 μm diameter and then fed to the entrance slit of the Echelle spectrometer (Aryelle 200 LTB). This spectrometer has a wavelength interval of 200 to 780 nm and pairs with an ICCD detector (camera iStar ICCD, Andor). Data acquisition and analysis were performed using Sophi software (LTB). The device is mapped on Fig. 55.1.

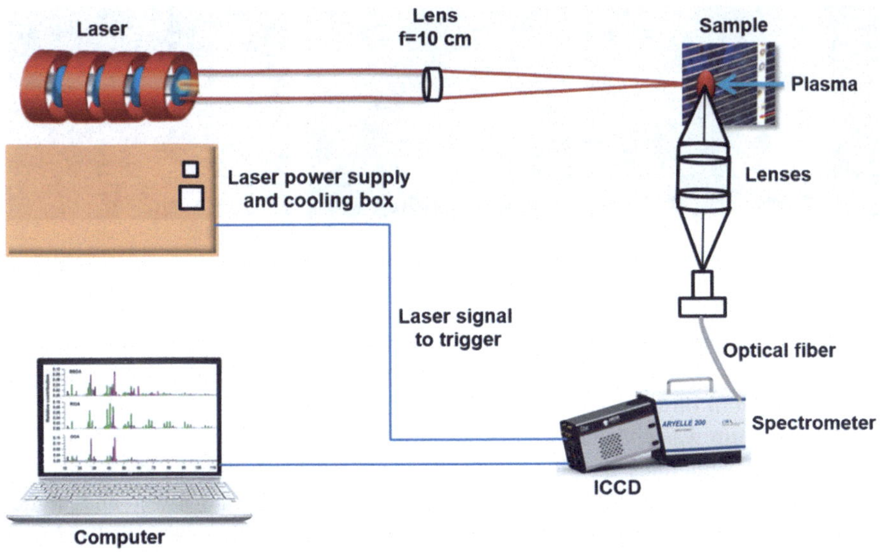

Fig. 55.1 Experimental setup

55.3 Results and Discussion

In our solar cell, we distinguish two different zones: the blue zone and the white zone (Fig. 55.2).

55.4 Crater Diameter Measurements

Using the optical microscope, we determined the diameter of the crater of the blue zone for 1, 5 and 10 shots (Fig. 55.3).

Fig. 55.2 Different zones of the used solar cell

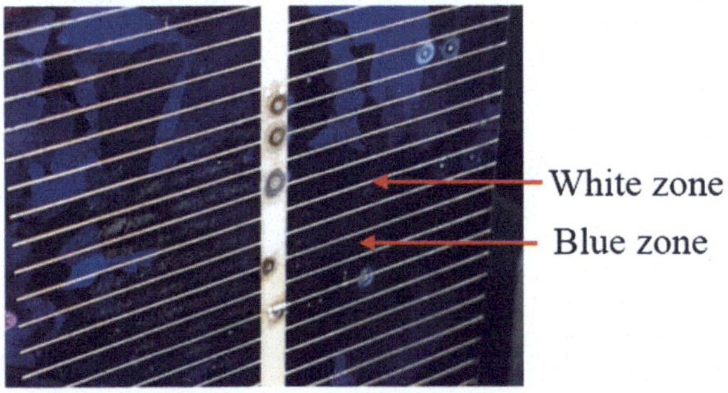

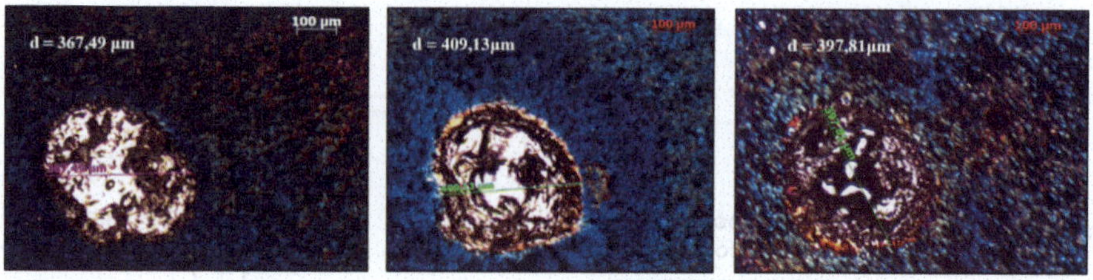

Fig. 55.3 Diameter of the crater for 1, 5 and 10 shots, respectively

After a gate and delay optimization, we found that the best signal to noise ratio was found for a gate and delay of 4 and 2 µs, respectively. Then we have recorded several spectrum on the two cited zone of the solar cell.

The spectrum was characterized by several atomic emission. On Table 55.1, we listed the relevant lines and their properties on the basic of Nist database. Alu represents the coefficient of Einstein, El and Eu represent the lower and higher energy levels, respectively. gl and gu represent the statistical weights of the lower and higher energy levels, respectively.

On Fig. 55.4. We present the evolution of emission intensities of Si, Ag, Na atomic lines as function number of laser shots. These spectral lines were chosen to measure plasma temperature, electron density and for the depth profiling study.

We notice that the intensity of the silicon line (Fig. 55.4 (a)) is very intense on the first shot (of the order of 11 × 104 a.u.), then begins to decrease until reaching a minimum intensity (of the order of 1 × 104 a.u) on the 6th shot. From the seventh shot, the intensity begins to increase until reaching the maximum value on the ninth shot (13 × 104 a.u.). This is due to the deposition of matter at the edge of the crater. At the 10th shot, the intensity decreases. The same phenomenon is observed in the silver line (Fig. 55.4 (b)); the intensity decreases gradually, soaps to the ninth shot, and then diminishes. For the sodium line (Fig. 55.4 (c)), it is detected only during the first five shots.

The white zone contains a very thin layer deposited on the silicon. The intensity of the silicon line (Fig. 55.5 (a)) during the first four shots is very low, which

Table 55.1 Atomic properties of the elements detected in the white and the blue zone

Elements	A_{lu} (10^7) (S^{-1})	E_l (eV)	E_u (eV)	g_l	g_u
Si I 288.158 nm	21.7	0.78	5.08	5	3
Ag I 328.068 nm	14	0	3.77	2	4
Na I 588.995 nm	6.16	0	2.10	2	4
Al I 309.270 nm	26	0.01	4.02	4	6
Ca I 442.673 nm	21.8	0	2.93	1	3
K I 769.896 nm	3.75	0	1.60	2	2
Pb I 405.780 nm	14	0	3.77	2	4

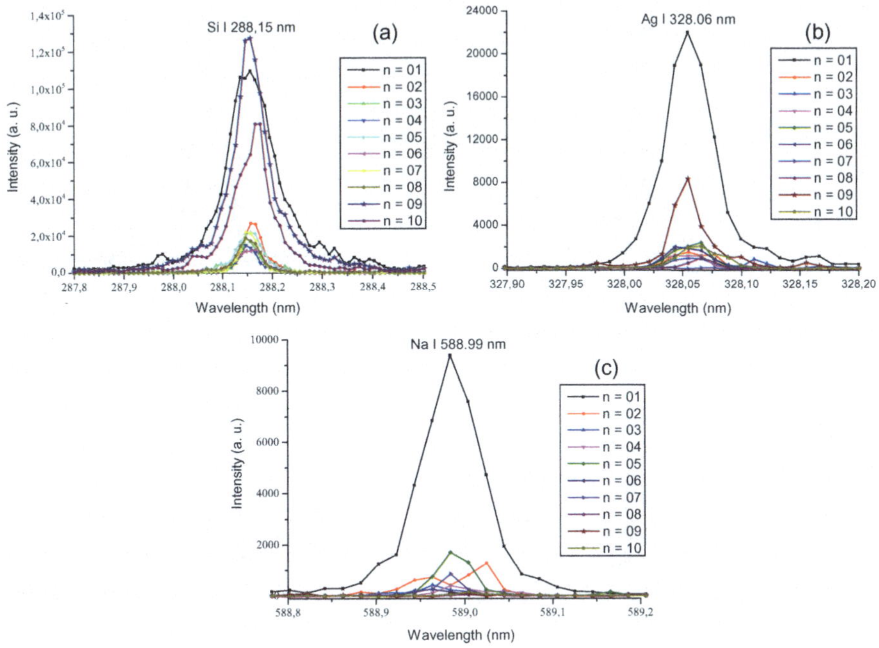

Fig. 55.4 The spectral signature of silicon **a**, silver **b** and sodium **c** lines as a function of the number of laser shots in the blue zone

assures us that we are still in the thin white layer. The deeper you dig, the more you reach the silicon layer and the intensity of the silicon line begins to increase. We observe the Silver during the first ten shots (Fig. 55.5 (b)), which shows that the white layer contains a lot of silver, The Aluminum, lead and calcium are also present during the first ten shots (Fig. 55.5 (c, d, f)).

Sodium is observed during the first ten shots (Fig. 55.5 (e)), which shows that sodium enters the white thin layer.

55.5 Comparison of Silicon Line Emission Between the Two Zones

In the blue zone, we notice that the silicon intensity is at maximum during the first 20 shots after it degrades with each shot (Fig. 55.6). The decrease in silicon intensity does not mean that we have a base concentration, but only that the ablated material decreases with the number of shots.

In the white zone, the depth profiling of silicon shows us that its initial concentration is low, and then increases with the number of shots (Fig. 55.6). This

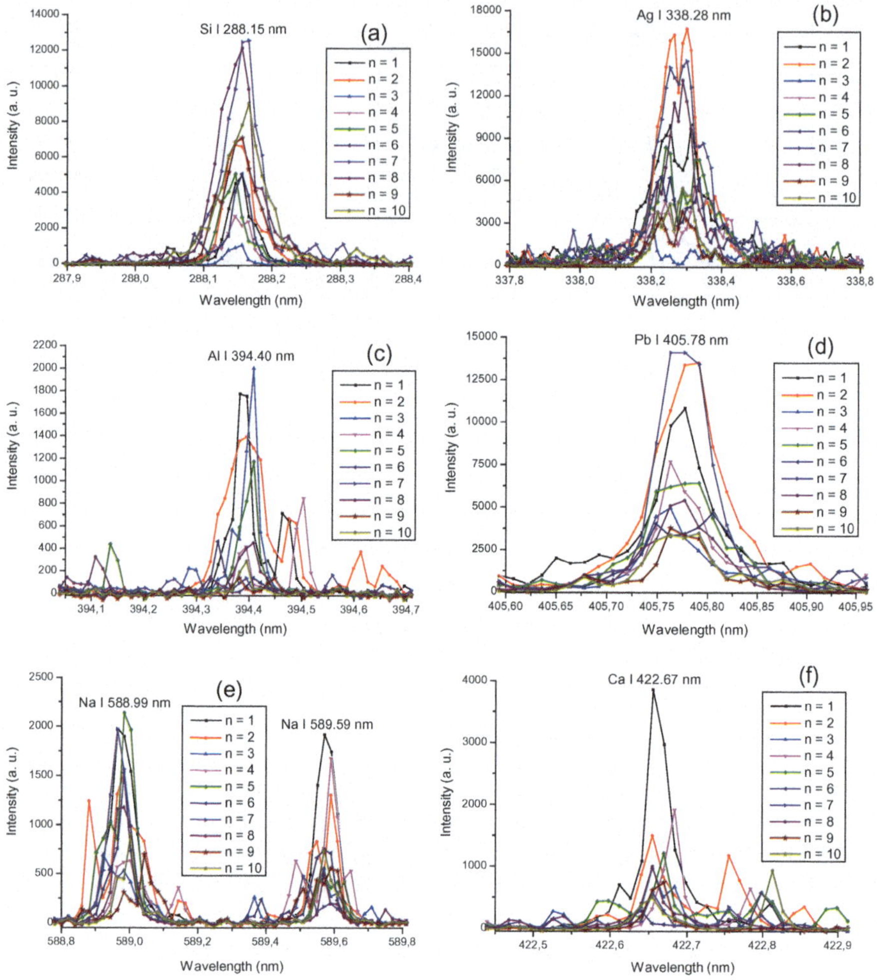

Fig. 55.5 The spectral signature of silicon **a**, silver **b**, alumina **c**, lead **d**, sodium **e** and calcium **f** lines as a function of the number of laser shots in the white zone

confirms the absence of silicon in the white thin layer. Its detection in this area is due to the migration of the silicon to the area where it is minority.

During the first sixty-five shots, we observe the same profiling of Silicon in both. After the intensity gradually decreases by going deeper (blue zone). However, in the white zone, the intensity increases because of the ablated volume, which forms at the edge of the crater.

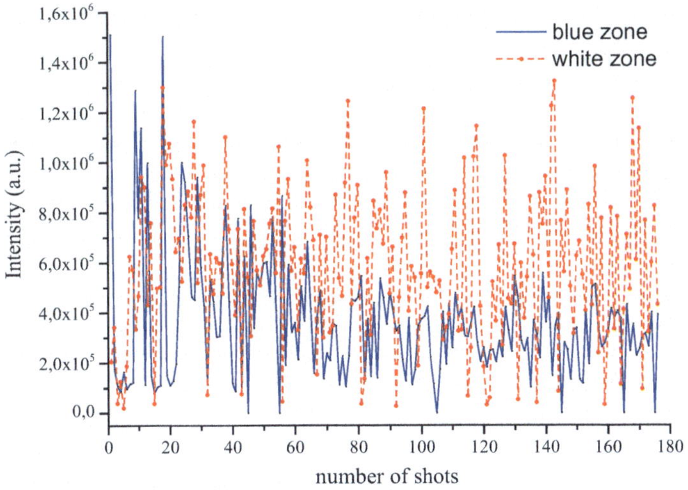

Fig. 55.6 Depth profiling of the Silicon line Si I 288.15 nm versus number of laser shots in the two zones

55.6 Conclusion

A spectroscopic study was done to examine photovoltaic cells based on polycrystalline silicon. The aim was to investigate the distribution of the detected elements in the volume of the sample. Spectroscopic analysis allowed as to detect seven elements in the white zone (Si, Ag, Pb, Al, Na, C, and Ca) and only four elements in the blue zone (Si, Ag, Na, K). Depth profiling enabled us to investigate the distribution of detected elements in the solar cell volume. This allowed us to compare the distribution and presence of the elements in the different layers of the photovoltaic cell. A further study with the LIBS procedure without "CF-LIBS" calibration is needed to understand the mechanisms and physical phenomena observed in this study.

References

1. E.R. Observ, Baromètre du solaire photovoltaïque. Système solaire (2004)
2. D. Sarti, R. Einhaus, Silicon feedstock for the multi-crystalline photovoltaic industry. Sol. Energy Mater. Sol. Cells **74**, 27–40 (2002)
3. A. Ciucci, M. Corsi, V. Rastelli, A. Salvetti, E. Tognoni, New procedure for quantitative elemental analysis by laser-induced plasma spectroscopy. Appl. Spectrosc. **53**, 960–964 (1999)
4. E. Axente, J. Hermann, G. Socol, L. Mercadier, S.A. Beldjilali, M. Cirisan, C. Luculescu, C. Ristoscu, I. Mihailescu, V. Craciun, Accurate analysis of indium-zinc oxide thin films via

laser-induced breakdown spectroscopy based on plasma modeling. J. Anal. At. Spectrom. **29**, 553–564 (2013)
5. J.E. Carranza, B.T. Fisher, G.D. Yoder, D.W. Dahn, On-line analysis of ambient air aerosols using laser-induced breakdown spectroscopy. Spectrochim. Acta B **56**, 851–864 (2001)
6. O. Samek, D.C.S. Beddows, J. Kaiser, S.V. Kukhlevsky, M. Liska, H.H. Telle, J. Young, Application of laser-induced breakdown spectroscopy to in situ analysis of liquid samples. Opt. Eng. **39**, 2248–2262 (2000)
7. S.A. Beldjilali, E. Axente, A. Belasri, T. Baba-Hamed, J. Hermann, Compositional analysis of drugs by laser-induced breakdown spectroscopy. J. Appl. Spectrosc. **83**(5), 472 (2017)
8. S.A. Beldjilali, W.L. Yip, J. Hermann, T. Baba-Hamed, A. Belasri, Investigation of plasmas produced by laser ablation using single- and double-pulses for food analysis demonstrated by probing potato skins. Anal. Bioanal. Chem. **400**, 2173–2183 (2011)
9. J.B. Sirven, Détection de métaux lourds dans les sols par spectroscopie d'émission sur plasma induit par laser (LIBS). Thèse de doctorat. L'université bordeaux 1 (Septembre 2006)
10. S. Messaoud Aberkane, M. Abdelhamid, K. Yahiaoui, C. Mahieddoune, S. Abdelli-Messaci, M.A. Harith, Depth profiling of alumina thin films using laser induced breakdown spectroscopy. Structural and morphological dependence. Thin Solid Films **653**, 293–300 (2018)
11. K. Yahiaoui, S. Abdelli-Messaci, S. Messaoud Aberkane, M. Siad, A. Kellou, Growth of γ-alumina thin films by pulsed laser deposition and plasma diagnostic. Appl. Phys. A **123**, 464–475 (2017)

Chapter 56
One-Dimensional Numerical Simulation of a Capacitively Coupled Oxygen Plasma Driven by a Dual Frequency Generator at Low Temperature

Z. Kechidi, A. Tahraoui, A. H. Belbachir, W. Adress, and N. Ouldcherchali

Abstract The Single and double frequency non-equilibrium capacitively coupled (CCP) radiofrequency plasma sources are commonly used in the laboratory for research and for a variety of processing and techniques. This study is carried out on the basis of a one-dimensional, self-consistent fluid model and corresponding governing equation are described in details in the present report. Our simulation results show that the ozone (O_3) is efficiently generated in the bulk of the discharge in case of helium-oxygen admixture as compared to our recent work [1] carried out on pure oxygen. They also show that the formation of the metastable singlet molecule ($O_2(a^1\Delta_g)$) and the atomic oxygen (O) are important as compared with pure oxygen plasma and which has a significant influence on the electron heating process. This is due to high rate of production of these species and low rate of destruction and recombination. The rate of production of negative oxygen (O^-) and positive (O_2^+) molecular ions are also important but less than the neutral atoms and molecules. The formation of positive (He^+), (He_2^+) and (O^+) ions are almost of the same order of magnitude which is about 10^{-2} less than the (O^-) and (O_2^+) ions. This can be explained by the fact that the electron-impact ionization of oxygen molecules can dominate over helium ionization due to the lower ionization threshold of the oxygen molecule (12.6 eV for O_2 vs. 24.6 eV for He).

Keywords Capacitively coupled plasma · Dual frequency · Species generation

Z. Kechidi (✉) · N. Ouldcherchali
LREA Laboratory, University of Médéa, Médéa, Algeria
e-mail: ziane.univm@gmail.com

A. Tahraoui
Quantum Electronics Laboratory, USTHB, Bab Ezzouar, Algeria

A. H. Belbachir
LAAR Laboratory, Faculty of Physics, USTO, Oran, Algeria

W. Adress
Technical College, Northern Technical University, Mosul, Iraq

A. Belasri and S. A. Beldjilali (eds.), *ICREEC 2019*, Springer Proceedings in Energy,
https://doi.org/10.1007/978-981-15-5444-5_56

56.1 Introduction

Low temperature oxygen plasma is of great interest for many applications in surfaces cleaning and treatments. It may also be mixed with other main feeding gases such as helium to etch a variety of materials, including polymerization and coating. They have been proven to be an effective plasma source due to their great capability to generate reactive species [2] widely used in research and industry for surface modification of polymers and fiber [3, 4] and biomedical applications [5–7]. Metastable singlet delta oxygen ($O_2(a^1\Delta_g)$) is recognized as very important [8, 9] in interaction of plasma with cells because it causes cellular damage and thus can be used for cancer treatment [10]. Moreover, it is used in various processes, such as fine chemical syntheses, wastewater treatment, sterilization [11], and in many biomedical applications. Ozone (O_3) is known worldwide as an environmental friendly, significantly powerful disinfectant and oxidant for a wide range of applications [12], including sterilization in hospitals and drinking water purification. It is increasingly being used to destroy bacteria effectively and render viruses inactive. Given the important applications of these species, this study is focused on issues of how to better understand and control the production of these species within the discharge bulk using a 1D numerical simulation of capacitively coupled RF plasma across a 10 mm electrode gap. The paper is structured as follows. The fluid model with detailed information on the governing equations as well as the used boundary conditions is presented in Sect. 56.2. The main results and discussion are detailed in section in Sect. 56.3. Finally, the conclusion of our findings is drawn in Sect. 56.4.

56.2 Description of the Model Discharge

56.2.1 Model Geometry

We consider a typical geometry of a capacitively coupled discharge in helium-oxygen admixture where the plasma is generated between two parallel electrodes driven by a dual rf power. The discharge is modeled in one-dimensional, with electrodes separation of d = 1 cm. The working gas is an admixture of helium (He) and oxygen (O_2) which is added to generate reactive oxygen species. The discharge model is based on a simplified set of the fluid equations described as bellow for electrons and 8 chemical species, solved self-consistently with Poisson's equation. The species accounted for in the model are electrons (e), the atomic oxygen (O), the molecular ozone (O_3), positive molecular oxygen ions (O_2^+), negative atomic oxygen ions (O^-), positive atomic oxygen ions (O^+) and singlet delta molecular oxygen ($O_2(a^1\Delta_g)$), singly charged helium positive ion (He^+) and helium molecular ion (He_2^+).

56.2.2 Governing Equations

The continuity equations for species k are given by:

$$\frac{\partial n_k}{\partial t} + \vec{\nabla}.\overrightarrow{\Gamma_k} = S_k(\in), \tag{56.1}$$

where, n_k is species (k) density, $\overrightarrow{\Gamma_k}$ the particle flux and $\in$ is the electrons mean energy. The right-hand side of Eq. (56.1) $S_k(\in)$ is the source term given by:

$$S_k(\in) = \sum\nolimits_l N_{k,l} R_l(\in), \tag{56.2}$$

where $N_{k,l}$ is the net number of particles of species k produced in one reaction event of type l (negative if particles are lost) and $R_l(\in)$ is the reaction rate.

By using the drift-diffusion approximation, the fluxes $\overrightarrow{\Gamma_k}$ of different species are given by:

For electrons:

$$\overrightarrow{\Gamma_e} = -n_e \mu_k(\in)\vec{E} - D_e(\in)\vec{\nabla} n_e, \tag{56.3}$$

where, $\overrightarrow{\Gamma_e}$, n_e, $\mu_k(\in)$, $D_e(\in)$ and $\vec{E}$ are the electron flux, density, mobility, diffusion coefficient and the electric field respectively.

For charged (+ positive or − negative) ions:

$$\vec{\Gamma_i} = -n_i \mu_i \vec{E} - D_i \vec{\nabla} n_i, \tag{56.4}$$

For neutrals [13]:

$$\overrightarrow{\Gamma_n} = -D_n \vec{\nabla} n_n, \tag{56.5}$$

Electrons flux due to secondary emission:

$$\overrightarrow{\Gamma_{\gamma_e}} = \sum\nolimits_\alpha \gamma_e \left|\overrightarrow{\Gamma_\alpha}\right|, \tag{56.6}$$

here, $\overrightarrow{\Gamma_{\gamma_e}}$ and $\overrightarrow{\Gamma_\alpha}$ are the flux of secondary electron emission and positive ions respectively.

Electrons energy equation:

The electrons energy equation is given by [14]

$$\frac{\partial(n_\in)}{\partial t} + \vec{\nabla}.\overrightarrow{\Gamma_\in} = -e\overrightarrow{\Gamma_e}.\vec{E} - Q_{el} - Q_{inel} \tag{56.7}$$

where, the energy flux $\overrightarrow{\Gamma_{\in}}$ is defined by [14–16]

$$\Gamma \in= -n \in \mu \in E - D \in \nabla n \in, \tag{56.8}$$

$n_{\in} = \in n_e$ represents the electron mean energy density, $\mu_{\in} = \frac{5}{3}\mu_e$ the mobility of energy, $D_{\in} = \frac{5}{3}D_e$ the diffusion coefficient of energy, $v_{th,\in} = \frac{5}{3}v_{th,e}$ the energy thermal velocity and $-e\overrightarrow{\Gamma_e}.\vec{E}$ the Joule heating term from the ambipolar electrostatic field. The second and third terms on the right hand side of Eq. (56.7) are the net gain and loss of electron power per unit volume due to elastic and inelastic collisions; and they are given by [16]

$$Q_{el} = 3\frac{m_e}{m_g}k_b v_m n_e (T_e - T_g) = \frac{m_e}{m_g}k_{el}n_e n_j\left(2 \in -3\frac{T_g k_b}{e}\right) \tag{56.9}$$

where v_m denotes the elastic collision frequency, m_e, m_g, T_e, T_g are the electron and mixture background gas masses and temperature respectively and k_{el} is the rate constant of momentum transfer for elastic collisions of electrons with He and (O_2) molecules.

$$Q_{inel} = \sum\nolimits_j \Delta \in_j k_j n_e n_j \tag{56.10}$$

where $\Delta \in_j$, k_j are the threshold energy and rate constant of the jth inelastic process, respectively.

For simplicity, we did not include the effects of ion and neutrals inertia in the present model, so that no energy equation solved for these species, however we have used the generalized Einstein relation to represent their diffusion

$$D_i = \frac{k_b T_i}{m_i v_i} = \frac{k_b T_i}{q_i}\mu_i, \tag{56.11}$$

where, D_i, v_i, T_i, k_b, m_i, q_i and μ_i are diffusion coefficient, ion-neutral collision frequency, temperature, Boltzmann constant, mass, charge and mobility of particles respectively.

The ions temperature is determined by using Wannier's formula [15]

$$T_i = T_g + \frac{m_i + m_g}{5m_i + 3m_g}\frac{m_g\left(\mu_i|\vec{E}|\right)^2}{k_b} \tag{56.12}$$

Finally we have closed the set of equation by using the Poisson's equation

$$\Delta\phi = -\frac{1}{\varepsilon_0}\sum\nolimits_i q_i z_i n_i \quad (56.13)$$

where, ϕ, ε_0, q_i, z_i and n_i are the electric potential, vacuum permittivity, species charge, charge number and the species density respectively.

The set of Eqs. 56.1–56.13 constitute a full system that describes the plasma ignition and track the species produced between the electrodes.

56.2.3 Boundary Conditions

Boundary conditions are required to be specified in solving and closing the transport equations. We have picked out from the literature the appropriate ones related to our model [14].

The boundary conditions for the charged species flux are,

$$\overrightarrow{\Gamma_p}.\vec{n} = \frac{1-r}{1+r}\left[(2a-1)sgn(q)n_p\mu_p\vec{E}.\vec{n} + \frac{1}{2}n_p v_{th,p}\right] \quad \text{at x} = 0, \text{d}, \quad (56.14)$$

where, r is the fraction of particles reflected by the surface, $sgn(q)$ is the electric sign of the charged species and a the switching function which is set to one if the drift velocity is directed toward the wall and to zero otherwise:

$$a = \begin{cases} 1, & sgn(q)\mu_s\vec{E}.\vec{n} > 0, \\ 0, & sgn(q)\mu_s\vec{E}.\vec{n} \le 0, \end{cases} \quad (56.15)$$

n_p and μ_p are the density and mobility for the positive ion species p respectively, $\vec{E}$ the electric field vector acting on the charged species, $\vec{n}$ is the normal vector pointing toward the wall and, $v_{th,p} = \sqrt{\frac{8k_B T_p}{\pi m_p}}$ is the thermal velocity. As far as for the negative ion species, we have used the Dirichlet boundary conditions at both electrodes.

For the neutral atoms and molecules, we have used the expression given in the reference [16],

$$\overrightarrow{\Gamma_n}.\vec{n} = \frac{1}{4}n_n v_{th,n}. \quad (56.16)$$

However, in the case of electrons, we have used the total electron flux's expression, which distinguishes between the electrons coming from the bulk and the electrons emitted by the surface of the walls, given by,

$$\begin{aligned}\overrightarrow{\Gamma_e}.\vec{n} = & \frac{1-r_e}{1+r_e}\left[-(2a_e-1)\mu_e(\in)n_p\vec{E}.\vec{n}+\frac{1}{2}(n_e-n_\gamma)v_{th,e}\right], \\ & -\frac{2}{1+r_e}(1-a_e)\sum\nolimits_p \gamma_{se}\overrightarrow{\Gamma_p}.\vec{n}\end{aligned} \tag{56.17}$$

where n_γ is given by,

$$n_\gamma = (1-a_e)\frac{\sum_p \gamma_{se}\overrightarrow{\Gamma_p}.\vec{n}}{\mu_e(\in)n_p\vec{E}.\vec{n}} \tag{56.18}$$

The electron energy flux at the electrodes is given by,

$$\begin{aligned}\overrightarrow{\Gamma_\in}.\vec{n} = & \frac{1-r_e}{1+r_e}\left[-(2a_e-1)\frac{5}{3}\mu_e(\in)n_\in\vec{E}.\vec{n}+\frac{2}{3}(n_\in-n_{\gamma,\in})v_{th,e}\right] \\ & -\frac{2}{1+r_e}(1-a_e)\in_{w,p}\sum\nolimits_p \gamma_{se}\overrightarrow{\Gamma_p}.\vec{n}\end{aligned} \tag{56.19}$$

where $n_\in = n_e\bar{\in}$ is the electron energy density and $n_{\gamma,\in}$ is given by the expression

$$n_{\gamma,\in} = (1-a_e)\in_{w,p}\frac{\sum_p \gamma_{se}\overrightarrow{\Gamma_p}.\vec{n}}{\frac{5}{3}\mu_e(\in)n_p\vec{E}.\vec{n}}\overrightarrow{\Gamma_\in}.\vec{n} \tag{56.20}$$

Finally, the boundary conditions on the electric potential are specified for both electrodes using the Dirichlet boundary conditions,

$$\phi = \begin{cases} V_{lf}cos(2\pi f_{lf}t) + V_{hf}cos(2\pi f_{hf}t), & at\ the\ powered\ electrode \\ 0, & at\ the\ grounded\ electrode\end{cases} \tag{56.21}$$

where V_{lf} and V_{hf} are the amplitudes of the lower and higher applied voltage respectively. The lower frequency f_{lf} and higher frequency f_{hf} were kept constant at 13.56 MHz and 27.12 MHz respectively.

56.3 Results and Discussions

The simulation is performed using time-dependent partial differential equation solver (PDE's) general form in COMSOL Multiphysics [17] in combination with MATLAB via Livelink with Matlab in order to rerun the model for several thousands of rf cycles to get a 'steady state' solution for all species and post-processing.

For a given power, the particle and power balance equations are solved simultaneously to determine the steady-state plasma composition [18]. In the following, we discuss the 1D simulation results for a capacitively coupled 95% He + 5% O_2 plasma with a discharge gap of 10 mm and voltage amplitude of 200 V. The pressure and temperature of the background gas are assumed to be 100 Pa and 300 K, respectively.

Figure 56.1 shows the evolution of the density profiles of ozone (O_3), the neutral oxygen atom and the molecular oxygen metastable $O_2\left(a^1\Delta_g\right)$ between the two electrodes. We found that the formation of these species is very important in case of helium-oxygen admixture as compared to our recent work [1] carried out on pure oxygen. This is due to high rate of production of these species and low rate of destruction and recombination.

Figure 56.2 illustrates the formation of (O^-) and (O_2^+) ions versus the versus inter-electrode gap distance. They are also important but less than the neutral atoms and molecules.

Figure 56.3, reveals that the formation of the positive (He^+), (He_2^+) and (O^+) are almost of the same order of magnitude which about 10^{-2} less than the (O^-) and (O_2^+) ions. This can be explained by the fact that the electron-impact ionization of oxygen molecules can dominate over helium ionization due to the lower ionization threshold of the oxygen molecule (12.6 eV for O_2 vs. 24.6 eV for He).

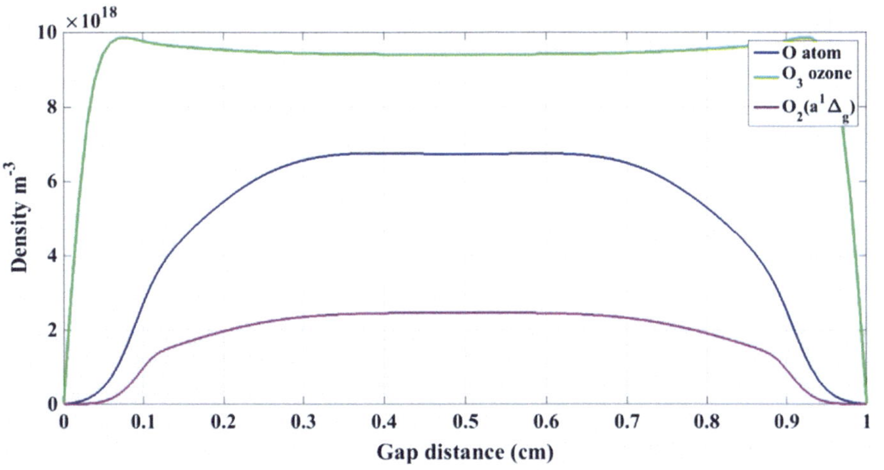

Fig. 56.1 Neutral (O) and (O_3) and metastable singlet delta state $O_2\left(a^1\Delta_g\right)$ molecule density profiles versus inter-electrode gap distance

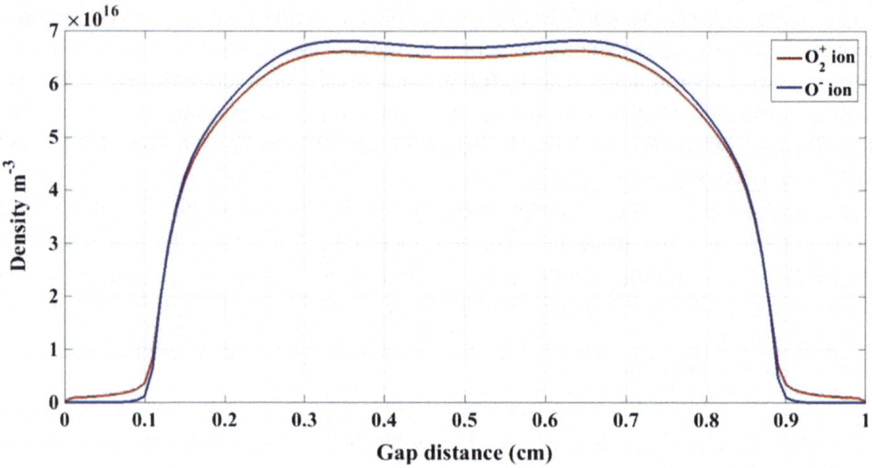

Fig. 56.2 Positive (O_2^+) and negative (O^-) density profiles versus inter-electrode gap distance

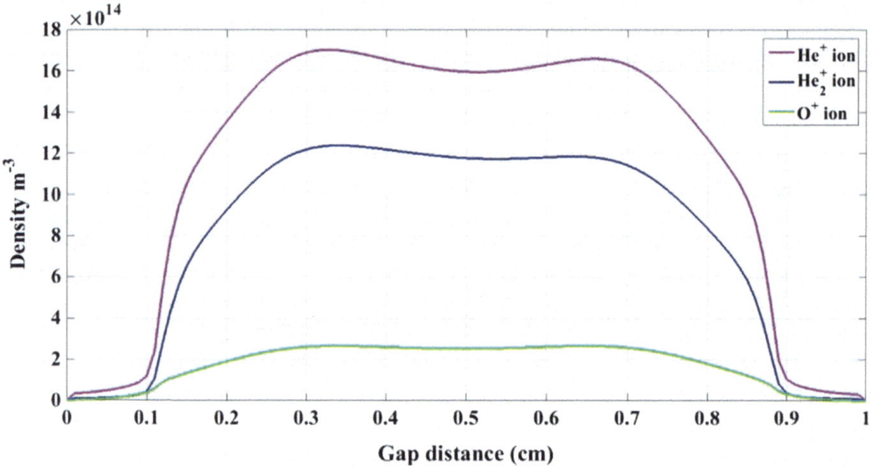

Fig. 56.3 Positive (He^+), (He_2^+) and (O^+) ions density profiles versus inter-electrode gap distance

56.4 Conclusion

In this work, we use a self-consistent one-dimensional fluid model to investigate the species production in a capacitively coupled helium-oxygen discharge. The densities of the active species, (O), (O_3), $O_2(a^1\Delta_g)$, (O^-), (O_2^+), (He^+), (He_2^+) and (O^+), have been determined as a function of the inter-electrode gap distance. We find that the ozone (O_3) is efficiently generated in the bulk of the discharge in case of helium-oxygen admixture as compared to our recent work [2] carried out on pure oxygen. They also show that the formation of the metastable singlet molecule $O_2(a^1\Delta_g)$ and the atomic oxygen (O) are important as compared with pure oxygen plasma and which has a significant influence on the electron heating process. This is due to high rate of production of these species and low rate of destruction and recombination. The rate of production of negative oxygen ion (O^-) and positive (O_2^+) ions are also important but less than the neutral atoms and molecules. The formation of positive (He^+), (He_2^+) and (O^+) are almost of the same order of magnitude which is about 10^{-2} less than the (O^-) and (O_2^+) ions. This can be explained by the fact that the electron-impact ionization of oxygen molecules can dominate over helium ionization due to the lower ionization threshold of the oxygen molecule (12.6 eV for O_2 vs. 24.6 eV for He). Comparison with experimental measurements and simulation investigation from literature for helium oxygen plasma admixture show good agreement. These simulation results yield insight into the dynamics of the helium-oxygen discharge and are used to analyze the basic physical mechanisms governing the formation of some important species and therefore may explain some experimental data.

References

1. Z. Kechidi, A. Tahraoui, Influence of external input parameters on species production in a dual frequency capacitively coupled radio-frequency oxygen plasma. Arab. J. Sci. Eng. **45**, 441 (2019)
2. S. Schröter, A. Wijaikhum, A.R. Gibson, A. West, H.L. Davies, N. Minesi, J. Dedrick, E. Wagenaars, N. De Oliveira, L. Nahon et al., Chemical kinetics in an atmospheric pressure helium plasma containing humidity. Phys. Chem. Chem. Phys. **20**(37), 24263–24286 (2018)
3. Y. Qiu, C. Zhang, Y.J. Hwang, B.L. Bures, M. McCord, The effect of atmospheric pressure helium plasma treatment on the surface and mechanical properties of ultrahigh-modulus polyethylene fibers. J. Adhes. Sci. Technol. **16**(1), 99–107 (2002)
4. M. Thiyagarajan, A. Sarani, C. Nicula, Optical emission spectroscopic diagnostics of a non-thermal atmospheric pressure helium-oxygen plasma jet for biomedical applications. J. Appl. Phys. **113**(23), 233302 (2013)
5. V. Léveillé, S. Coulombe, Atomic oxygen production and exploration of reaction mechanisms in a He-O2 atmospheric pressure glow discharge torch. Plasma Process. Polym. **3**(8), 587–596 (2006)
6. J. He, Y. Zhang, Generation of reactive oxygen species in helium–oxygen radio-frequency discharges at atmospheric pressure. IEEE Trans. Plasma Sci. **41**(10), 2979–2986 (2013)
7. H.M. Joh, J.Y. Choi, S.J. Kim, T.H. Chung, T.-H. Kang, Effect of additive oxygen gas on cellular response of lung cancer cells induced by atmospheric pressure helium plasma jet. Sci. Rep. **4**, 6638 (2014)
8. J.S. Sousa, G. Bauville, B. Lacour, V. Puech, M. Touzeau, J.-L. Ravanat, DNA oxidation by singlet delta oxygen produced by atmospheric pressure microdischarges. Appl. Phys. Lett. **97** (14), 141502 (2010)
9. Y. Inoue, R. Ono, Measurement of singlet delta oxygen in an atmospheric-pressure helium–oxygen plasma jet. J. Phys. D Appl. Phys. **50**(21), 214001 (2017)

10. G.Y. Park, Y.J. Hong, H.W. Lee, J.Y. Sim, J.K. Lee, A global model for the identification of the dominant reactions for atomic oxygen in He/O2 atmospheric-pressure plasmas. Plasma Process. Polym. **7**(3–4), 281–287 (2010)
11. G. Park, H. Lee, G. Kim, J.K. Lee, Global model of He/O2 and Ar/O2 atmospheric pressure glow discharges. Plasma Process. Polym. **5**(6), 569–576 (2008)
12. T. Vijayan, J.G. Patil, High concentration ozone generation in the laboratory for various applications. Int. J. Sci. Technol. Educ. Res. **1**(6), 132–142 (2010)
13. J. Waskoenig, K. Niemi, N. Knake, L.M. Graham, S. Reuter, V. Schulz-von der Gathen, T. Gans, Atomic oxygen formation in a radio-frequency driven micro-atmospheric pressure plasma jet. Plasma Sources Sci. Technol. **19**, 045018 (2010)
14. N. Balcon, *Atmospheric Pressure Radio Frequency Discharges, Diagnostic and Numerical Modeling* (Australian National University, Canberra, 2007)
15. G.J.M. Hagelaar, F.J. de Hoog, G.M.W. Kroesen, Boundary conditions in fluid models of gas discharges. Phys. Rev. E **62**, 1452–1454 (2000)
16. Y. Sakiyama, D.B. Graves, Neutral gas flow and ringshaped emission profile in non-thermal RF-excited plasma needle. Plasma Sources Sci. Technol. **18**(2), 025022 (2019)
17. Comsol.multiphysics (2014), http://www.comsol.com/multiphysics/
18. J.L. Walsh, D.-X. Liu, F. Iza, M.-Z. Rong, M.G. Kong, Contrasting characteristics of sub-microsecond pulsed atmospheric air and atmospheric pressure helium–oxygen glow discharges. J. Phys. D Appl. Phys. **43**(3), 032001 (2010)

Chapter 57
Discharge Parameters Effect on Joule Heating Phenomenon in O_2 DBD for Ozone Generation

Amar Benmoussa and Ahmed Belasri

Abstract The aim of the present work is to investigate the effect of gas heating distribution in the gap of dielectric barrier discharge DBD reactor in pure oxygen gas for ozone production. The fluid model combines the means physical processes in the DBD discharge for ozone generation, and the heat transport equation resolution were used for determining the gas temperature profile. The numerical findings of the model are able to predict the evolution of gas temperature in O_2 DBD reactor. In order to clarify the influence of the operating conditions of the discharge on the gas temperature, we study this instability phenomenon by varying of the applied voltage, the pressure, the frequency, and the pressure to optimize ozone generation. The results obtained from this study show clearly the rise in gas temperature is mainly depends to the high values of deposited power in DBD reactor. The increase of gas heating in the discharge can affects significantly the efficiency of ozone production.

Keywords Ozone production · Joule heating approximation · Modeling · DBD reactor

57.1 Introduction

A dielectric barrier discharge (DBD) was developed essentially for ozone production. In industry, ozone has a number of applications such as medical sterilization, treatment of water, food processing. Many intensive works were carried out in order to optimize the discharge parameters to increase the efficiency of ozone generation [1–6]. In the DBD, up 80% of discharge electrical energy is lost as heat [7, 8]. Although, the gas heating in discharges at atmospheric pressure can be increased with an increasing in energy density of discharge, which leads to a

A. Benmoussa (✉) · A. Belasri
LPPMCA, Faculté de Physique, Université des Sciences et de la Technologie d'Oran Mohamed Boudiaf USTO-MB, 31000 Oran, Algeria
e-mail: benmoussa_am@yahoo.fr

A. Belasri and S. A. Beldjilali (eds.), *ICREEC 2019*, Springer Proceedings in Energy,
https://doi.org/10.1007/978-981-15-5444-5_57

significant decreasing in the efficiency of ozone [9–15]. Due to the lack of published experimental data of gas temperature effect on ozone production, we focus our investigation on the influence of some discharge parameters on gas heating for improving the ozone efficiency [16]. In particular, ozone generation in DBD has shown other advantages such as longtime life and strong oxidization, but the rise in gas heating in the discharge is limited by the ozone dissociation process [17–22].

The gas temperature evolution along gap distance of DBD was determined by from heat transport equation solution at one-dimensional which the effect of the heat transfer to the dielectrics was neglected. We describe in detail the physical approach and heat transport equation in Sect. 57.2. In Sect. 57.3, the results of parametric study of gas heating in oxygen DBD are investigated and discussed.

57.2 Discharge Model

The physical model is a one-dimensional fluid model; this approach was used to execute the present numerical simulation as described in our previous work [23, 24]. The continuity and simplified momentum-transfer equations of positive ion density n_i (x), the electron density n_e (x), and the electric field E(x), are calculated self-consistently for charged particles coupled with Poisson's equation during the discharge pulse.

The system of equations describing the discharge:

For electrons,

$$\frac{\partial n_e}{\partial t} + \frac{\partial \varphi_e}{\partial t} = S_e \tag{57.1}$$

$$\varphi_e = -D_e \frac{\partial n_e}{\partial x} - \mu_e E n_e \tag{57.2}$$

For ions,

$$\frac{\partial n_i}{\partial t} + \frac{\partial \varphi_i}{\partial t} = S_i \tag{57.3}$$

$$\varphi_i = -D_i \frac{\partial n_i}{\partial x} - \mu_i E n_i \tag{57.4}$$

The Poisson's equation:

$$\nabla E = \frac{|e|}{\varepsilon_0}(n_i - n_e) \tag{57.5}$$

Where the indexes i and e refer to ions, and electrons, respectively, n_i and n_e are the ion and electron densities, v_e and v_i are the mean velocities,$\mu_{i,e}$ and $\varphi_{i,e}$ are the mobility, flux (for electron, and for ions).

The gas temperature evolution across the discharge gap of DBD reactor of oxygen (O_2) can be found by numerical resolution of heat transport equation in a one-dimensional approximation. We note also that the gas temperature equation was solved in Joule heating approximation, it's given as follows:

$$\frac{\partial^2 T_g(x)}{\partial x^2} + \frac{P(x)}{k} = 0 \tag{57.6}$$

Where $T_g(x)$, and P(x) are the gas spatial evolution of gas temperature in the gap distance, and the deposited power density, k is the thermal conductivity of pure oxygen.

57.3 Results and Discussion

An important study on the discharge DBD of Oxygen concerns the influence of the discharge parameters on the gas temperature distribution in plasma medium. The parametric investigation was carried out for parameters and operating conditions listed in Table 57.1.

Table 57.1 Operation conditions and discharge parameters

Discharge parameter	Symbol	Value
Gas discharge	O_2	–
Pressure	P	250 torr
Capacity of dielectric (F\cm^2)	C_d	0.23×10^{-9}
Applied voltage (sinusoidal) (V)	V	5000 V
Gap length	D	0.5 cm
Frequency	F	5×10^4 Hz
Thermal conductivity	k	26.4403 mW m^{-1} k^{-1} [25]

57.3.1 Parametric Study of Gas Temperature in O_2 DBD

In this subsection, we discuss the gas temperature peaks in oxygen DBD for different parameters of discharge such gas pressure, frequency and applied voltage as shown in Fig. 57.1(a), Fig. 57.1(b), and Fig. 57.2, respectively. In Fig. 57.1(a) we observe that grows of gas pressure in the discharge have a significant effect on gas temperature. However, an increasing in pressure induced more elastic collisions, which result in oxygen dissociation, this lead to an increasing in gas temperature in the DBD. The influence of the driven frequency on heating effect is also analyzed as displayed in Fig. 57.1(b). It shows clearly that the gas temperature takes important values when the driven frequency in discharge increases. The peak of gas temperature reaches a value of about 360 °K for frequency f = 30 kHz, and 470 °K for f = 60 kHz. We plotted in Fig. 57.2, the effect of sinusoidal applied voltage on temperature of gas.

The gas heating increases in discharge DBD when the amplitude of applied voltage increases. The high values of deposited power in oxygen DBD lead to important rise in gas temperature along gap discharge, and most of energy is dissipated to Joule heating effect. The peak of gas temperature reaches values of 309, 417, 525, and 640 °K for 4, 5, 6, and 7 kV, respectively.

The dependence of gas heating on the applied voltage waveforms and secondary electron coefficient emission (γ) is presented in Fig. 57.3(a) and (b) as a function of the gas pressure. Note here that the value of both rectangular and sinusoidal applied voltage waveforms is equal to 5 kV. The same variation is observed for gas temperature of discharge when coefficient emission (γ) varies from 0,01 and 0,3. The maximum of gas temperature profiles reaches a values of 418 °K for $\gamma = 0.01$, and 439 °K for $\gamma = 0.3$ at pressure of 250 torr with sine applied voltage waveform.

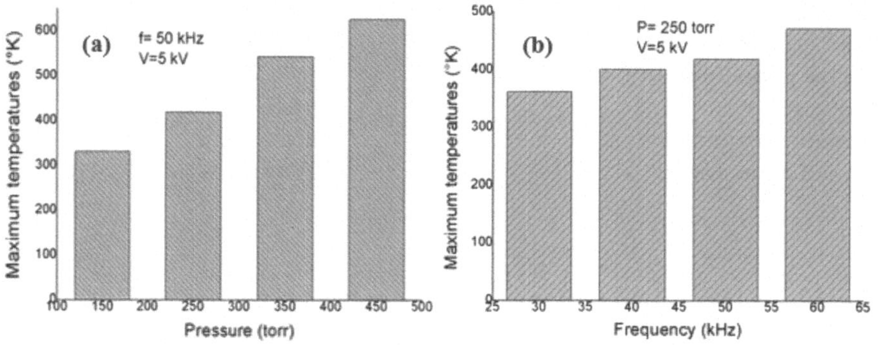

Fig. 57.1 Peaks of gas temperature as a function of: **a** Total gas pressure; **b** Frequency and total gas pressure p = 250 torr in DBD oxygen

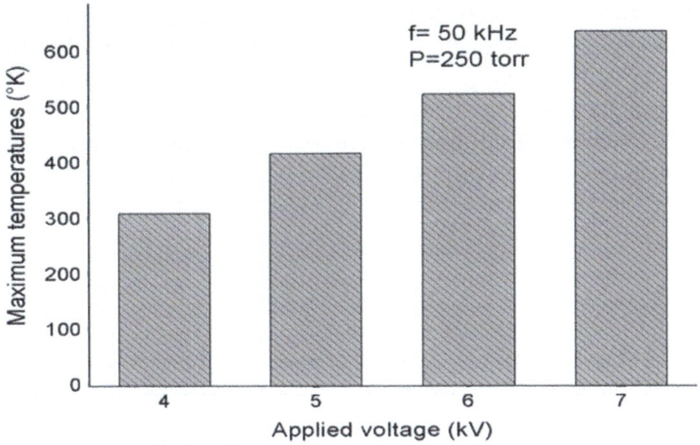

Fig. 57.2 Peaks of gas temperature as a function of frequency f = 50 kHz and total gas pressure p = 250 torr in DBD oxygen

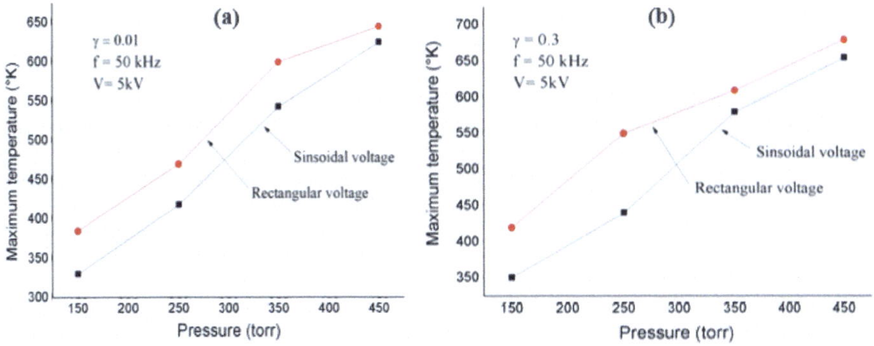

Fig. 57.3 Maximum of gas temperatures as a function of total gas pressure for sinusoidal and rectangular applied voltage for: **a** $\gamma = 0.01$, f = 50 kHz and V = 5 kV; **b** $\gamma = 0.3$, f = 50 kHz and V = 5 kV in DBD oxygen

57.4 Conclusion

This work presents a parametric study of gas heating phenomenon in oxygen DBD for ozone generation. The context of this investigation, based on a one-dimensional fluid coupled with heat transport equation in order to obtain the gas temperature in the DBD discharge of oxygen pure. The high value of gas temperature in the discharge is originating from Joule heating approximation due to the deposited power in dielectric barrier discharge DBD oxygen. The effect of discharge parameters appears clearly on gas heating evolution along the discharge. An increasing in the gas pressure, the frequency, and the applied voltage show a significant increase of gas temperature evolution in oxygen DBD. The effect of waveforms of applied voltage on gas heating is clearly observed. The DBD pulsed with rectangular applied voltage has a large values of gas temperature compared to that pulsed with sinusoidal applied voltage.

Finally, we conclude from these investigations that the choice of the optimum discharge conditions is necessary for effective ozone production.

References

1. M. Šimek, S. Pekàrek, V. Prukner, Influence of power modulation on ozone production using an AC surface dielectric barrier discharge in oxygen. Plasma Chem. Plasma Process. **30**, 607–617 (2010)
2. I.D. Chalmers, R.C. Baird, T. Kelly, Control of an ozone generator-theory and practice. Meas. Sci. Technol. **9**, 983–988 (1998)
3. K. Teranishi, N. Shimomura, S. Suzuki, H. Itoh, Development of dielectric barrier discharge-type ozone generator constructed with piezoelectric transformers: effect of dielectric electrode materials on ozone generation. Plasma Sources Sci. Technol. **18**, 045011 (2009). (10 p.)
4. W. Linsheng, P. Bangfa, L. Ming, Z. Yafang, H. Zhaoji, Dynamic characteristics of positive pulsed dielectric barrier discharge for ozone generation in air. Plasma Sci. Technol. **18**, 147 (2016)
5. N. Mastanaiah, P. Banerjee, J.A. Johnson, S. Roy, Examining the role of ozone in surface plasma sterilization using dielectric barrier discharge (DBD) plasma. Plasma Process. Polym. **10**, 1120–1133 (2013)
6. D.C. Seok, H.Y. Jeong, T. Lho, Y.H. Jeong, DBD parameter for production of high ozone concentration, in *31st ICPIG*, Granada, Spain, 14–19 July 2013
7. W.J.M. Samaranayake, Y. Miyahara, T. Namihira, S. Katsuki, R. Hackamland, H. Akiyama, Ozone production using pulsed dielectric barrier discharge in oxygen. IEEE Trans. Dielectr. Electr. Insul. **7**(6), 849–854 (2000)
8. T.-L. Sung, S. Teii, C.-M. Liu, R.-C. Hsiao, P.-C. Chen, Y.-H. Wu, C.-K. Yang, K. Teii, S. Ono, K. Ebihara, Effect of pulse power characteristics and gas flow rate on ozone production in a cylindrical dielectric barrier discharge ozonizer. Vacuum **90**, 65–69 (2013)
9. V.-I. Gibalov, G.-J. Pietsch, On the performance of ozone generators working with dielectric barrier discharges. Ozone: Sci. Eng. **28**, 119–124 (2006)
10. S. Zhang, W. Gaens, B. Gessel, S. Hofmann, E. Veldhuizen, A. Bogaerts, P. Bruggeman, Spatially resolved ozone densities and gas temperatures in a time modulated RF driven atmospheric pressure plasma jet: an analysis of the production and destruction mechanisms. J. Phys. D: Appl. Phys. **46**, 205202 (2013). (12 p.)
11. M. Benyamina, K. Khodja, F. Ghaleb, A. Belasri, Influence of the gas temperature in ozone production of mixture N2-O2. J. Chem. Chem. Eng. **6**, 391–395 (2012)
12. U. Kogelschatz, B. Eliasson, W. Egli, Dielectric-barrier discharges principle and applications. J. Phys. IV (Colloq.) **7**, 47–66 (1997)
13. K. Ohe, K. Kamiya, T. Kimura, Improvement of ozone yielding rate in atmospheric pressure barrier discharges using a time-modulated power supply. IEEE Trans. Plasma Sci. **27**, 1582–1587 (1999)
14. M. Šimek, M. Člupek, Efficiency of ozone production by pulsed positive corona discharge in synthetic air. J. Phys. D: Appl. Phys. **35**, 1171–1175 (2002)
15. L.-S. Wei, B.-F. Peng, M. Li, Y.-F. Zhang, A numerical study of species and electric field distributions in pulsed DBD in oxygen for ozone generation. Vacuum **125**, 123–132 (2016)

16. Y.-M. Sung, T. Sakoda, Optimum conditions for ozone formation in a micro dielectric barrier discharge. Surf. Coat. Technol. **197**, 148–153 (2005)
17. B. Eliasson, M. Hirth, U. Kogelschatz, Ozone synthesis from oxygen in dielectric barrier discharges. J. Phys. D Appl. Phys. **20**, 1421–1437 (1987)
18. T. Murata, Y. Okita, M. Noguchi, I. Takase, Basic parameters of coplanar discharge ozone generator. Ozone: Sci. Eng. **26**, 429–442 (2004)
19. S. Jodzis, J. Petryk, Gas temperature in an ozonizer. The computer modeling of an actual discharge system. IEEE Trans. Plasma Sci. **39**(11), 2020–2021 (2011)
20. S. Boonduang, P. Limsuwan, Effect of generating heat on ozone generation in dielectric cylinder-cylinder DBD ozone generator. Energy Power Eng. **5**, 523–527 (2013)
21. J. Kitayama, M. Kuzumoto, Theoretical and experimental study on ozone generation characteristics of an oxygen-fed ozone generator in silent discharge. J. Phys. D: Appl. Phys. **30**, 2453–2461 (1997)
22. K. Yanallah, F. Pontiga, A. Fernandez-Rueda, A. Castellanos, A. Belasri, Ozone generation by negative corona discharge: the effect of joule heating. J. Phys. D: Appl. Phys. **41**, 195206 (2008)
23. A. Benmoussa, A. Belasri, F. Ghaleb, Z. Harrache, Gas heating phenomenon in rare gas dielectric barrier discharge for excimer lamps. IEEE Trans. Plasma Sci. **42**(3), 706–711 (2014)
24. A. Benmoussa, A. Belasri, Z. Harrache, Numerical investigation of gas heating effect in dielectric barrier discharge for Ne-Xe excilamp. Curr. Appl. Phys. **17**, 479–483 (2017)
25. E.W. Lemmon, R.T. Jacobsen, Viscosity and thermal conductivity equations for nitrogen, oxygen, argon, and air. Int. J. Thermo Phys. **25**(1), 21–69 (2004)

Chapter 58
Effect of the Variation of the Electrode Geometrical Configuration on the Electric Wind Velocity Produced by an Electric Corona Discharge

M. Bouadi, K. Yanallah, M. R. Bouazza, and F. Pontiga

Abstract In this work, we determine precisely the electric wind velocity, produced by a direct current (DC) corona discharge in air, using three electrode geometrical configurations: 'wire-to-plate' (a), 'two wires-to-plate' (b) and 'three wires-to-plate' (c). Each electrode wire is subjected to the same high positive voltage while the plate is grounded. The electric wind velocity is determined through a mathematical model based on the resolution of Navier-Stokes equation, in which a source term consisting in the electro-hydrodynamic (EHD) force, already established by our group in the form of a simplified analytical expression, is used. The results found allow to compare the profile of the electric wind produced by the corona discharge for the three electrode geometrical configurations ((a), (b) and (c)).

Keywords (DC) Corona discharge · Plasma · Electric wind · Force electro-hydrodynamic (EHD)

58.1 Introduction

A typical corona reactor usually consists of a pair of electrodes, one of which (the corona or stressed electrode) has a sharp curvature, while the other electrode is less curved. The electric corona discharge takes place when a high potential difference is applied between these two electrodes, creating a strong electric field around the sharp electrode, and causing the gas breakdown. However, the ionization remains generally confined to the vicinity of the stressed electrode, since the intensity of the electric field decreases very rapidly away from this electrode. As a result, the

M. Bouadi (✉) · K. Yanallah · M. R. Bouazza
Dpt. de Physique, Faculté des Sciences de la Matière,
Laboratoire de Génie Electrique et des Plasmas, University of Tiaret, Tiaret, Algeria
e-mail: bouadi.univp12@gmail.com

F. Pontiga
Departamento de Física Aplicada II, Universidad de Sevilla, Seville, Spain

A. Belasri and S. A. Beldjilali (eds.), *ICREEC 2019*, Springer Proceedings in Energy,
https://doi.org/10.1007/978-981-15-5444-5_58

inter-electrode space can be divided into two zones: the ionization region and the drift region. In the ionization region, ions having the same polarity as the corona electrode are generated and injected into the drift region, where they drift under the action of the electric field, toward the opposite electrode. As the ions move between the electrodes, part of their momentum is transferred to the neutral gas molecules through 'ion-molecule' collisions. Thus, a macroscopic gas flow appears with velocities up to several meters per second. This gas flow is generally called 'electric wind', 'ionic wind' or even 'electro-hydrodynamic (EHD) flow'. In addition to electric corona discharges [1–5], this EHD gas flow is also specific to other types of electrical discharges, including dielectric barrier discharges [4, 6].

The existence of the ionic wind has been known for a long time [7]. However, it is still a subject of investigation because of its intrinsic complexity and its industrial interest. The ionic wind has an aspect of paramount importance which consists in the generation of an airflow without using mechanical parts. Various applications of the ionic wind have focused on the cooling of electronic components [8, 9], the boundary layer removal in aeronautics, where the main interest is to reduce the aerodynamic air friction force and, thus, to improve the aerodynamic properties of vehicles and/or the lift force [10]; electro-hydrodynamic micro-pumps [11] and the chemical and biological decontaminations in volume and on the surface (gas depollution, sterilization, and bio-films treatment) [12, 13].

After the first empirical model of the ionic wind proposed by Robinson [14], which links the velocity of the wind to the square root of the electric current, Béquin et al. [1] and Lacoste et al. [15] established similar models, and determined the spatial distribution of the ionic wind velocity for 'point-to-plane' positive and negative corona discharge. Kawamoto et al. [16] performed a two-dimensional numerical computation of the positive corona discharge, and then determined the ionic wind velocity. Similarly, El-Khabiry and Colver [17] described the two-dimensional (2D) dielectric barrier discharge in the stationary state. Loiseau et al. [18] have introduced a term of unsteady electrical force in the Navier-Stokes equation. Many other numerical studies of the ionic wind generated by corona discharges have been performed [19, 20], which have practical uses [21, 22].

It can be concluded from the analysis of these studies that many models require high computation time and pose serious problems of convergence in some cases, hence the need to develop additional approximations.

In the present work, the ionic wind produced by a direct current positive corona discharge in air will be modelled for the following electrode configurations: 'wire-to-plate' (a), 'two wires-to-plate' (b) and 'three wires-to-plate' (c). The objective is to study the effect of these electrode configurations on the structure of the ionic wind. The modelling consists in dividing the calculation tasks into two parts. The first part concerns the calculation of the electro-hydrodynamic force, since it is responsible for the generation of the electric wind and its structure in the inter-electrode space. For this, we rely on a simplified analytical expression proposed by Yanallah and al [23, 24], which allows the computation of the spatial distribution of the EHD force without any numerical effort. Finally, the second part consists in the resolution of the Navier-Stokes equation using the values of the EHD force obtained in the first part. The completion of the computational tasks leads to the precise determination of the ionic wind velocities for the three geometric configurations previously mentioned (a, b and c).

58.2 Mathematical Model

58.2.1 The Electro-Hydrodynamic (EHD) Force

The generation of the ionic wind is due to the presence of the EHD force acting on the gas. The charged particles generated in the electric corona discharge are accelerated by the electric field and gain kinetic energy and, therefore, momentum. Then, during collisions with neutral molecules of the gas, these charged particles transmit part of their momentum to these molecules, which induces a hydrodynamic gas flow. According to Boeuf and Pitchford's approach [25], the EHD force acting on the gas can be expressed as follows:

$$\mathbf{F} = e_0\left(N_p - N_n - N_e\right)\mathbf{E} \tag{58.1}$$

where e_0 is the elementary electric charge, the subscript (p) designates the positive ions, (n) the negative ions and (e) the electrons, N_i (i = e, p or n) is the charged particle density of the species i and $\mathbf{E}$ is the electric field. Using the expressions of the charged species flux, $\mathbf{J}_\mathrm{i} = \mu_i N_i \mathbf{E}$, where μ_i is the mobility, the EHD force takes the form:

$$\mathbf{F} = e_0\left(\frac{\mathbf{J}_p}{\mu_p} - \frac{\mathbf{J}_n}{\mu_n} - \frac{\mathbf{J}_e}{\mu_e}\right) \tag{58.2}$$

In the positive corona discharge, the ionization is confined within a thin layer around the stressed electrode. Inside this layer, the electron density is higher than the other charged species densities [26, 27]. Most of the momentum transfer from the charged species to the neutral molecules, however, takes place in the drift region, where the positive ion density is dominant. Therefore, the density of the EHD force takes the expression:

$$\mathbf{F} \approx e_0 N_p \mathbf{E} = e_0 \frac{\mathbf{J}_p}{\mu_p} \tag{58.3}$$

As a result of this approximation, the density of the EHD force can finally be written as follows [23]:

$$F_{EHD} = \frac{1}{\mu_p} j(0) \left[\frac{(d + r_0)(1 - \cos\sigma)}{\sqrt{(a \sin\sigma)^2 + (d + r_0)^2(1 - \cos\sigma)^2}} \right]^{9/2} \frac{\cosh\tau - \cos\sigma}{1 - \cos\sigma} \tag{58.4}$$

where μ_p is the electric mobility of positive ions, d the inter-electrode distance and r_0 the wire radius. σ and τ are the bipolar coordinates of any point in the fluid [28]. The value $j(0)$ is determined from the total electric current i [24].

58.2.2 Hydrodynamic Model of the Gas Flow

When the high voltage is applied to the electrodes, the plasma is created, and an interaction between this plasma and the neutral gas takes place. The gas movement or the EHD flow can be described by the Navier-Stokes and continuity equations, which represent the conservation of momentum and mass density. Here we assume that the flow is Newtonian and the air gas is incompressible. The EHD flow induced by the corona discharge will be in general turbulent [19, 29]. Thus, the usual time-averaging approach for the momentum and mass density conservation equations (called the 'Reynolds-Averaged Navier-Stokes' (RANS)) will be used in this work. Furthermore, neglecting the gravity and considering that the regime is stationary, these equations can be written in the following form:

$$\rho(\mathbf{V} \cdot \nabla)\mathbf{V} = -\nabla P + \lambda \nabla^2 \mathbf{V} + \nabla \cdot \tau_R + \mathbf{F} \tag{58.5}$$

$$\nabla \cdot \mathbf{V} = 0 \tag{58.6}$$

where $\mathbf{V}$ is the average gas velocity, λ the air dynamic viscosity, P the gas pressure, τ_R the Reynolds tensor and $\mathbf{F}$ the EHD force formulated by the analytic expression (58.4). These equations will be closed using the standard k-epsilon turbulence model [30]. The equations are integrated on each control volume in order to construct algebraic equations whose variables are scalar quantities, such as velocity, pressure, etc.

The electric wind produced within the 'wire(s)-to-plate' corona discharge reactor shown in Fig. 58.1 is the target to be modelled. In this reactor the EHD flow is confined between two identical parallel plates of length l and the distance between them is h.

The wires are located between the two plates, in the middle, and are arranged parallel to them, but perpendicular to the air flow. The distance between each wire, with radius r_0, and the grounded lower plate is d. Note that the wires (cases b and c) are parallel to each other with an inter-wire distance f. Finally, each wire is subjected to the same high positive voltage.

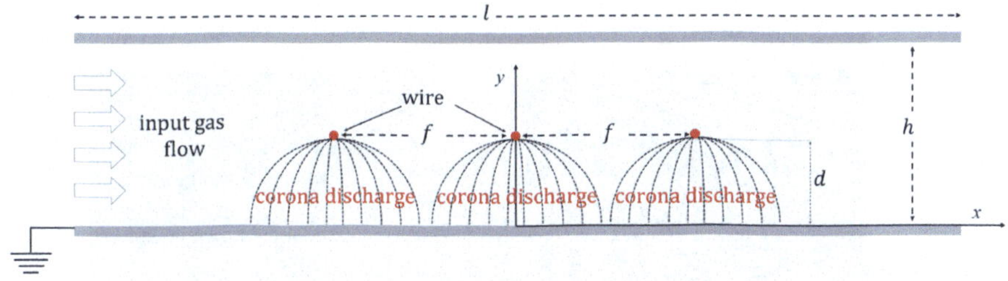

Fig. 58.1 Schematic illustration of the corona reactor

58.2.3 The Boundary Conditions

The resolution of the Navier-Stokes equation requires the definition of the boundary conditions for the velocity field: the no-slip condition is supposed to be satisfied at the internal surfaces of the two plates, and at the surface of each wire. The boundary condition at the gas inlet ($x = -7.5$ cm) assumes a unifom gas velocity V_0 and, at the gas outlet ($x = +7.5$ cm), the pressure is considered to be equal to the atmospheric pressure.

58.3 Results and Discussion

It is assumed that the problem is 2D and the length of the reactor channel is large compared with the inter-electrode space. The main geometrical parameters considered for the modelling are shown in Fig. 58.1.

In this work, the voltage applied to each wire is equal to $\phi = 10$ kV, which corresponds to an electric current intensity per unit of wire length of 24.70 μA/cm [31, 32]. The wire radius is $r_0 = 100$ μm, the inter-electrode distance is $d = 0.75$ cm, and the inter-wire distance is $f = 1$ cm. Finally, the length of the channel is $l = 15$ cm and the gas inlet velocity is taken equal to $V_0 = 2$ cm/s.

58.3.1 The EHD Force

Figure 58.2 shows the 2D spatial distribution of the magnitude of the EHD force. The inter-wire distance is large enough to neglect the interaction between the EHD forces of adjacent wires (cases b and c). Thus, since the EHD force has the same intensity around each wire for the three cases (a, b and c), Fig. 58.2 presents only the case (a). This figure reveals that the EHD force magnitude is very important around the wire. For this reason, just a small part of the inter-electrode space, close to the wire, is represented.

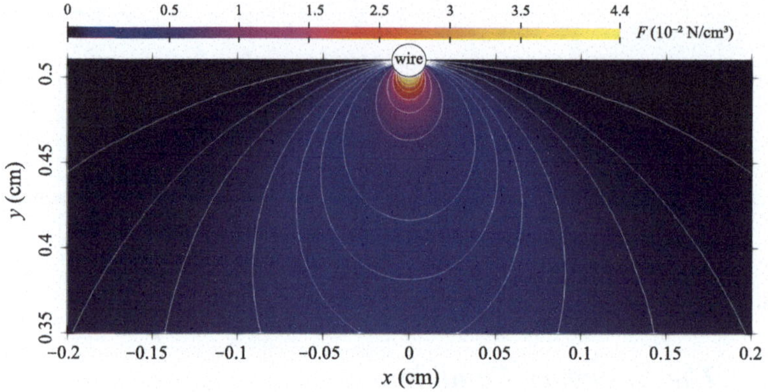

Fig. 58.2 2D-spatial distribution of the EHD force magnitude in the vicinity of the wire for r_0 = 100 μm, d = 0.75 cm and ϕ = +10 kV

During the corona discharge, the electrons produced via the ionization process in ionization region drift towards the wire(s) and the positive ions, produced in this area as well, move towards the plate. However, in the positive drift region the electric field is rather weak and therefore the gas ionization is negligible. So, this last region is electrically passive because it does not allow the formation of electronic avalanches. On the other hand, the production of negative ions via the attachment process is insignificant, as it is known when the corona discharge polarity is positive. Consequently, the contribution of negative ions in the EHD force is far less important. The inter-electrode space is then occupied, almost exclusively, by the positive ions. As a result, the total EHD force is generated by these positive ions: $F \approx F_p$ and it is, constantly, directed from the wire(s) to the plate [28].

58.3.2 The Gas Velocity

The profiles of the spatial distributions of the air velocity corresponding to the three electrode configurations (a, b and c) are presented in Fig. 58.3.

Figure 58.3 shows that, in the (case a), the maximum velocity is about 3.9 m/s, just above the ground plate. The air is accelerated by the EHD force towards the lower plate, where it is forced to change its direction. This leads to the formation of two large vortices, symmetrical with respect to the wire, which extend until reaching the two plates. The centre of each of these two vortices is located at the coordinates (x = ±1.028 cm, y = 0.75 cm). These high velocity values have been

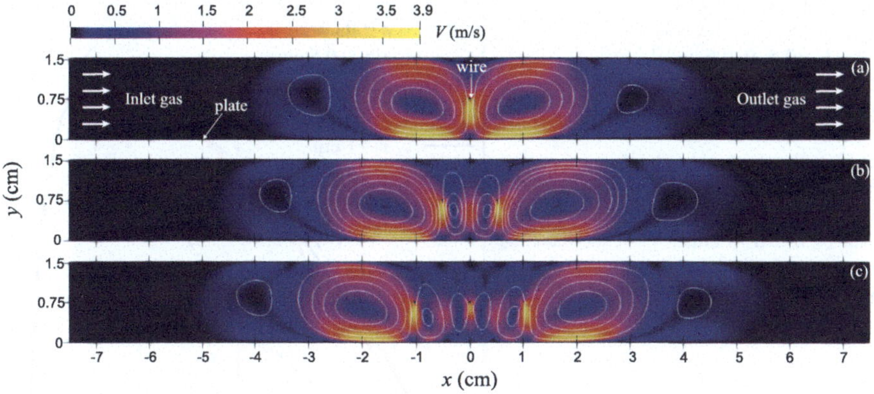

Fig. 58.3 2D-spatial distribution of the gas velocity magnitude corresponding to three electrode configurations: 'wire-to-plate' (a), 'two wires-to-plate' (b) and 'three wires-to-plate' (c); and for wire radius r_0 = 100 μm, inter-electrode distance d = 0.75 cm, inter-wire distance f = 1 cm and applied voltage ϕ = 10 kV

experimentally confirmed for different electrode configurations of the corona discharge [2]. In the other two cases (b and c), the general velocity profile is almost the same as in case (a). However, in (case b), two small vortices appear between the two big ones, which are confined between the two wires. This increases the distance between the centres of the two large vortices by 2 cm. Finally, in case (c), two additional vortices, smaller than those of case (b), are interposed in the same way, which almost doubles the separation distance between the two large vortices relative to case (a). Thus, the centre of each large vortex has moved a distance of the order of the inter-electrode space with the addition of each new wire.

Figure 58.4 illustrates the profile of the gas velocity along the y-axis for the three electrode configurations. The maximum values of the velocity always appear at a distance close to the wire(s). The general profile of the velocity, in case (b) and (c), is the same for the four positions of the wires ($x = -1, -0.5, 0.5$ and 1 cm), and the average maximum velocity is about 3.2 m/s. In case (c), $x = 0$, the gas velocity decreases by about 0.4 m/s with respect to case (a). Finally, the velocity gradient after the maximum velocity, in these two last cases, is slightly different. This is due to the formation of the small vortices which alter the air recirculation. The inset in Fig. 58.4 confirms, for all three cases, Robinson's relationship indicating that the maximum ionic wind velocity is proportional to the square root of the intensity of the electric current.

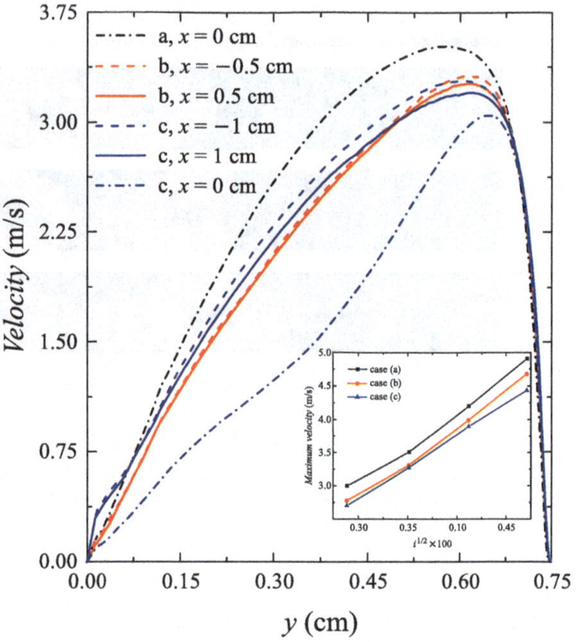

Fig. 58.4 Gas velocity magnitude along the y-axis at $x = -1, -0.5, 0, 0.5$ and 1 cm for the three electrode configurations: 'wire-to-plate' (a), 'two wires-to-plate' (b) and 'three wires-to-plate' (c); and for wire radius $r_0 = 100$ µm, inter-electrode distance $d = 0.75$ cm, inter-wire distance $f = 1$ cm and applied voltage $\phi = 10$ kV

58.4 Conclusion

The objective of the present work was to examine the effect of three electrode configurations (case (a), (b) and (c)) on the profile of the electric wind velocity produced by the positive corona discharge in the stationary regime. A simplified analytical expression of the EHD force was used, which allowed us, unlike many numerical models found in the literature, to compute the electric wind velocity with less numerical effort and a high accuracy. The injection of this EHD force, as an input parameter into the Navier-Stokes equation, has made it possible the determination of the 2D spatial distribution of the ionic wind velocity. As a main result, the general profile of the electric wind velocity remains globally the same for the three electrode configurations. Nevertheless, the distance between the two large vortices (case a) rises when adding more wires (b and c), due to the interposition of two small vortices (b) and four small vortices (c). Along the vertical planes containing the wires, the maximum velocities are located, in all cases, close to the wires. However, there is a slight decrease in these maximum velocities in cases (b) and (c) caused by the appearance of the small vortices that hinder the gas recirculation.

References

1. P. Béquin, K. Castor, J. Scholten, Electric wind characterisation in negative point-to-plane corona discharges in air. Eur. Phys. J. - Appl. Phys. **22**, 41–49 (2003)
2. D.F. Colas, A. Ferret, D.Z. Pai, D.A. Lacoste, C.O. Laux, Ionic wind generation by a wire-cylinder-plate corona discharge in air at atmospheric pressure. J. Appl. Phys. **108**, 103306 (2010)
3. C. Kim, D. Park, K.C. Noh, J. Hwang, Velocity and energy conversion efficiency characteristics of ionic wind generator in a multistage configuration. J. Electrost. **68**, 36–41 (2010)
4. E. Moreau, C. Louste, G. Touchard, Electric wind induced by sliding discharge in air at atmospheric pressure. J. Electrost. **66**, 107–114 (2008)
5. M. Rickard, D. Dunn-Rankin, F. Weinberg, F. Carleton, Characterization of ionic wind velocity. J. Electrost. **63**, 711–716 (2005)
6. J. Jolibois, N. Zouzou, E. Moreau, J.M. Tatibouët, Generation of surface DBD on rough dielectric: electrical properties, discharge-induced electric wind and generated chemical species. J. Electrost. **69**, 522–528 (2011)
7. M. Robinson, A history of the electric wind. Am. J. Phys. **30**, 366–372 (1962)
8. B. Kim, S. Lee, Y.S. Lee, K.H. Kang, Ion wind generation and the application to cooling. J. Electrost. **70**, 438–444 (2012)
9. M. Knap, J. Duga, T.C. Lui, Ionic wind generator on LED lighting application, in *20th International Workshop on Thermal Investigations of ICs and Systems* (2014), pp. 1–5. https://doi.org/10.1109/therminic.2014.6972502
10. L. Léger, E. Moreau, G. Artana, G. Touchard, Influence of a DC corona discharge on the airflow along an inclined flat plate. J. Electrost. **51–52**, 300–306 (2001)
11. E.D. Fylladitakis, M.P. Theodoridis, A.X. Moronis, Review on the history, research, and applications of electrohydrodynamics. IEEE Trans. Plasma Sci. **42**, 358–375 (2014)
12. E. Timmermann, F. Prehn, M. Schmidt, H. Höft, R. Brandenburg, M. Kettlitz, Indoor air purification by dielectric barrier discharge combined with ionic wind: physical and microbiological investigations. J. Phys. D: Appl. Phys. **51**, 164003 (2018)
13. T.P. Ryan, K.R. Stalder, Overview of current applications in plasma medicine. Proc. SPIE **10066**, 1006606 (2017). https://doi.org/10.1117/12.2255792
14. M. Robinson, Movement of air in the electric wind of the corona discharge. Trans. Am. Inst. Electr. Eng. Part Commun. Electron. **80**, 143–150 (1961)
15. D. Lacoste, D. Pai, C. Laux, Ion wind effects in a positive DC corona discharge in atmospheric pressure air, in *42nd AIAA Aerospace Sciences Meeting and Exhibit* (2004), p. 354
16. H. Kawamoto, H. Yasuda, S. Umezu, Flow distribution and pressure of air due to ionic wind in pin-to-plate corona discharge system. J. Electrostat. **64**, 400–407 (2006)
17. S. El-Khabiry, G.M. Colver, Drag reduction by dc corona discharge along an electrically conductive flat plate for small reynolds number flow. Phys. Fluids **9**, 587–599 (1997)
18. J.F. Loiseau, J. Batina, F. Noël, R. Peyrous, Hydrodynamical simulation of the electric wind generated by successive streamers in a point-to-plane reactor. J. Phys. Appl. Phys. **35**, 1020 (2002)
19. L. Zhao, K. Adamiak, Effects of EHD and external airflows on electric corona discharge in point-plane/mesh configurations. IEEE Trans. Ind. Appl. **45**, 16–21 (2009)
20. L. Zhao, K. Adamiak, Numerical simulation of the effect of EHD flow on corona discharge in compressed air. IEEE Trans. Ind. Appl. **49**, 298–304 (2013)
21. R.S. Islamov, An analytical model of the ionic wind in a regular ultracorona. J. Phys. Appl. Phys. **46**, 375204 (2013)
22. M. Molki, P. Damronglerd, Electrohydrodynamic enhancement of heat transfer for developing air flow in square ducts. Heat Transf. Eng. **27**, 35–45 (2006)

23. K. Yanallah, F. Pontiga, M.R. Bouazza, J.H. Chen, The effect of the electric wind on the spatial distribution of chemical species in the positive corona discharge. J. Phys. Appl. Phys. **50**, 335203 (2017)
24. M.R. Bouazza, K. Yanallah, F. Pontiga, J.H. Chen, A simplified formulation of wire-plate corona discharge in air: application to the ion wind simulation. J. Electrost. **92**, 54–65 (2018)
25. J.P. Boeuf, L.C. Pitchford, Electrohydrodynamic force and aerodynamic flow acceleration in surface dielectric barrier discharge. J. Appl. Phys. **97**, 103307 (2005)
26. J. Chen, J.H. Davidson, Electron density and energy distributions in the positive DC corona: interpretation for corona-enhanced chemical reactions. Plasma Chem. Plasma Process. **22**, 199–224 (2002)
27. K. Yanallah, F. Pontiga, A. Fernández-Rueda, A. Castellanos, Experimental investigation and numerical modelling of positive corona discharge: ozone generation. J. Phys. Appl. Phys. **42**, 065202 (2009)
28. K. Yanallah, F. Pontiga, J.H. Chen, A semi-analytical study of positive corona discharge in wire–plane electrode configuration. J. Phys. Appl. Phys. **46**, 345202 (2013)
29. S. Ould Ahmedou, M. Havet, Effect of process parameters on the EHD airflow. J. Electrost. **67**, 222–227 (2009)
30. H.K. Versteeg, W. Malalasekera, *An Introduction to Computational Fluid Dynamics: The Finite Volume Method* (Pearson Education, London, 2007)
31. W. Deutsch, Über die dichteverteilung unipolarer ionenströme. Ann. Phys. **408**, 588–612 (1933)
32. R.S. Sigmond, The unipolar corona space charge flow problem. J. Electrost. **18**, 249–272 (1986)

Chapter 59
Calculation of the Paschen Curve by Solving of the Self-sustaining Condition for Different Mixtures

Fatiha Ghaleb, Soumia Bendella, Wafà Benstaali, and Ahmed Belasri

Abstract The Electrical breakdown in a gas is the process of transition from the quasi-insulating state to a conducting state when a sufficiently intense electric field is applied. The value of the voltage associated with this transition is called breakdown voltage it corresponds to the first value of the voltage for which the discharge becomes autonomous or self-sustaining. The understanding of the breakdown process is necessary for the development of plasma applications. Our work consists in determining the breakdown voltage as a function of the inter-electrode distance pressure product which represents the Paschen curve in a plane geometry by an analytical resolution of the self-sustaining condition coupled to the equation of the Townsend first coefficient and the electric field. The results obtained from this resolution allowed us to describe and verify and of course to test the validity of our results with experimental measurements studied by several authors. This work was the object of analysis of the effect of certain parameters which are qualitatively in agreement with that of the experience.

Keywords Paschen curve · Self-sustaining condition · Mixture gases · Secondary emission

59.1 Introduction

The application of an electric field to a gas environment can result in complete or partial breakdown resulting in a conductive medium consisting of electrons and free ions. This conductive medium is generally referred to as plasma. Several factors such as field strength, gas type and pressure, gas impurities, initial electron number

F. Ghaleb (✉) · S. Bendella · A. Belasri
Laboratory of Plasma Physics, Conducting Materials and their Applications, Faculty of Physics, USTOMB, Oran, Algeria
e-mail: ghaleb.fatiha@gmail.com

W. Benstaali
University of Abdelhamid Ibn Badiss, Mostaganem, Algeria

A. Belasri and S. A. Beldjilali (eds.), *ICREEC 2019*, Springer Proceedings in Energy,
https://doi.org/10.1007/978-981-15-5444-5_59

and density, as well as electrode material and geometry influence the probability of the breakdown occurrence [1]. The process of transforming a neutral gas into a conductive state in a self-sustaining discharge is known as “breakdown” or “plasma ignition”. This phenomenon is related to several processes such as ionization of atoms by electron impact, displacing charged particles in an electric field, charge multiplication and secondary electron production at the cathode by ion impact [2]. Townsend’s breakdown theory describes how an effect of these microscopic processes can result in the ignition of a discharge between parallel electrodes in a low and high pressure environment [3]. For this purpose, we develop an analytical model to describe the electrical behavior of the breakdown for different gas mixtures.

59.2 Paschen Law

Indeed, the Paschen law, which represents the breakdown voltage of a pure gases in homogeneous field leaving obviously the condition of self-sustained Eq. (59.1) [4, 5], and of the expression of the first Townsend coefficient indicated by Eq. (59.2) [6, 7].

$$M = \exp(\alpha d) = 1 + 1/\gamma \tag{59.1}$$

$$\alpha/p = D_1 \exp\left[-D_2(p/E)^{0.5}\right] \tag{59.2}$$

Where d is the electrode separation, M is electronic multiplication, p is the gas pressure. This equation represents the general formula of the breakdown voltage of pure gases.

Since α is the number of ionizing collisions per unit length, E is electric field, the values of D_1 and D_2 positive constants, for the analytical model of the first Townsend ionization coefficient for noble gases, given in [8, 9].

In general, α depend also only on the electric field at this point and can be deduced by experimental measurements from Huxley and Crompton [10] or by numerical calculations from Ségur [11].

To have a comparison enters the various analytical results, one could calculate the breakdown voltage, by the relation of the electronic multiplication [12], according to the ionization coefficient [13].

The calculation principle of the breakdown voltage for a gas mixture is to determine the minimum voltage that must be applied between two electrodes, so that the average density of the charged particles present in the geometry reaches a charge density greater than its value initial. This criterion, although arbitrary, corresponds in fact to determining the moment when the condition of self-sustaining of the following equation is realized

$$M = \exp[(\alpha_1 + \alpha_2)d] = 1 + (\alpha_1 + \alpha_2)/(\alpha_1\gamma_1 + \alpha_2\gamma_2) \tag{59.3}$$

α_1 and α_2 are the ionization coefficient for first gas and second gas, the secondary emission is independent of the reduced electric field E/p, is defined by γ_1, γ_2 are the secondary Townsend coefficients (or secondary emission coefficient) for gas1, gas 2.

The expression of the electric field is [14]

$$E = V/d \tag{59.4}$$

V is breakdown voltage, in plane geometry, where we replace the ionization coefficient Eq. (59.2), we can deduce the breakdown voltage for a mixture of two gases at different percentages [15].

59.3 Results and Discussion

In this part we have tried to plot the Paschen curve for two mixtures of Ar10%-Ne and 50% argon in neon, by two different methods, the variation of the electronic multiplication and the secondary emission coefficient.

Figure 59.1, shows the variation of the breakdown voltage for a mixture Ar10%-Ne, obtained by resolution of self-sustaining equation in a plane geometry defined by parallel electrodes. These curves are compared with the experimental results of Capdeville [16]. For the analytical results we took into consideration two methods for determining the Paschen; the variation of the multiplication M of 10^3, 10^4 and 10^5 corresponds to the curves indicated by symbols square, circle and star respectively, the second method consists in calculating the Paschen voltage for a secondary emission coefficient equal to $\gamma_{Ar} = 3 \times 10^{-3}$ and $\gamma_{Ne} = 9 \times 10^{-2}$ for argon and neon. We note that the experimental calculation of Capdeville [15] is close to the Paschen curve for $M = 10^3$ at low product P.d, but at high product P.d the breakdown voltage is very high, we also see that the Paschen minimum is shifted towards low pressure-distance products.

Figure 59.2, represents the calculation of the Paschen curve as a function of the pressure-distance product for a mixture of 50% of the argon in the neon, compared with experimental measurements of Capdeville [16], the latter used pure iron electrodes. We distinguish more than the percentage of argon the higher the Paschen minimum moves to the left (low values of products P.d). It is noted that for this type of mixture, the curve of the breakdown voltage calculated and measured follows in a very similar way the theoretical form of the Paschen's law, a rapid decrease in the breakdown voltage to reach the Paschen minimum followed by growth when the product pressure distance increases. Similarly, for values of P.d greater than 2 torr.cm, it is clear that the addition of neon in the mixture reduces the breakdown voltage.

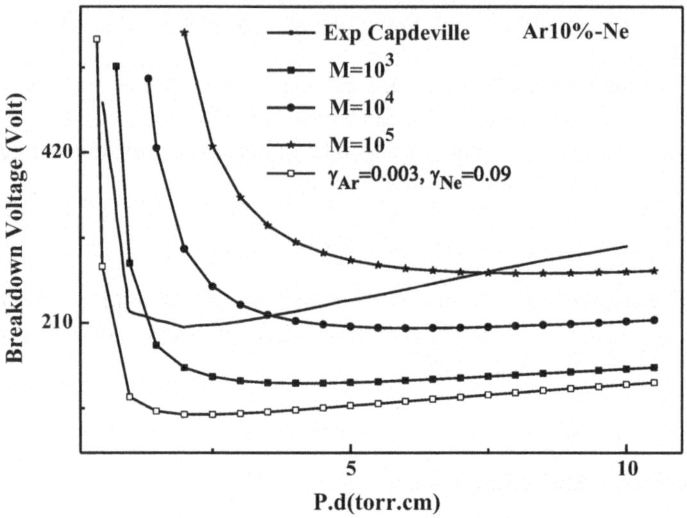

Fig. 59.1 Paschen voltage for Ar10%-Ne, for different multiplications, a secondary coefficient $\gamma_{Ar} = 3 \times 10^{-3}$ and $\gamma_{Ne} = 9 \times 10^{-2}$ for Ar, Ne, compared with experimental results [16]

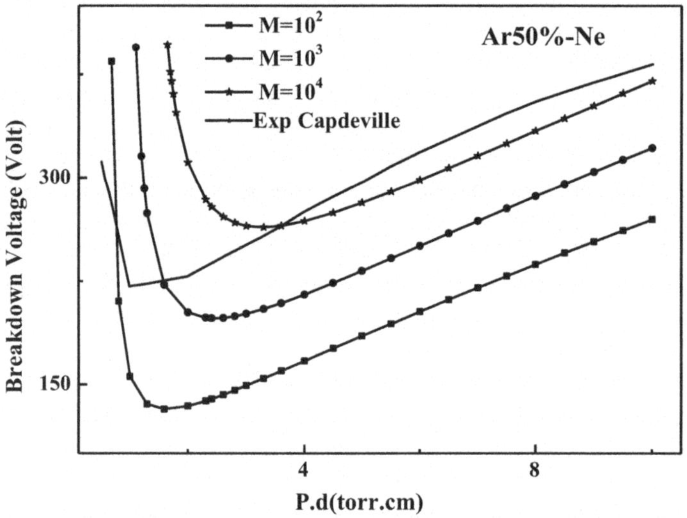

Fig. 59.2 Breakdown voltage as a function of product P.d, for a mixture of Ar50%-Ne, for different multiplications, compared with experimental results [16]

In Fig. 59.3, we presented the breakdown voltage as a function of the pressure product to the gas mixture and the inter-electrode distance, for a mixture of 50% xenon in helium. This mixture was used by Postel [17] in the micro-discharge, for the case of brass electrodes (a copper and zinc alloy, yellow in color, known for its

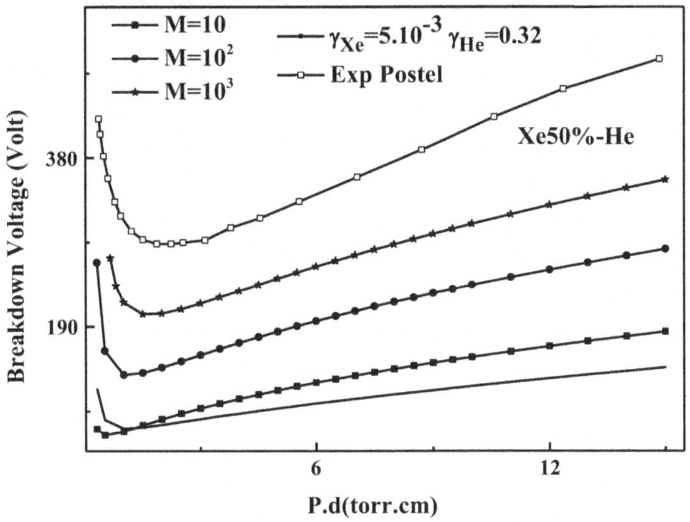

Fig. 59.3 Calculation of the Paschen curve as a function of the product P.d, for a mixture Xe50%-He, for different electronic multiplication, compared with experimental results [17]

ductility), these data are indicated in the figure by a square symbol curve (empty inside). We used in our analytical calculation which is represented by the curve without symbol, a secondary emission coefficient of xenon is equal to 5.10^{-3} and 0.32 for the helium, these values of the coefficients corresponding to the copper. For the illustrated curves by square symbols, circle and star, we took an electronic multiplication which is equal to 10, 10^2 and 10^3 respectively. From this comparison we can see that all the curves follow the same theoretical look of Paschen's law. Moreover, the curve of the Postel experiment [17] predicts higher voltages compared to theoretical estimates; this can be explained by the loss of electrons in the experiment, and the poor knowledge of the secondary emission coefficient of the brass material.

In the next two figures, (see Figs. 59.4 and 59.5), we took a mixture of argon and nitrogen. We calculated the Paschen voltage as a function of the pressure-distance product by the analytical approach taking into account the variation of the electronic multiplication, M = 10, 102 and 103, indicated by square, circle and star symbols respectively. The other analytical calculation gives the breakdown voltage for the case of a secondary emission coefficient equal to 0.03 for argon and 0.018, for the nitrogen gas, corresponding to the material of the magnesium oxide MgO electrode. (A conductive layer of magnesium oxide, its role is to lower the ignition voltages) [18]. these results are compared with work from the Klas experiment [19]. The results illustrate a Paschen minimum in the variable voltage range between 150 V and 180 V.

In Fig. 59.6 we determine the Paschen curve for a mixture of 50% argon in hydrogen. The results of our calculation are compared with that of Klas experiment,

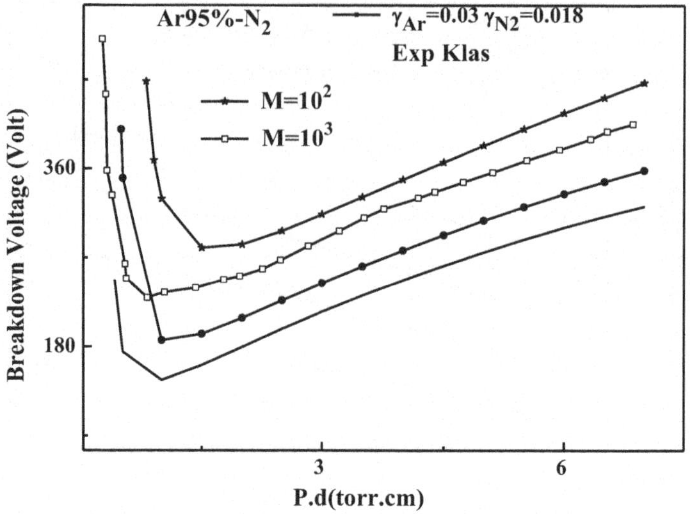

Fig. 59.4 Influence of the breakdown voltage as a function of the product P.d (in the unit of torr.cm) for the two mixtures Ar95%-N_2, compared to the Klas experiment calculation [19]

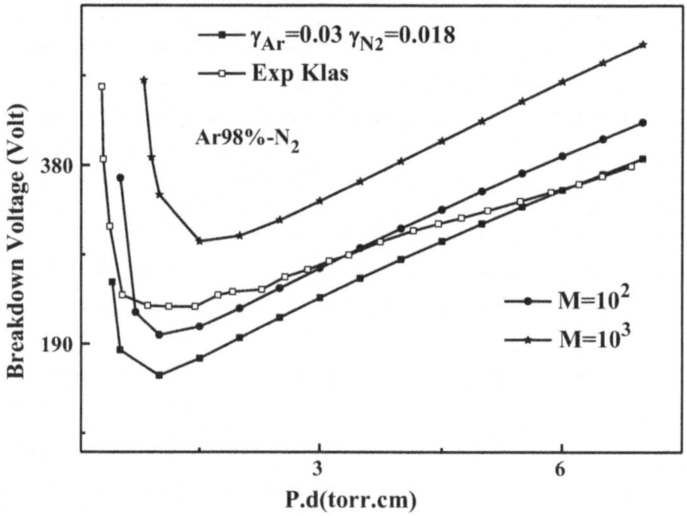

Fig. 59.5 Paschen voltage as a function of the product P.d, for an electronic multiplication varies from 10 to 10^3, in Ar98%-N_2 mixture, compared to the experimental results of Klas [19]

indicated by square symbols. The multiplication curves presented by solid circle and triangle symbols indicate the results where the electronic multiplication equals 102 and 103 [13] respectively. In the gamma curve, the secondary emission coefficient (second coefficient of Townsend) has been introduced, is indicated by a

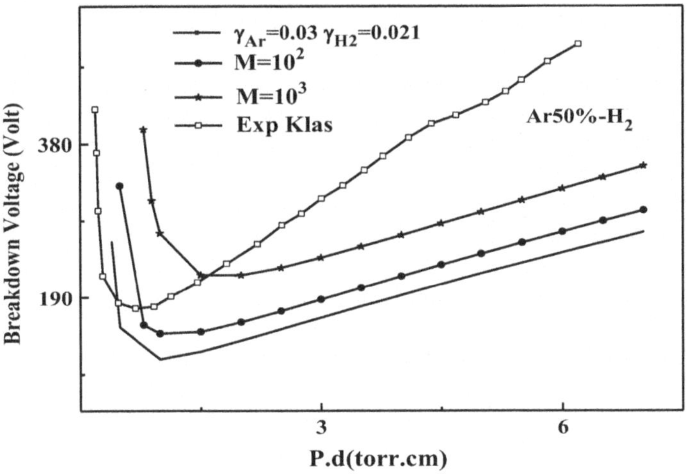

Fig. 59.6 Paschen curve for a secondary emission coefficient equal to 0.03 for Ar, 0.021 for H_2, for different multiplications, in mixture Ar50%-H_2, compared with Klas experience [19]

line, the secondary emission coefficient of argon and hydrogen are equal to 0.03 and 0.021. We note that although there are two different methods of solving the self-sustaining equation, the curves have the same pace, it is observed that the breakdown voltage for the results of experimental measurements of Klas [19], have the same pace. At low product P.d, the voltages drop, whereas they believe with the increase of this product. Regarding the Paschen minimum, it moves towards the low pressure-distance products.

59.4 Conclusion

The work that has just been reported in this paper is to provide a computational tool to predict electrical breakdown behavior and the Paschen curve for gas mixtures in different electrode configurations as well as different electronic multiplication in a uniform field, and a plane-plane geometry, these results allowed us to study the resolution of the condition of self-maintenance in a gas mixture for different percentage. The systematic comparison of the results of the analytical model with the experimental results, illustrates the role of secondary emission coefficient which is represented by the type of materials of the electrodes. It should be noted that according to the results found, the orders of magnitude of the breakdown voltage values obtained by the planar geometry are closer to the values predicted by the experiment.

References

1. J.M. Meek, J.D. Craggs, *Electrical Breakdown of Gases* (Wiley, New York, 1978)
2. F. Ghaleb, W. Benstâali, A. Belasri, Calculation of breakdown voltage in plasma display panels. J. Mater. Sci. Eng. C **28**, 791–794 (2008)
3. W.J.M. Brok, Modelling of transient phenomena in gas discharges. Ph.D. thesis, Technische Eindhoven, 2005
4. G. Auday, P. Guillot, J. Galy, H. Brunet, Experimental study of the effective secondary emission coefficient for rare gases and copper electrodes. J. Appl. Phys. **83**(11), 5917–5921 (1998)
5. M. Abdel-Salam, H. Anis, M. Dekker, High-Voltage Engineering. Technology & Industrial Arts, 2001
6. J.J. Mankowski, High voltage subnanosecond dielectric breakdown. Ph.D. thesis, Texas Tech University, Lubbock, 1997
7. D.A. Diver, *A Plasma Formulary for Physics, Technology and Astrophysics* (Wiley-VCH Verlag GmbH, Berlin, 2001)
8. J.R. Roth, *Industrial Plasma Engineering* (Institute of Physics Publishing, Bristol, 1995)
9. M.A. Hassouba, F.F. Elakshar, A.A. Garamoon, Measurements of the breakdown potentials for different cathode materials in the townsend discharge. Fizika A **11**(2), 81–90 (2002)
10. L.G.H. Huxley, R.W. Crompton, *The Diffusion and Drift of Electrons in Gases* (Wiley Interscience, New York, 1974)
11. P. Segur, M. Yousfi, J.P. Boeuf, in *Electrical Breakdown and Discharges in Gases*, ed. by E. E. Kunhardt, H.L. Luessen. NATO Advanced Study Institute, Series B: Physics, vol. 89a (Plenum, New York, 1983), p. 331
12. M. Levinshtein, J. Kostamovaara, S. Vainshtein, Breakdown phenomena in semiconductors and semiconductor devices. Int. J. High Speed Electron Syst. (IJHSES) **14**(4), 921–1114 (2005)
13. A.L. Ward, Calculations of cathode-fall characteristics. J. Appl. Phys. **33**(9), 2789–2794 (1962)
14. J.J. Cuomo, S.M. Rossnagel, W.D. Westwood, Handbook of Plasma Processing Technology (Editeur William Andrew Inc, USA, 1990)
15. F. Ghaleb, A. Belasri, Radiat. Eff. Defects Solids **167**(6), 377–383 (2012)
16. H. Capdeville, Mesures et Modélisation des Décharges Luminescentes a Basses Pressions: Comportement Electrique et Pulvérisation des Electrodes. Ph.D. thesis, Orléans, 2007
17. O.B. Postel, M.A. Cappelli, Vacuum emission and breakdown characteristics of a planar He–Xe Microdischarge. Appl. Phys. Lett. **76**(5), 544–546 (2000)
18. A. Benmoussa, A. Belasri, F. Ghaleb, Z. Harrache, Gas heating phenomenon in rare gas dielectric barrier discharge for excimer lamps. IEEE Trans. Plasma Sci. **42**(3), 706–711 (2014)
19. M. Klas, Š. Matejcik, B. Radjenovic, P. Papp, M.R. Radjenovic, The breakdown voltage characteristics, the effective secondary emission coefficient and the ionization coefficient of the argon-based mixtures. Nuclear Instrum. Methods Phys. Res. B **279**, 100–102 (2012)

Chapter 60
Recent Development in Diagnostic of Electronegative Plasmas Through Laser Pulse Induced Photo-Detachment

N. Oudini and A. Bendib

Abstract Laser pulse induced photo-detachment is an indirect measurement technique devoted for the estimation of plasma electronegativity, i.e. ratio of negative ion density in proportion to electron density. The technique combines the use of a laser pulse and a positively biased Langmuir probe. The laser pulse converts negative ion into electron-atom pair, while the Langmuir probe records the increase of electron saturation current due to the electron density increase induced by negative ion photo-detachment. The negative ion proportion is thereafter deduced from the recorded laser photo-detachment signal which is nothing but the temporal evolution of probe current measured from the instant when the laser pulse is fired until the probe current recovers to its nominal value (prior the laser pulse). As in any indirect measurement technique, one needs a model describing the relationship between observables (electron saturation current) and investigated quantities (electronegative plasma properties). Such model relies on a set of assumptions that are subject of discussion in the present review.

Keywords Plasma diagnostics · Negative ions · Laser pulse induced photo-detachment

60.1 Introduction

Laser pulse induced photo-detachment (LPD) has been introduced in 1979 by Bacal and Hamilton [1] to estimate the plasma electronegativity, i.e. ratio of negative ion density in proportion to electron density, in hydrogen discharge. The technique,

N. Oudini (✉)
Laboratoire des plasmas de décharges, Centre de Développement des Technologies Avancées, Cité du 20 Aout BP 17 Baba Hassen, 16081 Algiers, Algeria
e-mail: noureddine.oudini@gmail.com

A. Bendib
Laboratoire d'Electronique Quantique, Faculté de Physique, USTHB, El Alia BP 32 Bab Ezzouar, 16111 Algiers, Algeria

A. Belasri and S. A. Beldjilali (eds.), *ICREEC 2019*, Springer Proceedings in Energy,
https://doi.org/10.1007/978-981-15-5444-5_60

described schematically in Fig. 60.1a, combines the use of a laser pulse and a positively biased Langmuir probe. The laser pulse converts negative ion into electron-atom pair, while the Langmuir probe, biased positively, records the increase of electron saturation current due to the electron density increase induced by negative ion photo-detachment. The plasma electronegativity is estimated from the recorded laser photo-detachment signal. This signal, an example is illustrated in Fig. 60.1b, consists of the temporal evolution of probe current that is recorded from the instant when the laser pulse is fired until the probe current recovers to its nominal value (prior the laser pulse).

The analysis of laser photo-detachment signal relies on an analytical model that describes the relationship between plasma properties and the measured signal. For instance, Bacal and Hamilton assume [1] that the electron produced from negative ion photo-detachment thermalizes instantaneously with electron produced from other collisional process (background electron) in order to estimate plasma electronegativity. Unfortunately, this assumption is not verified and numerical simulations [2, 3] demonstrate that the aforesaid thermalization involves a time scale that might be longer than the duration of the measured signal. A more complex analytical model have been developed and verified through numerical experiments to better describe the relationship between the measured signal, the plasma properties and the laser pulse wavelength [4]. This model improves significantly the precision of the measured plasma electronegativity and extends the capability of the technique for measuring electron temperature and plasma density [4]. Experimental investigation [5] confirms that the laser photo-detachment signal is significantly affected by the laser wavelength as predicted in the previous numerical model [4]. Despite the recent progress in the understanding of the parameters affecting the aforesaid signal, the effects of a parasitical current, attributed by Kajita et al. [6] to

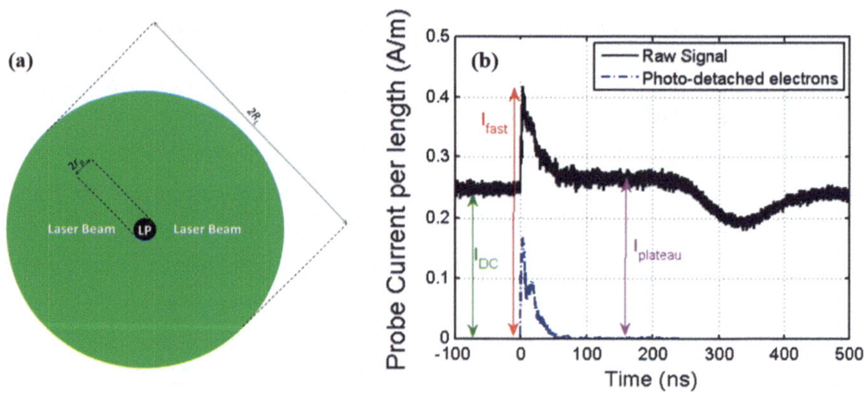

Fig. 60.1 **a** Schematic description of Laser pulse induced photo-detachment experiment where LP denotes the Langmuir probe. **b** Example of photo-detachment signal simulated for hydrogen plasma with $\alpha = 1$, positive ion density $10^{16}\ m^{-3}$, laser-beam radius $R_L = 3\ mm$, probe radius $r_p = 0.2\ mm$, photo-detachment frequency $f_{ph} = 5 \times 10^9\ s^{-1}$, laser photon energy $hv = 1.165\ eV$ and pulse duration $\tau = 5\ ns$ and a probe is biased at $+1\ V$ above the plasma potential

probe tip surface ablation, on the measured signal are neglected in all analytical models devoted to the analysis of laser photo-detachment signal. Interpreting the measured signal while neglecting the aforesaid parasitical current might lead to erroneous measurements. To surpass this limitation, Kajita et al. introduced [7], in 2005, a variant technique, so-called "eclipse photo-detachment", which involves a blocking wire to screen the Langmuir probe avoiding probe tip ablation. In this variant technique, the effects of the blocking wire use is assumed marginal on the measured signal. However, recent numerical investigation shows that the laser photo-detachment signal is distorted by the use of blocking wire [8]. Another alternative [9] consists on biasing the probe negatively such as to repel electron, including parasitical ones, and collect positive ions. In this case, the plasma properties are deduced from the change of positive ion Bohm velocity, sheath width and edge to center ratio. In this work, we present an overview of the recent progress on laser pulse induced photo-detachment and highlight the limits and prospects inherent to this technique.

60.2 Bacal and Hamilton Analytical Model and Its Limits

To analyze photo-detachment signals, Bacal and Hamilton [1] propose the following set of assumptions:

1. the probe collects thermal electron current;
2. the rise of electron saturation current is due exclusively to the creation of photo-detached electrons;
3. background electrons are assumed being not affected by the laser pulse and by phenomena induced by this pulse;
4. photo-detached electrons thermalize instantaneously with the background electrons, or at least this thermalization requires a time very small than the photo-detachment signal rise time (Fig. 60.1b).

Using these assumptions, one can show that plasma electronegativity can be estimated through

$$\alpha = \frac{I_{plateau} - I_{DC}}{I_{DC}} \tag{60.1}$$

where $I_{plateau}$ is probe current during plateau phase and I_{DC} is probe current prior laser pulse (see Fig. 60.1b).

It is important to mention, from one side, that Eq. (60.1) does not involve the probe current component I_{fast}. From the other side, some of assumptions proposed by Bacal and Hamilton [1] are motivated by simplicity reasons. Therefore, the validity of these latter might be questioned. For instance, it have been said by different authors [2, 10, 11] that a potential barrier, i.e. increase of plasma potential around the LP, forms after the laser pulse. This potential barrier affects the properties of background electron and therefore the assumption (iii) is not realistic. Indeed, Particle-In-Cell simulation [3] shows that background electron properties are significantly affected by the aforesaid potential barrier.

Figure 60.2 shows the spatiotemporal distribution of plasma potential (panel a), background electron density (panel b) and background electron temperature (panel c) after negative ion photo-detachment. Furthermore, the instantaneous thermalization of photo-detached and background electrons cannot be explained by any mechanism at low pressure plasmas. Indeed, Particle-In-Cell with Monte Carlo Collisions simulation [2] shows that the electron energy distribution function (EEDF) relaxes within a time duration that is not marginal in comparison to the photo-detachement signal duration. Indeed, photo-detached electron keeps mono-energetic characteristic of their EEDF, while the background EEDF tail extends due to Fermi acceleration enhanced by the propagation of the potential barrier inside and outside the illuminated region [2]. Figure 60.3 shows the EEDF calculated for different radial position ranges at t = 0.1 ns and t = 50 ns.

In view of results illustrated in Figs. 60.2 and 60.3, we consider that the use of Eq. (60.1) to diagnose plasma electronegativity might lead to erroneous estimation.

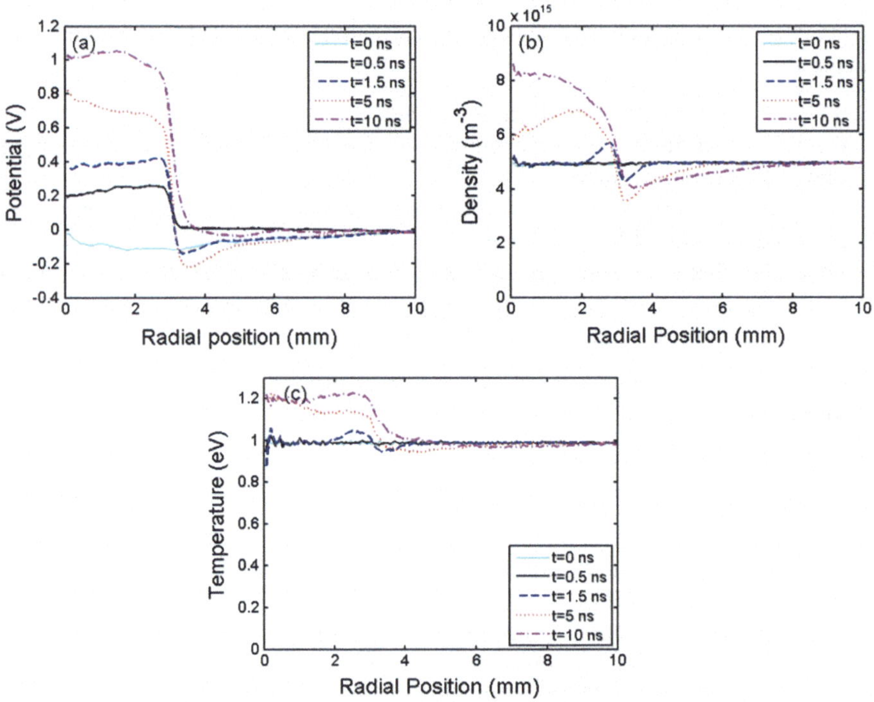

Fig. 60.2 Example of spatiotemporal distribution of plasma properties after negative ion photo-detachment for the conditions of Fig. 60.1b and a laser photon of 2.33 eV. **a** plasma potential, **b** background electron density and **c** background electron temperature

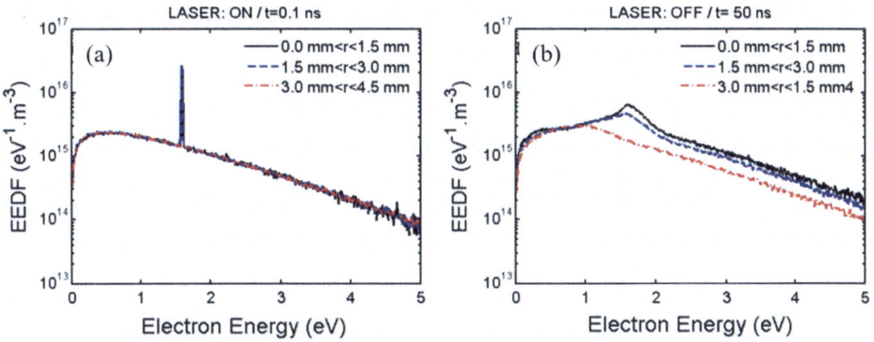

Fig. 60.3 Electron Energy Distribution Function (EEDF) calculated within different volume delimitated by the radius $r = 0$ mm, 1.5 mm, 3.0 mm and 4.5 mm for conditions summarized in Fig. 60.1. **a** $t = 0.5$ ns and **b** $t = 50$ ns

60.3 New Analytical Model and Its Limits

El Balghiti-Sube et al. [11] were first to describe ideal photo-detachment signal through the three component described in Fig. 60.1b. Both El Balghiti-Sube et al. [11] and Oudini et al. [4] agreed that the I_{fast}-I_{DC} components represent the contribution of non-thermal photo-detached electrons to probe current. However, these two latter works diverge on the interpretation of $I_{plateau}$. El Balghiti-Sube et al. [11] consider that the decrease of probe current from I_{fast} to $I_{plateau}$ is due to the depletion of photo-detached electron enhanced by losses on the probe, while background electrons are assumed unperturbed. In contrast, Particle-In-Cell simulation [4] demonstrate that, during the plateau phase, the contribution of photo-detached electron to probe current drops to zero, while the formed potential barrier focuses background electron flux entering the illuminated region towards the LP increasing the contribution of this electron group. Knowing these information, one can show that the different components of photo-detachment signal depend on several quantities, namely: laser wavelength $h\nu$, negative ion electron affinity E_A, background electron temperature T_e, the strength of the abovementioned potential barrier $\Delta\varphi$, the laser beam radius R_L and the probe radius r_p. The aforesaid relationship is summarized in Eq. (60.2).

$$\begin{cases} \exp\left(-\frac{e\Delta\varphi}{k_B T_e}\right) + \frac{\alpha}{\sqrt{\pi}}\sqrt{\frac{e(h\nu - E_A)}{k_B T_e}}\left(1 - \sqrt{\frac{e\Delta\varphi}{(h\nu - E_A)}}\right) = 1 \\ \frac{I_{fast} - I_{DC}}{I_{DC}} = \frac{\alpha}{\sqrt{\pi}}\sqrt{\frac{e(h\nu - E_A)}{k_B T_e}} \\ \frac{I_{plateau}}{I_{DC}} = \frac{R_L}{\sqrt{r_p^2 + R_L^2}}\left[\exp\left(\frac{e\Delta\varphi}{k_B T_e}\right) erfc\left(\sqrt{\frac{e\Delta\varphi}{k_B T_e}}\right) + 2\sqrt{\frac{e\Delta\varphi}{\pi k_B T_e}}\right] \end{cases} \tag{60.2}$$

Where e is the elementary charge and k_B is Boltzman constant.

Solving the set of Eq. (60.2) allows an accurate estimation of, amongst other, plasma electronegativity.

Despite the advances in the understanding of photo-detachment signal, it is important to mention that Eq. (60.2) neglects completely the presence of electron parasitical current that is attributed in literature to probe surface ablation [6, 7]. This parasitical current superimpose to the photo-detachment signal components, namely I_{fast} and $I_{plateau}$. Thus, these two components are measured with a non-marginal error and the use of Eq. (60.2), while neglecting the parasitical current, might leads, in practice, to erroneous measurement of plasma electronegativity. To overcome this limitation two variant techniques have been introduced. Oudini et al. [8] propose to bias the probe negatively such to collects positive ions and repels electrons, including the parasitical ones, from the probe. In this variant technique, the plasma properties such as plasma electronegativity and anisothermicity, i.e. ratio of electron temperature over negative ion temperature, are estimated on the basis of Bohm velocity, sheath width and edge to center ratio changes enhanced by the negative ion photo-detachment. Kajita et al. introduced a variant technique that adds a blocking wire to the experimental apparatus [7]. This blocking wire is placed such as to protect the probe tip from laser photons avoiding ablation. The recorded photo-detachment signal, in this variant technique, is assumed unperturbed by the presence of the blocking wire. However, recent numerical investigation shows that the presence of the blocking wire leads to the formation of a potential well, i.e. a potential pattern that impedes electron transport towards the probe, which distorts the photo-detachment signal [8].

60.4 Summary and Prospects

In view of what have been mentioned above, the comprehension of laser pulse induced photo-detachment combined with Langmuir probing gone through several evolutions. These evolutions improve the accuracy of the technique and extend it for measuring other plasma properties, such anisothermicity, in addition to plasma electronegativity. Despites these progresses an important issue persists on the use of this technique. Indeed, negative ion sources use a magnetic filter to enhance the volume negative ion production and/or impede electron transport to limit the co-extracted electron current. However, the effects of magnetic field on photo-detachment signal is completely neglected and not yet studied. Our future work will address this issue.

References

1. M. Bacal, G.W. Hamilton, Phys. Rev. Lett. **42**, 1538 (1979)
2. N. Oudini, F. Taccogna, A. Bendib, A. Aanesland, Phys. Plasmas **21**, 063515 (2014)
3. N. Oudini, N. Sirse, R. Benallal, F. Taccogna, A. Aanesland, A. Bendib, A.R. Ellingboe, Phys. Plasmas **22**, 073509 (2015)
4. N. Oudini, N. Sirse, F. Taccogna, A.R. Ellingboe, A. Bendib, Appl. Phys. Lett. **109**, 124101 (2016)
5. N. Sirse, N. Oudini, A. Bendib, A.R. Ellingboe, Plasma Sources Sci. Technol. **25**, 04LT01 (2016)
6. S. Kajita, S. Kado, A. Okamoto, S. Tanaka, Jpn. J. Appl. Phys. **44**, 8661 (2005)
7. S. Kajita, S. Kado, S. Tanaka, Plasma Sources Sci. Technol. **14**, 566 (2005)
8. N. Oudini, Plasma Sources Sci. Technol. **28**, 065016 (2019)
9. N. Oudini, N. Sirse, F. Taccogna, A.R. Ellingboe, A. Bendib, Phys. Plasmas **25**, 053510 (2018)
10. P. Devynck, J. Auvray, M. Bacal, P. Berlemont, J. Bruneteau, R. Leroy, R.A. Stern, Rev. Sci. Instrum. **60**, 2873 (1989)
11. F. El Balghiti-Sube, F.G. Baksht, M. Bacal, Rev. Sci. Instrum. **67**, 2221 (1996)

Chapter 61
Evaluation of the Efficiency of an Argon Coaxial Lamp

F. N. Haddou, Ph. Guillot, A. Belasri, T. Maho, and B. Caillier

Abstract This paper presents the experiments achieved in the DPHE laboratory in collaboration with the LPPMCA laboratory for the development of mercury-free fluorescent lamp concurrent light. Mercury emits UVs photons at a wavelength of 254 nm, but due to its high toxicity, mercury (Hg) has been classified by the World Health Organization (WHO) as one of the top ten chemicals posing a major risk to public health [1]. It is thus necessary to develop new free mercury UV light sources for decontamination of wastewater and water sterilization processing. Dielectric barrier discharge (DBD) interest to "lighting designers" who, for environmental considerations (energy saving and mercury disposal), would like to replace the mercury of discharge lamps [2, 3]. The excitation of rare gases by DBD provides a beneficial arrangement for the generation of radiation in the VUV, UV and IR range. The scope of this study concerns the domain of low-pressure, low-power fluorescent lamps. We focused on the electrical characteristic, optical emission spectroscopy of plasma produced in a dielectric barrier discharge by a high pulse generator (3 kV–100 kHz). The first purpose of these measurements is to analyze the operating range of the plasma lamp and optimize it in order to improve efficiencies.

Keywords DBD · Excimer lamp · Efficiency · Low pressure · Emission spectrum

---

F. N. Haddou (✉) · A. Belasri
Laboratoire de Physique des Plasmas, Matériaux Conducteurs et leurs Applications LPPMCA, Université des Sciences et de la Technologie d'Oran - Mohamed Boudiaf (USTO-MB), Oran, Algeria
e-mail: fatimanadia.hd@gmail.com

F. N. Haddou · Ph. Guillot · T. Maho · B. Caillier
Laboratoire Diagnostics des Plasmas Hors Équilibre (DPHE), Université de Toulouse, INU Champollion, Albi, France

A. Belasri and S. A. Beldjilali (eds.), *ICREEC 2019*, Springer Proceedings in Energy,
https://doi.org/10.1007/978-981-15-5444-5_61

61.1 Introduction

The objective of this exploratory study is to demonstrate that a light source with coaxial geometry and external electrodes can offer an undeniable advantage for the production of UV photons. The generation of VUV and UV radiation is always a requirement, especially for photocatalytic applications using phosphors. They are able to convert infrared radiation to visible [10] or the visible radiation into UV-C range [11, 12]. which is very interesting in the case of no intense UV photons are emitted.

Our lamp is a coaxial lamp with external electrodes that can be filled with a selected pressure of argon. This inert and cheap gas with industrial prospects, has already proven its worth in waste treatment [4].

This parametric study focuses on evaluating the luminous efficiency, maximal intensity and power of the DBD as a function of amplitude voltage to various parameters: pressure (p), inter-electrode distance (d) and frequencies. The measured optical spectra are used to determine the species responsible for UV emissions. Optimal operating conditions are thus determined.

The Sect. 61.1 provides the electrical and optical analyses, the Sect. 61.2 gives the information about the evolution of plasma between the inter electrode spacing followed by discussions of the results. The conclusions and the future prospects are discussed in Sect. 61.3.

61.2 Materials and Methods

61.2.1 Experimental Section

The cylindrical lamp made of Pyrex which is a dielectric material, has a height 120 mm and a diameter of 10 mm. Dielectric barrier discharge is initiated by the application of a pulse voltage between 600 V and 1000 V at a frequency in the range of 10–20 kHz. The metal electrodes (aluminum) are placed on the outer surface of the lamp. The UV-Visible spectra of the emitted light plasma and the corresponding luminous intensities were recorded by an optical fiber cable connected to a UV-Visible spectrometer Ocean Optics HR 2000+. The figures below show the current and voltage signals following the application of a square signal. Research has therefore focused on the study of the performance of lighted signs excited in electric pulse mode [5, 6]. In addition to that electrodes can be easily changed to study different electrode configurations (i.e. dimensions, position along the glass tube).

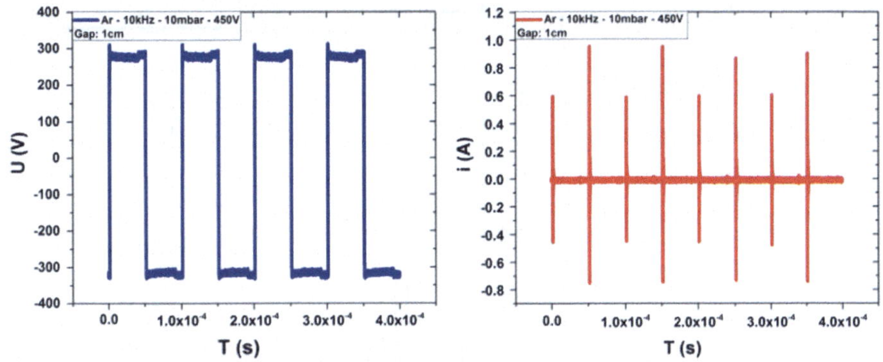

Fig. 61.1 Voltage measurement (left), current measurement (right), both as a function of time

Figure 61.1 shows the electrical voltage and current measurements taken from a HT pulse generator. A Pearson current probe acquires the current of the dielectric barrier discharge lamp and the voltage is acquired with a Tektronix P6015A probe. Studies have shown, in particular, that the luminous efficiency of the discharge increases when using a pulsed power supply [7, 8].

Based on electrical voltage and current measurements, the mean power delivered by the supply are respectively: where i the current of the discharge, u the applied voltage and T the period.

$$<P> = \frac{1}{T}\int_0^T u * idT \tag{61.1}$$

61.2.2 Plasma Composition

The spectrum below shows the active species in the argon-based lamp. A series of lines attributed to neutral argon, ranging from 696.5 to 950 nm originated from radiative transitions from the 3p54p excited levels (energies between 12.91 and 13.48 eV) to the 3p54s metastable and resonant levels (energies between 11.55 to 11.83 eV), from 300 nm to 400 nm for the presence of nitrogen and Hα at 656 nm, OH = 308 nm.

At 20 kHz, the spectrum shows a significant increase in nitrogen and this is caused by a strong ionization of the molecular states of nitrogen and an energy transfer from argon resonance states to a nitrogen and hydrogen [9].

The presence of impurities leads to a transfer of energy from the studied gas to these foreign bodies, thus affecting emission spectra and luminescence declines (Fig. 61.2).

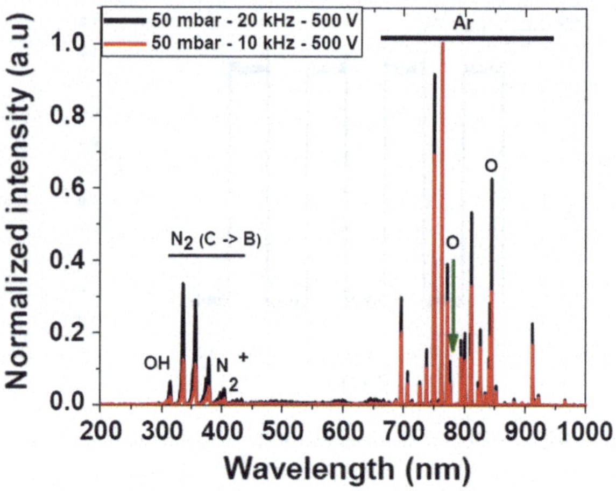

Fig. 61.2 The spectrum obtained at frequencies of 10–20 kHz respectively at different pressure and voltage values

61.3 Results and Discussions

The characterization of argon plasma by optical emission spectroscopy allowed to distinguish two very intense lines at 750.4 and 811.5 nm. These emissions correspond to transitions between the 3p5 4p and 3p5 4 s levels visible on the argon energy diagram below: from 2P1 to 1S2 for the 750.4 nm line and from 2P9 to 1S5 for the 811.5 nm line.

The results of Filipovic et al. [13], reported in several studies [14, 15] also concluded that the upper level of the 750.4 nm transition is mainly created by direct electronic impact on the fundamental state. From this observation, several authors [16, 17] used the argon line emitted at 750.4 nm to follow the electron density in their plasma (Fig. 61.3).

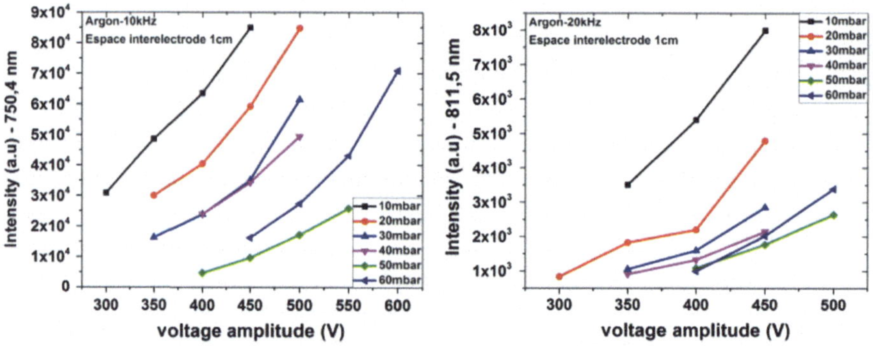

Fig. 61.3 Intensity for 750.4 nm (left) and 811.5 nm (right)

The results of Jelenak et al. [18] show that the 811.5 nm line was much more affected by collisional and radiative depopulation at both high density and high E/N argon discharges than the 750.5 nm line. Radiative decay of these levels to 1 s levels was responsible for the strongest line features in argon spectra, at 750.5 nm due to radiation from 2P1 and 2P5 levels, and at 811 nm due to transitions from 2P7 and 2P9 levels.

In addition to that, in this measurements, the follow-up of the maximum line Ar 763.5 nm was important for the estimation of metastable. These metastable species are considered as energy reservoirs due to the high energy they can store (11.74 eV for 3P0 and 11.55 eV for 3P2). Their relative long life is easily quenched by impurities.

The measurements of E, luminous efficiency are evaluated, provided that: I, the intensity of the discharge, the mean power is measured and t, the integration time was taken into account:

$$E = \frac{I}{t * <P>} \tag{61.2}$$

The follow-up of the most intense emission line 763.5 nm will give us the most information on the efficiency of the discharge. The heavier rare-gas species such as argon have lower ionization potential consequently the electron temperature can become lower and the electron density increases with increasing volumetric ratios of the heavier rare-gas species. When argon, with lower ionization energy is mixed into the nitrogen-rich plasma as in our case, in the case of impurities, it is found to lower the electron temperature with a resultant reduction in the degree of nitrogen dissociation [18].

Figure 61.4 shows that increasing the amplitude of the voltage leads to an increase in luminous efficiency. At the 20 kHz–10 mbar the luminous efficiency was better and at its maximum compared to 10 kHz. There was a process behind the emission of maximum line Ar 763.5 nm.

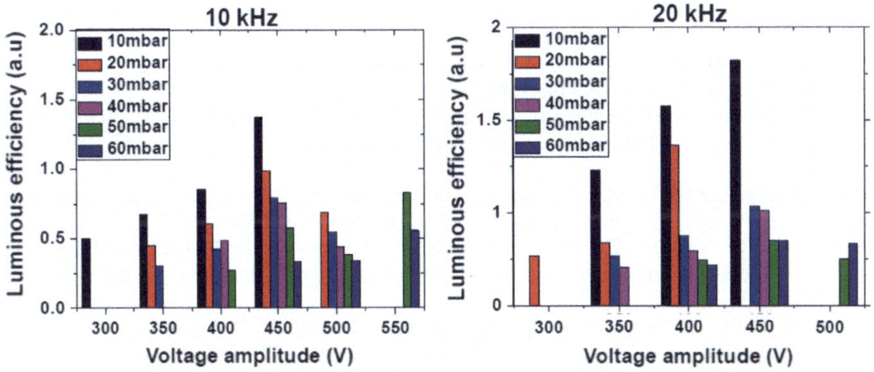

Fig. 61.4 Normalized efficiency as a function of voltage amplitude Ar 763.5 nm

61.4 Conclusion and Prospects

This work presents a first approach to study a homogeneous DBD for argon. Argon is relatively an inexpensive, chemically stable, safe gases that have proven as suitable filler base gases in mercury-free light sources. It is favorable for its interaction with nitrogen, there was an energy transfer between the metastable states of argon and nitrogen molecules.

The experimental results reported in this article show that under typical operating conditions 20 kHz–10 mbar–450 V the luminous efficiency is at his maximum compared to 10 kHz–450 V the discharge presents an optimum in terms of luminous efficiency functioning.

The decline of luminous efficiency was caused by the deactivation of metastable by collisions of molecular states of nitrogen which are favorable at 20 kHz, knowing that in our experiments the impurities (nitrogen) were uncontrolled.

In our following studies, we will work to have a high purity plasma with a new lamp, studies will continue on this context with the probability of adding ICCD cameras and using phosphorus.

In addition, the study of the emission spectrum mainly presents IR range and visible emissions. Spectroscopic studies were also carried out in order to identify the IR emission. The IR emission study allows us to understand the production process of metastable.

References

1. World Health Organization (2017), www.who.int/news-room/fact-sheets/detail/mercury-and-health
2. M. Obara, in *7th International Symposium on the Science & Technology of Light Sources*, Kyoto, Japan, 27–31 August 1995, pp. 149–158
3. T. Urakabe, S. Harada, T. Saikatsu, M. Karino, in *7th International Symposium on the Science & Technology of Light Sources*, Kyoto, Japan, 27–31 August 1995, p. 159
4. S. Horikoshi et al., Development of a Hg-free UV light source and its performance in photolytic and photocatalytic applications. Photochem. Photobiol. Sci. **18**(2), 328–335 (2019)
5. E. Robert et al., Light sources (2004), pp. 485–486
6. E. Robert et al., J. Phys. IV France **127**, 39–42 (2005)
7. F. Vollkommer, L. Hitzschke, Light Sources-8, Greifswald, Germany (1998), pp. 51–60
8. R.J. Carman, R.P. Mildren, B.K. Ward, D.M. Kane, J. Phys. D Appl. Phys. **37**, 2399–2407 (2004)
9. J.R. McNeely, G.S. Hurst, E.B. Wagner, M.G. Payne, Energy transfer from argon resonance states to nitrogen, hydrogen, and nitric oxide. J. Chem. Phys. **63**(6), 2717–2723 (1975)
10. A. Wnuk, M. Kaczkan, Z. Frukacz, I. Pracka, G. Chadeyron, M.-F. Joubert, M. Malinowski, Infra-red to visible up-conversion in holmium-doped materials. J. Alloys Compd. **341**, 353–357 (2002)
11. J.M.A. Caiut, S. Lechevallier, J. Dexpert-Ghys, B. Caillier, P. Guillot, UVC emitting phosphors obtained by spray pyrolysis. J. Lumin. **131**, 628–632 (2011)
12. E.L. Cates, M. Cho, J.H. Kim, Converting visible light into UVC: microbial inactivation by Pr3+ activated upconversion materials. Environ. Sci. Tech. **45**, 3680–3686 (2011)

13. D.M. Filipovic, B.P. Marinkovic, V. Pejcev, L. Vuskovic, Electron-impact excitation of argon: I. The 4 s' [1/2]1, 4p [1/2]0 and 4p' [1/2]0 states. J. Phys. B At. Mol. Opt. Phys. **33**, 677–691 (2000). https://doi.org/10.1088/0953-4075/33/11/308
14. V. Milosavljević, N. MacGearailt, P.J. Cullen, S. Daniels, M.M. Turner, Phase-resolved optical emission spectroscopy for an electron cyclotron resonance etcher. J. Appl. Phys. **113**, 163302 (2013). https://doi.org/10.1063/1.4802440
15. P.J. Cullen, V. Milosavljevi, Spectroscopic characterization of a radio-frequency argon plasma jet discharge in ambient air. Prog. Theor. Exp. Phys. **2015**, 0–17 (2015). https://doi.org/10.1093/ptep/ptv070
16. F. Jan, A.W. Khan, A. Saeed, M. Zakaullah, Mode transition in magnetic pole enhanced inductively coupled argon plasmas. Eur. Phys. J. D **66**, 5–11 (2012). https://doi.org/10.1140/epjd/e2012-20551-5
17. T.H. Chung, R. Kang, M.K. Bae, Optical emission diagnostics with electric probe measurements of inductively coupled Ar/O2/Ar-O2 plasmas. Phys. Plasmas **19**(11), 113502 (2013). https://doi.org/10.1063/1.4765357
18. Z.M. Jelenak, Z.B. Velikic, J.V. Bozin, Z.Lj. Petrovic, B.M. Jelenkovic, Institute of Physics, P.O. Box 57, I1000Belgrade, Yugoslavia (Received 11 November 1992) Electronic excitation of the 750- and 811-nm lines of argon

Chapter 62
Modeling of Dielectric Barrier Discharge in Pure Oxygen at Atmospheric Pressure

Mohammed Habib Allah Lahouel, Djilali Benyoucef, and Abdelatif Gadoum

Abstract A plasma producing between tow electrodes one of them or the tow of them covered by a solid dielectric or placed between them, called a Silent discharge or dielectric barrier discharge DBD. The presence of this dielectric it's to avoid the transition to arc. This process have a several applications, such as ozone generation, surface treatment, light source and other environmental industries. That for cold plasma characteristics, and the simple operation with atmospheric pressure. However, this paper propose contribution of modeling and simulation of dielectric barrier discharge in pure oxygen at atmospheric pressure using COMSOL Multiphysics to explain and understand physical behavior of various species created in plasma from O_2, by setting the mathematical and chemical model, courant and voltage characteristics during the break down was presented in results. In addition, the number density variations of species (charged species and neutral) across the gas in special scale is shown.

Keywords Discharge · Fluid model · Oxygen · Comsol multiphysics

62.1 Introduction

Electrode geometry of a dielectric barrier discharge DBD give specified characteristics to this discharge. Due to existing of dielectric barrier between electrodes, the formation of a sparks or the transaction of discharge to an arc is limited, and the plasma take a stable. The ignition of a DBD provide a variable voltage, with high frequency such as alternative or a pulsed, connected to one of electrodes and the other grounded [1]. The main objective of modeling a DBD is to serve the applications development and are based on the static behavior of plasma by considered the accumulation of charges on the dielectric surface [2]. This Modeling make place

M. H. A. Lahouel (✉) · D. Benyoucef · A. Gadoum
Laboratoire Génie Electrique et Energies Renouvelables, Hassiba Ben Bouali University, Chlef, Algeria
e-mail: habib221@live.com

A. Belasri and S. A. Beldjilali (eds.), *ICREEC 2019*, Springer Proceedings in Energy,
https://doi.org/10.1007/978-981-15-5444-5_62

of basic plasma operations and optimize current technologies which use plasmas, also predict the effectiveness of new applications. During the discharge there is a lot of reactive oxygen species generated in gas gap at atmospheric pressure. This species generated depend on plasma parameters, such as electron density and excitation temperature. [3] this proposed model include plasma chemistry, surface reactions, density of electrons and heavy species (ions, atom…), electric failed, electron temperature, for better understanding dbds at atmospheric pressure in O_2.

62.2 Numerical Model

62.2.1 Electron Transport

The electron density equation is written as:

$$\frac{\partial n_e}{\partial t} + \nabla \Gamma_e = R_e \tag{62.1}$$

$$\Gamma_e = -\mu_e E n_e - \nabla(D_e n_e) \tag{62.2}$$

With:

n_e: Electron density $(1/\mathrm{m}^3)$
R_e: Electron rate expression $(1/\mathrm{m}^3\,\mathrm{s})$
μ_e: Electron mobility $(\mathrm{m}^2/\mathrm{V\,s})$
E: Electric field (V/s)
D_e: Electron diffusivity $(\mathrm{m}^2/\mathrm{s})$
e: refers to electron.

Electron energy is given by:

$$\frac{\partial n_\varepsilon}{\partial t} + \nabla \Gamma_\varepsilon + E\,\Gamma_e = R_\varepsilon \tag{62.3}$$

$$\Gamma_\varepsilon = -\mu_\varepsilon E n_\varepsilon - \nabla(D_\varepsilon n_\varepsilon) \tag{62.4}$$

With:

n_ε: Electron energy density $(\mathrm{V/m}^3)$
R_ε: Energy loss or gain $(\mathrm{V}/(\mathrm{m}^3\,\mathrm{s}))$
μ_ε: Electron energy mobility $(\mathrm{m}^2/\mathrm{V\,s})$
D_ε: Electron energy diffusivity $(\mathrm{m}^2/\mathrm{s})$
ε: refers to energy.

Electron energy, $\varepsilon(V)$ computed by:

$$\varepsilon = \frac{n_\varepsilon}{n_e} \tag{62.5}$$

These sated equations above defined by the system stoichiometry and written with Townsend coefficients or rate coefficients.

The electron energy loss is obtained by summing the collisional energy loss over all reactions:

$$R_e = \sum\nolimits_{j=1}^{P} x_j k_j N_n n_e \Delta\varepsilon_j \tag{62.6}$$

With:

x_j: Mole fraction of the target species for reaction j;
k_j: Rate coefficient for reaction j (m^3/s);
N_n: Total neutral number density (1/m^3);
$\Delta\varepsilon_j$: Energy loss from reaction j (V);
and P refer to inelastic electron-neutral collisions;

The mean electron energy is exponentially related with the Townsend coefficients k_j. So, the Townsend coefficients can be written as:

$$k_j = A\varepsilon^\beta e^{-E/\varepsilon} \tag{62.7}$$

62.2.2 Electrostatics Equations

Electrostatic field:

$$-\nabla\, \varepsilon_0\, \varepsilon_r \nabla V = \rho \tag{62.8}$$

The space charge density:

$$\rho = q\left(\sum\nolimits_{k=1}^{N} Z_k n_k - n_e\right) \tag{62.9}$$

The constitutive relation:

$$D = \varepsilon_0 \varepsilon_r E \tag{62.10}$$

This relation (62.10) show the properties of the dielectric and the applicable material properties (relating the electric displacement D with the electric field E).

Boundary condition take onto account the accumulation of charges of the dielectric surface as it follow:

$$-n\,(D_1 - D_2) = \rho_s \tag{62.11}$$

With ρ_s is the solution of the following distributed ODE on the boundary:

$$\frac{d\rho_s}{dt} = n\,J_i + n\,J_e \tag{62.12}$$

Where the total ion and electron courant density describe by $n\,J_i$ and $n\,J_e$ respectively.

62.3 Plasma Model

This part of our work is an aggregation of possible chemical reactions in oxygen O_2 as a gas filled in discharge gap.

62.3.1 Electron Impact Reactions

The cross section set for electron collisions in O_2 used in this calculation is shown in Table 62.1 [4].

62.3.2 Other Chemical Reactions

The reactions used in our model token with their rate constant kf. Quantitatively the Arrhenius Equation determines the relationship between the rate constant proceeds and its temperature.

$$k^f = Ae^{-E_a/RT} \tag{62.13}$$

With:

A: Pre-exponential factor;
T: Temperature (kelvin);
Ea: Activation energy for the reaction (in the same units as R * T);
R: Universal gas constant.

Table 62.1 Phelps databases

Reaction	Formula
1	$e + O_2 \Rightarrow e + O_2$
2	$e + O_2 \Rightarrow O + O^-$
3	$e + O_2 \Rightarrow e +$ O2a1d
4	$e +$ O2a1d $\Rightarrow e + O_2$
5	$e + O_2 \Rightarrow e +$ O2b1s
6	$e +$ O2b1s $\Rightarrow e + O_2$
7	$e + O_2 \Rightarrow e +$ O245
8	$e +$ O245 $\Rightarrow e + O_2$
9	$e + O_2 \Rightarrow e + O + O$
10	$e + O_2 \Rightarrow e + O +$ O1d
11	$e + O_2 \Rightarrow e + O +$ O1s
12	$e + O_2 \Rightarrow e + e + O_2^+$
13	$e +$ O2a1d $\Rightarrow e +$ O2a1d
14	$e +$ O2a1d $\Rightarrow e + O + O$
15	$e +$ O2a1d $\Rightarrow 2e + O_2^+$
16	$e +$ O2b1s $\Rightarrow e +$ O2b1s
17	$e +$ O2b1s $\Rightarrow e + O + O$
18	$e +$ O2b1s $\Rightarrow 2e + O_2^+$
19	$e +$ O245 $\Rightarrow e + O + O$
20	$e +$ O245 $\Rightarrow 2e + O_2^+$
21	$e + O \Rightarrow e + O$
22	$e + O \Rightarrow e +$ O1d
23	$e +$ O1d $\Rightarrow e + O$
24	$e + O \Rightarrow e +$ O1s
25	$e +$ O1s $\Rightarrow e + O$
26	$e + O \Rightarrow 2e + O^+$
27	$e +$ O1d $\Rightarrow e +$ O1s
28	$e +$ O1d $\Rightarrow 2e + O^+$

62.3.3 Surface Reactions

Besides to the previous used chemical reactions, Table 62.2 set all the reactions between the existing species in gas gap and the wall.

Table 62.2 Surface reactions

N°	Reaction
1	$O_2^+ \Rightarrow O_2$
2	$O^+ \Rightarrow O$
3	$O_3^+ \Rightarrow O_3$
4	O2a1d => O_2
5	O2b1s => O_2
6	O245 => O_2
7	O1d =>O
8	O1s => O
9	$O_2^- \Rightarrow O_2$
10	$O^- \Rightarrow O$

62.4 Simulation Results

For this model, we consider a cylindrical geometry with a gap filled by oxygen.

Voltage used, is a sinusoidal function with high frequency f = 50 kHz (Fig. 62.1) as:

$$Vrf = 6000 * sin(2 * \pi * f * t) \tag{62.14}$$

Where t is time (μs), and Vrf applied voltage on the internal circle (V) and the external circle in related to the ground.

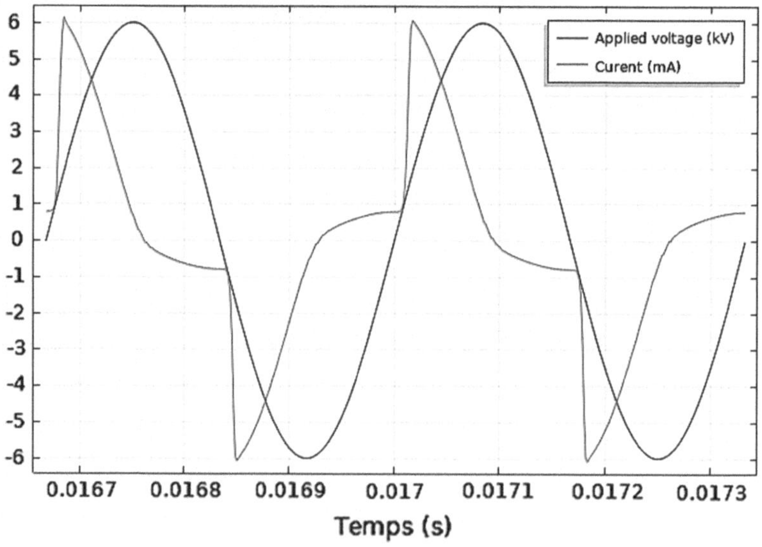

Fig. 62.1 Applied voltage and the result total discharge courant

The dielectric considered with relative permittivity $\varepsilon_r = 10$.

- Initial values:

The electrons number which uniformly distributed across the gas gap supposed $10^3 (1/\mathrm{m}^3)$ with the initial mean energy 4 V.

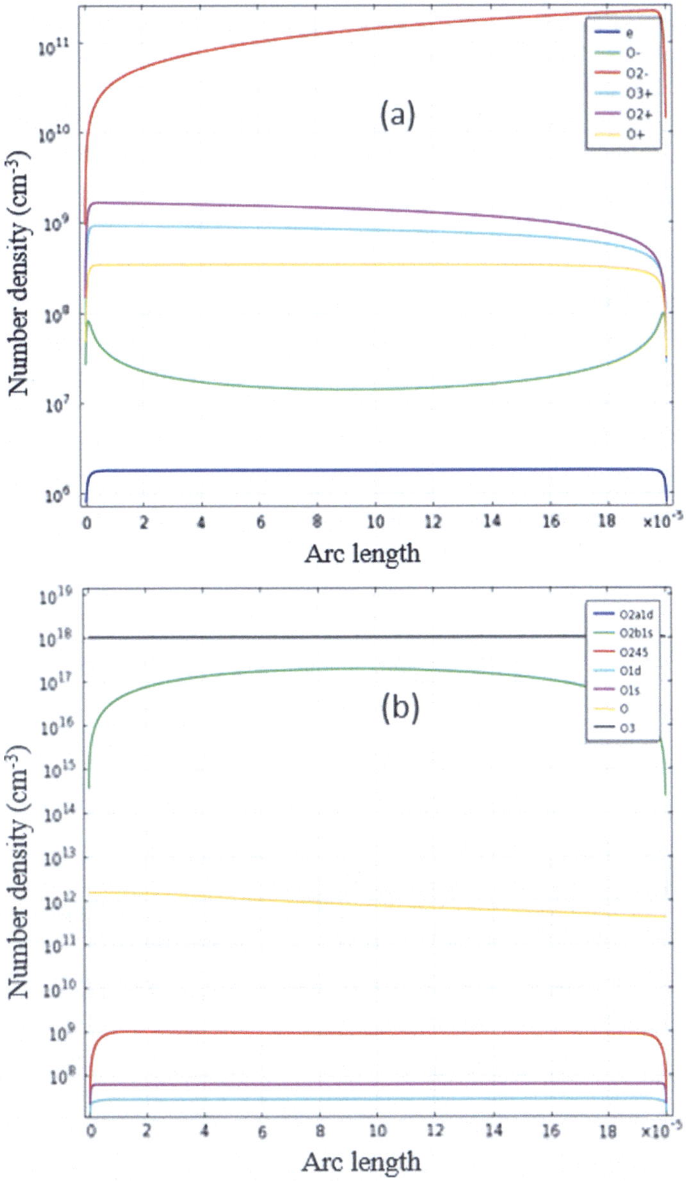

Fig. 62.2 **a** and **b** Ions and Excited Oxygen and ozone atoms number density

The number density of oxygen ions supposed also $10^3(1/m^3)$.

The Fig. 62.2 show the excited Oxygen and ozone atoms, and ion Oxygen number density in gap.

When the voltage applied to the inner circle, an electric field forms in the gap. A number of electrons rush toward acquiring enough energy to ionize some oxygen species creating new electrons, which they rush towards the dielectric coting, connected to the ground. Such as the voltage applied with the negative polarity, created ions accelerate to electrode surface. Therefore, for the reason of heaviness of Ions, their density increase near to grounded electrode in Fig. 62.2.

Figure 62.2 show also, that there is an increasing of densities of electron and ions and O_2^- ions specifically. With the increasing of applied voltage, there is creation of strong electric field that give enough energy to electron and they will accelerate across the gap, making remarkable ionization activities.

62.5 Conclusion

This paper present a model study of Oxygen discharge in atmospheric pressure with the existence of dielectric barrier based on plasma physics. Our model highlight the physics of the breakdown process by using a simple plasma chemistry model. The result of this model show:

- The creation of new elements during the discharge as Ozone O_3 for example;
- Most importantly the negative oxygen ions are a key element in producing plasma actuator's.

References

1. C. Riccardi, R. Barni, Chemical kinetics in air plasmas at atmospheric pressure, in *Chemical Kinetics*, ed. by V. Patel (InTech, London, 2012). ISBN 978-953-51-0132-1
2. S. Keller, P. Rajasekaran, N. Bibinov, P. Awakowicz, Characterization of transient discharges under atmospheric-pressure conditions applying nitrogen photoemission and current measurements. J. Phys. D: Appl. Phys. **45**(12), 125202 (2012)
3. D. Xiao, C. Cheng, J. Shen, Y. Lan, H. Xie, X. Shu, Y. Meng, J. Li, P.K. Chu, Characteristics of atmospheric-pressure non-thermal N2 and N2/O2 gas mixture plasma jet. J. Appl. Phys. **115**, 033303 (2014)
4. S.A. Lawton, A.V. Phelps, Excitation of the b1Σ+g state of O2 by low energy electrons. J. Chem. Phys. **69**, 1055 (1978)

Chapter 63
Effect of Dielectrics on Excilamp Discharge Characteristics

Wafà Benstâali, Fatiha Ghaleb, and Ahmed Belasri

Abstract In this paper, the characteristics of a xenon lamp are studied. In this lamp, the discharge is produced between two electrodes covered by dielectrics to protect them from the gas effects, to limit the current and to reduce the applied voltage. A one dimensional particle in cell model with Monte Carlo Collisions (PIC'MCC) was then used. The computer model allows to investigate two equivalent dielectric capacitances; 0.82 pF/cm^2, 230 pF/cm^2. The results show temporal variation of the different voltages in the simulation domain and the current density, spatiotemporal variation of electron density and electric field is also discussed. The discharge pulse is more important for 230 pF/cm^2 with less applied voltage (2.7 kV). The effect of dielectric capacitance is clearly seen on energy deposition into the different collisions. For higher capacitance, electrons use more energy to excite xenon atoms. This energy is used effectively into xenon metastable, which can contribute to an efficient discharge VUV emission. The energy deposited into high levels is not significantly affected.

Keywords Excilamp · Dielectric · Capacitance · Excitation · Energy · Model

63.1 Introduction

Excimer lamps are one of the most important ultraviolet emission sources. They are used in different domains; lighting, automobile, micro-electronics, pollution control, and recently in medical domain [1, 2]. This increase of interest is due to their several advantages. The most important are that they are environmentally friendly

W. Benstâali (✉)
Faculty of Sciences and Technology, Abdelhamid Ibn Badiss University, Mostaganem, Algeria
e-mail: Wafabenstaali7@yahoo.fr

W. Benstâali · F. Ghaleb · A. Belasri
Laboratoire de physique des plasmas, Matériaux Conducteurs et Leurs Applications, USTO-MB, Oran, Algeria

A. Belasri and S. A. Beldjilali (eds.), *ICREEC 2019*, Springer Proceedings in Energy,
https://doi.org/10.1007/978-981-15-5444-5_63

technology and are based on Dielectric Barrier Discharge (DBD) which provides non-equilibrium plasma at atmospheric pressure without transition to arc discharge.

The large domain of excilamp's applications requires their improvement, especially their luminous efficiency. Many experiences and theoretical works have been done to achieve this goal [3–12]. And in each one, dielectric characteristics have an important effect on discharge performances. A number of values of relative permittivity; ε_r and dielectric thickness; L_d were used in different works, let's quote Carman et al. [4] and Bogdanov et al. [5] who have used dielectrics with $\varepsilon_r = 3.7$ and $L_d = 2$ mm. Avtaeva et al. have used a dielectric with a relative permittivity equal to 4 and a thickness equal to 1 mm in [6]. Doanh et al. [7] have used dielectrics with $\varepsilon_r = 4$ and $L_d = 2$ mm. In these works, the dielectric capacitance is of about 1 pF/cm^2. Belznai et al. [8] have used a dielectric with $\varepsilon_r = 6.6$ and $L_d = 0.7$ mm for cylindrical geometry (a capacity of 8 pF/cm^2). Belasri et al. have used a dielectric capacitance equal to 3.27 pF/cm^2 [9]. Recently, Belasri et al., Khodja et al., and Benmoussa et al. have used a dielectric capacitance equal to 230 10^{-12} F/cm^2 [10–12]. In these works, the inter-electrodes distance is of about 3 to 5 mm. It is noted that the capacitance represents here the equivalent capacitance of the two dielectrics put on each electrode. Even the important difference in the value of dielectric capacitance, the behavior of excilamp's discharge has been well described. In this paper, we present a comparison between the two conditions using a one dimensional particle model. A description of its basics is presented in Sect. 2. The results obtained from our model are presented in Sect. 3.

63.2 Physical Model

Our basic model [13, 14] consists of a particle in cell method where electron and ion transport are described by following the trajectories of a large number of charged particles in the self-consistent electric field.

The motion of charged particles is described by Newton-Lorentz equations:

$$M\frac{dV}{dt} = F = qE \tag{63.1}$$

$$\frac{dX}{dt} = V \tag{63.2}$$

where m is the particle mass, V the velocity, x the position, F the applied force, q the charge, and E the electric field which is obtained form Poisson's equation:

$$\nabla^2\Phi = -\frac{\rho}{\varepsilon_0} \tag{63.3}$$

where Φ is the electric potential, ρ is the charge density, ε_0 is the vacuum permittivity.

A Monte Carlo collisions method is used to study particles collisions. When we consider the collision dynamics, a number of electron–neutral collision events are possible, including elastic scattering, excitation and ionization.

The energy deposited into ionization and excitation collisions can be calculated by the relations:

$$\Delta\varepsilon_{iz} = \varepsilon_2 + \varepsilon_{iz} \tag{63.4}$$

$$\Delta\varepsilon_{exc} = \varepsilon_{exc} \tag{63.5}$$

where $\Delta\ \varepsilon_{iz}$and $\Delta\varepsilon_{exc}$ represent the energy given into ionization and excitation respectively, ε_2 is the energy of electron produced by ionization, ε_{iz} and ε_{exc} are the ionization and excitation thresholds respectively.

63.3 Results and Discussion

In this section, we study a single discharge for a planar exilamp filled with xenon. The 1D PIC model is used in this paper to simulate a discharge pulse for a gap length d equal to 3 mm and a total gas pressure p of 400 Torr. The gas temperature is supposed to be constant and equal to 300 °K. The xenon secondary emission coefficient is 0.02. To study the effect of the equivalent dielectric capacitance, C_d, we compare the discharge characteristics for two capacitance values; 0.82 pF/cm^2 [4] and 230 pF/cm^2 [10]. Applied voltages of 5, and 2.7 kV are used for these capacitance values respectively.

The applied voltage through the discharge is given by the following formula:

$$V_{app} = Vd + Vg \tag{63.6}$$

The voltage across dielectrics at t = 0 is

$$Vg = \frac{C_d}{C_g + C_d} V_{app} = KV_{app} \tag{63.7}$$

In our conditions, K $\sim$ 0.73 for a dielectric capacitance of 0.82 pF/cm^2 (C_{d1}), and K $\sim$ 0.99 for a dielectric capacitance equal to 230 pF/cm^2 (C_{d2}).

Figure 63.1a and b represent temporal variation of current density and voltages on gas domain, Vg, and on dielectrics, Vd, for the two cases of dielectric capacitance. It is observed that voltages have the same behavior; at t = 0 s Vg is equal to $K^x V_{app}$. Due to memory charges, the dielectric voltage increases and the gas voltage decreases until canceled. However, for C_{d1}, it is observed that this period is characterized by two current pulses, the first pulse is small, about 13% of the principal pulse, which has a maximum value of about 0.7A/cm^2. Even the decay period of Vg takes 200 ns, each pulse lasts about 10 ns. For a capacitance of 230 pF/cm^2, the discharge is represented by a pulse which reaches a maximum value of 7 A/cm^2 with duration of about 100 ns.

Figure 63.2 shows the spatial distribution of electron density in the gas domain at several times, for C_{d1} (a) and C_{d2}(b). In each case, at the beginning, the electron density is small. The electron density maximum moves toward the cathode as the plasma expands. It increases to reach, at the moment of current maximum, a value of 4 10^{18} m^{-3} and 1.6 10^{20} m^{-3} for C_{d1} et C_{d2}. The distribution of the electric field at different times (the same times as electron density) is represented on Fig. 63.3, (a) and (b). At the moment when the current density is maximum, the electric field is on the order of 1.2×10^7 V/m for the first capacitance and 1.5×10^6 V/m, for C_{d2}.

In order to study the energy repartition in the discharge, we have calculated the energies dissipated by charged particles in different processes. The overall energy balance is obtained by summing the energy losses by each particle in the simulation in each category (elastic, excitation, ionization, loss on the wall). The energy lost into excitation and ionization is calculated by relations 63.4 and 63.5.

Figure 63.4, 63.5 and 63.6 represent the fractional energy deposited by electrons and ions into collisions over the total discharge. The major part of the discharge energy is used by electrons; about 80% and 85.5% for a dielectric capacitance equal to C_{d1} and C_{d2} respectively. The results of the two cases show also that the most of this energy contributes to xenon excitation (58.8% and 75% for C_{d1} and C_{d2} respectively), while ionization takes small energy compared with excitation (13.5% and 2.8% for C_{d1} and C_{d2} respectively). However, the energy used for ion heating is significant (20% and 14.5% for C_{d1} and C_{d2} respectively). The PIC model results indicate that there is only 1% to 3% of the total energy lost at walls. Energy lost by elastic collision is also low. A significant energy is spent for exciting xenon atom into Xe(1s5) states for C_{d2} (28%), these metastable uses less energy in the case of C_{d1} (15%). About 11% and 18% of the total energy is used into Xe(1s4) states (resonant), for C_{d1} and C_{d2} respectively.

This means that considerable energy is spent for exciting xenon atom into the first two lowest lying energy levels of xenon (about 26% and 46% for C_{d1} and C_{d2} respectively). The amount of energy used for exciting higher lying levels is

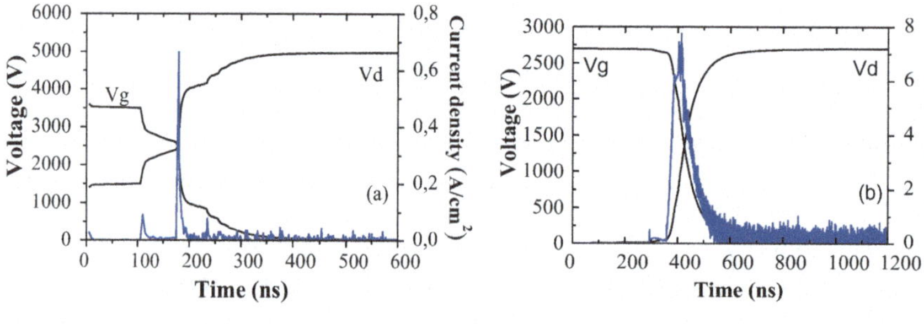

Fig. 63.1 Temporal variation of current density, gas and dielectric voltages, for a dielectric capacitance equal to **a** 0.82 pF/cm^2, **b** 230 pF/cm^2. The applied voltages are 5 and 2.7 kV respectively

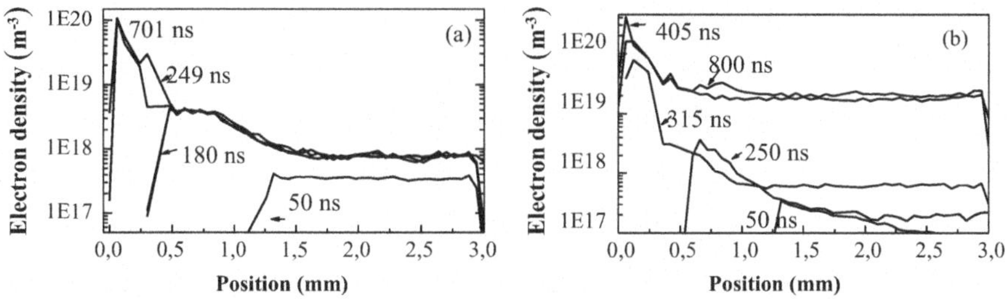

Fig. 63.2 Spatial distribution of electron density at different times of the discharge, for a dielectric capacitance equal to **a** 0.82 pF/cm^2, **b** 230 pF/cm^2

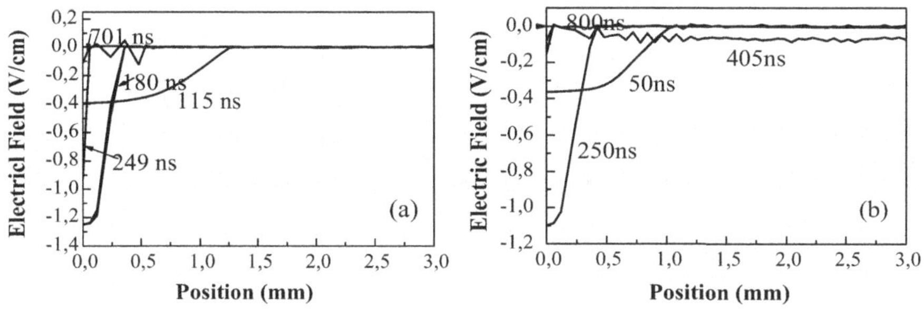

Fig. 63.3 Spatial distribution of electric field at different times of the discharge, for a dielectric capacitance equal to **a** 0.82 pF/cm^2, **b** 230 pF/cm^2

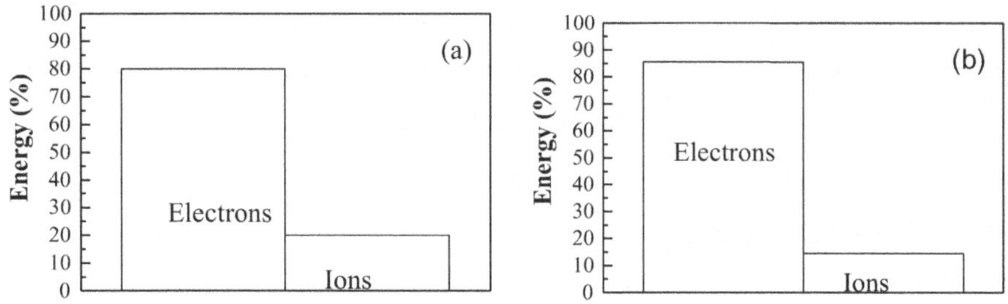

Fig. 63.4 Calculated fractional electrical energy deposition **a** for a dielectric capacitance equal to 0.82 pF/cm^2 with an applied voltage equal to 5 kV, **b** for a dielectric capacitance equal to 230 pF/cm^2 with an applied voltage equal to 2.7 kV

relatively large; ~28% % for Xe * (1s3 + 1s2 + 2p) (for the two capacitances), Xe * (2 s + 3d) use less than 5% of the energy input in the discharge. Many of the excited atoms at these levels drop back to the first two energy levels, taking part effectively at the VUV light generation process. It is noted that even electrons use approximately the same part of energy, the repartition is clearly different between

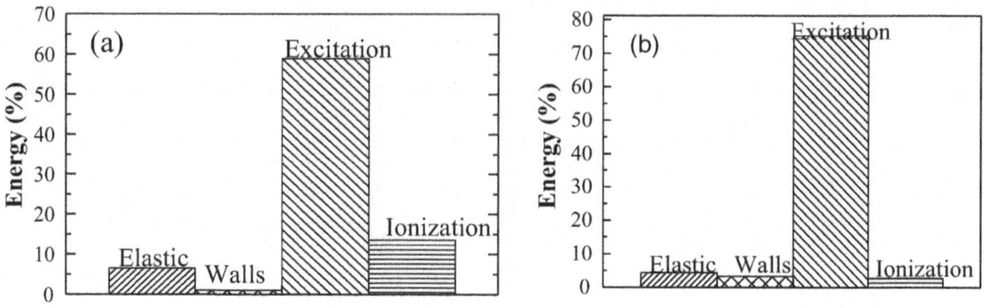

Fig. 63.5 Fractional energy deposited by electrons into different collisions; **a** for a dielectric capacitance equal to 0.82 pF/cm^2, **b** for a dielectric capacitance equal to 230 pF/cm^2

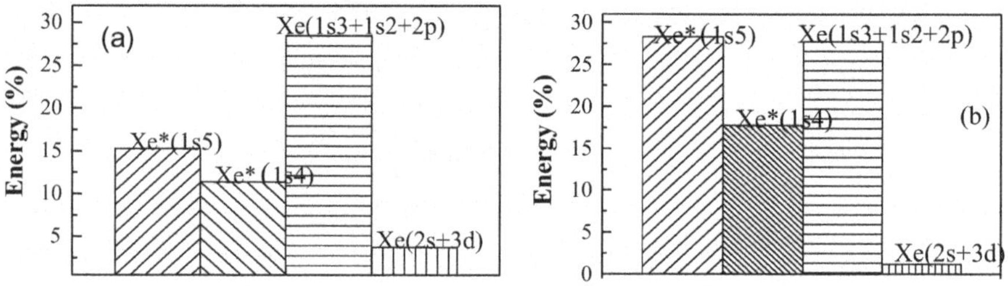

Fig. 63.6 Fractional energy deposited into different excited states of xenon; **a** for a dielectric capacitance equal to 0.82 pF/cm^2, **b** for a dielectric capacitance equal to 230 pF/cm^2

the two cases studied in this work. For a capacitance of 230 pF/cm^2, the total energy deposited by electrons into the first levels (1s4 and 1s5) is more important that in C_{d1}. There is no noticeable difference in the energy deposited in higher levels.

63.4 Conclusion

In this paper, a 1D particle in cell model has been used to study a xenon lamp discharge. The temporal variations of discharge voltage, dielectric voltage, discharge current, electric field are discussed to study the effect of dielectrics on excilamp characteristics. Two different dielectric capacitances were considered; 0.82 and 230 pF/cm^2 with the goal of providing insight for optimizing efficiencies. The results show that the conventional capacitance, C_{d1}, needs an applied voltage in the order of 5 kV; two pulses are observed. 2.7 kV is applied for a dielectric capacitance equal to 230 pF/cm^2, with a maximum of current ten times higher than for C_{d1}. The fractional energy deposited by electrons and ions into collisions over the total discharge is also calculated. The energy deposited into xenon excitation is influenced by the capacitance and voltage conditions; it reaches a value of 75% for

230 pF/cm^2. This is due essentially to the increase of the part of energy deposited into metastable states (28%), which contributes to an efficient discharge into excitation, and then in VUV emission. A dielectric capacitance equal to 230 pF/cm^2 allows a low applied voltage with higher efficiency.

References

1. U. Kogelschatz, Ultraviolet excimer radiation from non-equilibrium gas discharges and its application in photophysics, photochemistry and photobiology. J. Opt. Technol. **79**(8), 484–493 (2012)
2. D. Schitz, A. Ivankov, V. Pismennyi, Excimer lamp for dermatology. in *Conference Series*, vol. 1172 (2019), p. 012052
3. N.L.D. Bachir, A. Belasri, P. Guillot, B. Caillier, Radiative emissions in visible–IR of krypton excilamp: experimental and theoretical interpretations. Plasma Chem. Plasma Process. **39**, 1243 (2019)
4. R.J. Carman, R.P. Mildren, Computer modelling of a short-pulse excited dielectric barrier discharge xenon excimer lamp ($\lambda \sim$ 172 nm). J. Phys. D: Appl. Phys. **36**, 19–33 (2003)
5. E.A. Bogdanov, A.A. Kudryavtsev, R.R. Arslanbekov, V.I. Kolobov, Simulation of pulsed dielectric barrier discharge xenon excimer lamp. J. Phys. D Appl. Phys. **37**, 2987–2995 (2004)
6. S.V. Avtaeva, A.V. Skornyakov, Calculation of the characteristics of xenon excilamps using a one-dimensional hydrodynamic model. Russ. Phys. J. **53**(3), 257–262 (2010)
7. L.T. Doanh, S. Bhosle, G. Zissis, H. Piquet, Estimation of the light output power and efficiency of a XeCl dielectric barrier discharge exciplex lamp using one-dimensional drift–diffusion model for various voltage waveforms. IEEE Trans. Ind. Appl. **49**(1), 331–340 (2013)
8. S. Beleznai, G. Mihajlik, A. Agod, I. Maros, R. Juhasz, Z. Nemeth, L. Jakab, P. Richter, High-efficiency dielectric barrier Xe discharge lamp: theoretical and experimental investigations. J. Phys. D: Appl. Phys. **39**, 3777–3787 (2006)
9. A. Belasri, Z. Harrache, Electrical approach of homogenous high pressure Ne/Xe/HCl dielectric barrier discharge for XeCl (308 nm) lamp. Plasma Chem. Plasma Process. **31**, 787–798 (2011)
10. A. Belasri, Z. Harrache, Electrical and kinetical aspects of homogeneous dielectric-barrier discharge in xenon for excimer lamps. Phys. Plasmas **17**, 123501 (2010)
11. K. Khodja, A. Belasri, H. Loukil, Modeling of a Ne/Xe dielectric barrier discharge exilamp for improvement of VUV radiation production. Plasma Phys. Rep. **43**, 891–898 (2017)
12. A. Benmoussa, A. Belasri, Z. Harrache, Numerical investigation of gas heating in dielectric barrier discharge in Ne-Xe excilamp. Curr. Appl. Phys. **17**(4), 479–483 (2017)
13. W. Benstâali, A. Belasri, Discharge efficiency improvement in PDP cell via a 1D PIC-MCC method. IEEE Trans. Plasma Sci. **30**, 1460–1465 (2011)
14. W. Benstâali, Z. Harrache, A. Belasri, Study of a Ne/Xe pulsed planar dielectric barrier discharge characteristics and energy balance: simulation via 1D PIC/MCC method. Phys. Scr. **85**, 065502 (2012). 10 p.

Chapter 64
Spectroscopic Study of a Kr and Kr/Cl_2 Excilamps Under Sinusoidal and Pulsed Excitation

Nadjet Larbi Daho Bachir, Ahmed Belasri, Philippe Guillot, and Bruno Caillier

Abstract Recently considerable attention focused on the development of excimers and exciplexe lamp. These lamps are sources of spontaneous radiation based on the transitions of excited molecules called excimers when pure noble gas or a mixture of several noble gases is added. The excited molecules are considered to be exciplex when the gas mixture contains one or more halogens in addition to noble gases or mercury. The de-excitation of these molecules to their dissociative fundamental states induces a loss of energy in the form of UV, visible and IR radiation. The work is based on an experimental study of a pure krypton lamp and theoretical study of the Kr/Cl_2 plasma chemistry in terms of the homogenous model. The lamp is excited by a dielectric barrier discharge, the gas breakdown is obtained between two flat and identical electrodes of 5 cm $\times$ 5 cm surface area, these electrodes are insulated and separated by two dielectric plates of permittivity 4 and thickness 2 mm. The space between the two dielectrics is 2 mm and is filled with gas at pressure of 129 to 460 mbar. In order to carry out a parametric study, the influence of several parameters such as applied voltage, frequency and gas pressure was studied. A spectroscopic and kinetic analysis of a pure krypton and Kr/Cl_2 mixture excilamp excited by DBD is reported here. The interpretation of the experimental and theoretical results allowed to explain chemical processes responsible of the production of excimers inducing UV emission and to link these processes to the IR emissions that control the population of the excited metastable states of krypton. These states are indirectly responsible for UV emission.

Keywords Pure Krypton excilamp · Kr/Cl_2 gas mixture · UV emission · IR emission

---

N. Larbi Daho Bachir (✉) · A. Belasri
LPPMCA, University of Sciences and Technology of Oran Mohamed Boudiaf USTO-MB, Oran, Algeria
e-mail: ldbn@univ-oran1.dz

N. Larbi Daho Bachir
ISTA, University of Oran 1 Ahmed Ben Bella, Oran, Algeria

P. Guillot · B. Caillier
DPHE, Place de Verdun, 81012 Albi Cedex 9, France

A. Belasri and S. A. Beldjilali (eds.), *ICREEC 2019*, Springer Proceedings in Energy,
https://doi.org/10.1007/978-981-15-5444-5_64

64.1 Introduction

The lamp excited by DBD can have several types of configurations. In the literature, there are configurations of planar, coaxial or plan-plan geometry or other geometries that depend on the type of power supply used and the field of application required for these lamps [1, 2]. In our work we consider a flat discharge consisting of two metal electrode plates covered by a dielectric material. The insulating aspect of semiconductor materials introduced into a silent electrical discharge or dielectric barrier discharge requires a specific use of the applied voltage. In this type of discharge, it is not possible to supply the discharge with direct voltage because the insulating characteristic of the dielectrics leads the discharge to a total extinction just after the formation of the first avalanche and the accumulation of charges on the inner surface of the dielectrics, thus forming a wall that blocks the passage of the current [3]. The discharge will therefore eventually go out. However, the capacitive nature of dielectrics requires that these devices be powered by an alternative electrical source. In this paper we will study the kinetic parameters of the dielectric barrier discharge in a pure krypton discharge and in the Kr-Cl_2 mixture under two kinds of excitation: rectangular pulsed excitation and sinusoidal excitation. In particular, due to the pulsed and repetitive form of dielectric barrier discharges, the profile of the applied voltage plays a crucial role in determining the current profile that is often used to characterize the discharge. The shape of the excitation used closely influences the UV emission efficiency of the lamp [4].

However, in recent years many researchers have shown a growing interest in studying this gases and this is mainly due to recent advances in the application of radiation emitted by pure Kr or mixtures Kr-Cl_2 lamps in the medical field and in pollution control [5, 6]. The first work on gas discharges for mixing krypton and chlorine was thus carried out with the aim of developing UV/Visible lasers Green [7] was among the first to study the subject of mono-halide rare gas lasers and in his experimental work he evaluated some rate of reactions that may occur between rare gases such as Argon, Krypton and Xenon with halogens such as Fluore, Chlorine and Brome. Flannery's [8, 9] work on the recombination of rare gas molecular ions with halogen atomic ions in a gas discharge has made it possible to make very important progress in the study of the chemical kinetics of the mixture. He thus estimated the rates of recombination reactions that form the basis of kinetics in this type of mixture. Thus Flannery's work made it possible to understand the process of formation of the exciplex RgX^* from the rare gas atom Rg and the negative ion of the halogen X^- [10–12]. The three-body reaction rates given by Flannery are obtained after several experiments on several gas mixtures.

64.2 Kinetic Model

In a rare gas DBD lamp the electrical energy is efficiently converted into electro-magnetic energy by transforming the electrical energy of the discharge into the energy of movement of the electrons, which by colliding with heavy atoms contribute to chemical reactions [13].

64.2.1 *Kinetic Model for Kr/Cl_2 Gas Mixture*

In our case of the gas mixture between krypton and chlorine the process of formation of the excited exciplex molecule begins with an excitation neutral atoms to metastable and resonat excited sate and ionization of the rare gas atoms:

$$Kr + e \rightarrow Kr^{*} + e \tag{64.1}$$

$$Kr^{*} + e \rightarrow Kr^{+} + 2e \tag{64.2}$$

The second step is the formation of rare gas molecular ions:

$$Kr^{+} + Kr + Kr \rightarrow Kr_2^{+} + Kr \tag{64.3}$$

This process is often referred to as demirisation, which consists to formation of an ionic molecule of noble gas by reaction with three bodies. It is a fundamental process in the formation of the exciplex responsible for the characteristic UV emission of the lamp.

The formation of halogen ions occurs through the processes of the attachment reactions of the shape of:

$$Cl_2 + e \rightarrow Cl^{-} + Cl \tag{64.4}$$

For the formation of exciplex molecules, reaction processes are necessary and common to all types of rare gas and halogen mixtures: the three most important channels for exciplex formation are:

$$Kr^{+} + Cl^{-} + Kr \rightarrow KrCl^{*} + Kr \tag{64.5}$$

$$Kr_2^{+} + Cl^{-} \rightarrow KrCl^{*} + Kr \tag{64.6}$$

$$Kr^{*} + Cl_2 \rightarrow KrCl^{*} + Cl \tag{64.7}$$

The formed exciplex is not really stable and quickly dissociates into a neutral atom typically in a few nanoseconds. The energy due to the de-excitation of the excited molecules is released in the form of photon radiation in the ultraviolet range:

$$KrCl^* \rightarrow Kr + Cl + hv \quad (64.8)$$

Though, the efficient generation of UV photons depends on the form of excitation of the applied voltage, the pulse frequency, the concentration of chlorine in the gas and the gas pressure [14].

64.2.2 Kinetic Model for Pure Krypton

The kinetic model developed in case of pure Krypton excilamp describes the processes that accompany the excitation of a Kr gas. Since, the different excited and ionic atoms and molecules formed in inter-electrode space contribute to the formation of the excimer molecules responsible for emission in 148 nm range. The fundamental reactions which controls the radiation output of the dielectric barrier discharge for pure krypton gas. The chemical cycle leads to the creation of the excimers (Kr_2 * ($^1\Sigma u$), (Kr_2 * ($^3\Sigma u$)) whose de-excitation to the fundamental and dissociative state produces UV emissions in the 147 and 148 nm, respectively. the process begins with ionization and excitation of the Krypton atoms on different states of metastable, resonon and higher excited states. collisions between these different excited atoms and electrons as well as with other neutral atoms contribute to the formation of ionized krypton atoms and molecules which by three-body collisions induce a formation of the excited excimer molecules of krypton. The work in this article is a synthesis of several studies on excimer lamps for two types of gases (halogen-rare gas mixture and pure rare gas). The graphs presented in this section are the results of several studies on dielectric barrier discharges for the mixing of rare gases and halogen and work on pure rare gases. We present here a statistical, selective and comparative study in results of modeling work recognized and validated by several publications [15] and experimental work confirmed and recognized [16, 17].

64.3 Results

64.3.1 Efficiency Study for a Krypton-Chlorine Mixing Lamp

Where the lamp is alimented by a pulsed voltage, the results show a high efficacy for 222 nm, corresponding to a value of 8.4% for a total lamp efficacy of around 14.9%. Whereas, for sinusoidal excitation, the total efficiency of the discharge is

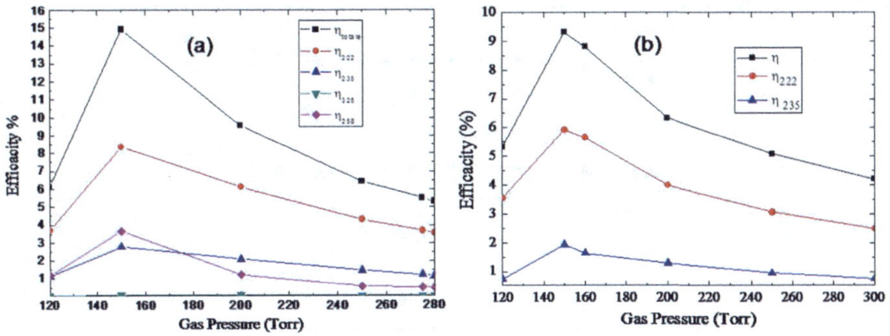

Fig. 64.1 Variation of the efficiencies for 222, 235, 325 and 258 nm emission for Kr/Cl$_2$ gas mixture as a function of the gas pressure under pulsed excitation in (a) and under sinusoidal excitation in (b)

around 9.6% and a partial efficiency for 222 nm is in the order of 6.4%. in contrast to sinusoidal excitation, under pulsed excitation the pulses are short with a rapidly increasing voltage over time, thus the atoms are rapidly excited and collide through harpoon processes faster for the formation of the KrCl * (B,C) exciplex. However, when the reaction rate is high the efficiency of the UV emission is high. Thus, the efficiency is in a maximum at a critical pressure of about 150 torr and then becomes low as the gas pressure increases. This is observed in both cases of pulsed excitation or sinusoidal excitation (Fig. 64.1).

64.3.2 Efficiency Study for a Pure Krypton Excilamp

The luminous efficacy behavior of the discharge in the case of pure Krypton discharge is completely different that in the krypton chlorine discharge. The lamp is in its maximum efficiency for a pressure of 450 torr (Fig. 64.2).

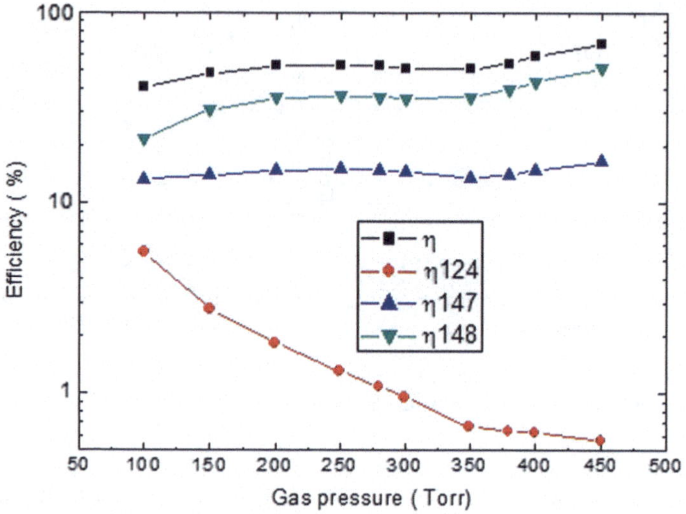

Fig. 64.2 Variation of efficiency with pressure of the gas for pure Krypton under sinusoidal excitation

64.4 Conclusion

The work carried out in this article consists in identifying and analyzing the dielectric barrier discharge system for plasma lamps. In this work, we have addressed, using digital discharge models, the kinetics of a high-pressure electric discharge for a noble gas halogen lamp (Kr-Cl_2) and for a pure krypton lamp. The use of two excitation modes, sinusoidal excitation and pulsed excitation, for the applied voltage was discussed to study the effect of the applied voltage waveform on the electrical and chemical parameters of the discharge. This allowed us to distinguish the best excitation mode to be used in the case of pure noble gas or rare gas and halogen mixture lamps in dielectric barrier discharges. A comparison between the efficiencies obtained for the different excitation modes: pulsed and sinusoidal shows that the use of pulsed voltage is the best way to improve the excilamps and this for critical and optimal working conditions on the gas pressure or on the voltage applied and the percentage of halogen in the case of gas mixtures in order to improve the radiative emission efficiency of the lamp.

References

1. K. Nassour, M. Brahami, S. Nemmich, N. Hammadi, N. Zouzou, A. Tilmatine, Comparative experimental study between surface and volume DBD ozone generator. Ozone-Sci. Eng. **38**, 70–76 (2016)
2. M.S. Tyagi, B.L. Meena, A.K. Sharma, R. Prakash, Analysis of discharge parameters in xenon-filled coaxial DBD tube. IEEE Trans. Plasma Sci. **39**(6), 1475–1481 (2011)
3. U.N. Pal, M. Kumar, H. Khatun, A.K. Sharma, Discharge characteristics of dielectric barrier discharge (DBD) based VUV/UV sources. in *International Symposium on "Vacuum Science and Technology", Journal of Physics: Conference Series*, vol. 114 (2008), p. 012065
4. S. Beleznai, G. Mihajlik, I. Maros, L. Balázs, P. Richter, High frequency excitation waveform for efficient operation of a xenon excimer dielectric barrier discharge lamp. J. Phys. D: Appl. Phys. **43**, 015203 (2010)
5. C. Heslin, D. Boehm, V. Milosavljevic, M. Laycock, P.J. Cullen, P. Bourke, Quantitative assessment of blood coagulation by cold atmospheric plasma. Plasma Med. **4**, 153–163 (2014)
6. H.J. Kim, H.I. Yong, S. Park, K. Kim, T.H. Kim, W. Choe, C. Jo, Effect of atmospheric pressure dielectric barrier discharge plasma on the biological activity of naringin. Food Chem. **160**, 241–245 (2014)
7. J.M. Green, A new generation of ultra-violet/visible gas lasers. Opt. Laser Technol. **10**(6), 289–300 (1978)
8. M.R. Fannery, T.P. Yang, App. Phys. Lett. **32**(5), 327 (1978)
9. M.R. Fannery, T.P. Yang, App. Phys. Lett. **33**(7), 574 (1978)
10. J.-Y. Zhang, I.W. Boyd, Multi-wavelength excimer ultraviolet sources from a mixture of krypton and iodine in a dielectric barrier discharge. Appl. Phys. B Lasers Opt. **71**, 177–179 (2000)
11. J.Y. Zhang, I.W. Boyd, Efficient excimer ultraviolet sources from a dielectric barrier discharge in raregas/halogen mixtures. Am. Inst. Phys. J. Appl. Phys. **80**(2), 633–638 (1996)
12. L.C. Ciobotaru, Investigation of monochromatization light effect at molecular/atomic level in electronegative-electropositive gas mixtures plasma. Int. J. Math. Comput. Phys. Electr. Comput. Eng. **6**, 1552–1556 (2012)
13. A. Belasri, N.L.D. Bachir, Z. Harrache, Plasma chemical and electrical modeling of a dielectric barrier discharge in Kr–Cl2 gas mixtures. Plasma Chem. Plasma Process. **33**(1), 131–146 (2013)
14. N.L.D. Bachir, A. Belasri, A simplified numerical study of the Kr/Cl2 plasma chemistry in dielectric barrier discharge. Plasma Sci. Technol. **15**(4), 343 (2013)
15. A. Belasri, Z. Harrache, Electrical and kinetical aspects of homogeneous dielectric-barrier discharge in xenon for excimer lamps. Am. Inst. Phys. Phys. Plasmas **17**, 123501 (2010)
16. N.L.D. Bachir, A. Belasri, P. Guillot, B. Caillier, Radiative emissions in visible–IR of krypton excilamp: experimental and theoretical interpretations. Plasma Chem. Plasma Process. **39**(5), 1243–1254 (2019)
17. B. Ren'an, S. Mingdong et al., Development of 146 nm vacuum UV light source. in *18th International Vacuum Congress*, 2012 Published by Elsevier under responsibility of Chinese Vacuum Society (CVS). Phys. Procedia **32**, 477–481(2012)

Chapter 65
Improvement in Electromechanical and Electrochemical of Low Carbon Steel Samples by PIII Treatment

M. M. Alim, R. Tadjine, A. Keffous, and M. Kechouane

Abstract In this work, nitriding by Plasma Immersion Ion Implantation (PIII) treatment of low carbon steel was performed in a plasma reactor with an inductive RF source (13.56 MHz) at low pressure in order to investigate the influence of the process conditions on the hardness and corrosion properties. A space charge sheath region is formed in front of the sample biased with a negative DC voltage. The positive ions accelerated within this sheath reach the sample with a high kinetic energy. The micro-hardness increased by $\sim 100\%$ for the nitrogen-implanted samples compared to the untreated one, the nitride phases (γ'-Fe_4N and ε-$Fe_{2\text{-}3}N$) are believed to participate to the hardening. A low corrosion rate is obtained, which indicates that nitriding by PIII improves the corrosion resistance behavior. These results are particularly interesting since they have been obtained for short treatment time (2 h) and relatively low bias voltages.

Keywords Plasma nitriding (PIII) · Low carbon steel · Micro-hardness · Corrosion resistance · DC bias voltage

---

M. M. Alim (✉) · R. Tadjine
Centre de Développement des Technologies Avancées,
Cité 20 Août 1956 B.P. 17, Alger, Algeria
e-mail: malim@cdta.dz

A. Keffous
Centre de Recherche en Technologie des Semi-conducteurs pour l'Energétique,
2 Bd Frantz Fanon B.P. 399, Alger, Algeria

M. Kechouane
Université des Sciences et des Technologies Houari Boumediene,
B.P. 32 El Alia, Bab Ezzouar, Alger, Algeria

A. Belasri and S. A. Beldjilali (eds.), *ICREEC 2019*, Springer Proceedings in Energy,
https://doi.org/10.1007/978-981-15-5444-5_65

65.1 Introduction

Nitriding is a surface treatment method that allows an improvement in hardness, wear and corrosion resistance of metallic materials [1]. This process can be produced by different methods such as gas, salt bath, plasma, and laser nitriding [1–3]. Due to the fact that these surface treatments (nitriding) are carried out at higher temperatures ($\sim$500 °C) and for a long time of treatment (20–80 h) [1], the deterioration of the surface, precipitation of carbides and grain growth might be possible, which may affect the mechanical properties of the material. Plasma Immersion Ion Implantation (PIII), which was introduced by Conrad et al. in 1988 [4], is the most distinguished technique used for nitriding purposes. Nowadays, this technique is used for the treatment of a large variety of materials [4–8]. PIII, which involves both the implantation and diffusion of nitrogen, is found to be effective in the temperature range of 250–500 °C for Carbon steels [7, 9]. In this technique, the sample to be nitrided is biased negatively with a high voltage [10–12] to create a space charge sheath region which accelerates the positive ions generated in the plasma towards the sample. The PIII-induced defects are responsible of modifications in some properties of the treated material, such as its hardness and corrosion resistance. These changes are highly dependent on the ion implantation conditions (power, pressure and negative bias voltage). This demonstrates the importance of optimizing the experimental parameters of this process. Low carbon steel is a metallic material employed for making several elements such as spindles of machines tools, bolts, shafts, keys and bigger gears [13]. In this study, we are interested in studying the effect of the self-heating mechanism induced by the application of a high negative DC voltage to the low carbon steel. Indeed, the DC voltage is known to be able to rapidly heat the samples, on the contrary to the pulsed polarization which necessitates more elevated bias voltages [14–16]. Thus, the DC polarization enables us to avoid an additional step of heating treatment which is generally used to diffuse the implanted ions. This work aims to improve the micro-hardness and corrosion resistance of the low carbon steel.

65.2 The Experimental Details

The experimental setup of the PIII system used for the nitriding treatment of the low carbon steel is shown in Fig. 65.1. The diameter of the spherical chamber is 930 mm, this chamber is connected to a pumping unit for evacuate the air up to a vacuum of 10^{-6} mbar. The pumping unit consists of a primary pump and a secondary oil diffusion pump to achieve a high order residual vacuum (10^{-6} mbar). To create the plasma a high-density plasma ($\sim 10^{12}$ cm^{-3}) [17–19], we used an inductive excitation system. The inductive excitation consists of a Pyrex tube surrounded by a Copper coil polarized by an (13.56 MHz) RF generator. A glass substrate holder with a length of 1.5 m has been used. To avoid the breakdown and arcs problem caused by the DC bias voltage [20], a quartz substrate holder shielded with a braid was necessitated. In order to polarize our samples, we used a negative high voltage power supply (HCN 2800–6500) was required. The chemical composition of the substrate material (Low carbon steels samples) of disc form (dimensions: 25 mm Ø and 8 mm thick) is; (in wt%) C: 0.38; Mn: 0.66; Si: 0.025; S: 0.0.16; P: 0.02; Mo: 0.02; Cr: 0.025; Ni: 0.02). The samples were polished to a surface finish using SiC emery papers (starting from 200 to 2400 grade). Subsequently, ultrasonic cleaning in a bath of acetone and alcohol for 10 min before treatment with PIII. The system was evacuated to a pressure (8.10^{-6} mbar), and the working pressure, from 3.10^{-2} to 12.10^{-2} mbar, was attained by introducing nitrogen into the reactor. Before the start of ion implantation, the native oxide layer on the top surface of the substrate was removed by sputtering with Argon gas for 20 min and a power of 200 W. A negative voltage ranging from 3.0 kV biased the substrate and the current varied from 6 to 10 mA. Then the Vickers micro-hardness of samples was calculated using the equation: $HV = 1854\ (F/d^2)$. At least, five measurements of the micro-hardness were taken for each test at different positions on the surface of the middle of each sample, and then the average values were taken as the average value of the micro-hardness of the sample. The measurements of the corrosion resistance of samples was carried on the electrochemical working station (PARSTAT 4000) in 3.5 wt% NaCl solution at room temperature (~25 °C) by DC polarization (Table 65.1).

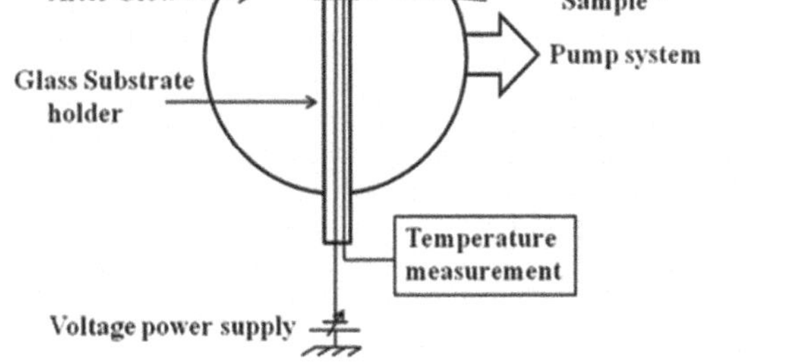

Fig. 65.1 The experimental setup

Table 65.1 The plasma nitriding parameters

Dimensions of inductively source	H = 0.17 m/D = 0.1 m
RF frequency	13.56 MHz
Injected power	500 W (at maximum)
Bias voltage	3.0 kV
Injected gas	N_2 (100%)
Pressure range	$2.10^{-2} - 8.10^{-2}$ mbar

65.3 Results and Discussion

65.3.1 Surface Hardening

Figure 65.2 shows the influence of the discharge power on the mico-hardness for 3.0 kV of negative bias voltage, 4.10^{-2} mbar of pressure and 120 min of treatment time. The micro-hardness increases from about 300 HV (at 160 W) to about 380 HV (at 320 W). This behavior can be explained by the increase of the discharge power which increases the density of ion species in the plasma, thus assisting the formation of the nitride phases γ'-F_4N and ε-$Fe_{2\text{-}3}N$. Gressman et al. [21] used ab initio calculations to confirm the presence of an elastically anisotropic nitride phase (γ'-Fe_4N) in carbon steels. This phase is believed be source of hardening as indicated by Yan et al. [22] who study the mechanical properties of the γ'-Fe_4N phase and found that this latter has a higher micro-hardness compared to that of the parent α-Fe phase.

Figure 65.3 shows the mico-hardness of the treated samples as a function of pressure, for a bias voltage of 3.0 kV, a discharge power of 160 W and 120 min of treatment time. The mico-hardness increases with pressure, and this trend is due to the increase of the nitrogen amount in the plasma (i.e. excited and neutral species).

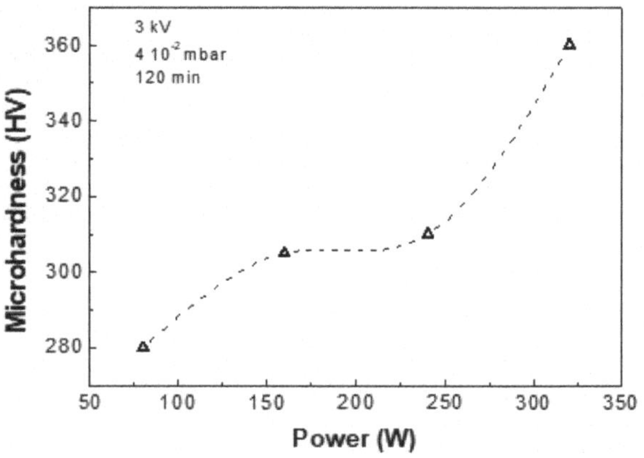

Fig. 65.2 Micro-hardness of the implanted lox carbon steel sample versus the injected power

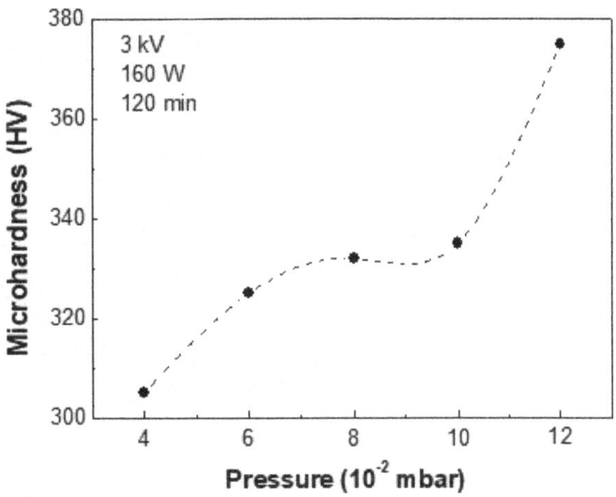

Fig. 65.3 Mico-hardness of the implanted low carbon steel sample versus the pressure

Thus, we can say that the ion density of species does not vary with the pressure. Consequently, ions are not the unique species that participate to the nitriding, neutral species are also suspected to contribute to the formation of nitride phases γ'-Fe_4N and ε-$Fe_{2\text{-}3}N$ as suggested by some authors [23, 24].

65.3.2 Potentiodynamic Polarization

Figure 65.4 shows the polarization curves potentiodynamic of untreated and treated low carbon steel samples by nitrogen with 60 min and 120 min under condition: 160 W, 10^{-1} mbar 3.0 kV into saline solution. We noted that the implanted sample with 120 min at lower current density leads to lower corrosion rate compared to the untreated one. The introduction of nitrogen gave more protection against corrosion for low carbon steel in an aqueous solution containing chlorine (Cl^-).

In the existing literature, negative high DC voltages ranging typically from 20 up to 100 kV were used [8]. Smith et al. [25] observed an increase in the pitting potential (i.e. the potential at which localized corrosion initiates and propagates) of ~600 mV for 304L stainless steel samples PIII-implanted with nitrogen at 20 kV. The work of Basu et al. [26] who studied the corrosion behaviour of nitrided steel by plasma in the DC pulsed glow discharge and found that at 3 h of treatment and at 560 °C, the corrosion current density is reduced by a factor of 0.25. The formation of the crystalline phases γ'-Fe_4N and ε-$Fe_{2\text{-}3}N$ to improve the corrosion resistance of treated samples.

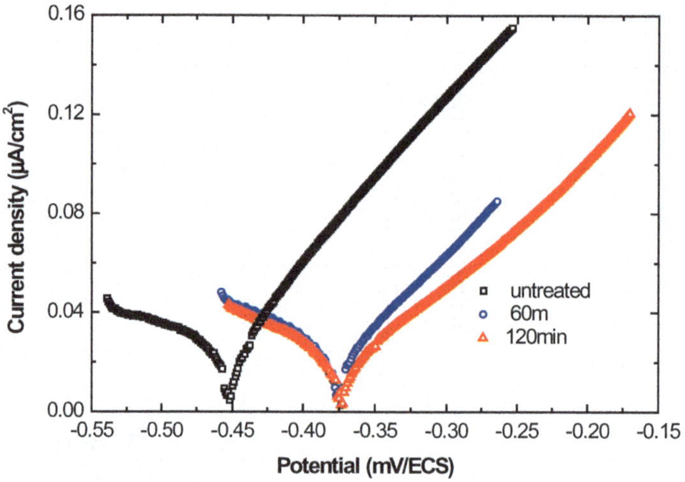

Fig. 65.4 Potentiodynamic polarization curves of untreated and treated samples, with different treatment times (0, 60 min, and 120 min)

65.4 Conclusion

The work presented in this paper is plasma immersion ion implantation of nitrogen into low carbon steel. In this study, plasma nitriding of the samples by self-induced heating mechanism is investigated under different operating conditions (injected power, pressure and negative bias voltage) in the aim of improving the corrosion resistance and micro-hardness of the studied material.

The micro-hardness increased by ~100% for the treated samples by nitriding compared to the untreated one. This study confirmed that both γ'-Fe_4N and ε-$Fe_{2\text{-}3}N$ phases participate to the hardening. The hardest surface (about 380 HV) is obtained for the experimental conditions: 3.0 kV, 320 W, 120 min and 4.10^{-2} mbar.

Plasma nitriding has improved the corrosion resistance of the treated samples. This result in corrosion resistance is mainly due to the presence of the γ'-Fe4N phase.

References

1. T. Balusamy, T.S.N. Sankara Narayanan, K. Ravichandran, I.S. Park, M.H. Lee, Mater. Charact. **85**, 38–47 (2013)
2. W. Jun, L. Yuanhua, F. Hongyuan, Z. Dezhi, P. Qian, S. Baoluo, J. Mater. Eng. Perform. **21**, 1708–1713 (2012)
3. Y. Zhao, B. Yu, L. Dong, H. Du, J. Xiao, Surf. Coat. Technol. **210**, 90–96 (2012)
4. J.R. Conrad, R.A. Dodd, F.J. Worzala, X. Qiu, Surf. Coat. Technol. **36**, 927 (1988)

5. M.M. Alim, N. Saoula, R. Tadjine, F. Hadj-Larbi, A. Keffous, M. Kechouane, Eur. Phys. J. Appl. Phys. **75**, 30801 (2016)
6. J.R. Conrad, F.J. Worzala, R.A. Dodd, Mater. Sci. Eng. **128**, 259–268 (1990)
7. A.M. Redsten, K. Sridharan, F.J. Worzala, J.R. Conrad, J. Mater. Process. Technol. **30**, 253–261 (1992)
8. A. Anders, *Handbook of Plasma Immersion Ion Implantation and Deposition* (Wiley, New York, 2000)
9. P. Saravanan, V.S. Raja, S. Mukherjee, Surf. Coat. Technol. **201**, 8131 (2007)
10. D. Gupta, Int. J. Adv. Technol. **2**, 471–490 (2011)
11. A. Anders, Surf. Coat. Technol. **156**, 3–12 (2002)
12. X.Y. Qian, N.W. Cheung, M.A. Lieberman, Nucl. Instrum. Method Phys. Res. **55**, 821–825 (1991)
13. K. Srithar, K. Ganesh Babu, Design of Machine Elements (2009), p. 504
14. M. Ueda, G.F. Gomes, E. Abramof, H. Reuther, Nucl. Instrum. Methods Phys. Res. B **206**, 749–753 (2003)
15. I.A. Rojas-Olmedo, R. López-Callejas, A. de la Piedad-Beneitez, R. Valencia-Alvarado, R. Peña-Eguiluz, A. Mercado-Cabrera, S.R. Barocio, A.E. Muñoz-Castro, B.G. Rodríguez-Méndez, Surf. Coat. Technol. **205**, 397–401 (2011)
16. L.A. Berni, M. Ueda, G.F. Gomes, A.F. Beloto, H. Reuther, J. Phys. D Appl. Phys. **33**, 1592–1595 (2000)
17. M.M. Alim, F. Hadj-Larbi, R. Tadjine, Int. J. Eng. Res. Afr. **1663–4144**(46), 1–6 (2020)
18. A. Anders, Surf. Coat. Technol. **200**, 1893–1906 (2005)
19. Z.J. Wang, X.B. Tian, C.Z. Gong, J.W. Shi, S.Q. Yang, K.Y.R. Fu, K.P. Chu, Surf. Coat. Technol. **206**, 5042–5045 (2012)
20. M.P. Fewell, J.M. Priest, M.J. Baldwin, G.A. Collin, K.T. Short, Surf. Coat. Technol. **131**, 284–290 (2000)
21. T. Gressmann, M. Wohlschlögel, S. Shang, U. Welzel, A. Leineweber, E.J. Mittemeijer, Z.-K. Liu, Acta Mater **55**, 5833–5843 (2007)
22. M.F. Yan, Y.Q. Wu, R.L. Liu, Appl. Surf. Sci. **255**, 8902–8906 (2009)
23. R. Asthana, A. Kumar, B. Dahotre, Materuials Processing and Manufacturing Science, (Academic Press is an imprint of Elsevier, 2005), p. 359
24. J.C. Díaz-Guillén, J.A. Díaz-Guillén, E.E. Granda-Gutiérrez, M.R. Díaz-Guillén, M.A. González-Albarrán, Int. J Electrochem Sci **8**, 973–982 (2013)
25. J. Alphonsa, G. Sinha, A. Kumar, G. Jhala, S.K. Tiwari, S. Gupta et al., Effect of frequency on theproperties of plasma nitrided AISI 4340 steel. J Metall Mat Sci **50**(2), 119–125 (2008)
26. A. Basu, J. Dutta Majumdarb, J. Alphonsa, S. Mukherjee, I. Manna, Mater. Lett. **62**, 3117 (2008)

Chapter 66
Distributions of Chemical Species Produced by Positive Corona Discharge Using Multi-wire Emitting Electrodes

M. R. Bouazza, K. Yanallah, M. Bouadi, and F. Pontiga

Abstract The objective of this work is to study the effect of changing the electrode configuration ('wire-plane' (a), 'two wires-plane' (b) and 'three wires-plane' (c)) on the spatial distribution of the neutral chemical species generated by a positive corona discharge, with a special focus on the effect of the electric wind on the distribution of these chemical species. This task is accomplished using a mathematical model that is based on the numerical resolution of Navier-Stokes equation coupled to the continuity equations of neutral species generated by the positive corona discharge. Regarding the chemical reactions, a model of plasma chemistry, which includes the most important chemical reactions occurring between electrons, atoms and molecules in the air, has been used. One of the main results is the comparison of the ozone density produced by the positive corona discharge for the three mentioned electrode configurations.

Keywords Positive corona discharge · Electric wind · Neutral chemical species

66.1 Introduction

In the positive corona discharge, when the anode is subjected to the high positive voltage, the seed electrons are accelerated by the electric field and, through collisions with the neutral particles, they generate other electrons and ions, which are accelerated by the electric field as well. During collisions of these ions with the neutral molecules of the gas, they transmit to them a part of their momentums. Thus, it appears a macroscopic gas flow named 'electric wind' or electro-hydrodynamic (EHD) flow whose maximum speed can reach several meters per second. Known for a long time [1], and because of its intrinsic complexity, this electric wind remains an

M. R. Bouazza (✉) · K. Yanallah · M. Bouadi
Dpt. de Physique, Faculté des Sciences de la Matière, Laboratoire de Génie Electrique et des Plasmas, University of Tiart, Tiaret, Algeria
e-mail: mohamed.bouazza@univ-tiaret.dz

F. Pontiga
Departamento de Física Aplicada II, Universidad de Sevilla, Seville, Spain

A. Belasri and S. A. Beldjilali (eds.), *ICREEC 2019*, Springer Proceedings in Energy,
https://doi.org/10.1007/978-981-15-5444-5_66

important subject of investigation and research, given its remarkable industrial interest. Among the industrial applications of the electric wind are mentioned the improvement of heat transfer [2], micro-pumps [3], the collection of dust particles [4], and the control of flows [5] (particularly in aeronautics).

In most of the previous studies interested in the chemical aspect of the corona discharge, the chemical species are supposed to be extracted from the corona reactor by imposing an external flow, without taking into account the effect of the electric wind. To our knowledge, the effect of the electric wind on the spatial distribution of chemical species generated by a positive corona discharge was simulated numerically for the first time by our group [6].

In the present work, we extend our previous study by considering three types of electrode configurations of the positive corona discharge which are 'wire-to-plate' (case a), 'two wires-to-plate' (case b) and 'three wires-to-plate' (case c), and we compare the effect of these three electrode configurations on the electric wind profile and the ozone distribution. The same strategy that we used before [7] will be adopted here to determine the air velocity for the three types of reactors. This strategy consists in injecting an analytical approximation of the EHD force, generated by the corona discharge, into the Navier-Stokes equation, which helps to determine the EHD flow velocity. This analytical approach reduces greatly the modelling effort since it is not necessary to solve the charged particle continuity equations and the Poisson equation, together, to determine the EHD force. In addition to the EHD flow velocity, which is required for solving the neutral continuity equations, the electron density is needed as well to determine the spatial distribution of the neutral particles. The electron density will be calculated by means of an analytical approximation [7] and, for comparison purposes, the plasma chemical kinetic model of Chen and Davidson [8], which has been validated experimentally, will be used.

66.2 The Neutral Species Model

In the present work, the positive corona discharge is supposed to be produced in air at atmospheric pressure and room temperature, between parallel 'wires-to-plate' electrodes, in a reactor formed by two parallel plates (see Fig. 66.1). The upper and the lower plates help to confine the gas flow between them. The distance between these two plates is $h = 1.51$ cm and the length of each one is $l = 15$ cm. The wires are centered between the two plates and parallel to them, but perpendicular to the gas flow. The radius of each wire and the distance between each wire and the lower plate are $r_0 = 100$ μm and $d = 0.75$ cm, respectively. The wires are subjected to high DC positive voltage and the lower plate is grounded. The electrical discharge is assumed to be uniform along each wire, which reduces the modelling geometry to two dimensions. For the configurations (case b) and (case c), the inter-wire distance is $f = 1$ cm.

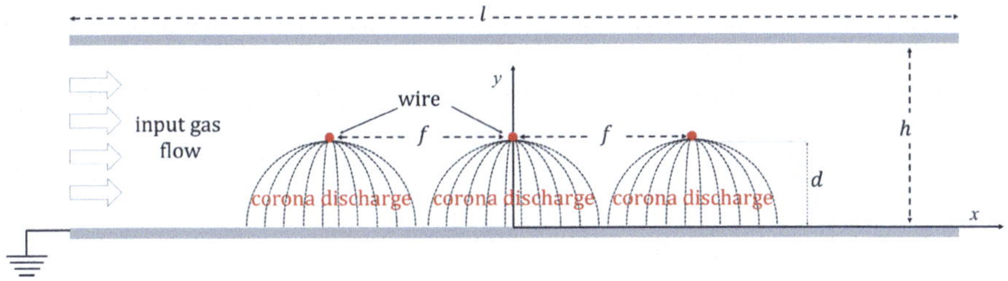

Fig. 66.1 Schematic illustration of the corona reactor

The spatial distributions of the neutral chemical species generated during the positive corona discharge are obtained by solving a set of continuity equations for each species. According to Soria et al. [9], the continuity equations in the stationary regime of the positive corona discharge are expressed as follows:

$$\nabla \cdot (N_j \mathbf{V} - D_j \nabla N_j) = S_j \tag{1}$$

where N_j, D_j and S_j are the neutral number density, the diffusion coefficient and the gain/loss rate of the jth neutral specie, respectively; and $\mathbf{V}$ is the gas velocity obtained from the solution of Navier-Stockes equation [10]. As indicated, and since this work focuses on the EHD flow and its effect on the distribution of neutral chemical species, a simplified chemical kinetics model for air, suggested by Chen and Davidson [8], will be considered. This model consists of 10 neutrals species (O, N, O_3, NO, NO_2, NO_3, N_2O, N_2O_5, O_2* and N_2*) and 24 reactions. The list of reactions and their rate constants were given in [11], and the values of the diffusion coefficients were taken from Chen [12]. Accurate knowledge of the distribution of the electron density is essential to determine the spatial distributions of neutral particles, since it is used to compute the electronic impact reaction rates in the continuity equations. So, the spatial distribution of the electron density in a 'wire-to-plate' positive corona discharge is taken from a previous analytical model [7]. Finally, let us point out that the resolution of the system of Eq. (1) is based on the finite volume method.

66.3 Results and Discussion

The results exposed in this section correspond to an applied voltage on each electrode wire ϕ = 10 kV, which gives rise to an electric current intensity per unit of wire length i = 24.70 μA/cm [13, 14], and a gas inlet velocity V_0 = 2 cm/s.

It is worth noting that the presented results corresponding to 'wire-to-plate' electrode configuration (case a) have been validated experimentally by Chen et al. [12], and by our simulation work [6].

Before describing the 2D spatial distribution of ozone, it is convenient to first present the profiles of the gas velocity and the atomic oxygen density generated by the corona discharge for the three electrode configurations mentioned before, since these two quantities have great influence on ozone.

66.3.1 Velocity Distribution

The velocity distributions for the three cases are presented in Fig. 66.2.

In the case (a) 'wire-plane', the EHD force accelerates the air up to a maximum velocity of 3.9 m/s and forms two large symmetrical vortices, rotating in opposite directions and extending until reaching the two plates. These high velocity values have been experimentally confirmed for different electrode configurations of the corona discharge [15]. In the case (b) 'two wires-to-plate', the gas velocity keeps the same overall profile, but the two large vortices, while keeping the same size and the same direction of rotation, are displaced along the x axis in opposite directions. These two large vortices are accompanied by the formation of two small ones between the two wires, each turning in opposite direction to that of the adjacent large vortex. Finally, in the configuration 'three wires-to-plate' (case c), while the velocity profile remains similar, two additional smaller vortices are interposed in the same way between the large vortices, which further increases the separation between them.

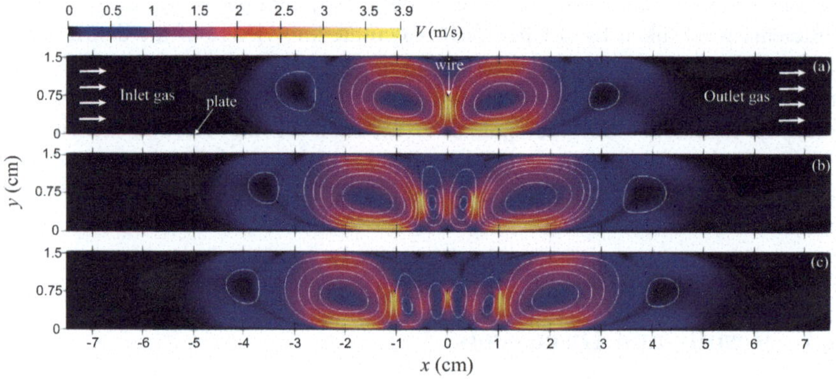

Fig. 66.2 2-D spatial distribution of air velocity amplitude and flow lines for three electrode configurations: 'wire-to-plate' (a), 'two wires-to-plate' (b) and 'three wires-to-plate' (c), and corresponding to a wire radius r_0 = 100 μm, an inter-electrode space d = 0.75 cm, an inter-wire distance f = 0.5 cm and an applied voltage ϕ = 10 kV

66.3.2 Spatial Distribution of Atomic Radicals

When the applied voltage is positive, free electrons in the inter-electrode space are accelerated by the electric field towards the wire (case a) or the wires (cases b and c). In the ionization region, very close to the wires, where the value of the reduced electric field (*E/N*, where *N* is the molecular density) exceeds 120 Townsend (Td), which represents the ionization threshold, inelastic collisions between the electrons and neutral molecules of the gas produce pairs of electron-positive ion. The new released electrons are accelerated as well by the electric field and there will be the formation of electronic avalanches. The positive ions drift towards the plate, while electrons rapidly go towards the wire initiating important chemical reactions. Among these reactions, the dissociation of the neutral molecules, such as O_2 and N_2, by the electronic impact leads to the formation of atomic species (or atomic radicals) like O (Fig. 66.3) and N:

$$e + O_2 \rightarrow O + O + e, \qquad k_1 = f(E/N) \tag{2}$$

$$e + N_2 \rightarrow N + N + e, \qquad k_2 = f(E/N) \tag{3}$$

The characteristic lifetime of the atomic oxygen (O) is very short, $\tau_O \approx (k_4\,[O_2]\,[N_2])^{-1} \sim 20\ \mu s$. This short lifetime duration can explain the rapid consumption of atomic oxygen, which contributes to ozone production:

$$O + O_2 + O_2 \rightarrow O_3 + O_2, \qquad k_3 = 6 \times 10^{-34} cm^6/s \tag{4}$$

$$O + O_2 + N_2 \rightarrow O_3 + N_2, \qquad k_4 = 5.6 \times 10^{-34} cm^6/s \tag{5}$$

Since the ionization layer is very small ($\sim$20 to 40 μm) [11], and the gas velocity must vanish on the surface of each electrode (no-slip boundary condition), the extraction of these species from the ionization zone is mainly based on the molecular diffusion. However, once these species diffuse a few hundred micrometres away from the positive electrode (the wire(s)), the ionic wind will affect their spatial distributions through the convection (Fig. 66.3).

66.3.3 Spatial Distribution of Ozone (O_3)

The 2D spatial distribution of ozone for the three electrode configurations is shown in Fig. 66.4.

The high concentration of ozone near the wire surface (ionization zone) is explained by the presence of a large amount of atomic oxygen O in this area, because it is through the consumption of O, reacting with O_2, that O_3 is formed.

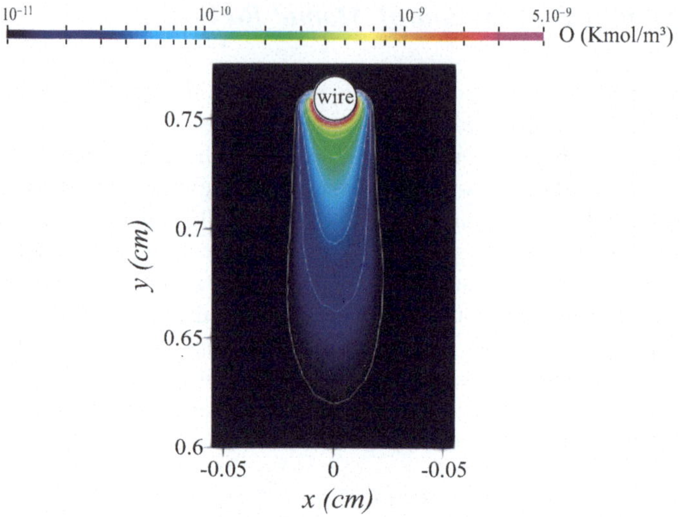

Fig. 66.3 2D-spatial distribution of oxygen atoms (O) near the wire, and corresponding to an input velocity $V_0 = 2$ cm/s, $r_0 = 100$ µm, $d = 0.75$ cm and $\phi = +10$ kV

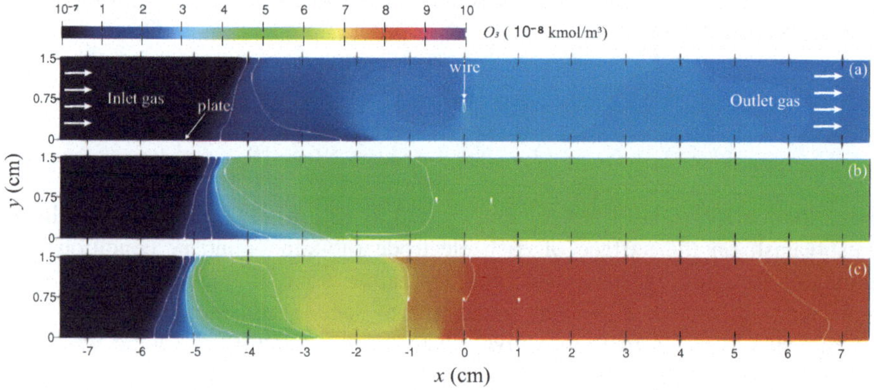

Fig. 66.4 2D-spatial distribution of ozone (O_3) for three electrodes configurations: 'wire-to-plate' (a), 'two wires-to-plate' (b) and 'three wires-to-plate' (c), and corresponding to a gas inlet velocity $V_0 = 2$ cm/s, a wire radius $r_0 = 100$ µm, an inter-electrode space $d = 0.75$ cm, an inter-wire distance $f = 0.5$ cm and an applied voltage $\phi = +10$ kV

Due to the negligible intensity of the ionic wind near the wire, ozone must escape from this region mainly by mollecular diffusion [6]. However, the vortices generated by the ionic wind homogenise very efficiently the distribution of ozone (Fig. 66.4). Even more, ozone spreads against the gas flow and approaches more and more the reactor inlet with the enlargement of the turbulence zone. Of course, the increasing number of wires is the origin of this enlargement. The main species responsible for the ozone destruction is the nitric oxide, which oxidizes ozone to form nitrogen dioxide according to the following reaction:

$$NO + O_3 \rightarrow NO_2 + O_2, \qquad k_5 = 2.0 \times 10^{-14} cm^3/s \tag{6}$$

However, and contrary to the atomic species, the lifetime of O_3 is very long, $\tau_{O3} \approx (k_5\,[NO])^{-1} \sim 10$ s. This lifetime is even longer than the residence time of the gas in the reactor ($\sim$7.5 s), according to the inlet velocity (2 cm/s). As a result, the ozone produced in the active region of the corona discharge can be transported far away from the wire with small losses. Finally, it should be noted that the maximum density of ozone rises with number of wires (Fig. 66.4). More precisely, at the reactor outlet and along the y axis, the maximum ozone density is about 2.71 $\times$ 10^{-8} kmol/m^3 for a single wire, while it nearly doubles for two wires, and almost triples for three wires.

66.4 Conclusion

In this work, we have obtained the spatial distributions of ozone (O_3) through the modelling of the positive corona discharge using multi wires-to-plate electrode configurations. A special emphasis has been placed on the role of the electric wind on the spatial distributions of ozone. The comparison between the three electrode configurations reveals that:

- The EHD flow generated by the positive corona discharge gives rise to a considerable enhancement to the chemical activity and to an intense recirculation of the gas. Furthermore, the number of vortices and the recirculation intensity of the gas increase with increasing the number of wires.
- The atomic species O, which is generated in the ionization zone, is less affected by the EHD flow, because of its very short lifetime. On the other hand, and unlike the results of other studies, where the spatial distributions of O_3 show a large density gradient almost exclusively downstream of the gas flow, this study demonstrates that, for the three configurations, high concentrations of the O_3 species in the presence of the EHD wind can even be found upstream of the gas flow, spreading progressively towards the reactor inlet when the number of wires increases.

References

1. M. Robinson, A history of the electric wind. Am. J. Phys. **30**, 366–372 (1962)
2. M. Molki, P. Damronglerd, Electrohydrodynamic enhancement of heat transfer for developing air flow in square ducts. Heat Transf. Eng. **27**, 35–45 (2006)
3. L. Zhao, K. Adamiak, EHD flow produced by electric corona discharge in gases: from fundamental studies to applications (a review). Part. Sci. Technol. **34**, 63–71 (2016)
4. J. Podliński, J. Dekowski, J. Mizeraczyk, D. Brocilo, J.-S. Chang, Electrohydrodynamic gas flow in a positive polarity wire-plate electrostatic precipitator and the related dust particle collection efficiency. J. Electrost. **64**, 259–262 (2006)
5. E. Moreau, G. Touchard, Enhancing the mechanical efficiency of electric wind in corona discharges. J. Electrost. **66**, 39–44 (2008)
6. K. Yanallah, F. Pontiga, M.R. Bouazza, J.H. Chen, The effect of the electric wind on the spatial distribution of chemical species in the positive corona discharge. J. Phys. Appl. Phys. **50**, 335203 (2017)
7. M.R. Bouazza, K. Yanallah, F. Pontiga, J.H. Chen, A simplified formulation of wire-plate corona discharge in air: application to the ion wind simulation. J. Electrost. **92**, 54–65 (2018)
8. J. Chen, J.H. Davidson, Ozone production in the positive DC corona discharge: model and comparison to experiments. Plasma Chem. Plasma Process. **22**, 495–522 (2002)
9. C. Soria, F. Pontiga, A. Castellanos, Plasma chemical and electrical modelling of a negative DC corona in pure oxygen. Plasma Sources Sci. Technol. **13**, 95 (2004)
10. H.K. Versteeg, W. Malalasekera, *An Introduction to Computational Fluid Dynamics: The Finite Volume Method* (Pearson Education, London, 2007)
11. K. Yanallah, F. Pontiga, J.H. Chen, A semi-analytical study of positive corona discharge in wire–plane electrode configuration. J. Phys. Appl. Phys. **46**, 345202 (2013)
12. J. Chen, Direct Current Corona-Enhanced Chemical Reactions. University of Minnesota, 2002
13. R.S. Sigmond, The unipolar corona space charge flow problem. J. Electrost. **18**, 249–272 (1986)
14. W. Deutsch, Über die dichteverteilung unipolarer ionenströme. Ann. Phys. **408**, 588–612 (1933)
15. D.F. Colas, A. Ferret, D.Z. Pai, D.A. Lacoste, C.O. Laux, Ionic wind generation by a wire-cylinder-plate corona discharge in air at atmospheric pressure. J. Appl. Phys. **108**, 103306 (2010)

Chapter 67
Parametric Study of Gas Heating Effect in Ne–Xe DBD Mixture, Application for Excimer Lamps

Amar Benmoussa and Ahmed Belasri

Abstract The effect of the discharge parameters on gas temperature in Ne–Xe dielectric barrier discharge DBD is investigated for operating conditions typical to excimer lamps functioning at high pressure. The gas heating effect in the DBD of gas mixture can affects the electrical characteristics of the excilamps, and consequently the homogeneity of excimer density. The gas temperature effect is originating from Joule heating approximation. The gas temperature development in the discharge was calculated by the heat transport equation resolution. This equation was coupled with the transport equation of the electrons and the ions of fluid model for a parallel-plate dielectric barrier discharge DBD. A parametric study of the influence of some discharge parameters such as the applied voltage, the total gas pressure, the dielectric capacitance, the xenon content in mixtures, and the secondary emission coefficient at the dielectric material is essential to see the effect of these discharge parameters on the rise of gas heating and consequently to the excilamps efficiency.

Keywords Gas heating effect · Excimer lamps · Ne–Xe DBD

67.1 Introduction

The technology application of dielectric barrier discharge (DBD) in eximer lamps have been an increased interest by scientific community [1], because it provide an efficiency to generate a high VUV and UV light sources in the excimer lamps. The dielectric discharge lamps convert electrical energy into radiation by transforming electrical energy into kinetic energy of moving electrons, which in turn is converted into electromagnetically radiation as a result of some kind of collision process with atoms of gas [2–6]. Over 80% of electrical energy is lost on discharge as heat Joule

A. Benmoussa (✉) · A. Belasri
LPPMCA, Faculté de Physique, Université des Sciences et de la Technologie d'Oran Mohamed Boudiaf USTO-MB, 31000 Oran, Algeria
e-mail: Benmoussa_am@yahoo.fr

A. Belasri and S. A. Beldjilali (eds.), *ICREEC 2019*, Springer Proceedings in Energy,
https://doi.org/10.1007/978-981-15-5444-5_67

effect in the DBD [7]. Recent experimental work showed that the development of the gas heating instability in high-pressure DBD is evident [8–11], and it has been observed that lower specific energy deposition would be beneficial for discharge stability by reducing the cathode sheath temperature [12–18]. This effect of gas heating development in the DBD has been the main goal of the present work, which represent a further work of our previous study of this kind of instability in Ne–Xe mixture [19]. The results obtained from this simulation gas temperature confirm that the choice of discharge parameters can plays an important role in gas heating. The calculation of gas temperature profiles are obtained by heat transport equation [20]. The model used in this work to describe the discharge behavior and our results of parametric study of gas heating are detailed in the next sections.

67.2 Description of the Physical Model

A one-dimensional fluid has been used to describe the behavior of Ne–Xe DBD. This model is based on continuity equations resolution for charged particles coupled to Poisson's equation for parallel plate geometry of DBD. We note that the present model is similar to that used in our previous works (See Refs. [21–23]).

The gas heating effect considered in this numerical investigation is originating from Joule heating approximation. The profile of gas temperature calculated in Ne–Xe DBD is determined by heat transport equation resolution. The formulation of heat transport equation is as follows:

$$\frac{\partial^2 T_g(x)}{\partial x^2} + \frac{P(x)}{k} = 0 \tag{1}$$

When T_g (x) represent the gas temperature distribution in discharge gap distance of DBD. P(x) and k are the dissipated power in the discharge and thermal conductivity, respectively. The value of thermal conductivity of Ne–Xe gas mixture was calculated by linear relationship mentioned in Ref. [24].

67.3 Results and Discussion

An important investigation of discharge parameters effect on gas temperature evolution in Ne–Xe DBD was carried out in order to obtain the operating conditions of DBD functioning and to improve the performance of excimer lamp. We note that the discharge conditions of Ne–Xe DBD used in this frame of work are based on recent literature, are considered as follows: the gas pressure (p) is taken equal to 250 Torr. The gap length (d) is 0.5 cm, the dielectric capacitance (C_d) is equal to 0.23 nF\cm^2. The gas mixtures Ne–Xe is taken for 5, 10, 40, and 50% of xenon in neon. The gas temperature (T) of both electrodes is equal to 300 °k, the peak value of applied voltage (Vsinus) is 5000 V, and the frequency (f) is fixed to 50 kHz.

In Fig. 67.1, we show the spatial variation of electric field in Ne–Xe DBD. We can see clearly that the electric field increases with an increasing in xenon concentration. The distortion of the electric field near the cathode sheath is caused by the positive charge developing at this region. It seems also that the increase of xenon concentration in Ne–Xe gas mixture leads to a significant values of electric field.

The gas temperature profiles of 5% Xe–Ne for sinusoidal and rectangular applied voltage were plotted in Fig. 67.2. We mentioned here that the input power in the DBD discharge is the same for both sinusoidal and rectangular applied voltage waveforms. The gas heating takes large values for rectangular applied voltage waveform this can explain by the production process of charged particles in the DBD. In the case of rectangular excitation waveform, the voltage polarity is so fast than the sinusoidal excitation waveform.

The spatial variations of the gas heating in the gap discharge of DBD operated with sinusoidal applied voltage for 5% xenon in neon and with different values of coefficient (γ) are plotted in Fig. 67.3. The gas temperature profile becomes important for high values of secondary electron emission process. We underline that the data of secondary electron emission used in this section was reported in literature (See Ref. [18]).

We plot in Fig. 67.4 the peak of gas temperature as a function of xenon percentage in the mixture for both rectangular and sinusoidal applied voltage. We can see that the maximum of the gas temperature increases linearly with xenon percentage in the mixture for rectangular and sinusoidal. This finding suggests that the gas heating take important values in the case of rectangular voltage when xenon percentage becomes important in Ne–Xe DBD.

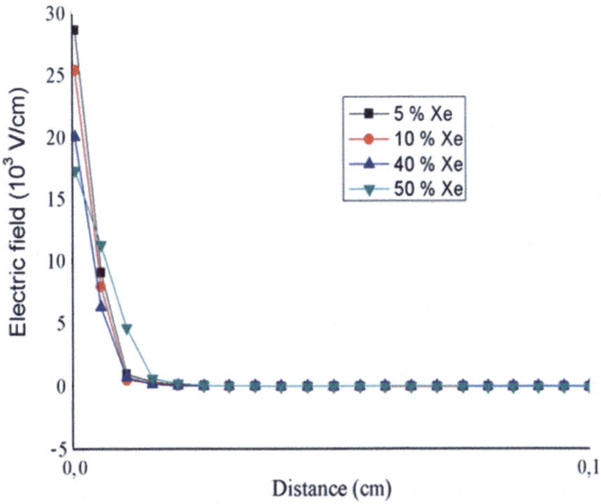

Fig. 67.1 The spatial variation of electric field in DBD for different concentrations of Xe in the mixture.

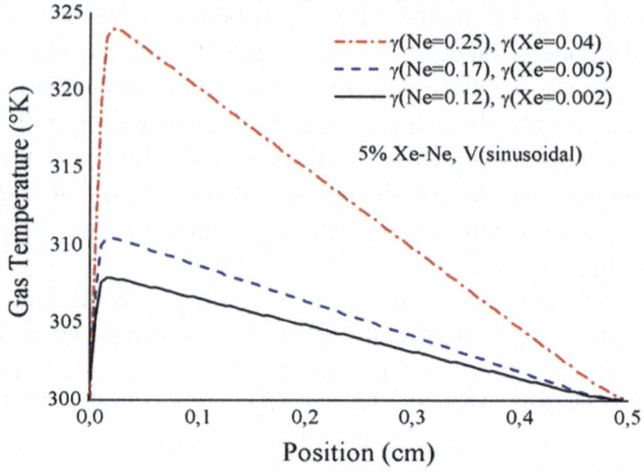

Fig. 67.2 The spatial variation of gas temperature of 5% Xe–Ne gas mixture in the DBD.

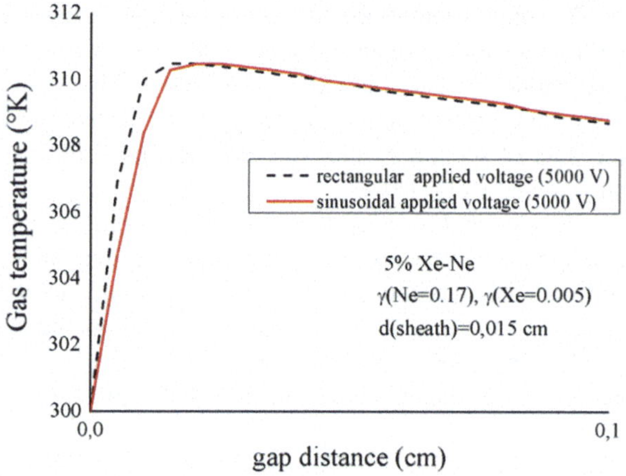

Fig. 67.3 The gas temperature profiles for many values of γ in the DBD.

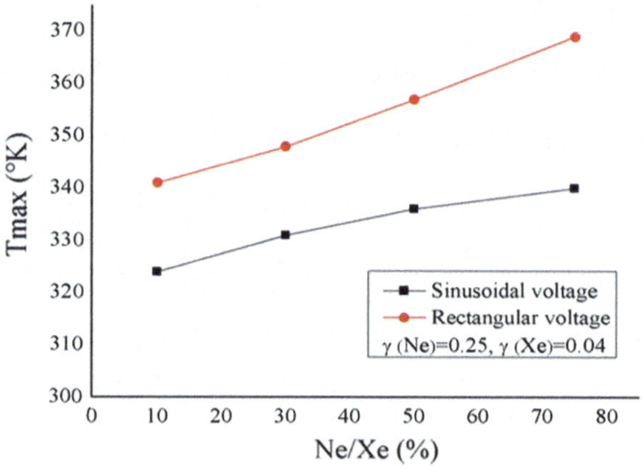

Fig. 67.4 The peaks of gas temperatures in DBD for sinus and rectangular voltages waveforms.

67.4 Conclusion

In the present work, a parametric investigation of gas heating in Xe–Ne DBD was carried out by using a one-dimensional fluid model coupled to heat transport equation. A significant increase of gas heating near the cathode region is owing to the high values of ionic power deposited in the discharge. This power is given by the product of the electric field and ion current density. The results show that the variation of some discharge parameters such as the shape of applied voltage, xenon percentage in Xe–Ne gas mixture, and secondary emission coefficient can play an important role to choice the optimum conditions of excimer lamp functioning.

Finally, the outcome of parametric study obtained by this simulation can allow us to a best understanding of the effect of discharge parameters on gas heating evolution which leads to an improvement in excilamps efficiency.

References

1. U. Kogelschatz, B. Eliasson, W. Egli, Dielectric-barrier discharges principle and applications. J. Phys. IV France **7**(C4), C4-47–C4-66 (1997)
2. R.P. Mildren, R.J. Carman, Enhanced performance of a dielectric barrier discharge lamp using short-pulsed excitation. J. Phys. D Appl. Phys. **34**(1), L1–L6 (2001)
3. S. Kanazawa, M. Kogoma, T. Moriwaki, S. Okazaki, J. Phys. D Appl. Phys. 21838 (1988)
4. P. Gulati, R. Prakash, U.N. Pal, M. Kumar, V. Vyas, Appl. Phys. Lett. **105**, 013505 (2014)
5. A. Belasri, Z. Harrache, Plasma Chem. Plasma Process. **31**, 787 (2011)
6. S. Bhosle, R. Diez, H. Piquet, D. Le Thanh, B. Rahmani, D. Buso, in *Excerpt from the Proceedings of the COMSOL Conference* (Hannover, 2008)
7. R.J. Carman, R.P. Mildren, J. Phys. D Appl. Phys. **36**, 19–33 (2003)
8. S. Beleznai, G. Mihajlik, A. Agod, I. Maros, R. Juhàsz, Z. Németh, L. Jakab, P. Richter, J. Phys. D. Appl. Phys. **39**, 3777–3787 (2006)
9. S. Baadj, Z. Harrache, A. Belasri, Plasma Phys. Rep. **39**, 1043–1054 (2013)
10. H. Akashi, A. Oda, Y. Sakai, 28th ICPIG, 15–20 July, Prague, Czech Republic (2007)
11. M. Boichenko, M.I. Lomaev, V.F. Tarasenko, Laser Phys. **18**, 738–748 (2008)
12. B. Eliasson, U. Kogelschatz, IEEE Trans. Plasma Sci. **19**(2), 309–323 (1991)
13. A. Belasri, K. Khodja, S. Bendella, Z. Harrache, J. Phys. D Appl. Phys. **43**(445202), 10 (2010)
14. J.-Y. Zhang, I.W. Boyd, J. Appl. Phys. **80**(2), 633–638 (1996)
15. J.P. Boeuf, Phys. Rev. **36**, 2782–2792 (1987)
16. I. Pérès, L.C. Pitchford, J. Appl. Phys. **78**, 774–782 (1995)
17. K.S. Moon, J. Lee, K.-W. Whang, J. Appl. Phys. **86**, 4049–4051 (1999)
18. G. Auday, P. Guillot, J. Galy, J. Appl. Phys. **88**, 4871–4874 (2000)
19. J.Y. Lim, J.S. Oh, B.D. Ko, J.W. Cho, S.O. Kang, G. Cho et al., J. Appl. Phys. **94**, 764–769 (2003)
20. K. Singh, N.K. Sood, Indian J. Pure Appl. Phys. **41**(2), 121–127 (2003)
21. A. Benmoussa, B. Larouci, A. Belasri, Numerical study of gas heating in ozone generated by DBD in oxygen. High Temp. Mater. Process. (N. Y.) **23**(2), 157–164 (2019)

22. A. Benmoussa, H. Si-Sabeur, Z. Harrache, A. Belasri, Study of gas heating in a xenon glow discharge. Plasma Devices Oper. **15**(4), 299–304 (2007)
23. A. Benmoussa, A. Belasri, F. Ghaleb, Z. Harrache, IEEE Trans. Plasma Sci. **42**, 706–711 (2014)
24. V. Vesovic, Int. J. Thermophys. **22**, 801–828 (2001)

Chapter 68
The Interaction of Solar Radiation with Earth's Atmosphere: Modeling the Total Atmospheric Transmittance by a Regression Function

M. Y. Boudjella, A. H. Belbachir, A. Dib, and M. Meftah

Abstract In this investigation, a numerical simulation has been performed based on the SMARTS2 model to study the influence of the zenith angle on the variation of the total atmospheric transmittances. This calculation has been made for ultraviolet and visible light in the range of [0.2, 0.39 μm] and [0.4, 0.75 μm], respectively by taking into account the absorption and the scattering of radiation by the atmospheric gas molecules as well as aerosols. In order to determine the direct dependence of the total atmospheric transmittance on the light wavelength, a regression approach has been used for four values of zenith angles [0°, 25°, 35°, 57°]. The result shows that total atmospheric transmittance can be modeled by a polynomial function of sixth order which describes fairly the atmospheric transmittance as a function of wave length and zenith angle. More importantly, the comparison between our present results and Modtran's results shows a good agreement with relative reduction ratio of transmission equals approximately to 1.

Keywords Transmittance · Solar radiation · Earth's atmosphere · Zenith angle

68.1 Introduction

The sun is seen as our fundamental natural source of energy, it emits solar radiations in isotropic ways where a part of these radiations reaches the ground after travelling through the space and the earth's atmosphere. The determination of the amount of solar radiation received at ground is important for study and deployment

M. Y. Boudjella (✉) · A. H. Belbachir · A. Dib
University of Sciences and Technology of Oran Mohamed Boudiaf USTO-MB,
BP 1505 El M'naouar, 31000 Oran, Algeria
e-mail: manalboudjella@gmail.com

M. Meftah
LATMOS, UVSQ, Université Paris Saclay, Sorbonne Université, CNRS, INSU, IPSL,
11 Boulevard d'Alembert, 78280 Guyancourt, France

A. Belasri and S. A. Beldjilali (eds.), *ICREEC 2019*, Springer Proceedings in Energy,
https://doi.org/10.1007/978-981-15-5444-5_68

of renewable energy. Electromagnetic solar radiation interacts with the earth's atmosphere constituents in many ways. Absorption and Rayleigh diffusion occur when the radiation interacts with the atmospheric gaseous molecules while Mie diffusion occurs generally with the atmospheric aerosols [2]. The total atmospheric transmittance is considered as the most important factor for both the estimation of the ground direct radiation and the atmospheric correction in determining the land surface parameters correctly [3]. The total atmospheric transmittance is a function of optical masses which is directly related to the zenithal angle. Its calculation requires the use of many parameters. In the present study, we investigate the influence on the atmospheric transmittance of two physical parameters, which are the light wavelength and the zenithal angle of the direction of light entering the atmosphere in the tropical region of the earth's atmosphere at the latitude (15° N). In this work, we use the values of the direct radiation received at the ground obtained by implementing of the known software SMARTS2 which stands for a Simple Models of Atmospheric Radiative Transfer of Sunshine [1].

68.2 Simulation Results and Discussion

The total atmospheric transmittance T_λ at a wavelength λ is given by [1]:

$$T_\lambda = T_{R\lambda} \times T_{O\lambda} \times T_{n\lambda} \times T_{g\lambda} \times T_{w\lambda} \times T_{a\lambda} \quad (1)$$

Where, $T_{R\lambda}$: is the transmittance due to Rayleigh scattering
$T_{O\lambda}, T_{n\lambda}, T_{g\lambda}, T_{w\lambda}$: Are the transmittances due to the absorption by Ozone, NO_2, mixed gazes and water vapor respectively.
$T_{a\lambda}$: is the transmittance due to the aerosol's extinction.
$T_{R\lambda}, T_{O\lambda}, T_{n\lambda}, T_{g\lambda}, T_{w\lambda}$ and $T_{a\lambda}$: are function of the optical masses m_k which is directly related the zenith angle by:

$$m_k = \left[cos(Z) + a_{k1} Z^{a_{k2}} (a_{k3} - Z)^{a_{k4}}\right]^{-1} \quad (2)$$

m_k: is the optical masse. It stands for m_R (Rayleigh), m_O (Ozone), m_N (NO_2), m_g (mixed gazes), m_W (water vapor) and m_a (aerosols).
Z: is the Zenith angle.
a_{k1}, a_{k2}, a_{k3} and a_{k4}: are the optical masse's coefficient [1].

The estimation of T_λ requires the use of many relationships and parameters such as ground temperature T_g, pressure P_g, visibility V … etc. In this work, the calculations was performed using MATLAB based on the SMATR2 model presented in [1] with sampling interval of 0.01 μm. We investigate the influence of zenithal angles Z on the total tropical atmospheric transmittance by keeping the atmospheric coefficients constant as indicated in Table 68.1 [3]. The transmission versus the wavelength characteristics in the range [0.2, 0.39 μm] and [0.39, 0.75 μm] are illustrated in Figs. 68.1 and 68.2, respectively. In Fig. 68.1 the simulation results show that the transmission is almost null in the wavelength range of [0.2, 0.28 μm] and starts increase monotonically in the range of [0.28, 0.55 μm]. As expected higher values of zenithal angles show lower transmission magnitudes.

Table 68.1 Tropical atmospheric properties at the latitude 15° North [1]

T_g (K)	P_g (mb)	V (Km)	Aerosol	NO_2 effective path length (atm-cm)	O_3 effective path length (atm-cm)
299.7	1013.0	23	Rural	2.11E−o4	0.2773

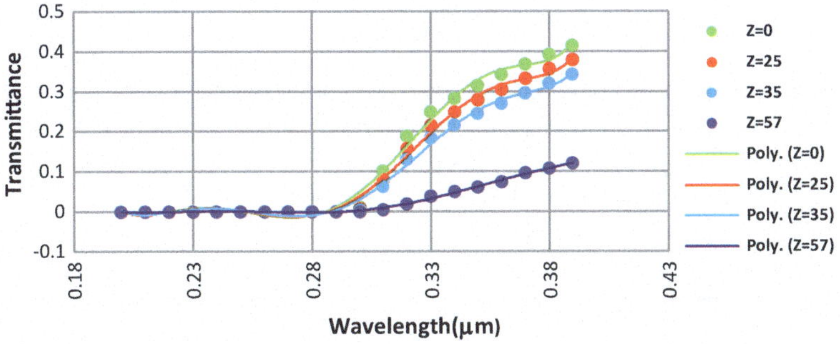

Fig. 68.1 Total atmospheric transmittances calculated based on Smarts2 Model in the ultraviolet range [0.2, 0.39 μm] at a sampling interval of 0.01 μm for various values of Z = 0°, 25°, 35°, 57°

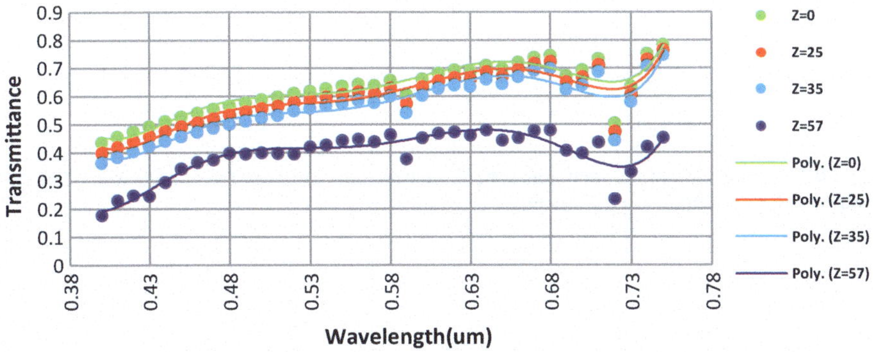

Fig. 68.2 Total atmospheric transmittance calculated based on Smarts2 Model in the ultraviolet range [0.2, 0.39 μm] at a sampling interval of 0.01 μm for (Smarts2 Model) in the visible range [0.4, 0.75 μm] at a sampling interval of 0.01 μm for various values of Z = 0°, 25°, 35°, 57°

68.2.1 Validation by Comparison to the Modtran's Transmittance

For the wavelength range from 0.2 to 0.75 μm (Fig. 68.3), a comparison between the calculated total transmittance T_c based on the Smarts2 model and the Modtran's transmittance $T_{(Modtran)}$ was made. In this case the incidence angle was taken to be normal (Z = 0). The values of $T_{Modtran}$ was extracted from the Modtran's website [4]. As shown in Fig. 68.3, For both Ultraviolet [0.2, 0.39 μm] and visible range [0.4, 0.75 μm], T_c and $T_{Modtran}$ in generally takes similar values. For the interval of the wavelength: $\lambda \in [0.2, 0.29\ \mu m]$, $T_{Modtran}$ and T_c are approximately equal to 0. For λ in the range of $[0.3, 0.75\ \mu m]$ they vary between 0.4 and 0.8. In order to compare them $T_{Modtran}$ and T_c, the relative reduction ratio is defined as:

$$R_r = \frac{T_c}{T_{Modtran}} \quad (3)$$

The average value of R_r is equal to 0.997. R_r varies between 0.94 and 1.08. The maximum values of T_c and $T_{Modtran}$ is observed at many points for example at $\lambda = 0.35, 0.55, 0.61\ \mu m$ ($R_r = 1$). This comparison shows that the developed code is correct and it can be used for furthermore study (Fig. 68.4).

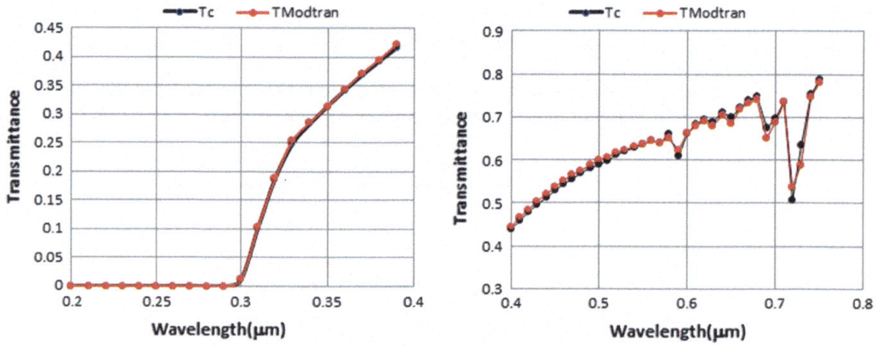

Fig. 68.3 Comparison of the total atmospheric transmittance between the present work (Smarts2 Model) and the Modtran's transmittance

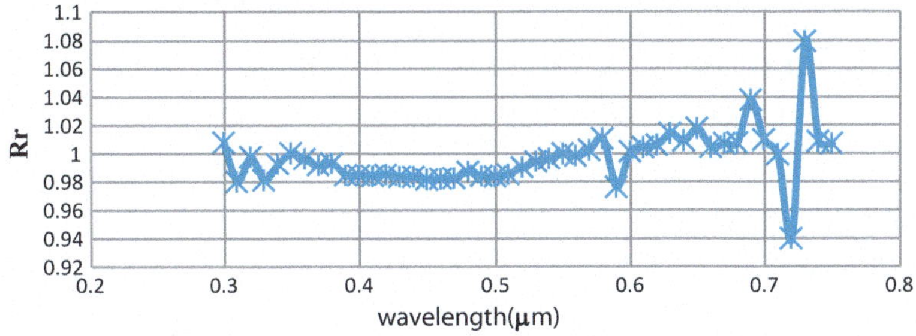

Fig. 68.4 The relative reduction ratio R_r between the calculated transmittance (Smarts2 Model) and Modtran's transmittance

68.2.2 The Total Atmospheric Transmittance Regression Relationships

In this investigation, Microsoft Excel was used to carry out a comparison between the atmospheric transmittances versus the zenithal angle for the wavelength ranges from 0 to 0.75 µm. During the analysis, we try to find the transmittances functions that fit with graphical with the data represented in Fig. 68.1. When z varies from 0° to 57°, the mathematical relationship between the transmission) and the wavelength can be extended to the general Eq. (4). Where a_1, a_2, a_3, a_4, a_5, a_6, b: are the regression polynomial coefficients, those relationships and their correlation factor R_Z^2 are given below a strong correlation is observed where $\left(0.85 \leq R_Z^2 \leq 0.99\right)$:

$$\mathrm{T_Z} = \mathrm{a_6}\lambda^6 - \mathrm{a_5}\lambda^5 + \mathrm{a_4}\lambda^4 - \mathrm{a_3}\lambda^3 + \mathrm{a_2}\lambda^2 - \mathrm{a_1}\lambda + \mathrm{b} \tag{4}$$

Ultraviolet Range

$$\begin{aligned}\mathrm{T_{UV-57}} = {} & 256471.51\lambda^6 - 457447.07\lambda^5 + 335353.64\lambda^4 - 129270.27\lambda^3 \\ & + 27633.12\lambda^2 - 3106.54\lambda + 143.55\end{aligned} \tag{5}$$

$R_{57}^2 = 0.9965$

$$\begin{aligned}\mathrm{T_{UV-35}} = {} & 652031.84\lambda^6 - 1146787.83\lambda^5 + 829142.14\lambda^4 - 315338.61\lambda^3 \\ & + 66540.70\lambda^2 + 7388.77\lambda + 337.45\end{aligned} \tag{6}$$

$R_{35}^2 = 0.99$

$$\begin{aligned}\mathrm{T_{UV-25}} = {} & 766784.66\lambda^6 - 1344781.01\lambda^5 + 969503.88\lambda^4 - 367666.60\lambda^3 \\ & + 77363.83\lambda^2 - 8566.90\lambda + 390.21\end{aligned} \tag{7}$$

$R_{25}^2 = 0.99$

$$T_{UV-0} = 878778.04\lambda^6 - 1537073.35\lambda^5 + 1105118.44\lambda^4 - 417948.48\lambda^3 + 87704.34\lambda^2 - 9685.83\lambda + 440.02 \quad (8)$$
$$R_0^2 = 0.99$$

Visible Range

$$T_{V-57} = 38357.8\lambda^6 - 129879.28\lambda^5 + 181557.47\lambda^4 - 134089.16\lambda^3 + 55185.95\lambda^2 - 11995.34\lambda + 1075.89 \quad (9)$$
$$R_{57}^2 = 0.87$$

$$T_{V-35} = 33906.97\lambda^6 - 114723.3\lambda^5 + 160242.60\lambda^4 - 118252.22\lambda^3 + 48618.09\lambda^2 - 10556.22\lambda + 945.79 \quad (10)$$
$$R_{35}^2 = 0.86$$

$$T_{V-25} = 32488.34\lambda^6 - 109900.99\lambda^5 + 153473.84\lambda^4 - 113231.22\lambda^3 + 46542.27\lambda^2 - 10102.76\lambda + 904.45 \quad (11)$$
$$R_{25}^2 = 0.86$$

$$T_{V-0} = 31053.00\lambda^6 - 105042.44\lambda^5 + 146632.89\lambda^4 - 108159.93\lambda^3 + 44447.19\lambda^2 - 9645.5\lambda + 863.8 \quad (12)$$
$$R_0^2 = 0.85$$

As shown in the Fig. 68.1, the regression analysis of the total atmospheric transmittances presents some anomalies for $\lambda \in [0.2, 0.3]$ where it takes negative values at some points for example for Z = 0: T_{UV-0} (0.2 μm, −0.00528) In order to compare the regressed transmittances to the calculated one point by point, for each value of Z, we have calculated the Ratio R_Z (the more the regression is smooth, R_Z have to be the smallest). R_Z is defined by the equation given below:

$$R_Z(\lambda) = \frac{\left|T_Z^c - T_Z^f\right|}{T_Z^c} \quad (13)$$

T_Z^f and T_Z^c are the calculated total transmittance and the deduced one from the regression formulas respectively for the 4 values of Z. Figure 68.5 shows the variation of R_Z versus wavelength from 0.31 to 0.39 μm and from 0.4 to 0.75 μm.

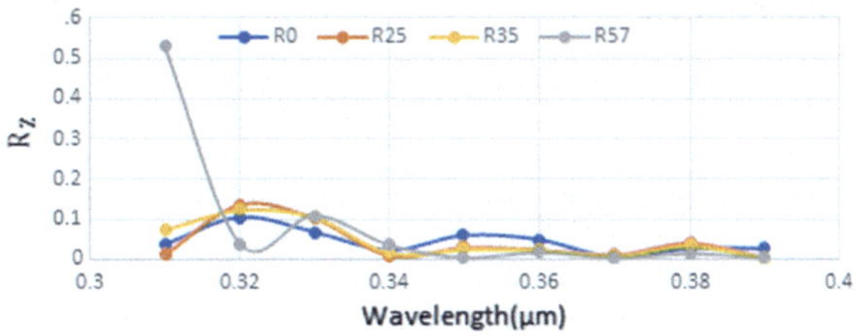

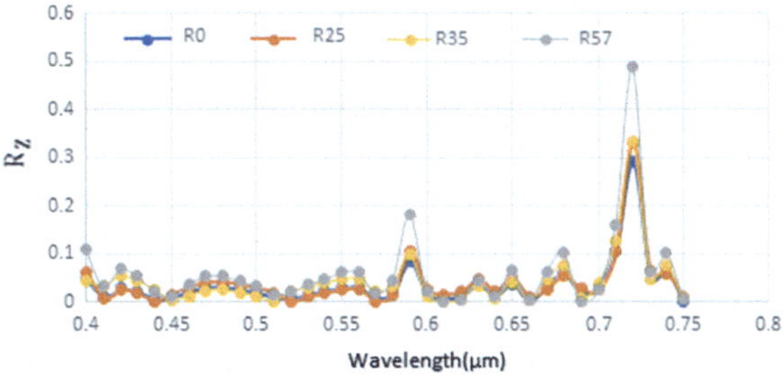

Fig. 68.5 Variation of the ratio R_Z versus wavelength for z = 0°, 25°, 35°, 57°: Left for the wavelength range from 0.31 to 0.39 μm, Right: from 0.4 to 0.75 μm

In the Ultraviolet range, a good fitting is observed for $\lambda \in [0.31, 0.39\ \mu\text{m}]$ where R_Z takes its minimum for $Z = 57°$ at the point (0.37, 0.003110326 μm); R_Z decreases at the points 0.31 μm and 0.32 μm and takes its maximum for Z = 57° (0.31 μm, 0.52966).

In the visible range, the regression was very smooth that can be explained by the values of R_{V-Z} which are in generally less than 0.1 in except for $Z = 57°$. R_{57} decreases at $\lambda = 0.59\ \mu\text{m}$ to be equal to 0.1835. The only limitation in the visible range was observed for $\lambda = 0.72\ \mu\text{m}$ where R_{V-Z} takes its maximum at $\lambda = 0.72\ \mu\text{m}$. The maximum values of R_Z vary between 0.281 to 0.486.

68.3 Conclusion

Through our work we made an attempt to study the effect of the wave length and of the zenith angle on the direct total atmospheric transmittance in the wavelength range from 0.2 to 0.75 μm. Four values of zenith angle were taken in consideration Z = 0°, 25°, 35°, 53°. The study shows that increasing the values of the zenith angle let the total atmospheric transmittance decrease as predicted. Moreover, a regression analysis of the values of the light wavelength and the zenith angle as well as the obtained values of the transmittance leads to a polynomial model of the sixth order for the total atmospheric transmittance which depends directly on the wavelength of light with a correlation factor R^2 which varies between 0.86 and 0.99. This polynomial model can be useful for rapid estimation of the direct radiation received at ground depending only on wavelength and zenith angle.

References

1. C. Guyemard, *SMARTS2, A Simple Model of the Atmospheric Radiative Transfer Sunshine: Algorithms and Performance*, 2nd edn. (Florida Solar Energy, Florida, 1999)
2. D.G. Andrews, *An Introduction to Atmospheric Physics*, 2nd edn. (Cambridge University Press, Cambridge, 2010)
3. L. Shi, Y. Qiu, J. Shi, J. Lemmetyinen, S. Zhao, Estimation of microwave atmospheric transmittance over China. IEEE Geosci. Remote Sens. J. **14**(12), 2210–2214 (2017)
4. Modtran Homepage, http://modtran.spectral.com/modtra_home. Accessed 15 Oct 2019

Chapter 69
Influence of Prolonged Exposure at Temperature on the Properties of a Thermoplastic Polycarbonate

S. Redjala, N. Aït Hocine, R. Ferhoum, and S. Azem

Abstract This work is devoted to studying the effects of temperature exposure on PC properties as a function of time. The samples underwent isothermal treatments at 80 °C, under air, in a thermo-regulated oven. Spectroscopic and physical-chemical analysis techniques highlight the structural modifications of the material. As a result, the maintenance generates physical and chemical aging phenomena and a crosslinking process which are causing a change in the characteristics of the aged material.

Keywords Polycarbonate · Thermal aging · Physical-chemical and thermal properties

69.1 Introduction

Polycarbonate (PC) is one of the technical polymers enjoying in wide application in various fields especially the automotive, aviation, space and building sectors. It is relatively preferred for its transparency and high resistance to shock justifying its use as riot shields, anti-vandalism glazing, protective screen machining machines, helmets, CDs, etc. Its high temperature resistance also makes it suitable for household appliances such as kettles, hair dryers and blenders. However, in the long term, this material undergoes aging mechanisms [1, 2] that limit its life service by causing changes in its physicochemical and mechanical properties [3–5].

The aim of this work is to carry out thermal aging of the PC at 80 °C under air and to highlight the effects of the exposure time on the physicochemical and thermal properties of the aged material.

S. Redjala (✉) · R. Ferhoum · S. Azem
LEC2M, Mouloud Mammeri University, BP 17, 15000 Tizi-Ouzou, Algeria
e-mail: radjalasonya@gmail.com

N. Aït Hocine
LaMé, INSA Centre Val de Loire, BP 3410 3 rue de la Chocolaterie, 41034 Blois, France

A. Belasri and S. A. Beldjilali (eds.), *ICREEC 2019*, Springer Proceedings in Energy,
https://doi.org/10.1007/978-981-15-5444-5_69

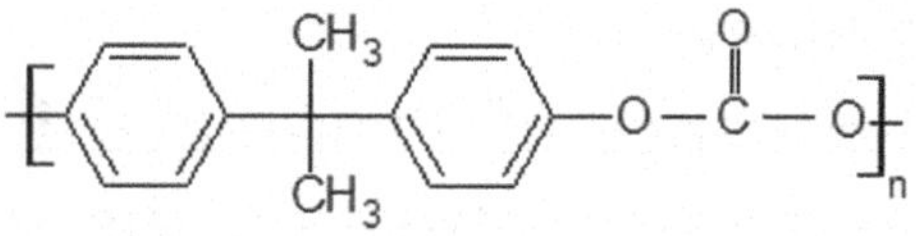

Fig. 69.1 Chemical formula of the PC

69.2 Study Material

The material used is a polycarbonate (PC) marketed by SABIC PLASTICS Algeria, in the plates form of 2 m × 1 m, with a thickness of 4.5 mm. The PC is generally an amorphous polymer but may have a nano-crystallinity, depending on how it is elaborate. It is transparent and has a good impact resistance and low density (ρ = 1.2 g/cm^3). Its glass transition temperature (Tg) is about 145 °C. Figure 69.1 shows its molecular structure.

Its analysis by Fourier Transform Infrared Spectroscopy (FTIR) reveals the main bonds that form the molecular architecture of this material, such as the C=C and C=O double bonds as well as the C-C and C-H mono-bonds [6].

69.3 Experimental Techniques

69.3.1 Aging Technique

In order to carry out the isothermal aging, we used an oven Memmert type controlled by an electronic regulator controlling the temperature maintenance. The samples, deposited on shelves, are heated to the processing temperature for appropriate times. The heating is done by convection of which an external air intake valve ensures the mixing of the air naturally, allowing a fast thermal transfer and a good homogeneity of the temperature.

69.3.2 Characterization Techniques

Scanning Electron Microscopy. The samples of virgin and aged polycarbonate were examined by means of an electron microscope TESCAN MIRA3 type with an acceleration voltage of 10 kV under secondary vacuum of 10-7 torr. Using a wrapper JEOL JFC-12003 type, we deposited two layers of gold on the surfaces to be analyzed in order to avoid the load effect, since the polycarbonate is an electrical insulator. The micrographs were obtained in secondary electron mode.

Thermogravimetric Analyzes. We used a PERKEN ELMER Diamond ATG/ATD device. The 10 mg sample is heated to 800 °C with a heating rate of 10 °C/min. This apparatus allows to record the thermal effects of the reactions or transformations which take place during the heating of the sample. Simultaneously, the sample is instantly weighed by an analytical balance, which allows a record of any change in the mass of the sample. In order to avoid the oxidation of the sample, the tests were carried out in a dynamic atmosphere of nitrogen.

X-Ray Diffraction. The studied material is an amorphous polymer but having a certain nano-crystallinity. To demonstrate any change in crystallinity after aging, this material was analyzed by X-ray diffraction. We used a Brucker D8 Advance X-ray diffractometer with $\theta - 2\theta$ assembly. It is equipped with a high-voltage 40 kV generator that accelerates the electrons in the copper-anode X-ray tube whose Kα line is 1.54 Å. The scanning interval was set from 5 to 60° with a step of 0.02° and an exposure time of one second per step. The sample of 1×1 cm^2 is placed on a hollow sample holder so that the surface to be analyzed is at the same level as the reference surface of the sample holder, in order to avoid line offsets.

69.4 Results and Discussion

69.4.1 Metallographic Observation

The microstructural observations of the virgin and aged PC at 80 °C for 216 h are shown in Fig. 69.2.

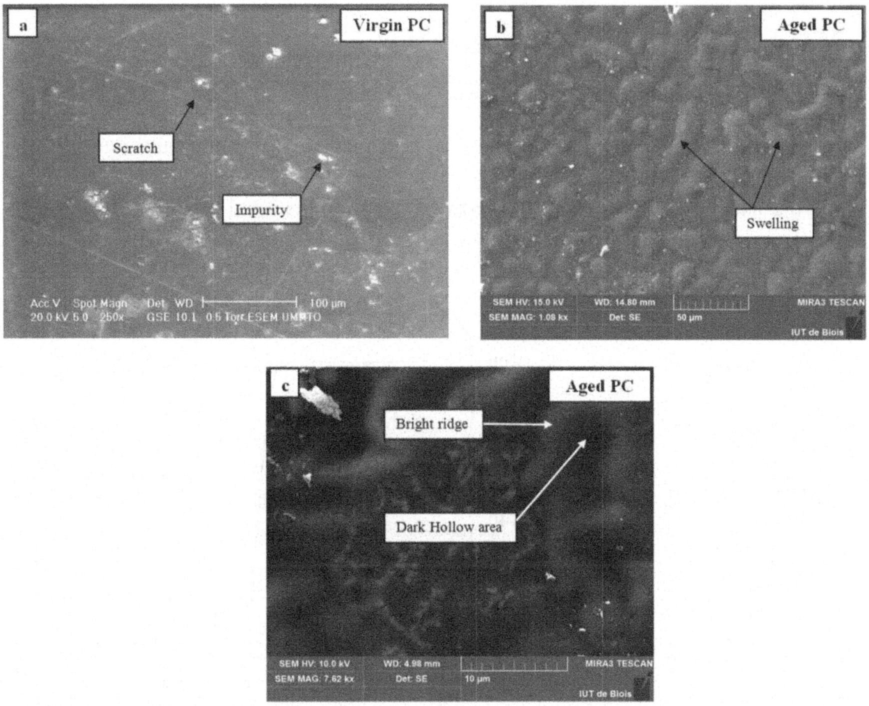

Fig. 69.2 SEM micrograph of the blank PC (a) and aged at 80 °C for 216 h (b, c)

The microscopic observation of the virgin material has shown fine superficial scratches as well as impurities due to handling during delivery (Fig. 69.2a). In contrast, Fig. 69.2b reveals a surface condition with swellings of different sizes that look like burn blisters. These probably contain gas such as carbon dioxide released during aging or more volume occupying substances. The observation at higher magnification (Fig. 69.2c) distinguishes dark hollow areas and relatively bright ridges.

69.4.2 Thermogravimetric Analysis

Thermogram of the blank PC shows a total thermal degradation occurring between 320 and 410 °C (Fig. 69.3). The degradation products are volatile such as carbon monoxide (CO), carbon dioxide (CO_2) and water vapor (H_2O) which explains the loss of mass of the material.

The beginning temperature of thermal degradation of the aged samples is shifted to the high temperatures for the three exposure times. This discrepancy is explained by an increase in the entanglement density of the molecular chains which requires a relatively greater energy to disentangle them [1].

69.4.3 XRD Analysis

Figure 69.4 shows the most intense diffraction peak of the blank and aged PC for 72, 144 and 216 h.

The intensities of the diffraction peaks of the samples aged at 80 °C decrease for the three aging times, compared to that of the virgin PC peak. This decrease is even

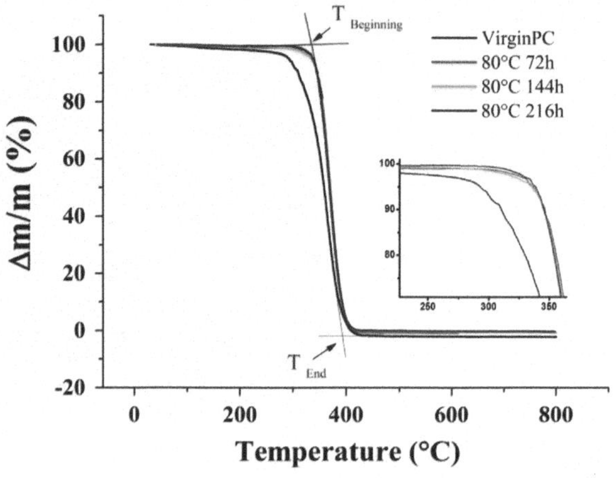

Fig. 69.3 Thermograms of the blank PC and aged at 80 °C for 72, 144 and 216 h

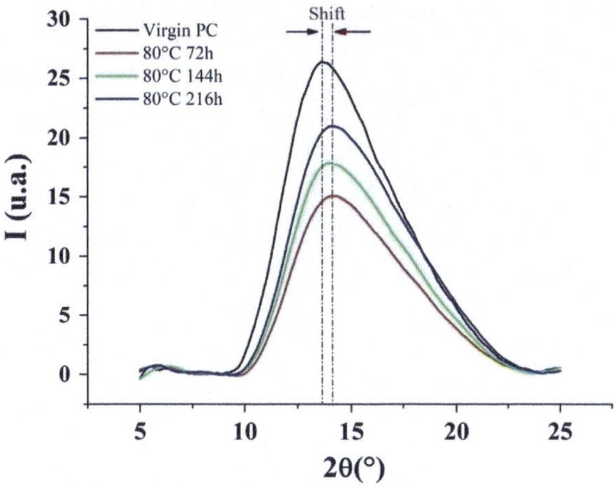

Fig. 69.4 XRD of virgin and aged PC at 80 °C

more important that the aging time is lower. A shift of the lines is also observed for the three peaks of the aged samples. Indeed, the thermo-oxidation phenomenon causes interruptions in intra and intermolecular bonds. It is at the origin of this redistribution at the level of the molecular chains of the aged material during aging time [7]. In addition, a line shift is observed towards the large angles, reflecting a tightening of the molecular chains in the crystallites.

69.5 Conclusion

The isothermal aging process leads to changes in the physicochemical and thermal properties of the polycarbonate. Thermal aging at 80 °C leads to:

- Thermo-oxidation of polycarbonate.
- The modification of the surface morphology of aged material.
- Increasing thermal stability
- Molecular rearrangement following the phenomenon of breakdown of chemical bonds

Thus, this study shows a deterioration of the properties of polycarbonate following aging in temperature.

References

1. C.H. Ho, Effects of time and temperature on physical aging of polycarbonate. Theoret. Appl. Fract. Mech. **39**(2), 107–116 (2003)
2. A. Boubakri, Influence of thermal aging on tensile and creep behavior of thermoplastic polyurethane. C. R. Mecanique **339**(10), 666–673 (2011)

3. M.M. Habib, Applications des méthodes de l'analyse thermique à l'étude du vieillissement des polymers. Thèse de doctorat, Université Blaise Pascal - Clermont-Ferrand II (2013)
4. J.H. Golden, The effect of thermal pretreatment on the strength of polycarbonate. J. Appl. Polym. Sci. **11**, 1571–1579 (1967)
5. L. Jiang, Aging induced ductile-brittle-ductile transition in bisphenol A polycarbonate. J. Polym. Res. **25**(2), 25–39 (2018)
6. E. Ghorbel, Characterization of thermo-mechanical and fracture behaviors of thermoplastic polymers. Mater. (Basel) **7**(1), 375–398 (2014)
7. K. Hareesh, Changes in the properties of Lexan polycarbonate by UV irradiation. Nucl. Instrum. Methods Phys. Res. Sect. B Beam Interact. Mater. Atoms **295**, 61–68 (2013)

Impact of Energy on the Environment

Chapter 70
Analysis of Pollution Caused by Road Traffic in the City of Oran

M. F. Lahlaf, O. Mosbah, A. Sahila, N. Zekri, and R. Bouamrane

Abstract A statistical analysis of simulated road traffic in the city of Oran is used in order to propose solutions for pollution reduction. There is a large number of vehicles circulating in Oran, leading often to traffic jams. These jams are an important source of air pollution because of the high fuel consumption. A characterization of the different phases of road traffic is required to predict and prevent the congestion phase. The SUMO code is used in this work to simulate the movement of cars on the most important roads and crossroads in Oran. A flow and velocity analysis clearly shows the existence of a critical vehicle density at which congestion phase transition appears. From the results analysis, it is concluded that at high speeds a car is polluting fast compared to average speeds. It has been noticed that the closer we get to traffic jams, higher is the pollution.

In a next step the effect of road signs and traffic lights would be examined to optimize the signaling as to reduce traffic jams and of course emissions of organic volatiles.

Keywords Road traffic · Car flows · SUMO · Air pollution · Congestion

70.1 Introduction

Traffic road has attracted scientists for several decades both analytically and numerically [1–4]. The city of Oran is the capital of Western Algeria and one of the most visited cities of this country especially in the summer period where many visitors want to spend their vacations in Oran's beaches, which leads to a large number (density) of vehicles circulate on the roads.

Vehicles (cars, buses and motorcycles) are the main means of urban transport in Algeria. They use hydrocarbons as fuels whose combustion products are known as

---

M. F. Lahlaf (✉) · O. Mosbah · A. Sahila · N. Zekri · R. Bouamrane
LEPM, Université des Sciences et de la Technologie d'Oran Mohamed Boudiaf USTO-MB, B.P 1505 El M'naouer, Oran, Algeria
e-mail: lahlaf_fethallah@yahoo.fr

A. Belasri and S. A. Beldjilali (eds.), *ICREEC 2019*, Springer Proceedings in Energy,
https://doi.org/10.1007/978-981-15-5444-5_70

anthropogenic volatile organic components (aVOC), and have an important impact on Ozone and secondary aerosol formation [5]. Among all aVOCs emitted from different industrial and other sources, 5% of them are emitted from road traffic [6]. This proportion is significantly larger in Oran, since these vehicles constitute the major means of transport. Emission of aVOC is enhanced in case of traffic congestion. Obviously this is the case since the density of vehicles is high. The exhaust gases released by cars affect not only public health but also flora and fauna. NO_x and CO_2 being pollutants produced by vehicles, we will study qualitatively these emissions according to road density and speed of cars. These emissions increase risks of mortality for drivers and any person living nearby roadways as confirmed by various health and environmental studies. On the other hand, three different traffic phases have been identified: free flow phase (where vehicles move freely), synchronized flow phase (where the velocity of each vehicle depends on the others), and the congestion phase (corresponding to jams) [7, 8].

Phase transitions imply generally a power-law behaviour near the threshold, and any quantity is expected to either vanish or diverge as $A \sim (T - T_c)^{\pm\delta}$, with δ the critical exponent, and T a control parameter (here the density of vehicles). Hence, T_c is the threshold density separating the two phases [9]. In particular, Appert-Rolland and Boisberranger [10] examined in 2013 the variation of the relaxation length in case of changing ways. However, there a very few works applying the theory of phases transitions [9] to traffic roads. Phase transitions theory indicates that disorder is maximum near the threshold, which shows the need of a statistical study of traffic road phase transitions.

This works focuses on the study of road traffic in Oran by a statistical approach to find a method avoiding the congestion phenomenon.

70.2 Method

The study of road traffic under its three regimes (fluid regime, synchronized regime and congestion) is of great interest for statistical physics as discussed above. Indeed, understanding phase transitions from one regime to another is crucial to avoid traffic jams and therefore pollute less.

Several analytic and numerical models have been proposed to study traffic road such as cellular automata, genetic algorithm, etc. (see for a review [11]). The main parameters used are the vehicles velocity, density and flux. However, the use of one of these models on a road network as that of Oran involves a large computation time and memory. In this work an open source code is used to account for the complex network in a continuous manner. We choose for this end the code SUMO (Simulation of Urban MObility), which is widely detailed in literature (see for example the Symposium of Greenshields in Juily 2008 [12] and the report no. 98-08 of the German Center for aeronautics [13]). This code reproduces all the road traffic physical quantities studied in literature, with the introduction of the road network of the city.

In our study, we imported into SUMO the road network of the city of Oran. We were then interested in road traffic around the district of "Les Amandiers", which is well known for its traffic jams. Every SUMO's parameter is set to its default value.

Before seeing the results on the exhaust gases, it is useful to mention that our simulation has provided a value of the critical density close to 0.6 and the congestion density close to 0.9, in agreement with many works (these densities are schematized in Fig. 70.1). The critical density corresponds to a maximum flow for the threshold transition to a synchronized flow. The Jam corresponds to the minimum flow in the congestion phase.

We also show the speed profile in the roads near "Les Amandiers" roundabout (Fig. 70.2). As expected, the vehicles speed is much larger in the third ringroad compared to the small road inside the district.

The emission of gases (CO_2 and NO_x) depends on the type of engine, its speed and the type of fuel. Different classes of vehicles can be used. As SUMO's goal is to simulate real-world traffic in large areas, the model of gas emission should cover the population of vehicles found on roads nowadays. This counts for passenger vehicles as well as for heavy duty vehicles, busses, motorcycles, etc. Some types of investigations require a distinction of regulative emission classes, e.g. the EURO-norm. Such a classification also helps in representing the population of vehicles over time, as most statistics on past and current vehicle fleets are represented this way. A second top-level requirement is that the emission model should match the resolution of the traffic simulation. It should be sensible to all vehicle (or traffic) state attributes that are available in the simulation. In case of a microscopic simulation, a vehicle's acceleration, speed, and the slope of the road beneath the vehicle are the major attributes to consider. On the contrary, the model is wanted to use only parameters that are offered by the traffic simulation model. Such a close connection to the traffic model implies the possibility to compute emission values for each simulated time step, usually 1 s. To achieve this, the emissions model must allow to compute emissions at such a time scale. In this work we restrict ourselves to a gasoline powered Euro norm 4 passenger model HBEFA3/PC_G_EU4 [14]. This means that SUMO will use the Handbook of Emission Factors for Road Transport (HBEFA version3) to calculate the amount of emitted pollutants [15]. Every emission class in the HBEFA describes a model for each emitted gas and is formulated using the velocity of the car.

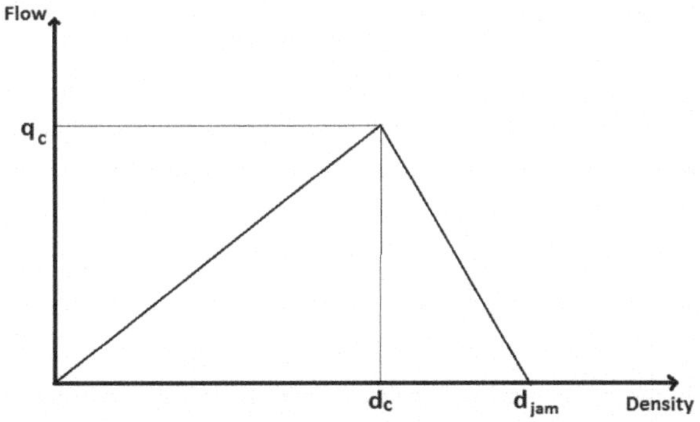

Fig. 70.1 Flow of cars according to density

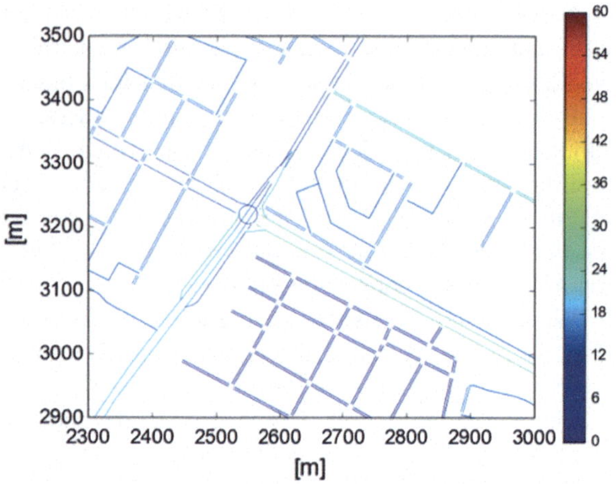

Fig. 70.2 Speed profile of cars in the region near the roundabout the district "Les Amandiers"

The random character of traffic and gas emission is due to the randomness in the vehicle's route. For the specifications of the roads, we imported a map of the "Les Amandiers" roundabout from the known OpenStreetMap website [16].

70.3 Results and Discussion

In this section, we will present results regarding NOx and CO2 emissions as functions of density and speed.

In Fig. 70.3 the effect of density of vehicles on CO_2 emission is presented. A cloud of events appears for low emissions. This cloud is widened as the density increases. As expected, it seems that average emission rate is enhanced as the density increases with a maximum emission around the critical density.

The same trends appear in Fig. 70.4 for the emission of NO_x, with an emission rate three order smaller than CO_2. Hence, the congestion phase induces a large pollutant emission. These results are consistent with literature [11].

Now let us examine the effect of vehicle speed on pollutants emission. Figure 70.5 shows a non-linear increase of CO_2 emission for speeds greater than 40 km/h, whereas the increase is linear for smaller speeds.

The maximum speed is around 100 km/h where the rate of CO_2 emission is about three times that for a 40 km/h speed. Note here that the maximum allowed car speed in this district is 50 km/h. The large increase of emission for high speed can be explained by the large fuel consumption at these speeds.

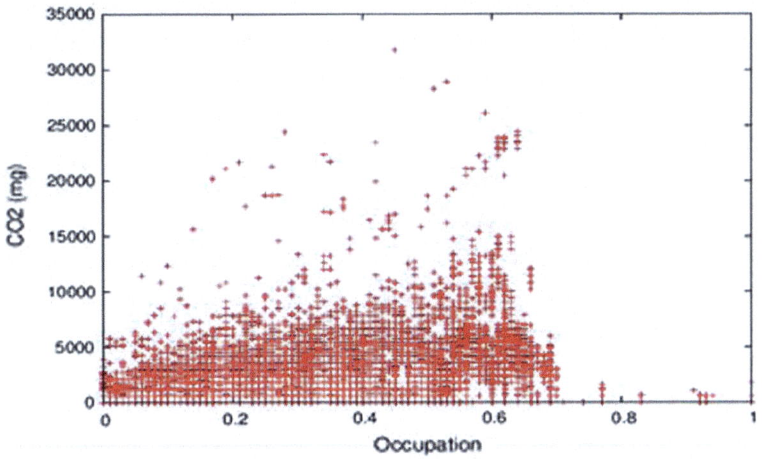

Fig. 70.3 CO_2 emission as a function of density

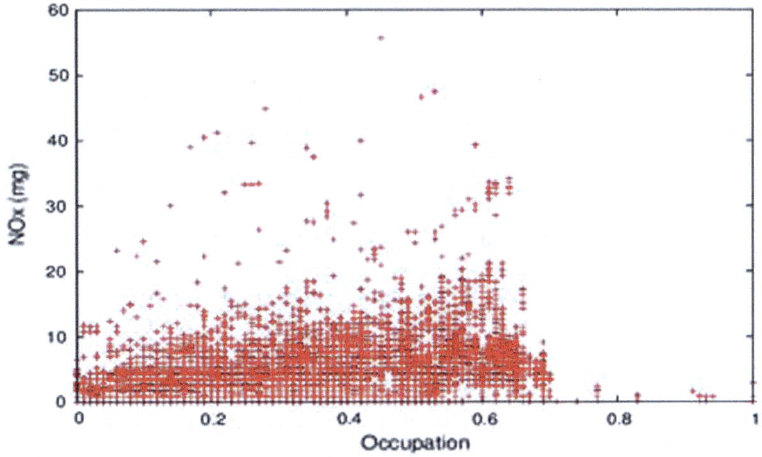

Fig. 70.4 NO_x emission as a function of density

The trend observed in Fig. 70.5 is also observed for NO_x emissions in Fig. 70.6. However, the rate of emission is three orders smaller than that of CO_2 emissions. In fact the emission process of various gases is strongly related to the combustion of the fuel. CO_2 production is much larger than NO_x.

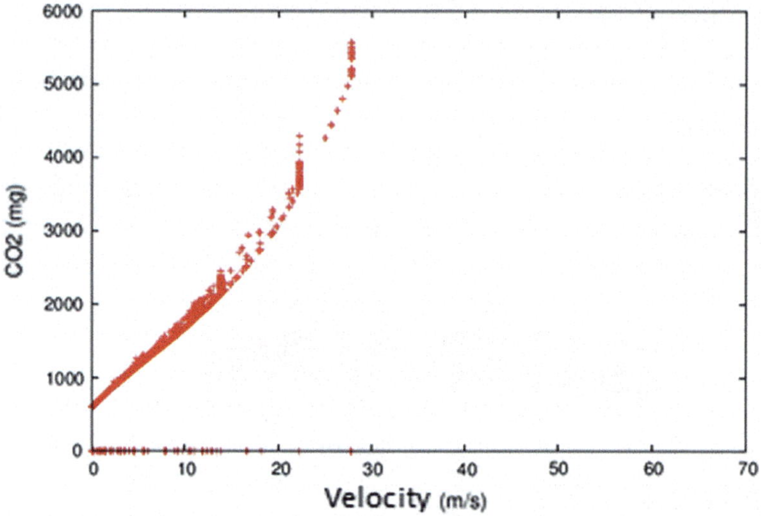

Fig. 70.5 CO_2 emission as a function of speed

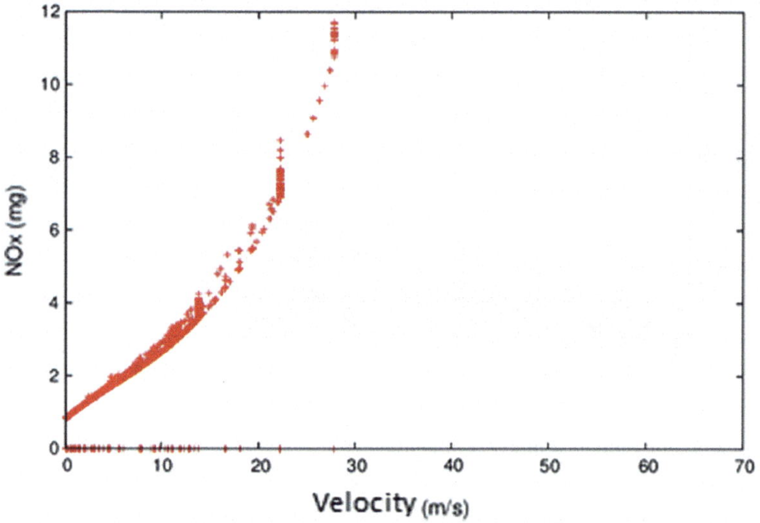

Fig. 70.6 NO_x emission as a function of speed

70.4 Conclusion

In this work, a simulation of the CO_2 and NO_x emission caused by traffic road of vehicles in the district of "les Amandiers" is realized in order to understand and control the behavior of car pollution. The Analysis of results concluded that at higher speeds correspond to higher pollution. The emission for the regulated speeds in the district (i.e. 50 km/h) reduces gas emission by third compared to that in highways (i.e. 100 km/h). On the other hand, the closer we get to traffic jams, higher is the pollution.

In a next step we intend to study the effect of road signs and traffic lights and see if one can reduce traffic jams and of course cars emissions.

References

1. S. Ferreira, A. Couto, Saf. Sci. **60**, 115–122 (2013)
2. H.S. Mahmassani, M. Saberi, A. Zockaie, Transp. Res. Part C **36**, 480 (2013)
3. D. Chowdhury, L. Santen, A. Schadschneider, Phys. Rep. **329**, 199 (2000)
4. N. Geroliminis, C.F. Daganzo, Existence of urban-scale macroscopic fundamental diagrams: some experimental findings. Transp. Res. Part B Methodol. **42**, 759 (2008)
5. G. Huang et al., Specification of anthropogenic emissions of non-metal VOC: a global gridded data set for 1970–2012. Atmos. Chem. Phys. **17**, 7683–7701 (2017)
6. K.R. Jain, Z. Cui, J.K. Domen, *Environmental Impact of Mining and Mineral Processing* (North Holland, Elsevier, 2016), pp. 53–157
7. B.S. Kerner et al., Phys. A **392**, 221–251 (2013)
8. T. Cheng, G. Tanaksaranond, C. Brunsdo, J. Haworth, Transp. Res. Part C **36**, 296 (2013)
9. H.E. Stanley, *Introduction to Phase Transitions and Critical Phenomena* (Clarendon Press, Oxford, 1971)
10. C. Appert-Rolland, J. Du Boisberranger, Transp. Res. Part C **34**, 162 (2013)
11. M. Treiber, A. Kesting, *Traffic Flow Dynamics* (Springer, Heidelberg, 2013)
12. Transportation Research Circular E-C149, *Proceedings du Greenshields Symposium de "75 Years of the Fundamental Diagram for Traffic Flow Theory" à Woods Hole*, Massachusetts, 8–10 Juillet 2008
13. S. Krauss, Microscopic modeling of traffic flow: investigation of collision free vehicle dynamics. Rapport 98-08 du Centre Allemand d'aéronautique, 2008
14. https://sumo.dlr.de/docs/Models/Emissions.html
15. Handbuch Emissionsfaktoren des Strassenverkehrs, Version 1.2, January 1999, INFRAS, im Auftrag Umweltbundesamt Berlin (Deutschland)
16. www.openstreetmap.org

Chapter 71
Environmentally Friendly Refrigerators Based on Electrocaloric Materials and Nanofluids

Brahim Kehileche, Younes Chiba, Noureddine Henini, and Abdelhalim Tlemçani

Abstract Electrocaloric cooling is an ecological refrigeration technology based on the electrocaloric effect. We don't use HCFCs, HFCs, CFCs, and NH_3, as a refrigerant. This paper reports an electrocaloric refrigeration with Al_2O_3/water as heat transfer nanofluid to vehiculate the fluxes between the CHEX and HHEX; the cold and hot heat exchanger; respectively, and $BaTiO_3$ as electrocaloric material near room temperature. The electrocaloric effect ECE is the adiabatic temperature change (ΔT) of a ferroelectric material during the application or removal of applied electric fields; during the polarization and depolarization processes, The mathematical model (Thermal, Electricity, and CFD) for electrocaloric refrigeration systems is composed with comsol, to optimize the electrocaloric effect ECE in electrocaloric refrigeration system. The results obtained are; the temperature span ΔT; cooling power for different frequency of cycles and coefficient of performance.

Keywords Electrocaloric effect · Nanofluid · Comsol

71.1 Introduction

Electrocaloric cooling device is actually as an alternative technology for the classical refrigeration based on HCFCs, HFCs, CFCs; near room temperature [1, 2]. The electrocaloric refrigeration systems based on electrocaloric effect (adiabatic/isothermal change in temperature ΔT/entropy ΔS; respectively, it is a reversible phenomenon for electrocaloric material) [3–5], and nanofluids as a refrigerant (Al_2O_3 suspended in water) [6–8]; the main advantages of alumina nanoparticles are their very low price and the absence of corrosion in thermal systems [9].

B. Kehileche (✉) · N. Henini · A. Tlemçani
Department of Electrical Engineering, Dr. Yahia Fares University, Medea, Algeria
e-mail: kehileche@yahoo.fr

Y. Chiba
Department of Mechanical Engineering, Dr. Yahia Fares University, Medea, Algeria

A. Belasri and S. A. Beldjilali (eds.), *ICREEC 2019*, Springer Proceedings in Energy,
https://doi.org/10.1007/978-981-15-5444-5_71

The performance of electrocaloric refrigeration system was then evaluated based on a simulation model with COMSOL Multiphysics, it is a challenging modelling and simulation because of the complex phenomenons and equations of the electrocaloric refrigeration systems made in parallel plates [10–12].

The temperature span ΔT; cooling power for different frequency of cycles and coefficient of performance COP for Al_2O_3/water nanofluids and $BaTiO_3$ are presented and discussed.

71.2 Electrocaloric Cooling Device

Figure 71.1 shows an electrocaloric cooling device, which is composed of electrocaloric material (regenerators made in parallel-plate, nanofluid and two heat exchanger (cold/hot).

Where; Length = 50 mm, width = 20 mm; number of ECM plates = 14

Nanofluid: Al_2O_3/water
ECM: $BaTiO_3$.

71.3 Mathematical Model

The mathematical model for electrocaloric refrigeration systems is composed of the following equations:

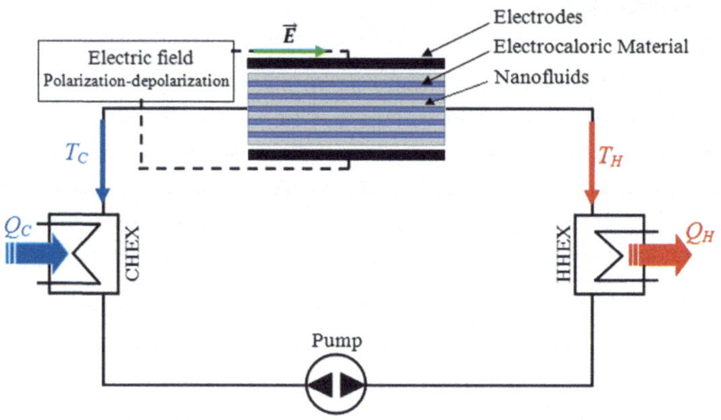

Fig. 71.1 Electrocaloric cooling device

Electrocaloric effect [2]

$$\left(\frac{\partial P}{\partial T}\right)_E = \left(\frac{\partial S}{\partial E}\right)_T \tag{1}$$

$$\Delta S = -\int_{E_2}^{E_1} \left(\frac{\partial P}{\partial T}\right)_E dE \tag{2}$$

$$\Delta T = \frac{T.\Delta S}{C_E} = -\int_{E_2}^{E_1} \frac{T}{C_E}\left(\frac{\partial P}{\partial T}\right)_E dE \tag{3}$$

The Navier-Stokes and heat transfer equations [1–5]:

$$\rho_f\left(\frac{dU}{dt} + (U \cdot \nabla)U\right) - \mu_f \nabla^2 + \nabla p = 0 \tag{4}$$

$$\nabla \cdot U = 0 \tag{5}$$

$$\rho_s c_{p,s} \frac{dT_s}{dt} - k_s \nabla^2 T_s = +\dot{Q}_{HT} \tag{6}$$

$$\rho_f c_{p,f}\left(\frac{dT_f}{dt} + (U \cdot \nabla)T_f\right) - k_f \nabla^2 T_f = -\dot{Q}_{HT} \tag{7}$$

The thermophysical properties of the nanofluids Al_2O_3 [10–12]:

$$\rho_{nf} = (1-\varphi)\rho_{bf} + \varphi\rho_{np} \tag{8}$$

$$\rho_{nf} C_{p,nf} = (1-\varphi)\left(\rho C_p\right)_{bf} + \varphi\left(\rho C_p\right)_{np} \tag{9}$$

$$k_{nf} = \frac{k_{np} + 2k_{bf} - 2\varphi\left(k_{bf} - k_{np}\right)}{k_{np} + 2k_{bf} + \varphi\left(k_{bf} - k_{np}\right)} \tag{10}$$

$$\mu_{nf} = \mu_{bf}(1 + 7.7\varphi) \tag{11}$$

71.4 Results

The problem is solved considering a ferroelectric material $BaTiO_3$ as electrocaloric material and Al_2O_3/water as heat transfer nanofluid. The velocity flow is 0.05 m/s.

During the polarization and depolarization processes, ΔT ($T_{hot} - T_{cold}$) in the two sides (hot exchanger and cold exchanger) of the electrocaloric refrigerator is shown in Figs. 71.2 and 71.3 as function of frequency and temperature; respectively at

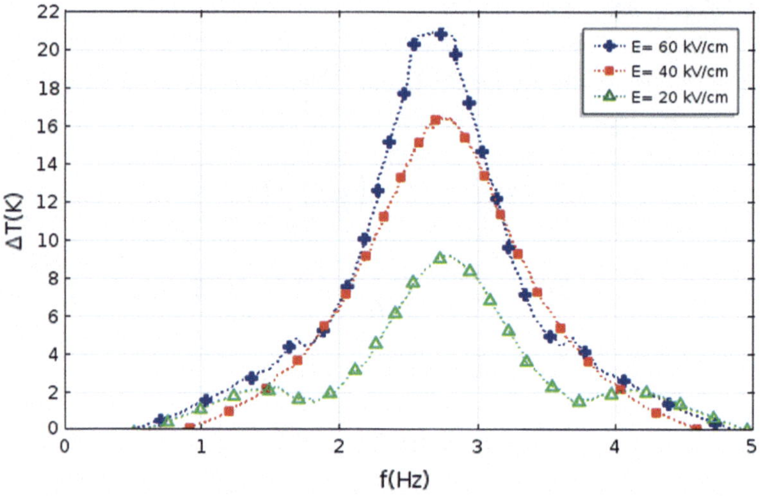

Fig. 71.2 Evolution of ΔT as function of cycle frequency at E = 20, 40 and 60 kV/cm

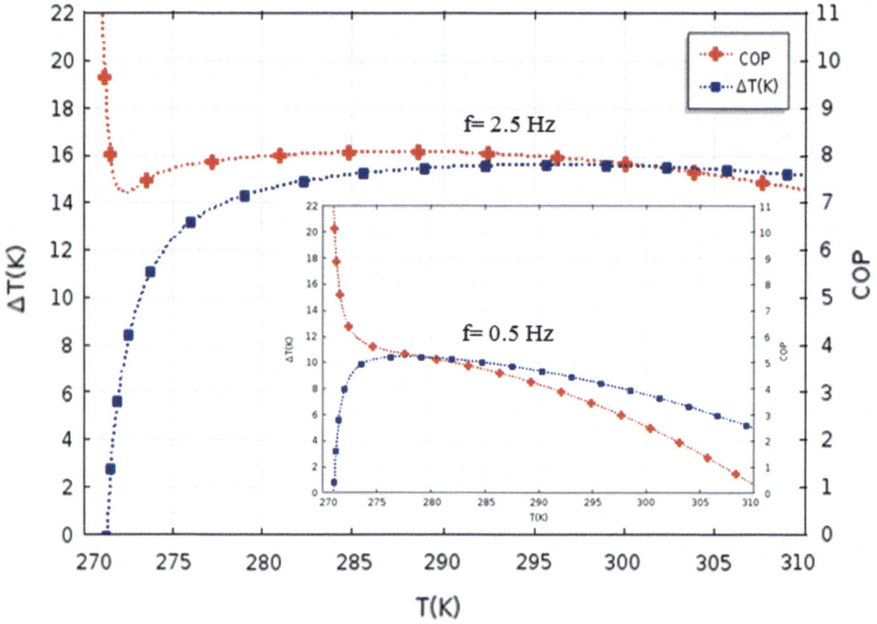

Fig. 71.3 Evolution of temperature span and COP as function of T[K] operation at frequency f = 0.5, 0.5 Hz; respectively

electric field E = [20; 40 and 60 kV/cm]. After a transient phase (Time = 60 s) the curves reach their steady state, the best result obtained in temperature span $\Delta T = 21$ K and coefficient of performance COP = 8 is f = 2.5 Hz at E = 60 kV/cm. the electric field is imposed E = [20; 40 and 60 kV/cm].

The variation of power between the electrocaloric material $BaTiO_3$ and the nanofluid Al_2O_3/water during polarization and depolarization with time and electric field is shown in Figs. 71.4 and 71.5; respectively. We note that, The curves given shows that after a transient phase (the first 60 s) the power reach a limit 500 W at t = 60 s. This can be justified by the fact that the regime becomes stable and the temperature span becomes constant $\Delta T = 21$ K. The best result obtained in power is f = 2.5 Hz.

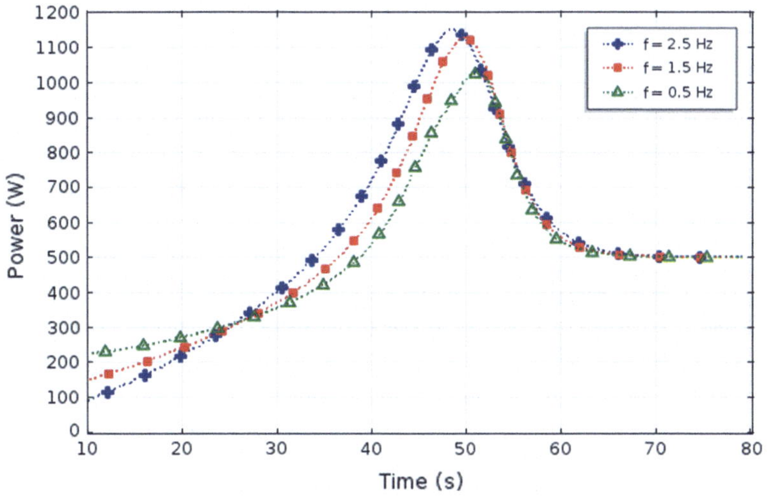

Fig. 71.4 Cooling power for different frequency as function of time

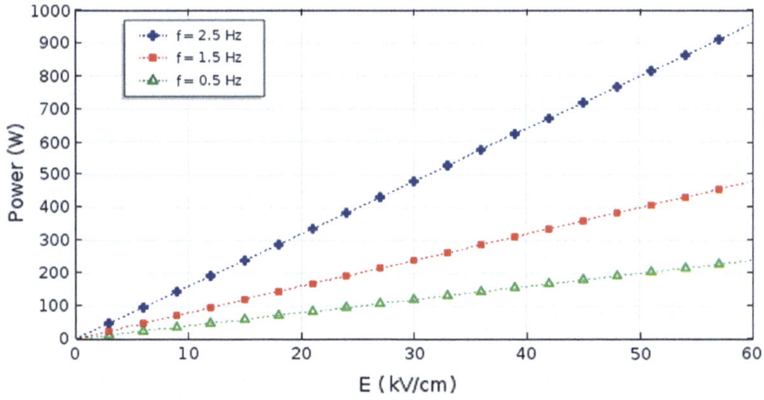

Fig. 71.5 Cooling power for different frequency as function of electric field

71.5 Conclusion

This paper to optimize the mathematical model for electrocaloric refrigeration systems made in parallel plates based on ferroelectrics materials $BaTiO_3$ (electrocaloric material) and nanofluids Al_2O_3/water, nanofluids improve the performance of electrocaloric cooling device; as shown in the temperature span ΔT; cooling power and coefficient of performance

Finally, the studies revealed that nanoparticles with a high density such as platinum or gold were the most expensive and the most efficient. However, the latter are among the most expensive nanoparticles, the interest of using these nanoparticles must be evaluated according to the needs of the interested industrialist.

References

1. M. Ozbolt, Electrocaloric refrigeration: thermodynamics, state of the art and future perspectives. Int. J. Refrig. **40**, 174–188 (2014)
2. T. Correia, Q. Zhang, *Electrocaloric Materials New Generation of Coolers*, 1st edn. (Springer, Heidelberg, 2014)
3. J.F. Scot, Electrocaloric materials. Annu. Rev. Mater. Res. **41**, 229–240 (2011)
4. C. Aprea, Electrocaloric refrigeration: an innovative, emerging, eco-friendly refrigeration technique. J. Phys. **796**(1), 012019 (2017)
5. V. Matjaz, Electrocaloric materials for future solid-state refrigeration technologies. Prog. Mater Sci. **57**, 980–1009 (2017)
6. Y. Chiba, Enhancements of thermal performances of an active magnetic refrigeration device based on nanofluids. Mechanika **23**(1), 31–38 (2017)
7. A. Greco, On the utilization of nanofluids as secondary fluid for heat transfer in a magnetocaloric cooler. Ital. J. Eng. Sci. **63**(1), 52–58 (2019)
8. H. Dhaiban, Numerical study of heat transfer enhancement in heat exchanger using Al_2O_3 nanofluids. J. Eng. **4**, 52–58 (2016)
9. I. Mugica, Exergy analysis of a parallel-plate active magnetic regenerator with nanofluids. Entropy **19**, 464 (2017)
10. C. Aprea, Is barocaloric an eco-friendly technology? ATEWI comparison with vapor compression under different operation modes. Climate **7**, 115 (2019)
11. C. Aprea, Enhancing the heat transfer in an active barocaloric cooling system using ethylene-glycol based nanofluids as secondary medium. Energies **12**, 2902 (2019)
12. A. Greco, Nanofluids as heat transfer fluids for high-efficiency caloric heat pumps. Ital. J. Eng. Sci. **63**(2–4), 251–256 (2019)

Chapter 72
Influence of Meteorological Parameters on the Performance of 10.5 kWp Photovoltaic System in Algerian Sahara

Abdelwehed Hamed, Djaaffar Rached, and Nordine Sahouane

Abstract This paper presents a performance evaluation of a grid connected photovoltaic system of 10.5 kWp installed on the roof of a building at the Renewable Energy Research Unit in the Saharan environment (URER/MS) in Adrar (Algeria) in the desert conditions. The monitoring data of the PV system over a 3 months period (26 March 2018 to 26 June 2018), are used to evaluate weekly and monthly energy yield, losses, conversion efficiency, performance ratio and capacity factor, which are the most and appropriate parameters widespread in the literature. This assessment results have been analyzed in detail in order to support the Algeria's national renewable energy program, particularly the choice of PV plants injected into the grid in the middle of the Sahara. This study shows that the maximum Array capture losses and thermal capture losses with a weekly average 2.39 h/day and 0.51 h/day respectively. The weekly average values of the performance ratio, capacity factor, efficiency of the PV module and system were 63.67%, 33.56%, 10.07% and 9.74%, respectively. The energy generated by PV system is 3.41 MWh and 96.63% of this energy is injected into the grid.

Keywords Grid-tied photovoltaic systems · Performance evaluation · Energy efficiency · Yield

A. Hamed (✉) · D. Rached
LPPMCA, Université des Sciences et de la Technologie d'Oran Mohamed Boudiaf
USTO-MB, Oran, Algeria
e-mail: hamed.abdelwehed@gmail.com

N. Sahouane
Unité de Recherche en Energies Renouvelables en Milieu Saharien URERMS,
Centre de Développement des Energies Renouvelables CDER, 01000 Adrar, Algeria

A. Belasri and S. A. Beldjilali (eds.), *ICREEC 2019*, Springer Proceedings in Energy,
https://doi.org/10.1007/978-981-15-5444-5_72

72.1 Introduction

In the latest years, many countries launch their green energy programs to replace usual energy sources with renewable ones such as biomass, wind, solar...Likewise Algeria initiates ambitious program to substitute fuel by clean green eco-energy, the goal is that 40% of energy sources will be renewable by 2030 [1]. In order to achieve this objective, the government execute several projects across the country [2], especially in the southern part due to their great solar energy potential. The climatic parameters such as ambient temperature, wind speed, relative humidity, solar radiation play important role in the operating performance of photovoltaic systems [3].

In this work, we have studied the different losses, the performances and the factors influencing the performance of grid-connected photovoltaic (PV) system with 10.5 kWp installed in the Saharan area of south Algeria (Adrar). The region of Adrar has a hot and dry desert climate BWh (B: arid, W: desert and h: hot arid) according to Köppen climate classification [4], which characterized by long, hot and dry summers and short, warm winters, recurrent sandstorm throughout the year, low humidity rate, rare rainfall and sunshine almost all the year long.

A similar study of this station was already done but, in another period, and with a system of 28 kWp [5].

72.2 Materials and Methods

72.2.1 Location of Study

All studies presented in this paper were executed in the field of Research Unit in Renewable Energy URERMS. Adrar is a region situated in the southern-west of Algeria (Latitude 27.88° N, Longitude −0.27° E, Altitude 262 m). This region is characterized by [6]:

- High ambient temperature in Summer.
- High solar irradiance potential.
- Large number days of clear and semi-clear sky.
- Few number days of dust storm.
- Low humidity rate.

72.2.2 The Grid-Tied System

The photovoltaic system of 28 kWp has 112 solar modules Mono-Crystalline (BJP-250SA with a power of 250 W), 14 modules are connected in series to form a branch. Three arrays (PV1 Array, PV2 Array and PV3 Array) are connected in parallel to form the PV system. The PV1 Array and PV2 Array which are composed of 3 branches each one connected in series, produces 10.5 kW each one, and PV3 Array containing only 02 branches connected in series and produces 7 kW. Each Array is connected to a 11 kW three-phase inverter of type SMA SMC 11000TLRP-10 [6], these inverters convert the DC current to AC 230–240 V, 50 Hz. The solar modules are mounted on metal frames supported by fixed concrete pillars at a 27° angle to the south (see Fig. 72.1).

72.2.3 Schematic of the Monitoring System

An SMA data logger (Sunny Web Box, multi-purpose data logger and automation device) collects and records the input and output voltage and current of the inverter (DC/AC), the solar radiation, the ambient and module temperature (see Fig. 72.2).

72.2.4 Performance PV Systems Evaluation Methodology

In this study, we are used the analysis process of the photovoltaic systems performances, given by the International Energy Agency Photovoltaic Power Systems (IEA) IEC 61724 Standard Norm, in order to evaluate the efficiency of photovoltaic system, the expressions of PR (Performance Ratio %) and CUF (Capacity Utilization Factor %) is given by Eq. (72.1) [6]:

$$PR = \frac{Y_F}{Y_R} \tag{72.1}$$

where Y_F is the Reference Yield in h/d and Y_R is Array Yield, h/d.

$$CUF = \frac{E_{AC}}{P_{PV,rated} * 24 * 365} \times 100 \tag{72.2}$$

where E_{AC} in the AC energy injected into the grid in kWh and $P_{PV,rated}$ is the Rated Power of PV system.

72.3 Assessment and Discussion

In order to be able to analyze the data well and to make a good study, we carried out the same study on two systems of different powers and during two different periods. The first data studied were from March 26, 2018 to May 07, 2018 with 42 PV

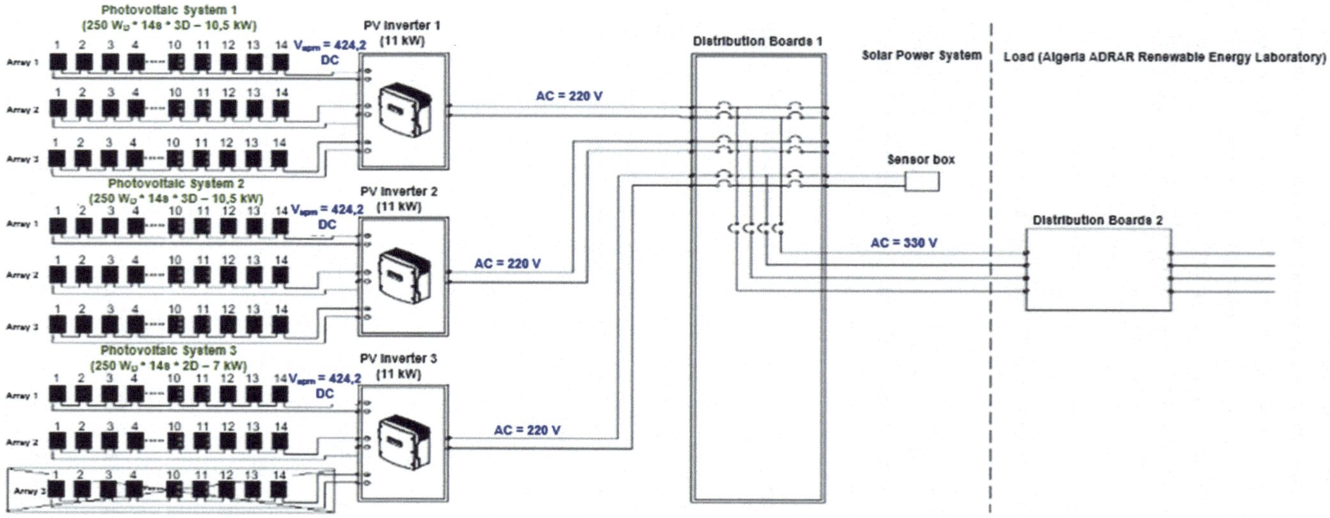

Fig. 72.1 Schematic of grid-connected photovoltaic system

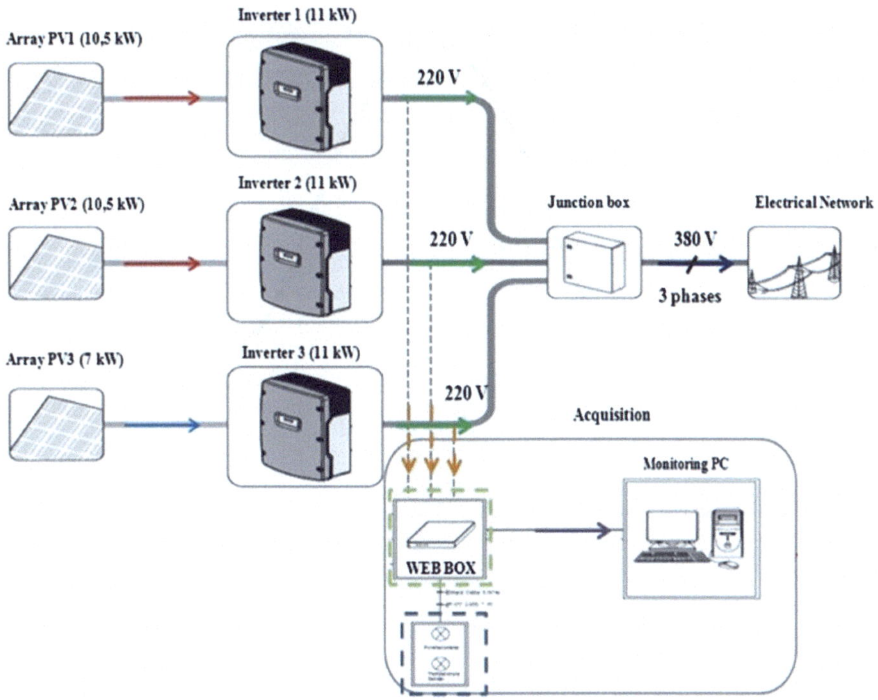

Fig. 72.2 Monitoring system diagram

panels, that is to say a power system of 10.5 kWc. The second part of the data studied was from May 08, 2018 to June 26, 2018, in which case we removed a branch of 14 panels and obtained a system with a power of 7.5 kWc with 28 panels of 250 W. Daily data was taken from sunrise to sunset for an interval of 5 min, the hours of operation of the system change from day to day. All of these steps allowed us to get real results from the average weekly and monthly PV system performance (Figs. 72.3 and 72.4).

The results show that the module temperature does not depend only on the ambient temperature and the solar irradiation, but additional factors can affect it, like the sand wind, defects in PV module [4], dust accumulation, open loop operation system, partial shading… etc. These factors also impact highly the behavior of the module.

72.3.1 *PV Array*

The first period (March 26, 2018 to May 07, 2018) 42 solar panels with a power of 10.5 kWc. During the first week of this period, we recorded an average minimum

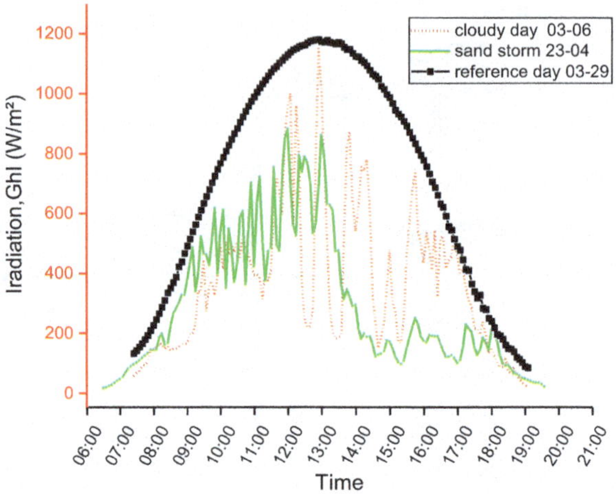

Fig. 72.3 Daily average irradiance, monthly ambient and module temperature

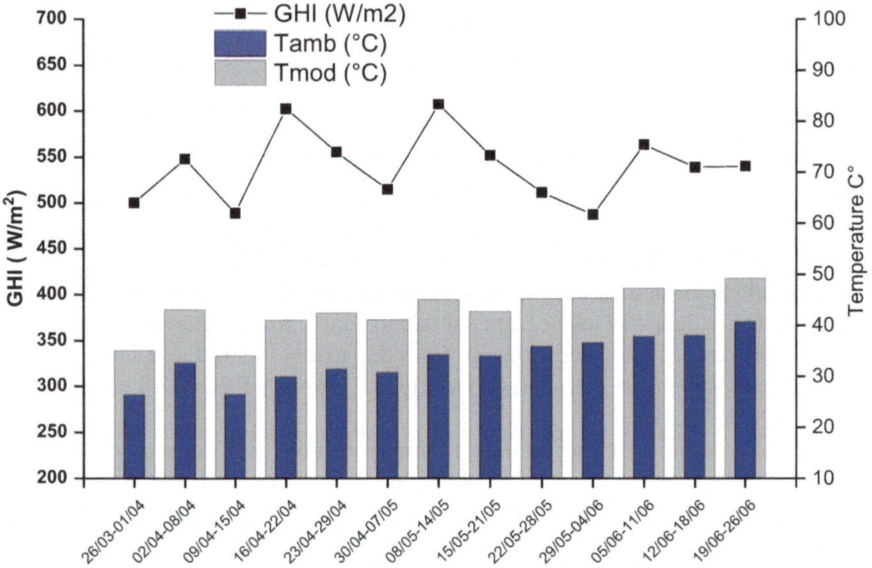

Fig. 72.4 Daily variation of irradiance average in different days. T_{amb}: ambient temperature in °C. T_{Mod}: module temperature in °C

value of 3.02 MWh of sunshine and during the 4th week, a maximum of 3.53 MWh was recorded. The second period (May 08, 2018 to June 26, 2018) 28 solar panels with a power of 7 kWp: During the 10th week of our study, we recorded a minimum value of 2.03 MWh of insolation and a maximum of

2.46 MWh during the 7th week. Otherwise the 6th and 13th week we recorded values of 3.76 MWh and 2.58 MWh respectively. These maximums are due because they are eight-day weeks. It is clear that the insolation is much greater for a field of 42 panels compared to a field of 28 panels.

The average weekly solar insolation that our system received during the entire period of our study (26-03 to 26-06-2018) is 2.769 MWh.

The energy produced can be affected by several parameters (meteorological, techniques, etc.). Indeed, the minimal energy produced was recorded during the 3rd week, caused by the instability of the network, which led to a repetitive shutdown of our system. The final average yield (RE, AR, Final) of the study period is 4.36 h/day, with a minimum of 1.87 h/day recorded during the 3rd week due to lower energy production, and a maximum of 6.07 h/day recorded during the 7th week (see Fig. 72.5) (Fig. 72.6).

72.3.2 Performance Ratio and Module Efficiency

The Capacity Utilization Factor (CUF) and Performance Ratio (PR) grade the performance of the PV system. The CUF mainly dependent on GHI (Global Horizontal Irradiance, W/m^2) of the location of the PV system while PR is a measure of the quality of PV systems independent of its location and it therefore often described as a quality factor, the module efficiency and hence it depends on the location. These parameters vary with weather condition of the location. The PR and CUF are represented in Fig. 72.7.

The efficiency of the inverter (n_{inv}) is almost constant throughout the period of this study, which means that climatic conditions do not affect the performance of

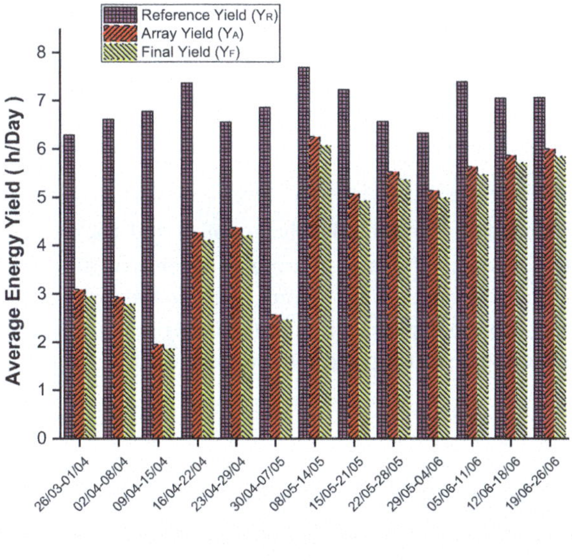

Fig. 72.5 Weekly average variation of reference Y_R, array Y_A and final yield Y_F

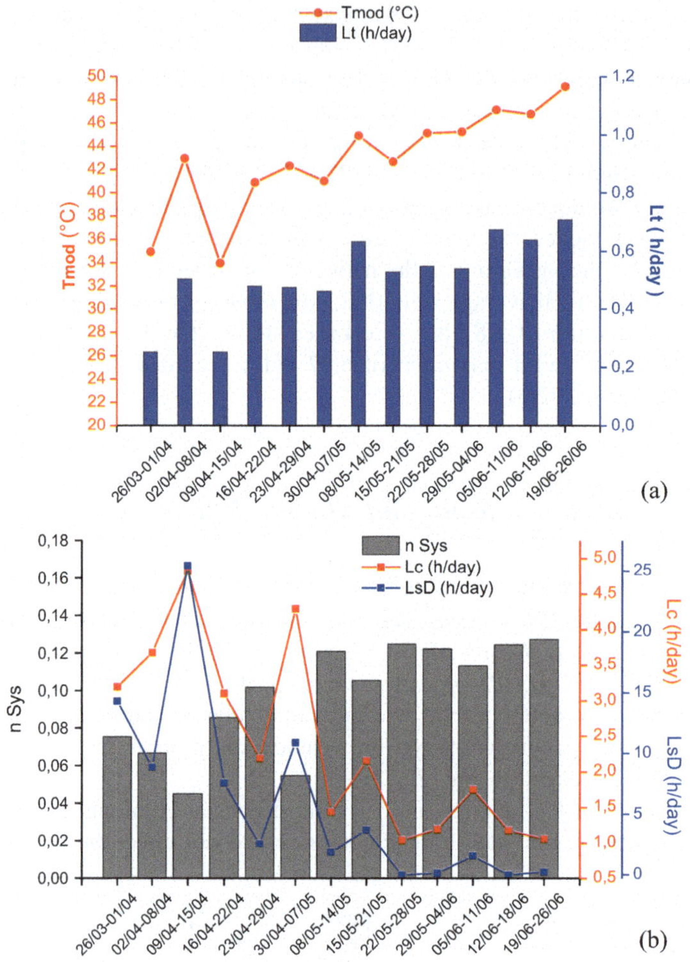

Fig. 72.6 Weekly average energy losses, module temperature and system disconnection losses over the monitored period. T_{Mod}: module temperature in °C. L_{TC}: thermal capture loss, h/d. L_C: array capture loss, h/d. L_{SD}: system disconnection loss%

the inverter. On the other hand, (n_{pv}) and (n_{Sys}) respectively the PV efficiency and the System efficiency, strongly depend on the various conditions. The weakest efficiency values were recorded during the 3rd week, which can be explained by the fact that the disconnection of the network was recurrent during this period and there were clouds during the two weeks (see Fig. 72.8).

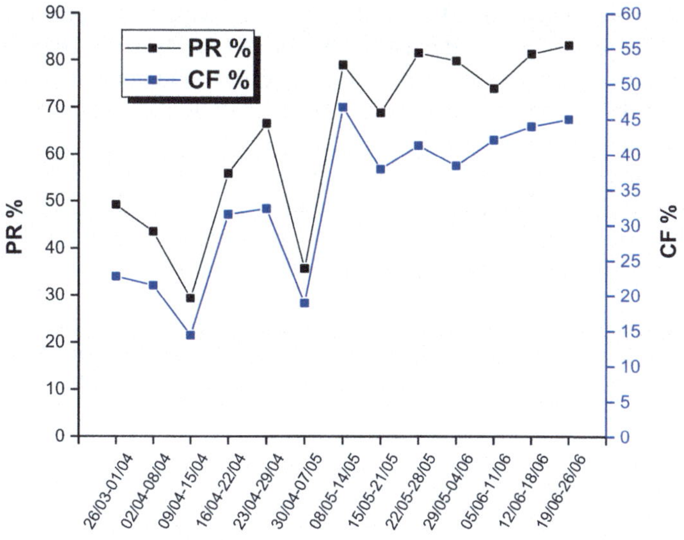

Fig. 72.7 Weekly average of CUF and PR

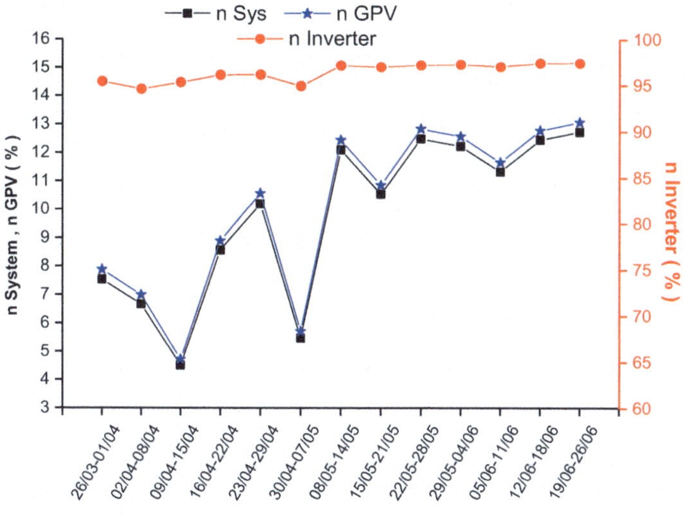

Fig. 72.8 Weekly average efficiency

72.4 Conclusion

The impact of the various parameters on the performance of the system has been analyzed and discussed. To be able to analyze the data well, we did the same study on two systems of different powers and during two different periods. The first data studied were from March 26, 2018 to May 07, 2018 with 42 PV panels (power of 10.5 kWc). The second part of the data studied was from May 08, 2018 to June 26, 2018, in which case we removed a branch of 14 panels and obtained a system with a power of 7.5 kWc with 28 panels of 250 W. This study focused on the losses and reasons that the system suffered.

Based on the experimental results, the most important parameters that dominated the yield and the production of the direct stations are: meteorological parameters (sand and dust wind, clouds ..), thermal parameters (temperatures) and the parameters of systems (network shutdown, inverters, wiring …).

Acknowledgements The authors are highly thankful to the authority of the URERMS, CDER for providing the support and facilities for the research work and giving access to the PV system data. We would like to thank all the members of electronic system team in UREMS.

References

1. A.B. Stambouli, Z. Khiat, S. Flazi, Y. Kitamura, A review on the renewable energy development in Algeria: current perspective, energy scenario and sustainability issues. Renew. Sustain. Energy Rev. **16**(7), 4445–4460 (2012)
2. A.B. Stambouli, Promotion of renewable energies in Algeria: strategies and perspectives. Renew. Sustain. Energy Rev. **15**(2), 1169–1181 (2011)
3. A. Bouraiou et al., Analysis and evaluation of the impact of climatic conditions on the photovoltaic modules performance in the desert environment. Energy Convers. Manag. **106**, 1345–1355 (2015)
4. M. Kottek, J. Grieser, C. Beck, B. Rudolf, F. Rubel, World map of the Köppen-Geiger climate classification updated. Meteorol. Zeitschrift **15**, 259–263 (2006)
5. A. Ziane, A. Necaibia, R. Dabou, M.D. Draou, M. Mostefaoui, N. Sahouane, Performance analysis of a grid connected photovoltaic station in the region of Adrar, in *5th International Conference on Electrical Engineering - Boumerdes (ICEE-B)* (2017)
6. N. Sahouane, R. Dabou, A. Ziane, A. Neçaibia, A. Bouraiou, A. Rouabhia, M. Blal, Energy and economic efficiency performance assessment of a 28 kWp photovoltaic grid-connected system under desertic weather conditions in Algerian Sahara. Renewable Energy **143**, 1318–1330 (2019)

Chapter 73
Thermal and Economical Study of a Direct Solar Dryer with Integration of Different Techniques of Heat Supply

Messaoud Sandali, Abdelghani Boubekri, and Djamel Mennouche

Abstract Solar drying has been identified as a promising alternative to sun drying for drying of fruit and vegetables in developing countries because of its minimal operational cost in terms of fuel. It is also a convenient alternative for the rural sector and other areas with scarce or irregular electricity supply. Studies conducted on solar drying have proved that it is a good alternative to sun drying for the production of high-quality dried products. This paper is focused on the thermal and the economical study of a direct solar dryer. The thermal study was performed using numerical simulation while the economic evaluation was done by computing life cycle cost (LCC) and life cycle benefit (LCB) of the solar dryer. Different techniques of heat supply such as: heat exchanger, porous medium and phase change material (PCM) were tested and studied throughout this paper. The performed economic analysis proves the feasibility of our solar dryer; however, the heat supply technique of heat exchanger with geothermal water was chosen as the most recommended one forasmuch to the short payback period (0.9 year) obtained when using this technique.

Keywords Direct solar dryer · Thermal study · Economical study · PCM · Heat exchanger · Porous medium

M. Sandali (✉) · A. Boubekri
Laboratoire de Développement des Energies Nouvelles et Renouvelables en Zones Arides (LENREZA), Université Kasdi Merbah Ouargla, 30000 Ouargla, Algeria
e-mail: sandalimessaoud@gmail.com

D. Mennouche
Faculté des Sciences Appliquées, Département de Génie des Procédées, Université Kasdi Merbah Ouargla, 30000 Ouargla, Algeria

A. Belasri and S. A. Beldjilali (eds.), *ICREEC 2019*, Springer Proceedings in Energy,
https://doi.org/10.1007/978-981-15-5444-5_73

73.1 Introduction

Solar drying has been identified as a promising alternative to sun drying for drying of fruit and vegetables in developing countries because of its minimal operational cost in terms of fuel. It is also a convenient alternative for the rural sector and other areas with scarce or irregular electricity supply. Studies conducted on solar drying have proved that it is a good alternative to sun drying for the production of high-quality dried products. Modeling of drying process is a valuable tool for prediction of performance of solar drying systems [1].

Several previous works which studied the use of different techniques of heat supply to improve the thermal performance of solar dryers such as Akmak and Yıldız [2] who studied the drying kinetics of seeded grape in solar dryer with PCM-based solar integrated collector. The system consists of an expanded-surface solar air collector, a solar air collector with phase change material and drying room. The phase change material has been used to perform the drying process even after sunset. It has been found that drying time decreases when drying air velocity increases and the drying process has occurred in decreasing drying period. Shalaby et al. [3] investigated a novel indirect solar dryer implementing phase change material (PCM) as energy storage medium. The system consists of PCM storage units in order to store the solar energy which can be used after sunset. It was found that with using PCM, the temperature of the drying air is higher than the ambient temperature by 2.5–7.5 °C after sunset for five hours. Jain et al. [4] studied the performance of an indirect solar crop dryer with phase change material. The system consists of a solar collector, packed bed energy storage material (paraffin wax), drying plenum with crop trays and natural ventilation system. After sunshine period, it was found that the temperature of the drying room was observed with 6 °C higher than the ambient temperature. Shalaby et al. [5] reviewed the previous works on solar drying systems which used the phase change material (PCM) as an energy storage medium. The results showed that the PCM reduces the heat losses and improves the thermal efficiency of drying systems. Srivastava et al. [6] investigated the using of Lauric acid as a phase change material to store solar energy. They studied the effect of inlet air velocity and inlet air temperature on the charge and discharge time, during the charge period, only the effect of inlet air velocity was taken into consideration. They concluded that thermal energy storage is the most effective use of solar energy to resolve the discrepancy between the energy supply and demand in solar heating applications. Dina et al. [7] evaluated the efficiency of solar dryer integrated with desiccants thermal energy storage for drying cocoa beans. It was found that during sunshine hours, the air temperature in drying chamber varied from 40 to 54 °C and higher than ambient temperature by 9–12 °C. Reyes et al. [8] studied Mushrooms dehydration in a hybrid-solar dryer using a phase change material. The system consists of a solar panel, paraffin wax and electric resistances. It was found that the use of solar energy reduces the electric energy consumption. Incorporating phase change material contributed significantly to improve the global thermal efficiency of the drying system. Agarwal et al. [9]

studied the shell and tube type latent heat storage (LHS) as a solar dryer. The heat transfer characteristics of the latent heat storage system have been evaluated during charging and discharging process using air as heat transfer fluid (HTF). The effects of temperature and flow rate of heat transfer fluid on the charging and discharging process of LHS have been studied. It was found that (HTF) is suitable to supply the hot air for drying of food product during non-sunshine hours or when the solar radiation is very low.

In other work, Reyes et al. [10] doubled the thermal conductivity of cans filled with paraffin wax by adding 5% w/w of aluminum wool. They designed a heat exchanger for solar energy accumulation, composed by 48 cans filled with 9.5 kg of paraffin wax mixed with aluminum wool. This application can be used in drying processes. The thermal conductivity of the mixture could be augmented by increasing the aluminum strips percentage. A new hybrid solar dryer was studied by Amer et al. [11]; the system consists of solar collector, reflector, heat exchanger with heat storage unit, and drying room. The heat energy was stored in water inside the heat storage unit during sun-shine time and with electric heaters. The results showed that the efficiency of the solar dryer was raised by recycling about 65% of the drying air. In order to improve the efficiency of a direct solar dryer, Sandali et al. [12] studied the effect of adding a porous medium as a sensible heat storage medium on the thermal performance of a direct solar dryer. Different kinds of materials with different thickness have been tested. The obtained results show that the drying air temperature is increased by 4 °C with integration of porous medium.

In this paper and as a first part, a combination mode is presented to show the variation of the thermal performance of the solar dryer when using two techniques of heat supply in combination. As a second part of this paper, an economic analysis of the direct solar dryer with using different techniques of heat supply is performed to show the feasibility and the viability of our dryer.

73.2 Physical Model

The geometry of the considered problem is presented in Figs. 73.1, 73.2 and 73.3. It is a direct solar dryer with integration of different techniques of heat supply which are: heat exchanger, porous medium and phase change material. The heat exchanger is supposed as connected with two sources of hot water; geothermal water and solar water heater. The material used as a porous medium is the gravel. The PCM used in this work is the paraffin. The geometrical dimensions of the solar dryer were taken as the real case of the prototype which is available at the level of the laboratory of development of new and renewable energies in arid zones (LENREZA, University of Ouargla, Algeria) (Tables 73.1, 73.2 and 73.3).

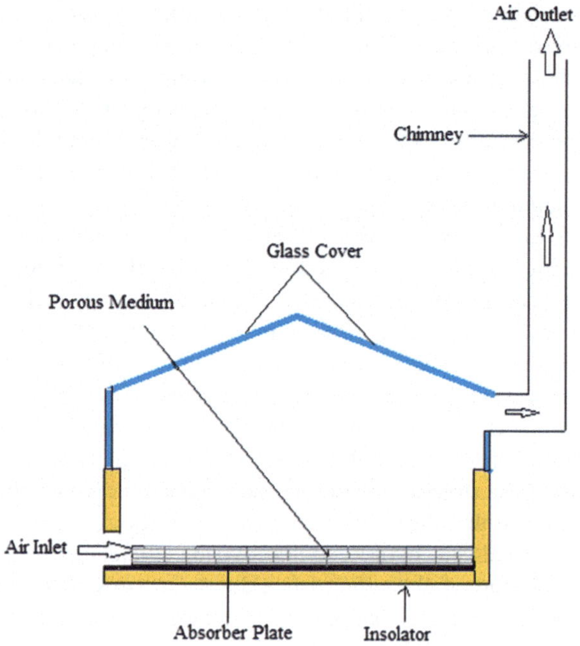

Fig. 73.1 Direct solar dryer with integration of porous medium

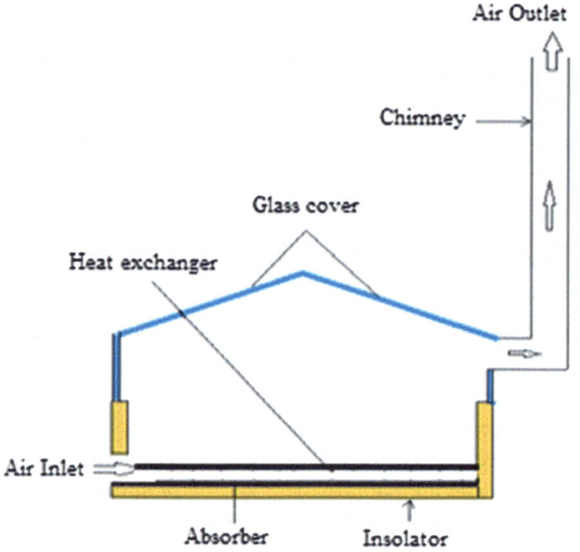

Fig. 73.2 Direct solar dryer with integration of heat exchanger

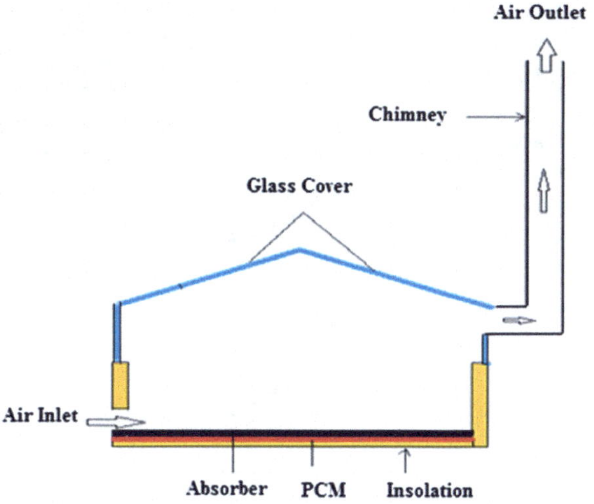

Fig. 73.3 Direct solar dryer with integration of PCM

Table 73.1 Geometrical dimensions of the direct solar dryer

Symbols	Values (m)
Wide	0.7
Length	1
High	0.04
Glass thickness	0.004
Absorber thickness	0.002
Insulation thickness	0.04
Chimney diameter	0.1
Chimney length	1

Table 73.2 Thermo-physical properties of PCM

Thermo-physical properties of PCM				
Melting temperature [K]	Latent heat [kJ/kg]	Density [kg/m^3]	Thermal conductivity [W/m K]	Specific heat [J/kg K]
326.5	266	780	0.21	2500

Table 73.3 Thermo-physical properties of porous medium

Material	Thermal conductivity (w/m-k)	Density (kg/m^{-3})	Specific heat (J/kg-k)
Gravel	2	2200	1100

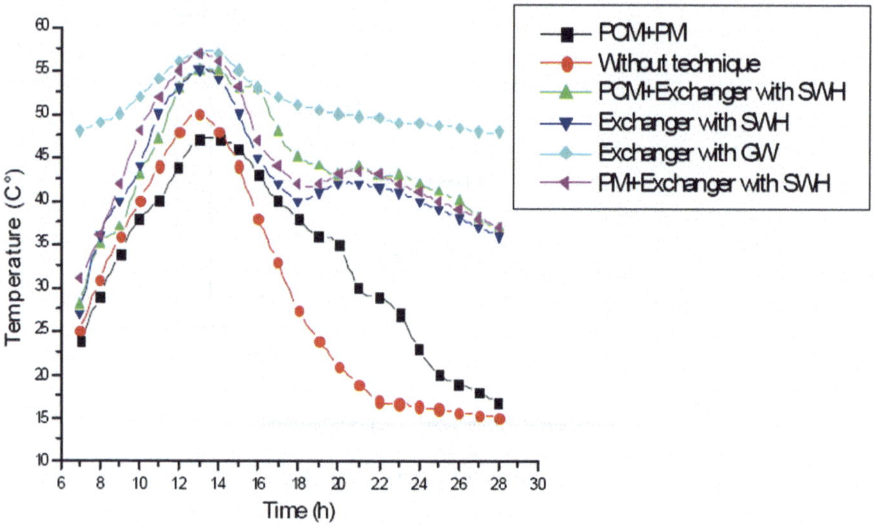

Fig. 73.4 Comparison between different techniques

73.3 Thermal Analysis and Comparison Between Different Techniques of Heat Supply

Figure 73.4 shows the evolution of the drying air temperature of the solar dryer with using different techniques. Between the tested combinations modes, it is appeared that the combination PCM + Heat exchanger with solar water heater is the best one inasmuch to the drying air temperature which this mode provides. However, between all the used techniques of heat supply in this study, it is noticed that the one of the heat exchanger with geothermal water is the more effective technique. By using heat exchanger with geothermal water, the lowest drying air temperature was found to be 46 °C, while the highest one was 58 °C. After sunset and throughout the night, the drying air temperature remains important and almost steady with an average value of 48 °C. The integration of the heat exchanger with geothermal water ensures the continuity of drying process at the night and even during cloudy days.

73.4 Economic Analysis of the Solar Dryer

The current practice today is to report the economy of a system or a process by its return time in how much time the investment or overinvestment is amortized considering the TB realized savings. These criteria of investment return time as well as the possession cost are determinative assessment criteria for the customer. This will allow us to discriminate the architectures of the dryer according to the customer's needs or specifications. Economic analysis of the dryer was carried out by computing its life cycle cost (LCC) and life cycle benefit (LCB).

73.4.1 *Life Cycle Cost (LCC)*

Life cycle cost (LCC) of the direct solar dryer is the sum of all the costs associated with a solar drying system over its lifetime in terms of money value at the present instant of time, and takes into account the time value of money. Economics of solar dryer was calculated through life cycle cost (LCC) analysis. The procedure of life cycle cost estimation as adopted by Barnwal and Tiwari [13], Singh et al. [14] and Sodha et al. [15] was used.

$$\mathrm{LCC} = \text{Initial cost of unit } (\mathrm{P_i}) + \mathrm{P_w}(\text{O \& M Costs including labor}) - \mathrm{P_w}\,(\mathrm{SV})$$

$$LCC = P_i + P_w \frac{X(1 - X^n)}{1 - X} - SV(1 + i)^n \tag{73.1}$$

where,

$$X = \frac{1 + e}{1 + i} = 0.945 \tag{73.2}$$

P_i = Initial investment (DA),
Pw = Operational and maintenance expenses, including replacement costs for damaged components (DA),
n = Life of the dryer (year),
P_w = Present worth of salvage value (SV) of the dryer at the end of life (DA),
e = Annual escalation in cost (4%),
i = Interest or discount rate (10%).

The table below presents the different cost values of the solar dryer system which allow us to calculate the economic attributes (Table 73.4).

After the numerical application on the correlation (73.1) we found that the life cycle cost of the direct solar dryer is changed from a case to another according to the technique of heat supply used:

$$\mathrm{LCC_{HE}} = 55000 + 421125(9.8) - 5500(0.239) = 4180707.5\,\mathrm{DA}$$

$$\mathrm{LCC_{SWH}} = 3239676.5\,\mathrm{DA}$$

$$\mathrm{LCC_{PCM}} = 3161526.1\,\mathrm{DA}$$

$$\mathrm{LCC_{PM}} = 3148094\,\mathrm{DA}$$

Table 73.4 Cost values of the direct solar dryer

Type of solar dryer	Coast
Direct solar dryer with heat exchanger cost (P_i)	15000 DA + 40000 DA = 55000 DA
Direct solar dryer with solar water heater cost (P_i)	60000 DA + 40000 DA = 115000 DA
Direct solar dryer with PCM cost (P_i)	500 DA/kg * 22 kg + 40000 DA = 51000 DA
Direct solar dryer with PM cost (P_i)	40000 DA
Present worth (P_w(SV))	10% of the solar dryer cost
Maintenance cost (P_w)	2.5% of the solar dryer cost
Number of functioning days in one year in case of using geothermal water heat exchanger	365 days
Number of functioning days in one year in case of using other techniques	275 days
Labor cost (P_w) (1000 DA * 275) or (1000 DA * 365)	275000 DA or 365000 DA
Life of the solar dryer	15 years
Solar dryer capacity	10 kg
Buying cost of tomatoes 15 DA/kg * 10 kg * 275 Buying cost of tomatoes 15 DA/kg * 10 kg * 365	41250 DA 54750 DA
Annual benefits 500 DA/kg * 3 kg * 275 Annual benefits 500 DA/kg * 3 kg * 365	412500 DA 547500 DA

73.4.2 Life Cycle Benefits (LCB)

The annual benefit was obtained by using total drying cycle of product. Thus, the total annual benefit from dried product was estimated as adopted by Barnwal and Tiwari [13], Singh et al. [14] and Sodha et al. [15].

$$LCB = R\frac{X(1 - X^n)}{(1 - X)} \tag{73.3}$$

where: R is the annual benefit.

LCB_{HE} = 5365500 DA
LCB = 4042500 DA

73.4.3 Payback Period

The payback period shows the length of time between cumulative net cash outflow recovered in the form of yearly net cash inflows. The payback period was calculated using the formula below [16].

$$\text{Payback period} = \frac{\text{Initial Investment}}{\text{Annual Net Undiscounted Benefits}} \tag{73.4}$$

The table below illustrates the payback period of the direct solar dryer with different techniques of heat supply:

The results presented in the Table 73.5 show that the payback period depends on two essential parameters; the initial cost of the solar dryer and the annual benefits which is depended on the dried product selling price. It appears that the payback period increased with increasing of the solar dryer cost and decreased with increasing of annual benefits. In our study, it is obvious that the short pay period which is 0.8 year is obtained when using the solar dryer with porous medium and that returns to the bare price of the used porous medium and which is found free of charge. While this technique (porous medium) was not recommended in thermal point of view because the daily operating time of the solar dryer obtained after using this technique was very small as compared with the other techniques and not enough to perform the drying process of such agricultural product. The second short payback period (0.9 year) was obtained with using geothermal water heat exchanger as heat supply technique. In addition to its short payback period, this kind of heat supply source is highly recommended in thermal point of view, it has given the highest drying air temperature compared with other techniques. The advantage of this technique is the possibility of performing the drying process in the whole year which gives an important value of the annual benefits. The payback period of the solar dryer with using PCM was obtained 1.31 year higher than the one of solar dryer with heat exchanger even though it costs less than it. This contradiction is explained by the role of the number of functioning day of the solar dryer which reverses its annual benefits; however, it is 275 days when using PCM and 365 days when using geothermal water heat exchanger. The highest payback period was obtained when using a solar water heater as a heat supply technique; it is 5.4 years and lower than the expected life of 15 years for the dryer.

Table 73.5 Payback period of solar dryer with different techniques

Technique	Payback period (year)
Direct solar dryer with geothermal water heat exchanger	0.9
Direct solar dryer with solar water heater	5.4
Direct solar dryer with PCM	1.31
Direct solar dryer with PM	0.8

73.5 Conclusion

Solar drying is one of the most fundamentals processes which allow increasing efficiently the storage time of the agro-alimentary products. This explains the recourse to the use of solar dryers, which have an undergone remarkable development in these recent years. In this context, the presented work concerns a simulation study aiming to improve the thermal performance of a direct solar dryer with using different techniques of heat supply and to evaluate economically this solar dryer. The techniques of heat supply used in this study are: heat exchanger with geothermal water, heat exchanger with solar water heater, storage of sensible heat in a porous medium and storage of latent heat with using phase change materials. Several numerical simulations using the finite volume method implemented on CFD Fluent software has been carried out.

Economic evaluation was done by computing life cycle cost (LCC) and life cycle benefit (LCB). The performed economic analysis proves the feasibility of our solar dryer; however, the heat supply technique of heat exchanger with geothermal water was chosen as the most recommended one forasmuch to the short payback period (0.9 year) obtained when using this technique.

Finally, among all the techniques of heat supply used in this study, it is noticed that the one of the heat exchanger with geothermal water is the most effective technique in thermally and economically point of view.

References

1. S. Poonia, A.K. Singh, P. Santra, D. Jain, Performance evaluation and cost economics of a low cost solar dryer for ber (*Zizyphus mauritiana*) fruit. Agric. Eng. Today **41**(1), 25–30 (2017)
2. G.C. Akmak, C. Yıldız, The drying kinetics of seeded grape in solar dryer with PCM-based solar integrated collector. Food Bioprod. Process. **89**, 103–108 (2011)
3. S.M. Shalaby, M.A. Bek, Experimental investigation of a novel indirect solar dryer implementing PCM as energy storage medium. Energy Convers. Manag. **83**, 1–8 (2014)
4. D. Jain, P. Tewari, Performance of indirect through pass natural convective solar crop dryer with phase change thermal energy storage. Renewable Energy **80**, 244–250 (2015)
5. S.M. Shalaby, M.A. Bek, A.A. El-Sebaii, Solar dryers with PCM as energy storage medium: a review. Renew. Sustain. Energy Rev. **33**, 110–116 (2014)
6. A.K. Srivastava, S.K. Shukla, S. Mishra, Evaluation of solar dryer/air heater performance and the accuracy of the result. Energy Procedia **57**, 2360–2369 (2014)
7. S.F. Dina, H. Ambarita, F.H. Napitupulu, H. Kawai, Study on effectiveness of continuous solar dryer integrated with desiccant thermal storage for drying cocoa beans. Case Stud. Therm. Eng. **5**, 32–40 (2015)
8. A. Reyes, A. Mahn, F. Vásquez, Mushrooms dehydration in a hybrid-solar dryer, using a phase change material. Energy Convers. Manag. **83**, 241–248 (2014)
9. A. Agarwal, R.M. Sarviya, An experimental investigation of shell and tube latent heat storage for solar dryer using paraffin wax as heat storage material. Eng. Sci. Technol. Int. J. **19**, 619–631 (2016)

10. A. Reyes, D. Negrete, A. Mahn, F.S. Olveda, Design and evaluation of a heat exchanger that uses paraffin wax and recycled materials as solar energy accumulator. Energy Convers. Manag. **88**, 391–398 (2014)
11. B.M.A. Amer, M.A. Hossain, K. Gottschalk, Design and performance evaluation of a new hybrid solar dryer for banana. Energy Convers. Manage. **51**, 813–820 (2010)
12. M. Sandali, A. Boubekri, D. Mennouche, Thermal behavior modeling of a cabinet direct solar dryer as influenced by sensible heat storage in a fractured porous medium, in *AIP Conference Proceedings*, vol. 1968 (2018), p. 020014. http://doi.org/10.1063/1.5039173
13. P. Barnwal, G.N. Tiwari, Grape drying by using hybrid photovoltaic–thermal (PV/T) greenhouse dryer: an experimental study. Sol. Energy **82**, 1131–1144 (2008)
14. D. Singh, A.K. Singh, S.P. Singh, S. Poonia, Economic analysis of parabolic solar concentrator based distillation unit. Indian J. Eco. Dev. **13**(3), 569–575 (2017)
15. M.S. Sodha, R. Chandra, K. Pathak, N.P. Singh, N.K. Bansal, Technoeconomic analysis of typical dryers. Energy Convers. Manage. **31**(6), 509–513 (1991)
16. S. Boughali, H. Benmoussa, B. Bouchekima, D. Mennouche, H. Bouguettaia, D. Bechki, Crop drying by indirect active hybrid solar – electrical dryer in the eastern Algerian Septentrional Sahara. Sol. Energy **83**, 2223–2232 (2009)

Chapter 74
Bioclimatic Architecture in the Ancient Village of Southern Algeria

Belkhir Hebbal, Yacine Marif, and Mohamed Mustapha Belhadj

Abstract South Algeria has a vast area with an arid climate. The summer season is characterized by the high-level use of conventional air in the residential sector. The stage of consumption in this sector is one of the major concerns expressed in the framework of the Algerian energy consumption model. The bioclimatic house standard is rapidly spreading across the world, it becomes necessary for Algeria to exploit this standard for facing the important increase in electricity consumption. In Algeria, fossil fuels are the largest source of electricity production. The environmental problems caused by the use of fossil fuels are well known: air pollution, greenhouse gases, and aerosol production. Ksar's architecture has been recognized as an example of adaptation in the desert climate in Algeria. This work presents useful ancient energy technology that has been used many years for the natural cooling of ancient buildings during the summer season in the South of Algeria. The investigation finds that two vernacular bioclimatic strategies can be translated to the present constructions: massive side walls and natural ventilation.

Keywords Southern regions · Ksar · Bioclimatic architecture · Passive cooling techniques

74.1 Introduction

In the last years, energy consumption in Algeria has increased with the rapid growth in the residential sector, especially in southern regions. According to the national agency for promotion and rationalization of energy, energy consumption in the residential sector in Algeria represents 42% of total energy consumption and the use of cooling systems in south regions accounted for more than 60% of the total building power consumption [1]. Furthermore, Ghedamsi et al. [2] in their investigation fined that the energy consumption can be increased to 179.78 TWh in 2040 in Algeria. The use of renewable energy resources and energy-efficient housing

B. Hebbal (✉) · Y. Marif · M. M. Belhadj
University of Ouargla, B.P. 511 Ghardaia Road, Ouargla, Algeria
e-mail: hebbalbelkhir@gmail.com

A. Belasri and S. A. Beldjilali (eds.), *ICREEC 2019*, Springer Proceedings in Energy,
https://doi.org/10.1007/978-981-15-5444-5_74

offer interesting opportunities for facing this important increase. Existing residential buildings are conventionally considered less energy efficient than traditional constructions. New materials have been introduced in construction without a proper study of their suitability for the climate in hot regions. This has resulted in extensive recourse to conventional air conditioning to guarantee human comfort. Numerous researches are recorded about the improvement of building thermal performance by introducing passive heating and cooling designs in Algeria (see Refs. [3–6]).

The use of ancient architecture technology offers real opportunity today to provide thermal comfort and to reduce the energy loads of the summer season. Extensive researches are recorded in many countries such as Spain\Iran\Qatar and United Arab Emirates [7–10], about optimization of energy consumption in newly-designed residential buildings by learning from the traditional construction. This paper presents an overview that has been carried out on the various techniques adopted in historic villages or Ksour as they are traditionally called in order to create an acceptable thermal comfort and improve people's quality of life in contemporary buildings. In the following sections, we focus on the case study location in Ouargla Ksar's which situated in the north of Ouargla new city.

74.2 Climatic and Bioclimatic Analysis of Ouargla City

Ouargla city is located in the southeast of Algeria (see Fig. 74.1). It is considered one oldest and comprehensive city of Algeria which saves the sites and historic vestiges until now.

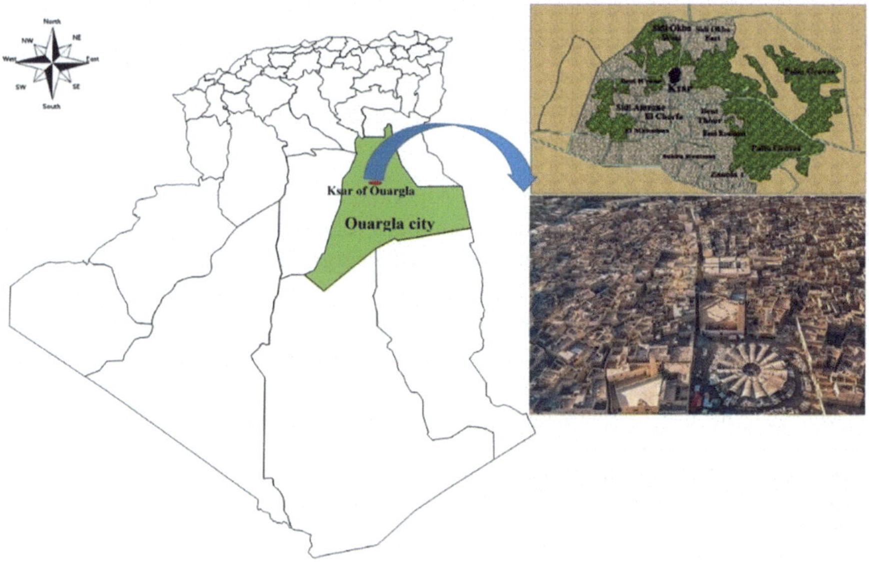

Fig. 74.1 Geographical location of Ouargla (right part) and location of the Ksar (left part)

74.3 Climatic Analysis

The climate of Ouargla is characterized by an arid climate. The aridity is expressed by very high temperatures in summer, very low rainfall and important evaporation due to the dryness of the air. The value of a 2.9 aridity index calculated according to De-Martonne formula confirms the aridity of this city.

The monthly average values of the meteorological parameters at Ouargla during 1990–2018 are presented in Fig. 74.2. The figure indicates that the mean monthly ambient temperature varies from a maximum of around 43.4 °C in the summer season (June to August) and a minimum of around 4.9 °C in the winter season (December to February). The mean daily relative humidity varies from a maximum of around 83.5% in December and a minimum of around 4.5% in July.

In this period, the mean monthly sunshine duration varies from a high of 322 h in July to 210 h in December and the prevailing winds are from Northeast and East but the most frequently coming from the North (see Fig. 74.3).

According to this analysis, the climate of Ouargla is characterized by an extremely hot and long summer, cold and dry winter. Four months are distinguished as being comfortable March, April, October, and November.

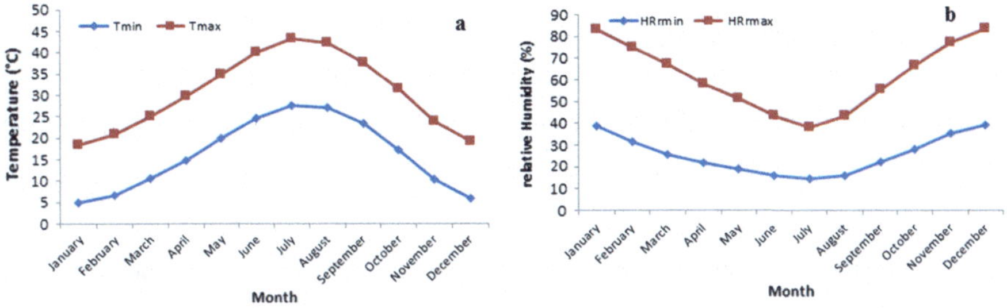

Fig. 74.2 **a** Maximum and minimum monthly temperature; **b** Maximum and minimum monthly relative humidity

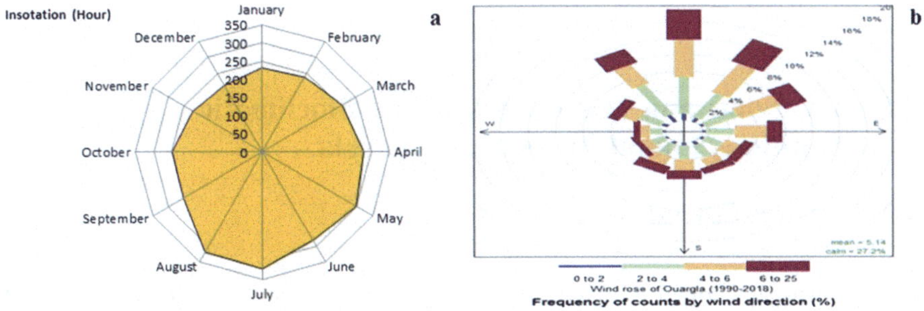

Fig. 74.3 **a** Monthly sunshine duration; **b** Prevailing winds

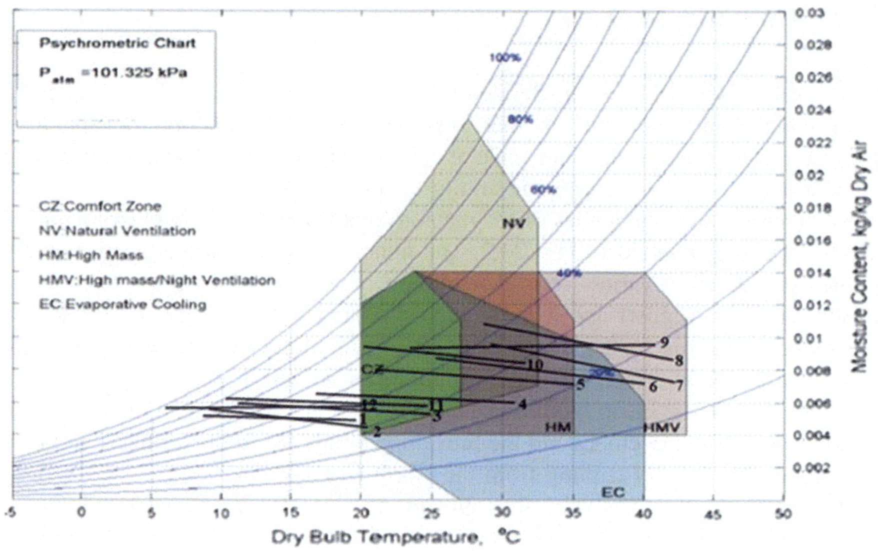

Fig. 74.4 Givoni's bioclimatic chart for Ouargla city

74.4 Bioclimatic Analysis

The bioclimatic charts facilitate the analysis of the climatic characteristics of a given location from the viewpoint of human comfort [11].

The interpretation of Givoni's Building Bioclimatic Chart (BBC) (see Fig. 74.4) applied to the city of Ouargla reveals two distinct periods: a cool period that solicits a passive heating, it concerns the months of November, December, January, and February, for this would require a good mastery of the sizes and orientation of the openings to bring the atmosphere closer to the comfort zone CZ. On the other hand, a hot and dry period during the months of June, July, August, and September which require a high thermal mass effect with night ventilation to get closer to the comfort zone CZ.

74.5 Overview and Description of the Bioclimatic Solutions Found in Ksar of Ouargla

74.5.1 Ksar Layout

The Ksar was developed in response to both the cultural and the climatic needs of its inhabitants. Large palm groves surrounded the ksar offers a better microclimate and protection for the hot winds (see Fig. 74.5 left). To reduce the surfaces exposed

Fig. 74.5 (Left) The Ksar surrounded by Palm groves, (right) Traditional narrow street

to the sun and sandstorms, the buildings were erected adjacent to each other and the streets are very narrow, irregular and generally covered [12]. The sun reaches the streets only at midday and the only building spaces that received a great amount of radiation were the roofs (see Fig. 74.5 right).

74.5.2 Massive Roof and Walls

The traditional houses are mostly made of locally available materials such as stones (Toub) and Timchemt which is a kind of traditional plaster with gray color, extracted from the limestone plateau one meter deep, calcined in partially buried ovens. It was stable in dry conditions. Toub mixed with Timchemt to form stronger walls. The roof and wall are mostly plastered using "Timchent". The thickness of the exterior walls kept around 0.5 m which leads to a high thermal mass of the building (see Fig. 74.6a). They absorb the maximum amount of heat during the day and prevent the warming of the indoor environment. At night by promoting natural ventilation, the walls recharge their mass of freshness by restoring the heat. The palm tree was used for ecological purposes: its wood and palm provide construction wood. It trunks offer planks of construction, poles, beams, and lintels (see Fig. 74.6b).

Fig. 74.6 Traditional massive roof and wall

74.5.3 Traditional House Architecture

Figure 74.7 shows an inner courtyard or "Imzargen" as called in the local language is the key element in traditional houses. It is clear that the inner courtyard is open from the top and room openings, this feature is important because they modify the indoor climate and contribute to natural ventilation. The sun's rays don't reach the courtyard until the afternoon and then the warm air rises, and convection is created which ventilates the rooms. It also serves daylight and provides privacy to the family. Furthermore, there are usually very few windows facing the street because almost all the openings and windows are facing the inner courtyard. The few windows that are facing the street are small and are fairly high up. The traditional house terraces are designed for people to enjoy the environment without leaving home.

Fig. 74.7 Traditional windows facing street and inner courtyard

74.6 Conclusion

The work reported here refers to highlights the importance of understanding the ancient architecture functioning, especially when the architectural and bioclimatic characteristics of the buildings can naturally improve indoor microclimate and comfort conditions naturally. The Ksar of Ouargla is a very interesting reference for adaptation to harsh climatic conditions and its traditional architecture brings good bioclimatic solutions using a combination of passive designs: vegetal, masse, shade, and ventilation. The passive design solutions allow buildings to adapt more appropriately to their local climates in order to create a favorable environment. On the other hand, modern architecture was found unsuitable for the climate of Ouargla. Therefore, it is imperative for architects and engineers to learn about ancient architecture.

References

1. APRUE Homepage, www.aprue.org.dz. Accessed 25 Aug 2019
2. N. Rébha Ghedamsi, A. Settou, A. Gouareh, N. Khamouli, B. Saifi, B. Recioui, Modeling and forecasting energy consumption for residential buildings in Algeria using bottom-up approach. Energy Build. **121**, 309–317 (2016)
3. S.M.A. Bekkouche, T. Benouaz, M.R. Yaiche, M.K. Cherier, M. Hamdani, F. Chellali, Introduction to control of solar gain and internal temperatures by thermal insulation, proper orientation and eaves. Energy Build. **43**, 2414–2421 (2011)
4. S.M.A. Bekkouche, T. Benouaz, M.K. Cherier, M. Hamdani, N. Benamrane, M.R. Yaiche, Thermal resistances of local building materials and their effect upon the interior temperature case of a building located in Ghardaïa region. Build. Mater. **52**, 59–70 (2014)
5. M. Benhammou, B. Draoui, Parametric study on thermal performance of earth-to-air heat exchanger used for cooling of buildings. Sustain. Energy Rev. **44**, 348–355 (2015)
6. M. Benhammou, B. Draoui, M. Hamouda, Improvement of the summer cooling induced by an earth-to-air heat exchanger integrated in a residential building under hot and arid climate. Appl. Energy **208**, 428–445 (2017)
7. I. Canas, S. Martin, Recovery of Spanish vernacular construction as a model of bioclimatic architecture. Build. Environ. **39**, 1477–1495 (2004)
8. M.S. Hatamipour, A. Abedi, Passive cooling systems in buildings: some useful experiences from ancient architecture for natural cooling in a hot and humid region. Energy Convers. Manag. **49**, 2317–2323 (2008)
9. A. Sayigh, A. Hamid Marafia, Vernacular and contemporary buildings in Qatar. Renew. Sustain. Energy Rev. **2**, 25–37 (1998)
10. K.A. Al-Sallal, L. Al-Rais, M. Bin Dalmouk, Designing a sustainable house in the desert of Abu Dhabi. Renew. Energy **49**, 80–84 (2013)
11. B. Givoni, Comfort, climate analysis and building design guidelines. Energy Build. **18**, 11–23 (1992)
12. M. Sebtia, D. Alkamaa, A. Bouchair, Assessment of the effect of modern transformation on the traditional settlement 'Ksar' of Ouargla in southern Algeria. Front. Archit. Res. **2**, 322–337 (2013)

Chapter 75
Cleaning Solar Panels Using the Travelling Wave Dielectrophoresis Method

Imene Nassima Mahi and Radjaa Messafeur

Abstract Once installed, photovoltaic panels tend to be forgotten, and covered with a thin layer of dust due to bad weather and pollution. To this can be added bird droppings, pollen, dead leaves or dead insects. All of these elements can have a long-term impact on the performance of the panels because they are likely to compromise the uptake of the sun's rays. Our work is essentially based on the removal of impurities deposited on the solar panels by introducing mobile wave conveyors where travelling-wave dielectrophoretic forces move the materials. In this paper, we studied the displacement of millimetric and nanometric size particles using three-phase and two-phase conveyors that are connected to the virtual instrument to follow the speed of displacement and we obtained good results or all the particles were moving off conveyors. The use of the two conveyors showed that the two-phase conveyor is more advantageous than the three-phase economy side in equipment. Both factors (voltage and frequency) are important for mobile wave creation, hence the ease of movement of particles. The advantage of this technology lies in the fact that the transport of the particles is ensured by the forces of the electric field.

Keywords Solar panels · Dielectrophoresis · Conveyors · Travelling waves · Displacement

75.1 Introduction

In 1839, Alexander Edmond Becquerel [1] discovered the photovoltaic effect that explains how electricity can be generated from sunlight. He claimed, "Light on an electrode immersed in a conductive solution would create an electric current".

I. N. Mahi (✉) · R. Messafeur
APELEC Laboratory, University of Djillali Liabes, Sidi Bel-Abbes, Algeria
e-mail: imenenassima@gmail.com

A. Belasri and S. A. Beldjilali (eds.), *ICREEC 2019*, Springer Proceedings in Energy,
https://doi.org/10.1007/978-981-15-5444-5_75

However, even after much research and development following this discovery, photovoltaic remained very inefficient. People mainly used solar cells to measure light. More than 100 years later, in 1941, Russell Ohl [2] invented the solar cell shortly after the invention of the transistor.

In solar photovoltaic, the efficiency of the overall system has increased thanks to numerous improvements in cell efficiency, system balance and general management and control [3]. However, the effect of environmental conditions, such as the deposition of foreign particles on the surface of modules of a photovoltaic solar generator [4], is one of the problems beyond the scope of these improvements. Such deposits reduce the transmission of light through the glass layer of the modules, which has an impact on the absorption of photons by solar cells [5, 6]. As deposits increase, conversion efficiency losses and, as a result, the energy yields of the modules and the matrix as a whole decrease progressively [7]. The dust removal technology in solar panels has attracted a lot of attention and investigation. When Clark et al. [8] have studied the strategy of attenuation of the dust that works on the lunar surface, they have suggested that there are probably two particle-charging mechanisms on the moon: 1. triboelectric loading, 2. photoemission of the electrons of the moon the surface of the particles by UV radiation.

Of all the known dust removal technologies, there has been a growing interest in electric curtain dust removal technology because of its high efficiency, clean, non-contact and no damage to solar panels. Therefore, it has become a hot spot in the dust collection area [9]. The fork electrodes are composed of a series of alternating, parallel and offset electrodes. The electric curtain is divided into two categories, one is the standing-wave electric curtain, which is connected to a single-phase power supply, and the other is the travelling-wave electric curtain, which is connected to a poly-phase power supply [10, 11]. The principle of travelling waves is as follows: if a dielectric particle is exposed to an electric field that oscillates with the period T, the induced multi-poles also oscillate with the same period. The dielectric particle is attracted either to field maxima or to field minima. On the other hand, the idea of travelling wave dielectrophoresis is to generate points of attraction that propagate continuously as much as the speed of the particle [12].

The purpose of this article is to present an experimental study to investigate the movement of millimetric and nanometric particles deposited on solar panels using a travelling wave conveyor (Fig. 75.1) (TWC) (two-phase and three-phase) to analyze their speed and their rate as a function of the variation of the applied voltage and the frequency of this same voltage.

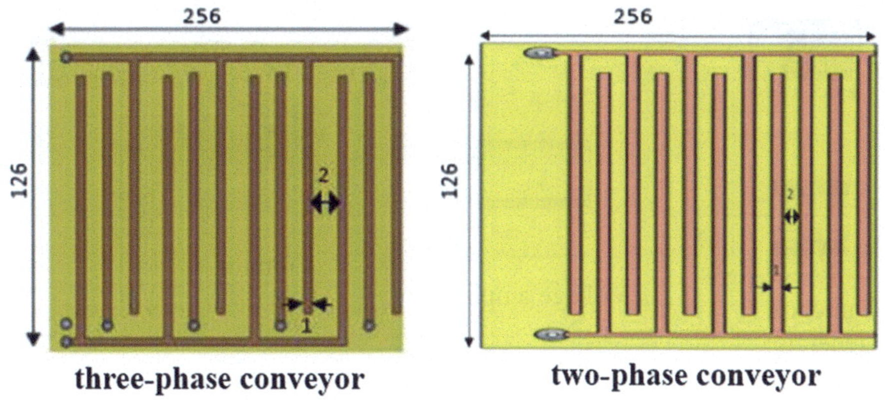

Fig. 75.1 The conveyors used

75.2 Materials and Methods

Moving materials in a travelling-wave conveyor depends on several factors. The objective of this work is the analysis of the particle displacement (Fig. 75.2) of 2 g PVC of sizes 100–150 μm and 10 g PA of sizes 1–2 mm using three-phase conveyors powered by high amplifiers voltage controlled by function generators [13] (Fig. 75.3).

The device is intended for the study of displacement of the powdery particles PVC and PA; in these experiments, a rectangular high voltage of variable amplitudes and frequencies feeds each phase of the conveyor; each conveyor is placed on a vibrator. Particles are collected in a bin on a scale and it is connected to the virtual instrument that has been developed in a LABVIEW Environment (National Instruments) [14].

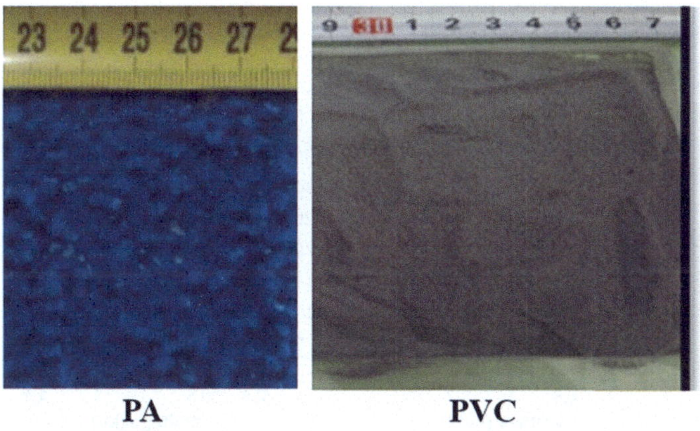

Fig. 75.2 Samples used

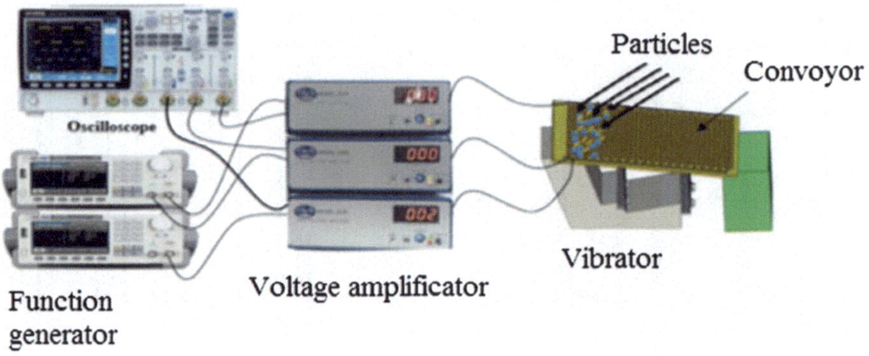

Fig. 75.3 Experimental device

75.3 Results and Discussions

By placing the particles PA on the fed conveyor, it was found that there was no movement. It was therefore forced to deposit the conveyor on a vibrator inclined by 30° and here is an example of displacement of PA and PVC as a function of time for a voltage of 1.4 kV and frequency of 50 Hz.

The results show that the measurement of the mass as a function of time makes it possible to know the speed of movement of the particles.

75.3.1 *Movement of PA Particles*

We have thus traced the evolution of PA particle displacement time as a function of frequency for the three selected voltage values (Fig. 75.4). The curves show that the more the voltage and the frequency increase the more the displacement is slow; this

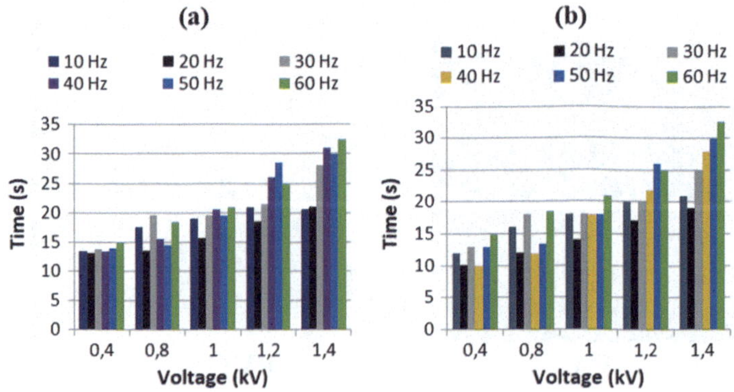

Fig. 75.4 Evolution of the travel time of PA particles **a** three-phase conveyor **b** two-phase conveyor

is due to the size of the particles. On the other hand, it seems that the polarization of the particles has an opposite effect on the millimetric granules, because there is a force of attraction between the electrodes and the granule, which has an opposite sign of charge with respect to the potential of the electrode.

Therefore, the easiest way to clean the solar panels of substances with a large size is to work with low frequencies, voltages and a vibrator.

75.3.2 Movement of PVC Particles

A second series of experiments was carried out for the micronized particles of PVC by varying the frequency from 20 Hz to 60 Hz as well as the voltage from 0.8 kV to 1.4 kV. For frequencies and voltages below 20 Hz and 0.8 kV respectively, there was no movement.

We plot the evolution of the displacement time of 500 mg of PVC as a function of frequency for the 4 chosen voltage values (Fig. 75.5). The curves show that the more the voltage and the frequency increase the more the displacement is fast, this is due to the size of the particles. Indeed, as soon as the conveyor is on, almost all the particles move in a single wave in the air rather than sliding on the surface of the conveyor. The remaining particles, which are more strongly attached to the conveyor, require a longer time to detach and move.

As shown in Fig. 75.5, the speed of movement of the particles is strongly affected by the tension because the release force applied to the particles is limited when the tension is minimal so the particle takes longer to peel off. We conclude the use of small frequencies and voltages is not a solution for cleaning solar panels of powder particles nanoscale.

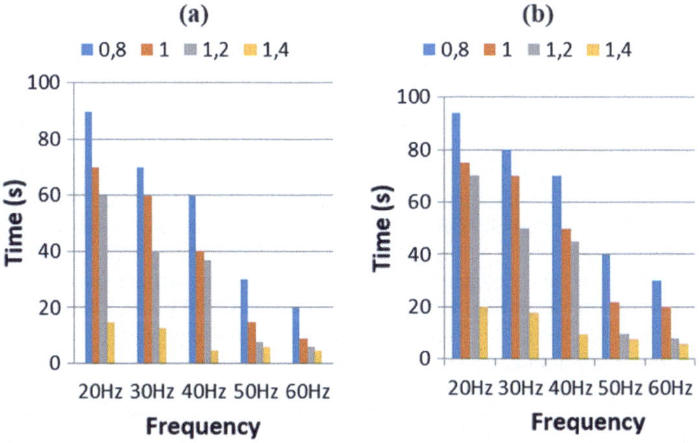

Fig. 75.5 Evolution of the travel time of PVC particles **a** three-phase conveyor **b** two-phase conveyor

75.4 Conclusion

The device concerns the cleaning of solar panels by moving millimetric and micronized particles. The purpose of this study was to analyze the feasibility of transporting millimeter PA and micronized PVC particles using travelling-wave conveyor technology (three-phase and two-phase) while considering factors that may influence the efficiency and the performance of this operation. The results obtained from this study are reliable at a frequency of 60 Hz and an amplitude of 1.4 kV for the rapid displacement of micronized particles, which is the inverse of the millimetric particles that requires only small voltages and frequencies.

The use of the two conveyors showed that the two-phase conveyor is more advantageous than the three-phase economy side in equipment. Both factors (voltage and frequency) are important for mobile wave creation, hence the ease of movement of particles. The advantage of this technology lies in the fact that the transport of the particles is ensured by the forces of the electric field.

References

1. P. Wolfe, *Solar Photovoltaic Projects in the Mainstream Power Market* (Routledge, New York, 2013)
2. R. Hantula, *How Do Solar Panels Work?* (Infobase Publishing, New York, 2010)
3. National Renewable Energy Laboratory (NREL), Solar Cell Record Efficiency Chart 2018 (National Renewable Energy Laboratory, Golden, 2018)
4. T. Ashley, E. Carrizosa, E. Fernández-Cara, Heliostat field cleaning scheduling for solar power tower plants: a heuristic approach. Appl. Energy **235**, 653–660 (2019)
5. H. Kawamoto, B. Guo, Improvement of an electrostatic cleaning system for removal of dust from solar panels. J. Electrostat. **1**, 28–33 (2018)
6. D. Deb, N.L. Brahmbhatt, Review of yield increase of solar panels through soiling prevention, and a proposed water-free automated cleaning solution. Renew. Sustain. Energy Rev. **82**, 3306–3313 (2018)
7. H.T. Ba, M.E. Cholette, R. Wang, P. Borghesani, L. Ma, T.A. Steinberg, Optimal condition-based cleaning of solar power collectors. Sol. Energy **157**, 762–777 (2017)
8. P.F.A. Minetto, J. Keller et al., Finding a dust mitigation strategy that works on lunar surface, in *The 38th Lunar and Planetary Science* (Texas, 2007)
9. H. Kawamoto, Electrostatic and electromagnetic cleaning of lunar dust adhere to spacesuits, in *Annual Meeting of LEAG, Lunar and Platenary Institute* (Houston, 2009)
10. G. Liu, J.S. Marshall, Particle transport by standing waves in electric curtain. J. Electrostat. **68**, 289–298 (2010)
11. G. Liu, S.Q. Li et al., Characteristics of the motion of martian dust simulant on the travelling-wave electric curtain. J. Eng. Thermal Phys. **32**, 2073–2075 (2011)
12. R. Hagedorn, G. Fuhr, T. Muller, J. Gimsa, Travelling wave dielectrophoresis of microparticles. Electrophoresis **19**, 49–54 (1992)
13. I.N. Mahi, R. Messafeur, A. Belgacem, Y. Bellebna, H. Louati, A. Tilmatine, New separation method of metal/plastic micronized particles using travelling wave conveyors. Int. J. Environ. Stud. **75**, 788–799 (2018)
14. Austin, National Instruments, LABVIEW, Measurement Manual, National Instruments (2000)

Chapter 76
Numerical and Statistical Approach of Aerosols Coagulation Process Affecting Global Climate

Joanna Dib and Djilali Ameur

Abstract Atmospheric aerosols stand for a complex mixture of microscopic solid or liquid droplets particles suspended in atmosphere, consisting of both natural and anthropogenic origin. The aim of this work is to study the atmospheric aerosols coagulation process greatly enhanced by the Van der Waals forces and monitored by the Brownian motion. We focus on the evolution of the mass concentration of particles, which is consistent with existing regulations, and we focus on the density change of the particles during the particle growth processes. We analyse an approach for solving Smoluchowski's coagulation equation employing the Monte Carlo probabilistic method based on the use of random numbers in repeated experiments. Additionally, several numerical simulations have been implemented and evaluated regarding their CPU times and their accuracy in terms of mass concentrations. The time step influence on the zeroth moment error is verified, as well as on the total number of simulated particles.

Keywords Atmospheric aerosols · Smoluchowski PDE equation · Coalescence · Monte Carlo

76.1 Introduction

Atmospheric aerosols stand for a complex mixture of microscopic solid or liquid droplets particles suspended in atmosphere, consisting of both natural and anthropogenic origin. They are made of soil, road dust, clay or from air pollution. In fact, aerosols have a profound effect on our lives in countless different ways and it depend on their size [4], it affects biosphere global climate and personal health, therefore environmental and aerosol health effects has been the subject of growing scientific research. Several studies have shown that, because of their small size, these particles can penetrate very deeply into the human body and link with

J. Dib (✉) · D. Ameur
Laboratory of Theoretical Physics, University Abou Bekr Belkaid, Tlemcen, Algeria
e-mail: joannadib2022@gmail.com

A. Belasri and S. A. Beldjilali (eds.), *ICREEC 2019*, Springer Proceedings in Energy,
https://doi.org/10.1007/978-981-15-5444-5_76

increased risk of asthma attacks and subsequently increased mortality. Furthermore, by absorbing and scattering of incoming solar radiation, aerosols may alter the ecosystems's radiative balance. Effects on the properties of clouds have also been brought to light. In fact, aerosol particles play a critical role in cloud formation. Many different types of atmospheric particulates can act as cloud condensation nuclei upon which water vapor condenses. Their number and composition can also affect the amount of water droplets contained within a cloud, lifetimes and precipitations.

Aerosol particles vary in size, concentration and distribution both spatially and temporally. Their concentration in atmosphere is variable due to the heterogeneity of chemical compositions and sources, interactions with clouds and solar and telluric radiation. One of the essential problems of aerosol physics is coalescence due to microphysical transformation processes. The dynamics of aerosols particles depends on various processes such as condensation/evaporation, followed by nucleation, coagulation, and finally deposition phenomena, but some additional phenomena must be taken into account such as the Van der Waals forces for coagulation. At the macroscopic level of aerosols, many studies [1, 2, 5, 9], experimentally highlight the significant increase in the coagulation rate of particles. It is in fact essential for the coagulation of nanoparticles, where the van der Waals forces act only on small particles.

The present paper is devoted to the analysis of a numerical approach for solving the Smoluchovski's equation for atmospheric aerosols simulation. We focus on the study of the number concentration of coagulated particles which decreases with the growth of the particle size.

76.2 Smoluchowski's Equations

The Smoluchowski equation is a population PDE equilibrium equation describing growth and division phenomena for nanoparticles population, characterized by its size density repartition. Since proposed in 1928, based on the Smoluchowski researches for coagulation process [7, 8], the SE has been widely applied in many chemical processes and environmental fields involving aerosol and cloud science, polymerization and many other newly emerging fields in nanoparticle engineering. The solution for SE remains a special issue and the analytical and numerical solution has been widely studied for many years [3, 6], where the determination, interpretation and quickened prediction of particles size distribution receive much more attention lately. It is to say that the collision rate per unit mass is due to the Brownian motion, which consists of the inter-particle mechanism modifying the particle size distribution, where the particles contact probability affects the particle agglomeration rate. By assumption, it follows that the collisions are binary and fluctuations in density are sufficiently small in order that collisions occur at random. Therefore elemental size distributions reveal an interesting pattern and the study of the numerical analysis of some coagulation fragmentation model behaviors makes a real challenge.

Then, the integral-differential SE has the unsteady-state form of:

$$\frac{\partial c}{\partial t}(i,t) = \frac{1}{2}\sum_{j=1}^{i-1} K(i-j,j)c(i-j,t)c(j,t) - \sum_{j=1}^{\infty} K(i,j)c(i,t)c(j,t) \quad (1)$$

$$K(i,j) = 1 \quad \text{and} \quad c(i,0) = c_i(0) = \begin{cases} 1 & if\ i = 1 \\ 0 & else \end{cases} \quad \forall i \in \aleph^*,\ t > 0 \quad (2)$$

System (1) is deduced for a set of particles 1, 2, 3, … or any other whole mass at time t, where c(i, t) and c(i, t) di denote respectively the particle size distribution number defined by a continuous distribution and the particle number concentration whose single particle mass lies between i and i + di. Two particles of mass i and j merge into a single particle, their probability of fusion is K(i, j) which stands for the coagulation Kernel, proportional to the density of these two particles. Moreover, we assume that the number and total mass are finite and the coagulation kernel is nonnegative and symmetric. The rate of agglomeration of aerosols of mass k by agglomeration of any aerosols pair of masses i and j, is represented by the first term on the right-hand side, and the second term accounts for the loss of aerosols of mass k by agglomeration with any other aerosols. The variance of c(i, t) is given by the melting of particles of masses i and j and by the disappearance of particles of sizes (i + j) following a fusion with another particle.

The particle size distribution mass is represented by $\check{c}(i,t)$, such that $\check{c}(i,t)$ di stands for the particle mass concentration with a single particle mass between i and i + di. Mass and number distributions are related by the single particle mass:

$$\check{c}(i,t) = i\, c(i,t) \quad (3)$$

In our study, the particle density is constant due to single component particles, therefore mass and volume distributions are equivalent. Therefore, we shall approximate the mass density, instead of approximating the density, which satisfies the following equation:

$$\frac{\partial \check{c}}{\partial t}(i,t) = \frac{1}{2}\sum_{j=1}^{i-1} \check{K}(i-j,j)\check{c}(i-j,t)\check{c}(j,t) - \sum_{j=1}^{\infty} \check{K}(i,j)\check{c}(i,t)\check{c}(j,t) \quad (4)$$

$$\check{c}(i,0) = \check{c}_i(0), \quad \text{where} \quad \check{c}(i,t) = i\,c(i,t) \quad \text{and} \quad \check{K}(i,j) = \frac{K(i,j)}{j} \quad (5)$$

An analytical solution is given by:

$$c(i,t) = \frac{\left(\frac{t}{2}\right)^{i-1}}{\left(1+\frac{t}{2}\right)^{i-1}} \quad (6)$$

A collision, in which aerosol particles, flowing in a carrier gas, stick to each other forming a single particle, is called coagulation. It is to say that the particle concentration and coagulation efficiency affect the particle population coagulation rate. This paper is mainly devoted to the study of zeroth moment, which corresponds to the number concentration of coagulated particles, and we compare approximate moment of order 0

$$M_{0,N}(t) = \sum_{k=1}^{N} \frac{1}{i_{N,n}(k)} \tag{7}$$

to the analytical moments of order 0 over the interval [0, T]:

$$M_p(t) = \sum_{i=1}^{\infty} i^p c(i,t) = \sum_{i=1}^{\infty} i^{p-1} \check{c}(i,t) \quad where \quad M_0(t) = \frac{2}{2+t} \tag{8}$$

76.3 Numerical Simulations

Numerical simulations were performed on an Intel (R) Core i7-3632 CPU 3.2 GHz computer with 4 GB memory. For all numerical and analytical solutions, the total time was up to unit. In the present study, the performance for the numerical simulations is evaluated employing the analytical method ability assessing criterion proposed for solving the SE due to Brownian coagulation. Therefore, we consider that the investigated analytical method and the reference numerical method show the same precision in solving SE if the absolute relative errors are less than 0.050.

The following table shows that considering a certain number of particles N, more the time step **Δt** decreases, more the error on moments of order 0 $M_{0,N}$ decreases (Table 76.1, Figs. 76.1 and 76.2).

Taking $\mathbf{\Delta t} = \frac{T}{200}$, the following table shows that more the number of particles N increases, more the error on zeroth moments $M_{0,N}$ decreases (Table 76.2 and Fig. 76.3).

Table 76.1 Evolution of the error on zeroth moment as a function of the number of particles N and the time step **Δt**

Time step	Error on $M_{0,N}$ for $\mathbf{N} = 5 \times 10^4$	Error on $M_{0,N}$ for $\mathbf{N} = 6 \times 10^4$
1/500	3.81×10^{-3}	8.63×10^{-4}
1/600	3.68×10^{-3}	7.05×10^{-4}
1/700	3.10×10^{-3}	4.55×10^{-4}
1/800	1.81×10^{-3}	3.64×10^{-4}
1/900	1.80×10^{-3}	2.37×10^{-4}
1/1000	1.47×10^{-3}	9.24×10^{-5}

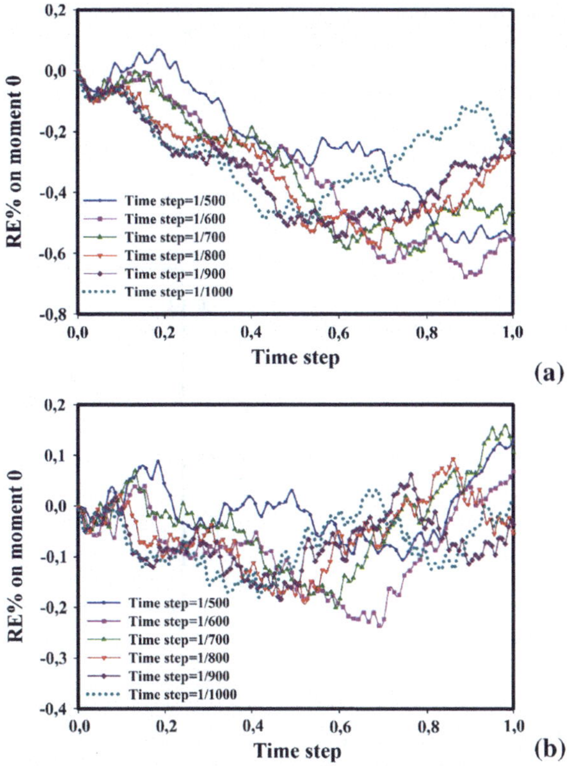

Fig. 76.1 Several relative error on zeroth moments corresponding to time steps varying between $\Delta t = \frac{T}{500}$ and $\Delta t = \frac{T}{1000}$ for particles number equal to $N = 5 \times 10^4$ (a) and $N = 6 \times 10^4$ (b)

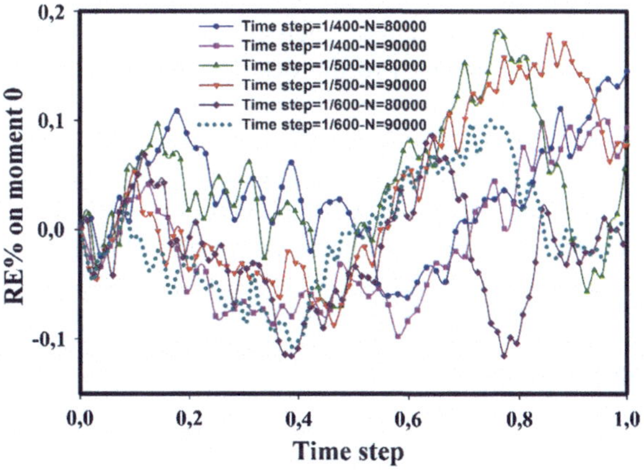

Fig. 76.2 Several relative error on zeroth moments corresponding to time steps varying between $\Delta t = \frac{T}{400}$ and $\Delta t = \frac{T}{600}$ for $N = 8 \times 10^4$ and $N = 9 \times 10^4$

Table 76.2 Variation of the error on zeroth moment and the relative error as function of the number of particles **N** and for the time step $\Delta t = \frac{T}{200}$

Particle number	Error on $M_{0,N}$	Relative error on $M_{0,N}$
$N = 6 \times 10^4$	1.48×10^{-3}	0.222
$N = 7 \times 10^4$	8.16×10^{-4}	0.122
$N = 8 \times 10^4$	7.66×10^{-4}	0.115
$N = 9 \times 10^4$	4.18×10^{-4}	0.062

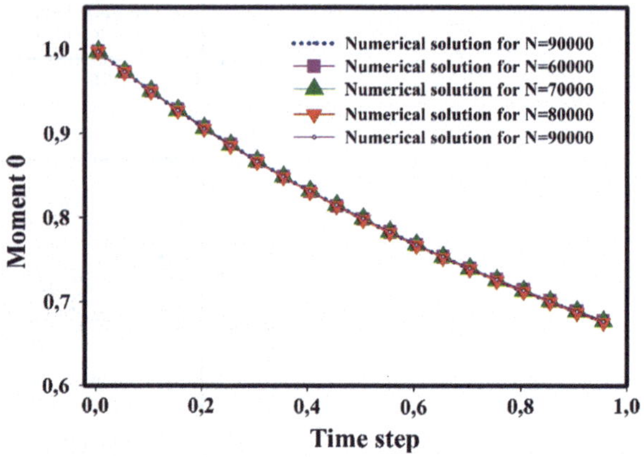

Fig. 76.3 Comparison between analytical and several numerical moments of order 0 corresponding to **N** varying between $N = 6 \times 10^4$ and $N = 9 \times 10^4$, for $\Delta t = \frac{T}{200}$

76.4 Conclusion

In this work, we develop and analyze a Monte Carlo method for atmospheric aerosols agglomeration phenomena. We are interested, in particular, in the numerical simulation of the discrete coagulation equations for aerosols. We write the equation verified by the mass distribution density. For the coagulation simulations, the numerical particles evolve by using random numbers for MC simulations. We prove the convergence of the MC numerical scheme in the case of the coagulation equation. All our numerical tests show that the numerical solutions calculated by MC algorithms converge to the exact solutions.

References

1. A. Alam, J.P. Shi, R.M. Harrison, Observations of new particle formation in urban air. J. Geophys. Res. **108**(D3), 4093 (2013). https://doi.org/10.1029/2001jd001417
2. D. Ameur, J. Dib, S. Belalia, I. Raspo, Statistical study for colloidal suspensions coagulation, in *ICEMAEP 2018 Proceedings* (2018)
3. H. Babovsky, On a Monte Carlo scheme for Smoluchowski's coagulation equation. Monte Carlo Methods Appl. **5**, 1–18 (1999)
4. EPA, Integrated science assessment for particulate matter. Technical report, U.S. Environmental Protection Agency, Washington D.C., USA, 2009

5. S.K. Friedlander, *Smoke, Dust and Haze Fundamentals of Aerosol Behavior*, 2nd edn. (Wiley, New York, 2000). WileyInterscience, New York, 1977, 333 p.
6. F.E. Kruis, A. Maisels, H. Fissan, Direct simulation monte carlo method for particle coagulation and aggregation. AlChe J. **46**, 1735–1742 (2000)
7. H. Müller, Zur Allgemeinen Theorie der Raschen Koagulation. Kolloidchem. Beih. **27**, 223–250 (1928)
8. M. Smoluchowski, Versuch einer Mathematischen Theorie der Koagulationskinetik Kolloider. Z. Phys. Chem. **92**, 129–168 (1917)
9. S. Twomey, *Atmospheric Aerosols* (Elsevier Scientific Publishing Co., New York, 1977)

Chapter 77
Study of the Magnetocaloric Effect in the New Compound $La_{0.67}Sr_{0.16}Ca_{0.17}MnO_3$: Application to Non-polluting Cooling

Asme Brahimi, Abdelkrim El Hasnaïne Merad, and Mohammed Benali Kanoun

Abstract Magnetic cooling based on magnetocaloric effect is one of the most important candidate applications as solutions for enjoying a non-polluting environment. In this work, a study of the magnetocaloric effect (EMC) in $La_{0.67}Sr_{0.16}$ $Ca_{0.17}MnO_3$ perovskite was made from an experimental measurement of magnetization as a function of temperature under low magnetic field and using phenomenological model. The results show 0.21 J/Kg K is the value of maximum magnetic entropy (ΔS_{max}) in this material perovskite near room temperature (Tc = 336 K). Furthermore it, from this phenomenological model, we were also able to calculate other properties. the values of the full-width at half-maximum (δT_{FWHM}), the maximum and the minimum specific heat were found to be, respectively, 10.19 K, 9.41 and −9.19 (J/(Kg K)) under 0.05 T field change. In addition, a relative cooling power (RCP) of about 2.12 J/Kg. Compared to other materials, $La_{0.67}Sr_{0.16}Ca_{0.17}MnO_3$ appears to be a promising candidate for magnetic cooling near room temperature.

Keywords Magnetic cooling · Magnetocaloric effect · Perovskite materials

---

A. Brahimi (✉)
Automatic Laboratory, A. Belkaid University, Box 119, 13000 Tlemcen, Algeria
e-mail: asmebrahimi@gmail.com

A. Brahimi · A. E. H. Merad
Solid State Physics Team, Theoretical Physics Laboratory, Faculty of Sciences, A. Belkaïd University, Box 119, 13000 Tlemcen, Algeria

M. B. Kanoun
Physics Department, College of Science, King Faisal University, P.O. Box 400, Al-Ahsa 31982, Saudi Arabia

A. Belasri and S. A. Beldjilali (eds.), *ICREEC 2019*, Springer Proceedings in Energy,
https://doi.org/10.1007/978-981-15-5444-5_77

77.1 Introduction

Refrigeration is one of the most important technologies used in everyday life (Food conservation, Beverage conservation, Health sector, air conditioning, …), where The number of residential refrigerators in 2013 was estimated at 1.4 billion units worldwide and the average annual consumption of all these cold appliances about 453 kWh, representing an annual electricity consumption of 634 TWh [1]. These refrigeration systems are based on the compression/expansion of the gas, which makes them polluting for the environment because they are a source of greenhouse gases emissions (hydrochlorofluarocarbons and hydrofluorocarbons). Magnetic cooling (MR) based on the magnetocaloric effect (MCE) can solve this problem, it has the advantage of replacing the gas with magnetic materials so it is environmentally friendly, Furthermore, the cooling efficiency in magnetic refrigerators is higher (the magnetic cooling efficiency can be reached up to 30–60% of a Carnot cycle, where as it is only 5–10% for conventional gas compression (CGC)) [2].

In this context, many experimental and theoretical studies have been carried out on magnetic materials, among the materials that have received particular attention are perovskite materials, due to their properties for example compared with the rare earth metals and their alloys, the perovskite-type oxides exhibit a considerably smaller magnetic hysteresis, a higher chemical stability, and a higher electrical resistivity that favors a lower value of eddy-current heating [3]. In this paper, the magnetocaloric properties of $La_{0.67}Sr_{0.3}Ca_{0.17}MnO_3$ compound in a low magnetic field were examined using a phenomenological model.

77.2 Phenomenological Model

To study magnetocaloric properties of the perovskite material $La_{0.67}Sr_{0.16}Ca_{0.17}MnO_3$, subjected to an applied magnetic field 0.05 T at different temperatures T, we have used a phenomenological model proposed by Hamad [4], in which the dependence of magnetization vs. temperature at a given magnetic field is given by:

$$M = \left(\frac{M_i - M_f}{2}\right) tanh(A(T_c - T)) + BT + C \tag{77.1}$$

The values of M_i and M_f represent the initial and final values of magnetization at ferromagnetic–paramagnetic (FM–PM) transition respectively (Fig. 77.1), B is the magnetization sensitivity dM/dT in the ferromagnetic (FM) region before transition, S_c the magnetization sensitivity dM/dT at Curie temperature (T_C), $A = \frac{2(B-S_c)}{M_i-M_f}$ and $C = \left(\frac{M_i+M_f}{2}\right) - BT_C$.

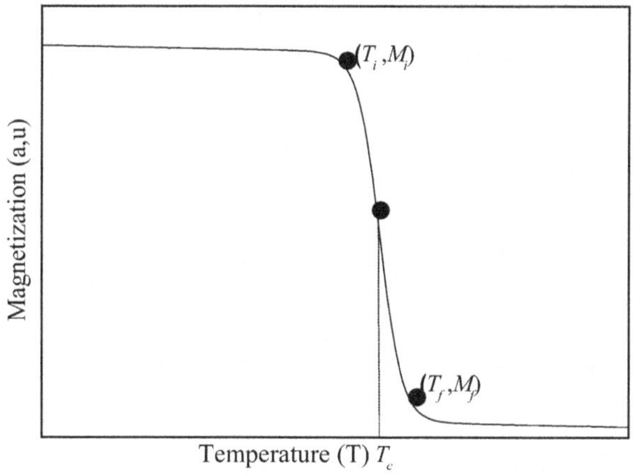

Fig. 77.1 Temperature dependence of magnetization in constant applied

The magnetic entropy change, under magnetic field from 0 to final value μ_0 H_{max}, is represented by:

$$\Delta S_M = \left(-A\left(\frac{M_i - M_f}{2}\right) sech^2(A(T_c - T)) + B\right) \tag{77.2}$$

Maximum entropy change obtained at T = T_C, and its value is evaluated by:

$$\Delta S_{max} = H_{max}\left(-A\left(\frac{M_i - M_f}{2}\right) + B\right) \tag{77.3}$$

A determination of the full-width at half-maximum δT_{FWHM} can be by the following equation:

$$\delta T_{FWHM} = \frac{2}{A} cosh^{-1}\left(\sqrt{\frac{2A(M_i - M_f)}{A(M_i - M_f) + 2B}}\right) \tag{77.4}$$

A magnetic cooling efficiency is evaluated by relative cooling power (RCP):

$$\begin{aligned} RCP &= -\Delta S_M(T, \mu_0 H_{max}) \times \delta T_{FWHM} \\ &= \left(M_i - M_f - 2\frac{B}{A}\right)\mu_0 H_{max} \times cosh^{-1}\left(\sqrt{\frac{2A(M_i - M_f)}{A(M_i - M_f) + 2B}}\right) \end{aligned} \tag{77.5}$$

The specific heat change is given by:

$$\Delta C_{P,H} = -TA^2(M_i - M_f) sech^2(A(T_c - T)) \tanh(A(T_c - T))\mu_0 H_{max} \tag{77.6}$$

Using this phenomenological model, the values of $-\Delta S_{max}$, δT_{FWHM}, RCP for $La_{0.67}Sr_{0.16}Ca_{0.17}MnO_3$ perovskite can be easily determined under magnetic field variation.

77.3 Model Application

From the experimental magnetization and the calculation of the magnetization sensitivity as a function of temperature, under a magnetic field 0.05 T [5]. The values of all the parameters M_i, M_f, T_C, B and S_C are determined (Table 77.1). Then, by applying these parameters in Eq. (77.1), we modeled the magnetization as a function of temperature, this modeling is presented in the Fig. 77.2. The curve obtained shows a good agreement between the experimental magnetization [5] and the calculated magnetization, where there is a transition from a ferromagnetic phase (FM) to a paramagnetic phase (PM) at a Curie temperature (T_C) equal to 336 K. It is believed that the double exchange mechanism between pairs of Mn^{3+} and Mn^{4+} ions is responsible for the ferromagnetic properties in these manganese oxides types [6].

To determine the magnetocaloric effect, it is necessary to determine the entropy variation. We have determined it using Eq. (77.2), and as shown in Fig. 77.3 the maximum entropy change ΔS_{max} reaches a peak of about 0.21 J/(kg K) at 0.05 T magnetic field variation.

Figure 77.4 shows the predicted values of the specific heat change as a function of temperature. Using Eqs. (77.4) and (77.5), the full width at half maximum (δT_{FWHM}), and relative cooling power (RCP) respectively at 0.05 T magnetic field variation for $La_{0.67}Sr_{0.16}Ca_{0.17}MnO_3$ are calculated and tabulated in Table 77.2.

These important values were compared with other samples in low applied magnetic field variations displayed in Table 77.3. The magnetocaloric properties of $La_{0.67}Sr_{0.16}Ca_{0.17}MnO_3$ are larger with magnetocaloric properties of $La_{0.7}Sr_{0.3}Mn_{1x}Al_xO_3$, $NdMnO_3$, $La_{0.5}Cd_{0.5}MnO_3$ and $Zn_{0.6}Ni_{0.4}Fe_2O_4$ as shown in Table 77.3 [7–10].

The results showed the possibility of applying this material in a non-polluting magnetic refrigeration technology.

Table 77.1 Model parameters for magnetocaloric samples $La_{0.67}Sr_{0.16}Ca_{0.17}MnO_3$ in an applied magnetic field of 0.05 T

Compound	M_i (emu/g)	M_f (emu/g)	T_c (K)	Sc (emu/(g k))	B (emu/(g k))
$La_{0.67}Sr_{0.16}Ca_{0.17}MnO_3$	49	1	336	−4.17	−0.01

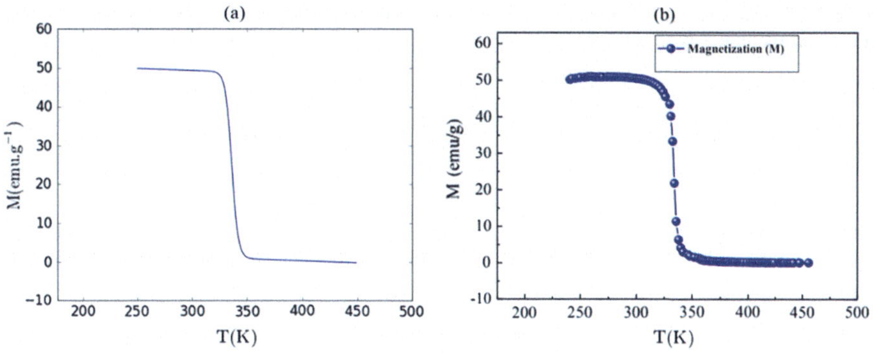

Fig. 77.2 Magnetization versus temperature in 0.05 T applied magnetic field for $La_{0.67}Sr_{0.16}Ca_{0.17}MnO_3$. **a** Represent modeled result **b** represent experimental result from Ref [5]

Fig. 77.3 Magnetic entropy variation versus temperature in 0.05 T applied magnetic field for $La_{0.67}Sr_{0.16}Ca_{0.17}MnO_3$

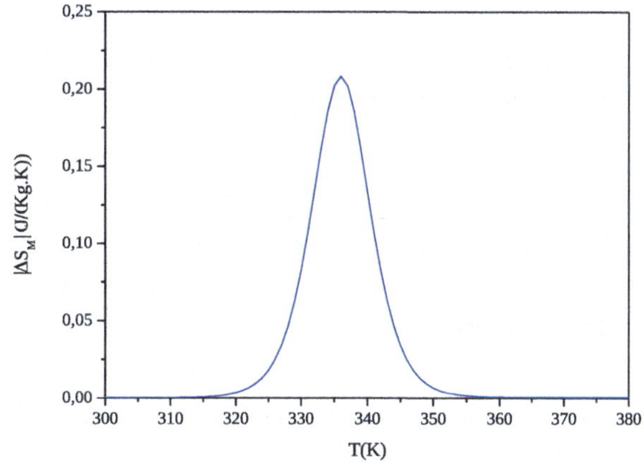

Fig. 77.4 Heat capacity change as a function of temperature in 0.05 T applied magnetic field for $La_{0.67}Sr_{0.16}Ca_{0.17}MnO_3$

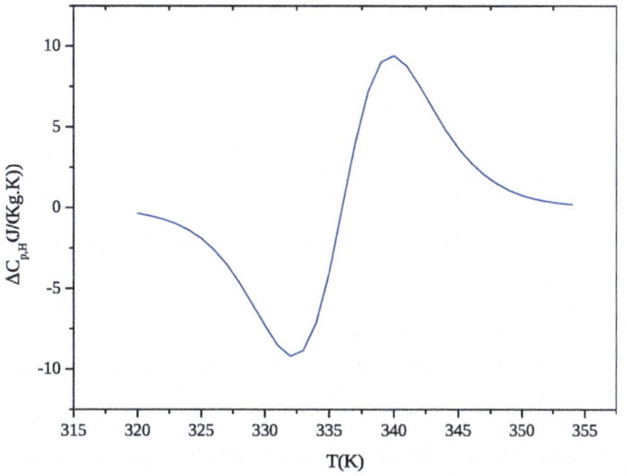

Table 77.2 The expected values of the magnetocaloric properties for $La_{0.67}Sr_{0.16}Ca_{0.17}MnO_3$ in an applied magnetic field 0.05 T

Compound	$\lvert\Delta S_{Max}\rvert/\Delta H$ (J/Kg K T)	δT_{FWHM} (K)	RCP (J/Kg)	$\delta C_{p,H(Max)}$ (J/(Kg K))	$\delta C_{p,H(Min)}$ (J/(Kg K))
$La_{0.67}Sr_{0.16}Ca_{0.17}MnO_3$	4.17	10.19	2.12	9.41	−9.19

Table 77.3 The expected values of the magnetocaloric properties for $La_{0.67}Sr_{0.16}Ca_{0.17}MnO_3$ in a applied magnetic field 0.05 T compared to the other compounds in different applied magnetic fields

Compound	ΔH (T)	$\lvert\Delta S_{Max}\rvert$ (J/Kg K)	δT_{FWHM} (K)	RCP (J/Kg)	$\delta C_{p,H}$ (Max) (J/(Kg K))	$\delta C_{p,H}$ (Min) (J/(Kg K))	Ref
$La_{0.67}Sr_{0.16}Ca_{0.17}MnO_3$	0.05	0.21	10.19	2.12	9.41	−9.19	This work
$La_{0.7}Sr_{0.3}MnAlO_3$	0.01	0.16	0.02	0.39	17.62	−17.784	[7]
$La_{0.7}Sr_{0.28}MnAl_{0.02}O_3$	0.01	0.08	0.05	0.37	9.21	−0.96	[7]
$La_{0.7}Sr_{0.25}MnAl_{0.05}O_3$	0.01	0.04	0.09	0.34	1.08	−1.12	[7]
$La_{0.7}Sr_{0.2}MnAl_{0.1}O_3$	0.01	0.02	0.14	0.32	0.35	−0.37	[7]
$La_{0.5}Cd_{0.5}MnO_3$	0.05	0.011	44.03	0.484	0.087	−0.076	[8]
$NdMnO_3$	1	0.0037	–	0.025	–	–	[9]
$Zn_{0.6}Ni_{0.4}Fe_2O_4$	1.7	0.025	–	–	–	–	[10]

77.4 Conclusion

In conclusion, the magnetocaloric effect of perovskite oxide $La_{0.67}Sr_{0.16}Ca_{0.17}MnO_3$ under an applied magnetic field of 0.05 T was found. The magnetization calculated on the basis of phenomenological model is in good agreement with the experimental magnetization. The maximum of magnetic change is 0.21 (J/Kg K) and it can be increased by increasing the magnetic field. The full-width at half-maximum (δT_{FWHM}) and relative cooling power (RCP) are well evaluated.

References

1. M.M. Vopson, Theory of giant-caloric effects in multiferroic materials. J. Phys. D Appl. Phys. **46**, 345304 (2013)
2. M.H. Phan, S.C. Yu, D. You-Wei, Review of magnetocaloric effect in manganite materials. J. Magn. Magn. Mater. **308**, 325–340 (2007)
3. Z. Wei, A. Chak-Tong, D. You-Wei, Review of magnetocaloric effect in perovskite-type oxides. Chin. Phys. B **22**, 057501 (2013)
4. M.A. Hamad, Prediction of thermomagnetic properties of $La_{0.67}Ca_{0.33}MnO_3$ and $La_{0.67}Sr_{0.33}MnO_3$. Phase Transit. **85**, 106–112 (2012)

5. Z.A. Mohamed, E. Tka, J. Dhahri, E.K. Hlil, Giant magnetic entropy change in manganese perovskite $La_{0.67}Sr_{0.16}Ca_{0.17}MnO_3$ near room temperature. J. Alloys Compounds (2014)
6. H.L. Ju, J. Gopalakrishnan, J.L. Peng, Q. Li, G.C. Xiong, T. Venkatesan, R.L. Greene, Dependence of giant magnetoresistance on oxygen stoichiometry and magnetization in polycrystalline $La_{0.67}Ba_{0.33}MnO_z$. Phys. Rev. B **51**, 6143 (1995)
7. P.T. Phong, L.V. Bau, L.C. Hoan, D.H. Manh, N. Phuc, I.-J. Lee, B-site aluminum doping effect on magnetic, magnetocaloric and electro-transport properties of $La_{0.7}Sr_{0.3}Mn_{1-x}Al_xO_3$. J. Alloys Compounds **645**, 243–249 (2015)
8. M.A. Hamad, Magnetocaloric effect in $La_{1-x}Cd_xMnO_3$. J. Supercond. Nov. Magn., 3459–3462 (2013)
9. R. Masrour, A. Jabar, A. Benyoussef, M. Hamedoun, E.K. Hlil, Monte Carlo simulation study of magnetocaloric effect in $NdMnO_3$ perovskite. J. Magn. Magn. Mater. **401**, 91–95 (2015)
10. O.M. Hemeda, N.Y. Mostafa, O.H. Abd Elkade, D.M. Hemeda, A. Tawfik, M. Mostafa, Electrical and morphological properties of magnetocaloric nano ZnNi ferrite. J. Magn. Magn. Mater. **394**, 96–104 (2015)

Chapter 78
Making a New Plaster-Foam Glass Composite Material Destined to Thermal Insulation in the Building

F. Benhaoua, N. Stiti, A. Ayadi, and M. Amrane

Abstract The protection of the environment through the reduction of solid waste and the reduction of greenhouse gases especially those emissions during the heating of buildings are objectives that can be achieved through the production of glass foam. In this study, we opted to prepare foam glass-gypsum composites destined to the coating ceilings and interior walls of buildings in order to expand employment of foam glass and to improve its performance including mechanical strength and facilitate their implementation. The foam glass is prepared by sintering at 800 °C with waste glass and calcium carbonate $CaCO_3$ in the form of a panel of 150 × 150 × 30 mm, the plaster-foam glass composites are prepared in the form of sandwiches panels or complexes. Physicochemical analysis techniques as SEM-EDS, porosity, density, thermal insulation and mechanical strength tests have been used for the characterization of these composites. The results obtained from thermal insulation and mechanical tests clearly show that the sandwich panel shows the best results with a thermal conductivity of 0.027 W/m°C and a compressive strength of 1.423 MPa.

Keywords Foam glass · Plaster · Composite · Thermal properties · Porous material

78.1 Introduction

The energy efficiency of buildings is one of the major concerns of social and economic development [1]. In Algeria, the building sector is the most energy-intensive sector with a consumption that is 42% higher than the national

F. Benhaoua (✉) · N. Stiti · A. Ayadi · M. Amrane
Research Unit: Materials, Processes and Environment, University of Boumerdes, Boumerdes, Algeria
e-mail: fayrouzbenhaoua@yahoo.fr

N. Stiti
University of Tizi-Ouzou, Tizi Ouzou, Algeria

A. Belasri and S. A. Beldjilali (eds.), *ICREEC 2019*, Springer Proceedings in Energy,
https://doi.org/10.1007/978-981-15-5444-5_78

overall energy consumption. Increasing the energy efficiency of buildings by thermal insulation is widely considered as a way to reduce up to 40% of energy demand related to heating and cooling of homes [2–4].

In this context, it is found that the foam glass is one of the most promising insulating materials and more requested in the construction, because of its power of thermal and acoustic insulation. This contributes to not only reduce energy consumption, but also to preserve natural resources, to reduce greenhouse gas emissions, improve a daily basis the comfort, and protect buildings against premature aging [5–7].

In that work, we seek to improve the propriety of the foam glass to increase its thermal insulation capacity, mechanical resistance and facilitate their implementation. To this end, we studied the characteristics of composite plaster-glass foam.

78.2 Methods and Materials

Foams glass (Fig. 78.1) were prepared from a 99 wt% cullet (waste glass) added to 01 wt% of foaming agent that is calcium carbonate $CaCO_3$, The resulting powder was well homogenized and pressed manually. The samples were thermally treated by sintering at a temperature of 800 °C for 10 min.

For the preparation of plaster-foam glass composites, a plaster was used for the construction; the characteristics of this plaster are given in Table 78.1.

The foams glass obtained are coated either a single layer of plaster with a thickness of 1 cm in order to prepare a complex panel plaster-foam glass (Fig. 78.2a) either are coated with two layers of 5 cm on the facades in order to prepare a sandwich panel plaster-foam glass (Fig. 78.2b).

Fig. 78.1 Panel of foam glass

Table 78.1 Characteristics physic for plaster

Characteristics	Results	Units
Fineness of grin	22	%
Normal consistency	40	%
Start of taking	4	min
End of taking	40	min
density	1.1	g/cm^3
Specific mass	2.6	g/cm^3

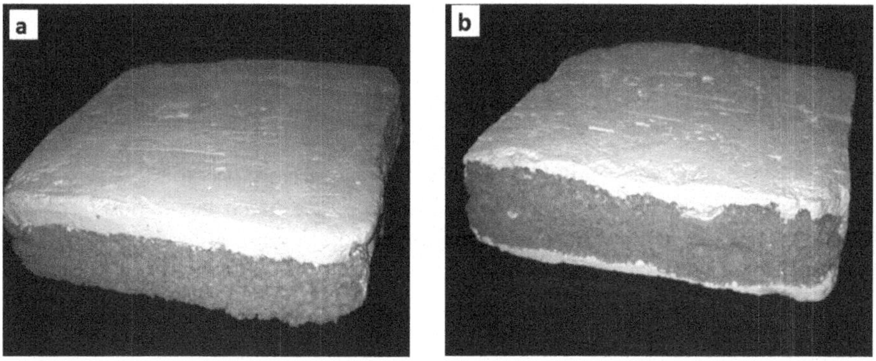

Fig. 78.2 Complex panel plaster-foam glass (a); sandwich panel plaster-foam glass (b)

The microstructure of the elaborated foam glass was analyzed by scanning electron microscopy (Philips XL 30 ESEM-type) on fragments of about 0.5 cm^2.

The density of samples with regular shape was calculated with mass and volume.

A thermal analyzer (Thermal Analyzer TCi) on samples of $50 \times 50 \times 20$ mm^3 determined measurements of thermal conductivity.

The tests of the compressive strength were made on $4 \times 4 \times 4$ cm^3 samples using a ZWICK ROELL machine.

78.3 Results and Discussion

78.3.1 Density Tests

Density measurements of complexes and sandwich plaster-foam glass composite were made on samples of a regular form according to the following formula:

$$\rho = m / V \quad (78.1)$$

where:

ρ: the density in g/cm^3;
m: the mass in g;
V: the volume in cm^3.

The results obtained show that elaborate composites are lighter materials with a density of 0.493 g/cm^3, for the complex and 0.511 g/cm^3, for the sandwich panel. These glass foams coated with plaster are denser than the glass foam panel alone with a value of 0.222 g/cm^3, this due to the high density of plaster compared to the foam glass.

78.3.2 Microstructural Characterization

Figure 78.3 shows the evolution of the microstructure of foam glass. This microstructure of foams glass shows the presence of a much larger pore size distribution, from a few micrometers 4.37 μm to several thousandths of a micrometer 1580 μm.

We also notice that the small pores fit between the large pores and play the role of a joint whose the majority of pores are communicating.

This microstructure also asserts that the shape of the small pores is uniform, in a spherical form; however, the big pores have some form.

78.3.3 Thermal Insulations Tests

Thermal insulation results (Fig. 78.4) shows that assembling plaster with foam glass in the form of a complex or a sandwich increases the thermal insulation capacity of foam glass from 0.4 for plaster alone to 0.027 W/m°C, for the plaster-foam glass sandwich. This coefficient of thermal conductivity value is the

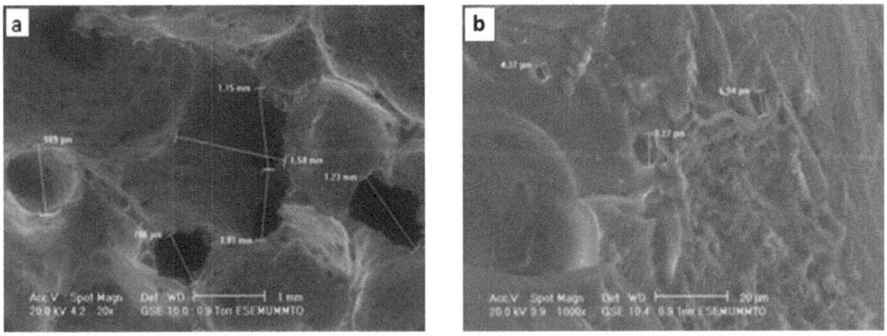

Fig. 78.3 SEM micrograph of foams glass; **a** 1 mm and **b** 20 μm

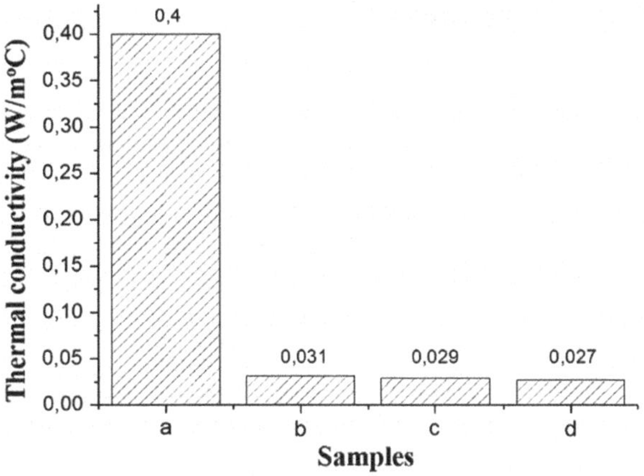

Fig. 78.4 Thermal conductivity of: **a** plaster; **b** foam glass; **c** Complex; **d** panel sandwich

lowest compared to the complex and the foam glass. This is interpreted by the suppression of the open porosity on the surface of the glass foams by the plaster, which has a very low porosity as indicated in Fig. 78.4.

78.3.4 Tests of Mechanical Resistance

Figure 78.5 shows that the compressive strength of foam glass increased with the number of plaster layers applied, from 0.59 to 0.88 MPa with the complex and at 1.423 MPa with the sandwich panel, this improvement remains weak in the face of the resistance plaster compression 10.95 MPa.

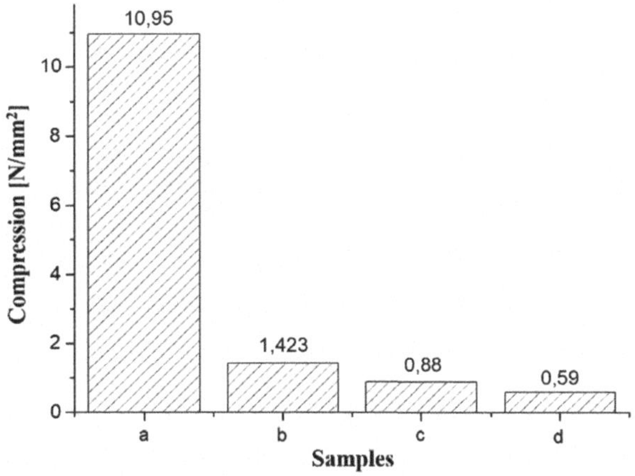

Fig. 78.5 Mechanical compressive resistance: **a** plaster; **b** foam glass; **c** complex; **d** panel sandwich

78.4 Conclusion

The manufacture of foams based on waste glass is a means of recycling. Which is part of an objective of environmental protection and maximum recovery of waste. Energy savings achieved with cullet result in a reduction of atmospheric pollution, especially carbon dioxide (CO_2) and a decrease in the price of the material.

Plaster is one of the building materials characterized by its anti-fire power and low thermal conductivity (0.4 W/m°C). The plaster-foam glass composite material achieves the desired quality by a considerable lightness ($\rho = 0.51$ g/cm^3) and a low thermal conductivity (0.027 W/m°C).

The use of plaster-foam glass sandwich panel in commercial or residential buildings to a great advantage. Whose consumption of energy for air conditioning or heating remarkably reduced, with a lower cost, more plaster skin facilitates the establishment of sandwich panel with a desired decorative appearance.

References

1. G. Huan, L. Hao, L. Libing, M. Lefu, S. Pengfei, X. Ziyan, L. Guocheng, A novel inorganic thermal insulation material utilizing perlite tailings. Energy Build. **190**, 25–33 (2019)
2. Le Plan d'Action en Matière d'Efficacité Énergétique, Ministère de l'Énergie d'Algerie
3. L. Aditya, T.M.I. Mahlia, B. Rismanchi, H.M. Ng, M.H. Hasan, H.S.C. Metselaar, O. Muraza, H.B. Aditiya, A review on insulation materials for energy conservation in buildings. Renew. Sustain. Energy Rev. **73**, 1352–1365 (2017)
4. T. Berger, A. Höltl, Thermal insulation of rental residential housing: do energy poor households benefit. A case study in Krems, Austria. Energy Policy **127**, 341–349 (2019)
5. V.E. Manevich, K.Yu. Subbotin, Foam glass and problems of energy conservation. Glass Ceram. **65**, 03–04 (2008)
6. H. Yin, M. Ma, J. Bai, Y. Li, S. Zhang, F. Wang, Fabrication of foam glass from iron tailings. Mater. Lett. **185**, 511–513 (2016)
7. H.W. Guo, X.F. Wang, Y.X. Gong, X.N. Liu, D.N. Gao, Improved mechanical property of foam glass composites toughened by glass fiber. Mater. Lett. **64**, 2725–2727 (2010)

Chapter 79
CO_2 Gas Sensors Based on Hydrophilic Vanadium Oxide Thin Film Coated QCM

Malika Berouaken, Chafiaa Yaddadene, Katia Chebout, Maha Ayat, Hamid Menari, Sabrina Belaid, and Noureddine Gabouze

Abstract Air pollution, which continues to increase, gives rise to considerable concern. It is generated by energy production, industrial activities, agriculture and transport. The main gases responsible for this pollution are sulfur dioxide (SO_2), nitrogen oxides (NOx) and carbon dioxide (CO_2). Nowadays, carbon dioxide is one of the main gases to detect. In the present work, vanadium oxide thin film has been deposited on quartz crystal microbalance (QCM) using vacuum thermal evaporating method followed by thermal treatment, for CO_2 gas sensor. The QCM covered with vanadium oxide film was heated at 200 °C for different times. The annealing time effect on the morphology and surface wettability of these structures has been investigated by using atomic force microscopy (AFM) and contact angle measurements (CA). The results show that elaborated surface exhibited roughness surface and Hydrophilic character. In addition, the exposing of this structure to CO_2 gas shows that the fabricated structure can be used as a CO_2 sensor at room temperature.

Keywords Vanadium oxide · Thin films · Quartz crystal microbalance · Gas sensor · CO_2

79.1 Introduction

Air pollution, which continues to increase, gives rise to considerable concern. It is generated by energy production, industrial activities, agriculture and transport. The main gases responsible for this pollution are sulfur dioxide (SO_2), nitrogen oxides (NOx) and carbon dioxide (CO_2). Nowadays, carbon dioxide is one of the main gases to detect. It is the largest greenhouse gas and is believed to be the main source

M. Berouaken (✉) · C. Yaddadene · K. Chebout · M. Ayat · H. Menari · S. Belaid · N. Gabouze
Division Couches Minces Surfaces et Interfaces, Centre de Recherche en Technologie des Semi-conducteurs pour l'Energétique, 2, Bd Frantz Fanon, BP 140 Alger 7-Merveilles, Algiers, Algeria
e-mail: BerouakenMalika@crtse.dz

A. Belasri and S. A. Beldjilali (eds.), *ICREEC 2019*, Springer Proceedings in Energy,
https://doi.org/10.1007/978-981-15-5444-5_79

of global warming. According to the World Meteorological Organization, CO_2 reached a record high in the atmosphere in 2018 with a concentration of about 410 ppm, a record level never seen in millions of years. While it is considered that the limit that should not exceed should be 350 ppm. Indeed, the detection and quantification of this gas is of great importance, given their heavy impact on humans and their environment.

Currently, there are several types of CO_2 sensors in the literature that are distinguished by the type of sensitive material and the transduction principle. Are found Metal oxide sensors [1, 2], polymer-based sensors [3, 4] and porous silicon sensors [5, 6]. Gas sensors based on metal oxides are considered interesting for their good characteristics: sensitivity, selectivity, speed, durability, low cost and simple detection methods. Nevertheless, the major problem of these sensors is the high operation temperature which requires high energy consumption and causes difficulties for their integration in electronic detection systems [7].

Alternatively, conductive polymers have attracted considerable interest as effective sensitive materials for gas sensors operating at room temperature. However, these sensors also have their own imperfections such as sensitivity to ambient temperature, sensitivity to moisture and pressure, and the problem of irreversibility or poisoning [8]. SiP has attracted a great attention as a material for the production of the sensitive layer of the sensors because of its large surface area, its cost and its operation at room temperature. However, these sensors are unstable because of the oxidation of the porous SiP layer in air. Therefore, the development of new detection device, working at low or at room temperature with high efficiency continues to be a very important research area on an industrial and academic scale.

One of the solutions proposed to resolve this problem is the use of quartz crystal microbalance (QCM) as a transducer. Indeed, QCM is one of the most promising sensor technologies. These gas sensors operate at room temperature even using metal oxide materials as a sensitive layer [9–11]. QCM sensors can directly detect the mass of adsorbed gas molecules on the detection layer, without a heating procedure, by changing the resonance frequency of the QCM. On the other hand, these sensors have good detection performance, high sensitivity, reproducibility, linearity, speed, and excellent stability over time. In addition, the cost of developing such sensors is inexpensive compared to other device commercialized. These performances have been proven by the rapid increase in the number of publications relating to detection applications based on QCM in recent years.

In this work, thermal evaporating process was used to deposit thin vanadium oxide layers onto the QCM in order to develop a new CO_2 detection device functioning at room temperature. The surface roughness and wettability of the films were characterized by atomic force microscopy and (AFM) and contact angle measurements (CA). The behavior of made structures in the presence of CO_2 has been studied at room temperature.

79.2 Experimental

The vanadium oxide thin films were coated onto QCM substrates (frequency of QCM was 6 MHz) from pure V_2O_5 powder (Aldrich 99.99%) by vacuum thermal evaporation method. The vanadium oxide is initially deposited in amorphous form, so thermal annealing is necessary in order to obtain a more stable oxidation phase of this oxide. Annealing of vanadium oxide thin films was done at 200 °C for 2 and 3 h. The surface of the film before and after thermal treatment was characterized by AFM using PHYWE compact AFM system and CA measurements with DGIDROP system (GBX scientific instruments) driven by Visiodrop software. The study of behaviour of these structures in the presence of CO_2 gas at room temperature was carried out by using homemade gas chamber, and the frequency change due to CO_2 adsorption/desorption was measured by a TF 930 frequency counter piloted by a computer via a USB interface.

79.3 Results and Discussion

79.3.1 AFM Analysis

Figure 79.1 shows AFM images of the vanadium oxide thin film before (Fig. 79.1a) and after thermal treatment at 200 °C for 2 h (Fig. 79.1b) and 3 h (Fig. 79.1c). As seen from AFM images with an increase in annealing time, grain growth is observed. The root mean square (RMS) of the all surfaces was calculated. It was found to be 6.1 nm for the film before annealing and after annealing at 200 °C for 2 and 3 h the RMS varying from 9.8 to 19.4 nm respectively. These results indicated that the annealing time had the effect of increasing the surface roughness, as consequence, increasing of the specific surface area.

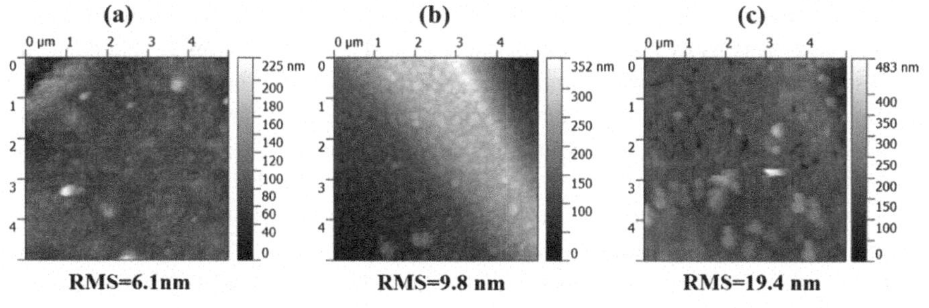

Fig. 79.1 AFM images of the QCM coated with Vanadium oxide film before (a) and after thermal treatment at 200 °C for 2 h (b) and 3 h (c)

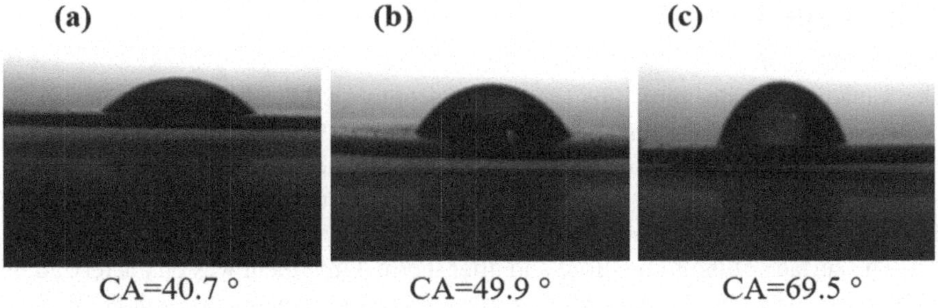

Fig. 79.2 Contact angle of the QCM coated with Vanadium oxide film before (a) and after thermal treatment at 200 °C for 2 h (b) and 3 h (c)

79.3.2 Contact Angle Measurement

Figure 79.2 shows the contact angle (CA) of a 3 µl drop of water deposited with a micro-syringe on QCM electrode surface coated with vanadium oxide before (Fig. 79.2a) and after annealing at 200 °C for 2 h (Fig. 79.2b) and 3 h (Fig. 79.2c). The measured values of the contact angle were in the order of 40.7°, 49.9° and 69.5° respectively. It can be seen that the contact angle increases with increasing annealing time. The observed increase in CA is probably due to the increase in surface roughness. Nevertheless, the annealing time does not significantly affect the surface wettability of the elaborated films since the CA values obtained are still in the range of hydrophilic surfaces (a surface having an AC value less than 90° is considered as a hydrophilic surface).

79.3.3 Sensor Response

Figure 79.3 illustrate the frequency shift (ΔF) of QCM sensors coated with vanadium oxide thin films before (a) and after annealing at 200 °C for 2 h (b) and 3 h (c), when exposed to the 40 mbar of CO_2 gas at room temperature. It can be seen that as the annealing time increases, the magnitude of the frequency shift increases, therefore the sensitivity of the sensor increases. This observation is related to an increase in the specific surface area of the sensor due to the increase in the surface roughness and probably it is related to the hydrophilicity character of the surface. As shown in the Table 79.1 when the roughness increases about 3 times, the sensitivity of the sensor increases approximately 2 times. In addition, the hydrophilic nature of the vanadium oxide thin film favorites the formation of more adsorbed species which increases the sensitivity of the sensor. At room temperature, the H_2O molecules from air (moisture) undergo a dissociative adsorption (hydroxyl

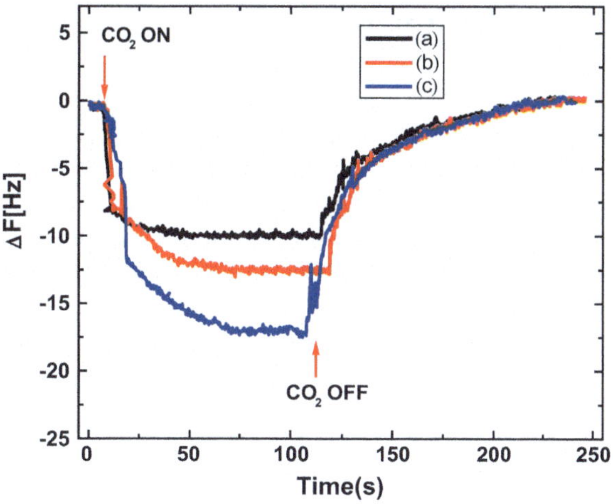

Fig. 79.3 Frequency change of QCM coated with vanadium oxide thin film when exposed to 40 mbar of CO_2 gas: before (a) and after annealing at 200 °C for 2 h (b) and 3 h (c)

Table 79.1 A frequency shift variation (ΔF) as a function of the surface roughness

Annealing time	Root mean square (nm)	ΔF (Hz)
Before annealing	6.1	10
200 °C/2 h	9.8	12.5
200 °C/3 h	19.4	17.5

OH^-, hydrogen H^+) on the surface of the vanadium oxide. The detection of CO_2 will take place by the reaction of the CO_2 molecules with the ions adsorbed (OH^-, H^+) on the surface. As shown schematically in Fig. 79.4, the higher specific surface area, more actives sites (OH^-, H^+) on the thin vanadium oxide surface are numerous, more the adsorption of CO_2 gas molecules is important. So the morphology of the sample heated at 200 °C for 3 h is favorable for CO_2 gas sensing application owing to its large surface area.

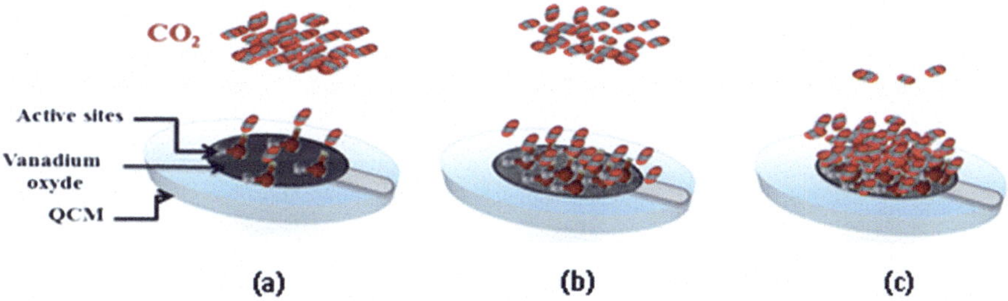

Fig. 79.4 Effect of the roughness and hydrophilic nature surface on the sensibility of vanadium oxide thin film before (a) and after annealing at 200 °C for 2 h (b) and 3 h (c)

79.4 Conclusion

In this work, CO_2 sensor working at room temperature was developed by using structure based on vanadium oxide as sensitive layer and QCM as transductor. Vanadium oxide thin film was deposited onto QCM substrate using thermal evaporation technique and heated at a temperature of 200 °C for 2 and 3 h. The effect of annealing time on the surface morphology and on the sensibility of the sensor was investigated. The results show that surface roughness is influenced by the annealing time unlike the wettability of the surface. In addition, the sensitivity of the sensor is closely related to the roughness and the wettability of the surface.

References

1. N.B. Tanvir, O. Yurchenko, E. Laubender, G. Urban, Investigation of low temperature effects on work function based CO_2 gas sensing of nanoparticulate CuO films. Sens. Actuators B Chem. **247**, 968–974 (2017)
2. F. Juang, W. Chern, B. Chen, Carbon dioxide gas sensing properties of $ZnSn(OH)_6$-ZnO nanocomposites with ZnO nanorod structures. Thin Solid Films **660**, 771–776 (2018)
3. S.A. Waghuley, S.M. Yenorkar, S.S. Yawale, S.P. Yawale, Application of chemically synthesized conducting polymer-polypyrrole as a carbon dioxide gas sensor. Sens. Actuators B Chem. **128**(2), 366–373 (2008)
4. F. Tebizi-Tighilt, F. Zane, N. Belhaneche-Bensemra, S. Belhousse, S. Sam, N. Gabouze, Electrochemical gas sensors based on polypyrrole-porous silicon. Appl. Surf. Sci. **269**, 180–183 (2013)
5. N. Gabouze, S. Belhousse, H. Cheraga, N. Ghellaib, Y. Ouadah, Y. Belkacema, A. Keffous, CO_2 and H_2 detection with a CHx/porous silicon-based sensor. Vacuum **80**, 986–989 (2006)
6. S. Naama, T. Hadjersi, A. Keffous, G. Nezzal, CO_2 gas sensor based on silicon nanowires modified with metal nanoparticles. Mater. Sci. Semiconductor Process. **38**, 367–372 (2015)
7. S. Wang, Y. Kang, L. Wang, H. Zhang, Y. Wang, Y. Wang, Organic/inorganic hybrid sensors: a review. Sens. Actuators B **182**, 467–481 (2013)
8. H. Bai, G. Shi, Gas sensors based on conducting polymers (Review). Sensors **7**(3), 267–307 (2007)
9. X. Wang, F. Cui, J. Lin, B. Ding, J. Yu, S.S. Al-Deyab, Functionalized nanoporous TiO_2 fibers on quartz crystal microbalance platform for formaldehyde sensor. Sens. Actuators B **171–172**, 658–665 (2012)
10. X. Li, X. Chen, Y. Yao, N. Li, X. Chen, High-stability quartz crystal microbalance ammonia sensor utilizing graphene oxide isolation layer. Sens. Actuators B **196**, 183–188 (2014)
11. T. Ito, Y. Fujii, N. Yamanishi, N. Asai, T. Shimizu, S. Shingubara, Electrodeposited ZnO thin film on twin sensor QCM for sensing of ethanol at room temperature. Procedia Eng. **168**, 411–414 (2016)

Chapter 80
Moth-Flame Optimization Algorithm for Solving Dynamic Economic Dispatch Considering Optimal Sizing of PV-ESS System

S. Sadoudi and M. Boudour

Abstract Photovoltaic power systems are known to be a clean energy source, their implementation into the electricity grid is the current trend in technologies and research. In this work dynamic economic dispatch issue incorporating an optimal sizing of a photovoltaic power system with battery storage system is solved by using Moth-flame optimization algorithm (MFO). Several simulations cases are processed while, the MFO algorithm is performed firstly over three units then six units test system. The obtained results are compared to other published works available in the literature to confirm the validity of the proposed method. In this paper, an optimal PV-ESS system have been sized through the proposed CDEED strategy. The reduction of emission gas and operating cost have been assumed in both of IEEE 3 unit and IEEE 6 unit systems. The Moth Flames Optimization (MFO) algorithm was employed as an optimization tool to minimize the objective function previously presented. The obtained results from the performed simulation of the 3 unit and 6 unit test system show the effectiveness of the proposed approach. As a further work, this issue will be extended by including the formulation of the Levelized Cost of Electricity (LCOE) considering the degradation rate and maintenance cost of PV panel, the battery lifetime and the cycles per day of charge discharge of the ESS.

Keywords CDEED · MFO · PV · ESS · IEEE 3 units · IEEE 6 units

80.1 Introduction

Fossil fuel power plants are the most used energy sources in electricity production, being a mature and controlled technology, fossil fuels such as gas and coal are considerably less expensive. However, through considering the limited natural

S. Sadoudi (✉) · M. Boudour
Laboratory of Electrical and Industrial Systems, USTHB, Algiers, Algeria
e-mail: slim.sa.93@gmail.com

A. Belasri and S. A. Beldjilali (eds.), *ICREEC 2019*, Springer Proceedings in Energy,
https://doi.org/10.1007/978-981-15-5444-5_80

reserve, the difficult access to oil fields issue [1], an effective dispatch of generation units have been investigated, in the aim of satisfying the power supply at the minimum operating cost.

High coal consumption in China leads to a high concentration of fine particles and a considerable deterioration in air quality. Among the main coal combustion sectors, industrial coal combustion, followed by coal combustion in power plants and the domestic sector [2]. Recently, the advancement and motivation of technology and scientific research have contributed to the promotion of renewable energies (RESs) for electricity production. In China, the policy on renewable energy and the reduction of pollutant emissions of atmosphere through electricity production is currently focused on investment and increasing of the installed capacity of renewable energy sources as wind, solar and biomass [3]. However, this kind of systems is often more expensive than conventional fossil-based [4]. Hence, a combined dynamic economic and emission dispatch (CDEED) strategy have been proposed in this paper in the aim to maximize the use of renewable energy systems. A holistic review for combined economic emission dispatch (CEED) have been presented in [5], through their investigation several optimization strategies have been discussed, showing their advantages and disadvantages. The CDEED issue consist on ensuring the optimal set-points of the output power generation of involved units, giving both of the lowest fuel cost operating and gas emission, while preserving the power balance and units capacities, considering the valve point effect. Exhaust gases due to the fuel combustion, the operating cost of units are closely related to the nature of used energy sources. Therefore, these objectives give rise to complexity based on the contradictory nature of conventional and renewable sources. In references [5, 6] a brief review of optimization techniques that have been proposed during the last years in purpose to solve the CDEED issue. Among them, a fuzzy logic controlled genetic algorithms (FLGA) have been used to deal with the 6 unit system under varying load demand [7], the niched pareto genetic algorithm (NPGA) also has been used to optimize the same 6 units system considering the valve point loading and transmission losses [8]. A modified harmony search method have been performed to gives the optimum cost and emission through 3 unit, 6 Unit, 10 Unit and 40 unit system at different power demand [9]. Over the last four years, new methods have often been requested to solve this problem such as the exchange market algorithm (EMA) [10], symbiotic organisms search (SOS) [11] and the coyote optimization [12]. During the recent emergence of the renewable energy sources, the CDEED issue integrates clean energy systems although they are expensive. The new strategy aims to maximise the integration of RESs [13, 14] such as wind, solar and hydro generators. Energy storage systems (ESS) have been used in [15] to amortize cost during the peak period by exploiting the energy stored during the lowest demand curve. The main contribution of this paper was the integration of PV-ESS source into the IEEE 3 and 6 units system, a strategy of conventional units reserve maximization have been included. A new optimization method inspired from the living style and the navigation method of moths in nature.

The rest of this paper is organized as follows. Section 80.2 is dedicated to the mathematical model and formulation of the proposed CDEED issue. Section 80.3 describes the MFO algorithm. Section 80.4 was consecrated to present the system's model of the two cases of studies, including the simulation results and discussions of the obtained results. Finally Sect. 80.5 concludes the paper.

80.2 Problem Formulation

The CDEED issue can be summarized as an optimization process with two contradictory objectives as described in Eqs. (80.1) and (80.2). The fuel costs and gas emissions are minimized by an appropriate distribution of the electricity production of the generating units.

$$Min \sum_{j=1}^{n} F_c(P_j) = Min \sum_{j=1}^{n} a_j P_j^2 + b_j P_j + c + \left| e_j \sin\left(f_j\left(P_j^{min} - P_j\right)\right)\right| \quad (80.1)$$

$$Min \sum_{j=1}^{n} F_e(P_j) = Min \sum_{j=1}^{n} 10^{-2}(\alpha_j P_j^2 + \beta_j P_j + \gamma) + \xi_j e^{(\delta_j P_j)} \quad (80.2)$$

where a_j, b_j, c_j, e_j and f_j are the fuel cost coefficients considering valve point effect of the unit j. α_j, β_j, γ_j, ξ_j and δ_j are the gas emissions coefficients of thermal units. The distributed generators (DG) units are composed of storage battery that are directly charged by the solar panels. Therefore the total cost of the PV-ESS system have been assumed as the photovoltaic system operating cost according to Eq. (80.3).

$$Min \sum_{j=1}^{n} C_{RESs}(P_j) = Min \sum_{j=1}^{n} \eta_j P_j \quad (80.3)$$

where η_j and P_j are respectively the PV-ESS cost coefficient and the generated output power of both PV-ESS systems. The high cost coefficient of PV-ESS unit significantly affects the participation of the renewable source to the main system. However, the clean aspect and the reserve preserving strategy have been incorporated in order to maximize the use of the PV-ESS system and their optimal sizing have been determined through the best solution of P_j. The reserve preserving strategy R_j consist on maximize the gap between the actual production of a conventional unit and its maximum capacity limit as shown in Eq. (80.4).

$$Max \sum_{j=1}^{n} R_J(P_j) = Max \sum_{j=1}^{n} \left(P_j^{max} - P_j\right) \quad (80.4)$$

All the objectives described above through Eqs. (80.1) to (80.4) have been assigned to the same scale and combined to a fitness function as follows.

$$Fit(P_j) = Min\,\omega\frac{\left(\sum_{j=1}^{n} F_c(P_j) + C_{RESs}\right) - C_{min}}{C_{max} - C_{min}} + (1-\omega)\frac{\sum_{j=1}^{n} F_e(P_j) - E_{min}}{E_{max} - E_{min}} + \frac{P_{min}}{R(P_{max} - P_{min})} \quad (80.5)$$

where ω is fixed at 0.5, it represent the weight factor, C_{min} and C_{max} are the minimal and maximal total units operating cost and E_{min}, E_{max} are the minimal and maximal emission exhaust of all units. Both of generation and load demand P_D must be in balance. Output power of each unit has to be bounded by upper and lower limits. The equality and inequality constraint have been satisfied as follow.

$$P_D - \left(\sum_{j=1}^{n} P_j + C_{RESs}\right) = 0 \quad (80.6)$$

$$P_{min} \le P_j \le P_{max} \quad (80.7)$$

80.3 Moth-Flame Optimization Algorithm

In 2015, Seyedali Mirjalili have been inspired from the living style and the navigation method of moths in nature to develop for the first time an optimization algorithm based on the convergence of the moth towards the flame [16]. This method, guarantees a good exploration of the research space around the optimal location. The main flowchart of the MFO algorithm have been presented in the Fig. 80.1. In this work, moth are represented by the output generation of each unit and the flame is the fitness function cited bellow in Eq. (80.5).

80.4 Simulation and Results

The MFO algorithm has been implemented in MATLAB to solve the CDEED issue and the optimal sizing of the PV-ESS. Firstly, an IEEE 3 UNIT test system have been used through the simulation. Two cases of study have been explored, the case I aims to validate the dispatching tool. Case II have been proposed to compare and prove the impact of the optimal PV-ESS system sizing. Finally, a second test have been investigated through the IEEE 6 UNIT system. Obtained results are presented as follows, the convergence characteristic of the MFO algorithm for both cases and systems are presented in Figs. 80.2 and 80.3. The operating cost, exhausted gas and total reserve of the IEEE 3 unit system are summarized in Table 80.1, Table 80.2 is dedicated to the IEEE 6 unit system.

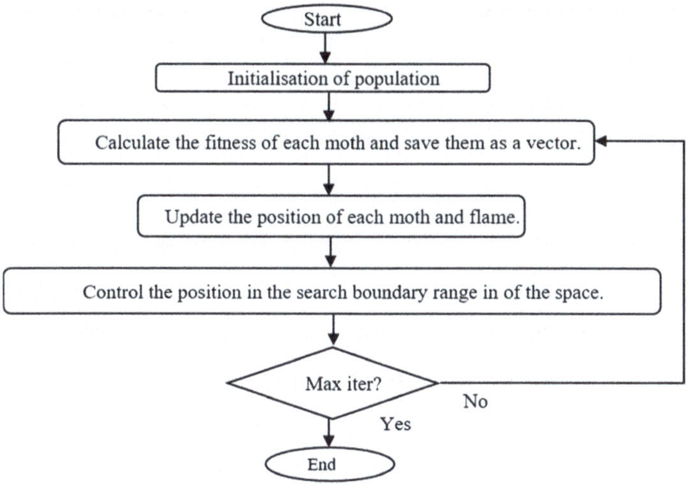

Fig. 80.1 The Moth-Flame Optimization flowchart algorithm

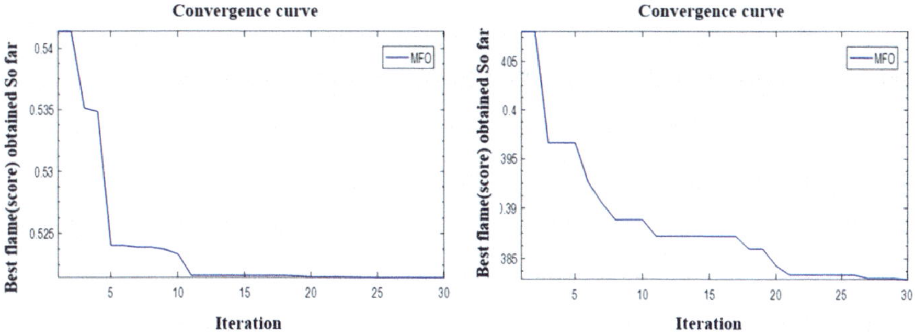

Fig. 80.2 The convergence characteristic of the MFO algorithm for IEEE 3 UNIT

The obtained results show the validity and robustness of the proposed method. The convergence of the proposed algorithm through the best solution at each iteration have been presented in Figs. 80.1 and 80.2, the maximum number of iteration is 50 in both cases of each system, the number of search agents is 20 in the three unit system and 40 in the six unit system. Through Tables 80.1 and 80.2, both of gas exhaust and main reserve of the systems have been improved by including renewable energy source. The cost of conventional units have been also reduced slightly. However, the total operating cost of both 3 unit and 6 unit systems have

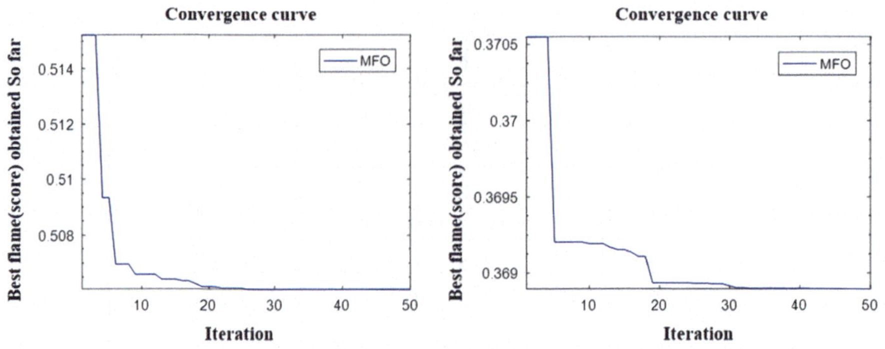

Fig. 80.3 The convergence characteristic of the MFO algorithm for IEEE 6 UNIT

Table 80.1 Simulation results for the IEEE 3 UNIT at 850 MW

	Case I		Case II		
	Total	Conventional	Total	Conventional	PV-ESS
Cost ($/h)	8251.53	8251.53	9846.22	7354.43	2491.78
Emission (kg/h)	115.85	115.85	85.438	85.43	0
Reserve (MW)	605.19		704.67		

Table 80.2 Simulation results for the IEEE 6 UNIT at 900 MW considering losses

	Case I		Case II		
	Total	Conventional	Total	Conventional	PV
COST ($/h)	49856.45	49856.45	57239.53	46544.91	10694.61
GES (Kg/h)	746.04	746.04	638.25	638.25	0.00
Reserve (MW)	563.99		635.29		

been affected due to the incorporation of PV-ESS. The proposed PV-ESS sizing have been considered through a combined economic and emission dispatch strategy.

80.5 Conclusion

In this paper, an optimal PV-ESS system have been sized through the proposed CDEED strategy. The reduction of emission gas and operating cost have been assumed in both of IEEE 3 unit and IEEE 6 unit systems. The Moth Flames Optimization (MFO) algorithm was employed as an optimization tool to minimize the objective function previously presented. The obtained results from the performed simulation of the 3 unit and 6 unit test system show the effectiveness of the proposed approach. As a further work, this issue will be extended by including the formulation of the Levelized Cost of Electricity (LCOE) considering the degradation rate and maintenance cost of PV panel, the battery lifetime and the cycles per day of charge discharge of the ESS.

References

1. S. Shafiee, E. Topal, When will fossil fuel reserves be diminished? Energy Policy **37**(1), 181–189 (2009)
2. Q. Ma et al., Impacts of coal burning on ambient PM 2.5 pollution in China. Atmos. Chem. Phys. **17**(7), 4477–4491 (2017)
3. T. Qi, X. Zhang, V.J. Karplus, The energy and CO_2 emissions impact of renewable energy development in China. Energy Policy **68**, 60–69 (2014)
4. Y. Zhang, Y. Wang, X. Wang, Greenware: greening cloud-scale data centers to maximize the use of renewable energy. in *ACM/IFIP/USENIX International Conference on Distributed Systems Platforms and Open Distributed Processing* (Springer, Heidelberg, 2011)
5. F.P. Mahdi et al., A holistic review on optimization strategies for combined economic emission dispatch problem. Renew. Sustain. Energy Rev. **81**, 3006–3020 (2018)
6. T.K. Panigrahi, A.K. Sahoo, A. Behera, A review on application of various heuristic techniques to combined economic and emission dispatch in a modern power system scenario. Energy Procedia **138**, 458–463 (2017)
7. Y.H. Song et al., Environmental/economic dispatch using fuzzy logic controlled genetic algorithms. IEEE Proc. Gener. Transm. Distrib. **144**(4), 377–382 (1997)
8. M.A. Abido, A niched Pareto genetic algorithm for multiobjective environmental/economic dispatch. Int. J. Electr. Power Energy Syst. **25**(2), 97–105 (2003)
9. A modified harmony search method for environmental/economic load dispatch of real-world power systems
10. A. Rajan, T. Malakar, Optimum economic and emission dispatch using exchange market algorithm. Int. J. Electr. Power Energy Syst. **82**, 545–560 (2016)
11. M.K. Dosoglu et al., Symbiotic organisms search optimization algorithm for economic/emission dispatch problem in power systems. Neural Comput. Appl. **29**(3), 721–737 (2018)
12. U. Güvenç, E. Kaymaz, Economic dispatch integrated wind power using coyote optimization algorithm, in *2019 7th International Istanbul Smart Grids and Cities Congress and Fair (ICSG)* (IEEE, 2019)
13. S. Sadoudi, M. Boudour, N.E.Y. Kouba, Optimal combined dynamic economic and emission dispatch including wind and photovoltaic power systems, in *2018 International Conference on Electrical Sciences and Technologies in Maghreb (CISTEM)* (IEEE, 2018)
14. S. Sadoudi, M. Boudour, N.E.Y. Kouba, Gravitational search algorithm for solving equal combined dynamic economic-emission dispatch problems in presence of renewable energy sources, in *2018 International Conference on Applied Smart Systems (ICASS)* (IEEE, 2018)
15. S.P. da Rocha, et al., Two-stage optimization combining PSO and TOPSIS for allocation of energy storage in electric power systems, in *2018 IEEE Congress on Evolutionary Computation (CEC)* (IEEE, 2018)
16. S. Mirjalili, Moth-flame optimization algorithm: a novel nature-inspired heuristic paradigm. Knowl. Based Syst. **89**, 228–249 (2015)

Chapter 81
CO_2 Capture from Natural Gas with Ionic Liquid

Nesrine Amiri, Zoubida Lounis, and Hassiba Benyounes

Abstract Carbon dioxide capture is a global concern because of its effect on climate change, especially as regards to global warming. Absorption using physical and chemical solvents is the commonly used method to capture CO_2. Ionic liquids (ILs) as advanced solvents have been regarded as appropriate candidates for CO_2 capture because of their advantages such as non-volatility potentially that makes ILs "green" solvents, negligible vapor pressure and high CO_2 solubility.

This work aimed to study the feasibility of CO_2 separation from a gas containing methane using the low viscosity ionic liquid 1-hexyl-3-methylimidazolium tetra-cyanoborate [hmim][TCB] as an alternative to conventional ones by simulation using Aspen plus V.8.0. Thermophysical properties were calculated using empirical correlations, and the experimental data for CO_2/CH_4 in presence of [hmim][TCB] were fitted with the NRTL activity coefficient model to determine the binary interaction parameters. The results show that lower solvent flow rate and lower energy consumption are required for the absorption with [hmim][TCB] than with organic solvents, especially for gas streams with moderate acid gas content.

Keywords CO_2 capture · Ionic liquid · Physical absorption

N. Amiri (✉) · Z. Lounis
Laboratoire d'Ingénierie de la Sécurité Industrielle et du Développement Durable, Université d'Oran 2, B.P 1015 El M'naouer, 31000 Oran, Algérie
e-mail: nesrine19941@gmail.com

N. Amiri · H. Benyounes
Laboratoire Physico-chimie des matériaux, catalyse et environnement LPCMCE, Faculté de Chimie, Université des Sciences et de la Technologie d'Oran Mohamed Boudiaf USTO-MB, B.P 1505 El M'naouer, 31000 Oran, Algérie

A. Belasri and S. A. Beldjilali (eds.), *ICREEC 2019*, Springer Proceedings in Energy,
https://doi.org/10.1007/978-981-15-5444-5_81

81.1 Introduction

The CO_2 absorption can be conducted by solvents which chemically react with the acid gas, or by physical solvents, where the acid gas is just dissolved in the liquid. Chemical solvents, such as monoethanolamine (MEA) are frequently preferred because of their capability of reducing the concentration of CO_2 to ppm levels. However, the MEA solvent has a high heat of reaction with CO_2 that leads to higher energy consumption during CO_2 recovery. The use of physical absorbents (e.g. Selexol) overcome the drawback of high-energy consumption of the regeneration step, although there are several drawbacks for physical solvents, such as the high affinity to heavy hydrocarbons which will be removed with the CO_2 resulting into hydrocarbon losses. A new generation of green solvents called "ionic liquids" are being studied in this work as promising sorbents to replace the conventional ones.

In this work, the IL 1-hexyl-3-methylimidazolium tetracyanoborate ([hmim][TCB]) has been selected because of its high CO_2 solubility, and especially because of the availability of experimental data on the solubility of CO_2 and CH_4 in this ionic liquid. The objective of this work is to study the feasibility of separating CO_2 from a gas mixture containing CO_2 and CH_4, as a model system representing natural gas, using the ionic liquid 1-hexyl-3-methylimidazolium tetracyanoborate [hmim][TCB].

81.2 Methodology

For the feasibility study of CO_2 absorption using IL [hmim][TCB], a binary gas mixture containing CH_4-CO_2 has been chosen as a model system. The experimental data on the properties of the IL were collected from literature such as the gas

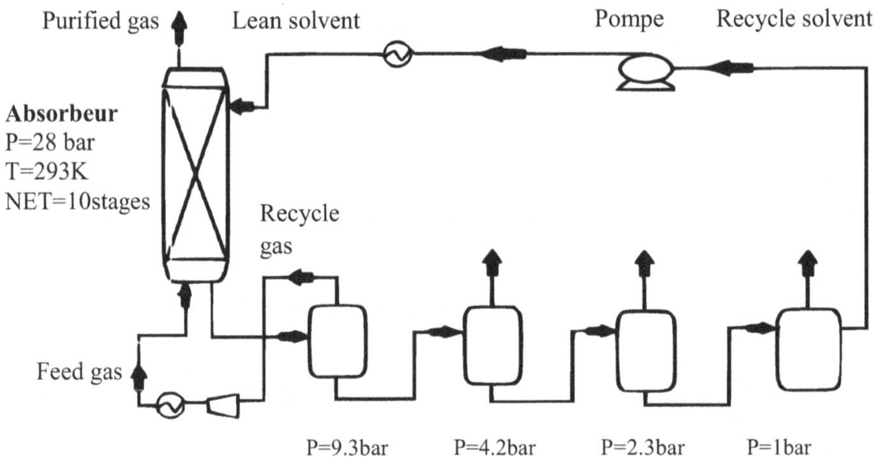

Fig. 81.1 CO_2 absorption unit based on Selexol process

solubilities. The physical and thermodynamic properties were then calculated to define the IL [hmim][TCB] in Aspen plus database. The NRTL model was used to describe the thermodynamic behavior of CO_2 and CH_4 in the presence of [hmim] [TCB], and the binary interaction parameters were fitted using experimental gas—liquid equilibrium. The general flowsheet for CO_2 capture by DEPG is presented in (Fig. 81.1) [1]. A comparative study between the proposed process using the IL as a solvent and the benchmark model using Selexol was carried out by simulation to determine the most efficient process to separate CO_2 from methane at different concentrations.

Table 81.1 Properties of [hmim][TCB]

Properties	Values and units
Chemical name	1-**h**exyl-3-**m**ethyl**im**idazolium tetra**c**yano**b**orate "**HMIM TCB**"
Structure	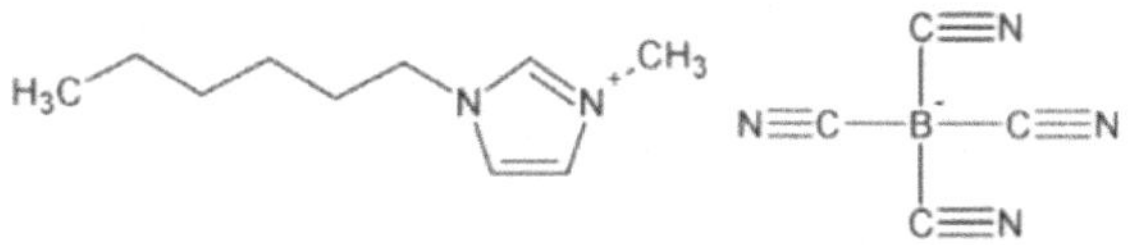
Molecular weight	282.16 g/mol
Critical properties	T_b=975.58 K T_c=1193.45 K P_c=1.9 MPa V_c=1000.62 cm^3/mol ω=1.462 Z_c=0.192

Temperature-dependent properties of [hmim][TCB] [2]

Properties	Correlations	ASPEN Model name
Vapor pressure (bar)	$lnP_i^{*j} = C_{1i} + \frac{C_{2i}}{T + C_{3i}} + C_{4i}T + C_{5i}lnT + C_{6i}T^{C_{7i}}$	PLXANT
Viscosity (N.s/m²)	$ln\eta_i^{*j} = C_{1i} + \frac{C_{2i}}{T} + C_{3i}lnT$	MULAND
Enthalpy of vaporization (kJ/kmole)	$\Delta_{vap}H_i^*(T) = \Delta_{vap}H_1^*(T_1)\left(\frac{1 - T/T_{ci}}{1 - T_1/T_{ci}}\right)^{a+b(1-T/T_{ci})}$	DHVLWT

CO_2 separation from a gas mixture of CO_2 and CH_4 using a physical absorption has been simulated in the Aspen Plus v8.0.

81.3 Results and Discussion

81.3.1 Thermodynamic Modeling and Calculation of Physico-Chemical Properties of hmim TCB

The database of Aspen plus do not contain the component 1-hexyl-3-methylimidazolium tetracyanoborate, so its properties should be implemented. The properties of this ionic liquid are briefly presented in Table 81.1, and the temperature-dependent properties have been calculated using empirical correlations [2], which are also presented with corresponding Aspen Physical Property model names.

The gas-Liquid equilibrium of CH_4-CO_2 system in presence of "DEPG" was represented by PC-SAFT equation of state model. And the equilibrium of CO_2 and CH_4 in presence of [hmim][TCB] was modeled with the NRTL activity coefficient model, where the binary interaction parameters of CO_2 + [hmim][TCB] and CH_4 + [hmim][TCB] were fitted to the experimental data [3]. The function to be minimized in this case is indicated in Eq. (81.1), and the results of regression of binary parameters are shown in Table 81.2.

$$RMS = \sqrt{\frac{1}{n}\left(\sum\left(T_{exp} - T_{reg}\right)^2 + \sum\left(P_{exp} - P_{reg}\right)^2 + \sum\left(x_{exp} - x_{reg}\right)^2\right)} \quad (81.1)$$

81.3.2 Simulation of CO_2 Absorption Process and Sensitivity Analysis

The simulation of the absorption process (Fig. 81.1) of CO_2 from an equimolar gas mixture of 100 kmol/h containing CO_2 and CH_4 has been performed using two different solvents DEPG and [hmim][TCB].

A sensitivity analysis of operating parameters of the absorption column has been carried out first for absorption using DEPG as solvent in order to set the optimal operating conditions to get a high recovery ratio of 99.94% in CO_2. These

Table 81.2 Regression parameters of NRTL model for CO_2 and CH_4 systems with [hmim][TCB]

Binary parameters	CO_2	CH_4
a_{ij}	−1.7989	6.0544
a_{ji}	1.2122	−0.8782
b_{ij}	0.4162	6.6179
b_{ji}	7.7214	5.7014
α_{ij}	0.3000	0.3000
RMS	1.935	0.176

Table 81.3 Operating parameters of CO_2 absorption column using DEPG and IL

Properties	Temperature (K)	Pressure (bar)	Number of stages	Mole flow rate (kmol/h)	
Feed gas	293	28	10	CO_2	50
				CH_4	50
Solvent	293	28		DEPG	160
				[hmim][TCB]	101.42

parameters are presented in Table 81.3. Then these parameters were used to get the same recovery ratio of CO_2 using [hmim][TCB] as solvent.

New studies were conducted setting the same operating conditions such as: temperature, pressure and number of theoretical stages to separate gas mixture containing CO_2 and CH_4 at different concentrations in CO_2 varying between 5% and 15% in mole, which have been chosen as model system representing natural gas. The obtained results of the absorption column are illustrated in Fig. 81.2 for each case of organic solvent DEPG and IL [hmim][TCB]. The results indicate that CO_2 capture with DEPG solvent consume high amounts of solvent, that increases when the CO_2 concentration in feed gas decreases. While in the case of IL solvent this behavior was inversed.

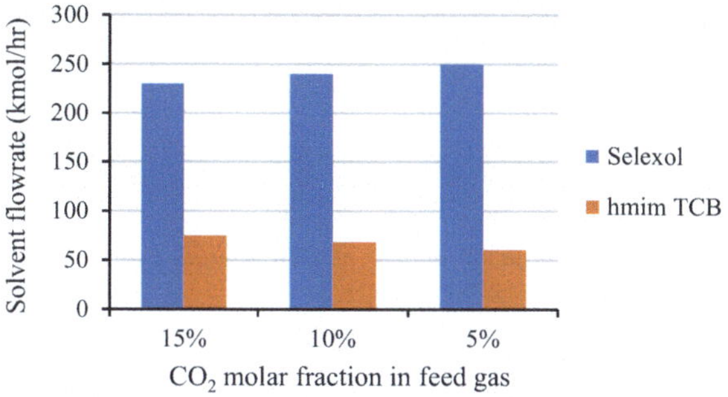

Fig. 81.2 Amount of solvent required to recover CO_2 in the absorber at deferent concentrations of CO_2

Table 81.4 Simulation results for the regeneration system

Solvent type	Selexol (DEPG)			[hmim][TCB]		
CO_2 mol fraction in feed gas (%)	15	10	5	15	10	5
Recovery ratio of CO_2 (%)	58.28	47.52	33.42	67.45	55.67	31.45
Energy consumed (cal/sec)	18566.2	66938.2	12917.9	2524.3	1594.8	532.9

The solvent can be regenerated by reducing the pressure in four flash drums (PSA process) as presented above in Fig. 81.1 for each solvent case. The used flash drums operate at pressures of 9.3 bar; 4.2 bar; 2.3 bar and 1 bar respectively. The simulation results of the regeneration system are summarized in Table 81.4.

81.4 Conclusion

The CO_2 separation using the ionic liquid [hmim][TCB] as alternative solvent to conventional ones was investigated by simulation. The results show that the solvent flow rate used in the case of [hmim][TCB] over DEPG is reduced by more than 3 for recovery ratio of 99.94% in the absorption column. The energy consumption of such an IL-based process is significantly low. e.g. The energy required to regenerate the IL is reduced by more than 16% compared to DEPG. That is still more less energy consumed than chemical absorption because the heat of desorption of acid gases from a physical solvent is lower than that required to desorb them from chemical solvents.

Consequently, the IL-based CO_2 capture process can be recommended as a very promising technology in CO_2 separation from gases moderately concentered in CO_2 such as natural gas.

In addition to the non-volatility property of [hmim][TCB] and the non-toxicity, it requires a low energy for their regeneration, which means that this type of solvent is very environmentally friendly.

References

1. C.H. Yu, C.H. Huang, C.S. Tan, A review of CO_2 capture by absorption and adsorption. Aerosol Air Qual. Res. **12**(5), 745–769 (2012)
2. B.E. Poling, J.M. Prausnitz, J.P. O'Connell, *The Properties of Gases and Liquids* (McGraw-Hill, New York, 2001)
3. M.T. Mota-Martinez, M. Althuluth, M.C. Kroon, C.J. Peters, Solubility of carbon dioxide in the low-viscosity ionic liquid 1-hexyl-3-methylimidazolium tetracyanoborate. Fluid Phase Equilib. **25**(332), 35–39 (2012)

Chapter 82
Study of the Influence of Thermal Insulation on the Energy Consumption of Buildings in Different Climates in Algeria

O. Sotehi, A. Chaker, and K. Mahdi

Abstract The aim of our work is to conduct a study on the use of thermal insulation in the building sector. A solar prototype of the solar village built in Boussaâda is chosen to carry out simulation calculations using the Trnsys 17 software, in three different climates of Algeria (Algiers, Constantine and Ouargla). The solar prototype studied is equipped with devices allowing the use of passive solar energy with improved insulation. The simulation results of the passive systems show that the use of thermal insulation allows significant energy savings estimated at 34.48, 30.44 and 21.49% respectively for the cities of Algiers, Constantine and Ouargla.

The passive system simulation results show that the use of these devices, such as Trombe walls and glazed surfaces, allows significant energy savings (although considered insufficient). A reduction in heating needs of 2.12, 1.7 and 2.64 times respectively for the cities of Algiers, Constantine and Ouargla is obtained.

Keywords Passive solar systems · Boussaâda solar prototype · Thermal insulation · Low energy consumption building · Passivhauss standard

---

O. Sotehi (✉)
Laboratoire Energie et Environnement, Université Salah Boubnider Constantine 3, Constantine, Algeria
e-mail: sotehioualid@gmail.com

A. Chaker · K. Mahdi
Laboratoire Physique Energétique, Université Les Fréres Mentouri Constantine 1, Constantine, Algeria

A. Belasri and S. A. Beldjilali (eds.), *ICREEC 2019*, Springer Proceedings in Energy,
https://doi.org/10.1007/978-981-15-5444-5_82

82.1 Introduction

Final energy consumption in the building sector of developed countries represents 20 to 40% of the total energy produced. This consumption is mainly due to the use of heating and air conditioning processes [1, 2]. In Algeria, the analysis of final energy consumption shows that the building sector is the most energy consumers.

According to statistics carried out by the Ministry of Energy in 2012, the building sector consumes more than 42% of national production in primary energy, which makes this sector the most energy-consuming [3]. Commercial electricity consumption increased from 6 MTep in 1970 to 40 MTep in 2005 [4, 5].

On the other hand, Algeria's renewable energy resources are enormous and un-exploitable for the moment [6]. Algeria has a potential of 169.44 TWh per year in solar thermal, 13.9 TWh per year in solar photovoltaic and 35 TWh per year in wind energy. As a result, the Algerian government is paying special attention to the development of this sector by launching several projects with a target of reaching 10% of electricity production by 2020 from renewable sources and 20% by 2030 [4, 5].

However, making the building more efficient is the most appropriate solution to solve these problems. This solution can be considered by two possibilities [1]: (A) reduce the energy consumption of new buildings by 40 to 50%. (B) The reduction of 15 to 25% of the energy consumption of existing buildings.

In order to expect a low level of energy consumption, special attention must be paid to the envelope. The latter is the main part of the building.

Several techniques are used in the building to reduce its energy consumption. Edwin Rodriguez-Ubinas et al. [8] gives these different techniques.

In this sense, we propose to study the influence of thermal insulation used in building. The solar prototype studied is equipped with passive solar energy and improved insulation. Three cities are chosen, namely: Algiers (zone A) characterized by a Mediterranean climate, Constantine (Zone B) characterized by a semi-arid climate and Ouargla (zone D1) characterized by an arid climate.

82.2 Description of the Studied Prototypes

The solar village is built under the direction of the Polytechnic School of Architecture and Urbanism (EPAU) in the city of Boussaâda which is a semi-arid zone [7]. The solar village has three houses: 2 houses on one level and 1 house on 2 levels (ground floor + 1). In this study, we will evaluate the operation of a single house described later.

This prototype is designed with a compact and central space facing south with a large bay window in the family living room and two rooms equipped with small windows for ventilation and lighting and a trombe wall to provide heating in cold weather (Fig. 82.1).

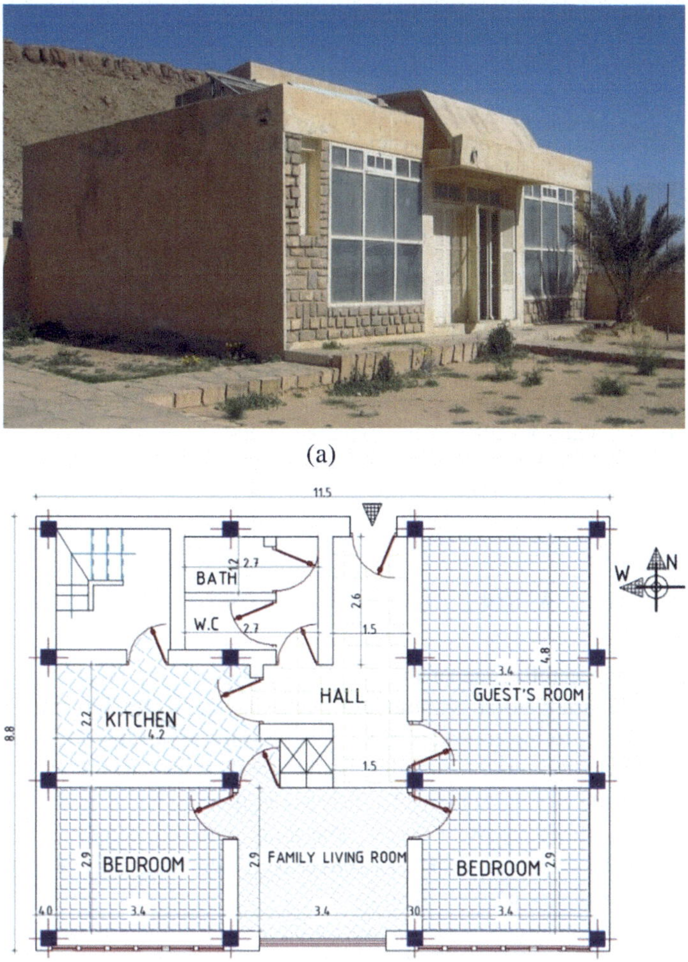

(a)

(b)

Fig. 82.1 **a** View of the South side, **b** Plan view

The characteristics of the solar prototype are given in Table 82.1.

82.3 Results and Discussion

In order to show the influence of certain passive techniques on the energy behavior of the solar prototype, we will study the influence of the glass surface and the trombe wall and the thermal insulation.

Table 82.1 Characteristics of the solar prototype construction

Wall/windows/ door	Composition	U value (W/m² · K)	Surface (m²)
North wall	Exterior plaster, 2 cm, Parpaing, 25 cm, Expanded cork insulation, 5 cm, Parpaing, 10 cm, Interior plaster, 2 cm	0.583	29.40
East/West wall (without insulation)	Exterior plaster, 2 cm, Parpaing, 40 cm, Interior plaster, 2 cm	1.51	23.80
East/West wall (with insulation)	Exterior plaster, 2 cm, Parpaing, 25 cm, Expanded cork insulation, 5 cm, Parpaing, 10 cm, Interior plaster, 2 cm	0.583	23.80
South wall	Exterior plaster, 2 cm, Stone, 40 cm, Interior plaster, 2 cm	0.668	20.10
Trombe wall	Exterior plaster, 2 cm, Parpaing, 25 cm, Interior plaster, 2 cm, Air layer, 10 cm, Glass, 0.5 cm	2.17	9.25 (one wall)

82.3.1 Effect of Adding Trombe Walls and Glazed Surfaces

In order to show the influence of the passive devices on the heating needs of the solar house, we simulated it without the two Trombe walls and with a glazing surface of the South living room equal to 2 m^2. Figure 82.2 illustrates the variation of the heating requirements of the solar house for the two cases with and without solar devices for the city of Algiers. An increase in heating needs is obtained for the case of a house without passive devices. The total annual heating requirements are 3619.82 kWh with an increase of 2.12 times. It is clear that passive solar devices reduce heating requirements significantly.

An increase in heating needs is also obtained for the city of Constantine (Fig. 82.2) and estimated at 6136.96 kWh or 1.74 times increase.

For the city of Ouargla, the annual heating requirement for solar homes without solar devices is 2119.39 kWh, with an increase of 3.07 compared to a solar house with solar devices, under the same climate (Fig. 82.2). The intensity of the solar radiation of this climate zone has a considerable influence on the operation of the Trombe walls and on the glass surface. The increase in heating needs is considerable.

82.3.2 Improvement of Thermal Insulation

Figure 82.3 illustrates the variation in the heating and cooling requirements of the solar house in different climates of selected cities with an insulated and uninsulated

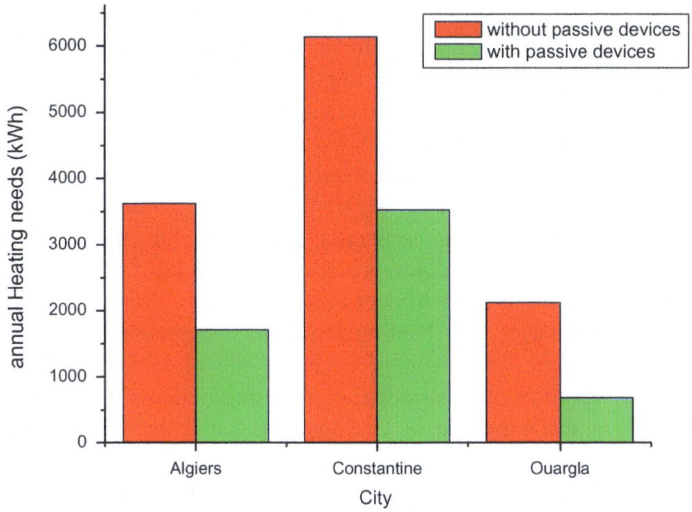

Fig. 82.2 Variation of heating requirements with and without passive devices

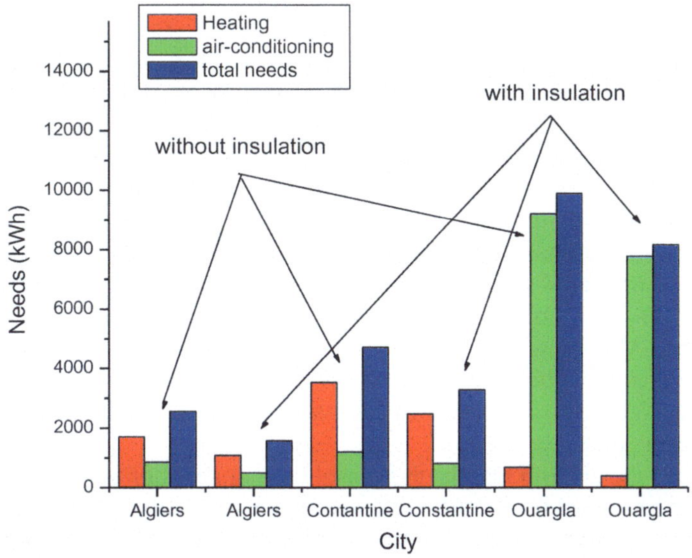

Fig. 82.3 Heating and cooling requirements for solar house with and without insulation

solar house (East and West wall with and without insulation as shown in Table 82.1).

For the house without insulation and for the city of Algiers, the annual heating needs of 1706.47 kWh are obtained. While for air conditioning, an estimated annual total of 847.61 kWh is obtained. Heating needs are 2.01 times more than the need for air conditioning. According to the Passivhauss standard, a building is considered passive if the annual consumption of heating and cooling does not exceed 15 kWh/m^2.year. For a total area of the solar house of 87.75 m^2, the consumption is 29.10 kWh/m^2.year.

Regarding the city of Constantine, annual heating needs are 3522.17 kWh (2.06 times more compared to the city of Algiers). While the annual cooling needs are 1197.33 kWh (1.41 times more compared to the city of Algiers). The annual consumption is 4719.5 kWh, which represents 53.78 kWh/m^2.year. A considerable increase compared to the city of Algiers is obtained.

For the city of Ouargla, the heating requirements obtained are 684.85 kWh. Compared to other cities, the heating needs of the city of Ouargla are lower, because of the availability of solar radiation, resulting in better operation of passive devices, as well as the reduction of heat losses through the walls. For air conditioning needs, a total annual air conditioning of 9207.51 kWh. The annual consumption of the solar house is 9892.36 kWh is obtained, which represents 112.73 kWh/m^2.year. Structural and other improvements can reduce the thermal needs of the solar house. For this we will show the influence of thermal insulation on the thermal needs of the solar house. The East and West walls are replaced by other insulated walls (see characteristics in Table 82.1).

A decrease in heating needs of the solar house in the city of Algiers is obtained. The annual heating requirements are 1075.31 kWh, a reduction of 36.98%. While cooling needs are estimated at 495.84 kWh, a decrease of 41.50%. The total needs of the house decreased by 38.48% (17.90 kWh/m^2.year). The influence of thermal insulation is important and can significantly reduce the thermal needs of the solar house. A house with low energy consumption can be obtained.

Regarding the city of Constantine, the simulation of the solar house shows that the total heating requirements are estimated at 2470.22 kWh, a decrease of 29.86% compared to the same solar house under the same climate but without thermal insulation. For air conditioning, a total annual air conditioning needs of 812.30 kWh. A decrease of 32.15% is obtained compared to the same solar house without thermal insulation. The total annual needs of the solar house are estimated at 3282.52 kWh, with a decrease of 30.44%. The annual requirements represent 37.40 kWh/m^2.year.

Regarding the city of Ouargla, the annual heating needs of the solar house with insulation are estimated at 389.05 kWh, a reduction of 43.19%. While the needs for air conditioning with thermal insulation are estimated at 7468.46 kWh, with a decrease of 18.88%. The total needs of the house without insulation are estimated at 7765.68 kWh or 88.49 kWh/m^2.year (a decrease of 21.49%).

82.4 Conclusion

A solar prototype of the solar village built in Boussaâda incorporating two trombe walls and a large glazed area is used for the calculation. The influence of the climate is characterized by the climate change of the climate, the city of Constantine characterized by a semi-arid climate and the city of Arid climate.

The analysis of the heating and cooling requirements of the solar prototype for the three cities shows that these needs are low for the city of Algiers compared to the other two cities. The air conditioning needs for the city of Ouargla are excessively high.

The contribution of several passive processes is studied and we can draw the following conclusions:

- The use of trombe walls and glazed surfaces allows a great reduction in heating needs. An increase of these needs estimated at 2.12, 1.74 and 3.07 times respectively for the cities of Algiers, Constantine and Ouargla is obtained for a prototype not equipped with these devices.
- The use of thermal insulation for the East and West walls reduces the total consumption of the prototype by 38.48, 30.44 and 21.49% respectively for the cities of Algiers, Constantine and Ouargla.

The study of the influence of certain passive devices on the thermal behavior of the solar prototype shows that it is possible to significantly reduce the thermal needs by passive techniques. A low energy house can be obtained just by using passive techniques.

References

1. N.N.A. Bakar, Energy efficiency index as an indicator for measuring building energy performance: a review. Renew. Sustain. Energy Rev. **44**, 1–11 (2015)
2. S. Stevanović, Optimization of passive solar design strategies: a review. Renew. Sustain. Energy Rev. **25**, 177–196 (2013)
3. APrue, Consommation Energétique Finale de l'Algérie: Chiffres clés Année 2005. Ministère de l'Energie et des Mines, Agence Nationale pour la Promotion et la Rationalisation de l'Utilisation de l'Energie (2007)
4. A.B. Stambouli, Promotion of renewable energies in Algeria: Strategies and perspectives. Renew. Sustain. Energy Rev. **15**, 1169–1181 (2011)
5. A.B. Stambouli, Algerian renewable energy assessment: the challenge of sustainability. Energy Policy **39**, 4507–4519 (2011)
6. Ministère de l'énergie algérien, http://www.memalgeria.org/francais/index.php?page=potenties
7. M. Boudiaf, M. Capderou, M. Janowski, Étude thermique des trois prototypes du village solaire intégré, CRAU, Algerie (1983)
8. E. Rodriguez-Ubinas, Passive design strategies and performance of Net Energy Plus Houses. Energy Build. **83**, 10–22 (2014)

Chapter 83
Geothermal Energy: A New Source of Energy for Heating Networks

Mohammed Mekki, Kheira Ouzza, and Amel Aattache

Abstract Geothermal is a word that refers to both the science that studies the internal thermal phenomena of the terrestrial globe, and the technology that aims to exploit it. To capture the geothermal energy and apply it directly to the buildings is made use of a network of tubes in which circulates a heat transfer fluid which allows by coupling it to a heat pump, the heat exchange with the ground. This technology is still not widespread in Algeria; hence, the objective of this study is to allow its development by providing a methodology. In order to validate this methodology, a calculation of an installation of vertical probes was conducted to meet the hot and cold needs of a family home. For its design, we will take into account the contribution of solar energy passive and the use of thermal insulation along the envelope to reduce the energy requirements to allow the geothermal installation; it will then be dimensioned for the supply of heating, cooling and hot water. This design includes geothermal drilling, heat pump equipment and radiant floor installation.

Keywords Renewable energies · Geothermal energy · Air conditioner · Heat exchange · Heater

83.1 Introduction

Algeria is launching a green energy dynamic by launching an ambitious program for the development of renewable energies and energy efficiency. This vision of the Algerian government is based on a strategy focused on the development of

M. Mekki (✉)
LM2SC, Département de Génie Civil, Université des Sciences et de la Technologie d'Oran Mohamed Boudiaf, USTO-MB, B.P. 1505, El M'nouar, Oran, Algeria
e-mail: mohammed.mekki@univ-usto.dz

K. Ouzza · A. Aattache
Département de Génie Civil, Université des Sciences et de la Technologie d'Oran Mohamed Boudiaf, USTO-MB, B.P. 1505, El M'nouar, Oran, Algeria

A. Belasri and S. A. Beldjilali (eds.), *ICREEC 2019*, Springer Proceedings in Energy,
https://doi.org/10.1007/978-981-15-5444-5_83

inexhaustible resources such as solar and wind energy and their use to diversify energy sources. Except that geothermal energy remains an unexploited potential in Algeria. Apart from balneotherapy, which is practically the only use, the real possibilities of this resource are unknown.

In a world where technology and renewable energy are constantly catching up, the increase in energy demand in the cities is similar, heating and cooling are no exception, so why not go that way? Why not follow this technology, these new innovative methods? Is the use of geothermal energy a solution for Algeria?

Geothermal systems using heat exchangers in the form of buried piping systems are gaining popularity among residential and commercial building owners. The use of soil to draw, store or dispose of energy greatly improves the efficiency of equipment for heating and cooling. For example, geothermal heat pumps have relatively constant source input temperatures throughout the year, even during periods of extreme heat or cold. This is a significant advantage over a conventional heat pump that uses outdoor air as a heat exchange media. In addition, the energy released during cooling is found somehow stored in the soil whose thermal inertia is very high. This same energy is subsequently recovered during the cold season to heat the building. This type of installation, commonly known as a closed-loop geothermal system, can be perceived as a huge reservoir that allows us to store energy when we want to get rid of it and then reuse it when necessary. Since the most expensive component for this type of project is the geothermal heat exchanger, particular attention will have to be paid to the type of heat exchanger used and its size in order to ensure the viability of the installation [1].

This article discusses the geothermal installation with vertical probes of a fictitious family home designed for doing this work. The geothermal installation will be sized for the supply of heating, cooling and domestic hot water (DHW).

The objective of this study is the design of a family home with low energy consumption, to achieve optimal comfort conditions for occupants throughout the year with minimal use of fossil fuels.

For the design of our very basic energy geothermal installation that will cover the heating, cooling and hot water demands, we will have to choose the appropriate heat pump model that will be determined in response to these requests as well as to the number of boreholes necessary geothermal and their depth.

In addition, the thermodynamic cycles of the heat pump will be studied according to two scenarios: one when the external climatic conditions are extreme; and the other considering average outside temperatures.

83.2 Building Sector in Energy Consumption in Algeria

Algeria is launching a green energy dynamic by launching an ambitious program for the development of renewable energies and energy efficiency. This vision of the Algerian government is based on a strategy focused on the development of inexhaustible resources such as solar energy and their use to diversify energy sources.

Thanks to the combination of initiatives and intelligence, Algeria is embarking on a new sustainable energy era.

The program consists of installing a renewable power of nearly 22,000 MW between 2011 and 2030 of which 12,000 MW will be dedicated to cover the national demand for electricity and 10,000 MW for export.

Thanks to this program, renewable energies are at the heart of Algeria's energy and economic policies: by 2030, around 40% of the electricity produced for national consumption will be of renewable origin.

83.3 Energy Consumption

Energy consumption in the household sector in Algeria, was probably multiplied by five between the year 1980 and 2012. Electricity, natural gas, fuel, coal, wood and even batteries are the energy vectors of the sector domesticated. The different types of energy serve us globally for four different uses. Heating accounts for the largest amount of energy: about 60% of domestic energy. Lighting, household appliances, audio-visual equipment and air conditioning account for nearly 20% Energy.

Knowing that the final energy consumption per capita is 0.48 tep in 1990, its evolution will increase from 0.71 toe in 2000 to 1.35 toe in 2010 (Fig. 83.1).

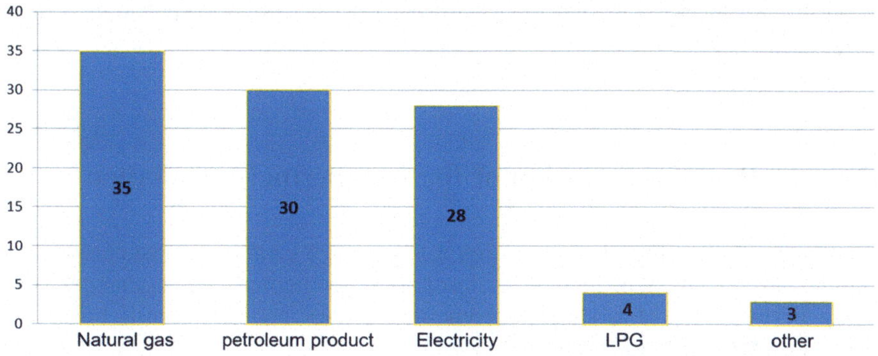

Fig. 83.1 The macro button chooses the correct format automatically

83.4 Geothermal Energy

Geothermal is a word of Greek origin, derived from "geos" which means earth, and from "thermos" which means heat: the heat of the Earth [2, 3].

It is used interchangeably to designate both the science that studies the internal thermal phenomena of the planet and all the industrial processes that attempt to exploit this heat to produce electrical energy and/or the useful heat for human being [4].

From the center to the surface, the globe is made up of three successive layers of decreasing temperature:

- The core, solid inside and liquid outside. Its temperature can reach 4200 °C.
- The mantle that surrounds it, with temperatures ranging from 3000 to 1000 °C. Plastic texture in the center, it becomes solid on the surface.
- The crust, which corresponds to the surface envelope, its temperature can range from 1.000 °C (contact with the mantle) up to 15–20 °C on the Earth's surface. Its thickness varies from 5 to 20 km in the depths of the ocean, and from 30 to 70 km under the continents. The origin of the internal heat of the earth refers to:
- The disintegration of radioactive isotopes present in the cortex and the mantle, mainly uranium 235, uranium 238, thorium 282 and potassium 40.
- The initial heat that was released during the formation of the planet.
- The movements between the different layers constituting the earth, mainly between the mantle and the nucleus.
- The crystallization of the nucleus. The outer core (liquid) crystallizes continuously, and in the transition zone with the nucleus the internal (solid) heat is released.

83.5 Methodology: Design of the Geothermal Installation

For the air conditioning of the housing, a low energy geothermal installation was chosen. This facility will include a closed-loop geothermal heat exchanger consisting of two vertical U-shaped geothermal probes housed in drill holes not exceeding 150 to 200 m depth [4].

Geothermal probes are associated with a heat pump, which must be reversible to meet heating and cooling needs.

For the system for conveying heat or cold to different rooms of the building, a radiant floor installation was chosen because it is the most efficient.

On the other hand, there is a storage tank for domestic hot water, with a support electric resistance.

The steps to be followed for sizing the geothermal installation are shown in Fig. 83.2.

83.6 Housing Situation

The house can be located in the Wilaya of Oran. The geographical coordinates of the municipality are: latitude: 35° 42′ 40″ N; longitude: 2° 59′ 39″ W and average altitude: 88 m above sea level.

Oran is a Mediterranean city par excellence, it is located at the edge of the southern bank of the Allégro-Provençal basin, it is located in north-west of Algeria to 432 km west of the capital Algiers.

The city lies at the end of a bay open to the north on the Gulf of Oran. The relief of its climate enjoys a mixture of 06 distinct natural components, one distinguishes its coastal border which extends from the mountains of Arzew to the sandy beach of Andalusian, to add to that a set of high hills bordering all the cliffs from Arzew to Canastel, on the west side there is the low coastal plain of Bousfer. The east side of Oran also characterizes the natural component specific to Oran with on one side the Oran-Gdyel plateau and on the other the Arzew saltworks.

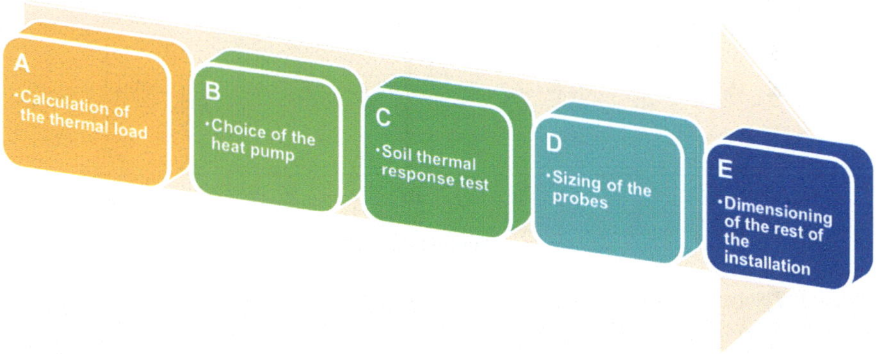

Fig. 83.2 The steps to follow for the sizing of the geothermal installation

83.7 Weather

The thermal amplitude is high with a difference of about 15 °C between the summer months and the winter months which reflects a Mediterranean climate [5].

The absolute maximum temperatures are between the months of July and August and are around 38.5 °C. On the other hand, absolute minimum temperatures are in January and are around 2 °C, although values of 0 °C can be reached.

83.8 Estimation of the Temperature of the Ground

The temperature gradient of the ground, at a certain depth, is constant and is not affected by climatic variations, as is the case in the areas closest to the surface. This phenomenon can be noticed by looking at Fig. 83.3 [6, 7].

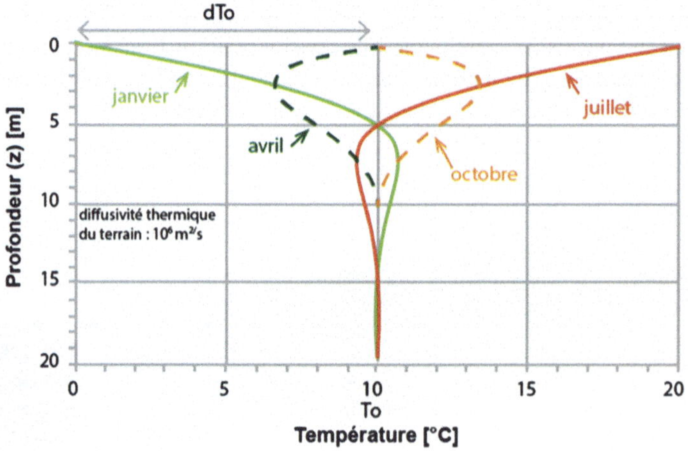

Fig. 83.3 Temperature variation as a function of depth [6, 7]

Table 83.1 Dimensions of the rooms on the ground floor and on the first floor

Ground floor		1st floor	
Stay	Surface (m^2)	Stay	Surface (m^2)
Hall	20	Distributor	8.75
bedroom	35	Bathroom 1	18
Living room	18	Bathroom 2	18
Dining room	16	Bath	5
Kitchen	18	Master bathroom	18
Utility room	1.875	Dresser	3
Sanitary	1.875	Main bathroom	6.25
		Room (useful surface)	10.5

83.9 Housing Design

The house consists of two levels of 100 m^2 each and a non-air conditioned basement. The height of each floor is 2.7 m. Below the layout of the rooms on the ground floor and 1st floor (Table 83.1).

83.10 Calculation of the Thermal Load of the Accommodation

For the calculation, we will have to take into account certain phenomena such as geometry and solar radiation, weather conditions, constructive characteristics of the building and the influence of activity. As some of these phenomena are very variable or difficult to quantify, the study of thermal loads will have to be based on considerations and hypotheses [8].

During calculations, the most unfavorable values will be taken, that is to say those that affect the climate of the house (depending on the time of year). On the other hand, we will have to be careful because, in this project, it is better to oversize the installation and meet its needs than not to meet it completely. However in summer, the thermal load contribution phenomena will be taken into account.

83.11 Calculation of the Maximum Heating Demand

During the winter months, heating is necessary to keep the rooms in the desired comfort conditions. For the study of this request, the thermal load by transmission and the thermal load by ventilation and infiltration were taken into account. The demand for each room was analyzed individually so that the total demand will be the sum of the latter [4].

Thermal Load by Transmission. The heat is transmitted through the enclosures (walls and glazing) to unheated rooms or to the outside. The thermal load per transmission is calculated as follows:

$$\dot{Q}_t = C_0 \,.\, C_i \,.\, U \,.\, S \,.\, \Delta T \tag{83.1}$$

where:

$\dot{Q}_t$: Thermal load by transmission, [W].
C_0: Coefficient d'orientation.
C_i: Intermittence coefficient of the installation.
U: Overall coefficient of thermal transmission, [W/ms^{2}K].
S: Surface of the wall exposed to the temperature difference, [m^2].
ΔT: Temperature difference between outside and inside, [K].

Thermal Load by Ventilation or Infiltration of Outside Air. A poor seal in the building causes air infiltration, ventilation and air renewal also means an increase in the thermal load. This is determined as follows:

$$\dot{Q}_v = V \,.\, N \,.\, 0{,}34 \,.\, \Delta T \tag{83.2}$$

where:

$\dot{Q}_v$: Thermal load by ventilation or infiltration of outside air [W].
V: Volume of the room to be heated, [m^3].
N: Hourly renewal rate, [h^{-1}].
0,34: Specific heat of air as a function of volume, [w h m^3 k].
ΔT: Différence de température entre l'extérieur et l'intérieur, [K].

The total heat demand $\dot{Q}_c$ is the sum of the thermal load by transmission ($\dot{Q}_t$) and the thermal load by ventilation or infiltration of outside air $\dot{Q}_v$.

According to this, the total heating demand obtained for the worst possible conditions is 15907 kW.

Table 83.2 Average seasonal thermal demand

Season	$\frac{change\ air}{h}$	$T_{interior}$ (°C)	$T_{exterior}$ (°C)	T_{ground} (°C)	Global irradiation $\frac{kWh}{m^2}$. day	Thermal demand (W)
Fall	1	22	12.4	9	–	5136
Winter	1	22	3.5	6	–	9296
Spring	1	24	9.5	9	5.6	−1.295
Summer	1	24	21.13	12	7.59	−7.528

The result obtained after calculating the heat losses through the walls of the house is 482 W. However, the result obtained after calculating the heat losses by ventilation in the dining room is 470 W.

The total heat loss of the dining room will be the sum of the losses by transmission (482 W) and by ventilation (470 W), or 952 W.

The average seasonal thermal demand is calculated in the same way as in the previous cases. Fall and winter will be calculated taking into account the heating demand, spring and summer, the cooling demand (the passive solar energy gains are taken into account).

Table 83.2 shows the parameters considered and the average thermal demand. Note that in spring and summer the result is negative because it is cooling.

83.12 Choice of Heat Pump

The geothermal heat pump must be chosen so as to be able to supply the maximum peaks of thermal power calculated in point 7. These requirements are 15.907 kW for heating and 14.844 kW for cooling. On the other hand, it must be a reversible heat pump to cope with both heating and cooling.

In a heat pump, the coolant is responsible for transporting heat from the cold point to the hot point. In the closed compression and expansion cycle in which the heat pump operates, the refrigerant will undergo two phase changes, one in the evaporator and the other in the condenser. The outside temperature varies throughout the year and so does the thermal demand, so the working temperatures of the refrigerant and the pressures at which phase changes occur will vary appropriate to the intended range of use.

83.13 Field Thermal Response Test

The thermal conductivity of the ground is easy to determine from the tabulated values, according to the type of rocks which compose the ground, and the values obtained during laboratory tests with samples of soundings. But none of these takes into account the natural conditions of the earth or the effects of the flow of groundwater. For this, it is necessary to resort to the thermal response test of the terrain.

The thermal response test (TRT) is an experimental test allowing to calculate the real temperature of the ground along the geothermal drilling and the effective thermal conductivity, taking into account the different geological strata crossed, the influence of water flow underground, geothermal probes, etc.

The test consists of circulating a heat exchanger fluid through a probe until the inlet and outlet temperatures stabilize. Then a constant amount of heat is injected over an extended period of about 50 or 60 h.

During the test, a recording of the most significant parameters is carried out. When the temperature of the fluid remains constant, the effective conductivity of the soil can be determined.

Since the means necessary to carry out this test are not available for carrying out this project, the thermal conductivity will be determined as an approximation from the tabulated values. As the geological profile of the area is not available, it will be assumed that the depth at which the geothermal probes are housed will be formed by a single stratum, mainly granite.

83.14 Sizing of the Probes

For the geothermal heat exchange system, we opted for vertical geothermal probes. In this project, two surveys will be carried out, so that a minimum distance of 2 m from the housing and a minimum distance between the probes of 6 m are respected, all over a distance of 10 m.

The diameter of the probes is chosen so that the geothermal area is easily introduced later. It is generally between 140 and 150 mm.

The drilling technique will depend on the type of geological layer drilled. In this case, the earth is mainly granite and therefore hard, compact and stable. Therefore, the most suitable technique is pneumatic rotary percussion with a hammer in the background.

We chose a geothermal heat exchanger made of cross-linked polyethylene (PEX-a) instead of a geothermal heat exchanger made of polyethylene (PE 100) because it offers great advantages in installation. On the one hand, PEX-a supports a higher working temperature range and an improvement in physical, chemical and mechanical performance even if the most important factor remains its useful life, more than 50 years.

Once the drilling is completed and the probes inserted, the empty space between the pipes and the neighboring soil must be filled by injecting a suspension of bentonite with cement to prevent water leaks. This suspension of bentonite and cement is injected from bottom to top into the borehole by means of an additional tube in order to avoid air pockets and to ensure good thermal contact between the heat transfer fluid and the soil.

Table 83.3 Data for two sounds to the depth 76 m

	Data exchanger		Operating data				
	Number of surveys	Depth (m)	COP/ *EER	Pc/Pf	Pa (kw)	Sal T (°C)	Debit (m^3/h)
Heater	2	76	4.46[a]	15.92	3581	0.57	3.68
Refrigeration	2	76	5.8[a]	21.37	3.73	34	3.51

[a]This is not the COP and the EER in normal operation of the pump

In summary, the results obtained for heating and cooling to the depth of 76 m are presented in Table 83.3.

The COP and EER obtained as program output data are the relationship between the heating power or the cooling power, respectively, and the electric power absorbed by the compressor for the conditions detailed in Table 83.3. However, the actual speed of the pump is far from these conditions and will vary depending on the thermal demand.

83.15 Design of the Radiant Floor Installation

For the heating system of the house, we will opt for a radiant floor installation. This type of installation was chosen because it is the one that best fits the optimal temperature profile of the human body. In addition, since the thermal emitter is uniform (it occupies the entire surface of the room) we will therefore avoid hot and cold zones caused by other types of emitters such as radiators.

83.16 Conclusion

In this study, we have shown the various possibilities of exploiting low and very low energy geothermal energy and in particular the heat exchange through vertical probes, stressing the importance of a good thermal insulation as well as that of the sizing of the appropriate heat pump.

Indeed, even if the cost of this installation seems expensive, it will be quickly depreciated by the time that will show us this installation is largely more profitable.

Finally, we can not finish this study, which proves to be a first in this country, without hoping for continuity in this line of research, without concealing our great interest in thermoactive structures such as soles or piles exchangers.

References

1. J. Bundschuh, D. Chandrasekharam, *Geothermal Energy Resources for Developing Countries* (Balkema, Rotterdam, 2002)
2. V. Trillat-Berdal, Intégration énergétique dans les bâtiments par l'utilisation combinée de l'énergie solaire et de la géothermie basse température. Ph.D. thésis, Université de Savoie-Chambery, 2006
3. H. Brandl, Energy foundations and other thermo-active ground structures. Géotechnique **56**(2), 81–122 (2006)

4. H.Y. Zeng, N.R. Diao, Z.H. Fang, Heat transfer analysis of boreholes in vertical ground heat exchangers. Int. J. Heat Mass Transf. **46**(23), 4467–4481 (2003)
5. O. Fruchier, F. Thiery, T. Talbert, J. Nou, D. Gachon, Etude d'un mix-énergétique solaire/ geothermie a l'échelle 1, in *Colloque sur l'Enseignement des Technologies et des Sciences de l'Information et des Systemes,* Grenoble, mars 2010
6. C. Greg Egg, J. Orio, D. Carl, *Modern Geothermal HVAC, Engineering and Control Applications* (McGraw-Hill Education, New York, 2013)
7. APRUE, *programme de développement de l'efficacité énergétique à l'horizon 2030,* Edition 2015
8. J.-P. Borel, Exposé pour la journée professionnelle de l'énergie, forum habitat et jardin, 28 février 2008

Chapter 84
Green Capacited Vehicles Routing Optimization with Constraints Rule-Based Approach

Latifa Dekhici, Khaled Guerraiche, and Khaled Belkadi

Abstract A large quantity of air pollution is caused by the combustion of fuel. The Green Capacited Vehicle Routing Problem (GCVRP) is an extension of Vehicle Touring (VRP) that aims to minimize fuel or CO_2 emissions or fuel consumption instead of total distance or with it in case of bi-objective optimization. The objective of the present work is to resolve the GCVRP with heterogeneous vehicles through a rule-based approach combined to an oriented object design. This technique using an existing environment that is Optaplanner. The modularity of the environment separates design and heuristics such as tabu search in java, rules and objective function in drools language, heuristics' setting in xml from data and results in excel files. Moreover, it appears very useful to the resolution of the Green Capacited Vehicles Routing Problem. In this paper, the resolution was investigated on a customized Belgic dataset taken from an existing benchmark. The instance contains 50 cities. The final project will allow the reusability and the flexibility to add easily soft and hard constraints depending on the case study.

Keywords Vehicles routing problem · CO_2 emission · Heuristics · Maximal constraints satisfaction

84.1 Introduction

Transportation area comprise daily shippers of millions of tons of supplies. Unfortunately, despite its necessity to the world life, it is one of the great facts of the existing pollutant particles in the air, particularly carbon dioxide (CO_2) [4]. From 1990 to 2007 as an example, the global CO_2 emissions due this sector have been bred by 45%, and are expected to remain to multiply by 2030 by around 40% [7].

L. Dekhici (✉) · K. Belkadi
Département d'Informatique, Université des Sciences et de la Technologie d'Oran USTO-MB, Oran, Algeria
e-mail: latifa.dekhici@univ-usto.dz

L. Dekhici · K. Guerraiche
Laboratoire LABGEM, Ecole Supérieure En Génie Electrique et Energétique, Oran, Algeria

A. Belasri and S. A. Beldjilali (eds.), *ICREEC 2019*, Springer Proceedings in Energy,
https://doi.org/10.1007/978-981-15-5444-5_84

Nowadays, the vehicle routing cost has not to be calculated according to distances but also pollutants emission. This work aims to minimize the emission of carbon dioxide from the heterogeneous charged vehicles and to respect some constraints. The optimization of vehicle routing problem is resolved using a rules-based approach combined to an oriented object design. That is why, definitions of the problem constraints and purposes retained are given in the second section. In the third section, Optaplanner environment and the approach used are presented. In Sect. 84.4, the dataset is described with results. Finally, a brief conclusion is given.

84.2 Green Capacited Vehicles Routing Problem

84.2.1 VRP and GVRP

The vehicle routing problem (VRP) is a combinatorial optimization and integer programming problem that deals with the optimal set of roads for a fleet of vehicles to cross in order to deliver to a given set of customers. It generalizes the well-known travelling salesman problem (TSP). The road network can be described using a graph where the arcs are roads and vertices are junctions between them. The arcs may be directed or undirected due to the possible presence of one-way streets or different costs in each direction. Each arc has an associated cost that is generally its length or travel time, which may be dependent on vehicle type. The classic Bjorklund [6] defines Green Transportation as a vehicle delivery service that, paralleled to competing shipper services that work for the similar purpose, reduces damaging effects on the natural environment and on living beings health. Recently, the green transportation concept for sustainable development has soared due to governmental regulations and customers preference for green products. Consequently, transportation companies are reviewing their processes to take into account such concern. In some cases, transforming the traditional logistics systems to be environmentally friendly shall even lower the costs enabling it to meet classic logistics objectives. Moreover, the green transportation system could be completed by determining emission factors and quantifying trucks emissions to integrate them into the logistics systems networks.

84.2.2 Constraints

In this work, we admit the following hypothesis: Each good must be shipped to its customers. Each customer must receive its good; each vehicle has specific CO_2

emission according to its consumption of fuel per kilometer; CO_2 emission depends on the weight of the goods transported and finally there are a heterogeneous fleet of vehicles in term of fuels and capacities.

84.2.3 Emission of CO_2

Total emission emitted per tons of CO_2 depends on the number of vehicles, the average distance they travel, and the extent to which they are loaded. It is difficult to do an exact estimation because of the uncertain effects on climate change. When an engine is started below its normal operating temperature, it uses fuel inefficiently, and the amount of pollution produced is higher than when it is hot [5]. In other hypothesis, Elbouzekri et al. [1] using genetics metaheuristics used another equation in the fuel optimization of Vehicle routing problem in order to calculate transportation cost between two points i and j. These observations lead to the basic relationship used in the calculation method (Eq. 84.1):

$$E(q,d) = d.\left[\left(\frac{e_{fl} - e_{el}}{Q}\right)q + e_{el}\right] \tag{84.1}$$

Where: E(q,d) the CO_2 emission from a vehicles in kg/km with a load q and distance d;efl is the CO_2 emissions of a fully loaded (by weight) vehicle; eel is the CO_2 emissions of an empty vehicle and Q is the volume capacity of a vehicle.

84.3 Constraint Programming and Heuristics Approach

The approach used in this paper is based on constraints rules in drools language. To model the problem, an existing oriented object model implemented in a java environment in Optaplanner [2] is modified. The settings of time limitation and other heuristics parameters is done on an XML file while reading Data is via a gaps-separated files. In this section, some notions related to this approach is described.

84.3.1 Constraint Programming

Constraint programming is a powerful paradigm for solving combinatorial search problems that draws on a wide range of techniques from artificial intelligence, computer science, databases, programming languages and operations research: It is based on the concept of constraints and decision variables, which limit the values for variables. Its advantage lies in the ability to represent problems in natural way, with variables that correspond to problem entities, leading to better understanding of the solution and relatively simple formulation of the problem.

84.3.2 Optaplanner

OptaPlanner is a constraint satisfaction solver that optimizes business resource planning. Use cases include Job Scheduling, Employee Rostering and more. OptaPlanner, released under the Apache Software License is a lightweight, open source embeddable planning engine. It enables Java programmers to apply plain domain objects and reuse existing code to solve optimization problems. OptaPlanner combines optimization heuristics and metaheuristics (such as Tabu Search) with score calculation.

Optimization using OptaPlanner requires to perform several steps:

- Model the planning problem: as classes that implements the interface solution.
- Define rules to calculate score: there are several types of scores Simple, hard Soft, Hard Medium Soft and Benadble (or with priorities levels). There are several ways to implement such score function: Easy Java, Incremental Java and Drools language.
- Configure a Solver: Optaplanner is generally used to solve problems in two stages. First, a solution is initialized using construction heuristics, whereas in the second stage, metaheuristic algorithms are used to optimize the solution. The configuration allows the setting of optimization algorithms as First Fit and Tabu Search and their parameters. It also describes problem domain.
- Load a problem dataset: to define a way of persistence.
- Solve and Get the best solution.

Annotation. Regarding the dataset of the planning problem, we can sort the domain classes in three categories:

Unrelated classes. Are not used by any of the score constraints. From a planning standpoint, this data is obsolete.
A problem fact class. Which is used by the score constraints, but does not change during planning (as long as the problem stays the same). For example: Customer, Depot. All the properties of a problem fact class are problem properties.
A planning entity class. Which is used by the score constraints and change during planning. For example: visits. The properties that change during planning are planning variables. The other properties are problem properties (previousStandstill, assigned tour).

Domain. The most important domain classes are customer and vehicle. The standstill is an interface of the two classes behaviors, while solving the problem (Fig. 84.1). In this work, we add to the existing design, the characteristics of each vehicle according to the fuel formula 84.1. Other classes also act as Air Distance, Depot and Location.

Constraints, Rules and Score. Drools Language [2] is chosen to define rules. Drools rule engine, providing custom implementation of Rete algorithm towards object-oriented systems. It offers an alternative way to Java-based score calculation, by providing its own declarative language. An example of hard score is the arriving after the due time.

Solver Configuration. Several xml configuration files are proposed in this work to compare methods and times limitation criteria. Files defines:

- Problem domain paths: we define solution class, entities classes…paths.
- Score director factory: We choose in this section a type score called bendable that sort hard and soft constraints in levels, numbers of hard and soft levels and drools rules definition file.
- Termination criteria. In each xml file a second spent limit is configured.
- Metaheuristics and constructive Heuristics. In this work First fit algorithm provides a straightforward approach to solution initialization. Then solver is configured to include Tabu Search, Hill Climbing, Simulated Annealing, or Hill Climbing with Late Acceptance.

Data Loading. Many ways are possible to load data and fill instance of problem domain. Java persistence API can be used in the case of real utilization with DAO modules. Second Optaplanner allow user to load data from excel files. In this work, the extension of the basic Optaplanner problem (homogenous vehicles) involves improvement of java source code and data files in order to add many fields as emisson in mg CO_2/km for heterogeneous vehicles. The new files have the extension *.vrp. So, the improvement of the approach concerns many steps in the Optaplanner basic problem such as the reading of data, the class model in java and especially the constraints in drools.

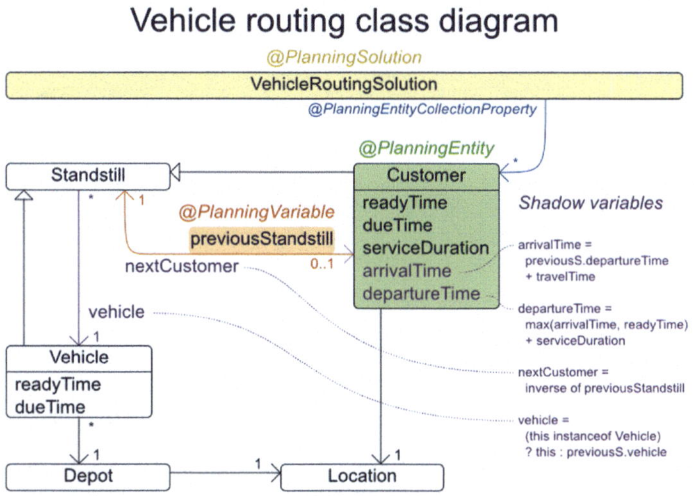

Fig. 84.1 The standard Optaplanner TWVRP class diagram

84.4 Data and Results

After extensive experimentation involving the solution of several instances using the different heuristics (local search, Tabu search and constructive heuristics…) and different times limit the best overall parameter settings of the Solver file is provided. Data were taken from existing instances of VRP Benchmark [3]. The Belgic dataset contains 49 cites (addresses) and 1 depot. Information according the transportation and fuel cost of each vehicles were added to fit a green capacited vehicles routing problem Table 84.1.

The results obtained showed the convergence towards solutions minimizing CO_2 emission and minimizing the abuse of other constraints. Figure 84.2 shows the state of the routing and the measures of emission in gr. in the final step of the resolution. We can see beside each truck a fraction (as 55/125) that expresses the load per the capacity (125 kg) of the truck, a colored line that characterizes the path to assigned customers and a dotted line that denotes the path to the depot.

Table 84.1 CO_2 emission improvement using the approach

	CO_2 emission
Initial state	122880 g 288 mg CO_2
Final state	108452 g 192 mg CO_2

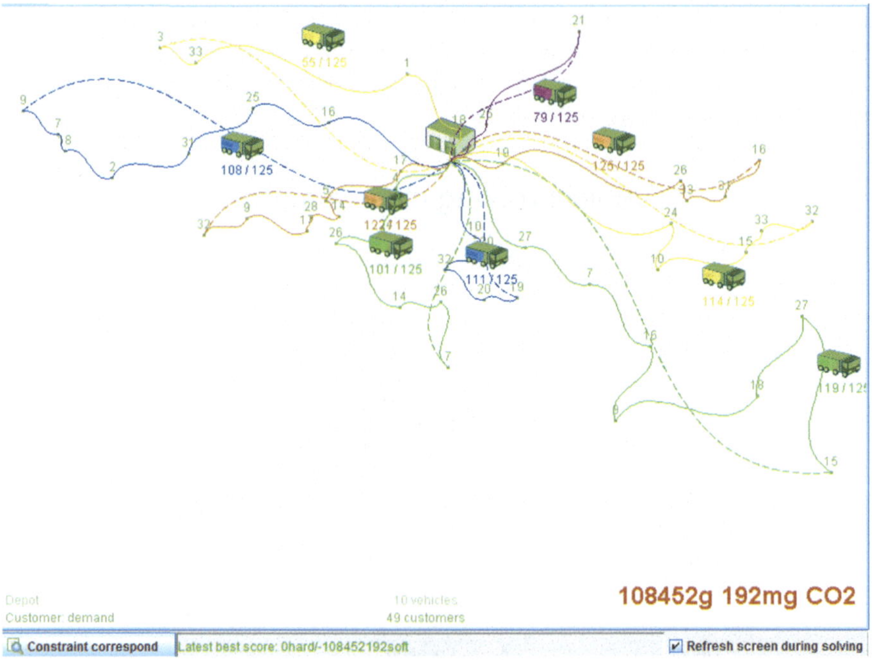

Fig. 84.2 The Fleet Routing state after optimization

84.5 Conclusion

In this paper, an approach that combine object-oriented modeling with rules based was tried to resolve a GCVRP. Results on a 50 adresses Belgic instance has shown how the CO_2 emission could be reduced without abuse of hard constraint. However, more great random tests must be done in the same way with [1] to be compared. Others constraints as those related to time windows must also be tested in a real case study.

References

1. A. ElBouzekri, A. ElHilali Alaoui, Y. Benadada, A genetic algorithm for optimizing the amount of emissions of greenhouse GAZ for capacitated vehicle routing problem in green transportation. Int. J. Soft Comput. **8**, 406–415 (2013)
2. http://Optaplanner.org
3. http://vrp-rep.org
4. T. Bektas, G. Laporte, The pollution Routing problem. Transp. Res. Part II Methodol. **45**(8), 1232–1250 (2011). https://doi.org/10.1016/j.trb.2011.02.004
5. A. Palmer, The development of an integrated routing and carbon dioxide emissions model for goods vehicles. PhD thesis, School of Management. Cranfield, University Cranfield, UK, 2007
6. M. Bjorklund, Influence from the business environment on environmental purchasing — drivers and hinders of purchasing green transportation services. J. Purchasing Supply Manag. **17**, 11–22 (2011)
7. ITF. "Reducing transport greenhouse Gas emissions: Trends and Data". Leipzig, Germany. Bakground for the 2010 International Transport Forum (2010)

Chapter 85
Promoting Biomimicry Towards, Sustainable Strategies for Energy-Efficient Building Design

Afef Ouis and Nassira Benhassine

Abstract The elevating problems of climate change throughout the world along with the indiscriminate consumption of natural resources are creating a sense of urgency and environmental awareness, which call for fundamental changes in many industries, and the building sector is no exception. In this questioning of our model of economic growth driven by consumption and based on a greedy fossil fuel industry. Buildings consume the larger part of worldwide energy and are therefore major contributors to global greenhouse gas emissions. As a result, the construction sector has been identified as the sector with the largest potential to reduce energy demand and decrease GHG emissions. Over the past years, the need to fight the impacts of climate change imposes on designers and architects to improve existing techniques and adopt new strategies, technologies and methods that enable to fulfill the current challenges in terms of environmental, societal and economic considerations. Thus, the motivation arose to investigate new design ideas for buildings that could help solve such problems. We suggest in this research to question biomimicry, a new field that studies and emulates the forms, functions, and process found in nature, to solve human challenges. In this regard, the main challenge is how architects could use biomimicry as a tool for energy efficiencies design to attain buildings that respond adaptively to climate change and the environment? To sum up, our research aim is to demonstrate how to use biomimicry as an approach to enhance sustainable solutions for energy-efficient building design.

Keywords Biomimicry · Building design · Energy efficiency · Environment · Sustainability

A. Ouis (✉) · N. Benhassine
Laboratory of Bioclimatic Architecture and Environment (ABE),
Faculty of Architecture and Urban Planning,
University of Constantine 3 Salah Boubnider, Constantine, Algeria
e-mail: afef.ouis@univ-constantine3.dz

A. Belasri and S. A. Beldjilali (eds.), *ICREEC 2019*, Springer Proceedings in Energy,
https://doi.org/10.1007/978-981-15-5444-5_85

85.1 The Challenge of Sustainability in Building Design

The path to sustainability in the built environment is transitional and embraces improving existing techniques and presenting new innovative solutions [1]. To this end, building design is becoming more difficult and complicated, at the same time, it acquires importance and a growing need to accomplish the crucial aspects of sustainable development [2], and it is currently the most pressing, complex and challenging agenda facing architects. Moreover, buildings consume the largest part of global energy, accounting for 40% of total energy consumption. Over the coming years, carbon emissions from the built environment are expected to increase faster than in other sectors [3].

On the other hand, in the 1970s the first oil shock increased the environmental awareness that carried out new requirements for energy conservation and functional strategies [4]. In the last decades, two diverse techniques have been implemented to improve and lower energy in building design, passive strategies, and active technologies [2]. In fact, the shape of the building: walls, windows, floors, and roofs play a major role to collect, reflect, store and distribute the energy. Passive strategies such as sun shading, insulation, atrium, natural ventilation, and the Trombe wall do not implicate the usage of electrical and mechanical devices [5–7]. unlike active systems which aim to enhance indoor comfort and reduce the building's impact on the environment, through systems or structures like solar panels, wind turbine, and heat pump to produce and supply of energy [2].

This matter plays a leading role in the context of changing climate many researchers investigate different ways to tackle this issue and seek novel methods to respond to the environment adaptively. In this paper, we explore Biomimicry, a new field that studies and emulates the forms, processes, and strategies found in nature to look into new solutions in architecture. Indeed, biomimicry has gained certain visibility and growing interest, in both research settings and design practice. According to Lenau et al. there is a growing interest in biomimetic research [8]. Finally, the primary aim of this study is to review the different approaches of biomimicry and present existing biomimetic strategies aiming to achieve higher levels of sustainability and energy-efficiency in building design.

85.2 Biomimicry Overview

85.2.1 History and Terminology

The influence of ideas derived from nature has always been pervasive through the history of humanity. Nevertheless, biomimetics as a scientific domain of research was introduced by Otto Herbert Schmitt, an American biophysicist during the 1950s, to explain the approach of drawing inspiration from biology to technology [9]. He is credited with having used the term biomimetics, a derivative of Greek words bios (life) and mimesis (imitate) [10]. Formally, he didn't define the word biomimetics but he frequently employed it to describe his work [9, 11]. Three decades later, the biologist and environmentalist Janine Benyus published a book entitled: 'Biomimicry: Innovation Inspired by Nature' which was the starting point for a new paradigm termed biomimicry [12].

Recent work within the International Organization of Standardization ISO [13], about biomimetics, has led to the following definitions and terminology:

- Biomimetics: '*Interdisciplinary cooperation of biology and technology or other fields of innovation with the goal of solving practical problems through the function analysis of biological systems, their abstraction into models and the transfer into and application of these models to the solution*' [8, 13].
- Biomimicry: '*Philosophy and interdisciplinary design approaches taking nature as a model to meet the challenges of sustainable development in a social, environmental, and economic*' [8, 13].

It can be seen that the definition proposed for the term biomimetics supports its methodological aspect, while that of biomimicry emphasizes the philosophical aspect and the integration of sustainable development as the primary motivation in the biomimetic methodology [14].

85.2.2 Biomimicry Approaches

Through a comparative literature review, biomimicry approaches in the design process consistently fall into two categories with various terminologies (summarized in Table 85.1):

As explained by Fayemi et al., in the solution-based approach new biomimetic design for technical implementation is born from the fundamental biological research. After their abstraction, the promising principals from the biological model can be applied in technology [21]. However, the problem-based approach attempts to solve a technical problem, designers look to potential organisms or ecosystems for solutions with initial goals and parameters to implement and test the design [22] (Fig. 85.1).

Table 85.1 Biomimicry approaches

Solution-based	Problem-based
Biology influencing design Zari [15]	Design referencing biology Zari [15]
Bottom up Speck et al. [16, 17]	Top down Speck et al. [16, 17]
Biology to design Baumeister et al. [18]	Challenge to biology Baumeister et al. [18]
Solution-driven Helms et al. [19]	Problem-driven Helms et al. [19]
Solution- based Badarnah and Kadri [20]	Problem-based Badarnah and Kadri [20]
Biology push ISO: 18458 [13]	Technology pull ISO: 18458 [13]

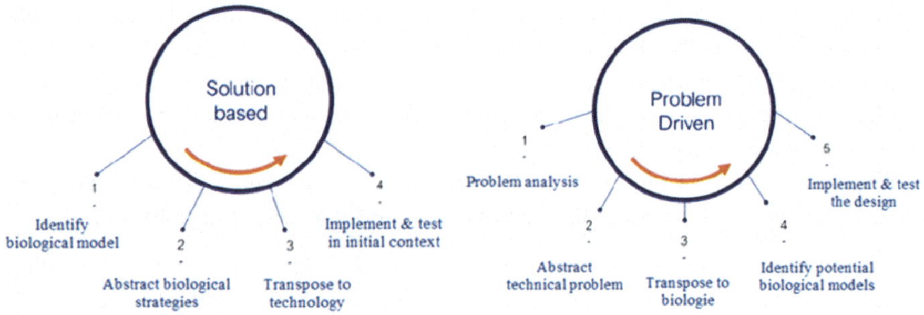

Fig. 85.1 Biomimicry approaches [23]

85.3 State of the Art in Application

85.3.1 Biomimetic Case Study

With a view to clarifying the design process through biomimicry approaches introduced in Sect. 85.2, we use this section to demonstrate its application.

Multidisciplinary cooperation between biologists, engineers, and architects at the Institute of Building Structures and Structural Design (ITKE), University of Stuttgart, Germany has led to the invention of a self-adapting shading system Flectofin®, by drawing inspiration from the elastic deformation of the perch of the Bird-of-paradise flower *'Strelitzia reginae'* as shown in (Fig. 85.2) [24, 25]. This biological system has been successfully analyzed, the kinematic mechanism of the flower was abstracted, then the biomimetic principles needed was simulated for architectural purposes [24, 26]. Indeed, the entire system replaces the technical hinges and depends only on the reversible material deformation (Fig. 85.3) in order to reduce maintenance expenses and energy consumption frequently related to interactive shading systems [24].

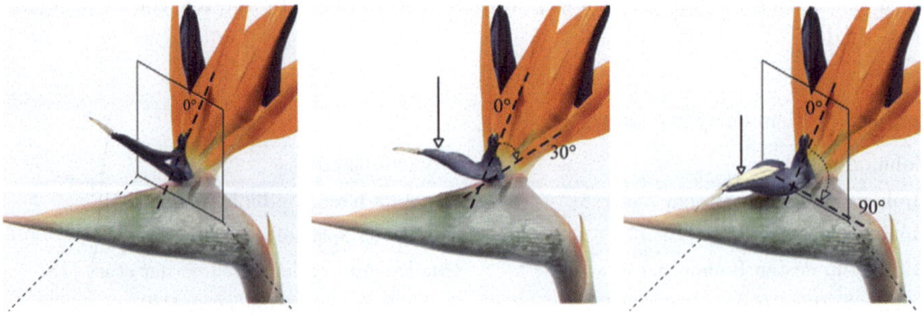

Fig. 85.2 The elastic deformation system of the flower (*Strelitzia reginae*) [25]

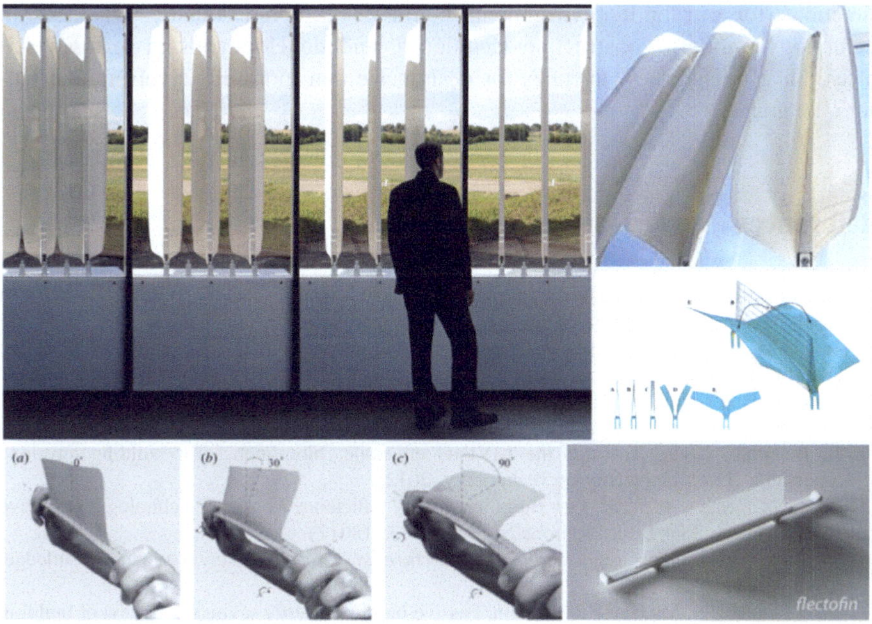

Fig. 85.3 The Flectofin® facade shading system [26]

Examples could also be found in research projects in the Architectural Association School of Architecture, the Austrian Institute of Technology, and the MIT Media Lab among many others. Biomimetic projects in practice could be seen in some of the works of HOK, Michael Pawlyn, Vincent Callebaut and Atelier One, also among others.

85.4 Conclusion

Over the last decades, several research studies on biomimetics for architectural applications have emerged, to investigate the ability to reduce energy demand and decrease GHG emissions. However, the first attempts of biomimetics in the building sector were mostly dedicated to building envelopes, building systems or experimental architecture.

The examples of biomimetic strategies, technologies, and concepts examined in this study suggest that buildings and urban environments should be expected to become more adaptable, resilient, and active contributors to climate restoration. By looking at the living world, beneficial solutions could be extracted and new directions for our built environments could be explored. Finally, advances in technology, modern construction techniques, and developments in material sciences, have opened new avenues for design; we can more effectively mimic nature's language.

References

1. R. Looman, Design strategy for the integration of climate-responsive building elements in dwellings, in *CIB World Building Congress*, South Africa (2007), pp. 1106–1114
2. R.C.G.M. Loonen, M. Trčka, D. Cóstola, J.L.M. Hensen, Climate adaptive building shells: state-of-the-art and future challenges. Renew. Sustain. Energy Rev. **25**, 483–493 (2013)
3. N. Chayaamor-Heil, N. Hannachi-Belkadi, Towards a platform of investigative tools for biomimicry as a new approach for energy-efficient building design. Buildings **7**(1), 19 (2017)
4. L. Badarnah Kadri, Towards the LIVING envelope: biomimetics for building envelope adaptation (Delft University of Technology, 2012)
5. X. Li, C. Shen, C.W.F. Yu, Building energy efficiency: passive technology or active technology? Indoor Built Environ. **26**(6), 729–732 (2017)
6. C. Porteous, K. MacGregor, *Solar Architecture in Cool Climates*, 1st edn. (Routledge, London, 2005)
7. S.B. Sadineni, S. Madala, R.F. Boehm, Passive building energy savings: a review of building envelope components. Renew. Sustain. Energy Rev. **15**(8), 3617–3631 (2011)
8. T.A. Lenau, A.-L. Metze, T. Hesselberg, Paradigms for biologically inspired design, in *Bioinspiration, Biomimetics, and Bioreplication VIII*, Denver, USA (2018), pp. 1–20
9. S. Jacobs, Biomimetics: a simple foundation will lead to new insight about process. Int. J. Des. Nat. Ecodyn. **9**(2), 83–94 (2014)
10. Y. Bar-Cohen, *Biomimetics: Biologically Inspired Technologies*, 1st edn. (CRC/Taylor & Francis, Boca Raton, 2006)
11. A. Iouguina, J.W. Dawson, B. Hallgrimsson, G. Smart, Biologically informed disciplines: a comparative analysis of bionics, biomimetics, biomimicry, and bio-inspiration among others. Int. J. Des. Nat. Ecodyn. **9**(3), 197–205 (2014)
12. J.M. Benyus, *Biomimicry: Innovation Inspired by Nature* (William Morrow Paperbacks, New York, 2002)
13. ISO: 18458, Biomimetics—Terminology, Concepts and Methodology (ISO, 2015)
14. P.E. Fayemi, Innovation par la conception bio-inspirée: proposition d'un modèle structurant les méthodes biomimétiques et formalisation d'un outil de transfert de connaissances (2016)
15. M.P. Zari, Biomimetic approaches to architectural design for increased sustainability, in *The SB07 NZ Sustainable Building Conference*, New Zealand (2007), pp. 1–10
16. T. Speck, O. Speck, Process sequences in biomimetic research, in *Design and Nature IV*, vol. I (Algarve, Portugal, 2008), pp. 3–11
17. O. Speck, D. Speck, R. Horn, J. Gantner, K.P. Sedlbauer, Biomimetic bio-inspired biomorph sustainable? An attempt to classify and clarify biology-derived technical developments. Bioinspir. Biomim. **12**(1), 011004 (2017)
18. D. Baumeister, R. Tocke, J. Dwyer, S. Ritter, J. Benyus, J. Smith, *Biomimicry Resource Handbook: A Seed Bank of Best Practices*, 2014th edn. (CreateSpace Independent Publishing Platform, Missoula, 2014)
19. M. Helms, S.S. Vattam, A.K. Goel, Biologically inspired design: process and products. Des. Stud. **30**(5), 606–622 (2009)
20. L. Badarnah, U. Kadri, A methodology for the generation of biomimetic design concepts. Archit. Sci. Rev. **58**(2), 120–133 (2015)

21. P.E. Fayemi, K. Wanieck, C. Zollfrank, N. Maranzana, A. Aoussat, Biomimetics: process, tools and practic. Bioinspir. Biomim. **12**(1), 011002 (2017)
22. M.P. Zari, Ecosystem services analysis for the design of regenerative urban built environments. Build. Res. Inf. **40**, 54–64 (2012)
23. A. Letard, N. Maranzana, K. Raskin, A. Aoussat, Design et Biomimetisme: quel rôle pour le designer? (2018), p. 10
24. J. Lienhard et al., Flectofin: a hingeless flapping mechanism inspired by nature. Bioinspir. Biomim. **6**(4), 045001 (2011)
25. J. Knippers, T. Speck, Design and construction principles in nature and architecture. Bioinspir. Biomim. **7**(1), 015002 (2012)
26. A. Team, Flectofin hingeless louver system. AskNature, https://asknature.org/idea/flectofin-hingeless-louver-system/. Accessed 17 May 2019

The manufacturer's authorised representative in the EU is Springer Nature Customer Service Centre GmbH, Europaplatz 3, 69115 Heidelberg, Germany. If you have any concerns regarding our products, please contact ProductSafety@springernature.com

Printed and bound by CPI Group (UK) Ltd, Croydon, CR0 4YY
05/10/2026
02228126-0020